Kappa Distributions

Kappa Distributions

Theory and Applications in Plasmas

Edited by

George Livadiotis

Southwest Research Institute,
San Antonio, Texas, United States

Elsevier
Radarweg 29, PO Box 211, 1000 AE Amsterdam, Netherlands
The Boulevard, Langford Lane, Kidlington, Oxford OX5 1GB, United Kingdom
50 Hampshire Street, 5th Floor, Cambridge, MA 02139, United States

Notices
Knowledge and best practice in this field are constantly changing. As new research and experience broaden our understanding, changes in research methods, professional practices, or medical treatment may become necessary.

Practitioners and researchers must always rely on their own experience and knowledge in evaluating and using any information, methods, compounds, or experiments described herein. In using such information or methods they should be mindful of their own safety and the safety of others, including parties for whom they have a professional responsibility.

To the fullest extent of the law, neither the Publisher nor the authors, contributors, or editors, assume any liability for any injury and/or damage to persons or property as a matter of products liability, negligence or otherwise, or from any use or operation of any methods, products, instructions, or ideas contained in the material herein.

Library of Congress Cataloging-in-Publication Data
A catalog record for this book is available from the Library of Congress

British Library Cataloguing-in-Publication Data
A catalogue record for this book is available from the British Library

ISBN: 978-0-12-804638-8

For information on all Elsevier publications visit our website at https://www.elsevier.com/books-and-journals

Working together
to grow libraries in
developing countries

www.elsevier.com • www.bookaid.org

Publisher: Nikki Levy
Acquisition Editor: Marisa LaFleur
Editorial Project Manager: Marisa LaFleur
Production Project Manager: Paul Prasad Chandramohan
Cover Designer: Victoria Pearson Esser

Typeset by TNQ Books and Journals

Contents

CONTRIBUTORS... vii

PREFACE xi

PART 1 • Theory and Formalism

CHAPTER 1 **Statistical Background of Kappa Distributions: Connection With Nonextensive Statistical Mechanics** 3
G. Livadiotis

CHAPTER 2 **Entropy Associated With Kappa Distributions** 65
G. Livadiotis

CHAPTER 3 **Phase Space Kappa Distributions With Potential Energy**... 105
G. Livadiotis

CHAPTER 4 **Formulae of Kappa Distributions: Toolbox** 177
G. Livadiotis

PART 2 • Plasma Physics

CHAPTER 5 **Basic Plasma Parameters Described by Kappa Distributions** 249
G. Livadiotis

CHAPTER 6 **Superstatistics: Superposition of Maxwell–Boltzmann Distributions** 313
C. Beck, E.G.D. Cohen

CHAPTER 7 **Linear Kinetic Waves in Plasmas Described by Kappa Distributions** 329
A.F. Viñas, R. Gaelzer, P.S. Moya, R. Mace, J.A. Araneda

CHAPTER 8   Nonlinear Wave–Particle Interaction and Electron Kappa Distribution 363
P.H. Yoon, G. Livadiotis

CHAPTER 9   Solitary Waves in Plasmas Described by Kappa Distributions 399
G.S. Lakhina, S. Singh

PART 3 • Applications in Space Plasmas

CHAPTER 10   Ion Distributions in Space Plasmas 421
G. Livadiotis, D.J. McComas

CHAPTER 11   Electron Distributions in Space Plasmas 465
V. Pierrard, N. Meyer-Vernet

CHAPTER 12   The Kappa-Shaped Particle Spectra in Planetary Magnetospheres 481
K. Dialynas, C.P. Paranicas, J.F. Carbary, M. Kane, S.M. Krimigis, B.H. Mauk

CHAPTER 13   Kappa Distributions and the Solar Spectra: Theory and Observations 523
E. Dzifčáková, J. Dudík

CHAPTER 14   Importance of Kappa Distributions to Solar Radio Bursts 549
I.H. Cairns, B. Li, J.M. Schmidt

CHAPTER 15   Common Spectrum of Particles Accelerated in the Heliosphere: Observations and a Mechanism 569
L.A. Fisk, G. Gloeckler

CHAPTER 16   Formation of Kappa Distributions at Quasiperpendicular Shock Waves 609
G.P. Zank

CHAPTER 17   Electron Kappa Distributions in Astrophysical Nebulae 633
D.C. Nicholls, M.A. Dopita, R.S. Sutherland, L.J. Kewley

APPENDIX A: ABBREVIATIONS 657
APPENDIX B: MAIN SYMBOLS 661
REFERENCES 665
INDEX 711

Contributors

J.A. Araneda
University of Concepción, Concepción, Chile

C. Beck
Queen Mary University of London, London, United Kingdom

I.H. Cairns
University of Sydney, Sydney, NSW, Australia

J.F. Carbary
Johns Hopkins University Applied Physics Laboratory, Laurel, MD, United States

E.G.D. Cohen
Rockefeller University, New York, NY, United States
University of Iowa, Iowa City, IA, United States

K. Dialynas
Academy of Athens, Athens, Greece

M.A. Dopita
Australian National University, Canberra, ACT, Australia
King Abdulaziz University, Jeddah, Saudi Arabia
University of Hawaii at Manoa, Honolulu, HI, United States

J. Dudík
Astronomical Institute of the Czech Academy of Sciences, Ondřejov, Czech Republic

E. Dzifčáková
Astronomical Institute of the Czech Academy of Sciences, Ondřejov, Czech Republic

L.A. Fisk
University of Michigan, Ann Arbor, MI, United States

R. Gaelzer
Federal University of Rio Grande do Sul, Porto Alegre, Brazil

G. Gloeckler
University of Michigan, Ann Arbor, MI, United States

M. Kane
Harford Research Institute, Bel Air, MD, United States

L.J. Kewley
Australian National University, Canberra, ACT, Australia
University of Hawaii at Manoa, Honolulu, HI, United States

S.M. Krimigis
Academy of Athens, Athens, Greece
Johns Hopkins University Applied Physics Laboratory, Laurel, MD, United States

G.S. Lakhina
Indian Institute of Geomagnetism, New Panvel (W), Navi Mumbai, India

B. Li
University of Sydney, Sydney, NSW, Australia

G. Livadiotis
Southwest Research Institute, San Antonio, TX, United States

R. Mace
University of KwaZulu-Natal, Durban, South Africa

B.H. Mauk
Johns Hopkins University Applied Physics Laboratory, Laurel, MD, United States

D.J. McComas
Princeton University, Princeton, NJ, United States

N. Meyer-Vernet
LESIA, Observatoire de Paris, CNRS, PSL Research University, UPMC, Sorbonne University, Paris Diderot, Paris, France

P.S. Moya
University of Chile, Santiago, Chile

D.C. Nicholls
Australian National University, Canberra, ACT, Australia

C.P. Paranicas
Johns Hopkins University Applied Physics Laboratory, Laurel, MD, United States

V. Pierrard
Royal Belgian Institute for Space Aeronomy, Brussels, Belgium
Université Catholique de Louvain, Louvain-La-Neuve, Belgium

J.M. Schmidt
University of Sydney, Sydney, NSW, Australia

S. Singh
Indian Institute of Geomagnetism, New Panvel (W), Navi Mumbai, India

R.S. Sutherland
Australian National University, Canberra, ACT, Australia

A.F. Viñas
NASA Goddard Space Flight Center, Greenbelt, MD, United States

P.H. Yoon
University of Maryland, College Park, MD, United States
Kyung Hee University, Yongin, Korea

G.P. Zank
University of Alabama in Huntsville, Huntsville, AL, United States

Preface

God created everything by number, weight, and measure.
Sir Isaac Newton

Following Gibbs, we consider "an ensemble of mechanical systems identical in nature and subject to forces determined by identical laws, but distributed in phase in any continuous manner....The number of systems of an ensemble which fall within th[is] extension will be represented by the [phase-space normalization] integral..., [where] no systems are supposed to be created or destroyed..." (Gibbs, 1902). According to Gibbs, a statistical ensemble is an idealization consisting of a large number of mechanical *analogs*, each of which represents a possible state that the real system may reside. Further, in the canonical statistical ensemble, the energy is not known exactly but the number of particles is fixed.

The study of thermodynamics is concerned with particle systems, which can be considered as stationary. A system is in a stationary state when all of its observables are independent of time. The characteristic phase space probability distribution parameterized by these observables is also time independent. Thermal equilibrium—the concept that any flow of heat (thermal conduction, thermal radiation) is in balance—is a stationary state of the system, but this is not unique. Any other stationary state of the system should be equally capable to describe the system. For instance, the internal energy of the system should be independent of the selection of the specific stationary state (relativity principle for statistical mechanics, Chapter 1).

Statistical mechanics is frequently used to determine the average behavior of a particle system when this resides at thermal equilibrium. When a particle system is at thermal equilibrium (typical behavior of earthy gases, e.g., the air), the particles are distributed in a specific way: There are many particles with small velocities and very few with large velocities. It is possible to write a mathematical equation describing how many particles are found at each velocity; this mathematical expression is given by the "Maxwellian distribution." However, space plasmas are particle systems distributed such that there are more high-velocity particles than there should be if the space plasma were in equilibrium. The mathematical equation used to describe the particle velocity in space plasmas is called the "kappa distribution."

Empirical kappa distributions have become increasingly widespread across space and plasma physics. Space plasmas from the solar wind to planetary magnetospheres and the outer heliosphere are systems out of thermal equilibrium, better described by the generalized formula of a kappa distribution rather than a Maxwellian, where the higher-velocity particles are well described when kappa has some finite small value. In general, the larger the kappa parameter, the closer the plasma is to thermal equilibrium. When the kappa reaches infinity, the plasma is exactly at thermal equilibrium and the distribution of space plasma is reduced to a Maxwellian. In this way, the kappa parameter is a novel thermal observable (such as temperature, density, pressure, etc.), which can label a stationary state and define a measure of the "thermodynamic distance" of a system from thermal equilibrium.

A breakthrough in the field came with the connection of kappa distributions to the solid background of nonextensive statistical mechanics. The kappa distribution maximizes the entropy of nonextensive statistical mechanics under the constraints of the canonical ensemble. This entropic formulation had been investigated by several authors in 1970s (e.g., Daróczy, 1970; Sharma and Taneja, 1975; Dial, 1982), but its final form and exploitation in statistical mechanics was succeeded later by Tsallis (1988).

Understanding the statistical origin of kappa distributions was the cornerstone of further developments of these distributions, by means of the (1) Foundation theory, (2) Plasma formalism, and (3) Space plasma applications. Some examples are (1) the concept of temperature and thermal pressure; (2) the physical meaning of the kappa parameter, that is, its role in the kappa spectrum arrangement, its connection with correlations, and degrees of freedom; (3) the multiparticle and multispecies joint kappa distribution; (4) the generalization to phase space kappa distribution of a Hamiltonian with nonzero potentials; (5) the *Sackur–Tetrode* entropy associated with kappa distributions.

The book reviews the theoretical developments of kappa distributions, their implications, and applications in plasma mechanisms and processes, and how these affect the underpinnings of our understanding of astrophysics, space and plasma physics, and statistical mechanics—thermodynamics.

The book is separated into three major parts: (A) theoretical methods; (B) analytical methods in plasmas; (C) applications in space plasmas.

(A) *Theory and formalism*: The first part of the book focuses on basic aspects of the theory of kappa distributions. The book starts from the connection of kappa distributions with a solid statistical background, the nonextensive statistical mechanics, then derives the entropy associated with kappa distributions, and develops the kappa distributions in cases with nonzero potentials. Finally, a toolbox of the important formulae of kappa distributions is provided.

(B) *Plasma physics*: Kappa distributions have become increasingly widespread across plasma physics with the publications rate following, remarkably, an exponential growth. This part is devoted to analytical methods related to kappa distributions on various basic plasma topics, spanning among others, the Debye shielding, polytropes, superstatistics (superposition of Maxwell—Boltzmann distributions), linear waves and the nonlinear approach, turbulence, solitons and double layers.

(C) *Applications in space plasmas*: This part is devoted to several important applications of theoretical and analytical developments in space plasmas from all over the heliosphere and beyond, including the ion and electron kappa distributions; the effect of kappa distributions in planetary magnetospheres, solar spectra, and radio bursts; shock waves; and beyond the heliosphere, the astrophysical nebulae. In addition, several mechanisms exist for generating these distributions in space and other plasmas, where some of them are described in detail in this book: polytropes (Chapter 5), superstatistics (Chapter 6), turbulence (Chapter 8), effect of pickup ions (Chapter 10), pump acceleration mechanism (Chapter 15), effect of shock waves (Chapter 16).

The book of kappa distributions is ideal for space, plasma, and statistical physicists; geophysicists especially of the upper atmosphere; Earth and planetary scientists; and astrophysicists. However, we claim, by no means, this book to be the alpha and omega of kappa distributions; many other topics could fill all three parts. Nevertheless, the book is, indeed, a necessary and sufficient assistant for accomplishing basic research in the mentioned physical subjects. It should be used for future space and plasma physics analyses that seek to apply kappa distributions in data analyses, simulations, modeling, or other theoretical work. Usage of the involved concepts and equations guarantees results that remain firmly grounded on the foundation of nonextensive statistical mechanics.

I must certainly thank all the authors and coauthors of this book. The whole authoring, reviewing, and editing process took about 2years. Parts B and C were given to be authored by leading experts on space and plasma physics. Each leading author and their teams of coauthors have numerous collection of exceptional representative publications related to their chapter.

I will always be grateful to the ever memorable professor Nikolaos Voglis (1948—2007) for inspiring me the concept of nonextensive statistical mechanics. I would also like to thank professor Xenophon Moussas (National and Kapodistrian University of Athens) for introducing me the usage of kappa distributions in space physics, as well as professor David McComas (Princeton University) for the numerous fruitful collaborations on the topic of kappa distributions. Finally, I could not be less grateful to my wife Eliana for all her spiritual support, especially the last 2 years when the book was in preparation. Τέλος, τω εν Τριάδι Θεώ δόξα.

George Livadiotis, PhD,
Senior Scientist
Space Science and Engineering
Southwest Research Institute

PART 1
Theory and Formalism

1. Statistical Background of Kappa Distributions: Connection
   With Nonextensive Statistical Mechanics   3
   G. Livadiotis
2. Entropy Associated With Kappa Distributions   65
   G. Livadiotis
3. Phase Space Kappa Distributions With Potential
   Energy   105
   G. Livadiotis
4. Formulae of Kappa Distributions: Toolbox   177
   G. Livadiotis

CHAPTER 1

Statistical Background of Kappa Distributions: Connection With Nonextensive Statistical Mechanics

G. Livadiotis
Southwest Research Institute, San Antonio, TX, United States

Chapter Outline

1.1 Summary 5
1.2 Introduction 5
1.3 Mathematical Motivation 14
1.4 Nonextensive Statistical Mechanics, in Brief! 16
  1.4.1 General Aspects 16
  1.4.2 q-Deformed Functions 17
    1.4.2.1 q-Unity 17
    1.4.2.2 q-Exponential 17
    1.4.2.3 q-Logarithm 17
    1.4.2.4 q-Hyperbolic 18
    1.4.2.5 q-Gamma 18
  1.4.3 Ordinary and Escort Probability Distributions 18
  1.4.4 Tsallis Entropy 20
  1.4.5 The Physical Temperature 21
1.5 Entropy Maximization 22
  1.5.1 Discrete Description 22
  1.5.2 Continuous Description 28
1.6 Connection of Kappa Distributions With Nonextensive Statistical Mechanics 29
  1.6.1 Derivation 29
  1.6.2 Historical Comments 34
1.7 Structure of the Kappa Distribution 37
  1.7.1 The Base of the Kappa Distribution 37
  1.7.2 The Exponent of the Kappa Distribution 39
1.8 The Concept of Temperature 41
  1.8.1 The Definition of Temperature Out of Equilibrium and the Concept of Physical Temperature 41

Kappa Distributions. http://dx.doi.org/10.1016/B978-0-12-804638-8.00001-2

1.8.2 Mean Kinetic Energy Defines Temperature 42
1.8.3 Misleading Considerations About Temperature 43
   1.8.3.1 *The Misleading Temperature-Like Parameter T_κ 43*
   1.8.3.2 *The Misleading Dependence of the Temperature on the Kappa Index 44*
   1.8.3.3 *The Misleading "Nonequilibrium Temperature" 45*
   1.8.3.4 *The Misleading "Equilibrium Temperature" 46*
   1.8.3.5 *The Most Frequent Speed 46*
   1.8.3.6 *The Divergent Temperature at Antiequilibrium 48*
   1.8.3.7 *The Thermal Pressure 49*
   1.8.3.8 *The Debye Length 49*
1.8.4 Relativity Principle for Statistical Mechanics 50
1.9 The Concept of the Kappa (or *q*) Index 51
1.9.1 General Aspects 51
1.9.2 Dependence of the Kappa Index on the Number of Correlated Particles: Introduction of the Invariant Kappa Index κ_0 52
1.9.3 Formulation of the *N*-Particle Kappa Distributions 56
1.9.4 Negative Kappa Index 57
1.9.5 Misleading Considerations About the Kappa (or q) Index 59
   1.9.5.1 *Kappa Index Sets an Upper Limit on the Total Number of Particles 59*
   1.9.5.2 *Correlation: Independent of the Total Number of Particles 60*
   1.9.5.3 *The Problem of Divergence 61*
1.10 Concluding Remarks 62
1.11 Science Questions for Future Research 63

1.1 Summary

Classical particle systems reside at thermal equilibrium with their velocity distribution function stabilized into a Maxwell distribution. On the contrary, collisionless and correlated particle systems, such as space plasmas, are characterized by a non-Maxwellian behavior, typically described by the kappa distributions or combinations thereof. Empirical kappa distributions have become increasingly widespread across space and plasma physics. However, a breakthrough in the field came with the connection of kappa distributions with the solid background of nonextensive statistical mechanics. Understanding the statistical origin of kappa distributions is the cornerstone of further theoretical developments and applications, which, among others, involve (1) the physical meaning of temperature, thermal pressure, and other thermodynamic parameters; (2) the physical meaning of the kappa index and its connection to the degrees of freedom and their correlation; (3) the Sackur–Tetrode entropy for kappa distributions; (4) the multiparticle description of kappa distributions; and (5) the kappa distribution of a Hamiltonian with a nonzero radial or angular potential. With the results provided in this study, the full strength and capability of nonextensive statistical mechanics are available for the physics community to analyze and understand the kappa-like properties of the various particle and energy distributions observed in geophysical, space, astrophysical and other plasmas.

Science Question: Is there a connection of kappa distributions with statistical mechanics?

Keywords: Correlations; Entropy; Nonextensive statistical mechanics; Temperature.

1.2 Introduction

Numerous analyses have established the theory of kappa distributions and provided a plethora of different applications in geophysical, space, astrophysical, or any other types of plasmas exhibiting non-Maxwellian behavior. Fig. 1.1 shows the number of publications related to kappa or Lorentzian distributions and their statistical background, the nonextensive statistical mechanics.

Kappa distributions were introduced about half a century ago to describe magnetospheric electron data by Olbert and its PhD students and colleagues (e.g., Binsack, 1966; Olbert, 1968; Vasyliūnas, 1968); it should be noted that Binsack (1966) was the first to publish the usage of kappa distributions, but he acknowledged that the kappa function was actually "introduced by Prof. Olbert of MIT in his studies of IMP-1."

The kappa distributions were employed to describe numerous space plasma populations in: (1) the *inner heliosphere*, including solar wind (e.g., Collier et al., 1996; Maksimovic et al., 1997, 2005; Pierrard et al., 1999; Mann et al., 2002; Marsch, 2006; Zouganelis, 2008; Štverák et al., 2009; Livadiotis and McComas,

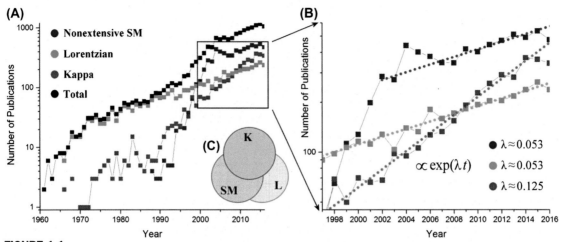

FIGURE 1.1

(A) Number of publications cataloged by Google Scholar since 1960 and related with kappa (*red*) or Lorentzian (*green*) distributions, and the nonextensive statistical mechanics (*blue*); their total number is also shown (*black*). (B) A magnification of the last 20 years showing the exponential growth rates λ (on a semilog scale). (C) The inset panel explains the selection criteria: (1) (L) without (SM) or (K); (2) (SM) without (K); and (3) any (K), where we symbolize: Kappa distribution (K), Lorentzian distributions (L), and nonextensive statistical mechanics (SM). The total number of publications is ~15,000, from which ~7000 (SM), ~3500 (K), and ~4500 (L). Interestingly, the growth publication rate of the kappa distributions is more than doubled the rates of the Lorentzian distributions and the nonextensive statistical mechanics.

2010a, 2011a, 2013a,c; Yoon, 2014; Pierrard and Pieters, 2015; Pavlos et al., 2016), solar spectra (e.g., Chapter 13; Dzifčáková and Dudík, 2013; Dzifčáková et al., 2015), solar corona (e.g., Owocki and Scudder, 1983; Vocks et al., 2008; Lee et al., 2013; Cranmer, 2014), solar energetic particles (e.g., Xiao et al., 2008; Laming et al., 2013), corotating interaction regions (e.g., Chotoo et al., 2000), and solar flares—related (e.g., Mann et al., 2009; Livadiotis and McComas, 2013b; Bian et al., 2014; Jeffrey et al., 2016); (2) the *planetary magnetospheres*, including magnetosheath (e.g., Binsack, 1966; Olbert, 1968; Vasyliūnas, 1968; Formisano et al., 1973; Ogasawara et al., 2013), near magnetopause (e.g., Ogasawara et al., 2015), magnetotail (e.g., Grabbe, 2000), ring current (e.g., Pisarenko et al., 2002a), plasma sheet (e.g., Christon, 1987; Wang et al., 2003; Kletzing et al., 2003), magnetospheric substorms (e.g., Hapgood et al., 2011), magnetospheres of giant planets (Chapter 12), such as Jovian (e.g., Collier and Hamilton, 1995; Mauk et al., 2004), Saturnian (e.g., Schippers et al., 2008; Dialynas et al., 2009; Livi et al., 2014; Carbary et al., 2014), Uranian (e.g., Mauk et al., 1987), Neptunian (Krimigis et al., 1989), magnetospheres of planetary moons, such as Io (e.g., Moncuquet et al., 2002) and Enceladus (e.g., Jurac et al., 2002), or cometary magnetospheres (e.g., Broiles et al., 2016a,b); (3) the *outer heliosphere and the inner heliosheath* (e.g., Decker and Krimigis, 2003; Decker et al., 2005; Heerikhuisen et al., 2008, 2010, 2014, 2015; Zank et al., 2010; Livadiotis et al., 2011, 2012, 2013; Livadiotis and McComas, 2011a,b, 2012, 2013a,c,d; Livadiotis, 2014a; Fuselier et al., 2014; Zirnstein and McComas,

2015); (4) *beyond the heliosphere*, including HII regions (e.g., Nicholls et al., 2012), planetary nebula (e.g., Nicholls et al., 2012; Zhang et al., 2014), and supernova magnetospheres (e.g., Raymond et al., 2010); or (5) *other space plasma—related analyses* (e.g., Milovanov and Zelenyi, 2000; Saito et al., 2000; Du, 2004; Yoon et al., 2006; Raadu and Shafiq, 2007; Livadiotis, 2009, 2014a, 2015a,b,c,e, 2016a,b,d; Tribeche et al., 2009; Hellberg et al., 2009; Livadiotis and McComas, 2009, 2010a,b,c, 2011b, 2014a; Baluku et al., 2010; Le Roux et al., 2010; Eslami et al., 2011; Kourakis et al., 2012; Randol and Christian, 2014, 2016; Varotsos et al., 2014; Fisk and Gloeckler, 2014; Liu et al., 2015; Viñas et al., 2015; Ourabah et al., 2015; Dos Santos et al., 2016; Nicolaou and Livadiotis, 2016). Maxwell distributions have also been used in space science (e.g., Hammond et al., 1996), especially due to their simplicity; for example, they are often used to fit the "core" of the observed distributions, that is, the part of the distribution around its maximum.

Table 1.1 contains the representative values of density n, temperature T, and kappa index κ of ~ 40 different space plasmas. The kappa indices were collected from the results of one or more published analyses. The values of the triplet (n,T,κ) and their processing are shown in Figs. 1.2 and 1.3. We also calculate the measure $M = 1/(\kappa - 0.5)$, an alternative of the kappa index κ. This is a measure of how far the system resides from thermal equilibrium (Livadiotis and McComas, 2010a,b, 2011b).

Fig. 1.2 illustrates the parameter values of space plasmas of Table 1.1. Panel (A) plots the parameters in an (n,T) diagram and uses a color map to illustrate the values of the kappa index κ and the measure $M = 1/(\kappa - 0.5)$. Panel (B) shows a 3-D scatter map of the triplet values (n,T,M), while panel (C) depicts the histogram of the values of M. In Figs. 1.2A and B, we observe that hotter and denser space plasmas reside closer to thermal equilibrium (i.e., larger κ, smaller M). The histogram of M values in Fig. 1.2C shows that the most frequent values are $M \sim 0.25$ and $M \sim 0.45$, corresponding to $\kappa \sim 4.5$ and $\kappa \sim 2.5$. The value of $\kappa \sim 4.5$ coincides with that of the most frequent solar wind value (Gloeckler and Geiss, 1998). The value of $\kappa \sim 2.5$ is frequently observed in space plasmas characterizing a special state called "escape state" (e.g., Chapters 2 and 8; Livadiotis and McComas, 2010a, 2013d; Yoon, 2014).

Fig. 1.3 shows that the measure M is negatively correlated with both the density and the temperature. We would like to find out when this correlation is maximized, that is, when do we consider only the density or the temperature, or when do we use some combination of both? More precisely, the correlation between M and $\log(nT^v)$ is examined for various values of the exponent v. A linear relation is fitted for (a) $v = 0$, (b) $v = 0.6$, and (c) $v = 1$. Then we seek for the maximum correlation. In particular, panel (d) shows the modified coefficient $R \equiv 1 - r^2$ (where r denotes the Pearson's correlation coefficient), which is plotted as a function of the exponent v. The modified correlation R is minimized, and the correlation r is maximized, for $v = 0.55$. The error derived from the correlation maximization method (Livadiotis and McComas, 2013c) is $\delta v = 0.38$. Therefore, the fitting for the optimal exponent $v = 0.6$ (in Fig. 1.3B) gives

Table 1.1 Space Plasmas Out of Thermal Equilibrium and Characteristic Values of n, T, and κ

Space Plasma	$\log n(\mathrm{m}^{-3})$	$\log T(\mathrm{K})$	κ	$M \equiv 1/(\kappa - 0.5)$	References
CIR	6.85 ± 0.19	5.59 ± 0.21	2.76 ± 0.18	0.44 ± 0.04	Chotoo et al. (2000)
SEP	6.0 ± 1.0	10.0 ± 0.5	6.0 ± 1.0	0.18 ± 0.03	Xiao et al. (2008)
ST e⁻ (Ulysses)	6.0 ± 0.3	5.3 ± 0.3	3.5 ± 1.5	0.33 ± 0.17	Zouganelis (2008)
Lower solar corona, e⁻	15.0 ± 1.0	6.5 ± 0.5	17 ± 7	0.061 ± 0.026	Cranmer (2014)
Outer solar corona, e⁻	12.0 ± 1.0	6.5 ± 0.5	3.0 ± 1.0	0.40 ± 0.16	Lee et al. (2013)
Slow SW He⁺ (Wind)	6.0 ± 1.0	6.0 ± 0.5	1.96 ± 0.13	0.69 ± 0.06	Collier et al. (1996)
Fast SW He⁺ (Wind)	6.0 ± 1.0	6.0 ± 0.5	2.65 ± 0.27	0.47 ± 0.06	»
Slow SW e⁻ (Ulysses)	5.7 ± 0.3	5.7 ± 0.3	2.71 ± 0.56	0.45 ± 0.12	Maksimovic et al. (1997b)
Fast SW e⁻ (Ulysses)	5.7 ± 0.3	5.7 ± 0.3	1.90 ± 0.08	0.71 ± 0.04	»
Slow SW e⁻ (Helios)	7.5 ± 0.5	5.6 ± 0.3	7.0 ± 1.0	0.154 ± 0.024	Štverák et al. (2009)
Slow SW e⁻ (Cluster)	6.6 ± 0.6	4.9 ± 0.3	5.0 ± 1.0	0.22 ± 0.05	»
Slow SW e⁻ (Ulysses)	5.7 ± 0.5	5.0 ± 0.3	2.4 ± 0.1	0.526 ± 0.028	»
Fast SW e⁻ (Helios)	7.5 ± 0.5	5.6 ± 0.3	6.0 ± 0.7	0.182 ± 0.023	»
Fast SW e⁻ (Cluster)	6.6 ± 0.6	4.9 ± 0.3	5.00 ± 0.10	0.222 ± 0.005	»
Fast SW e⁻ (Ulysses)	5.7 ± 0.5	5.0 ± 0.3	2.30 ± 0.10	0.56 ± 0.03	»
Fast SWe⁻ (H/W/U)	6.5 ± 1.0	5.0 ± 0.5	5.0 ± 1.5	0.22 ± 0.07	Maksimovic et al. (2005)
Ambient SW	4.0 ± 0.5	5.0 ± 0.5	~1.50 ± 0.10	1.00 ± 0.10	Fisk and Gloeckler (2006)
Quiet SWe⁻ (Wind)	4.9 ± 0.3	6.6 ± 0.6	3.0 ± 0.5	0.40 ± 0.08	Lin et al. (1996)
Outer heliosphere	4.0 ± 0.5	4 ± 0.5	1.63 ± 0.05	0.89 ± 0.04	Decker et al. (2005)
IH	4.15 ± 0.25	6.0 ± 0.3	1.75 ± 0.10	0.80 ± 0.06	Livadiotis et al. (2011)
MSh	7.0 ± 0.5	6.62 ± 0.19	3.5 ± 1.1	0.33 ± 0.12	Ogasawara et al. (2013)
MSh, no upstream waves	7.0 ± 0.5	6.62 ± 0.19	2.0 ± 0.5	0.67 ± 0.22	Formisano et al. (1973)
MSp substorms	7.81 ± 0.20	6.08 ± 0.20	4.0 ± 1.0	0.29 ± 0.08	Hapgood et al. (2011)
Plasma sheet H⁺ e⁻	5.7 ± 0.8	6.3 ± 0.7	5.30 ± 0.10	0.208 ± 0.004	Christon et al. (1989)
Plasma sheet e⁻	5.0 ± 0.9	6.8 ± 0.8	4.0 ± 1.0	0.29 ± 0.08	Kletzing et al. (2003)
Ring current	6.3 ± 1.0	7.5 ± 0.5	6.0 ± 0.5	0.182 ± 0.017	Pisarenko et al. (2002a)
Magnetotail	5.0 ± 1.0	5.5 ± 0.5	6.5 ± 0.5	0.167 ± 0.014	Grabbe (2000)
Jupiter's MSp	7.3 ± 2.0	8.1 ± 0.4	4.5 ± 1.5	0.25 ± 0.09	Collier and Hamilton (1995)

Saturnian MSp e⁻	6.7 ± 0.3	5.3 ± 0.3	1.85 ± 0.10	0.741 ± 0.055	Schippers et al. (2008)
Saturnian MSp e⁻	5.5 ± 0.3	7.3 ± 0.3	4.30 ± 0.10	0.263 ± 0.007	≫
Saturnian MSp H⁺	6.0 ± 0.5	8.0 ± 1.0	3.5 ± 1.0	0.33 ± 0.11	Dialynas et al. (2009)
Saturnian MSp O⁺	6.0 ± 0.5	8.0 ± 0.5	5.0 ± 1.5	0.22 ± 0.08	≫
MSp, Enceladus, OH cloud	9.0 ± 0.5	9.3 ± 0.5	2.90 ± 0.10	0.417 ± 0.017	Jurac et al. (2002)
Uranian MSp	6.0 ± 0.5	9.0 ± 0.3	3.25 ± 0.75	0.36 ± 0.10	Mauk et al. (1987)
HI regions	10.0 ± 1.0	4.0 ± 0.5	2.8 ± 0.4	0.44 ± 0.08	Zhang et al. (2014)
HII regions	10.0 ± 1.0	4.0 ± 0.5	15 ± 5	0.069 ± 0.024	Nicholls et al. (2012)
HII regions, e⁻	10.0 ± 1.0	4.0 ± 0.5	12 ± 7	0.09 ± 0.05	Binette et al. (2012)
Planetary nebulae	10.0 ± 1.0	4.0 ± 0.5	100 ± 50	0.010 ± 0.005	Nicholls et al. (2012)
Supernova	5.5 ± 0.5	8.42 ± 0.27	2.73 ± 0.88	0.45 ± 0.18	Raymond et al. (2010)

(1) If it is not mentioned otherwise, the parameters characterize the proton plasma.
(2) The name of the planet for magnetospheric and relevant plasmas is avoided only for Earth.
(3) Abbreviations: *CIR*, Corotating Interaction Region; *IH*, Inner Heliosheath; *MSh*, Magnetosheath; *MSp*, Magnetosphere; *SEP*, Solar Energetic Particle; *ST*, Suprathermal; *SW*, Solar Wind; *H/W/U*, Helios/Wind/Ulysses.

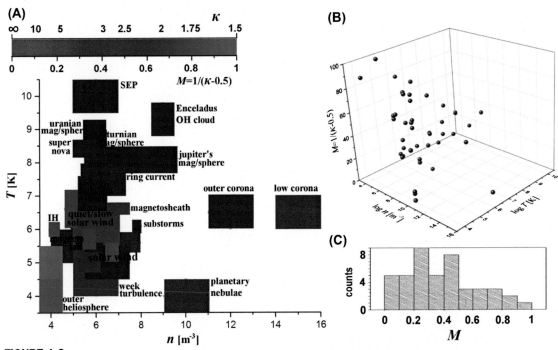

FIGURE 1.2

Examples of space plasmas with their representative values of density n, temperature T, and kappa index κ (Table 1.1). (A) A color map based on the measure $M = 1/(\kappa - 0.5)$ in the n-T plane. This measure M takes values from $M = 0$ (thermal equilibrium) to $M = 1$ ("antiequilibrium," the furthest state from thermal equilibrium). (B) 3-D scatter map of the values of (n, T, M). (C) Histogram of the values of M. Adopted from Livadiotis (2015a).

$M \sim 2.27 - 0.19 \cdot \log(nT^\nu)$. This is written as $1 - M \sim 0.19 \cdot \log(nT^\nu / 5 \times 10^6)$ or $nT^\nu \cong 5 \times 10^6 \cdot 10^{5.26(1-M)}$. If we consider the value $\nu \sim 0.5$, the product nT^ν can be thought as the "electron thermal flux," given by $J \equiv n\theta_e \sim nT^{1/2}$, where we find $M \sim 2.22 - 0.19 \cdot \log(nT^{1/2})$ or $J = J_0 e^{-a_j M}$ with $J_0 \sim 2 \times 10^{15}$ s^{-1}m^{-2} and $a_j \sim 12$. Also, near the error bars is the exponent value of $\nu = 1$ that corresponds to thermal pressure $nT = P/k_B$. In this case, the fitting gives $M \sim 2.43 - 0.16 \cdot \log(nT)$, which is written as $1 - M \sim 0.16 \cdot \log\left(P/k_B / 0.87 \times 10^9\right)$ or $P/k_B = 0.87 \times 10^9 \cdot 10^{5.88(1-M)}$, i.e., $P = P_0 e^{-a_p M}$ with $P_0 \sim 0.91 \times 10^{-8}$ Pa and $a_p \sim 13.54$. Thus the lowest and highest thermal pressure of a space plasma that is stabilized on a single kappa distribution are, on average, given by $P/k_B = 0.87 \times 10^9$ or $P \cong 0.12$ pdyne·cm^{-2}, and $P/k_B = 1.54 \times 10^{15}$ or $P \cong 2.13 \times 10^5$ pdyne·cm^{-2}, respectively. (Note that we used the pressure units of $k_B \cdot$m$^{-3} \cdot$K, Pa, and pdyne·cm^{-2}.) The previously described relations quantified the existence of negatively correlated trends between M (or the positive correlated trends between and the kappa index) and the density−temperature quantity nT^ν.

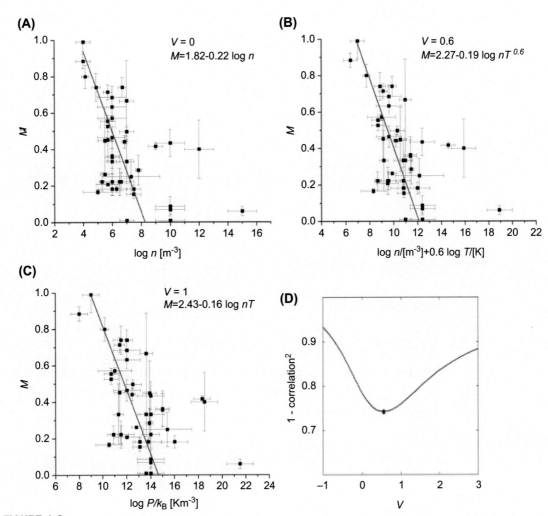

FIGURE 1.3
Fitting and correlation between $M = 1/(\kappa - 0.5)$ and $\log(nT^{\nu})$ are examined for various values of the exponent ν, (A) $\nu = 0$, (B) $\nu = 0.6$, and (C) $\nu = 1$. Panel (D) shows the modified correlation coefficient ($\equiv 1 - \text{correlation}^2$) and its minimization at $\nu = 0.6$ (corresponding to a maximization of correlation). Notes: panels (A–C) apply a fitting that takes into account the errors of both the variables (e.g., see the fitting method in Frisch et al., 2013); panel (D) applies a correlation maximization method (as defined in Livadiotis and McComas, 2013c). Adopted from Livadiotis (2015a).

The standard formulation of the particle kinetic energy or velocity is given by

$$P(\varepsilon_K; \kappa_0, T) = \frac{(k_B T)^{-\frac{1}{2}d_K}}{\Gamma\left(\frac{d_K}{2}\right)} \cdot \frac{\Gamma(\kappa)\kappa^{-\kappa}}{\Gamma\left(\kappa - \frac{d_K}{2}\right)\left(\kappa - \frac{d_K}{2}\right)^{-\left(\kappa - \frac{d_K}{2}\right)}}$$

$$\times \left(1 + \frac{1}{\kappa} \cdot \frac{\varepsilon_K - \langle\varepsilon_K\rangle}{k_B T}\right)^{-\kappa - 1} \varepsilon_K^{\frac{1}{2}d_K - 1}, \tag{1.1a}$$

$$P(\vec{u}; \kappa, T) = \pi^{-\frac{1}{2}d_K}\theta^{-d_K} \cdot \frac{\Gamma(\kappa)\kappa^{-\kappa}}{\Gamma\left(\kappa - \frac{d_K}{2}\right)\left(\kappa - \frac{d_K}{2}\right)^{-\left(\kappa - \frac{d_K}{2}\right)}}$$

$$\times \left[1 + \frac{1}{\kappa} \cdot \frac{\frac{1}{2}m(\vec{u} - \vec{u}_b)^2 - \frac{d_K}{2}k_B T}{k_B T}\right]^{-\kappa - 1}, \tag{1.1b}$$

with normalization

$$\int_{-\infty}^{\infty} P(\vec{u}; \kappa_0, T)d\vec{u} = 1, \quad \int_0^{\infty} P(\varepsilon_K; \kappa_0, T)d\varepsilon_K = 1. \tag{1.1c}$$

The bases of the distributions can be written in terms of the kinetic energy $\varepsilon_K(\vec{u}) = \frac{1}{2}m(\vec{u} - \vec{u}_b)^2$ instead of the meanless kinetic energy $\varepsilon_K - \langle\varepsilon_K\rangle = \frac{1}{2}m(\vec{u} - \vec{u}_b)^2 - \frac{d_K}{2}k_B T$,

$$P(\varepsilon_K; \kappa_0, T) = \frac{\left[\left(\kappa - \frac{d_K}{2}\right)k_B T\right]^{-\frac{1}{2}d_K}}{B\left(\frac{d_K}{2}, \kappa - \frac{d_K}{2} + 1\right)} \cdot \left(1 + \frac{1}{\kappa - \frac{d_K}{2}} \cdot \frac{\varepsilon_K}{k_B T}\right)^{-\kappa - 1} \varepsilon_K^{\frac{1}{2}d_K - 1}, \tag{1.2a}$$

$$P(\vec{u}; \kappa, T) = \left[\pi\left(\kappa - \frac{d_K}{2}\right)\theta^2\right]^{-\frac{1}{2}d_K} \cdot \frac{\Gamma(\kappa + 1)}{\Gamma\left(\kappa - \frac{d_K}{2} + 1\right)} \cdot \left[1 + \frac{1}{\kappa - \frac{d_K}{2}} \cdot \frac{(\vec{u} - \vec{u}_b)^2}{\theta^2}\right]^{-\kappa - 1}. \tag{1.2b}$$

where the Beta function is defined by $B(x,y) \equiv \Gamma(x) \cdot \Gamma(y)/\Gamma(x+y)$.

In terms of the invariant kappa index $\kappa_0 \equiv \kappa - \frac{d_K}{2}$, the distributions become

$$P(\varepsilon_K; \kappa_0, T) = \frac{(\kappa_0 k_B T)^{-\frac{1}{2}d_K}}{B\left(\frac{d_K}{2}, \kappa_0 + 1\right)} \cdot \left(1 + \frac{1}{\kappa_0} \cdot \frac{\varepsilon_K}{k_B T}\right)^{-\kappa_0 - 1 - \frac{1}{2}d_K} \varepsilon_K^{\frac{1}{2}d_K - 1}, \tag{1.3a}$$

$$P(\vec{u};\kappa_0,T) = \left(\pi\,\kappa_0\theta^2\right)^{-\frac{1}{2}d_K} \cdot \frac{\Gamma\left(\kappa_0 + 1 + \frac{d_K}{2}\right)}{\Gamma(\kappa_0 + 1)} \cdot \left[1 + \frac{1}{\kappa_0}\cdot\frac{(\vec{u} - \vec{u}_b)^2}{\theta^2}\right]^{-\kappa_0 - 1 - \frac{1}{2}d_K}.$$

(1.3b)

At thermal equilibrium, these distributions become

$$P(\varepsilon_K;\kappa_0\to\infty,T) = \frac{(k_B T)^{-\frac{1}{2}d_K}}{\Gamma\left(\frac{d_K}{2}\right)}\cdot\exp\left(-\frac{\varepsilon_K}{k_B T}\right),$$

(1.4a)

$$P(\vec{u};\kappa_0\to\infty,T) = \left(\pi\,\theta^2\right)^{-\frac{1}{2}d_K}\cdot\exp\left[-\frac{\frac{1}{2}m(\vec{u}-\vec{u}_b)^2}{k_B T}\right].$$

(1.4b)

where $\theta \equiv \sqrt{2k_B T/m}$ is the thermal speed of a particle with mass m, that is, the particle temperature T expressed in speed units; both the particle $\vec{u}$ and bulk $\vec{u}_b \equiv \langle\vec{u}\rangle$ velocities are in an inertial reference frame. Note that the kinetic energy is expressed in the co-moving system of the particle flow with bulk velocity $\vec{u}_b$, i.e., $\varepsilon_K = \frac{1}{2}m(\vec{u} - \vec{u}_b)^2$.

This chapter presents the connection between the basic theory and formalism of kappa distributions and the statistical framework of nonextensive statistical mechanics. In Section 1.3, we provide a brief mathematical motivation for generating and using the kappa distribution, that is, the deformation of the exponential canonical distribution and the logarithmic entropic form. In Section 1.4, we briefly review the basic aspects of nonextensive statistical mechanics, emphasizing on the Tsallis entropic formulation. In Section 1.5, we develop the particle canonical distribution function derived from the maximization of the Tsallis entropy under the constraints of canonical ensemble. The derivation is shown for both discrete and continuous datasets. Then in Section 1.6, we proceed in the connection of kappa distributions with nonextensive statistical mechanics, providing also important historical facts and comments. In Section 1.7, we analyze the structure of the kappa distribution, focusing on its base and exponent. In Section 1.8, we determine the physical meaning of the concept of temperature for stationary states out of the specific stationary state called "thermal equilibrium," utilizing the main classical definitions of temperature. Then we specify eight misleading considerations regarding the temperature and provide their resolution. We state the relativity principle for statistical mechanics, which concerns the constancy of the mean kinetic energy along the stationary states. In Section 1.9, we analyze and discuss the concept of the kappa index, emphasizing on the dependence of the kappa index on the number of correlated particles, which introduces the notion of the invariant kappa index κ_0. We describe the formulation of the multiparticle kappa distributions and the negative kappa distribution. We also specify three misleading considerations about the kappa (or q) index and provide their resolution. Finally, the concluding remarks are given in Section 1.10, while three general science questions for future analyses are posed in Section 1.11.

1.3 Mathematical Motivation

Thermal equilibrium is a special stationary state. It is the state where any flow of heat (e.g. thermal conduction, thermal radiation) is in balance. Systems at thermal equilibrium have two special statistical features: (1) they have their particle velocities described by stationary distribution functions, and (2) these stationary functions are Maxwellian distributions (or combination thereof).

Space plasmas reside also in stationary states. However, the classical Maxwell distributions are extremely rare in these plasmas, while their vast majority is described by non-Maxwellian distributions; they reside in stationary states but out of thermal equilibrium. *How can these exotic stationary states be described?*

First, Jacob Bernoulli (1683; see O'Connor and Robertson, 2016) defined the limit

$$e \equiv \lim_{\kappa \to \infty} \left(1 + \frac{1}{\kappa}\right)^{\kappa}, \quad e^{-1} \equiv \lim_{\kappa \to \infty} \left(1 + \frac{1}{\kappa}\right)^{-\kappa}. \tag{1.5}$$

This limit led Euler to the definition of the exponential function (e.g., see Maor, 1994), which can be expressed by

$$e^{x} = \lim_{\kappa \to \infty} \left(1 - \frac{1}{\kappa} \cdot x\right)^{-\kappa}, \quad e^{-x} = \lim_{\kappa \to \infty} \left(1 + \frac{1}{\kappa} \cdot x\right)^{-\kappa}, \tag{1.6}$$

and its inverse, the natural logarithm,

$$\ln x = \lim_{\kappa \to \infty} \left[\kappa \cdot \left(1 - x^{-\frac{1}{\kappa}}\right)\right]. \tag{1.7}$$

Note that the kappa index can be switched to a q-index defined by $\kappa = (q-1)^{-1}$ (Section 1.6), where the limit of $\kappa \to +\infty$ corresponds to $q \to 1^{+}$. Then the previous relations can be given by

$$e \equiv \lim_{q \to 1} q^{\frac{1}{q-1}}, \quad e^{\pm x} = \lim_{q \to 1} [1 \pm (1-q)x]^{\frac{1}{1-q}}, \quad \ln x \equiv \lim_{q \to 1} \left(\frac{1 - x^{1-q}}{q-1}\right). \tag{1.8}$$

The two basic formalisms of statistical mechanics are the entropy and the canonical distribution (under the constraints of the canonical ensemble). In the classical Boltzmann–Gibbs (BG) statistical mechanics, the canonical distribution is given by the exponential distribution of energy, while the entropy is given by the known probabilistic functional that involves the natural logarithm; thus we need only two functions, the exponential and its inverse, the natural logarithm. If we use the Euler's limit definitions for exponential and logarithm, we have:

- Euler limit for BG (exponential) canonical distribution

$$p(\varepsilon) = \frac{e^{-\frac{\varepsilon}{k_B T}}}{\sum_\varepsilon e^{-\frac{\varepsilon}{k_B T}}} = \lim_{\kappa \to \infty} \left[\frac{\left(1 + \frac{1}{\kappa} \cdot \frac{\varepsilon}{k_B T}\right)^{-\kappa}}{\sum_\varepsilon \left(1 + \frac{1}{\kappa} \cdot \frac{\varepsilon}{k_B T}\right)^{-\kappa}} \right]. \tag{1.9}$$

- Euler limit for BG (logarithmic) entropy

$$S \equiv \sum_p p \cdot \ln(p) = \lim_{\kappa \to \infty} \left[\kappa \cdot \left(1 - \sum_p p^{1+\frac{1}{\kappa}}\right) \right]. \tag{1.10}$$

We observe that the particle systems at the specific stationary state of thermal equilibrium are described by the following distribution and entropic forms:

$$p(\varepsilon; \kappa) = \frac{\left(1 + \frac{1}{\kappa} \cdot \frac{\varepsilon}{k_B T}\right)^{-\kappa}}{\sum_\varepsilon \left(1 + \frac{1}{\kappa} \cdot \frac{\varepsilon}{k_B T}\right)^{-\kappa}}, \quad S([p]; \kappa) = \kappa \cdot \left(1 - \sum_p p^{1+\frac{1}{\kappa}}\right), \tag{1.11}$$

but only for a specific kappa value, that is, $\kappa \to \infty$.

The question that arises is the following: why should the stationary states of particle systems describe only with the infinity kappa value? Or can the finite kappa values describe the stationary states observed in space plasmas (see Fig. 1.4)?

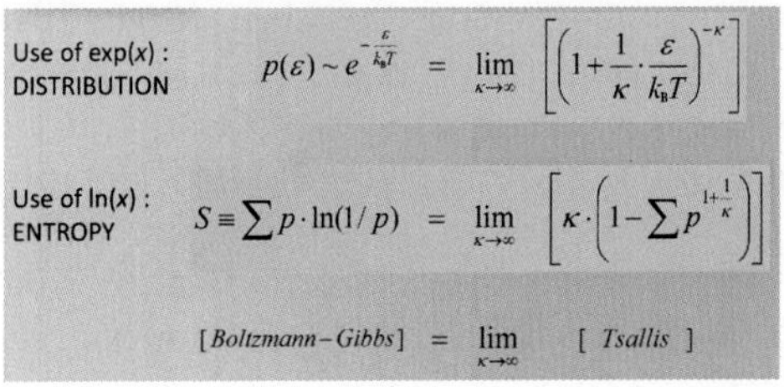

Use of exp(x): DISTRIBUTION	$p(\varepsilon) \sim e^{-\frac{\varepsilon}{k_B T}}$	$= \lim_{\kappa \to \infty}$	$\left[\left(1 + \frac{1}{\kappa} \cdot \frac{\varepsilon}{k_B T}\right)^{-\kappa}\right]$
Use of ln(x): ENTROPY	$S \equiv \sum p \cdot \ln(1/p)$	$= \lim_{\kappa \to \infty}$	$\left[\kappa \cdot \left(1 - \sum p^{1+\frac{1}{\kappa}}\right)\right]$
	[Boltzmann–Gibbs]	$= \lim_{\kappa \to \infty}$	[Tsallis]

FIGURE 1.4
Boltzmann–Gibbs exponential distribution and logarithmic entropy can be written as limit cases of the deformed exponential and deformed logarithm functions, respectively (see Section 1.4.1), which are the distribution and entropy of Tsallis nonextensive statistical mechanics.

1.4 Nonextensive Statistical Mechanics, in Brief!
1.4.1 General Aspects

The classical BG statistical mechanics has stood the test of time for describing classical systems at thermal equilibrium. In contrast, nonextensive statistical mechanics has offered a theoretical basis for describing and analyzing complex systems out of thermal equilibrium (e.g., see Borges et al., 2002; and refs. therein).

Nonextensive statistical mechanics was introduced by Tsallis (1988) as a possible generalization of the classical framework of BG statistical mechanics. The main suggestion was to generalize the formulation of entropy to a mono-parametrical function of the probability distribution. The entropic parameter, the so-called q-index, can attain any value in $(-\infty, +\infty)$, but it is usually restricted to finite intervals, depending on the examined system. In any case, at $q \to 1$, the entropy formulation converges to the old, known BG entropy, and the whole statistical theory recovers to the BG statistical mechanics. The maximization of this new type of entropy under the constraint of internal energy had derived the probability distribution of energy within the framework of canonical ensemble, also called the canonical distribution.

The nonextensive version of the canonical distribution is given by the so-called "q-deformed exponential" (or simply q-exponential) distribution (e.g., Silva et al., 1998; Yamano, 2002), that is, a distribution governed by the flexible parameter q that recovers the BG exponential distribution at the limit of $q \to 1$. The q-exponential was considered an anomalous distribution (Abe, 2002) from the point of view of the standard BG exponential distribution. Surprisingly, however, q-exponential distributions are observed frequently in nature, and it is now widely accepted that these distributions constitute a suitable generalization of the BG exponential distribution, rather than describing a kind of rare or anomalous behavior. Applications of the q-exponential distribution can be found in a wide variety of topics: *sociology—sociometry*: e.g., Internet (Abe and Suzuki, 2003); citation networks of scientific papers (Tsallis and de Albuquerque, 2000); urban agglomeration (Malacarne et al., 2001); linguistics (Montemurro, 2001); economy (Borland, 2002); *biology*: biochemistry (Andricioaei and Straub, 1996; Tsallis et al., 1999); ecology (Livadiotis et al., 2015, 2016); *statistics* (Habeck et al., 2005; Livadiotis, 2008, 2012, 2016b); *physics*: e.g., nonlinear dynamics (Robledo, 1999; Borges et al., 2002); condensed matter (Hasegawa, 2005); earthquakes (Sotolongo-Costa et al., 2000; Sotolongo-Costa and Posadas, 2004; Silva et al., 2006; Varotsos et al., 2014); turbulence (Beck et al., 2001); physical chemistry (Livadiotis, 2009); and space physics/astrophysics (Tsallis et al., 2003; Du, 2004; Sakagami and Taruya, 2004) (a more extended bibliography of q-deformed exponential distributions can be found in Swinney and Tsallis, 2004; Gell-Mann and Tsallis, 2004; Tsallis, 2009a,b; for a complete bibliography on "nonextensive statistical mechanics and thermodynamics," see http://tsallis.cat.cbpf.br/TEMUCO.pdf).

The connection of kappa distributions with nonextensive statistical mechanics has already been examined by several authors (e.g., Treumann, 1997; Milovanov and Zelenyi, 2000; Leubner, 2002; Livadiotis and McComas, 2009). The empirical kappa distribution and the Tsallis-like Maxwellian distribution of velocities are accidentally of the same form under the transformation of indices: $q = 1 + 1/\kappa$. Livadiotis and McComas (2009) showed that the consistent connection of the theory and formalism of kappa distributions with nonextensive statistical mechanics is based on four fundamental physical notions and concepts: (1) q-deformed functions, (2) escort probability distribution (Beck and Schlogl, 1993); (3) Tsallis entropy (Tsallis, 1988), and (4) physical temperature (Abe, 1999, 2001; Rama, 2000). We briefly review these four concepts as follows.

1.4.2 q-Deformed Functions

Any well-defined generalization of a function $F(x)$ into some other function $F(x;q)$ that is parameterized by the q-index and recovers to $F(x)$ for $q \rightarrow 1$ is practically considered a "q-deformed function." The q-deformed functions that will be used in this book are given as follows.

1.4.2.1 q-Unity

The q-deformed "unity function" (Livadiotis and McComas, 2009) is defined as

$$1_q(x) = [1 + (1 - q) \cdot x]_+ = \left(1 - \frac{1}{\kappa} \cdot x\right)_+. \qquad (1.12)$$

The subscript "+" denotes the Tsallis cut-off condition:

$$x_+^\gamma = \begin{cases} x^\gamma & \text{if } x > 0, \\ 0 & \text{if } x \leq 0, \ \forall \gamma. \end{cases} \qquad (1.13)$$

If the exponent of the distribution is positive, then the condition can be written also as: $x_+ = x$, if $x \geq 0$ and $x_+ = 0$, if $x \leq 0$. In other words, $\exp_q(x)$ becomes zero if its base is non-positive.

1.4.2.2 q-Exponential

The q-exponential function can be defined (either in terms of q or of κ) (Silva et al., 1998; Yamano, 2002) as follows:

$$\exp_q(x) = [1 + (1 - q) \cdot x]_+^{-\frac{1}{q-1}} = \left(1 - \frac{1}{\kappa} \cdot x\right)_+^{-\kappa}. \qquad (1.14a)$$

1.4.2.3 q-Logarithm

The inverse of the q-exponential function is called the q-logarithm function,

$$\ln_q(x) = \frac{x^{1-q} - 1}{1 - q} = \kappa \cdot \left(1 - x^{-\frac{1}{\kappa}}\right). \qquad (1.14b)$$

1.4.2.4 q-Hyperbolic

The q-hyperbolic functions of sine, cosine, and tangent (or cotangent), can be defined using the q-exponential function (Borges, 1998),

$$\sinh_q(x) = \frac{1}{2}\cdot\left[\exp_q(x) - \exp_q(-x)\right], \quad \cosh_q(x) = \frac{1}{2}\cdot\left[\exp_q(x) + \exp_q(-x)\right],$$

$$(1.15a)$$

$$\tanh_q(x) = \left[\coth_q(x)\right]^{-1} = \frac{\exp_q(x) - \exp_q(-x)}{\exp_q(x) + \exp_q(-x)}.$$

$$(1.15b)$$

1.4.2.5 q-Gamma

The q-gamma function $\Gamma_q(x)$ was first introduced by Suyari (2006) and Niven and Suyari (2009). Here we utilize the definition of Livadiotis and McComas (2009, 2011b). This is expressed either in terms of the entropic q-index or the kappa index, $\kappa = 1/(q - 1)$,

$$\Gamma_q(a) \equiv \int_0^\infty \exp_q^q(-\gamma)\gamma^{a-1}\ d\gamma, \quad \text{or}$$

$$(1.16a)$$

$$\Gamma_q(a) \equiv \int_0^\infty [1 - (1-q)\gamma]_+^{\frac{q}{1-q}}\gamma^{a-1}\ d\gamma, \quad \Gamma_\kappa(a) \equiv \int_0^\infty \left(1 + \frac{1}{\kappa}\cdot\gamma\right)_+^{-\kappa-1}\gamma^{a-1}\ d\gamma,$$

$$(1.16b)$$

and it is equal to

$$\Gamma_\kappa(a) = \Gamma(a)\cdot\kappa^{a-1}\cdot\frac{\Gamma(\kappa + 1 - a)}{\Gamma(\kappa)} = \kappa^a\cdot\frac{\Gamma(a)\cdot\Gamma(\kappa + 1 - a)}{\Gamma(\kappa + 1)}$$

$$= \kappa^a\cdot B(a, \kappa + 1 - a).$$

$$(1.16c)$$

1.4.3 Ordinary and Escort Probability Distributions

Consider a system described by a discrete energy spectrum $\{\varepsilon_k\}_{k=1}^W$, which is associated with a discrete probability distribution $\{p_k\}_{k=1}^W$. The formalism of nonextensive statistical mechanics is interwoven with the concept of escort probabilities (Tsallis, 1999; Gell-Mann and Tsallis, 2004; Tsallis, 2009b). The escort probability distribution $\{P_k\}_{k=1}^W$ is constructed from the ordinary probability distribution $\{p_k\}_{k=1}^W$, as $P_k \propto p_k^q$, $\forall k = 1, \ldots, W$, coinciding thus with $\{p_k\}_{k=1}^W$ for $q \to 1$ (Beck and Schlogl, 1993). In fact, there is a duality between the ordinary $\{p_k\}_{k=1}^W$ and escort probabilities $\{P_k\}_{k=1}^W$, such as $P_k = P_k(\{p_k\}_{k=1}^W; q)$ and $p_k = p_k\left(\{P_k\}_{k=1}^W; \frac{1}{q}\right)$, $\forall k = 1, \ldots, W$, expressed by

$$P_k = \frac{p_k^q}{\sum_{k=1}^{W} p_k^q} \Leftrightarrow p_k = \frac{P_k^{\frac{1}{q}}}{\sum_{k=1}^{W} P_k^{\frac{1}{q}}}, \text{ for } \forall k = 1, ..., W. \tag{1.17}$$

Within the framework of Tsallis statistics, the interpretation for the internal energy U is given by the escort expectation value of energy $\langle \varepsilon \rangle_{esc}$, that is,

$$U = \langle \varepsilon \rangle_{esc} = \sum_{k=1}^{W} P_k \varepsilon_k = \frac{\sum_{k=1}^{W} p_k^q \varepsilon_k}{\sum_{k=1}^{W} p_k^q}, \tag{1.18a}$$

where the symbol $\langle \ \rangle_{esc}$ denotes the escort expectation value. Note that the escort expectation value can be thought, either as a new type of expectation functional when expressed in terms of the ordinary distribution or as the standard type of expectation functional when expressed in terms of the escort distribution. Nevertheless, the escort distribution is the one associated with the observed and measured statistical moments. Hereafter, the subscript "*esc*" will be ignored since the expectation value will be taken associated with the escort distributions by default, unless it is stated otherwise. Also, the subscript "*ord*" will indicate that the mean value was derived using the ordinary distribution.

In general, any statistical moment is always derived in association with the escort probability distribution. Then, the mean of a function of energy $f(\varepsilon)$ is given by

$$\langle f \rangle = \sum_{k=1}^{W} P_k f(\varepsilon_k) = \frac{\sum_{k=1}^{W} p_k^q f(\varepsilon_k)}{\sum_{k=1}^{W} p_k^q}. \tag{1.18b}$$

In the continuous description, the ordinary $p(\varepsilon)$ and escort $P(\varepsilon)$ probability distributions are related to each other by the relations

$$P(\varepsilon) = p(\varepsilon)^q \bigg/ \int_0^\infty p(\varepsilon)^q d\varepsilon \Leftrightarrow p(\varepsilon) = P(\varepsilon)^{\frac{1}{q}} \bigg/ \int_0^\infty P(\varepsilon)^{\frac{1}{q}} d\varepsilon, \tag{1.19a}$$

or, considering the density of energy states $g_E(\varepsilon)$,

$$P(\varepsilon) = p(\varepsilon)^q \bigg/ \int_0^\infty p(\varepsilon)^q g_E(\varepsilon) d\varepsilon \Leftrightarrow p(\varepsilon) = P(\varepsilon)^{\frac{1}{q}} \bigg/ \int_0^\infty P(\varepsilon)^{\frac{1}{q}} g_E(\varepsilon) d\varepsilon. \tag{1.19b}$$

Then, the mean energy $U = \langle \varepsilon \rangle$ and the mean function of energy $f(\varepsilon)$ are given by

$$U = \langle \varepsilon \rangle = \frac{\int_0^\infty \varepsilon \cdot P(\varepsilon) g_E(\varepsilon) d\varepsilon}{\int_0^\infty P(\varepsilon) g_E(\varepsilon) d\varepsilon}, \langle f(\varepsilon) \rangle = \frac{\int_0^\infty f(\varepsilon) \cdot P(\varepsilon; q) g_E(\varepsilon) d\varepsilon}{\int_0^\infty P(\varepsilon; q) g_E(\varepsilon) d\varepsilon}. \tag{1.20}$$

(Note that the density of states can be sometimes incorporated in the distribution formula.) Finally, it has to be stressed that the ordinary and escort distributions have differences in their meaning and usage. Both distributions are characteristic of the system; mathematically, the one leads to the other via the duality relation (see Eqs. 1.17, 1.19a, and 1.19b); physically, however, they are fundamentally different, as the ordinary distribution is rather an auxiliary mathematical tool related to the information arrangement and dynamics, while the escort distribution is related to the observed stationary states and measurements (statistical moments).

1.4.4 Tsallis Entropy

Nonextensive statistical mechanics generalizes the classical formulation of BG entropy using the notion of the canonical distribution via the formalism of escort probabilities. For the discrete energy spectrum $\{\varepsilon_k\}_{k=1}^{W}$ and the associated distribution $\{p_k\}_{k=1}^{W}$, the nonextensive entropy is given by

$$S(q) = \frac{1}{q-1} \cdot \left(1 - \sum_{k=1}^{W} p_k^q\right), \text{ or } S(q) = \frac{1}{q-1} \cdot \left[1 - \left(\sum_{k=1}^{W} P_k^{\frac{1}{q}}\right)^{-q}\right], \quad (1.21a)$$

or in terms of the kappa index

$$S(\kappa) = \kappa \cdot \left(1 - \sum_{k=1}^{W} p_k^{\frac{\kappa+1}{\kappa}}\right), \text{ or } S(\kappa) = \kappa \cdot \left[1 - \left(\sum_{k=1}^{W} P_k^{\frac{\kappa}{\kappa+1}}\right)^{-\frac{\kappa+1}{\kappa}}\right], \quad (1.21b)$$

leading to the BG formulation for $\kappa \to \infty$ (or $q = 1$)

$$S(\kappa \to \infty) = -\sum_{k=1}^{W} p_k \ln(p_k). \quad (1.21c)$$

(Note that all the previous entropic formulations are given in units of the Boltzmann's constant k_B).

Nonextensive statistical mechanics involves a consistent and solid mathematical framework. In particular, the entropy fulfills all the required mathematical conditions (Tsallis, 2009b):

1. *Nonnegativity*: the entropy is either positive or zero (when one of the possibilities equals unity).
2. *Maximization at equidistribution*: S is maximized when $p_k = 1/W$, $\forall k = 1, ..., W$, and $\kappa > 0$.
3. *Nonadditivity*: given two sets A and B of independent discrete probability distributions $\{p_i^A\}_{i=1}^{W}$ and $\{p_j^B\}_{j=1}^{W}$, then

$$S(A+B) = S(A) + S(B) - \frac{1}{\kappa} \cdot S(A) \cdot S(B). \quad (1.22)$$

4. *Experimental robustness*: it is also stabilized under fluctuations of the relevant probability distribution (Abe, 2002).

5. *Uniqueness*: it is the only entropic form that fulfills the generalized information measure (Santos, 1997; Abe, 2000).

Nonextensive statistics have been studied in detail. Indeed, notions as such of the q-Fourier transformation and the relevant central theorem (Moyano et al., 2006; Umarov et al., 2008), the consistent q-generalization of basic mathematical functions, such as trigonometric and hyperbolic functions (Borges, 1998), gamma functions (Suyari, 2006; Niven and Suyari, 2009; Livadiotis and McComas, 2009), hyperbolic functions (Borges, 1998), the related Langevin function (Chakrabarti and Chandrashekar, 2010; Livadiotis, 2016a), and numerous other theorems and properties lead to a well-established foundation of the nonextensive statistical mechanics.

Finally, it must be clear that there is not any single or absolute universal functional form for the entropy. On the contrary, the entropy is determined by the detailed particle dynamics of each system. The general characteristic is that the entropy can always be determined as a superposition of Tsallis entropy $S(\kappa)$ on the kappa (or q) index, while the canonical probability distribution as superposition of the kappa (or q-exponential) distributions (see Chapter 4 and the concept of spectral statistics in Tsallis, 2009b).

1.4.5 The Physical Temperature

Once the exact characterization of the statistical mechanics that justifies the kappa distribution is specified, then the exact definition of temperature can also be determined. Having interpreted the kappa distribution as the q-Maxwellian probability distribution, Livadiotis and McComas (2009) showed how the exact definition of temperature is given by the so-called physical temperature T_{phys}.

In classical BG statistical mechanics, temperature is primarily defined on one of three ways: (1) thermodynamics: the thermodynamic definition $T_S \equiv (\partial S / \partial U)^{-1}$ (with S and U standing for the classical BG entropy and internal energy, respectively) (e.g., see Tsallis, 1999; Milovanov and Zelenyi, 2000); (2) kinetic theory: the kinetic temperature T_K, determined by the second statistical moment of the probability distribution of velocities; and (3) statistics: the Lagrangian temperature T_L, defined by the second Lagrangian multiplier that corresponds to the constraint of internal energy in the canonical ensemble. All of these three definitions coincide in equilibrium $T_S = T_K = T_L$, but they are typically different when the system is out of equilibrium.

In nonextensive statistical mechanics, the thermodynamic definition of temperature is generalized to the physical temperature T_{phys}. All the three definitions coincide in equilibrium $T_{phys} = T_K = T_L$. In contrast to the BG formalism, the nonextensive approach maintains the equality between thermodynamic and kinetic definitions, even when the system is relaxing into stationary states out of equilibrium, i.e., $T_{phys} = T_K$. In this way, the kinetic temperature T_K, which is used in the majority of space plasmas analyses, even in the primary works on kappa distributions (Binsack, 1966; Olbert, 1968; Vasyliūnas, 1968), has been provided with a solid statistical foundation within the concept of physical temperature T_{phys} and the formalism of Tsallis statistical mechanics. In contrast to these more recent developments, the use of BG statistical mechanics in space physics is highly problematic since it provides neither a reliable derivation of kappa distribution nor a well-defined temperature out of equilibrium (note that this is also symbolized by T_{phys} or T_q, but we adopt the simpler notation of T since it is indicating to the actual temperature of the system, and it is independent of the kappa or q indices).

1.5 Entropy Maximization
1.5.1 Discrete Description

In order to obtain the stationary probability distribution $\{p_k\}_{k=1}^{W}$ associated with a conservative physical system of energy spectrum $\{\varepsilon_k\}_{k=1}^{W}$, we follow along the famous Gibbs path (1902), where the entropy $S = S(\{p_k\}_{k=1}^{W})$ is maximized (under constraints). The maximum is derived from $\vec{\nabla}_p\, S(\{p_k\}_{k=1}^{W}) = 0$, where

$$\vec{\nabla}_p \equiv \left(\frac{\partial}{\partial p_1}, \frac{\partial}{\partial p_2}, \cdots, \frac{\partial}{\partial p_W} \right), \tag{1.23}$$

is the gradient in the W-dimensional probability space $\vec{p} \equiv (p_1, p_2, ..., p_W) \in \{p_1 \in [0,1]\} \otimes \{p_2 \in [0,1]\} \otimes \cdots \{p_W \in [0,1]\} \subseteq \Re^W$. Hence, we have

$$\frac{\partial}{\partial p_j} S(\{p_k\}_{k=1}^{W}) = 0, \ \forall j = 1, ..., W. \tag{1.24}$$

On the other hand, the "variables" $\{p_k\}_{k=1}^{W}$ are not independent because of the two constraints:

  1. Normalization of the probability distribution,

$$\sum_{k=1}^{W} p_k = 1. \tag{1.25a}$$

  2. Fixed internal energy,

$$\sum_{k=1}^{W} p_k\, \varepsilon_k = U. \tag{1.25b}$$

The Lagrange method involves maximizing the functional form

$$G(\{p_k\}_{k=1}^{W}) = S(\{p_k\}_{k=1}^{W}) + \sum_{m=1}^{M} \lambda_m\, B_m(\{p_k\}_{k=1}^{W}), \tag{1.26}$$

instead of directly maximizing the entropy $S(\{p_k\}_{k=1}^{W})$, in the case we have constraints, such as $B_m(\{p_k\}_{k=1}^{W}) = b_m, \ \forall m = 1, ..., M$. The unknown Lagrange multipliers $\{\lambda_m\}_{m=1}^{M}$ are supposed to be linearly expressed in terms of the constraints known values $\{b_m\}_{m=1}^{M}$. Thus we have

$$G\left(\{p_k\}_{k=1}^{W}\right) = S\left(\{p_k\}_{k=1}^{W}\right) + \lambda_1 \sum_{k=1}^{W} p_k + \lambda_2 \sum_{k=1}^{W} p_k \ \varepsilon_k, \tag{1.27}$$

where the maximization follows by

$$\frac{\partial}{\partial p_j} G\left(\{p_k\}_{k=1}^{W}\right) = 0, \ \forall j = 1, ..., W. \tag{1.28}$$

Within the framework of non-extensive statistical mechanics, the interpretation for the internal energy U is given by the escort expectation value of energy, that is, the mean kinetic energy associated with the escort distribution

$$U = \langle \varepsilon \rangle_{esc} = \sum_{k=1}^{W} P_k \ \varepsilon_k = \frac{\sum_{k=1}^{W} p_k^q \ \varepsilon_k}{\sum_{k=1}^{W} p_k^q}. \tag{1.29}$$

Therefore, we maximize the functional form

$$G\left(\{p_k\}_{k=1}^{W};q\right) = S\left(\{p_k\}_{k=1}^{W};q\right) + \lambda_1 \sum_{k=1}^{W} p_k + \lambda_2 \sum_{k=1}^{W} P_k\left(\{p_{k'}\}_{k'=1}^{W};q\right)\varepsilon_k, \tag{1.30}$$

with the entropy S now given by the discrete form

$$S\left(\{p_k\}_{k=1}^{W};q\right) = \frac{1}{q-1} \cdot \left(1 - \sum_{k=1}^{W} p_k^q\right), \tag{1.31}$$

that recovers the BG entropy $S\left(\{p_k\}_{k=1}^{W};q \to 1\right) = -\sum_{k=1}^{W} p_k \ln(p_k)$, while the constrain (2) is referred to the internal energy U as interpreted by the escort expectation value of the energy spectrum $\{\varepsilon_k\}_{k=1}^{W}$, Eq. (1.29). (The Boltzmann's constant k_B is ignored during the entropy maximization for the sake of simplicity, that is, choosing suitable units so that $k_B = 1$; however, it is restored directly after the maximization.)

The maximization of the functional $G\left(\{p_k\}_{k=1}^{W};q\right)$ gives

$$-\frac{q}{q-1} \cdot p_j^{q-1} + \lambda_1 + \lambda_2 \sum_{k=1}^{W} \varepsilon_k \frac{\partial}{\partial p_j} P_k\left(\{p_{k'}\}_{k'=1}^{W}\right), \text{with}$$

$$\frac{\partial}{\partial p_j} P_k\left(\{p_{k'}\}_{k'=1}^{W}\right) = \frac{\partial}{\partial p_j} \left[\frac{p_k^q}{\sum_{k'=1}^{W} p_{k'}^q}\right] = \frac{q p_j^{q-1}}{\phi_q} \cdot (\delta_{kj} - P_k), \text{ or}$$

$$-\frac{q}{q-1} \cdot p_j^{q-1} + \lambda_1 + \lambda_2 \frac{q p_j^{q-1}}{\phi_q} \cdot (\varepsilon_j - U),$$

that leads to

$$p_j = \frac{1}{\widetilde{Z}_q}[1 - (1-q)\beta(\varepsilon_j - U)]^{\frac{1}{1-q}}, \tag{1.32}$$

where we set $\widetilde{Z}_q \equiv \left(\lambda_1 \frac{q-1}{q}\right)^{-\frac{1}{1-q}}$ and $\beta_L \equiv -\lambda_2$. The quantity $\beta \equiv 1/(k_B T)$, given by

$$\beta \equiv \beta_L/\phi_q \Leftrightarrow T \equiv T_L \cdot \phi_q, \quad \text{with} \quad \phi_q \equiv \sum_{k=1}^{W} p_k^q, \tag{1.33}$$

defines the inverse of the physical temperature T, (Section 1.4.5); hereafter, it will be noted either with subscript "phys," or simply with T, since it represents the actual temperature of the system; $\beta_L \equiv 1/(k_B T_L)$ defines the inverse of the Lagrangian temperature T_L.

The physical temperature T characterizes the actual temperature of the system and thus does not depend on the q-index, while the Lagrangian "temperature" T_L is a thermal parameter that depends on the actual temperature T and the q-index $T_L = T_L(T;q)$. At the classical limit of $q \to 1$, T_L coincides with the system's temperature (in Section 1.8, we will discuss the concept of temperature in more detail).

In addition, the distribution, Eq. (1.30), can be written as

$$p_j = \frac{1}{\widetilde{Z}_q} \exp_q[-\beta(\varepsilon_j - U)], \tag{1.34}$$

where $\exp_q(x) \equiv [1 + (1-q)x]_+^{\frac{1}{1-q}}$ denotes the q-deformed exponential function. Note that the nonextensive version of the partition function has to be settled as $Z_q = \widetilde{Z}_q \exp_q(-\beta U)$ in order for the relations that connect statistical mechanics and thermodynamics to be valid (for details, see Tsallis, 1999, 2009b; Gell-Mann and Tsallis, 2004; Livadiotis, 2016d). The subscript "+" denotes the cut-off condition in Eq. (1.13).

The ordinary and escort probability distributions can be rewritten as

$$p_k \equiv p(\varepsilon_k; q) = \frac{\exp_q\left[-\frac{1}{1_q(\beta U)} \cdot \beta \varepsilon_k\right]}{\sum_{k'=1}^{W} \exp_q\left[-\frac{1}{1_q(\beta U)} \cdot \beta \varepsilon_{k'}\right]},$$

$$P_k \equiv P(\varepsilon_k; q) = \frac{\exp_q^q\left[-\frac{1}{1_q(\beta U)} \cdot \beta \varepsilon_k\right]}{\sum_{k'=1}^{W} \exp_q^q\left[-\frac{1}{1_q(\beta U)} \cdot \beta \varepsilon_{k'}\right]}, \tag{1.35a}$$

where we use the notation of the q-deformed "unity function," Eq. (1.12).

In the case of the continuous energy spectrum, the probability distributions are written as

$$p(\varepsilon; q) = \frac{\exp_q\left[-\dfrac{1}{1_q(\beta U)}\cdot\beta\varepsilon\right]}{\int_0^\infty \exp_q\left[-\dfrac{1}{1_q(\beta U)}\cdot\beta\varepsilon\right]d\varepsilon}, \quad P(\varepsilon; q) = \frac{\exp_q^q\left[-\dfrac{1}{1_q(\beta U)}\cdot\beta\varepsilon\right]}{\int_0^\infty \exp_q^q\left[-\dfrac{1}{1_q(\beta U)}\cdot\beta\varepsilon\right]d\varepsilon},$$

(1.35b)

while, considering the density of energy states $g_E(\varepsilon)$, the distributions become

$$p(\varepsilon; q)g_E(\varepsilon) = \frac{\exp_q\left[-\dfrac{1}{1_q(\beta U)}\cdot\beta\varepsilon\right]g_E(\varepsilon)}{\int_0^\infty \exp_q\left[-\dfrac{1}{1_q(\beta U)}\cdot\beta\varepsilon\right]g_E(\varepsilon)d\varepsilon},$$

(1.35c)

$$P(\varepsilon; q)g_E(\varepsilon) = \frac{\exp_q^q\left[-\dfrac{1}{1_q(\beta U)}\cdot\beta\varepsilon\right]g_E(\varepsilon)}{\int_0^\infty \exp_q^q\left[-\dfrac{1}{1_q(\beta U)}\cdot\beta\varepsilon\right]g_E(\varepsilon)d\varepsilon},$$

with normalization

$$\int_0^\infty p(\varepsilon; q)g_E(\varepsilon)d\varepsilon = 1, \quad \int_0^\infty P(\varepsilon; q)g_E(\varepsilon)d\varepsilon = 1.$$

(1.35d)

Note that if we use the identity,

$$\exp_q\left[-\frac{1}{1_q(\beta U)}\cdot\beta\varepsilon\right] = \left[1+\frac{q-1}{1-(q-1)\beta U}\cdot\beta\varepsilon\right]^{-\frac{1}{q-1}}$$

$$= [1+(1-q)\cdot\beta U]^{\frac{1}{q-1}}\cdot[1-(1-q)\cdot\beta(\varepsilon-U)]^{-\frac{1}{q-1}}, \quad (1.36)$$

we may rewrite the distributions as

$$p(\varepsilon; q)g_E(\varepsilon) \propto \frac{\exp_q[\beta(\varepsilon-U)]}{\exp_q(-\beta U)}\cdot\varepsilon^{a-1}, \quad P(\varepsilon; q)g_E(\varepsilon) \propto \frac{\exp_q^q[\beta(\varepsilon-U)]}{\exp_q^q(-\beta U)}\cdot\varepsilon^{a-1}. \quad (1.37)$$

We utilize the classical case of a power law density of energy states $g_E(\varepsilon) \propto \varepsilon^{a-1}$ with $a = \frac{f}{2}$, where f denotes the degrees of freedom for each of the particles. For 3-D monatomic particles, $f = 3$, $a = \frac{3}{2}$, and thus $g_E(\varepsilon) \sim \varepsilon^{\frac{1}{2}}$. Let's ignore the normalization constant for simplicity; then, we have

$$p(\varepsilon; q)g_E(\varepsilon) \propto \exp_q\left[-\frac{1}{1_q(\beta U)}\cdot\beta\varepsilon\right]\cdot\varepsilon^{a-1}, \quad P(\varepsilon; q)g_E(\varepsilon) \propto \exp_q^q\left[-\frac{1}{1_q(\beta U)}\cdot\beta\varepsilon\right]\cdot\varepsilon^{a-1}.$$

(1.38)

Next we express the physical temperature T, or $\beta = (k_B T)^{-1}$, in terms of the kinetic temperature T_K, which is defined through the internal energy. This is given by the mean energy (or, in the absence of a potential energy, this is given by the mean kinetic energy).

The mean energy $\langle \varepsilon \rangle$ and the mean function of energy $f(\varepsilon)$ are given by

$$\langle \varepsilon \rangle = \frac{\int_0^\infty \varepsilon \cdot P(\varepsilon; q) g_E(\varepsilon) d\varepsilon}{\int_0^\infty P(\varepsilon; q) g_E(\varepsilon) d\varepsilon} = \frac{\int_0^\infty \exp_q^q\left[-\frac{1}{1_q(\beta U)} \cdot \beta \varepsilon \right] \cdot \varepsilon^a d\varepsilon}{\int_0^\infty \exp_q^q\left[-\frac{1}{1_q(\beta U)} \cdot \beta \varepsilon \right] \cdot \varepsilon^{a-1} d\varepsilon}, \quad (1.39a)$$

$$\langle f(\varepsilon) \rangle = \frac{\int_0^\infty f(\varepsilon) \cdot P(\varepsilon; q) g_E(\varepsilon) d\varepsilon}{\int_0^\infty P(\varepsilon; q) g_E(\varepsilon) d\varepsilon} = \frac{\int_0^\infty f(\varepsilon) \cdot \exp_q^q\left[-\frac{1}{1_q(\beta U)} \cdot \beta \varepsilon \right] \cdot \varepsilon^{a-1} d\varepsilon}{\int_0^\infty \exp_q^q\left[-\frac{1}{1_q(\beta U)} \cdot \beta \varepsilon \right] \cdot \varepsilon^{a-1} d\varepsilon},$$

$$(1.39b)$$

(compared with Eq. 1.20). We then calculate the internal energy, that is, the mean energy $U = \langle \varepsilon \rangle$,

$$U = \frac{1_q(\beta U)}{\beta} \cdot \frac{\int_0^\infty \exp_q^q(-x) x^a \, dx}{\int_0^\infty \exp_q^q(-x) x^{a-1} \, dx} = \frac{1_q(\beta U)}{\beta} \cdot \frac{\Gamma_q(a+1)}{\Gamma_q(a)} = \frac{1_q(\beta U)}{\beta} \cdot \frac{a}{1_q(a)}$$

$$\Rightarrow \frac{1_q(a)}{a} = \frac{1_q(\beta U)}{\beta U} \Rightarrow \beta U = a,$$

$$(1.40a)$$

where we utilized the q-deformed gamma function $\Gamma_q(a)$. Therefore, we end up with $U = a k_B T$. $\quad (1.40b)$

Hence, for the escort probability distribution, the parameter T is independent of the q-index, coinciding with the kinetically defined temperature T_K,

$$T_K = T. \quad (1.41)$$

Then we substitute into the ordinary and escort distributions,

$$p(\varepsilon; q) g_E(\varepsilon) \propto \exp_q\left[-\frac{1}{1_q(a)} \cdot \frac{\varepsilon}{k_B T_K} \right] \cdot \varepsilon^{a-1} = \left(1 + \frac{1}{\frac{1}{q-1} - a} \cdot \frac{\varepsilon}{k_B T_K} \right)^{-\frac{1}{q-1}} \cdot \varepsilon^{a-1},$$

$$(1.42a)$$

$$P(\varepsilon;q)g_E(\varepsilon) \propto \exp_q\left[-\frac{1}{1_q(a)}\cdot\frac{\varepsilon}{k_BT_K}\right]^q\cdot\varepsilon^{a-1} = \left(1+\frac{1}{\frac{1}{q-1}-a}\cdot\frac{\varepsilon}{k_BT_K}\right)^{-\frac{1}{q-1}-1}\cdot\varepsilon^{a-1}.$$

$$(1.42b)$$

Namely, the distributions can be expressed in terms of the kinetic temperature T_K and the index $\left(\frac{1}{q-1}\right)$. We can easily verify the relation between ordinary and escort distributions:

$$P(\varepsilon;q) \propto \exp_q\left[-\frac{1}{1_q(a)}\cdot\frac{\varepsilon}{k_BT_K}\right]^q = \left(1+\frac{1}{\frac{1}{q-1}-a}\cdot\frac{\varepsilon}{k_BT_K}\right)^{-\frac{1}{q-1}-1}$$

$$= \left[\left(1+\frac{1}{\frac{1}{q-1}-a}\cdot\frac{\varepsilon}{k_BT_K}\right)^{-\frac{1}{q-1}}\right]^q \propto p(\varepsilon;q)^q.$$

$$(1.43)$$

The escort distribution of particle energy $P(\varepsilon)$ generalizes the classical Boltzmannian exponential distribution of energy, while the escort distribution of particle velocity $P(\vec{u})$ generalizes the Maxwell distribution of velocity,

$$P(\varepsilon;q) \propto \left(1+\frac{1}{\frac{1}{q-1}-\frac{f}{2}}\cdot\frac{\varepsilon}{k_BT}\right)^{-\frac{1}{q-1}-1}, \quad P(\vec{u};q) \propto \left[1+\frac{1}{\frac{1}{q-1}-\frac{f}{2}}\cdot\frac{\varepsilon_K(\vec{u})}{k_BT}\right]^{-\frac{1}{q-1}-1},$$

$$(1.44a)$$

where we set $a = \frac{f}{2}$ and kept the simpler notation of the physical temperature T; $\varepsilon_K(\vec{u}) = \frac{1}{2}m(\vec{u}-\vec{u}_b)^2$ is the particle kinetic energy; $P(\varepsilon)$ and $P(\vec{u})$ are usually called q-exponential distributions of energy and of velocity; sometimes they may be also called q-Boltzmannian and q-Maxwellian distributions. Using the notion of kappa index $\kappa = 1/(q-1)$, the q-exponential distributions can be transformed to their equivalent, the kappa distribution, written in terms of energy or velocities as follows:

$$P(\varepsilon;\kappa) \propto \left(1+\frac{1}{\kappa-\frac{f}{2}}\cdot\frac{\varepsilon}{k_BT}\right)^{-\kappa-1}, \quad P(\vec{u};\kappa) \propto \left[1+\frac{1}{\kappa-\frac{f}{2}}\cdot\frac{\varepsilon_K(\vec{u})}{k_BT}\right]^{-\kappa-1}. \quad (1.44b)$$

1.5.2 Continuous Description

The entropy is a functional of the ordinary probability distribution in the velocity space $p(\vec{u})$,

$$S[p(\vec{u})] = \frac{1 - \phi_q[p(\vec{u})]}{q - 1}, \tag{1.45a}$$

where the argument ϕ_q is the following probability functional,

$$\phi_q[p(\vec{u})] \equiv \int_{-\infty}^{\infty} [p(\vec{u}) \cdot \sigma_u^f]^q \frac{du_1 \, du_2 \dots du_f}{\sigma_u^f}, \tag{1.45b}$$

σ_u is the smallest speed scale parameter characteristic of the system (Chapter 2) so that the quantity

$$d\Omega \equiv \frac{du_1 \, du_2 \dots du_f}{\sigma_u^f}, \tag{1.46}$$

gives the number of microstates in the f-dimensional phase space. (Note that the probability distribution $p(\vec{u})$ scales as σ_u^{-f}, or $p(\vec{u}) \cdot \sigma_u^f$ is dimensionless.)

In the canonical ensemble, the entropy is maximized under the constraint of normalization $\mu[p(\vec{u})] = 1$, where

$$\mu[p(\vec{u})] \equiv \int_{-\infty}^{\infty} \left[p(\vec{u}) \cdot \sigma_u^f \right] \frac{du_1 \, du_2 \dots du_f}{\sigma_u^f}, \tag{1.47}$$

and the constraint of the mean energy $E[p(\vec{u})] = 0$, where

$$E[p(\vec{u})] \equiv \frac{1}{\phi_q} \cdot \int_{-\infty}^{\infty} [p(\vec{u}) \cdot \sigma_u^f]^q \cdot [\varepsilon_K(\vec{u}) - U] \frac{du_1 \, du_2 \dots du_f}{\sigma_u^f}, \tag{1.48}$$

where we use again the particle kinetic energy $\varepsilon_K(\vec{u}) = \frac{1}{2}m(\vec{u} - \vec{u}_b)^2$ and the particle mean kinetic energy $\langle \varepsilon_K \rangle = \frac{1}{2}m\langle(\vec{u} - \vec{u}_b)^2\rangle$, that is, the internal energy U in the absence of a potential energy. Given the two constraints, the entropy is maximized using the Lagrange method, that is, by using the two Lagrange multipliers λ_1 and λ_2 to maximize the functional $G[p(\vec{u})] \equiv S[p(\vec{u})] + \lambda_1 \cdot \mu[p(\vec{u})] + \lambda_2 \cdot E[p(\vec{u})]$, i.e.,

$$\delta S = -\frac{q}{q-1} \cdot \int_{-\infty}^{\infty} [p(\vec{u}) \cdot \sigma_u^f]^{q-1} \cdot [\delta p(\vec{u}) \cdot \sigma_u^f] \frac{du_1 \, du_2 \dots du_f}{\sigma_u^f}, \tag{1.49}$$

$$\delta\mu \cong \int_{-\infty}^{\infty} [\delta p(\vec{u}) \cdot \sigma_u^f] \frac{du_1 \, du_2 \dots du_f}{\sigma_u^f}, \tag{1.50}$$

$$\delta E \cong \frac{q}{\phi_q} \cdot \int_{-\infty}^{\infty} [p(\vec{u}) \cdot \sigma_u^f]^{q-1} [\varepsilon(\vec{u}) - U] \cdot [\delta p(\vec{u}) \cdot \sigma_u^f] \frac{du_1 du_2 \dots du_f}{\sigma_u^f}, \text{ hence,} \tag{1.51}$$

$$\delta G = \delta S + \lambda_1 \delta \mu + \lambda_2 \delta E = \int_{-\infty}^{\infty} I(\vec{u}) \cdot \left[\delta p(\vec{u}) \cdot \sigma_u^f \right] \frac{du_1 du_2 \ldots du_f}{\sigma_u^f} = 0. \quad (1.52)$$

In order for the integral, Eq. (1.52), to be zero for every fluctuating distribution $p(\vec{u})$, must $I(\vec{u}) = 0$, where

$$I(\vec{u}) \equiv \lambda_1 - \frac{q}{q-1} \cdot \left[p(\vec{u}) \cdot \sigma_u^f \right]^{q-1} \cdot \left[1 + (1-q) \cdot \lambda_2 \cdot \frac{\varepsilon_K(\vec{u}) - U}{\phi_q} \right] \quad (1.53)$$

The maximization leads to the canonical probability distribution of velocities,

$$p(\vec{u}) \sim \left[1 - (1-q) \cdot (-\lambda_2) \cdot \frac{\varepsilon_K(\vec{u}) - U}{\phi_q} \right]^{\frac{1}{1-q}}. \quad (1.54)$$

The second Lagrangian multiplier λ_2 represents the negative inverse of the Lagrangian temperature T_L, i.e., $\lambda_2 \equiv -\beta_L$ with $\beta_L \equiv 1/(k_B T_L)$, while it is related to the actual temperature T of the system, i.e., $\beta = \beta_L/\phi_q$ and $\beta = 1/(k_B T)$. Hence, we observe that the argument ϕ_q has a double role; while it is necessary for determining the entropy, Eq. (1.45b), it also connects the Lagrangian temperature to the actual temperature T,

$$T_L = T/\phi_q. \quad (1.55)$$

The mean kinetic energy $U = \langle \varepsilon \rangle$ is given by

$$\langle \varepsilon \rangle = \frac{1}{2} m \cdot \left\langle (\vec{u} - \vec{u}_b)^2 \right\rangle = \frac{f}{2} k_B T \text{ or } \left\langle (\vec{u} - \vec{u}_b)^2 \right\rangle = \frac{f}{2} \cdot \theta^2, \quad (1.56)$$

where $\theta = \sqrt{2k_B T/m}$ is the thermal speed; thus the ordinary $p(\vec{u})$ and escort $P(\vec{u})$ distribution functions are written as

$$p(\vec{u}) \sim \left[1 + (q-1) \cdot \frac{\varepsilon_K(\vec{u}) - U}{k_B T} \right]^{\frac{1}{1-q}}, \quad P(\vec{u}) \sim \left[1 + (q-1) \cdot \frac{\varepsilon_K(\vec{u}) - U}{k_B T} \right]^{\frac{q}{1-q}}. \quad (1.57)$$

The escort distribution $P(\vec{u})$ is expressed by the q-Maxwellian distribution, the generalization of the classical Maxwellian distribution. Using the notion of kappa index $\kappa = 1/(q-1)$, the q-Maxwellian distribution is transformed to its equivalent, the kappa distribution of velocities, as shown in Eq. (1.44b).

1.6 Connection of Kappa Distributions With Nonextensive Statistical Mechanics

1.6.1 Derivation

We have already seen that the q-Maxwellian distribution, given by the escort distribution $P(\vec{u})$, is transformed to its equivalent, the kappa distribution of velocities, under the transformation of the kappa and q indices $\kappa = 1/(q-1)$ see Eqs. (1.44a) and (1.44b),

$$
P(\overrightarrow{u};q) \propto \left[1 + \frac{1}{\frac{1}{q-1} - \frac{f}{2}} \cdot \frac{\varepsilon_K(\overrightarrow{u})}{k_BT}\right]^{-\frac{1}{q-1}-1} , \quad P(\overrightarrow{u};\kappa) \propto \left[1 + \frac{1}{\kappa - \frac{f}{2}} \cdot \frac{\varepsilon_K(\overrightarrow{u})}{k_BT}\right]^{-\kappa-1} .
$$

$$(1.58)$$

One may think that we should also examine the ordinary distribution. Would it be wrong to consider the ordinary, instead of the escort, distribution? We recall that the constraint of the mean energy (internal energy) in Eqs. (1.29) and (1.48) is given with respect to the escort distribution, and the same holds for any statistical moment (Section 1.4.3).

However, if the statistical moments were carried out by the ordinary probability distribution, then the mean of a function of energy $f(\varepsilon)$ should be given by

$$
\langle f(\varepsilon)\rangle = \frac{\int_0^\infty f(\varepsilon)\cdot p(\varepsilon;q)g_E(\varepsilon)d\varepsilon}{\int_0^\infty p(\varepsilon;q)g_E(\varepsilon)d\varepsilon} = \frac{\int_0^\infty f(\varepsilon)\cdot \exp_q\left[-\frac{1}{1_q(\beta U)}\cdot\beta\varepsilon\right]\cdot\varepsilon^{a-1}d\varepsilon}{\int_0^\infty \exp_q\left[-\frac{1}{1_q(\beta U)}\cdot\beta\varepsilon\right]\cdot\varepsilon^{a-1}d\varepsilon},
$$

$$(1.59a)$$

considering again the classical energy density of states $g_E(\varepsilon) \propto \varepsilon^{a-1}$. Then the mean energy should be

$$
U_{ord} = \langle\varepsilon\rangle = \frac{\int_0^\infty \varepsilon\cdot p(\varepsilon;q)g_E(\varepsilon)d\varepsilon}{\int_0^\infty p(\varepsilon;q)g_E(\varepsilon)d\varepsilon} = \frac{\int_0^\infty \exp_q^q\left[-\frac{1}{1_q(\beta U)}\cdot\beta\varepsilon\right]\cdot\varepsilon^a d\varepsilon}{\int_0^\infty \exp_q^q\left[-\frac{1}{1_q(\beta U)}\cdot\beta\varepsilon\right]\cdot\varepsilon^{a-1}d\varepsilon},
$$

$$(1.59b)$$

(we used again the subscript "*ord* " to indicate that association with the ordinary distribution) we note that the mean energy U was involved in the expressions of the ordinary and escort distribution, but it was derived using the escort distribution (e.g., Eq. 1.39a). Then Eq. (1.59b) is written as

$$
U_{ord} = \frac{\int_0^\infty \left[1 + \frac{q-1}{1-(q-1)\beta U}\cdot\beta\varepsilon\right]^{-\frac{1}{q-1}}\cdot\varepsilon^a d\varepsilon}{\int_0^\infty \left[1 + \frac{q-1}{1-(q-1)\beta U}\cdot\beta\varepsilon\right]^{-\frac{1}{q-1}}\cdot\varepsilon^{a-1}d\varepsilon}.
$$

$$(1.60)$$

We calculate the involved integral I as follows:

$$I = \int_0^\infty \left(1 + \frac{\frac{1}{q-1}}{\frac{1}{q-1} - \beta U} \cdot \beta\varepsilon\right)^{-\frac{1}{q-1}} \cdot e^f \, d\varepsilon = \beta^{-f-1}$$

$$\times \int_0^\infty \left[1 + \frac{1}{\frac{1}{q-1} - 1} \cdot \left(\frac{\frac{1}{q-1} - 1}{\frac{1}{q-1} - \beta U}\right) \cdot x\right]^{-\left(\frac{1}{q-1} - 1\right) - 1} \cdot x^f \, dx$$

$$= \beta^{-f-1} \cdot \left(\frac{\frac{1}{q-1} - 1}{\frac{1}{q-1} - \beta U}\right)^{-f-1} \cdot \int_0^\infty \left[1 + \frac{1}{\frac{1}{q-1} - 1} \cdot y\right]^{-\left(\frac{1}{q-1} - 1\right) - 1} \cdot y^f \, dy$$

$$= \beta^{-f-1} \cdot \left(\frac{\frac{1}{q-1} - 1}{\frac{1}{q-1} - \beta U}\right)^{-f-1} \cdot \Gamma_{\frac{1}{q-1} - 1}(f + 1)$$

(1.61)

where we have used the notion of the q-gamma distribution (Section 1.4.2),

$$\Gamma_{\frac{1}{q-1} - 1}(f + 1) = \Gamma(f + 1) \cdot \left(\frac{1}{q-1} - 1\right)^f \cdot \frac{\Gamma\left(\frac{1}{q-1} - f - 1\right)}{\Gamma\left(\frac{1}{q-1} - 1\right)}.$$

(1.62)

Hence, we have

$$I = \beta^{-f-1} \left(\frac{1}{q-1} - \beta U\right)^{f+1} \cdot \frac{\Gamma(f + 1) \cdot \Gamma\left(\frac{1}{q-1} - f - 1\right)}{\Gamma\left(\frac{1}{q-1}\right)}, \text{ and}$$

(1.63)

$$\frac{I(a)}{I(a-1)} = \beta^{-1}\left(\frac{1}{q-1} - \beta U\right) \cdot \frac{\Gamma(a+1) \cdot \Gamma\left(\frac{1}{q-1} - a - 1\right)}{\Gamma(a) \cdot \Gamma\left(\frac{1}{q-1} - a\right)}$$

$$= \frac{a}{\frac{1}{q-1} - a - 1} \cdot \frac{\frac{1}{q-1} - \beta U}{\beta}, \tag{1.64}$$

leading to

$$\beta U_{ord} = a \cdot \frac{\frac{1}{q-1} - \beta U}{\frac{1}{q-1} - a - 1}, \quad \text{or}, \quad \beta \cdot (U_{ord} - U) \cdot \frac{\frac{1}{q-1} - a - 1}{\frac{1}{q-1} - 1} + \beta U$$

$$= a \cdot \frac{\frac{1}{q-1}}{\frac{1}{q-1} - 1}. \tag{1.65}$$

The system has to be characterized by the same internal energy, independently of the probability distribution that is being considered, namely,

$$U_{ord} = U \Rightarrow \beta U = a \cdot \frac{\frac{1}{q-1}}{\frac{1}{q-1} - 1}. \tag{1.66}$$

However, how can this relation be true, if the mean kinetic energy in Eq. (1.40a) was found to be $\beta U = a$? The answer is that the (inverse) temperature β is different for the two cases, namely, when the escort or the ordinary distribution is used. Indeed, given the kinetic definition of temperature T_K, we obtain:

- Ordinary distribution:

$$U = a \cdot k_B T \cdot \frac{\frac{1}{q-1}}{\frac{1}{q-1} - 1} \equiv a \cdot k_B T_K \Rightarrow T = T_K \cdot \frac{\frac{1}{q-1} - 1}{\frac{1}{q-1}}. \tag{1.67a}$$

- Escort distribution:

$$U = a \cdot k_B T \equiv a \cdot k_B T_K \Rightarrow T = T_K. \tag{1.67b}$$

Then, we substitute T_K into the ordinary and escorts distribution,

$$p(\varepsilon; q) \propto \left[1 + \frac{(q-1)(2-q)}{1-(q-1)(a+1)} \cdot \frac{\varepsilon}{k_B T}\right]^{-\frac{1}{q-1}}$$

$$= \left[1 + \frac{1}{\left(\frac{1}{q-1}-1\right)-a} \cdot \frac{\varepsilon}{k_B T_K}\right]^{-\left(\frac{1}{q-1}-1\right)-1}, \qquad (1.68a)$$

$$P(\varepsilon; q) \propto \left[1 + \frac{q-1}{1-a(q-1)} \cdot \frac{\varepsilon}{k_B T}\right]^{-\frac{q}{q-1}} = \left[1 + \frac{1}{\left(\frac{1}{q-1}\right)-a} \cdot \frac{\varepsilon}{k_B T_K}\right]^{-\left(\frac{1}{q-1}\right)-1}. $$

$$(1.68b)$$

Namely, the ordinary distribution can be expressed in terms of the kinetic temperature T_K, which coincides with the actual temperature of the system (as we will see in Section 1.8.2) and the index $\left(\frac{1}{q-1}-1\right)$.

The two distributions, ordinary and escort, become equivalent under the transformations:

$$\left(\frac{1}{q-1}\right)_{ord} - 1 = \left(\frac{1}{q-1}\right)_{esc}, \quad T_{ord} \cdot \left(\frac{1}{q-1}\right)_{ord} = T_{esc} \cdot \left(\frac{1}{q-1}\right)_{esc}. \qquad (1.69)$$

We recall, however, that in their meaning and usage, the ordinary distribution is *not* equivalent to the escort distribution (Section 1.4.3). Both distributions are characteristic of the system, but the ordinary distribution is related to the information arrangement and dynamics, while the escort distribution is related to the observed stationary states and measurements (statistical moments); mathematically, the one leads to the other through the duality relation in which the mean energy is taken via the escort distribution. Namely,

- Duality relation:

$$p\big(\varepsilon; U = \langle\varepsilon\rangle_{esc}\big) \underset{1/q}{\overset{q}{\rightleftarrows}} P\big(\varepsilon; U = \langle\varepsilon\rangle_{esc}\big). \qquad (1.70a)$$

- Equivalence:

$$p\big(\varepsilon; U = \langle\varepsilon\rangle_{ord}\big) = P\big(\varepsilon; U = \langle\varepsilon\rangle_{esc}\big). \qquad (1.70b)$$

The escort distribution in Eq. (1.70a) and both the equivalent distributions in Eq. (1.70b) represent the kappa distribution, according to the transformations,

$$
\left(\frac{1}{q-1}\right)_{ord} - 1 = \left(\frac{1}{q-1}\right)_{esc} \equiv \kappa \ ,
$$

$$
T_{ord} \cdot \left(\frac{1}{q-1}\right)_{ord} = T_{esc} \cdot \left(\frac{1}{q-1}\right)_{esc} \equiv T_K \cdot \kappa,
$$

(1.71)

leading to

$$
P(\varepsilon;\kappa) \propto \left(1 + \frac{1}{\kappa - a} \cdot \frac{\varepsilon}{k_B T_K}\right)^{-\kappa-1} , \ a = \frac{1}{2}f,
$$

(1.72)

with f noting the degrees of freedom.

1.6.2 Historical Comments

We have seen that the empirically introduced kappa distribution and the Tsallis q-exponential distribution have accidentally the same functional forms under a transformation of the kappa and q indices.

The origin of the kappa distribution in nonextensive statistical mechanics has already been examined by several authors (e.g., see Milovanov and Zelenyi, 2000; Leubner, 2002). Historically, however, the first evidence of a connection between the theory of kappa distributions and the framework of nonextensive statistical mechanics was given by Treumann (1997) in the appendix of his investigation regarding the superdiffusion near the magnetopause. (It is particularly impressive how that author captured the statistical origin of kappa distributions just nine years after the introduction of nonextensive statistical mechanics.)

A detailed derivation of the kappa distribution through nonextensive statistical mechanics was referred to in the analysis of Milovanov and Zelenyi (2000) and Leubner (2002). They showed that the kappa distribution constitutes the canonical probability distribution by maximizing the Tsallis entropy under the constraints of canonical ensemble. However, in regards to the second constraint, the one of internal energy, they did not consider the escort expectation value. In this case, after maximizing the Tsallis entropy S, one finds

$$
p(\varepsilon;q;T_*) \propto \left[1 - (q-1) \cdot \frac{\varepsilon}{k_B T_*}\right]_+^{\frac{1}{q-1}} ,
$$

(1.73a)

instead of the already derived (Section 1.5) ordinary probability distribution

$$
p(\varepsilon;q;T) \propto \left[1 - (1-q) \cdot \frac{\varepsilon - U}{k_B T}\right]_+^{\frac{1}{1-q}} ,
$$

(1.73b)

In fact, Eq. (1.73a) was the first derived type of q-exponential distribution (Tsallis, 1988). However, this result was facing some fundamental problems: in particular, the canonical distribution of energy ε was not invariant for an arbitrary selection of the ground level of energy, and the internal energy U was not extensive as it should be for uncorrelated distributions. These inconsistencies, as well as other problems, were corrected by the later work of Tsallis et al. (1998) that considers the escort–expectation values.

Although nonextensive statistical mechanics was correctly revised, the old version had kept being used in theories and applications to space physics. A decade later, Livadiotis and McComas (2009) showed that the connection between kappa distributions and the modern nonextensive statistics holds again, under the transformation $\kappa = 1/(q - 1)$. As it was shown, the first kind of kappa distribution is connected with the earliest and incomplete version of nonextensive statistical mechanics, while the second kind is related to the modern and consistent version of nonextensive statistical mechanics.

In particular, Milovanov and Zelenyi (2000) and Leubner (2002, 2004a,b) used Eq. (1.73a) to find the first kind of kappa distribution with exponent $-\kappa$, using the transformation

$$\kappa \equiv \frac{1}{1-q}, \text{ or}, q \equiv 1 - \frac{1}{\kappa}. \tag{1.74a}$$

The second kind of kappa distribution, with exponent $-(\kappa + 1)$, is derived using the transformation

$$\kappa \equiv \frac{1}{q-1}, \text{ or }, q \equiv 1 + \frac{1}{\kappa}. \tag{1.74b}$$

It must be noted that Leubner (2002) had mentioned the second kind of kappa distribution, though described it as a "reduced" form of the first kind and without the temperature interpretation.

Under the previously transformations, Eqs. (1.74a) and (1.74b), the canonical distributions, Eqs. (1.73a) and (1.73b) become, respectively, the following two kinds of kappa distributions:

$$P^{(1)}(\varepsilon) \sim \left[1 + \frac{1}{\kappa_*} \cdot \frac{\varepsilon}{k_B T_*}\right]^{-\kappa_*}, P^{(2)}(\varepsilon) \sim \left[1 + \frac{1}{\kappa} \cdot \frac{\varepsilon - U}{k_B T}\right]^{-\kappa-1}. \tag{1.75}$$

During the last half century, these were the main two empirical models of kappa distributions used to describe space plasma particle populations.

The mean kinetic energy depends, by definition, only on the temperature T (see Section 1.8.2). The first kind has an exponent $-\kappa$, while the second kind has an exponent $-\kappa-1$. The two formulations become equivalent under the transformation given by (1) $\kappa_* = \kappa + 1$, and (2) $T_* = T \cdot \left(\kappa - \frac{1}{2}f\right) \Big/ (\kappa + 1)$, including the (kinetic and potential) degrees of freedom $f = 2U/(k_\mathrm{B}T)$; their substantial difference is in the notion of temperature and the connection to statistical mechanics.

Most important, the second kind of kappa distribution is the only formulation that includes correctly the notion of temperature. The Maxwell's kinetic definition of temperature acquires a physical meaning as soon as it coincides with the Clausius's thermodynamic definition. This is the actual temperature of a system. It is unique and independent of the kappa index; thus it characterizes the system at any kappa index, even for infinity (thermal equilibrium). Unfortunately, it is sometimes confused with statements like "temperature for a kappa distribution," "temperature at thermal equilibrium," or "Maxwellian temperature" (see Section 1.8.3). Livadiotis and McComas (2009, 2010a) completed the connection of kappa distributions with the modern nonextensive statistical mechanics using the concept of physical temperature that unifies the thermodynamic and kinetic definitions of temperature (see also: Livadiotis, 2014a, 2015a, 2016d) (Fig. 1.5). Thereafter, the connection of kappa distributions with nonextensive statistical mechanics revealed the robust physical meaning of temperature, thermal pressure, and other thermodynamic parameters. The properties of kappa distributions and the proven tools of nonextensive statistical mechanics have been successfully applied to a variety of space plasmas throughout the heliosphere. These analyses led to the determination of the thermodynamic variables of these plasmas, as well as the understanding of the underpinning plasma processes.

Having connected the empirical kappa distributions with a solid background of statistical mechanics, it is straightforward now to generalize the distribution from the description of velocities to the description of the Hamiltonian. More specifically, we are able to handle the existence of a nonnegligible potential energy using the mathematical formulations and the physical theory of kappa distributions.

$$P(\overrightarrow{r}, \overrightarrow{u}; \kappa, T) \sim \left[1 + \frac{1}{\kappa} \cdot \frac{H(\overrightarrow{r}, \overrightarrow{u}) - \langle H \rangle}{k_\mathrm{B}T}\right]^{-\kappa - 1}, \tag{1.76}$$

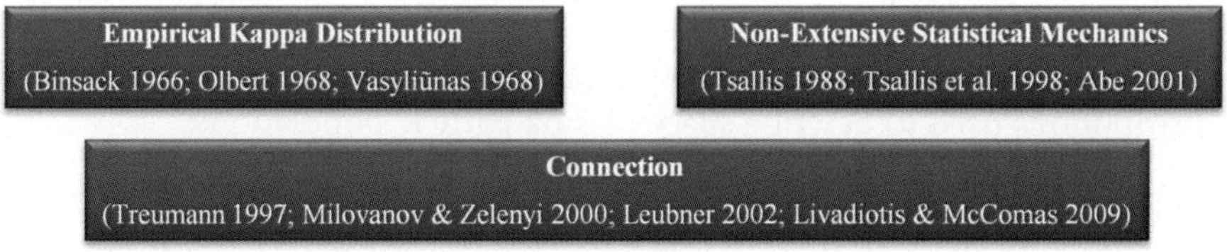

FIGURE 1.5
Connection of empirical kappa distributions with nonextensive statistical mechanics.

Empirical Kappa Distribution	Non-Extensive Statistical Mechanics
(Binsack 1966; Olbert 1968; Vasyliūnas 1968)	(Tsallis 1988; Tsallis et al. 1998; Abe 2001)

Connection
(Treumann 1997; Milovanov & Zelenyi 2000; Leubner 2002; Livadiotis & McComas 2009)

where $\langle H \rangle$ is the ensemble phase space average of the Hamiltonian function $H(\vec{r}, \vec{u})$.

Kappa distributions were mostly involved for the statistical description of the velocity or kinetic energy of particles, but not of the potential energy. This is expected since the empirical kappa distribution originated as a mathematical function generalizing the Maxwellian distribution. However, using the non-extensive statistical mechanics, it was possible to develop the kappa distribution of a Hamiltonian function in the presence of a nonzero potential energy.

In general, the connection of the kappa distribution with the solid background of nonextensive statistical mechanics put the foundations for an analysis on the behavior of systems described by kappa distributions. Among many others, some examples are the following:

1. the kappa distribution of the Hamiltonian function in the presence of a potential energy; this constitutes a phase space distribution, given by the sum of the (velocity-dependent) kinetic energy and the (position-dependent) potential energy (from which we can construct the two marginal distributions of velocities and of positions);
2. the concept of temperature (and thermal pressure) was shown to be consistently defined for systems, characterized by stationary states out of thermal equilibrium and of the classical BG statistical framework, typically described by kappa distributions and combinations thereof;
3. the physical meaning of the kappa (or q-) index that labels the kappa (or the q-exponential) distributions, and its connection with the correlation between particle energies; and
4. the derivation of the Sackur–Tetrode entropic formulation for particle systems described by kappa distributions.

1.7 Structure of the Kappa Distribution

Here we are describing the base and exponent of kappa distributions. For the sake of simplicity, we are ignoring the normalization constant and using the proportionality operator " $\propto$ " instead of the equal " $=$."

1.7.1 The Base of the Kappa Distribution

It is important to understand that not only kappa distributions, but any canonical distribution that describes the energy of a particle system must be expressed in terms of the difference between that energy and its mean value; in other words, the zero level of energy must be set at its mean value, and the distribution is just a function (F) of that difference, normalized by the temperature,

$$P(\varepsilon) \propto F\left(\frac{\varepsilon - \langle \varepsilon \rangle}{k_{\mathrm{B}} T}\right). \tag{1.77}$$

The classical BG energy distribution can be trivially transformed to this form because of the factorization property of the exponential function,

$$P(\varepsilon) \propto \exp\left(-\frac{\varepsilon - \langle\varepsilon\rangle}{k_B T}\right) = \frac{\exp\left(-\frac{\varepsilon}{k_B T}\right)}{\exp\left(-\frac{\langle\varepsilon\rangle}{k_B T}\right)}. \tag{1.78}$$

The kappa distribution can follow the above scheme, but with a slight change on the kappa index in the base $\kappa \to \kappa - \frac{\langle\varepsilon\rangle}{k_B T}$,

$$\left(1 + \frac{1}{\kappa}\cdot\frac{\varepsilon - \langle\varepsilon\rangle}{k_B T}\right)^{-\kappa-1} = \frac{\left(1 + \frac{1}{\kappa - \frac{\langle\varepsilon\rangle}{k_B T}}\cdot\frac{\varepsilon}{k_B T}\right)^{-\kappa-1}}{\left(1 + \frac{1}{\kappa - \frac{\langle\varepsilon\rangle}{k_B T}}\cdot\frac{\langle\varepsilon\rangle}{k_B T}\right)^{-\kappa-1}}. \tag{1.79}$$

As we will see in Section 1.9.2, this auxiliary index equals the invariant kappa index $\kappa \to \kappa_0 \equiv \kappa - \frac{\langle\varepsilon\rangle}{k_B T}$,

$$\left(1 + \frac{1}{\kappa}\cdot\frac{\varepsilon - \langle\varepsilon\rangle}{k_B T}\right)^{-\kappa-1} \propto \left(1 + \frac{1}{\kappa_0}\cdot\frac{\varepsilon}{k_B T}\right)^{-\kappa_0-1-\frac{1}{2}f}, \tag{1.80}$$

(where $\frac{1}{2}f \equiv \frac{\langle\varepsilon\rangle}{k_B T}$). The same behavior characterizes the escort q-exponential distribution since this is equivalent to the kappa distribution.

$$\left[1 - (1 - q)\cdot\frac{\varepsilon - \langle\varepsilon\rangle}{k_B T}\right]^{\frac{q}{1-q}} = \frac{\left[1 - (1 - q)\cdot\frac{\varepsilon}{k_B T + (1 - q)\langle\varepsilon\rangle}\right]^{\frac{q}{1-q}}}{\left[1 - (1 - q)\cdot\frac{\langle\varepsilon\rangle}{k_B T + (1 - q)\langle\varepsilon\rangle}\right]^{\frac{q}{1-q}}}. \tag{1.81}$$

Nevertheless, the statistical mechanics community is more used to an auxiliary "temperature" instead of an auxiliary q-index, i.e., $k_B T \to k_B T_* \equiv k_B T + (1 - q)\langle\varepsilon\rangle$,

$$\left[1 - (1 - q)\cdot\frac{\varepsilon - \langle\varepsilon\rangle}{k_B T}\right]^{\frac{q}{1-q}} \propto \left[1 - (1 - q)\cdot\frac{\varepsilon}{k_B T_*}\right]^{\frac{q}{1-q}}. \tag{1.82}$$

The difference between those two manipulations is that the auxiliary temperature, which is not a temperature but a temperature-like parameter, is computational-wise useful, but still physically meaningless. On the other hand,

$$\left(1 + \frac{1}{\kappa} \cdot \frac{\varepsilon - \langle \varepsilon \rangle}{k_B T}\right)^{-\kappa - 1} \propto \left(1 + \frac{1}{\kappa_0} \cdot \frac{\varepsilon}{k_B T}\right)^{-\kappa_0 - 1 - \frac{1}{2}f} \text{, where } \kappa \to \kappa_0 \equiv \kappa - \tfrac{1}{2}f$$

$$\Updownarrow$$

$$\kappa = 1/(q-1)$$

$$\Downarrow$$

$$\left[1 - (1-q) \cdot \frac{\varepsilon - \langle \varepsilon \rangle}{k_B T}\right]^{\frac{q}{1-q}} \propto \left[1 - (1-q) \cdot \frac{\varepsilon}{k_B T_*}\right]^{\frac{f}{1-q}} \text{, where } T \to T_* \equiv T + (1-q)\langle \varepsilon \rangle / k_B$$

FIGURE 1.6
The manipulations of auxiliary kappa index (upper) and auxiliary temperature (lower). The auxiliary kappa index κ_0 is invariant of the degrees of freedom and thus is more fundamental than the standard kappa index κ. On the other hand, the auxiliary temperature is a useful temperature-like parameter, but without any physical meaning.

the auxiliary kappa index κ_0 is more fundamental than the standard kappa index κ, as it is invariant of the degrees of freedom (a property that is shown in Section 1.9.2) (Fig. 1.6).

1.7.2 The Exponent of the Kappa Distribution

We have seen how the base of kappa distributions can change forms according to

$$P(\varepsilon) \propto \left(1 + \frac{1}{\kappa} \cdot \frac{\varepsilon - \langle \varepsilon \rangle}{k_B T}\right)^{-\kappa - 1} \propto \left(1 + \frac{1}{\kappa - \frac{\langle \varepsilon \rangle}{k_B T}} \cdot \frac{\varepsilon}{k_B T}\right)^{-\kappa - 1} = \left(1 + \frac{1}{\kappa - \frac{f}{2}} \cdot \frac{\varepsilon}{k_B T}\right)^{-\kappa - 1}. \tag{1.83}$$

A frequently asked intriguing question is about the exponent, whether this can be different. The kappa distribution form can be thought to be generalized according to

$$P(\varepsilon) \propto \left(1 + \frac{1}{\kappa - \frac{f}{2}} \cdot \frac{\varepsilon}{k_B T}\right)^{-\kappa - 1} \overset{?}{\to} P(\varepsilon) \propto \left(1 + \frac{1}{\kappa - \frac{f}{2}} \cdot \frac{\varepsilon}{k_B T}\right)^{-\kappa - a}. \tag{1.84}$$

As a mathematical model, it is useful. However, it is not as useful as a physical model because in order the temperature to attain a physical meaning needs to be connected with the mean kinetic energy. The latter is derived as follows:

$$\langle \varepsilon \rangle = \frac{\displaystyle\int_0^\infty \varepsilon \cdot P(\varepsilon) g_E(\varepsilon) d\varepsilon}{\displaystyle\int_0^\infty \varepsilon \cdot P(\varepsilon) g_E(\varepsilon) d\varepsilon} = \frac{\displaystyle\int_0^\infty \left(1 + \frac{1}{\kappa - \frac{f}{2}} \cdot \frac{\varepsilon}{k_B T}\right)^{-\kappa - a} \cdot \varepsilon^{\frac{f}{2}} d\varepsilon}{\displaystyle\int_0^\infty \left(1 + \frac{1}{\kappa - \frac{f}{2}} \cdot \frac{\varepsilon}{k_B T}\right)^{-\kappa - a} \cdot \varepsilon^{\frac{f}{2} - 1} d\varepsilon}, \tag{1.85}$$

where we used the classical energy states $g_E(\varepsilon) \propto \varepsilon^{\frac{f}{2}-1}$. We find

$$\langle \varepsilon \rangle \equiv \frac{f}{2}k_B T_K = \frac{\kappa - \frac{f}{2}}{\kappa + a - \frac{f}{2}} \cdot \frac{f}{2}k_B T \quad \text{or} \quad T_K = \frac{\kappa - \frac{f}{2}}{\kappa + a - \frac{f}{2}} \cdot T. \tag{1.86}$$

Thus the parameter T does *not* coincide with the actual temperature T_K. Substituting the parameter T in terms of the temperature T_K, we obtain

$$P(\varepsilon) \propto \left[1 + \frac{1}{(\kappa + a - 1) - \frac{f}{2}} \cdot \frac{\varepsilon}{k_B T_K} \right]^{-(\kappa + a - 1) - 1}. \tag{1.87}$$

We now compare the two kappa distributions below; *do they have any difference?*

$$P(\varepsilon) \propto \left(1 + \frac{1}{\kappa - \frac{f}{2}} \cdot \frac{\varepsilon}{k_B T_K} \right)^{-\kappa - 1} \quad , \quad \frac{f}{2} < \kappa \leq \infty,$$

$$\updownarrow \tag{1.88}$$

$$P(\varepsilon) \propto \left[1 + \frac{1}{(\kappa + a - 1) - \frac{f}{2}} \cdot \frac{\varepsilon}{k_B T_K} \right]^{-(\kappa + a - 1) - 1} \quad , \quad \frac{f}{2} < \kappa + a - 1 \leq \infty.$$

We just need to rename the kappa index $\kappa + a - 1 \to \kappa$ so that

$$P(\varepsilon) \propto \left(1 + \frac{1}{\kappa - \frac{f}{2}} \cdot \frac{\varepsilon}{k_B T} \right)^{-\kappa - a}$$

$$= \left[1 + \frac{1}{(\kappa + a - 1) - \frac{f}{2}} \cdot \frac{\varepsilon}{k_B T_K} \right]^{-(\kappa + a - 1) - 1} \quad \xrightarrow{\kappa + a - 1 \to \kappa} \quad \left(1 + \frac{1}{\kappa - \frac{f}{2}} \cdot \frac{\varepsilon}{k_B T_K} \right)^{-\kappa - 1}. \tag{1.89}$$

Any mathematical model suitable for describing the distribution function of particle energy acquires a physically meaningful notion of temperature within the framework of thermodynamics. In this way, the abstract kappa distribution model with exponent $-\kappa-a$, once described correctly with the concept of temperature, ends up again on the old, known escort's exponent of $-\kappa-1$.

1.8 The Concept of Temperature

1.8.1 The Definition of Temperature Out of Equilibrium and the Concept of Physical Temperature

The definition of temperature is controversial whenever the classical weak interactions scenario of the classical BG statistical mechanics is no longer valid. Over the last three decades, several different concepts of "nonequilibrium temperatures" have been examined. For a classical gas in equilibrium, the definition of the kinetic temperature T_K emerges from the equipartition of the internal energy $U \equiv \frac{f}{2} k_B T_K$, where f are the total degrees of freedom. This definition is often adopted for systems in nonequilibrium (e.g., Chapman and Cowling, 1990; Fort et al., 1999).

Alternatively, a completely different definition of nonequilibrium temperature is possible in terms of a nonequilibrium entropy by analogy to an equilibrium expression (e.g., Luzzi et al., 1997), namely,

$$T_S \equiv \left(\frac{\partial S}{\partial U}\right)^{-1}, \tag{1.90a}$$

which constitutes the thermodynamic definition of temperature. However, it was claimed that away from equilibrium, the phase space probability distribution $P(\vec{r}, \vec{u})$ is typically fractal (Hoover, 2001; Hoover et al., 2004; Hoover and Hoover, 2008). Hence, the BG entropy, that is, the phase space average logarithm of $P(\vec{r}, \vec{u})$, diverges. Thus the existence of a nonequilibrium temperature, which is based on Eq. (1.90a) and the BG entropic formulation, appears to be doubtful.

In 1988, Tsallis introduced the generalized formulation of entropy S (Chapter 2). Eventually, it was shown that Tsallis entropy can successfully describe complex systems that are either out of equilibrium or characterized by the presence of long-range interactions (Tsallis, 1999). This was achieved under specific values of the entropic index q (different from $q = 1$, which recovers the BG entropy). Still, when the Tsallis generalized entropy is utilized in Eq. (1.90a), it is dubious that a thermometer immersed in a complex system will measure the quantity $(\partial S/\partial U)^{-1}$. In contrast to this quantity, the definition of the temperature given in Eq. (1.63) is generalized to the physical temperature T (Abe, 1999, 2001; Rama, 2000),

$$T_{phys} = \left(\frac{\partial S}{\partial U}\right)^{-1} \cdot [1 + (1-q) \cdot S/k_B]. \tag{1.90b}$$

The physical temperature T_{phys} is obtained in accordance with the generalized zero-th law (Abe, 2001; Abe et al., 2001; Wang et al., 2002; Toral, 2003) and it serves the role of the kinetic definition of temperature within the framework of nonextensive statistical mechanics (Livadiotis and McComas, 2009). Therefore, all the advantages of a kinetically defined temperature, in contrast to other configurational definitions (Hoover and Hoover, 2008), can be ascribed to T_{phys}. Several inconsistencies concerning the kinetic definition in regards to the zero-th law of thermodynamics (Baranyai, 2000a, 2000b) are fully recovered since the origin of T_{phys} establishes the generalized zero-th law.

1.8.2 Mean Kinetic Energy Defines Temperature

The temperature is a well-understood concept in physics when it characterizes particle systems at thermal equilibrium. This is based on the equivalence of the two fundamental definitions of temperature, that is, (1) the kinetic definition of Maxwell (1866), and (2) the thermodynamic definition of Clausius (1862). The same equivalence is shown to exist for particle systems out of thermal equilibrium when residing in stationary states described by kappa distributions (Livadiotis and McComas, 2009, 2010a, 2011b, 2013a). The mean particle kinetic energy is given by $\langle \varepsilon_K \rangle \equiv \frac{1}{2} d_K k_B T$. Then the kinetic definition is given by $T_K \equiv \langle \varepsilon_K \rangle \cdot 2/(d_K k_B)$. On the other hand, the thermodynamic definition of Clausius is given by the notion of "physical temperature"

$T_{phys} \equiv (\partial S/\partial U)^{-1} \cdot \left[1 - \frac{1}{\kappa} \cdot S \Big/ k_B \right]$, which is consistent with the zero-th law of

thermodynamics (Abe, 2001) (S is the entropy and U is the internal mean energy of the system).

In the history of the kappa distribution, the temperature of plasmas has been calculated from the mean kinetic energy. However, this method was true only for systems at thermal equilibrium, i.e., described by Maxwellian distributions. The connection with nonextensive statistical mechanics (Livadiotis and McComas, 2009, 2010a) confirmed the equivalence of the two definitions of temperature. Therefore, the mean kinetic energy (or the variance of the velocities) describes truly the temperature. It is only now that we are allowed to calculate the outcome of mixing two nonequilibrium plasmas, similarly to the mixing of classical gases, following simple calorimetry rules. For example, we may ask, what would be the resulting temperature of mixing equal volumes of some solar corona plasma and of some laboratory plasma (Fig. 1.7).

Finally, we mention the concept of thermal pressure for stationary states out of thermal equilibrium, which is tightly related to the temperature. Given the system's volume V or number density $n = 1/V$, the kinetic definition of the pressure is given by $P \equiv (2/f) \cdot n \cdot U$, which coincides with its thermodynamic

definition, that is, $P \equiv (\partial S/\partial V) \cdot T \Big/ \left(1 - \frac{1}{\kappa} \cdot S/k_B \right)$ (Abe, 2001). Similar to the case

of temperature definitions, the equivalence of the kinetic and thermodynamic definitions of pressure also provides a common and unique definition of the nonequilibrium pressure of systems. The classical ideal gas state equation

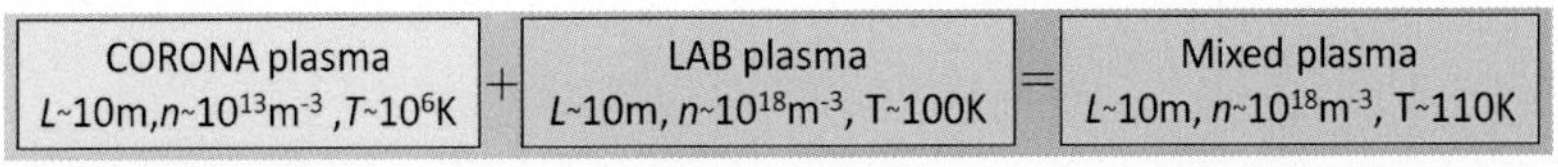

CORONA plasma $L\sim10m, n\sim10^{13}m^{-3}, T\sim10^6K$	+	LAB plasma $L\sim10m, n\sim10^{18}m^{-3}, T\sim100K$	=	Mixed plasma $L\sim10m, n\sim10^{18}m^{-3}, T\sim110K$

FIGURE 1.7
The connection of kappa distributions with nonextensive statistical mechanics led to the equivalence of the two main definitions of temperatures, the kinetic and the thermodynamic ones. Hence, the temperature is a consistent and physically meaningful quantity, not only for systems at thermal equilibrium described by Maxwellian distributions, but even for systems out of thermal equilibrium described by kappa distributions. One of the important applications of this equivalence is that space and laboratory plasmas can be understood under the same unique thermodynamics. For example, we do know how to calculate the temperature of mixing equal volumes of some type of space plasma (e.g., outer corona proton plasma) with some type of laboratory proton plasma. The temperature of solar corona can be calculated using the kinetic definition, while the temperature of a laboratory plasma can be measured using simply a thermometer (that makes use of the thermodynamic definition). Thus the resulting temperature follows from simple calorimetry rules. Taken from Livadiotis (2015a).

$P = n \cdot k_B \cdot T$ still holds for any dimensionality (f) and any nonequilibrium stationary state (κ).

1.8.3 Misleading Considerations About Temperature

During the history of kappa distributions, there were several confusing and incorrect misuses of the notion of temperature. Some of these misinterpretations of temperature can be found even today. We mention the most common ones as follows.

1.8.3.1 The Misleading Temperature-Like Parameter T_κ
1.8.3.1.1 Misinterpretation

The standard formulation of the particle velocity or kinetic energy is

$$P(\vec{u}) \propto \left[1 + \frac{1}{\kappa - \frac{d_K}{2}} \cdot \frac{(\vec{u} - \vec{u}_b)^2}{\theta^2} \right]^{-\kappa-1}, \quad P(\varepsilon_K) \propto \left(1 + \frac{1}{\kappa - \frac{d_K}{2}} \cdot \frac{\varepsilon_K}{k_B T} \right)^{-\kappa-1} \varepsilon_K^{\frac{1}{2}d_K-1},$$

$$(1.91)$$

where $\theta \equiv \sqrt{2k_B T/m}$ is the thermal speed. However, there are cases where a different formulation of a kappa distribution is used, causing confusion about the notion temperature,

$$P(\vec{u}) \propto \left[1 + \frac{1}{\kappa} \cdot \frac{(\vec{u} - \vec{u}_b)^2}{\theta_\kappa^2} \right]^{-\kappa-1}, \quad P(\varepsilon_K) \propto \left(1 + \frac{1}{\kappa} \cdot \frac{\varepsilon_K}{k_B T_\kappa} \right)^{-\kappa-1} \varepsilon_K^{\frac{1}{2}d_K-1}, \qquad (1.92)$$

where $\theta_\kappa \equiv \sqrt{2k_B T_\kappa/m}$ and the quantity T_κ are incorrectly interpreted as the actual temperature of the particle system. Inevitably, the mean kinetic energy is found to be dependent on the kappa index. Indeed, comparing Eqs. (1.91) and (1.92), we derive the relation $\left(\kappa - \frac{d_K}{2} \right) \cdot k_B T = \kappa \cdot k_B T_\kappa$; then, substituting T in $\langle \varepsilon_K \rangle \equiv \frac{1}{2} d_K k_B T$, we obtain

$$\langle \varepsilon_K \rangle \;=\; \frac{1}{2} d_K \cdot k_B T_\kappa \cdot \frac{\kappa}{\kappa - \dfrac{d_K}{2}}.$$

(1.93)

1.8.3.1.2 Resolution

The misinterpretation can be avoided by considering that the mean kinetic energy, by definition, does *not* depend on any thermodynamic variable other than the actual temperature $T \equiv \langle \varepsilon_K \rangle \cdot 2/(d_K \cdot k_B)$,

$$\langle \varepsilon_K \rangle \;=\; \frac{1}{2} d_K \cdot k_B T_\kappa \cdot \frac{\kappa}{\kappa - \dfrac{d_K}{2}} \equiv \frac{1}{2} d_K \cdot k_B T.$$

(1.94)

Hence, T_κ (and not T) does depend on the kappa index, i.e., $T_\kappa = T_\kappa(T, \kappa)$, $T \neq T(T_\kappa, \kappa)$.

1.8.3.2 The Misleading Dependence of the Temperature on the Kappa Index

1.8.3.2.1 Misinterpretation

The temperature is not a fundamental independent parameter, but instead, it is considered dependent on the kappa index.

1.8.3.2.2 Resolution

Since their introduction (Binsack, 1966; Olbert, 1968; Vasyliunas, 1968), kappa distributions involved a temperature that was calculated via the mean kinetic energy. The truth is that this method was known to be meaningful only for systems at thermal equilibrium, i.e., particles described by Maxwellian distributions. However, the examined space plasma system was residing in a stationary state out of thermal equilibrium, described by the non-Maxwellian behavior of kappa distributions. Nevertheless, that was acceptable back then since there was no basic theory to consider regarding the physical concept of temperature, neither was a statistical theory known to be used as a background to these distributions. The connection of kappa distributions with nonextensive statistical mechanics had been developed years later, also confirming the equivalence of the two definitions of temperature for kappa distributions (Livadiotis and McComas, 2009).

The mean kinetic energy (or the variance of the velocities) describes truly the temperature, as it is equivalent to the thermodynamic definition of temperature. The thermodynamic definition is consistent with the zero-th law of thermodynamics (Abe, 2001), "two bodies that are in equilibrium with a third are in equilibrium with each other," to cover stationary states out of thermal equilibrium: "two bodies that are in equilibrium or the same non-equilibrium stationary state with a third, are in the same stationary state with each other" (Livadiotis and McComas, 2010a). In particular, starting from first thermodynamic principles with the only assumption the stationarity of the system (that is, any stationary state other than the classical thermal equilibrium), Abe (2001) was able to show the remarkable result that the most generalized theoretical framework that is aligned to the concept of temperature and the zero-th law of thermodynamics is the kappa distributions and the Tsallis nonextensive statistical mechanics. The notion of temperature used is independent of any other parameter, such as the kappa index or the entropy. In this way, the zero-th law of thermodynamic is valid and equivalent for all the observers that belong to states of different kappa indices (Livadiotis and McComas, 2009, 2010a).

Having understood the independence of temperature on other thermodynamic parameters, we now need to interpret correctly the observations of correlations between temperature and the kappa index. Let's start with a similar example, the case of plasmas at thermal equilibrium. The plasma particles are described by Maxwell distributions, dependent only on the temperature T and the (number) density n. What is the relation of T and n? Trivially, these are two independent parameters. Nevertheless, on the streamlines of a plasma flow, we frequently observe various relations between n and T, typically described by a polytropic relation $T = A \cdot n^{a-1}$, with a noting the polytropic index. However, each streamline has different constant A, while the polytropic index a may characterize the specific streamline or a package of different streamlines. More often, we say that a certain value of the polytropic index characterizes the whole plasma a_*, but what we mean in reality is that there is a distribution of different polytropic indices with the most frequent one given by a_* (Chapter 5; Nicolaou et al., 2014; Livadiotis and Desai, 2016; Nicolaou and Livadiotis, 2017).

There are other descriptions of generalized polytropic relations, $T = f(n)$ (Livadiotis and McComas, 2012; Livadiotis, 2016c). Generalized polytropic relations may also include the kappa index for space plasmas out of thermal equilibrium, i.e., $T = f(n, \kappa)$. In fact, very often, the dependence on the density is ignored, and published analyses are dealing with the more interesting relation of $\kappa - T$. For example, see the positively correlated kappa index and temperature values in the inner heliosheath (Livadiotis et al., 2011) and in the magnetospheres of Earth (Ogasawara et al., 2015), Jupiter (Collier and Hamilton, 1995), and Saturn (Dialynas et al., 2009).

The observed relations between the temperature and the kappa index by no means imply any universal dependence between those two thermodynamic parameters, in the same way that the observed relations between the temperature and the density do not imply any universal dependence between them. Care must be shown to distinguish the local relations that depend on certain plasma streamlines or the whole plasma, and not misinterpret these with any universal behavior that indicates to the foundations of physical laws.

1.8.3.3 The Misleading "Nonequilibrium Temperature"

1.8.3.3.1 Misinterpretation

Another similar misinterpretation is given by the statement that T_κ characterizes the "nonequilibrium temperature," i.e., the temperature of a nonequilibrium system described by a kappa distribution, in contrast to T that expresses the old, known "Maxwellian" temperature, i.e., the temperature of an equilibrium system described by a Maxwell distribution.

1.8.3.3.2 Resolution

In other words, a new type of temperature is thought to exist (T_κ), different from the Maxwellian temperature (T), while the two types are supposed to be identical at thermal equilibrium ($\kappa \to \infty$). However, there is only one consistent type or characterization of temperature that emerges from the equivalence between the thermodynamic and kinetic definitions of temperature. The thermodynamically defined temperature is what an ideal thermometer would indicate. This is the physical temperature T and not the auxiliary temperature-like parameter T_κ (for more details, see Livadiotis and McComas, 2009, 2010a, 2012, 2013a, Livadiotis, 2009, 2014a, 2015a, 2016d).

1.8.3.4 The Misleading "Equilibrium Temperature"
1.8.3.4.1 Misinterpretation

There is also the impression that T is the temperature that would characterize the system if this was at thermal equilibrium.

1.8.3.4.2 Resolution

However, the temperature is a thermodynamic variable independent of the kappa index, and thus it is invariant under the system's transitions to different kappa indices. In other words, the same system will have the same temperature for any kappa index, even when this is infinity and the distribution becomes Maxwellian. As it was correctly stated in Hellberg et al., (2009), "...clearly, from the definition of temperature, all distributions with the same mean energy per particle have the same temperature." It is worth noting that only the second statistical moment of velocities is related to the mean kinetic energy and the kinetic definition of temperature, and thus no other statistical moment is independent of the kappa index (see Fig. 1.2 in Livadiotis, 2014a).

1.8.3.5 The Most Frequent Speed
1.8.3.5.1 Misinterpretation

Let the speed distribution for a system of $d_K = 3$ degrees of freedom per particle,

$$P_w(w; \kappa_0, \theta) = \frac{2\kappa^{-\frac{1}{2}d_K}\theta^{-d_K}}{B\left(\dfrac{d_K}{2}, \kappa_0 + 1\right)} \cdot \left(1 + \frac{1}{\kappa_0}\cdot\frac{w^2}{\theta^2}\right)^{-\kappa_0 - 1 - \frac{1}{2}d_K} w^{d_K - 1} \quad \text{with } w \equiv |\vec{u} - \vec{u}_b|,$$

$$(1.95)$$

where we have set $P_w(w) \equiv P(w)g_V(w)$ for the isotropic distribution $P(w) = P(\vec{u})$, with $g_V(w) = B_{d_K} \cdot w^{d_K - 1}$; $B_f = 2\pi^{\frac{1}{2}f}/\Gamma\left(\frac{1}{2}f\right)$ is the surface of the f-dimensional unit sphere. In the case of $d_K = 3$ degrees of freedom per particle, the distribution becomes

$$P_w(w; \kappa_0, \theta) = \frac{2\kappa_0^{-\frac{3}{2}}\theta^{-3}}{B\left(\frac{3}{2}, \kappa_0 + 1\right)} \cdot \left(1 + \frac{1}{\kappa_0} \cdot \frac{w^2}{\theta^2}\right)^{-\kappa_0 - \frac{5}{2}} w^2, \tag{1.96}$$

while in terms of θ_κ, this is written as

$$P_w(w; \kappa, \theta_\kappa) = \frac{2\kappa^{-\frac{3}{2}}\theta_\kappa^{-3}}{B\left(\frac{3}{2}, \kappa - \frac{1}{2}\right)} \cdot \left(1 + \frac{1}{\kappa} \cdot \frac{w^2}{\theta_\kappa^2}\right)^{-\kappa - 1} w^2. \tag{1.97}$$

The speed that maximizes the distribution (the most frequent speed) is given exactly by $w_{max} = \theta_\kappa \equiv \sqrt{2k_B T_\kappa/m}$. This coincidence may be interpreted as an evidence that θ_κ has a fundamental role in characterizing the distribution, e.g., use it instead of the temperature or its equivalent, the thermal speed θ (Scudder and Karimabadi, 2013). According to this idea, the independent parameters are the maximum speed and the kappa index (instead of the temperature and the kappa index),

$$P_w(w; \kappa, w_{max}) = \frac{2\kappa^{-\frac{3}{2}}w_{max}^{-3}}{B\left(\frac{3}{2}, \kappa - \frac{1}{2}\right)} \cdot \left(1 + \frac{1}{\kappa} \cdot \frac{w^2}{w_{max}^2}\right)^{-\kappa - 1} w^2. \tag{1.98}$$

1.8.3.5.2 Resolution

Why should we utilize the most frequent speed and not the most frequent square root of speed or some other power or more general function of it? If we argue that these functions may lack physical meaning, then how about using the kinetic energy: the most frequent kinetic energy would have been equally important but it is *not* equal to T_κ; instead, we obtain $\varepsilon_{K\,max} = T_\kappa \cdot \kappa/(\kappa + 2)$.

Even if there is special theoretical reason to seek for the most frequent speed, its coincidence to θ_κ holds only for the case of particles of three degrees of freedom each. For all the other cases, we have

$$w_{max}(\kappa, \theta_\kappa) = \sqrt{\frac{\kappa}{\kappa - \frac{d_K - 3}{2}}} \cdot \sqrt{1 + \frac{d_K - 3}{2}} \cdot \theta_\kappa, \text{ or,}$$

$$w_{max}(\kappa_0, \theta) = \sqrt{\frac{\kappa_0}{\kappa_0 + \frac{3}{2}}} \cdot \sqrt{\frac{d_K - 1}{2}} \cdot \theta, \tag{1.99}$$

that is, in terms of the d_K-dimensional kappa index or the invariant kappa index $\kappa_0 = \kappa - \frac{1}{2}d_K$ (see Section 1.9.2). Thus, in general, $w_{max} \neq \theta_\kappa$, but is rather a function of $w_{max}(\kappa, \theta_\kappa)$. As we can see, the two following distributions have significant differences,

$$P_w(w; \kappa, \theta_\kappa) = \frac{2\kappa^{-\frac{1}{2}d_K}\theta_\kappa^{-d_K}}{B\left(\frac{d_K}{2}, \kappa + 1 - \frac{d_K}{2}\right)} \cdot \left(1 + \frac{1}{\kappa} \cdot \frac{w^2}{\theta_\kappa^2}\right)^{-\kappa-1} w^{d_K-1}, \tag{1.100a}$$

$$P_w(w; \kappa, w_{\max}) = \frac{2\left(\frac{\kappa - \frac{d_K-3}{2}}{1 + \frac{d_K-3}{2}}\right)^{-\frac{1}{2}d_K} w_{\max}^{-d_K}}{B\left(\frac{d_K}{2}, \kappa - \frac{1}{2} - \frac{d_K-3}{2}\right)} \cdot \left(1 + \frac{1 + \frac{d_K-3}{2}}{\kappa - \frac{d_K-3}{2}} \cdot \frac{w^2}{w_{\max}^2}\right)^{-\kappa-1} w^{d_K-1}. \tag{1.100b}$$

(Before a suggestion regarding a redefinition of the kappa distribution functional form according to Eq. (1.100b), we should examine the case of $d_K = 1$ degrees of freedom per particle; it should be enough for discouraging us.)

Aside from the previous comments, it is again the argument of the absence of a consisting temperature (and thermodynamics); T_κ is not a physically meaningful temperature according to its thermodynamic definition, while T surely is. The kappa distribution is a stationary distribution in which the zero-th law of thermodynamics can apply (Abe, 2001; Livadiotis and McComas, 2010a); therefore, independently of what is the mechanism that is generating it, it will automatically acquire a temperature.

1.8.3.6 The Divergent Temperature at Antiequilibrium
1.8.3.6.1 Misinterpretation

Another mistaken impression is that the temperature can be defined for all the kappa indices, but it diverges at $\kappa \to \frac{3}{2}$ or less (in the case of $d_K = 3$ degrees of freedom per particle), that is, *the antiequilibrium state*. In general d_K dimensionalities, the divergence occurs for $\kappa \to \frac{d_K}{2}$.

1.8.3.6.2 Resolution

Indeed, Eq. (1.94) gives

$$T = T_\kappa \cdot \frac{\kappa}{\kappa - \frac{d_K}{2}}, \tag{1.101a}$$

from which we observe that the temperature could diverge at $\kappa \to \frac{d_K}{2}$ if T_κ was a constant parameter, independent of the kappa index, but it is not. It is the temperature that is independent of the kappa index, and Eq. (1.101a) is actually written as

$$T_\kappa = T \cdot \frac{\kappa - \frac{d_K}{2}}{\kappa}. \tag{1.101b}$$

Thus the only true statement is that the nontemperature parameter T_κ tends to zero as $\kappa \to \frac{d_K}{2}$. The actual temperature T is invariant of the transition among kappa indices, even if this is the antiequilibrium at $\kappa \to \frac{d_K}{2}$.

1.8.3.7 The Thermal Pressure

1.8.3.7.1 Misinterpretation

The misinterpretations of temperature may lead to incorrect or misused formulations of other related thermodynamic variables, such as the thermal pressure.

1.8.3.7.2 Resolution

The thermal pressure is given by the well-known relation of ideal gases $p = n\,k_B T$ (e.g., Abe, 2001; Livadiotis and McComas, 2012), but if T_κ is taken as temperature, then the pressure may be handled as dependent on the kappa index as $p = p(\kappa) = n\,k_B T_\kappa \cdot \kappa / \left(\kappa - \frac{d_K}{2} \right)$ (e.g., Heerikhuisen et al., 2008; Fahr and Siewert, 2013). It must be remembered, though, that $p \neq p(\kappa)$, $T \neq T(\kappa)$, and $T_\kappa = T_\kappa(\kappa)$.

1.8.3.8 The Debye Length

1.8.3.8.1 Misinterpretation

The outstanding work of Treumann (1999b) and Treumann et al. (2004) included derivations of various thermodynamic quantities using kappa distributions within the framework of grand canonical ensemble. Specifically, the mean kinetic energy (for a 3-D particle) was found to be

$$\langle \varepsilon_K \rangle = \frac{3}{2} \cdot \frac{1}{\beta} \cdot \frac{\kappa}{\kappa - \frac{3}{2}} \cdot \left(1 - \frac{\beta\mu}{\kappa} \right), \tag{1.102}$$

where μ denotes here the chemical potential. However, the mean kinetic energy constitutes the kinetic definition of temperature $\langle \varepsilon_K \rangle \equiv \frac{3}{2} k_B T$. Then, by comparing this with Eq. (1.102), we obtain an expression of β in terms of temperature and kappa index,

$$\frac{1}{\beta} \cdot \left(1 - \frac{\beta\mu}{\kappa} \right) \cdot \frac{\kappa}{\kappa - \frac{3}{2}} = k_B T, \text{ or, } \frac{1}{\beta} = \frac{\mu}{\kappa} + \frac{\kappa - \frac{3}{2}}{\kappa} k_B T. \tag{1.103}$$

Note that the argument "β" represents the second Lagrangian coefficient. In the early years of nonextensive statistical mechanics, it was thought to be the inverse of the actual temperature $1/\beta \to \beta^{-1} = k_B T$ (Tsallis, 1988). Later, it was realized that the inverse $\beta_L^{-1} = k_B T_L$ does not coincide with the actual temperature

(Tsallis et al., 1998; Livadiotis and McComas, 2009, 2013a) (it is simply a parameter called Lagrangian temperature T_L that coincides with the actual temperature only at thermal equilibrium, $\kappa \to \infty$) (e.g., see Livadiotis, 2014a).

Comparing $\langle \varepsilon_K \rangle \equiv \frac{1}{2} d_K k_B T$ with Eq. (1.102), we derived the relation between the Lagrangian β and the actual temperature T, Eq. (1.103), which can shed light to the functional form of the Debye length. Treumann et al. (2004) showed that the Debye length is given by

$$\lambda_D^2 = \left(\varepsilon \, e^{-2} n^{-1} \right) \cdot \frac{\kappa}{\kappa - \frac{1}{2}} \cdot \frac{1}{\beta} \cdot \left(1 - \frac{\beta \mu}{\kappa} \right). \tag{1.104}$$

This equation is against all the previously or thereafter published papers where they derive a positive correlation of the Debye length with the kappa index (e.g., Chapters 5 and 8; Bryant, 1996; Rubab and Murtaza, 2006; Gougam and Tribeche, 2011; Livadiotis and McComas, 2014a).

1.8.3.8.2 Resolution

The argument β is not independent of the kappa index. Indeed, substituting $1/\beta$ from Eq. (1.103) into Eq. (1.104), we obtain the actual dependence of λ_D on kappa index,

$$\lambda_D^2 = \left(\varepsilon \, e^{-2} n^{-1} \right) \cdot \frac{\kappa}{\kappa - \frac{1}{2}} \cdot \frac{\kappa - \frac{3}{2}}{\kappa} \cdot k_B T = \left(\varepsilon \, e^{-2} n^{-1} \right) \cdot k_B T \cdot \frac{\kappa - \frac{3}{2}}{\kappa - \frac{1}{2}}, \text{ or}$$

$$\lambda_D^2 = \lambda_{D\infty}^2 \cdot \frac{\kappa - \frac{3}{2}}{\kappa - \frac{1}{2}}, \tag{1.105}$$

where we used the actual temperature in the formulation of the Debye length at equilibrium $\lambda_{D\infty}^2 = \left(\varepsilon \, e^{-2} n^{-1} \right) \cdot k_B T$. Hence, we ended up with the well-known dependence of the Debye length on the kappa index.

1.8.4 Relativity Principle for Statistical Mechanics

The equivalence of the kinetic and thermodynamic definitions of temperature $(T_K = T_{phys})$ implies that the isothermal variations of the kappa index (i.e., with constant temperature T_{phys}) will not vary the mean kinetic energy (because $U = \langle \varepsilon \rangle \equiv \frac{1}{2} f \, k_B T_K = \frac{1}{2} f \, k_B T_{phys}$). However, the kappa index characterizes and arranges the stationary states into the κ-spectrum. Then, the isothermal transitions of the system across stationary states will not vary the mean kinetic energy.

Therefore, the internal energy is independent of the specific kappa index, that is, of the specific stationary state.

The stationary state at thermal equilibrium is neither unique nor has any special properties that can be identified by measuring the mean kinetic energy of the particle system. The same holds for any other stationary state (out of thermal equilibrium) that acquires temperature and entropy and is described by a kappa distribution. The previously mentioned considerations lead to the statement of the relativity principle for statistical mechanics: "*a particle system is identified with the same mean kinetic energy for all the possible stationary states that can attain.*"

1.9 The Concept of the Kappa (or *q*) Index
1.9.1 General Aspects

The physical meaning of the kappa index is interwoven with the statistical correlation between the energy of the particles. It has been shown that a simple relation exists between the correlation (Pearson's correlation coefficient R) and the kappa index $R = \frac{3}{2}/\kappa$ (Chapter 5; Livadiotis and McComas, 2011b) (for $N = 1$, particle and $d = 3$ degrees of freedom per particle). The largest value of the kappa index is infinity, corresponding to the system residing at thermal equilibrium. This is a special stationary state for which the particles are characterized by zero correlation. The whole structure of classical statistical mechanics is based on this ideal property that the energies of any two particles are uncorrelated (i.e., zero covariance). On the other hand, the smallest possible kappa value is $\frac{3}{2}$ and corresponds to the furthest state from thermal equilibrium, a state called antiequilibrium (Livadiotis and McComas, 2013a). As the system approaching this state, more particles lose their energy becoming motionless; for this reason, this state is sometimes mentioned the "q-frozen state" to emphasize that by approaching this state, the system behaves very much like when the temperature decreases approaching the absolute zero (Chapter 10; Livadiotis and McComas, 2010a).

As it was shown (Livadiotis and McComas, 2011b; Livadiotis, 2015c), the kappa index depends on the degrees of freedom with the simple linear relation,

$$\kappa(f) = \kappa_0 + \frac{1}{2}f, \; f = d \cdot N, \tag{1.106}$$

where f are the total kinetic degrees of freedom from all the N particles of d degrees each. The kappa index κ_0, called "invariant kappa index," classifies the different stationary states independently of the dimensionality, the degrees of freedom, or the number of particles.

The kappa index arranges the stationary states ("kappa spectrum") from thermal equilibrium to the furthest state, the antiequilibrium (e.g., Chapter 15; Fisk and Gloeckler, 2006, 2014). The "escape state" (at $\kappa \sim 2.5$) separates the kappa spectrum in two halves, the near- and far-equilibrium regions. This separation is frequently observed in space plasmas (e.g., Chapter 10; Livadiotis and McComas, 2010a,c, 2013d). A measure of the "thermodynamic distance" of a stationary state from thermal equilibrium ("nonextensivity") is defined by

$$M \equiv \alpha/(\kappa_0 + \alpha), \tag{1.107}$$

for any value of α, e.g., $\alpha = 2$ (Livadiotis and McComas, 2010a,b) or $\alpha = 1$ (Livadiotis and McComas, 2011b) (bringing the escape state exactly at the middle of the kappa spectrum).

1.9.2 Dependence of the Kappa Index on the Number of Correlated Particles: Introduction of the Invariant Kappa Index κ_0

The N-particle phase space kappa distribution is given by

$$P(\vec{r}_1, \vec{r}_2, ..., \vec{r}_N; \vec{u}_1, \vec{u}_2, ..., \vec{u}_N)$$
$$\propto \left[1 + \frac{1}{\kappa} \cdot \frac{H(\vec{r}_1, \vec{r}_2, ..., \vec{r}_N; \vec{u}_1, \vec{u}_2, ..., \vec{u}_N) - \langle H \rangle}{k_B T} \right]^{-\kappa-1}, \tag{1.108}$$

where $H(\vec{r}_1, \vec{r}_2, ..., \vec{r}_N; \vec{u}_1, \vec{u}_2, ..., \vec{u}_N)$ is the N-particle Hamiltonian, and $\langle H \rangle$ is the ensemble average of the Hamiltonian over the phase space. If there is no potential energy, the Hamiltonian is simply expressed by the kinetic energy $H(\vec{u}_1, \vec{u}_2, ..., \vec{u}_N) = \sum_{i=1}^{N} \varepsilon_{K\,i} = \frac{1}{2} m \cdot \sum_{i=1}^{N} (\vec{u}_i - \vec{u}_b)^2$, with $\langle H \rangle = \sum_{i=1}^{N} \langle \varepsilon_{K\,i} \rangle = \frac{1}{2} m \cdot \sum_{i=1}^{N} \langle (\vec{u}_i - \vec{u}_b)^2 \rangle = \frac{1}{2} d_K N$, thus

$$P(\vec{u}_1, \vec{u}_2, ..., \vec{u}_N) \propto \left[1 + \frac{1}{\kappa} \cdot \frac{H(\vec{u}_1, \vec{u}_2, ..., \vec{u}_N) - \langle H \rangle}{k_B T} \right]^{-\kappa-1}$$
$$= \left\{ 1 + \frac{1}{\kappa} \cdot \frac{1}{\theta^2} \left[\sum_{i=1}^{N} (\vec{u}_i - \vec{u}_b)^2 - \frac{1}{2} d_K N \right] \right\}^{-\kappa-1}. \tag{1.109}$$

Therefore, the N-particle kappa distribution of velocities is written as (Livadiotis and McComas, 2011b)

$$P(\vec{u}_1, \vec{u}_2, ..., \vec{u}_N) \propto \left[1 + \frac{1}{\kappa - \frac{1}{2} d_K N} \cdot \frac{1}{\theta^2} \sum_{i=1}^{N} (\vec{u}_i - \vec{u}_b)^2 \right]^{-\kappa-1}. \tag{1.110}$$

Consider the N-particle kappa distribution of velocities in Eq. (1.110). If this is integrated over the d-dimensional velocity space of the Nth particle, the result is the $(N-1)$ particle kappa distribution of velocities. If, then, this is integrated over the $(N-1)$th particle velocity, the $(N-2)$ particle kappa distribution will be derived, and by continuing the integration over all $(N-1)$ particle velocities, we end up to the one-particle distribution. Namely,

$$P(\vec{u}_1, \vec{u}_2, ..., \vec{u}_{N-2}, \vec{u}_{N-1}, \vec{u}_N) \sim \left[1 + \frac{1}{\kappa - \frac{1}{2}d_K N} \cdot \frac{1}{\theta^2} \sum_{i=1}^{N} (\vec{u}_i - \vec{u}_b)^2 \right]^{-\kappa-1}$$

$\Downarrow$ integration over $\vec{u}_N$

$$P(\vec{u}_1, \vec{u}_2, ..., \vec{u}_{N-2}, \vec{u}_{N-1}) \sim \left[1 + \frac{1}{\kappa - \frac{1}{2}d_K N} \cdot \frac{1}{\theta^2} \sum_{i=1}^{N-1} (\vec{u}_i - \vec{u}_b)^2 \right]^{-\kappa-1+\frac{1}{2}d_K}$$

$\Downarrow$ integration over $\vec{u}_{N-1}$

$$P(\vec{u}_1, \vec{u}_2, ..., \vec{u}_{N-2}) \sim \left[1 + \frac{1}{\kappa - \frac{1}{2}d_K N} \cdot \frac{1}{\theta^2} \sum_{i=1}^{N-2} (\vec{u}_i - \vec{u}_b)^2 \right]^{-\kappa-1+\frac{1}{2}2d_K}$$

$\Downarrow$ integration over $\vec{u}_{N-2}$

$\vdots$

$\Downarrow$ integration over $\vec{u}_2$

$$P(\vec{u}_1) \sim \left[1 + \frac{1}{\kappa - \frac{1}{2}d_K N} \cdot \frac{1}{\theta^2} (\vec{u}_1 - \vec{u}_b)^2 \right]^{-\kappa-1+\frac{1}{2}d_K(N-1)} .$$

$$(1.111)$$

Nevertheless, the one-particle kappa distribution has a different formulation, that is

$$P(\vec{u}_1) \sim \left[1 + \frac{1}{\kappa - \frac{1}{2}d_K} \cdot \frac{1}{\theta^2} (\vec{u}_1 - \vec{u}_b)^2 \right]^{-\kappa-1} , \qquad (1.112)$$

and thus we wonder how these two different formulations can become one,

$$P(\vec{u}_1) \propto \left[1 + \frac{1}{\kappa - \frac{1}{2}d_K N} \cdot \frac{1}{\theta^2} (\vec{u}_1 - \vec{u}_b)^2 \right]^{-\kappa-1+\frac{1}{2}d_K(N-1)} \quad ?$$

$\Leftrightarrow$

$$P(\vec{u}_1) \propto \left[1 + \frac{1}{\kappa - \frac{1}{2}d_K} \cdot \frac{1}{\theta^2} (\vec{u}_1 - \vec{u}_b)^2 \right]^{-\kappa-1} .$$

The problem can be solved if we realize that the kappa index is *not* an invariant quantity, but instead, it depends on the degrees of freedom $\kappa = \kappa(f)$, so that

$$P(\vec{u}_1) \propto \left[1 + \frac{1}{\kappa(d_K N) - \frac{1}{2}d_K N} \cdot \frac{1}{\theta^2}(\vec{u}_1 - \vec{u}_b)^2 \right]^{-\left[\kappa(d_K N) - \frac{1}{2}d_K N\right] - 1 - \frac{1}{2}d_K}$$

$$\updownarrow \tag{1.113}$$

$$P(\vec{u}_1) \propto \left[1 + \frac{1}{\kappa(d_K) - \frac{1}{2}d_K} \cdot \frac{1}{\theta^2}(\vec{u}_1 - \vec{u}_b)^2 \right]^{-\left[\kappa(d_K) - \frac{1}{2}d_K\right] - 1 - \frac{1}{2}d_K}$$

Hence, the following dependence of the kappa index is shown

$$\kappa(f) - \frac{1}{2}f = \kappa(d_K N) - \frac{1}{2}d_K N = \kappa(d_K) - \frac{1}{2}d_K \equiv \kappa_0. \tag{1.114}$$

The kinetic degrees of freedom for the one-particle's phase space are given by $d_K \equiv 2\langle \varepsilon_K \rangle / (k_B T)$. As mentioned, κ_0 indicates the kappa index that is invariant of the (kinetic) degrees of freedom of the system; in contrast to κ_0, the well-known kappa index κ depends on the particle kinetic degrees of freedom d_K and the number of correlated particles N. Therefore, the quantity κ_0 is a modified kappa index that is independent on the dimensionality. Its range of values is $(0, \infty)$, with $\kappa_0 \to \infty$ corresponding to thermal equilibrium and $\kappa_0 \to 0$ to the furthest state from thermal equilibrium, also called antiequilibrium.

Using this invariant kappa index, the one-particle distribution is written with a unified way,

$$P(\vec{u}_1) \propto \left[1 + \frac{1}{\kappa_0} \cdot \frac{1}{\theta^2}(\vec{u}_1 - \vec{u}_b)^2 \right]^{-\kappa_0 - 1 - \frac{1}{2}d_K}, \tag{1.115}$$

while the N-particle distribution becomes

$$P(\vec{u}_1, \vec{u}_2, ..., \vec{u}_N) \propto \left[1 + \frac{1}{\kappa_0} \cdot \frac{1}{\theta^2} \sum_{i=1}^{N}(\vec{u}_i - \vec{u}_b)^2 \right]^{-\kappa_0 - 1 - \frac{1}{2}d N}. \tag{1.116}$$

Fig. 1.8 summarizes the described paradox and its resolution.

The previous developments led naturally to a new property of kappa distributions: the degeneration of the kappa index. The invariant kappa index κ_0 indicates an "absolute" value of kappa, corresponding to zero dimensionality or degrees of freedom. This does not mean that the system has zero degrees of freedom, the absolute vacuum; it rather means that the system may have any degrees of freedom, kinetic or potential, but our measurement of kappa is free of those degrees.

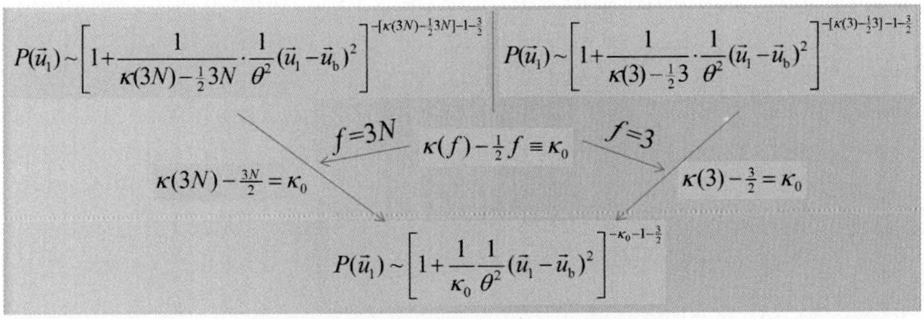

$$P(\vec{u}_1) \sim \left[1 + \frac{1}{\kappa(3N) - \frac{1}{2}3N} \cdot \frac{1}{\theta^2} (\vec{u}_1 - \vec{u}_b)^2 \right]^{-[\kappa(3N) - \frac{1}{2}3N] - 1 - \frac{3}{2}} \qquad P(\vec{u}_1) \sim \left[1 + \frac{1}{\kappa(3) - \frac{1}{2}3} \cdot \frac{1}{\theta^2} (\vec{u}_1 - \vec{u}_b)^2 \right]^{-[\kappa(3) - \frac{1}{2}3] - 1 - \frac{3}{2}}$$

$$f = 3N \qquad \kappa(f) - \tfrac{1}{2} f \equiv \kappa_0 \qquad f = 3$$

$$\kappa(3N) - \frac{3N}{2} = \kappa_0 \qquad\qquad \kappa(3) - \frac{3}{2} = \kappa_0$$

$$P(\vec{u}_1) \sim \left[1 + \frac{1}{\kappa_0} \frac{1}{\theta^2} (\vec{u}_1 - \vec{u}_b)^2 \right]^{-\kappa_0 - 1 - \frac{3}{2}}$$

FIGURE 1.8
The dependence of the kappa index on the degrees of freedom f. Taken from Livadiotis (2015c).

Let us assume that the particle system has zero potential energy. Even then, the kinetic energy is unavoidably nonzero, leading to the already described dependence of the kappa index: $\kappa(f) = \kappa_0 + \frac{1}{2}f$, with $f = N \cdot d_K$. The system could be investigated as a whole in a 3-D velocity space; alternatively, it could be examined in 1-D or 2-D subspaces, for instance, measuring only the parallel or the perpendicular velocities with respect to a certain direction (e.g., of some external magnetic field). In all these cases, the kappa index varies according to $\kappa(d_K) = \kappa_0 + \frac{1}{2}d_K$ when using the one-particle, d_K-D kappa distribution; if it were possible to investigate the system using only single points of the velocity space, then we would have $\kappa(0) = \kappa_0$.

The kinetic degrees of freedom may be constant as soon as the number of correlated particles N and the per-particle degrees of freedom d_K are constant. But the potential degrees of freedom may vary, depending on the interactions acting on the particle system. The process of varying the potential energy has no effect on the *invariant* kappa index κ_0, but it does have on the total kappa index

$$\kappa(d_K) = \kappa_0 + \frac{1}{2}d_K \quad \text{(no potential)}$$
$$\Rightarrow \kappa(d_K, d_\Phi) = \kappa_0 + \frac{1}{2}d_K \pm \frac{1}{2}d_\Phi \quad \text{(with potential } \Phi\text{).} \tag{1.117a}$$

The one-half multiplying the kinetic degrees of freedom is related to the kinetic energy expressed by the *square* velocity; the one-half multiplying the potential degrees is set by definition. More precisely, the potential degrees depend on what type of positional function is the potential; for example, the power law potential $\Phi \propto \pm r^{\pm b}$ leads to

$$\kappa(d_K) = \kappa_0 + \frac{1}{2}d_K \quad \text{(no potential)}$$
$$\Rightarrow \kappa(d_K, d_r) = \kappa_0 + \frac{1}{2}d_K \pm \frac{1}{b}d_r \left(\text{with potential } \Phi \propto \pm r^{\pm b}\right), \tag{1.117b}$$

where d_r is the dimensions of the position vector $\vec{r}$ (compare with Eq. (1.106); see also: Chapter 3; Livadiotis, 2015b,c). The deformation of kappa index, shown in Eq. (1.117b), intrigues us as regards the nature of the invariant kappa index. Is it associated with some hidden internal degrees of freedom, $\kappa_0 = d_0$?

There are several various methods to calculate the kappa index from a set of data following a kappa distribution (e.g., see Chapter 10): (1) fitting the distribution function, where the kappa index is one of the free parameters to be determined (e.g., see Livadiotis and McComas, 2013c; Nicolaou and Livadiotis, 2016); (2) determine the value that maximizes the distribution, which is a function of kappa; and (3) determine the quasilinear slope (on a log–log scale) in the high-energy region (suprathermal tail) that is either exactly or trivially related to the kappa index (e.g., see Table 1.1 in Livadiotis and McComas, 2009; Livadiotis et al., 2011).

Once the kappa index is observed, it is necessary to be reduced to its invariant measure in order to be compatible to other observations that they may use modeling with different dimensionality. Assuming that the per-particle degrees are $d_K = 3$ and that a 3-D distribution was used to determine the kappa index, then $\kappa_0 = \kappa_3 - \frac{3}{2}$; also, κ_3 is the total kappa index. However, if a potential was in effect with potential degrees d_Φ, then $\kappa - \frac{1}{2}d_K \neq constant$, and the once invariant kappa index is "further degenerated," $\kappa_0(\Phi = 0) = \kappa_0 + \frac{1}{2}d_\Phi$.

1.9.3 Formulation of the *N*-Particle Kappa Distributions

In the absence of a potential energy, the particle energy is given simply by their kinetic energy. The Hamiltonian is simply given by

$$H(\vec{u}_1, ..., \vec{u}_N) = \frac{1}{2}m \cdot \sum_{i=1}^{N} (\vec{u}_i - \vec{u}_b)^2, \tag{1.118}$$

and the *N*-particle kappa distribution becomes

$$P(\vec{u}_1, \vec{u}_2, ..., \vec{u}_N) \propto \left[1 + \frac{1}{\kappa - \frac{1}{2}d_K N} \cdot \frac{1}{\theta^2} \sum_{i=1}^{N} (\vec{u}_i - \vec{u}_b)^2 \right]^{-\kappa-1}, \tag{1.119}$$

or, in terms of the invariant kappa index

$$P(\vec{u}_1, ..., \vec{u}_N) \propto \left[1 + \frac{1}{\kappa_0} \cdot \frac{1}{\theta^2} \sum_{i=1}^{N} (\vec{u}_i - \vec{u}_b)^2 \right]^{-\kappa_0-1-\frac{1}{2}d_K N}. \tag{1.120}$$

In the presence of a potential energy $\Phi(\vec{r}_1, ..., \vec{r}_N)$, with Hamiltonian

$$H(\vec{r}_1, \vec{u}_1, ..., \vec{r}_N, \vec{u}_N) = \frac{1}{2}m \cdot \sum_{i=1}^{N} (\vec{u}_i - \vec{u}_b)^2 + \Phi(\vec{r}_1, ..., \vec{r}_N), \tag{1.121}$$

the *N*-particle kappa distribution becomes

$$P(\vec{r}_1, \vec{u}_1, ..., \vec{r}_N, \vec{u}_N) \propto \left[1 + \frac{1}{\kappa - \frac{1}{2}d N} \cdot \frac{H(\vec{r}_1, \vec{u}_1, ..., \vec{r}_N, \vec{u}_N)}{k_B T} \right]^{-\kappa-1}, \tag{1.122}$$

or, in terms of the invariant kappa index

$$P(\vec{r}_1, \vec{u}_1, \ldots, \vec{r}_N, \vec{u}_N) \propto \left[1 + \frac{1}{\kappa_0} \cdot \frac{H(\vec{r}_1, \vec{u}_1, \ldots, \vec{r}_N, \vec{u}_N)}{k_B T}\right]^{-\kappa_0 - 1 - \frac{1}{2}d\,N}$$

$$= \left\{1 + \frac{1}{\kappa_0} \cdot \left[\frac{1}{\theta^2} \sum_{i=1}^{N} (\vec{u}_i - \vec{u}_b)^2 + \frac{1}{k_B T} \cdot \Phi(\vec{r}_1, \ldots, \vec{r}_N)\right]\right\}^{-\kappa_0 - 1 - \frac{1}{2}d\,N},$$

$$(1.123)$$

where we set d as the kinetic and potential degrees of freedom per particle,

$$\frac{1}{2}d \equiv \frac{1}{N} \cdot \frac{\langle H \rangle}{k_B T} = \frac{1}{N} \cdot \frac{\sum_{i=1}^{N} \langle (\vec{u}_i - \vec{u}_b)^2 \rangle}{\theta^2} + \frac{1}{N} \cdot \frac{\langle \Phi \rangle}{k_B T} = \frac{1}{2}d_K + \frac{1}{2}d_\Phi, \qquad (1.124a)$$

$$\frac{1}{2}d_K = \frac{1}{N} \cdot \frac{\sum_{i=1}^{N} \langle (\vec{u}_i - \vec{u}_b)^2 \rangle}{\theta^2} \quad \text{and} \quad \frac{1}{2}d_\Phi \equiv \frac{1}{N} \cdot \frac{\langle \Phi \rangle}{k_B T}. \qquad (1.124b)$$

1.9.4 Negative Kappa Index

The kappa distribution for negative kappa indices is called the "negative kappa distribution" (while the standard kappa distribution with positive kappa index may be referred to as "positive kappa distribution"). We obtain the negative kappa distribution by setting $\kappa_0^+ \to -\kappa_0^-$ in the formalism of the positive kappa distribution; hence, the new kappa index κ_0^- will be still positive, i.e.,

$$P(\vec{u}; \kappa_0^+, T) \propto \left[1 + \frac{1}{\kappa_0^+} \cdot \frac{\frac{1}{2}m(\vec{u} - \vec{u}_b)^2}{k_B T}\right]^{-\kappa_0^+ - 1 - \frac{1}{2}d_K} \xrightarrow{\kappa_0^+ = -\kappa_0^-}$$

$$(1.125)$$

$$P(\vec{u}; \kappa_0^-, T) \propto \left[1 - \frac{1}{\kappa_0^-} \cdot \frac{\frac{1}{2}m(\vec{u} - \vec{u}_b)^2}{k_B T}\right]_+^{\kappa_0^- - 1 - \frac{1}{2}d_K}.$$

In similar for the positive kappa distribution, $\theta \equiv \sqrt{2k_B T/m}$ is the thermal speed of particles of mass m and temperature T. Hereafter we keep the notation $\kappa_0^- = \kappa_0$ for simplicity. Note the cut-off condition indicated by the subscript "+," indicating that the distribution becomes zero anytime its base is negative (Eq. 1.13). Hence, in Eq. (1.125), if $\varepsilon_K < \kappa_0 k_B T$, then $P(\varepsilon_K; \kappa_0, T) = 0$, i.e., the kinetic energy has a maximum value determined by the value of the kappa index and temperature.

Next, we derive the statistical moments of the negative kappa distribution, Eq. (1.125). The mean kinetic energy is given by

$$\langle \varepsilon_K \rangle = \mu_1/\mu_0, \text{ where } \mu_m \equiv \int_0^{\kappa_0 k_B T} (\kappa_0 k_B T - \varepsilon_K)^{\kappa_0 - 1 - \frac{1}{2}d_K} \varepsilon_K^{\frac{1}{2}d_K - 1 + m} d\varepsilon_K. \qquad (1.126)$$

Setting $\kappa_0 k_B T - \varepsilon_K = \kappa_0 k_B T / (1 + t)$, we obtain

$$\mu_m \equiv (\kappa_0 k_B T)^{\kappa_0 - 1 + m} \int_0^\infty (1 + t)^{-\kappa_0 - m} t^{\frac{1}{2} d_K - 1 + m} dt. \tag{1.127}$$

For $t \to \infty$, the integrand becomes $t^{\frac{1}{2} d_K - \kappa_0 - 1}$ and the integral converges for $\kappa_0 > \frac{d_K}{2}$. Then, we find

$$\mu_m \equiv (\kappa_0 k_B T)^{\kappa_0 - 1 + m} \cdot (\kappa_0 + m - 1)^{-\frac{1}{2} d_K - m} \cdot \Gamma_{\kappa_0 + m - 1}\left(\frac{d_K}{2} + m\right), \tag{1.128}$$

where we use the notion of q-gamma function $\Gamma_q(x)$ or $\Gamma_\kappa(x)$ (Section 1.4.2). Then we have

$$\Gamma_{\kappa_0 + m - 1}\left(\frac{d_K}{2} + m\right) = (\kappa_0 + m - 1)^{\frac{1}{2} d_K + m} \cdot \frac{\Gamma\left(\frac{d_K}{2} + m\right) \cdot \Gamma\left(\kappa_0 - \frac{d_K}{2}\right)}{\Gamma(m + \kappa_0)}. \tag{1.129}$$

Hence,

$$\mu_m \equiv (\kappa_0 k_B T)^{\kappa_0 - 1 + m} \cdot \Gamma\left(\frac{d_K}{2} + m\right) \cdot \Gamma\left(\kappa_0 - \frac{d_K}{2}\right) \Big/ \Gamma(\kappa_0 + m), \tag{1.130}$$

and thus the mean kinetic energy is

$$\langle \varepsilon_K \rangle = \mu_1 / \mu_0 = \frac{1}{2} d_K \cdot k_B T. \tag{1.131}$$

The same result was found in the case of the positive kappa distribution, Eq. (1.3b) (Livadiotis and McComas, 2009, 2010a, 2013a). Therefore, the equipartition theorem, as well as the notion of temperature, is valid for both positive and negative kappa indices (Livadiotis and McComas, 2010a, 2012, 2013a).

The normalization constant is $\mu_0 \equiv (\kappa_0 k_B T)^{\kappa_0 - 1} \Gamma\left(\frac{d_K}{2}\right) \Gamma\left(\kappa_0 - \frac{d_K}{2}\right) \Big/ \Gamma(\kappa_0)$; thus the d_K-dimensional kappa distribution of the velocity in Eq. (1.125) becomes

$$P(\vec{u}; \kappa_0, T) = \left(\pi \kappa_0 \theta^2\right)^{-\frac{1}{2} d_K} \cdot \frac{\Gamma(\kappa_0)}{\Gamma\left(\kappa_0 - \frac{d_K}{2}\right)} \cdot \left[1 - \frac{1}{\kappa_0} \cdot \frac{(\vec{u} - \vec{u}_b)^2}{\theta^2}\right]_+^{\kappa_0 - 1 - \frac{1}{2} d_K}, \tag{1.132a}$$

that is, in terms of the kinetic energy, Eq. (1.3a),

$$P(\varepsilon_K; \kappa_0, T) = \frac{(\kappa_0 k_B T)^{-\frac{1}{2} d_K}}{B\left(\frac{d_K}{2}, \kappa_0 - \frac{d_K}{2}\right)} \cdot \left(1 - \frac{1}{\kappa_0} \cdot \frac{\varepsilon_K}{k_B T}\right)_+^{\kappa_0 - 1 - \frac{1}{2} d_K} \varepsilon_K^{\frac{1}{2} d_K - 1}. \tag{1.132b}$$

Notes: (1) The distributions (1.132a,b) converge for $\kappa_0 > \frac{d_K}{2}$. This can be shown as follows: using the transformation $X + 1 = [1 - \varepsilon_K/(\kappa_0 k_B T)]^{-1}$, the normalization of the distribution, Eq. (1.132b), requires the convergence of the integral $\int_0^\infty (X + 1)^{-\kappa_0} X^{\frac{1}{2}d_K - 1} \, dX$, which holds only if the integrand "falls" with $X^{-\delta}$, $\delta > 1$, for $X \to \infty$. This is true if $\kappa_0 > \frac{d_K}{2}$. (2) The kappa index could have a different symbol than the positive kappa index κ_0. Indeed, there is no reason for the positive and negative kappa indices to be (absolutely) equal. However, we use the same notation for both in order to keep the paper as simple as possible. (3) Finally, it has to be noted that the negative kappa distribution was first given in the work of Leubner (2004). However, this formulation is suffering by misinterpretation of the temperature and kappa index, which is similar to all the early statistical interpretations of kappa distributions (for details, see p. 204−205 in Livadiotis and McComas, 2013a).

Note that in Chapter 3, we will develop the multiparticle negative kappa distribution in combination with a nonzero potential energy and examine the degeneration of the kappa index.

1.9.5 Misleading Considerations About the Kappa (or q) Index

1.9.5.1 Kappa Index Sets an Upper Limit on the Total Number of Particles

Given the nonnegative value of the invariant kappa index, Eq. (1.106) may written as

$$\frac{1}{2}d \cdot N = \kappa - \kappa_0 \leq \kappa, \tag{1.133a}$$

which gives the misleading impression that the kappa index bounds the system's number of particles with the existence of an upper limit N_*, given by

$$N \leq N_* \equiv \frac{2}{d} \cdot \kappa. \tag{1.133b}$$

If Eq. (1.133b) were true, it will definitely be a knockout to the foundations of nonextensive statistical mechanics (Plastino and Rocca, 2017). But it is not!

1. In order for the kappa index to operate as an upper limit in Eq. (1.133b), it should be a constant parameter independent of N. However, the kappa index depends on the number of particles; in fact, Eq. (1.133a) reads exactly this functional dependence $\kappa(N) = \kappa_0 + \frac{d}{2}N$. Therefore, the number of particles N described by a kappa distribution can be any large number independently of the value of the kappa index $\kappa(N)$ that will be always larger. The measure of the thermodynamic distance of a stationary state from thermal equilibrium ("nonextensivity") should be invariant of the number of particles and degrees of freedom, and thus it is expressed in terms of the invariant kappa index κ_0, which is (as shown by Eq. 1.107).

2. There is, indeed, an upper limit on the number of particles that are described by the kappa distribution N, but this concerns the number of the correlated particles and not the number of all the particles in the system. Note that the multiparticle kappa distribution (see as follows) describes the energies of the correlated particles. Plasmas are systems with local correlations among their particles, manifested by correlation lengths such as the correlation length λ_D (Chapters 2 and 5; Livadiotis and McComas, 2014a). The number of correlated particles cannot exceed the number of particles in a Debye sphere, N_D, that is, the spherical volume with radius equal to a Debye length. Therefore, the thermodynamic limit is not affected by the condition $N < N_D$ since this is a limitation of the correlation number of particles and not of the system's total number of particles.

1.9.5.2 Correlation: Independent of the Total Number of Particles

The origin of the vastly different statistical behavior between classical particle systems and space plasmas is the manifestation of local correlations between the plasma particles. The stronger the correlation, the furthest the plasma resides from thermal equilibrium (Livadiotis and McComas, 2011b, 2013a). While correlations shift plasmas away from thermal equilibrium, collisions destroy correlations, recovering plasmas back at thermal equilibrium (Livadiotis and McComas, 2013b; Livadiotis, 2017a). Certainly, there may be various and different mechanisms of creating kappa distributions in space and other plasmas, e.g., the presence of weak turbulence (Yoon, 2014) and pickup ions (Livadiotis and McComas, 2010a, 2011a). The reason that kappa distributions exist and sustain themselves in space plasmas is the presence and preservation of correlations in a collisionless environment.

The correlation of the energies of two particles is found (Chapters 5 and 10; Abe, 1999; Livadiotis and McComas, 2011b; Livadiotis, 2015c)

$$R(\kappa_0; d) = \frac{\frac{1}{2}d}{\kappa_0 + \frac{1}{2}d}. \tag{1.134a}$$

The correlation between the energies of two particles of different degrees of freedom, d_1 and d_2, is

$$R(\kappa_0; d_1, d_2) = \sqrt{\frac{\frac{1}{2}d_1}{\kappa_0 + \frac{1}{2}d_1}} \cdot \sqrt{\frac{\frac{1}{2}d_2}{\kappa_0 + \frac{1}{2}d_2}} = \sqrt{R(\kappa_0; d_1) \cdot R(\kappa_0; d_2)}. \tag{1.134b}$$

In terms of the total kappa index, given by $\kappa = \kappa_0 + \frac{1}{2}d\,N$, which depends on the degrees of freedom $f = d\,N$, the correlation is written as

$$R(\kappa; d) = \frac{\frac{1}{2}d}{\kappa - \frac{1}{2}d\,(N-1)}. \tag{1.135a}$$

or in terms of the q-index,

$$R(q;d) = \frac{(q-1)d}{2+(1-q)d(N-1)}. \tag{1.135b}$$

This result was first found by Abe (1999). One of the strange consequences of this result is that the correlation between two particles decreases with the number of (correlated) particles; in fact, when $N \to \infty$, the correlation vanishes, $R \to 0$. But the correlation between two particles should not be dependent on the number of other particles. The worst paradox is that we find again that the value of the kappa index must be large. Namely, for $0 \le R(\kappa;d) \le 1$, we obtain $\frac{1}{2}d\,N \le \kappa$. On the other hand, for $-1 \le R(\kappa;d) \le 0$, we have $\kappa \le \frac{1}{2}d(N-2)$, i.e., the kappa index cannot be infinity, which is necessary property of nonextensive statistical mechanics to include systems that recover at thermal equilibrium and BG statistical mechanics.

1.9.5.3 The Problem of Divergence

One of the major issues in the formalism of nonextensive statistical mechanics is the multiple integrations of the canonical distribution (in terms of the kinetic energy or the velocities). As it was shown in Section 3, each of the integrations contributes to the exponent by adding $\frac{1}{2}d$. Hence, $(N-1)$ multiple integrations of the velocity of the $(N-1)$ particles increase the exponent by $\frac{1}{2}d\cdot(N-1)$; then the whole exponent becomes $-\left(\kappa - \frac{1}{2}d\cdot N\right) - 1 - \frac{1}{2}d$, and the whole one-particle distribution is

$$P(\vec{u}_1) \sim \left[1 + \frac{1}{\kappa - \frac{1}{2}d\,N} \cdot \frac{1}{\theta^2}(\vec{u}_1 - \vec{u}_b)^2 \right]^{-\left(\kappa - \frac{1}{2}d\,N\right) - 1 - \frac{1}{2}d}. \tag{1.136}$$

The integral of the mean kinetic energy (second statistical moment of velocity at the co-moving frame) must converge (so that the temperature is to be defined). For this, the asymptotic behavior at infinity of $u_1^2 P(\vec{u}_1)u_1^{d-1}$ must be falling faster than a power law of u_1^{-1}. We have

$$u_1^2 P(\vec{u}_1)u_1^{d-1} \sim u_1^2 \left[1 + \frac{1}{\kappa - \frac{1}{2}d\,N} \cdot \frac{1}{\theta^2}(\vec{u}_1 - \vec{u}_b)^2 \right]^{-\left(\kappa - \frac{1}{2}d\,N\right) - 1 - \frac{1}{2}d} \tag{1.137}$$

$$u_1^{d-1} \xrightarrow{u_1 \to \infty} u_1^2 u_1^{-2\left(\kappa - \frac{1}{2}d\,N\right) - 2 - d} u_1^{d-1} \sim u_1^{-2\left(\kappa - \frac{1}{2}d\,N\right) - 1}.$$

or $-2(\kappa - \frac{1}{2}d\,N) - 1 < -1$, from which we obtain the range of the allowable values of the kappa or q-indices, that is,

$$\kappa > \frac{1}{2}d\,N \text{ or } q < 1 + 1 \Big/ \left(\frac{1}{2}d\,N\right), \tag{1.138}$$

while for indices out of this range, $\kappa \leq \frac{1}{2}dN$ or $q \geq 1 + 1/(\frac{1}{2}dN)$, the integrals diverge and the distributions cannot be normalized.

Given the large number of particles, practically $N \to \infty$, we obviously find that $\kappa \to \infty$ or $q \to 1$, which corresponds to the limit of nonextensive statistical mechanics to the classical BG statistical mechanics. Therefore, in contrast to our experience from space and other laboratory plasma observation, here we find that *nonextensive statistical mechanics is not applicable* (!) in the continuous phase spaces with Hamiltonian $H(\vec{u}) = \varepsilon_K(\vec{u}) = \frac{1}{2}m(\vec{u} - \vec{u}_b)^2$ or $H(\vec{r}, \vec{u}) = \varepsilon_K(\vec{u}) + \Phi(\vec{r})$.

This paradox is easily resolved when we take into account the dependence of the kappa index on the degrees of freedom. In this case, we have $\kappa = \kappa_0 + \frac{1}{2}dN$; hence, the restriction (1.138) now becomes $\kappa_0 > 0$ (it is interesting to note that the paradox can be resolved only if the kappa index has this specific relation to the degrees of freedom. In addition, the number N corresponds to the correlated number of particles instead of the total number of particles in the system).

1.10 Concluding Remarks

The kappa distributions were empirical statistical models of the velocities of particle populations in space plasmas. A breakthrough in the field of space physics came with the connection of the observed kappa distributions with the solid statistical framework of nonextensive statistical mechanics, in contrast to the Maxwell distributions, which are connected with the classical BG statistical mechanics. Since then, the kappa distributions have become increasingly widespread across space physics with the number of relevant publications following, remarkably, an exponential growth rate. In this chapter, we presented the formulation and characteristics of the connection between the basic theory and formalism of kappa distributions and the statistical framework of non-extensive statistical mechanics for both the discrete and continuous descriptions of data.

The main developments are the following:

- The maximization of the Tsallis entropy under the constraints of the canonical ensemble to derive the canonical distribution: this is given in terms of the q-exponential function that replaces the classical Boltzmannian or Maxwellian exponential distribution function;

- The equivalence of the ordinary and escort q-exponential distributions with the main kinds of kappa distributions.

- The only exponent of kappa distributions that is compatible with the kinetically defined temperature is $-\kappa-1$. Even if we consider some other arbitrary exponent, e.g., $-\kappa-1-\alpha$, after substituting the correct notion of temperature taken from the mean kinetic energy, we will end up again with the same exponent, $-\kappa-1$.

- The physical meaning of the temperature for stationary states out of thermal equilibrium was established, showing the equivalence of the thermodynamic and kinetic definitions of temperature. The concept of temperature was exclusively known for systems at thermal equilibrium, the stationary state described by a Maxwellian distribution of velocities and the BG statistical mechanics.

- The physical meaning of the kappa index is interwoven with the statistical correlation between the energy of the particles. The kappa index classifies the different stationary states independently of the dimensionality, the degrees of freedom, or the number of particles.

- The kappa index depends on the degrees of freedom with $\kappa(f) = \kappa_0 + \frac{1}{2}f$, where $f = d \cdot N$ are the total kinetic degrees of freedom from all the particles. The kappa index κ_0, called "invariant kappa index," classifies the different stationary states independently of the dimensionality, the degrees of freedom, or the number of particles. The concept of the invariant kappa index allows the construction of the multiparticle kappa distributions and the negative kappa distribution.

- Several misleading considerations regarding the temperature and the kappa index were explained and resolved.

- We stated the relativity principle for statistical mechanics, which concerns the constancy of the mean kinetic energy along the stationary states.

Now that the connection is complete between empirically-derived kappa distributions and nonextensive statistical mechanics, the full strength and capability of the tools of this statistical background are available for the space and plasma physics community to analyze and understand the properties of the particle energy kappa distributions observed in space.

1.11 Science Questions for Future Research

Future analyses and observations need to address the following questions:

1. Why systems have positive or negative correlation between T and κ?
2. What is the origin of the degrees of freedom connected to κ_0?
3. How is the entropy maximized for anisotropic energy constraint?

CHAPTER 2

Entropy Associated With Kappa Distributions

G. Livadiotis
Southwest Research Institute, San Antonio, TX, United States

Chapter Outline

2.1 Summary 66
2.2 Introduction 66
2.3 The Role and Impact of Scale Parameters in the Entropic Formulation 67
  2.3.1 The Units' Paradox 67
  2.3.2 The Length Scale, σ_r 72
  2.3.3 The Speed Scale, σ_u 75
  2.3.4 Impact on Entropy 76
2.4 Derivation of the Entropic Formula for Velocity Kappa Distributions 78
  2.4.1 The Argument ϕ_q 78
  2.4.2 The Entropy 82
    2.4.2.1 Formula 82
    2.4.2.2 Thermodynamic Limits 84
2.5 Entropy for Isothermal Transitions Between Stationary States 87
  2.5.1 Derivation 87
  2.5.2 Survey 89
  2.5.3 Spontaneous Entropic Procedures 94
2.6 The Discrete Dynamics of Transitions Between Stationary States 96
  2.6.1 Discrete Dynamics 96

2.6.2 The Discrete Map of Stationary States Transitions 97
2.6.3 Numerical Application of the Discrete Transitions of Stationary States 98
2.6.4 The Five Stages of Stationary State Transitions 101
  2.6.4.1 The Intermediate Transitions Toward Antiequilibrium 101
  2.6.4.2 The Isentropic Switching and Its Importance 101
  2.6.4.3 The Double Role of Antiequilibrium 102
  2.6.4.4 The Final Transitions Toward Equilibrium 102
  2.6.4.5 Back to Far-Equilibrium Region 102
2.7 Concluding Remarks 103
2.8 Science Questions for Future Research 103
Acknowledgment 103

Kappa Distributions. http://dx.doi.org/10.1016/B978-0-12-804638-8.00002-4

2.1 Summary

The kappa distribution is the outcome of the maximization of entropy under the constraints of canonical ensemble. However, there is no systematic analysis focusing on the respective entropy itself. Indeed, while the kappa distributions are exclusively used to describe plasma populations, when it comes to their entropy, the Boltzmann Gibbs entropic formulation is employed. This chapter presents the theory, formulations, and properties of the entropy corresponding to kappa distributions. These developments will allow the researcher to calculate, formulate, and study the entropy, processes, and transitions, and improve the understanding of thermodynamics, in general, of the particle populations in space, geophysical, laboratory, or other plasmas, which are described by kappa distributions or combinations thereof.

Science Question: What is the entropy of kappa distributions telling us about plasma populations?

Keywords: Correlation length; Debye length; Entropy; Quantization constant; Speed scale; Temperature.

2.2 Introduction

Classical particle systems reside at thermal equilibrium with their velocity/energy distribution function stabilized into a Maxwell/Boltzmann form. On the contrary, collisionless and correlated particle systems, such as space and geophysical plasmas, are out of thermal equilibrium, whose particle populations are typically described by kappa distributions (or combinations thereof).

Since their introduction (Binsack, 1966; Olbert, 1968; Vasyliūnas, 1968), empirical kappa distributions become increasingly widespread across space and plasma physics. The implications and applications of kappa distributions in plasma physics were based on both (1) data analyses, and (2) analytical plasma methods, which aim to study the effects of kappa distributions on various basic plasma topics, spanning linear/nonlinear plasma waves, solitons, shock waves, dusty plasmas, etc. A watershed in the field came with the connection of kappa distributions to the statistical framework of nonextensive statistical mechanics (Chapter 1; Livadiotis and McComas, 2009). Understanding the statistical origin of kappa distributions was the cornerstone of important theoretical developments and applications. One of these developments is the formulation of entropy for particles described by kappa distributions.

The kappa distribution is the outcome of the maximization of entropy under the constraints of canonical ensemble. While there were numerous publications related to the kappa distributions in plasmas, there is no systematic analysis focusing on the respective entropy itself. Indeed, the kappa distributions are exclusively used to describe plasma populations, but when it comes to their entropy, the Boltzmann Gibbs entropic formulation is employed. However, this can lead to physically meaningless results because we use the entropic formulation, which is valid only for particles at thermal equilibrium (e.g., particles with no correlations) in order to describe particles out of thermal equilibrium (e.g., particles with correlations).

The appropriate entropic formulation is the one involved in the framework of nonextensive statistical mechanics (Tsallis, 1988). This is derived as follows: the entropy is a functional of the particle velocity distribution. First, the entropy is expressed in the continuous description of the velocity distribution (Livadiotis, 2014a). Then the kappa distribution, which is the specific function found to maximize this entropy, is substituted to the entropic functional. The derived entropy characterizes the particle systems described by kappa distributions.

This chapter presents the derivation and properties of the nonextensive entropic formulation for kappa distributions. The chapter is structured as follows: in Section 2.3, we examine the speed and length scale parameters involved in the entropic formulation, their physical meaning, and properties. We present the paradox of units that is connected with the entropic equation in the continuous description and how this is resolved by using the scale parameters. In Section 2.4, we first derive the entropic argument ϕ_q, which is related with the entropic formula but also with the ratio of the Lagrangian and actual temperature of the system. Then, we derive the entropic formula for the kappa distribution of velocities. We study its properties and thermodynamic limits. In Section 2.5, we derive and study a special entropic measure that depends only on the kappa index and characterizes the transition of the system among stationary states. We also provide a survey of the relevant literature regarding the observed kappa indices and compare those to the characteristics of the presented entropic measure. In Section 2.6, we present the discrete dynamics of transitions between stationary states, derive analytically the characteristic discrete dynamical map of stationary states transitions, develop numerical applications of this map, and examine the five stages of stationary state transitions. Finally, the concluding remarks are given in Section 2.7, while three general science questions for future analyses are posed in Section 2.8.

2.3 The Role and Impact of Scale Parameters in the Entropic Formulation

2.3.1 The Units' Paradox

Let the kappa distribution of the kinetic energy, $\varepsilon_K(\vec{u}) = \frac{1}{2}m(\vec{u} - \vec{u}_b)^2$ (in the co-moving system),

$$P(\varepsilon_K; \kappa, T) \propto \left(1 + \frac{1}{\kappa} \cdot \frac{\varepsilon_K - \langle \varepsilon_K \rangle}{k_B T}\right)^{-\kappa-1} \propto \left(1 + \frac{1}{\kappa - \frac{\langle \varepsilon_K \rangle}{k_B T}} \cdot \frac{\varepsilon_K}{k_B T}\right)^{-\kappa-1}, \text{ or,}$$

(2.1a)

$$P(\varepsilon_K; \kappa_0, T) \propto \left(1 + \frac{1}{\kappa_0} \cdot \frac{\varepsilon_K}{k_B T}\right)^{-\kappa_0 - 1 - \frac{1}{2}d_K},$$

or of the velocities,

$$P(\vec{u};\kappa_0,\theta) \propto \left[1 + \frac{1}{\kappa_0}\cdot\frac{(\vec{u}-\vec{u}_b)^2}{\theta^2}\right]^{-\kappa_0-1-\frac{1}{2}d_K}, \tag{2.1b}$$

where $\theta \equiv \sqrt{2k_BT/m}$ is the thermal speed of particle of mass m, that is, the particle temperature T expressed in speed units; both the particle $\vec{u}$ and bulk $\vec{u}_b$ velocities are in an inertial reference frame; $d_K \equiv 2\langle\varepsilon_K\rangle/(k_BT)$ denotes the kinetic degrees of freedom, while the kappa index depends on the degrees of freedom so that the difference $\kappa_0 \equiv \kappa - \frac{1}{2}d_K$ is independent of the degrees defining the invariant kappa index κ_0. Nonetheless, the distribution may be given in terms of the dimension-dependent kappa index κ and the kappa-dependent thermal speed θ_κ,

$$P(\vec{u};\kappa,\theta) \propto \left[1 + \frac{1}{\kappa}\cdot\frac{(\vec{u}-\vec{u}_b)^2}{\theta_\kappa^2}\right]^{-\kappa-1}, \theta_\kappa \equiv \sqrt{\left(\kappa - \frac{1}{2}d_K\right)\Big/\kappa}\cdot\theta$$

$$= \sqrt{\kappa_0\Big/\left(\kappa_0 + \frac{1}{2}d_K\right)}\cdot\theta. \tag{2.2}$$

In the presence of a potential energy $\Phi(\vec{r})$, the distribution of Hamiltonian $H(\vec{r},\vec{u}) = \varepsilon_K(\vec{u}) + \Phi(\vec{r})$ is used for describing the particle system instead of the kinetic energy (Chapter 3),

$$P(\vec{r},\vec{u};\kappa,T) \propto \left[1 + \frac{1}{\kappa}\cdot\frac{H(\vec{r},\vec{u})-\langle H\rangle}{k_BT}\right]^{-\kappa-1} \propto \left[1 + \frac{1}{\kappa-\frac{f}{2}}\cdot\frac{H(\vec{r},\vec{u})}{k_BT}\right]^{-\kappa-1}, \text{or}$$

$$P(\vec{r},\vec{u};\kappa_0,T) \propto \left[1 + \frac{1}{\kappa_0}\cdot\frac{H(\vec{r},\vec{u})}{k_BT}\right]^{-\kappa_0-1-\frac{1}{2}f}, \tag{2.3}$$

where $f \equiv 2\langle H\rangle/(k_BT)$ provides the sum of kinetic d_K and potential $d_\Phi \equiv 2\langle|\Phi|\rangle/(k_BT)$ degrees of freedom, while the invariant kappa index is now defined by $\kappa_0 \equiv \kappa - \frac{1}{2}f$. Note that in the case of central power law potentials, i.e., $\Phi(r) \propto \pm r^{\pm b}$, the potential degrees of freedom are connected with the positional dimensionality d_r, i.e., $\frac{1}{2}d_\Phi = \frac{1}{b}d_r$.

Let us first focus on the distribution of the velocities. The derivation of entropy involves calculating a functional of the velocity probability distribution $p(\vec{u})$, that is, $S(p(\vec{u}))$. However, the distribution is not dimensionless, i.e., the units of the distribution $p(\vec{u})$ (or of its normalization constant) are the inverse velocity volume. This is problematic because of the logarithm or the power law of the distribution included in the entropic formulations:

- classical Boltzmann–Gibbs entropy,

$$S(p(\overrightarrow{u})) = -k_{\mathrm{B}} \cdot \int_{-\infty}^{+\infty} p(\overrightarrow{u}) \ln p(\overrightarrow{u}) d\overrightarrow{u};$$ (2.4a)

- nonextensive Tsallis entropy,

$$S(p(\overrightarrow{u}); q) = k_{\mathrm{B}} \cdot (q-1)^{-1} \cdot \left[1 - \int_{-\infty}^{+\infty} p(\overrightarrow{u})^q d\overrightarrow{u} \right],$$ (2.4b)

where the entropic index q and the kappa index κ are related with $\kappa = (q-1)^{-1}$ (Chapter 1; Livadiotis and McComas, 2009). The nonextensive entropy recovers the classical Boltzmann–Gibbs formulation at thermal equilibrium ($\kappa \to \infty$ or $q \to 1$). Recall that $d\overrightarrow{u}$ denotes the infinitesimal d_{K}-dimensional velocity volume, avoiding writing $d^{d_{\mathrm{K}}}\overrightarrow{u}$ for simplicity. The L_q-normed Lebesgue integral (Livadiotis, 2007, 2008) of the probability distribution $\int p(\overrightarrow{u})^q d\overrightarrow{u}$ defines the entropic argument ϕ_q that we will examine further as follows.

The units of the velocity distribution density p is inverse velocity volume; this can be readily derived from its normalization, that is, $1 = \int_{-\infty}^{\infty} p(\overrightarrow{u}) d\overrightarrow{u}$, leading to $[p] = [u]^{-d_{\mathrm{K}}}$, where we used the symbol [X] for the units of X. However, the units of the distribution $p(\overrightarrow{u})$ in Eq. (2.4b) would depend on the value of the q-index; i.e., $[p] = [u]^{-d_{\mathrm{K}}/q}$; in other words, the entropy in Eq. (2.4b) would be dependent on the speed units!

Also problematic is the classical entropic formulation of Eq. (2.4a) because the logarithm must refer to a dimensionless number. Nonetheless, this is easily resolved because the units affect the entropy only by some constant, while physically meaningful are the variations of the entropy,

$$S(p(\overrightarrow{u})/[p]) = -k_{\mathrm{B}} \cdot \int_{-\infty}^{+\infty} \frac{p(\overrightarrow{u})}{[p]} \ln \left[\frac{p(\overrightarrow{u})}{[p]} \right] \frac{d\overrightarrow{u}}{[u]^{d_{\mathrm{K}}}}$$

$$= k_{\mathrm{B}} \cdot \ln[p] - k_{\mathrm{B}} \cdot \int_{-\infty}^{+\infty} p(\overrightarrow{u}) \ln p(\overrightarrow{u}) d\overrightarrow{u} = k_{\mathrm{B}} \cdot \ln[p] + S(p(\overrightarrow{u})).$$ (2.5)

The origin of the units' paradox is the misuse of the probability density as simple dimensionless probability. This can be true only in the discrete case where both entropic formulations are consistently defined because the involved discrete distribution $\{p_k\}$ is a dimensionless probability instead of a probability density:

- classical Boltzmann—Gibbs entropy of a system described by a Maxwell velocity distribution,

$$S(p_k) = -k_B \cdot \sum_k p_k \ln p_k, \tag{2.6a}$$

- Tsallis entropy of a system described by a kappa distribution,

$$S(p_k; q) = k_B \cdot (q - 1)^{-1} \cdot \left[1 - \sum_k p_k^q \right]. \tag{2.6b}$$

In order to resolve the units paradox, we consider a considerably small speed scale σ_u, characteristic of the system; at this small scale, the distribution density can be considered approximately constant, so the probability in the narrow interval $(u_i, u_i + \sigma_u)$, $i : 1, 2, \ldots, d_K$, can be approached by $p(\vec{u}) \cdot \sigma_u^{d_K}$.

The same units' paradox appears when considering the distribution of Hamiltonian:

$$S(p(\vec{r}, \vec{u}); q) = k_B \cdot (q - 1)^{-1} \cdot \left[1 - \int_{-\infty}^{+\infty} p(\vec{r}, \vec{u})^q d\vec{r} d\vec{u} \right]; \tag{2.7}$$

Indeed, the units of the distribution $p(\vec{r}, \vec{u})$ in Eq. (2.7) depend on the q-index; i.e., $[p] = \left([r]^{-d_r} [u]^{-d_K} \right)^{1/q}$, or $[p] = ([r][u])^{-d/q}$ when $d_r = d_K \equiv d$. In other words, the entropy is now dependent on the length and speed units! The paradox is resolved by considering the probability in the narrow phase—space interval $(r_i, r_i + \sigma_r) \times (u_j, u_j + \sigma_u)$, $i : 1, \ldots, d_r$; $j : 1, \ldots, d_K$ that can be approached by $p(\vec{r}, \vec{u}) \cdot \sigma_r^{d_r} \sigma_u^{d_K}$; for $d_r = d_K \equiv d$, the probability becomes $p(\vec{r}, \vec{u}) \cdot (\sigma_r \sigma_u)^d$, with $\sigma_r \cdot m \sigma_u = h$ is the phase—space scale, typically given by the Planck constant (see Chapter 5, Section 5.9).

Using the speed scale σ_u, we redefine the distribution of velocities:

$$1 = \int_{-\infty}^{\infty} p(\vec{u}) \sigma_u^{d_K} d\Omega, \tag{2.8a}$$

where $d\Omega$ is the number of microstates included in a system's velocity space,

$$d\Omega \equiv \frac{d\vec{u}}{\sigma_u^{d_K}} = \frac{du_1 du_2 \cdots du_{d_K}}{\sigma_u^{d_K}} = \frac{du_1}{\sigma_u} \frac{du_2}{\sigma_u} \cdots \frac{du_{d_K}}{\sigma_u}. \tag{2.8b}$$

Note that $p(\vec{u})$ is a probability density; the quantity $\widetilde{p}(\vec{u}) \equiv p(\vec{u}) \cdot \sigma_u^{d_K}$ gives the dimensionless probability $\widetilde{p}(\vec{u})$ in the narrow interval $[u_i - u_{bi}, u_i - u_{bi} + \sigma_u]$ (of the ith velocity component). For example, the velocity kappa distribution is

$$
\widetilde{P}(\vec{u}; \kappa_0, \theta) \equiv P(\vec{u}; \kappa_0, \theta) \cdot \sigma_u^{d_K} = \left[\pi \kappa_0 (\theta / \sigma_u)^2 \right]^{-\frac{1}{2}d_K} \cdot \frac{\Gamma\left(\kappa_0 + 1 + \dfrac{d_K}{2}\right)}{\Gamma(\kappa_0 + 1)}
$$
$$
\times \left[1 + \frac{1}{\kappa_0} \cdot \frac{(\vec{u} - \vec{u}_b)^2}{\theta^2} \right]^{-\kappa_0 - 1 - \frac{1}{2}d_K}.
$$
(2.9)

The normalization of the dimensionless distribution is

$$
\int \widetilde{p}(\vec{u}) \frac{d\vec{u}}{\sigma_u^{d_K}} = \int \left[p(\vec{u}) \sigma_u^{d_K} \right] \frac{d\vec{u}}{\sigma_u^{d_K}} = \int p(\vec{u}) d\vec{u}.
$$
(2.10)

Therefore, the integration of any functional of the probability distribution $f(p(\vec{u}))$ is given by

$$
\int_{-\infty}^{\infty} f\left[p(\vec{u}) \sigma_u^{d_K} \right] \frac{d\vec{u}}{\sigma_u^{d_K}}.
$$
(2.11)

In the case of entropy, we need to calculate the power law functional $f(x) = x^q$, namely,

$$
S = (q-1)^{-1} \cdot \left\{ 1 - \int_{-\infty}^{\infty} \left[p(\vec{u}) \sigma_u^{d_K} \right]^q \frac{d\vec{u}}{\sigma_u^{d_K}} \right\}.
$$
(2.12a)

Nonextensive statistics generalize the classical Boltzmann–Gibbs statistical mechanics, not only by means of the entropic formulation, but also by the notion of the canonical distribution. This is characterized by the dual formalism of ordinary $p(\vec{u})$ and escort $P(\vec{u})$ probability distributions, related with $P(\vec{u}) \propto p(\vec{u})^q$. While the observable distribution is given by the escort distribution, the entropy is simpler when expressed in terms of the ordinary distribution. Nonetheless, as we will see further as follows in detail, the entropy can be also expressed in terms of the escort distribution,

$$
S = (q-1)^{-1} \cdot \left\{ 1 - \left\{ \int_{-\infty}^{+\infty} \left[P(\vec{u}) \cdot \sigma_u^{d_K} \right]^{\frac{1}{q}} \frac{d\vec{u}}{\sigma_u^{d_K}} \right\}^{-q} \right\}.
$$
(2.12b)

Using the length σ_r and speed scale σ_u, we redefine the phase–space distribution:

$$
\int \widetilde{p}(\vec{r}, \vec{u}) \frac{d\vec{r} d\vec{u}}{\sigma_r^{d_r} \sigma_u^{d_K}} = \int \left[p(\vec{r}, \vec{u}) \sigma_r^{d_r} \sigma_u^{d_K} \right] \frac{d\vec{r} d\vec{u}}{\sigma_r^{d_r} \sigma_u^{d_K}} = \int p(\vec{u}) d\vec{u}.
$$
(2.13)

Hence, the integration of a functional of the probability distribution is

$$\int_{-\infty}^{\infty} f\left[p(\overrightarrow{r}, \overrightarrow{u})\sigma_{\mathrm{r}}^{d_{\mathrm{r}}}\sigma_{\mathrm{u}}^{d_{\mathrm{K}}}\right] \frac{d\overrightarrow{r}\,d\overrightarrow{u}}{\sigma_{\mathrm{r}}^{d_{\mathrm{r}}}\sigma_{\mathrm{u}}^{d_{\mathrm{K}}}}, \text{ or } \int_{-\infty}^{\infty} f\left[p(\overrightarrow{r}, \overrightarrow{u})(\sigma_{\mathrm{r}}\sigma_{\mathrm{u}})^{d}\right]\frac{d\overrightarrow{r}\,d\overrightarrow{u}}{(\sigma_{\mathrm{r}}\sigma_{\mathrm{u}})^{d}} \text{ if } d_{\mathrm{r}} = d_{\mathrm{K}} \equiv d.$$

$$(2.14)$$

For example, the entropy is given by

$$S = (q-1)^{-1} \cdot \left\{ 1 - \int_{-\infty}^{\infty}\left[p(\overrightarrow{r}, \overrightarrow{u})\sigma_{\mathrm{r}}^{d_{\mathrm{r}}}\sigma_{\mathrm{u}}^{d_{\mathrm{K}}}\right]^{q}\frac{d\overrightarrow{r}\,d\overrightarrow{u}}{\sigma_{\mathrm{r}}^{d_{\mathrm{r}}}\sigma_{\mathrm{u}}^{d_{\mathrm{K}}}} \right\}, \qquad (2.15a)$$

or given the escort distribution

$$S = (q-1)^{-1} \cdot \left\{ 1 - \left\{ \int_{-\infty}^{+\infty}\left[P(\overrightarrow{r}, \overrightarrow{u})\sigma_{\mathrm{r}}^{d_{\mathrm{r}}}\sigma_{\mathrm{u}}^{d_{\mathrm{K}}}\right]^{\frac{1}{q}}\frac{d\overrightarrow{r}\,d\overrightarrow{u}}{\sigma_{\mathrm{r}}^{d_{\mathrm{r}}}\sigma_{\mathrm{u}}^{d_{\mathrm{K}}}} \right\}^{-q} \right\}. \qquad (2.15b)$$

We recall that the description using the entropic q-index can be simply transformed into the more familiar kappa index, $\kappa = (q-1)^{-1}$, or the more fundamental, invariant kappa index, $\kappa_0 = \kappa - \frac{1}{2}f = (q-1)^{-1} - \frac{1}{2}f$.

We have seen that the speed scale solves the units' paradox. According to Tsallis (Tsallis et al., 1995; Prato and Tsallis, 1999; Tsallis, 1999, 2009b), the σ_{u} parameter depends on the characteristics of the given problem. Next we examine the interpretation of the speed scales.

2.3.2 The Length Scale, σ_{r}

We investigate the statistical implications in particle systems with local correlations, e.g., plasma particles within the scale of the Debye length. The statistical treatment of the phase space spanned by correlation clusters (e.g., Debye spheres) is fundamentally different from that of the classical case, which assumes uncorrelated particles. However, it can be thought as in the classical case, where the essentially uncorrelated clusters replace the uncorrelated particles in Boltzmann–Gibbs statistics.

The number of microstates included in a system's phase space that is clustered in groups of N_{C} correlated particles is different from that of a classical system of uncorrelated particles (at thermal equilibrium). For classical systems, the number of microstates in an infinitesimal phase–space volume of N-particle $2d\cdot N$-dimensional phase–space is given by

$$d\Omega_N = m^{d\cdot N}d\overrightarrow{r}_1 d\overrightarrow{u}_1 \cdots d\overrightarrow{r}_N d\overrightarrow{u}_N \big/ (N!h^{d\cdot N}). \qquad (2.16a)$$

The infinitesimal one-particle microstates are given by

$$d\Omega_1 = \frac{d\overrightarrow{r}\,d\overrightarrow{u}}{N!^{\frac{1}{N}}(h/m)^{d}} = \frac{(d\overrightarrow{r}/V)d\overrightarrow{u}}{(n/e)\cdot(h/m)^{d}}, \qquad (2.16b)$$

where we used the approximation $N!^{\frac{1}{N}} \cong N/e$; V is the volume of the particle system (see Chapter 3, Section 3.4). Note that the infinitesimal d-dimensional volume $d\vec{r}$ (short for $d^d\vec{r}$) is divided by the whole volume V because the distribution is independent of particle position. By inserting the smallest possible speed scale σ_u, the one-particle microstates can be obtained by $d\Omega_1 \equiv (d\vec{r}/V)(d\vec{u}/\sigma_u^d)$, i.e., each velocity component is divided by σ_u (note that the infinitesimal volume is normalized to the total volume V, because the distribution is independent of the position — zero potential energy). Hence,

$$d\Omega_1 \equiv (d\vec{r}/V)\left(d\vec{u}/\sigma_u^d\right) \cong \frac{(d\vec{r}/V)(d\vec{u}/\sigma_u^d)}{(n/e)\cdot[h/(m\sigma_u)]^d} \cong \frac{(d\vec{r}/V)(d\vec{u}/\sigma_u^d)}{[h/(a_d m\sigma_u b)]^d}, \qquad (2.17)$$

where we substituted the interparticle distance b, given by $1 = (B_d/d)n\cdot b^d$, leading to

$$h \approx m\sigma_u b \qquad (2.18a)$$

(note that we ignored $a_d \equiv (eB_d/d)^{1/d}$. Then, setting $\sigma_r \equiv h/(m\sigma_u)$, we obtain that the characteristic length scale is the interparticle distance,

$$h \equiv m\sigma_u\sigma_r \Rightarrow \sigma_r \approx b \qquad (2.18b)$$

Systems with local correlations among particles form clusters of correlated particles; then the number of microstates is given by

$$d\Omega_N = m^{d\cdot N} d\vec{r}_1 d\vec{u}_1 \cdots d\vec{r}_N d\vec{u}_N \big/ \left(c_N h^{d\cdot N}\right), \qquad (2.19)$$

where the combinations of correlated particles determines the argument c_N. Namely, the microstates are obtained by dividing the phase–space volume of N nonidentical particles with the factor of $c_N h^{d\cdot N}$ instead of $N! h^{d\cdot N}$.

The argument c_N stands for the maximized combinations of having N particles within M clusters. Any possible combination is given by the M-variable function of $\{N_{C,m}\}_{m=1}^M$, that is, $C_M^N(N_{C,1}, N_{C,2}, ..., N_{C,M})$, where $N_{C,m}$ is the number of correlated particles in the mth uncorrelated cluster. We have

$$C_M^N(N_{C,1}, N_{C,2}, ..., N_{C,M}) = N! \Big/ \prod_{m=1}^M N_{C,m}!, N = \sum_{m=1}^M N_{C,m}. \qquad (2.20)$$

This function can be easily shown that is maximized when $N_{C,m}$ is constant for any m. Indeed, by maximizing the function $F(N_{C,1}, N_{C,2}, ...N_{C,M}) = \ln C_M^N(N_{C,1}, N_{C,2}, ..., N_{C,M}) - \lambda\sum_{m=1}^M N_{C,m}$, we find $\ln N_{C,m} = \lambda$, or $N_{C,m}$: constant for any m. Therefore, $N_{C,m} \equiv N_C = N/M$. Substituting in Eq. (2.20), we find

$$c_N \equiv C_M^N(N_{C,m} = N_C) = N! \big/ (N_C!)^{N/N_C}, N_C \gg 1, \qquad (2.21a)$$

hence, we calculate

$$c_N^{1/N} \equiv C_M^N(N_{C,m} = N_C) = (N!)^{1/N} \big/ (N_C!)^{1/N_C} \cong N/N_C, N_C \gg 1. \qquad (2.21b)$$

Note that we can give a general expression for c_N in regard to the local correlation among the particles:

$$c_N = \begin{cases} N! & \text{for uncorrelated particles,} \\ (N/N_C)^N & \text{for particles correlated by clusters,} \\ 1 & \text{for fully correlated particles.} \end{cases} \qquad (2.21c)$$

Then the one-particle microstates are given by

$$d\Omega_1 = \frac{d\vec{r}\,d\vec{u}}{c_N^{1/N}(h/m)^d} = \frac{(d\vec{r}/V)d\vec{u}}{(n/N_C)\cdot(h/m)^d}. \qquad (2.22)$$

Again, by inserting the smallest possible speed scale σ_u, the one-particle microstates can be obtained by $d\Omega_1 \equiv (d\vec{r}/V)(d\vec{u}/\sigma_u^d)$, hence,

$$d\Omega_1 \equiv (d\vec{r}/V)\left(d\vec{u}/\sigma_u^d\right) \cong \frac{(d\vec{r}/V)(d\vec{u}/\sigma_u^d)}{(n/N_C)\cdot[h/(m\sigma_u)]^d} = \frac{(d\vec{r}/V)(d\vec{u}/\sigma_u^d)}{\left[h/\left((B_d/d)^{1/d}m\sigma_u\ell_C\right)\right]^d}, \qquad (2.23)$$

where we substituted $N_C = (B_d/d)n\cdot\ell_C^d$, while we ignored $a_d' \equiv (B_d/d)^{1/d}$. Therefore, we derive

$$h \approx m\sigma_u\ell_C. \qquad (2.24a)$$

The importance of this relation is the interpretation of the correlation length as the smallest possible length scale in the case of particle systems with local correlations. Indeed, setting $\sigma_r \equiv h/(m\sigma_u)$, we obtain

$$h \equiv m\sigma_u\sigma_r \Rightarrow \sigma_r \approx \ell_C. \qquad (2.24b)$$

These results show that in this case of local correlations among plasma particles (collisionless space plasmas, Chapter 5; Livadiotis and McComas, 2013b, 2014a, 2014b; Livadiotis, 2015d; 2017a), the characteristic length scale is given by the correlation length $\sigma_r \approx \ell_C$;

$$N_C \approx n\cdot\ell_C^d, \sigma_r \approx \ell_C. \qquad (2.25)$$

However, in the case of uncorrelated particles (e.g., classical gas), the characteristic length scale may be represented by the interparticle distance, $\sigma_r \approx b \propto n^{-\frac{1}{d}}$. The absence of local correlations can be interpreted as $N_C \approx 1$; this is readily shown when comparing $N_C \approx n\cdot\ell_C^d$, $\sigma_r \approx \ell_C$, and $1 \approx n\cdot b^d$ (a d-dimensional cubic volume was considered for simplicity). Namely,

$$N_C \approx 1 \approx n\cdot b^d, \sigma_r \approx b \propto n^{-\frac{1}{d}} \qquad (2.26)$$

Therefore, the length scale σ_r is given by the correlation length ℓ_C or the interparticle distance b.

The smallest correlation length in plasmas is the Debye length (e.g., Livadiotis and McComas, 2014a), $\ell_C \approx \lambda_D$, where a maximum number of N_D particles can be correlated within a Debye sphere, the sphere of radius equal with the Debye length, $N_C \approx N_D \propto n\cdot\lambda_D^d$. Yet the particles within a Debye sphere may not all be significantly correlated; a smaller number of particles $N_C < N_D$ may be characterized by some correlation, defining an effective correlation length smaller than the Debye length $\ell_C \approx (N_C/n)^{\frac{1}{d}} < \lambda_D$. Thus, the number of the correlated particles cannot be considered identical to the Debye particles number N_D, and the correlation ratio $\chi_C \equiv N_C/N_D$ is rather another independent parameter. Then the length scale may be written as $\sigma_r = \ell_C$, where the effective correlation length may have any value within the lower limit of interparticle distance b and the upper limit of the Debye length λ_D,

$$\sigma_r = \begin{cases} \ell_C \sim \lambda_D \propto \left[\dfrac{\kappa_0}{\kappa_0 + 1} (T/n) \right]^{\frac{1}{2}}, & \text{for Debye spheres full of correlated particles,} \\ b < \ell_C < \lambda_D \\ b \propto n^{-\frac{1}{d}}, & \text{for uncorrelated particles.} \end{cases}$$

(2.27)

(For the kappa dependence of the Debye length, see: Chapters 5 and 8; Bryant, 1996; Rubab and Murtaza, 2006; Gougam and Tribeche, 2011; Livadiotis and McComas, 2014a).

2.3.3 The Speed Scale, σ_u

The speed scale is connected with the phase–space cell, the Planck constant h, namely, $\sigma_u \approx h/(m\sigma_r)$, with either $\sigma_r \approx \ell_C$ (e.g., λ_D) or $\sigma_r \approx b$. This is the fundamental interpretation of speed scale. Yet it may be equal to some characteristic speeds of the particle velocity distribution, e.g., proportional to the thermal speed or other similar thermal parameters.

Possible speed scales are the thermal speed θ or the most frequent particle speed w_{max}. The distribution of speed $w \equiv |\vec{u} - \vec{u}_b|$ is given by

$$P(w; \kappa_0, \theta) \propto \left[1 + \frac{1}{\kappa_0} \cdot \frac{w^2}{\theta^2} \right]^{-\kappa_0 - 1 - \frac{1}{2} d_K} w^{d_K - 1},$$

(2.28)

which is maximized for

$$w_{max} = \frac{1}{2} (d_K - 1) \cdot \sqrt{\kappa_0 \Big/ \left(\kappa_0 + \frac{3}{2} \right)} \cdot \theta.$$

(2.29a)

For $d_K = 3$, this becomes

$$w_{max} = \sqrt{\kappa_0 \Big/ \left(\kappa_0 + \frac{3}{2} \right)} \cdot \theta = \sqrt{\left(\kappa - \frac{3}{2} \right) \Big/ \kappa} \cdot \theta = \theta_\kappa.$$

(2.29b)

Note that θ_κ is not the most frequent particle speed, in general, but it coincides with it only for $d_K = 3$ (see paragraph 1.8.3.5 in Chapter 1).

The role of the thermal speed θ is more fundamental than that of w_{max} because the former is actually the temperature expressed in speed units $\theta \equiv \sqrt{2k_B T/m}$, while the latter is a physical characteristic deducible from the temperature and other parameters of the particle speed distributions. Given a temperature-like physical quantity T_*, one may derive the respective thermal speed by $\theta_* \equiv \sqrt{2k_B T_*/m}$. Then, finding a speed scale $\sigma_u \approx \theta_*$ is equivalent to finding a temperature scale, i.e.,

$$T_* \approx \sigma_T \equiv \sigma_u^2 m/(2k_B) \text{ or } \theta_* \equiv \sqrt{2k_B T_*/m} \approx \sigma_u. \tag{2.30}$$

In Section 2.5, we will apply Eq. (2.30) to set the temperature scale equal to the actual temperature of the system or to its Lagrangian temperature (Chapter 1; Livadiotis, 2014a), that is, to set the speed scale equal to the thermal speed or the Lagrangian thermal speed.

2.3.4 Impact on Entropy

Given Eq. (2.24a), we find for particle systems with local correlations

$$\sigma \equiv \sigma_u/\theta \approx h/(m\theta \ell_C). \tag{2.31a}$$

For quasineutral plasmas of ions and electrons, the dimensionless parameter is given by

$$\sigma = (4\pi/3)^{-1/3} h \Big/ \Big(\sqrt{m_i m_e \theta_i \theta_e} \, \ell_C \Big), \text{ with } \ell_C \sim \lambda_D, \tag{2.31b}$$

that can be written as

$$\sigma \cong 6^{\frac{1}{3}} \pi^{\frac{2}{3}} 2^{-\frac{1}{2}} k_B^{-1} \varepsilon_0^{-\frac{1}{2}} q_e (m_i m_e)^{-\frac{1}{4}} h n^{\frac{1}{2}} (T_i T_e)^{-\frac{3}{4}} (T_i + T_e)^{\frac{1}{2}}, \tag{2.31c}$$

with electron and ion mass, m_e, m_i, and temperature, T_e, T_i; ε_0 is the vacuum permittivity. In this chapter, we denote the elementary electric charge with q_e to avoid confusion with the Euler's number e. The phase—space quantization is represented here by the Planck constant h (or its reduced value, $\hbar = h/(2\pi)$), though evidence is shown for a different quantization constant in space plasmas, about 12 orders of magnitude larger, $\hbar_* \sim 10^{12} \cdot \hbar$ (Chapter 5, Section 5.9; Livadiotis and McComas, 2013b, 2014b,c; Livadiotis, 2015d, 2016c, 2017a; Livadiotis and Desai, 2016).

At thermal equilibrium, the entropy is given by the Sackur—Tetrode equation

$$S/N = d \cdot k_B \ln(\sqrt{\pi e}/\sigma), \tag{2.32}$$

where the nonnegativity of entropy implies the condition $\sigma \leq \sqrt{\pi e}$. We show this classical formula as follows. Substituting $p(\vec{u})$ in Eq. (2.4a) with the Maxwell distribution

$$P(\vec{u}; \theta) = \pi^{-\frac{1}{2}d} \theta^{-d} \exp\left[-\frac{(\vec{u} - \vec{u}_b)^2}{\theta^2} \right], \theta \equiv \sqrt{2k_B T/m}, \tag{2.33}$$

we obtain

$$
\begin{aligned}
S(N = 1)/k_{\mathrm B} &= -\int_{-\infty}^{+\infty}\left[P(\vec{u};\theta)\sigma_{\mathrm u}^{d}\right]\ln\left[P(\vec{u};\theta)\sigma_{\mathrm u}^{d}\right]\frac{d\vec{u}}{\sigma_{\mathrm u}^{d}}\\
&= -\int_{-\infty}^{+\infty}P(\vec{u};\theta)\ln\left[P(\vec{u};\theta)\theta^{d}\right]d\vec{u} - d\cdot\ln\sigma\\
&= \int_{-\infty}^{+\infty}P(\vec{u};\theta)\frac{(\vec{u}-\vec{u}_{b})^{2}}{\theta^{2}}d\vec{u} + \frac{1}{2}d\cdot\ln\pi - d\cdot\ln\sigma\\
&= \frac{1}{2}B_{d}\cdot\pi^{-\frac{1}{2}d}\Gamma\left(\frac{d}{2}+1\right) + \frac{1}{2}d\cdot\ln\pi - d\cdot\ln\sigma = d\cdot\ln\left(\sqrt{\pi e}/\sigma\right),
\end{aligned}
$$

(2.34)

where $B_{d} = 2\pi^{\frac{1}{2}d}\big/\Gamma\left(\frac{d}{2}\right)$ stands for the d-dimensional spherical surface. The Boltzmann–Gibbs entropy is extensive, i.e., $S(N) = N\cdot S(N = 1)$; hence we end up with Eq. (2.32).

Independently of how complicated is the general entropy formulation out of thermal equilibrium (e.g., for finite kappa indices), it has to be analytically definable at thermal equilibrium. There, the generalized entropy must be reduced to the Sackur–Tetrode equation that requires the condition $\sigma < \sqrt{\pi e}$ to be fulfilled. The limit of $\sigma \sim \sqrt{\pi e}$ has different meaning for systems with or without correlations:

- For systems with no correlations, the limit $\sigma \sim \sqrt{\pi e}$ means the system approaches "quantum degeneracy," beyond which the nonquantum approach of statistical mechanics is no longer valid. This concept is usually described by comparing the thermal de Broglie wavelength $\lambda_{\mathrm{Bg}} \equiv h\cdot(2\pi k_{\mathrm B}mT)^{-\frac{1}{2}}$ and the interparticle distance $b \sim n^{-\frac{1}{d}}$. The passage of the system into the quantum regime implies $\lambda_{\mathrm{Bg}} < b$.
- For systems with correlations, the limit $\sigma \sim \sqrt{\pi e}$ means the system transitions to "correlation degeneracy," beyond which the description through correlation clusters (e.g., Debye spheres) no longer applies. Then the correlation clusters dissolve and the entropy increases, leading to plasma states characterized by insignificant correlations among the particles.

Consider space plasmas with Debye spheres full of correlated particles, $N_{\mathrm C} \sim N_{\mathrm D}$, according to 2.27, for simplicity. The correlation degeneracy can be caused by increasing the dimensionless parameter $\sigma \propto \sqrt{n}/T$, that is, by increasing the density and/or decreasing the temperature. It would be interesting to calculate the increase of entropy generated by this destruction of particle correlations. However, the change of the length scale from the correlation length $\ell_{\mathrm C} \sim \lambda_{\mathrm D}$ to the smaller scale of the interparticle distance b decreases rather than increases the entropy; the deviation of the entropy is negative because $\lambda_{\mathrm D} > b$, i.e.,

$$
\begin{aligned}
\Delta S/N &= (S_{\mathrm{no-C}} - S_{\mathrm{Corr}})/N = d\cdot k_{\mathrm B}\cdot\ln(\sigma_{\mathrm{Corr}}/\sigma_{\mathrm{no-C}})\\
&\approx d\cdot k_{\mathrm B}\cdot\ln(b/\lambda_{\mathrm D}) \approx -\frac{1}{6}d\cdot k_{\mathrm B}\cdot\ln N_{\mathrm D} < 0.
\end{aligned}
$$

(2.35a)

Nonetheless, a new quantization constant characterizes the phase–space of plasmas with local correlations; this is similar to the classical Planck constant, but ~ 12 orders of magnitude larger, $h_*/h \approx 10^{12}$ (Chapter 5, Section 5.9; Livadiotis and McComas, 2013b, 2014b, 2014c; Livadiotis, 2015d, 2016c, 2016e; Livadiotis and Desai, 2016). However, the Planck constant characterizes isolated particles or uncorrelated particles systems; thus the deviation of the entropy is positive because $N_D << 10^{36}$, i.e.,

$$\Delta S/N \approx \frac{1}{2}d \cdot k_B \cdot \ln\left(\frac{h_*}{h}N_D^{-\frac{1}{3}}\right) \approx \frac{1}{6}k_B \cdot d \cdot \ln 10 \cdot (36 - \log N_D) > 0. \qquad (2.35b)$$

2.4 Derivation of the Entropic Formula for Velocity Kappa Distributions

2.4.1 The Argument ϕ_q

Nonextensive statistical mechanics were introduced by Tsallis (1988) as a possible generalization of the classical framework of Boltzmann–Gibbs statistical mechanics. The main suggestion was to generalize the formulation of entropy, depending on a parameter q. The maximization of this new type of entropy under the constraint of fixed internal energy leads to the probability distribution of energy within the framework of canonical ensemble. The characteristic exponent of the canonical distribution was $1/(q-1)$, and it has been connected with a type of kappa distributions under the transformation $\kappa = -1/(q-1)$. However, the earliest nonextensive statistical mechanics were suffering from some fundamental issues. In particular, the canonical distribution of energy was not invariant under arbitrary selections of the zero-level energy, while the internal energy was not extensive as it should be for uncorrelated distributions. These two inconsistencies as well as various other problems were finally solved a decade later by Livadiotis and McComas (2009). Thereafter, the characteristic exponent of the canonical distribution is $-q/(q-1)$, which is connected with the primary type of kappa distributions under the transformation $\kappa = 1/(q-1)$ (for more details, see Chapter 1).

There are several differences between the structures of the old and the modern nonextensive statistical mechanics. The latter generalizes the notion of the canonical distribution via the formalism of escort probabilities. The ordinary $p(\overrightarrow{r}, \overrightarrow{u})$ and escort $P(\overrightarrow{r}, \overrightarrow{u})$ probability distributions (Beck and Schlogl, 1993) are related as follows:

$$P(\overrightarrow{r}, \overrightarrow{u}) = \frac{p(\overrightarrow{r}, \overrightarrow{u})^q}{\int p(\overrightarrow{r}, \overrightarrow{u})^q d\overrightarrow{r}\, d\overrightarrow{u}} \Leftrightarrow p(\overrightarrow{r}, \overrightarrow{u}) = \frac{P(\overrightarrow{r}, \overrightarrow{u})^{1/q}}{\int P(\overrightarrow{r}, \overrightarrow{u})^{1/q} d\overrightarrow{r}\, d\overrightarrow{u}}. \qquad (2.36)$$

The kappa distribution for a d_K-dimensional system is given by the escort distribution

$$P(\vec{u}) \propto \left[1 + \frac{1}{\kappa_0} \cdot \frac{(\vec{u} - \vec{u}_b)^2}{\theta^2}\right]^{-\kappa_0 - 1 - \frac{1}{2}d_K} \text{, or } P(\vec{u}) \propto \left[1 + \frac{1}{\kappa - \frac{d_K}{2}} \cdot \frac{(\vec{u} - \vec{u}_b)^2}{\theta^2}\right]^{-\kappa - 1},$$

$$(2.37a)$$

while the respective ordinary probability distribution is

$$p(\vec{u}) \propto \left[1 + \frac{1}{\kappa_0} \cdot \frac{(\vec{u} - \vec{u}_b)^2}{\theta^2}\right]^{-\kappa_0 - \frac{1}{2}d_K} \text{, or } p(\vec{u}) \propto \left[1 + \frac{1}{\kappa - \frac{d_K}{2}} \cdot \frac{(\vec{u} - \vec{u}_b)^2}{\theta^2}\right]^{-\kappa},$$

$$(2.37b)$$

which is useful for deriving the entropic formula. Both distributions are expressed either in terms of the invariant kappa index κ_0 or the dimensionality dependent kappa index $\kappa(f) = \kappa_0 + \frac{f}{2}$, where we are restricted here to zero potential energy, $f = d_K$.

The expression of the entropy is related with the argument ϕ_q. In the continuous description, the argument ϕ_q is expressed in terms of a speed scale parameter σ_u, or equivalently, in terms of a dimensionless scale parameter $\sigma = \sigma_u/\theta$. Then, given the f-dimensional velocity space, we have

$$\phi_q \equiv \int_{-\infty}^{+\infty} \left[p(\vec{u}) \cdot \sigma_u^f\right]^q \frac{du_1 du_2 \ldots du_f}{\sigma_u^f} = \sigma_u^\ell \cdot \int_{-\infty}^{+\infty} p(\vec{u})^{1+\frac{1}{\kappa}} du_1 du_2 \ldots du_f. \quad (2.38)$$

Note that with the presence of the speed scale parameter σ_u, the probability distribution $p(\vec{u})$ scales as σ_u^{-f}, while $p(\vec{u}) \cdot \sigma_u^f$ is dimensionless.

The ordinary probability distribution, Eq. (2.37b), is normalized as follows:

$$p(\vec{u}) = \pi^{-\frac{f}{2}} \left(\kappa - \frac{f}{2}\right)^{-\frac{f}{2}} \frac{\Gamma(\kappa)}{\Gamma\left(\kappa - \frac{f}{2}\right)} \cdot \theta^{-f} \cdot \left[1 + \frac{1}{\kappa - \frac{f}{2}} \cdot \frac{(\vec{u} - \vec{u}_b)^2}{\theta^2}\right]^{-\kappa}. \quad (2.39)$$

Without loss of generality, we set $\vec{u}_b = 0$; thus $p(\vec{u}) = p(u)$ and

$$\phi_q = \sigma_u^{f/\kappa} B_f \cdot \int_{-\infty}^{+\infty} p(u)^{1+1/\kappa} u^{f-1} du. \quad (2.40)$$

Then by substituting $p(u)$ to Eq. (2.40), we obtain

$$\phi_q = \sigma^{\frac{\kappa - \frac{f}{2}}{\kappa}} \cdot \left[\pi^{\frac{f}{2}} \left(\kappa - \frac{f}{2}\right)^{\frac{f}{2}} \Gamma\left(\kappa - \frac{f}{2}\right) \Big/ \Gamma(\kappa)\right]^{-\frac{1}{\kappa}} \quad (2.41)$$

where we used the dimensionless scale parameter $\sigma = \sigma_u/\theta$.

Another way to derive the argument ϕ_q is by using directly the escort distribution function $P(\vec{u})$. This is a more convenient method because the distribution that describes the particles of the system and its statistical moments is the escort $P(\vec{u})$

and not the ordinary $p(\overrightarrow{u})$ distribution. The duality of ordinary/escort distributions is given by the following scheme (similar to Eq. (2.36)),

$$P(\overrightarrow{u}) \cdot \sigma_{\mathrm{u}}^{f} \equiv \frac{\left[p(\overrightarrow{u}) \cdot \sigma_{\mathrm{u}}^{f}\right]^{q}}{\displaystyle\int_{-\infty}^{+\infty} \left[p(\overrightarrow{u}) \cdot \sigma_{\mathrm{u}}^{f}\right]^{q} \dfrac{du_1 du_2 \ldots du_f}{\sigma_{\mathrm{u}}^{f}}}$$

$$\updownarrow \tag{2.42}$$

$$p(\overrightarrow{u}) \cdot \sigma_{\mathrm{u}}^{f} \equiv \frac{\left[P(\overrightarrow{u}) \cdot \sigma_{\mathrm{u}}^{f}\right]^{\frac{1}{q}}}{\displaystyle\int_{-\infty}^{+\infty} \left[P(\overrightarrow{u}) \cdot \sigma_{\mathrm{u}}^{f}\right]^{\frac{1}{q}} \dfrac{du_1 du_2 \ldots du_f}{\sigma_{\mathrm{u}}^{f}}},$$

from where we obtain that

$$\phi_q \equiv \int_{-\infty}^{+\infty} \left[p(\overrightarrow{u}) \cdot \sigma_{\mathrm{u}}^{f}\right]^{q} \frac{du_1 du_2 \ldots du_f}{\sigma_{\mathrm{u}}^{f}} = \left\{ \int_{-\infty}^{+\infty} \left[P(\overrightarrow{u}) \cdot \sigma_{\mathrm{u}}^{f}\right]^{\frac{1}{q}} \frac{du_1 du_2 \ldots du_f}{\sigma_{\mathrm{u}}^{f}} \right\}^{-q},$$

$$\tag{2.43a}$$

or, in terms of the kappa index,

$$\phi_q = \left\{ \int_{-\infty}^{+\infty} \left[P(\overrightarrow{u}) \cdot \sigma_{\mathrm{u}}^{f}\right]^{\frac{\kappa}{\kappa+1}} \frac{du_1 du_2 \ldots du_f}{\sigma_{\mathrm{u}}^{f}} \right\}^{-\frac{\kappa+1}{\kappa}}. \tag{2.43b}$$

The normalized escort distribution, Eq. (2.37a), is given by

$$P(\overrightarrow{u}) = \pi^{-\frac{f}{2}} \left(\kappa - \frac{f}{2}\right)^{-\frac{f}{2}} \frac{\Gamma(\kappa+1)}{\Gamma\left(\kappa - \dfrac{f}{2} + 1\right)} \cdot \theta^{-f} \cdot \left[1 + \frac{1}{\kappa - \dfrac{f}{2}} \cdot \frac{(\overrightarrow{u} - \overrightarrow{u}_b)^2}{\theta^2}\right]^{-\kappa-1}. \tag{2.44}$$

Again, we set $\overrightarrow{u}_b = 0$ for simplicity so that $P(\overrightarrow{u}) = P(u)$. Then the argument ϕ_q is deduced by substituting $P(u)$ into

$$\phi_q = \sigma_{\mathrm{u}}^{f/\kappa} B_f^{-\frac{\kappa+1}{\kappa}} \cdot \left\{ \int_{-\infty}^{+\infty} P(u)^{\frac{\kappa}{\kappa+1}} u^{f-1} du \right\}^{-\frac{\kappa+1}{\kappa}}. \tag{2.45}$$

Next, we express the argument ϕ_q in terms of the invariant kappa index κ_0. Given the number of correlated particles N_C and the degrees of freedom per particle d, then $f = d \cdot N_C$,

$$\phi_q = \left[\left(\frac{\sigma^2}{\pi e}\right)^{\frac{1}{2}d \cdot N_C} \cdot \frac{\Gamma\left(\kappa_0 + \dfrac{d}{2} \cdot N_C\right)}{\left(\kappa_0 + \dfrac{d}{2} \cdot N_C\right)^{\kappa_0 + \frac{1}{2}d \cdot N_C}} \cdot \frac{\kappa_0^{\kappa_0}}{\Gamma(\kappa_0)} \cdot e^{\frac{1}{2}d \cdot N_C} \right]^{-\frac{1}{\kappa_0 + \frac{1}{2}d \cdot N_C}}, \text{ or} \tag{2.46a}$$

$$
\phi_q = \left(\frac{\sigma^2}{\pi e}\right)^{1 - \frac{\kappa_0}{\kappa_0 + \frac{1}{2}d \cdot N_C}} \left[\frac{\Gamma\left(\kappa_0 + \frac{d}{2}\cdot N_C\right)\kappa_0^{\kappa_0} e^{\frac{1}{2}d \cdot N_C}}{\left(\kappa_0 + \frac{d}{2}\cdot N_C\right)^{\kappa_0 + \frac{1}{2}d \cdot N_C} \Gamma(\kappa_0)}\right]^{\frac{1}{\kappa_0 + \frac{1}{2}d \cdot N_C}}, \tag{2.46b}
$$

that can be rewritten as

$$
\phi_q = \left[\left(\frac{\sigma^2}{\pi e}\right)^{\frac{1}{2}d \cdot N_C} \sqrt{\frac{\kappa_0}{\kappa_0 + \frac{d}{2}\cdot N_C}} \cdot \frac{g\left(\kappa_0 + \frac{d}{2}\cdot N_C\right)}{g(\kappa_0)}\right]^{\frac{1}{\kappa_0 + \frac{1}{2}d \cdot N_C}}, g(x) \equiv \frac{\Gamma(x)}{\sqrt{2\pi/x}(x/e)^x}. \tag{2.46c}
$$

Using the Stirling's approximation,

$$
\Gamma(n) = \sqrt{\frac{2\pi}{n}} \cdot \left(\frac{n}{e}\right)^n \cdot E_n, \tag{2.47a}
$$

with the error E_n converging to 1 for $n \to \infty$ (e.g., Robbins, 1955),

$$
E_n \cong 1 + \frac{1}{12}\cdot\frac{1}{n} + \frac{1}{288}\cdot\frac{1}{n^2} - \frac{139}{51840}\cdot\frac{1}{n^3} + \cdots, \frac{1}{12n+1} < \ln E_n < \frac{1}{12n}, \tag{2.47b}
$$

(e.g., for $n = 8$ the error is about 1%), we may approximate the argument ϕ_q by

$$
\phi_q \cong \left[\left(\frac{\sigma^2}{\pi e}\right)^{\frac{1}{2}d \cdot N_C} \cdot g(\kappa_0)^{-1} \cdot \sqrt{\frac{\kappa_0}{\kappa_0 + \frac{d}{2}\cdot N_C}}\right]^{\frac{1}{\kappa_0 + \frac{1}{2}d \cdot N_C}}, \tag{2.48}
$$

$$
g(x) = E_x, 8 < \kappa_0 + \frac{1}{2}d \cdot N_C.
$$

The Stirling's approximation is accurate enough for $8 < \kappa_0 + \frac{1}{2}d \cdot N_C$, that is for all the values of κ_0 and $N_C \geq 10$ plotted in Fig. 2.1 ϕ_q may be further approximated for $\kappa_0 \ll \frac{1}{2}d \cdot N_C$, where Eq. (2.48) gives

$$
\phi_q \cong \left(\frac{\sigma^2}{\pi e}\right) \cdot \left[\left(\frac{\pi e}{\sigma^2}\right)^{\kappa_0} g(\kappa_0)^{-1} \cdot \sqrt{\frac{\kappa_0}{\frac{d}{2}\cdot N_C}}\right]^{\frac{1}{\frac{1}{2}d \cdot N_C}}, \text{ or} \tag{2.49a}
$$

$$
\phi_q \cong \left(\frac{\sigma^2}{\pi e}\right) \cdot \left\{1 + \frac{\kappa_0}{\frac{1}{2}d \cdot N_C} \ln\left[\left(\frac{\pi e}{\sigma^2}\right)g(\kappa_0)^{-\frac{1}{\kappa_0}} \left(\frac{\kappa_0}{\frac{d}{2}\cdot N_C}\right)^{\frac{1}{2\kappa_0}}\right]\right\}, \tag{2.49b}
$$

while at the limit of $\kappa_0 \ll \frac{1}{2}d \cdot N_C \to \infty$, Eq. (2.49b) becomes

$$
\phi_q \to \left(\frac{\sigma^2}{\pi e}\right). \tag{2.49c}
$$

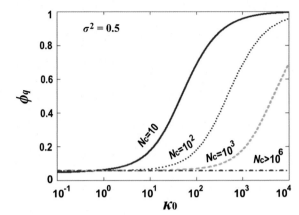

FIGURE 2.1
Dependence of ϕ_q on the invariant kappa index κ_0 for $\sigma^2 = 0.5$ and various numbers of correlated particles N_C. Adapted from Livadiotis (2014a).

The deviation of ϕ_q from the constant value of $\cong \sigma^2/(\pi e)$ is larger once the condition is not valid, i.e., when the kappa index is larger than N_C. This is shown in Fig. 2.1, where ϕ_q is plotted as a function of the invariant kappa index κ_0 and for various N_C.

The importance of the argument ϕ_q is not only its direct relation to the entropic formulation, but also its role to connect the actual temperature with the Lagrangian temperature-like parameter. The latter is a parameter related to the maximization of the entropy within the framework of the canonical ensemble and leads to a well-defined temperature only at thermal equilibrium. The Maxwellian temperature was thought to be defined and determined by the temperature that the system would have if it was residing at thermal equilibrium. Now we understand that such a definition is inconsistent because the temperature of the system is independent of the value of the kappa index. As it is shown by Livadiotis (2014a), the Maxwellian temperature is not an actual temperature, but a general definition of a temperature-like parameter that coincides with the actual temperature at thermal equilibrium. Therefore, the Lagrangian is a Maxwellian "temperature." The argument ϕ_q is given by the ratio of the actual temperature T to the Lagrangian temperature-like parameter T_L.

2.4.2 The Entropy

2.4.2.1 Formula

The entropy is a functional of the ordinary probability distribution $p(\vec{u})$ through the argument ϕ_q,

$$S(p(\vec{u}))/k_B = \frac{1 - \phi_q(p(\vec{u}))}{q - 1} = \kappa \cdot \left[1 - \phi_q(p(\vec{u}))\right]. \tag{2.50}$$

Substituting ϕ_q from Eq. (2.46b), we obtain

$$
S(\sigma; \kappa_0; N_C)/k_B = \left(\kappa_0 + \frac{d}{2} \cdot N_C\right)
$$

$$
\times \left\{ 1 - \left[\left(\frac{\sigma^2}{\pi e}\right)^{\frac{1}{2}d \cdot N_C} \cdot \frac{\Gamma\left(\kappa_0 + \frac{d}{2} \cdot N_C\right)}{\left(\kappa_0 + \frac{d}{2} \cdot N_C\right)^{\kappa_0 + \frac{1}{2}d \cdot N_C}} \cdot \frac{\kappa_0^{\kappa_0}}{\Gamma(\kappa_0)} \cdot e^{\frac{1}{2}d \cdot N_C}\right]^{\frac{1}{\kappa_0 + \frac{1}{2}d \cdot N_C}}\right\}, \text{or}
$$

$$(2.51a)$$

$$
S(\sigma; \kappa_0; N_C)/k_B = \left(\kappa_0 + \frac{d}{2} \cdot N_C\right)
$$

$$
\times \left\{ 1 - \left(\frac{\sigma^2}{\pi e}\right)^{1 - \frac{\kappa_0}{\kappa_0 + \frac{1}{2}d \cdot N_C}} \left[\frac{\Gamma\left(\kappa_0 + \frac{d}{2} \cdot N_C\right)\kappa_0^{\kappa_0} e^{\frac{1}{2}d \cdot N_C}}{\left(\kappa_0 + \frac{d}{2} \cdot N_C\right)^{\kappa_0 + \frac{1}{2}d \cdot N_C} \Gamma(\kappa_0)}\right]^{\frac{1}{\kappa_0 + \frac{1}{2}d \cdot N_C}}\right\},
$$

$$(2.51b)$$

which gives the entropy of a cluster of N_C correlated particles.

In plasmas, the Debye shielding introduces local correlations among particles, and the Debye length λ_D is the smallest possible correlation length (Chapter 5). The Debye sphere is the smallest possible correlation cluster with a number of particles given by $N_D = \frac{B_d}{d} n \cdot \lambda_D^d$. However, not all the particles within a Debye sphere are equally correlated; we may consider that only a number of N_C particles out of N_D Debye particles are correlated. In this general configuration, N particles are separated to $M = N/N_D$ clusters, each of which has N_D particles, N_C correlated and N_D uncorrelated particles; the system has entropy

$$
S_{tot}(N)/N = (1/N_D) \cdot [S(\kappa_0; N_C) + (N_D - N_C) \cdot S(\kappa_0 \to \infty; N = 1)], \text{or,}
$$

$$(2.52a)$$

per particle entropy

$$
\frac{S_{tot}}{Nk_B} = \frac{\kappa_0 + \frac{d}{2} \cdot N_C}{N_D} \cdot \left\{ 1 - \left(\frac{\sigma^2}{\pi e}\right)^{1 - \frac{\kappa_0}{\kappa_0 + \frac{1}{2}d \cdot N_C}} \left[\frac{\Gamma\left(\kappa_0 + \frac{d}{2} \cdot N_C\right)\kappa_0^{\kappa_0} e^{\frac{1}{2}d \cdot N_C}}{\left(\kappa_0 + \frac{d}{2} \cdot N_C\right)^{\kappa_0 + \frac{1}{2}d \cdot N_C} \Gamma(\kappa_0)}\right]^{\frac{1}{\kappa_0 + \frac{1}{2}d \cdot N_C}}\right\}
$$

$$
+ \frac{d}{2} \cdot \left(1 - \frac{N_C}{N_D}\right) \cdot \ln\left(\frac{\pi e}{\sigma^2}\right).
$$

$$(2.52b)$$

Thereafter, we may consider the simple case that there are $M = N/N_C$ correlation clusters in the system, where each cluster has N_C correlated particles.

2.4.2.2 Thermodynamic Limits

Next we derive the entropy at the two extreme thermodynamic limits of (1) a large number of correlated particles but smaller than the kappa index, $\frac{1}{2}d \cdot N_C \ll \kappa_0$; and (2) kappa index smaller than the large number of correlated particles, $\kappa_0 \ll \frac{1}{2}d \cdot N_C$:

1. For $\frac{1}{2}d \cdot N_C \ll \kappa_0$. The argument ϕ_q, given by Eq. (2.48), becomes

$$\phi_q \cong \exp\left\{ \frac{1}{\kappa_0 + \frac{1}{2}d \cdot N_C} \ln\left[\left(\frac{\sigma^2}{\pi e}\right)^{\frac{1}{2}d \cdot N_C} \cdot g(\kappa_0)^{-1} \cdot \sqrt{\frac{\kappa_0}{\kappa_0 + \frac{d}{2} \cdot N_C}} \right] \right\}$$

$$\cong 1 + \frac{1}{\kappa} \ln\left[\left(\frac{\sigma^2}{\pi e}\right)^{\frac{1}{2}d \cdot N_C} \right],$$
(2.53)

where $g(\kappa_0) \to 1$ and $\sqrt{\kappa_0 \big/ \left(\kappa_0 + \frac{d}{2} \cdot N_C\right)} \to 1$. Hence, the entropy becomes

$$S(N_C) \equiv k_B \cdot \kappa (1 - \phi_q) \cong \frac{1}{2}d \cdot N_C k_B \cdot \ln\left(\frac{\pi e}{\sigma^2}\right), \quad \text{or} \quad \frac{S(N)}{N k_B} = \frac{S(N_C)}{N_C k_B} \cong \frac{1}{2}d \cdot \ln\left(\frac{\pi e}{\sigma^2}\right),$$
(2.54)

that is, the entropic formula at thermal equilibrium, given in Eq. (2.32) and shown in Eq. (2.34) within the Boltzmann–Gibbs statistical framework. Note that we use the relation

$$S(N)/N = S(N_C)/N_C,$$
(2.55)

because the $M = N/N_C$ correlation clusters are uncorrelated to each other so that the total entropy among the clusters is additive $S(N) = M \cdot S(N_C)$.

We notice that for small values of entropy, we have

$$\frac{S(N)}{N k_B} \cong \frac{1}{2}d \cdot \left(1 - \frac{\sigma^2}{\pi e}\right), \quad \text{for } S \cong 0,$$
(2.56a)

because in order to attain $S \cong 0$, we must have $\sigma^2 \cong \pi e$, i.e.,

$$\ln\left(\frac{\pi e}{\sigma^2}\right) \cong \frac{\pi e}{\sigma^2} \cdot \left(1 - \frac{\sigma^2}{\pi e}\right) \cong 1 - \frac{\sigma^2}{\pi e}.$$
(2.56b)

2. For $\kappa_0 \ll \frac{1}{2}d \cdot N_C$, the argument ϕ_q is given by Eq. (2.49a,b), or by Eq. (2.49c) for $N_C \to \infty$. Hence,

$$
S \equiv k_B \cdot \kappa (1 - \phi_q) \cong k_B \cdot \frac{d}{2} \cdot N_C \left\{ 1 - \left(\frac{\sigma^2}{\pi e} \right) - \frac{\kappa_0}{\frac{1}{2}d \cdot N_C} \cdot \left(\frac{\sigma^2}{\pi e} \right) \cdot \ln \left[\left(\frac{\pi e}{\sigma^2} \right) g(\kappa_0)^{-\frac{1}{\kappa_0}} \left(\frac{\kappa_0}{\frac{d}{2} \cdot N_C} \right)^{\frac{1}{2\kappa_0}} \right] \right\},
$$

(2.57a)

or

$$
\frac{S(N)}{N k_B} = \frac{S(N_C)}{N_C k_B} \cong \frac{d}{2} \cdot \left\{ 1 - \left(\frac{\sigma^2}{\pi e} \right) - \frac{\kappa_0}{\frac{1}{2}d \cdot N_C} \cdot \left(\frac{\sigma^2}{\pi e} \right) \cdot \ln \left[\left(\frac{\pi e}{\sigma^2} \right) g(\kappa_0)^{-\frac{1}{\kappa_0}} \left(\frac{\kappa_0}{\frac{d}{2} \cdot N_C} \right)^{\frac{1}{2\kappa_0}} \right] \right\}
$$

$$
\cong \frac{d}{2} \cdot \left(1 - \frac{\sigma^2}{\pi e} \right).
$$

(2.57b)

The results at the two thermodynamic limits, (1) $\frac{1}{2}d \cdot N_C \ll \kappa_0$, and (2) $\kappa_0 \ll \frac{1}{2}d \cdot N_C$, can also be interpreted as those in the two limiting stationary states of thermal (1) equilibrium, and (2) antiequilibrium, i.e.,

$$
\frac{S(\sigma; \kappa_0; N_C)}{\frac{d}{2} \cdot N_C k_B} = \left(1 + \frac{\kappa_0}{\frac{d}{2} \cdot N_C} \right)
$$

$$
\times \left\{ 1 - \left(\frac{\sigma^2}{\pi e} \right)^{1 + \frac{1}{\frac{1}{2}d \cdot N_C}} \cdot \left[\frac{\Gamma \left(\kappa_0 + \frac{d}{2} \cdot N_C \right)}{\left(\kappa_0 + \frac{d}{2} \cdot N_C \right)^{\kappa_0 + \frac{1}{2}d \cdot N_C} \Gamma(\kappa_0)} \cdot \frac{\kappa_0^{\kappa_0}}{\Gamma(\kappa_0)} \cdot e^{\frac{1}{2}d \cdot N_C} \right]^{\frac{1}{\kappa_0 + \frac{1}{2}d \cdot N_C}} \right\}
$$

$$
= \begin{cases} \ln \left(\frac{\pi e}{\sigma^2} \right) & \text{for } \dfrac{\kappa_0}{\frac{d}{2} \cdot N_C} \gg 1, \text{ Equilibrium,} \\[3mm] 1 - \dfrac{\sigma^2}{\pi e} & \text{for } \dfrac{\kappa_0}{\frac{d}{2} \cdot N_C} \ll 1, \text{ Antiequilibrium.} \end{cases}
$$

(2.58)

As shown in Eqs. (2.56a,b), the entropy, if small, is identical for the two extreme cases of thermal (1) equilibrium, and (2) antiequilibrium.

Fig. 2.2 depicts the entropy $S(\sigma; \kappa_0; N_C)$ given in Eq. (2.58) for $d = 3$ and various values of the kappa index κ_0, the number of correlated particles N_C, and the modified dimensionless scale parameter $\tilde{\sigma} \equiv \sigma^2/(\pi e)$. Panels (a) and (b) depict

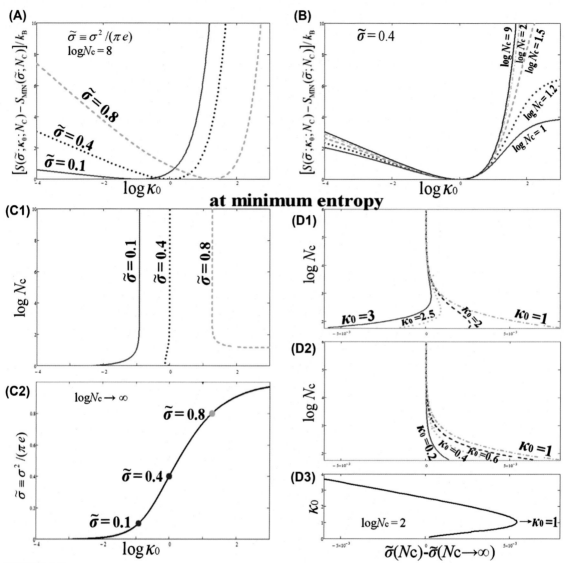

FIGURE 2.2

Entropy $S(\sigma;\kappa_0;N_C)$ as a function of the kappa index (on semilog scale) for various values of (A) $\tilde{\sigma} \equiv \sigma^2/(\pi e)$, and (B) $\log N_C$. The minimum of entropy is studied in panels (C$_1$), (C$_2$), (D$_1$), (D$_2$), and (D$_3$).

the difference of the entropy $S(\sigma;\kappa_0;N_C)$ from its minimum (that is, setting the entropic minimum to zero), $S(\sigma;\kappa_0;N_C)-S_{MIN}(\sigma;N_C)$; panel (A) plots the entropy for $\log N_C = 8$ and $\tilde{\sigma} = 0.1, 0.4, 0.8$, while panel (b) plots the entropy for $\tilde{\sigma} = 0.4$ and $\log N_C = 1, 1.2, 1.5, 2, 9$. Panels (C$_1$), (C$_2$), (D$_1$), (D$_2$), and (D$_3$) show plots corresponding to the minimum entropy. In (C$_1$), the kappa index κ_0 that is corresponding to the minimum entropy is shown as function of $\log N_C$ for $\tilde{\sigma} = 0.1, 0.4,$ and 0.8. The orientation of this panel is chosen to match the respective minima of panel (A). In (C_2), the kappa index κ_0 is plotted for $N_C \to \infty$ as a function of $\tilde{\sigma}$. Finally, panels (D_1) and (D_2) plot $\tilde{\sigma}$ as a function of $\log N_C$ for various kappa indices with the value of $\tilde{\sigma}$ standardized to that at $N_C \to \infty$; we observe that the largest deviation of $\tilde{\sigma}(N_C \ll \infty)$ from $\tilde{\sigma}(N_C \to \infty)$ occurs for $\kappa_0 \approx 1$. This is illustrated in (d_3), where the deviation $\tilde{\sigma}(N_C = 2) - \tilde{\sigma}(N_C \to \infty)$ is plotted as a function of κ_0.

Setting $n = N/V$ and $U = \frac{3}{2} N k_B T$, we can derive the generalization of the Sackur–Tetrode entropy (Sackur, 1911; Tetrode, 1912), which in the extreme states of equilibrium and antiequilibrium becomes:

$$\frac{S(U; V; N)}{\frac{3}{2} \cdot N k_B} = \begin{cases} \ln\left(c_1 \cdot \dfrac{V^{\frac{2}{3}} U}{N^{\frac{5}{3}}}\right), & \text{Equilibrium,} \\[2mm] \ln\left(c_2 \cdot \dfrac{V\, U^2}{N^3}\right), & \text{near Equilibrium,} \\[2mm] 1 - \left(c_2 \cdot \dfrac{V\, U^2}{N^3}\right)^{-1}, & \text{Antiequilibrium,} \end{cases} \tag{2.59a}$$

where

$$c_2 \equiv \left[\frac{9(36\pi)^{\frac{1}{3}}}{4e} \cdot \frac{q_e^2 \hbar_*^2}{\varepsilon_0 (m_i m_e)^{\frac{1}{2}}}\right]^{-1} \cong 21.9 \times 10^{40}, c_1 \equiv \left[\frac{3\pi}{e^{\frac{5}{3}}} \cdot \frac{\hbar^2}{(m_i m_e)^{\frac{1}{2}}}\right]^{-1} \cong 0.2 \times 10^{40},$$

$$\tag{2.59b}$$

(with all the units in SI).

2.5 Entropy for Isothermal Transitions Between Stationary States

2.5.1 Derivation

A spectrum-like arrangement of stationary states can be detectable within the framework of Tsallis statistical mechanics (Chapters 5, 10), where the values of the kappa index are in the interval $\kappa_0 \in (0, \infty]$ or $\kappa \in \left(\frac{3}{2}, \infty\right]$ (for three degrees of freedom); the corresponding q-indices are in the interval $q \in [1, \frac{5}{3})$. The first boundary value, i.e., $\kappa_0 \to \infty$ (or, $\kappa \to \infty$, $q = 1$), corresponds to thermal equilibrium, while the second boundary value defines the furthest stationary state from equilibrium, the antiequilibrium, i.e., $\kappa_0 \to 0$ (or, $\kappa \to \frac{3}{2}$, $q \to \frac{5}{3}$, in a 3-D system). In the following investigation, aside of the invariant kappa index, we also may use the kappa index for $d = 3$ degrees of freedom $\kappa = \kappa_0 + \frac{3}{2}$, the entropic q-index $q \equiv 1 + 1/\kappa$ (Chapter 1; Livadiotis and McComas, 2009), and the metastability measure $M_q = 4/(2\kappa + 1)$ (Chapter 10; Livadiotis and McComas, 2010b).

Here we are interested to find an entropic measure that can characterize the system in each of these stationary states. As we have already discussed, the temperature is a parameter that does not affect the value of the kappa index. Namely, by simply varying the temperature, while retaining a fixed value of kappa, the system should remain in the same stationary state. On the other hand, by varying the kappa index (or equivalently, the q-index or the M_q measure), the system can transit through various stationary states without any variation of the temperature (isothermal procedure). Then we wish to find an entropy expression, which is dependent on the kappa index and independent of the temperature; this is the purpose of this section. This entropy will be used to describe the variations of the kappa index for fixed temperature, namely, the isothermal transitions between the stationary states.

Livadiotis and McComas (2009) derived the argument ϕ_q in the case of one-particle description, $N_C = 1$, with $d = 3$ degrees of freedom and using a formalism expressed in terms of the entropic q-index. Both ordinary and escort distributions are isotropic $p(\vec{w}) = p(w)$ (where we use the velocity in the commoving reference frame, $\vec{w} \equiv \vec{u} - \vec{u}_b$); thus from Eq. (2.38), we obtain the argument ϕ_q by

$$\phi_q \equiv \int_0^\infty \left[p(w) \cdot \sigma_u^f \right]^q \cdot \frac{4\pi w^2 dw}{\sigma_u^f}. \tag{2.60}$$

The normalized ordinary distribution is

$$p(w; q; \theta) = \theta^{-3} \cdot \frac{1}{2\pi} \cdot \frac{1}{1_q\left(\frac{3}{2}\right)^{\frac{3}{2}} \tilde{\Gamma}_q\left(\frac{3}{2}\right)} \cdot \exp_q\left[-\frac{1}{1_q\left(\frac{3}{2}\right)} \cdot \frac{w^2}{\theta^2} \right]; \tag{2.61}$$

thus the extracted argument ϕ_q is given by

$$\phi_q(q; \sigma) = 1_q\left(\frac{3}{2}\right)^{1_q\left(\frac{1}{2}\right)} \left[2\pi \Gamma_q\left(\frac{3}{2}\right) \right]^{1-q} \cdot \sigma^{3(q-1)}. \tag{2.62a}$$

This may be written in terms of kappa index as follows:

$$\phi_q(\kappa_0; \sigma) = \left[\left(\frac{\sigma^2}{\pi e}\right)^{\frac{3}{2}} \frac{\kappa^{-\kappa}\Gamma(\kappa)}{\kappa_0^{-\kappa_0}\Gamma(\kappa_0)} e^{\frac{3}{2}} \right]^{\frac{1}{\kappa}}, \tag{2.62b}$$

where we have the standard 3-D kappa index related with the invariant kappa index by $\kappa = \kappa_0 + \frac{3}{2}$.

The functions $\tilde{\Gamma}_q(a)$ and $\Gamma_q(a)$ are the q-deformed gamma functions of the first and second kind. These were defined and studied in Livadiotis and McComas (2009; see also Chapter 1), namely,

$$\tilde{\Gamma}_q(a) \equiv \int_0^\infty \exp_q(-\gamma)\gamma^{a-1} d\gamma = \int_0^\infty [1 - (1-q)\gamma]_+^{\frac{1}{1-q}} \gamma^{a-1} d\gamma, \tag{2.63a}$$

$$\Gamma_q(a) \equiv \int_0^\infty \exp_q(-\gamma)^q \gamma^{a-1} d\gamma = \int_0^\infty [1 - (1-q)\gamma]_+^{\frac{q}{1-q}} \gamma^{a-1} d\gamma, \qquad (2.63b)$$

where $\exp_q(x) \equiv [1 + (1-q)x]_+^{\frac{1}{1-q}}$ denotes the q-deformed exponential function (e.g., Silva et al., 1998; Yamano, 2002), while the subscript "+" denotes the cut-off condition $[x]_+ = x$, if $x \geq 0$ and $[x]_+ = 0$, and if $x \leq 0$, in accordance to Eq. (1.13) in Chapter 1. Both kinds of q-deformed gamma functions recover the ordinary gamma function $\Gamma(a)$ for $q \to 1$. The q-deformed "unit function," is defined by $1_q(u) \equiv 1 + (1-q)u$ (see 1.4.2 in Chapter 1; Livadiotis and McComas, 2009). Also, we may use the kappa index for the q-deformation symbolism, i.e., $\Gamma_q(a) = \Gamma_\kappa(a)$. For $a = \frac{3}{2}$, we have

$$\Gamma_q\left(\frac{3}{2}\right) = \tilde{\Gamma}_q\left(\frac{3}{2}\right) \cdot 1_q\left(\frac{3}{2}\right) = \frac{\sqrt{\pi}}{2} \cdot (q-1)^{-\frac{1}{2}} \cdot \frac{\Gamma\left(\dfrac{1}{q-1} - \dfrac{1}{2}\right)}{\Gamma\left(\dfrac{1}{q-1}\right)}, \text{ or} \qquad (2.63c)$$

$$\Gamma_\kappa\left(\frac{3}{2}\right) = \frac{\sqrt{\pi}}{2} \cdot \kappa^{\frac{1}{2}} \cdot \frac{\Gamma\left(\kappa - \dfrac{1}{2}\right)}{\Gamma(\kappa)}.$$

Note that the argument ϕ_q in Eq. (2.62b) can be also derived from Eq. (2.46c), set for the special case of $N_C = 1$ particle and $d = 3$ degrees of freedom. We also recall that $\kappa = \kappa_0 + \frac{d}{2} \cdot N_C = \kappa_0 + \frac{3}{2}$.

2.5.2 Survey

We address the physical meaning of the speed scale σ_u and its relation with the thermal speed parameter $\theta \equiv \sqrt{2k_B T/m}$. In Subsection 2.3.2, we have seen how the speed scale can be connected to the length scale σ_r, which can be the correlation length, e.g., the Debye length, or the interparticle distance, while in Subsection 2.3.3, the speed scale was interpreted by means of thermal speed and other temperature-related parameters. In particular, instead of relating σ_u with a speed scale, we equivalently relate $\sigma_u^2 m/(2k_B)$ with a temperature scale:

$$\sigma_T \equiv \sigma_u^2 m/(2k_B). \qquad (2.64)$$

Further, given the three main definitions of temperature (Chapter 1; Livadiotis and McComas, 2009, 2010a), we can potentially have three different interpretations of the temperature scale σ_T. Namely, we deal with the following temperature or temperature-like parameters: (1) the thermodynamic (called also physical temperature) T, (2) the kinetic, T_K, and (3) the Lagrangian, T_L.

In the classical case when systems are assumed to be at thermal equilibrium, all three definitions are equivalent, i.e., $T = T_K = T_L$, and thus only one trivial interpretation appears: $\sigma_T = T = T_K = T_L$, or $\sigma_u^2 m/(2k_B) = T$, namely, the dimensionless quantity is simply given by $\sigma \equiv \sigma_u/\theta = 1$. However, for stationary states out of thermal equilibrium, the equality holds only between two definitions, $T = T_K$ and T_L. Hence, there are two possible interpretations: (1)

$\sigma_T = T = T_K$, that is, $\sigma_u = \theta \equiv \sqrt{2k_B T/m}$ or $\sigma \equiv \sigma_u/\theta = 1$. This is the same as in the case of equilibrium; and (2) $\sigma_T = T_L$, that is, $\sigma_u = \theta_L \equiv \sqrt{2k_B T_L/m} = \theta / \sqrt{\phi_q}$ or $\sigma \equiv \sigma_u/\theta = 1 / \sqrt{\phi_q}$; we recall that T is independent parameter, while T_L is dependent, $T_L = T_L(T;q)$. The connecting function is the argument ϕ_q. Hence, we have two possible interpretations:

$$(1)\ \sigma_T = T = T_K \text{ or } \sigma \equiv \sigma_u/\theta = 1, \quad (2)\ \sigma_T = T_L \text{ or } \sigma \equiv \sigma_u/\theta = 1 / \sqrt{\phi_q}.$$

$$(2.65)$$

Let us start examining interpretation (1). Substituting $\sigma = 1$ in the derived argument ϕ_q given in Eq. (2.62a) or Eq. (2.62b), we find, respectively,

$$\phi_q(q; \sigma = 1) = 1_q\left(\frac{3}{2}\right)^{1_q\left(\frac{1}{2}\right)} \cdot \left[2\pi\Gamma_q\left(\frac{3}{2}\right)\right]^{1-q}, \text{ or}$$

$$(2.66)$$

$$\phi_q(\kappa_0; \sigma = 1) = \left[\pi^{-\frac{3}{2}} \frac{\kappa^{-\kappa}\Gamma(\kappa)}{\kappa_0^{-\kappa_0}\Gamma(\kappa_0)}\right]^{\frac{1}{\kappa}}.$$

In this way, the argument ϕ_q depends only on the q-index that characterizes a particular stationary state, and thus the same holds for the entropy S, i.e.,

$$S(q) = \frac{1 - 1_q\left(\frac{3}{2}\right)^{1_q\left(\frac{1}{2}\right)} \cdot \left[2\pi\Gamma_q\left(\frac{3}{2}\right)\right]^{1-q}}{q - 1}, \text{ or } S(\kappa_0) = \kappa \cdot \left\{1 - \left[\pi^{-\frac{3}{2}} \frac{\kappa^{-\kappa}\Gamma(\kappa)}{\kappa_0^{-\kappa_0}\Gamma(\kappa_0)}\right]^{\frac{1}{\kappa}}\right\}.$$

$$(2.67a)$$

For the extreme states at equilibrium ($q = 1$, $\kappa_0 \to \infty$) and antiequilibrium ($q \to \frac{5}{3}$, $\kappa_0 \to 0$), the entropic formulae in Eq. (2.67a) give

$$S_{EQ} \equiv S(q = 1) = \frac{3}{2} \cdot \ln(\pi e), S_{aEQ} \equiv S\left(q \to \frac{5}{3}\right) = \frac{3}{2}.$$

$$(2.67b)$$

In the whole interval $q \in [1, \frac{5}{3})$, and as q decreases (or kappa increases), the entropy increases monotonically from its minimum value S_{aEQ} to its maximum S_{EQ}.

In the second interpretation (2), we use the only temperature-like quantity emerging from statistical mechanics, the inverse of the second Lagrangian multiplier, that is to say $\sigma = 1 / \sqrt{\phi_q}$, concluding in

$$\phi_q(q) = \left\{1_q\left(\frac{3}{2}\right)^{1_q\left(\frac{1}{2}\right)} \cdot \left[2\pi\Gamma_q\left(\frac{3}{2}\right)\right]^{1-q}\right\}^{\frac{1}{1_q\left(-\frac{3}{2}\right)}}, \text{ or}$$

$$(2.68a)$$

$$\phi_q(\kappa_0) = \left[\pi^{-\frac{3}{2}} \frac{\kappa^{-\kappa}\Gamma(\kappa)}{\kappa_0^{-\kappa_0}\Gamma(\kappa_0)}\right]^{\frac{1}{\kappa+\frac{3}{2}}}$$

and

$$S(q) = \frac{1 - \left\{ 1_q \left(\frac{3}{2}\right)^{1_q\left(\frac{1}{2}\right)} \cdot \left[2\pi \Gamma_q \left(\frac{3}{2}\right) \right]^{1-q} \right\}^{\frac{1}{1_q\left(-\frac{3}{2}\right)}}}{q - 1}, \text{ or} \tag{2.68b}$$

$$S(\kappa_0) = \kappa \cdot \left\{ 1 - \left[\pi^{-\frac{3}{2}} \frac{\kappa^{-\kappa} \Gamma(\kappa)}{\kappa_0^{-\kappa_0} \Gamma(\kappa_0)} \right]^{\frac{1}{\kappa + \frac{3}{2}}} \right\}.$$

The main characteristic of this entropy is that it is not increasing monotonically over the whole interval $q \in [1, \frac{5}{3})$. In particular, as the q-index decreases (or the kappa index increases), from antiequilibrium to equilibrium, the entropy decreases from $S_{aEQ} = 1.5$ until a minimum at $q_{Fund} \cong 1.6125$ (corresponding to $\kappa_{Fund} \cong 1.6327$ - see Chapter 16) with entropy $S_{Fund} \cong 1.2856$. The minimum value of entropy defines a "fundamental" stationary state. The new interesting features that characterize only the second interpretation of entropy (e.g., the nonmonotonicity, the cavity, and the deceleration branch in addition to the acceleration one) appear because the fundamental does not coincide with antiequilibrium, as in the first interpretation.

After the entropic minimum and as the q-index continues to decrease, the entropy increases until equilibrium at S_{EQ}. Both the entropic behaviors of interpretations (1) and (2) are plotted in Fig. 2.3; the first interpretation (1) is represented by the blue dash line, while the nonmonotonic entropy expression that follows the second interpretation (2) is represented by the red solid line.

Under spontaneous procedures that increase entropy, interpretation (1) leads to a monotonically increasing kappa index (decreasing q-index). On the other hand, interpretation (2) suggests a more complicated behavior: starting from the fundamental state, i.e., with minimum entropy S_{Fund}, spontaneous entropy increases can now produce either a monotonically increasing kappa index

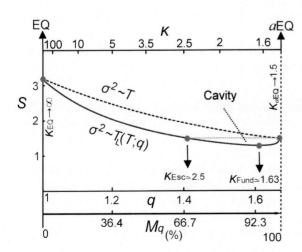

FIGURE 2.3
The entropy S is depicted with respect to the q- and κ-indices and the measure $M_q = 4/(2\kappa + 1) = 1/\left(\frac{1}{2}\kappa_0 + 1\right)$. The entropy is shown for both interpretations (1) and (2). The cavity is the nonmonotonic part of the entropy; $\kappa_{Esc} \geq \kappa > \frac{3}{2}$ in interpretation (2) and can explain the observed kappa indices, $\kappa_{Esc} \cong 2.5$ and $\kappa_{Fund} \cong 1.63$, and their dynamic relation with thermal equilibrium and antiequilibrium. Adapted from Livadiotis and McComas (2010a).

(decreasing q-index) along the left-hand side branch of $\kappa > \kappa_{Fund}$ ($q < q_{Fund}$), or a monotonically decreasing kappa index (increasing q-index) along the right-hand side branch of $\kappa < \kappa_{Fund}$ ($q > q_{Fund}$). Ultimately, the increasing kappa index (decreasing q-index) will reach equilibrium, $\kappa \to \infty$ ($q \to 1$), and the decreasing kappa index (increasing q-index) will approach antiequilibrium $\kappa \to \frac{3}{2}$ ($q \to \frac{5}{3}$). Thus, as S increases, we have: According to interpretation (1), the kappa index can continuously increase without any barrier from $\kappa = \kappa_{Fund}$ to $\kappa \to \infty$. According to interpretation (2), along the left-hand side branch we have the same behavior of the increasing kappa index, but along the right-hand side branch the decreasing kappa index approaches antiequilibrium; then, any further augments of entropy can occur only due to isentropic procedures that allow transitions between the two branches and can divert the kappa index from $\kappa \to \frac{3}{2}$ to $\kappa = \kappa_{Esc}$, where the kappa index can keep increasing along the left-hand branch. The critical point at $\kappa_{Esc} \cong 2.5$ ($q_{Esc} \cong 1.41$) separates the interval $\kappa \in \left(\frac{3}{2}, \infty\right]$ or $q \in \left[1, \frac{5}{3}\right)$ into two parts, namely, $\kappa \in (\kappa_{Esc}, \infty]$ or $q \in [1, q_{Esc})$, in which the entropy S is a bijective function of the kappa index, $\kappa \in \left(\frac{3}{2}, \kappa_{Esc}\right]$ and $q \in \left[q_{Esc}, \frac{5}{3}\right)$, in which the entropy S is only a surjective function since the one-to-one correspondence is lost. As mentioned, the cavity that is enclosed in the subinterval $\kappa \in \left(\frac{3}{2}, \kappa_{Esc}\right]$ or $q \in \left[q_{Esc}, \frac{5}{3}\right)$ can host isentropic procedures.

It is remarkable that under interpretation (2), both the predicted critical values of $\kappa_{Esc} \cong 2.5$ and $\kappa_{Fund} \cong 1.63$ are observed in space plasmas: the former constitutes a separator between the stationary states near equilibrium and those near antiequilibrium (Chapter 10; Livadiotis and McComas, 2010a, 2013a); the latter value minimizes the entropy and coincides with that estimated by Decker et al. (2005). External factors that can decrease the entropy of the system, move it into a different stationary state, closer to the fundamental state. In the case of solar wind, for example, newly formed PickUp Ions (PUIs) may play just such a role because their motion is highly ordered (Livadiotis and McComas, 2011a, 2011b; see also Livadiotis et al., 2012, 2013). This motion is dictated by the relative orientation of the solar wind velocity vector and the interplanetary magnetic field, which become increasingly perpendicular on average as one moves out through the heliosphere.

After ionization, the PUIs gyrate about the solar wind magnetic field with their guiding center moving at the solar wind velocity, forming a ring velocity distribution. Very quickly, these ions are scattered by self-generated Alfvénic fluctuations, forming spherical shell distributions, centered approximately at the solar wind velocity with a radius of the solar wind speed (e.g., Sagdeev et al., 1986; Lee and Ip, 1987). The hybrid probability distribution, which is a normalized sum of both the kappa distribution of the original stationary state and the spherical shell distribution, describes a more organized state. Therefore, this hybrid state should have less entropy than the stationary state assigned by the original kappa distribution. As the PUIs co-move with the solar wind, they undergo adiabatic cooling, which reduces the radius of the shell (Zilbersher and Gedalin, 1997). In this way, PUIs move from the shell distribution and merge with the original particles into a new kappa distribution. Thus the hybrid state transits to a new stationary state with a different kappa and lower entropy. The departure of the PUIs from the spherical distribution to the solar wind new stationary state can re-increase the entropy. However, this procedure is much slower so that the whole transition of the original to the new stationary state can be characterized by a decrease of entropy. In this model, the hybrid state is transient, tending finally to a new stationary state with lower entropy than the original stationary state, closer to the fundamental state.

Following this reasoning, in the distant solar wind of the outer heliosphere, where pickup protons represent an increasing fraction of the population, the entropy can become smaller and the specific stationary state with the lowest entropy becomes achievable. If interpretation (1) applies, then the effect of PUIs leads to antiequilibrium and internal increases of entropy would direct the system again back to equilibrium. On the other hand, if interpretation (2) applies, then the effect of PUIs leads to the fundamental state. In this case, the internal increases of entropy would direct the system both toward equilibrium and antiequilibrium (see next subsection).

Finally, we stress that something must be playing the role of decreasing the entropy of stationary states in space plasmas and driving the outer heliospheric states back toward the fundamental. The influence of PUIs is the most likely source of this reordering of the plasma and that this procedure, and the competition between this reordering and the natural progression of increasing entropy, likely play a critical element in defining space plasmas. Nevertheless, other mechanisms may also act and stabilize the particles kappa in the far-equilibrium region.

While interpretation (1) was used to describe diffusion processes within the framework of Tsallis statistical mechanics (e.g., Tsallis et al., 1995; Prato and Tsallis, 1999), there are a number of good reasons to believe that interpretation (2) is also physically meaningful and may have important implications in at least some physical systems, including space plasmas:

1. The critical value of the kappa index $\kappa_{aEQ} \rightarrow 1.5$, which is a limiting value in both interpretations, should be difficult to observe in the interpretation (1) as any increase in entropy drives systems away from this particular stationary state. In contrast, Fisk and Gloeckler (2006) claim this to be almost "ubiquitously" observed in space plasmas, suggested that systems might collect at this value, rather than moving away from it. Interpretation (2) provides a natural explanation for just this sort of collection at $\kappa_{aEQ} \rightarrow 1.5$, as increases in entropy lead to this value along the deceleration branch.

2. The critical values of the kappa index that arise from interpretation (2) have been routinely observed in space plasmas: $\kappa_{Fund} \cong 1.63$ and

$\kappa_{Esc} \cong 2.5$, in addition to $\kappa_{EQ} \to \infty$ and $\kappa_{aEQ} \to 1.5$ that arise from interpretation (1). These critical values of the kappa index are also related to certain dynamical behavior: starting from (or near) the fundamental state, $\kappa_{Fund} \cong 1.63$, the system transits to various stationary states before finally reaching equilibrium $\kappa_{EQ} \to \infty$.

3. Significantly different dynamical behavior characterizes the near- and far-equilibrium stationary states. The separator $\kappa_{Esc} \cong 2.5$ is the escape point of stationary states from the nonmonotonic entropic cavity of $\kappa_{Esc} \geq \kappa > \frac{3}{2}$ to the monotonic entropic behavior of $\infty \geq \kappa > \kappa_{Esc}$. Inside the cavity, the system can approach this escape point either directly, through a specific stochastic acceleration process, or indirectly, interceding antiequilibrium, $\kappa_{aEQ} \to 1.5$ through an opposite stochastic deceleration. This exposed dynamical scenario is also supported by observations: while the extreme states of equilibrium and antiequilibrium are both observed in the inner heliosphere, the fundamental state $\kappa_{Fund} \cong 1.63$ was detected in the distant solar wind of the outer heliosphere, where the influence of PUIs in minimizing entropy, for example, should be stronger. In the presence of internal irreversible procedures that increase entropy, and in the absence of any external procedure that decreases entropy, plasmas are expected to be characterized by stationary states that lie in the near-equilibrium region, $\infty \geq \kappa > \kappa_{Esc}$ (e.g., inner heliosphere); however, when the internal procedures that increase entropy are rare, plasmas can be characterized by stationary states that lie in the far-equilibrium region, $\kappa_{Esc} \geq \kappa > \frac{3}{2}$ (e.g., inner heliosheath) (for example, see Livadiotis et al., 2011, 2012, 2013; Livadiotis and McComas, 2013d).

4. The previously mentioned critical values were also found in other plasmas, or plasma-related, slowly driven systems that emit energy bursts, as it is shown in Chapter 10.

Next, we examine the dynamics of the transition of stationary states and show the detailed paths by which the transition of stationary states evolves toward equilibrium.

2.5.3 Spontaneous Entropic Procedures

Having expressed the entropy of stationary states as a function of the q-index, we are ready to examine how the spontaneous entropic variations will affect the values of this index. (In this section we choose to use only the q-index, but the exact analysis could be done with any of the equivalent indices of kappa or M_q). Any internal procedure is expected to either increase (irreversible procedures) or preserve (reversible procedures) the entropy of the system. As soon as the system is in stationary states outside the cavity, i.e., $q \in [1, q_{Esc})$, a specific value of entropy corresponds to a specific value of q-index (one-to-one relation). Augments of entropy lead the system to transit through stationary states of smaller q-index toward equilibrium. On the other hand, when the system is in stationary states that are located inside the cavity, i.e., $q \in [q_{Esc}, \frac{5}{3})$, a specific value of entropy corresponds to two values of q-index. Then, isentropic transitions might also take place.

As we observed in Fig. 2.3, the depicted entropy $S(q)$ is a nonmonotonic function inside the cavity that is for values of entropy $S_{Fund} \leq S \leq 1.5$ and q-index $q_{Esc} \leq q < \frac{5}{3}$. Thus the entropy is characterized by two monotonic branches: the acceleration branch for $q_{Esc} \leq q \leq q_{Fund}$ with entropy $S_A(q)$ (A-branch) and the deceleration branch for $q_{Fund} \leq q < \frac{5}{3}$ with entropy $S_D(q)$ (D-branch). Notice that the A-branch extends out of the cavity all the way to equilibrium ($q \to 1$). As q increases, the entropy on the A-branch, $S_A(q)$, decreases, approaching the minimum value $S_A(q) = S_{Fund}$ with index $q = q_{Fund}$, while on the D-branch, the entropy $S_D(q)$ increases, deviating from the minimum value S_{Fund} and the index q_{Fund}, and approaching $S_D(q) \to 1.5$, or $q \to \frac{5}{3}$. In other words, as soon as the entropy increases, the A-branch $S_A(q)$ is leading to a decrease of q (decrease of M_q, increase of κ), while the D-branch $S_D(q)$ is leading to an increase of q (increase of M_q, decrease of κ). Therefore, along the A-branch, spontaneous entropic increments lead to smaller q-indices and the stationary states transit gradually to equilibrium ($M_q = 0\%$). This procedure is related to an isothermal shifting of the energy probability distribution to higher energy values (e.g., see Chapter 10, Figure 10.4) (stochastic acceleration). On the other hand, along the D-branch, spontaneous entropic increments lead to larger q-indices, and thus the relevant transition of stationary states lead gradually to antiequilibrium ($M_q \to 100\%$). This procedure is related to an opposite isothermal shifting of the energy probability distribution toward lower energy values (stochastic deceleration).

The cavity enclosed by both the A- and D-branches for $S_{Fund} \leq S \leq 1.5$ is of great interest. Even though the system can transit to stationary states of higher entropy, a specific isentropic transition should also be possible. This can be realized by the two solutions of the equation $S(q) = \text{constant} \equiv S \leq 1.5$, namely, $q_A = S_A^{-1}(S)$ (along the A-branch) and $q_D = S_D^{-1}(S)$ (along the D-branch). Then the isentropic switching of the system between the two stationary states of deviation measures q_A and q_D is possible as soon as the entropy is $S_{Fund} \leq S \leq 1.5$, namely, it is included within the cavity (or equivalently, the metastability measure is $M_{q,Esc} \leq M_q < 100\%$, or the q-, κ-indices are $q_{Esc} \leq q < \frac{5}{3}$, $\kappa_{Esc} \geq \kappa > \frac{3}{2}$). The isentropic switching of the system can be represented by the dual function $D(q)$, where by duality, we mean the property $D^2(x) = D(D(x)) = x$, for $x \in \Re$. Namely,

$$q_D = D(q_A), q_A = D(q_D), \text{with}$$

$$D(q) = \begin{cases} (S_A^{-1} \circ S)(q) & \text{if } q \geq q_{Fund} (D - \text{branch}), \\ (S_D^{-1} \circ S)(q) & \text{if } q \leq q_{Fund} (A - \text{branch}). \end{cases} \tag{2.69}$$

We stress the fact that the isentropic switching is not a procedure that takes place instantly, which would require a huge instantaneous transfer of energy from (or to) the "far tail" particles with a kappa index close to antiequilibrium to (or from) the bulk of the distribution at lower energies with a kappa index close to the escape state. This incorrect impression might appear, since the exact meaning of Eq. (2.69) is to describe transitions only between stationary states, and it ignores all the intermediate nonstationary states that occur between them. In this way, the deformation of distributions during the isentropic switching is a finite time process whose end points are described by Eq. (2.69) as the initial and final stationary states of this switching.

The transition of the system through various stationary states along the A- and D-branches results in the extreme stationary states at equilibrium and anti-equilibrium, respectively. However, the isentropic switching allows all stationary states to eventually reach equilibrium because all the stationary states that are accumulated near antiequilibrium are eventually diverted to the value $M_{q,\text{Esc}} \cong 67.795\%$ ($q_{\text{Esc}} \cong 1.4082$, $\kappa_{\text{Esc}} \cong 2.5$), namely, to the separator of the stationary states near equilibrium and antiequilibrium states. As the entropy keeps increasing out of the cavity, the transition takes place along the A-branch. Therefore, the A-branch for $M_q < M_{q,\text{Esc}}$ (outside the cavity) is the ultimate fate followed by all stated, regardless of whether they followed by the A- or D-branch inside the cavity. This dynamical behavior of the transient stationary states is the subject of Section 2.6.

2.6 The Discrete Dynamics of Transitions Between Stationary States

2.6.1 Discrete Dynamics

The description of stationary states out of thermal equilibrium within the framework of nonextensive statistical mechanics is a steady theory. The stationary probability distribution, that is, the kappa distribution, is extracted by maximizing the entropy under the constraints of canonical ensemble. However, this procedure does not examine the dynamical behavior of the system's arrival at (or departure from) these Tsallis-like stationary states that fill the interval

$\kappa \in \left(\frac{3}{2}, \infty\right]$ (or, $q \in [1, \frac{5}{3})$, $M_q \in [0\%, 100\%)$). In the case where the stationary state

is the specific one at thermal equilibrium, then the dynamical behavior of systems approaching (or leaving) equilibrium is given by the well-known Vlasov–Maxwell differential equations (specified by the Parker's equation for the solar wind). The extracted solution describes the time-dependent probability distribution of particles in the phase space, $p(\vec{x}; \vec{w}; t)$, under the initial condition of a Maxwellian $p(\vec{x}; \vec{w}; t = 0) = P(\vec{w}; \kappa \to \infty; \theta)$. Obviously, the solution $p(\vec{x}; \vec{w}; t)$ refers either to the system's departure from equilibrium, for $t > 0$, or to its arrival to equilibrium, for $t < 0$. In a similar way, one can study the dynamical behavior of the system's approaching (or leaving) stationary states out of equilibrium. In this case, the initial probability distribution is given by a kappa distribution, namely, $p(\vec{x}; \vec{w}; t = 0) = P(\vec{w}; \kappa; \theta)$ for any permissible

value of the κ-index in the interval $\kappa \in \left(\frac{3}{2}, \infty\right]$. Interestingly, among others, Prested et al. (2008), Heerikhuisen et al. (2008, 2010, 2014), and Schwadron et al. (2009), used as initial probability distribution the specific kappa distribution with $\kappa = \kappa_{\text{Fund}} \cong 1.63$; this coincides with the kappa value measured by Decker et al. (2005). This kappa is considered here as the starting point of the transient stationary states, that is, the kappa index for which the entropy is minimized (fundamental state).

In this approach, we consider that the system transits through a number N_{ss} of alternating stationary states. Each stationary state is assigned by a kappa distribution of a specific κ-index (q-index or measure M_q). Hence, the evolution of the system during its transition into these N_{ss} stationary states is interpreted by the set of kappa distributions $\{P(\overrightarrow{w};\kappa_n;\theta)\}_{n=0}^{N_{ss}}$ in which the index n counts the various states, with $n=0$ and $n=N_{ss}$, corresponding to the initial and final stationary state, respectively. In this way, the continuous time that is included in Vlasov's, Parker's, or other stochastic equations, $p(\overrightarrow{x};\overrightarrow{w};t)$ is now replaced by the index $n=0, 1, 2, ..., N_{ss}$ that has the role of the discrete time.

Precisely speaking, a differential equation with respect to time t, as well as its solution in the time interval $t \in [0, t_f]$, can be always "discretized" by considering a finite step of time δt so that $t = n \cdot \delta t$, with $n = 0 \Leftrightarrow t = 0, n = Nss \Leftrightarrow t = t_f = N_{ss} \cdot \delta t$. Then the differential equation degenerates to a difference equation (e.g., Elaydi, 2005; Livadiotis, 2005; Livadiotis and Moussas, 2007) that is a recurrent relation (a map, multidimensional in the generic case) of the $(n + 1)$-th arguments and parameters of the distribution (e.g., the κ-index) to the n-th respective ones ($\kappa_{n+1} = \kappa_{n+1}(\kappa_n)$). Specifically, the discrete approach neglects the times of the sequential departures (and arrivals) of the system at each one of the N_{ss} metastable stationary states, while focusing instead on the sequential stationary states themselves. Most important is the fact that at each discrete time, $n = 0, 1, 2, ..., N_{ss}$, the attained stationary state is a kappa distribution where the κ-index is the only flexible parameter so that all information about the dynamics is included in the discrete time dependence of kappa indices $\{\kappa_n\}_{n=0}^{N_{ss}}$ (or equivalently, $\{q_n\}_{n=0}^{N_{ss}}, \{\kappa_{0n}\}_{n=0}^{N_{ss}}, \{M_{q,n}\}_{n=0}^{N_{ss}}$, etc.).

2.6.2 The Discrete Map of Stationary States Transitions

We demonstrate the transition of stationary states toward equilibrium and antiequilibrium states by examining the simplest case where all of the internal procedures that the system is subject to lead to equal augments of entropy δS, that is

$$S_{n+1} \equiv S_n + \delta S, \forall n = 0, 1, 2, ..., N_{ss} - 1. \tag{2.70a}$$

For our presentations, we choose the entropic index q, but the other indices and measures can be readily deduced, e.g., $\kappa = (q-1)^{-1}$, $\kappa_0 = \kappa - \frac{3}{2}$, $M_q = \left(1 + \frac{1}{2}\kappa_0\right)^{-1}$, etc. Then the relevant values of the q-index $\{q_n\}_{n=0}^{N_{ss}}$ are given by

$$q_{n+1} = f(q_n), \text{with } f(q) \equiv \begin{cases} S_D^{-1}[S(q) + \delta S] & \text{if } q \geq q_B (D - \text{branch}), \\ S_A^{-1}[S(q) + \delta S] & \text{if } q \leq q_B (A - \text{branch}). \end{cases} \tag{2.70b}$$

The transition of stationary states toward equilibrium along the A-branch and toward antiequilibrium along the D-branch can be shown in a special

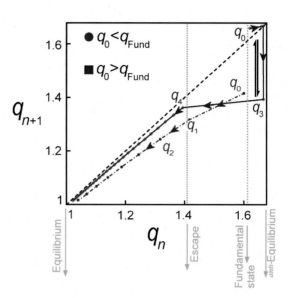

FIGURE 2.4
The phase space portrait for transient stationary states along the A-branch with $q_0 = 1.6 < q_{Fund}$ and (constant) entropic step $\delta S = 0.2$ (*red dash–dot line*), and along the D-branch with $q_0 = 1.62 > q_{Fund}$ and $\delta S = 0.06$ (*blue solid line*). In this case, when the values of the q-index become sufficiently close to antiequilibrium, that is in the n_{aEQ}-th iteration ($n_{aEQ} = 3$), then, in the next ($n_{aEQ} + 1$)-th iteration, the isentropic mechanism shifts the value of the q-index out and near the edge of the cavity, namely, $q_{n_{aEQ}} + 1 \cong q_{ESC}$. The isentropic switching is possible for any state within the cavity (*black vertical arrows*). Adapted from Livadiotis and McComas (2010a).

diagram called discrete phase space of the map (2.70b). This is the plot of all the pairs (q_n, q_{n+1}) of the map $q_n \rightarrow q_{n+1} = f(q_n)$, $\forall n = 0,1,2,\ldots$. The sequence of the q-index values q_0, q_1, q_2, and so on, is formed by the iterations of the map, Eq. (2.70b). Each of the iterations $n = 0,1,2,\ldots$ is geometrically represented by a horizontal line segment from the graph of the map $f(q)$ to the bisector, followed by a vertical line segment from the bisector back to the map. In Fig. 2.4, we present the phase space portrait for transient stationary states along the A-branch, i.e., the initial value of q-index is $q_0 < q_{Fund}$ (near-equilibrium region, see Chapter 10), where we choose $q_0 = 1.6$, and along the D-branch, i.e., the initial value of q-index is $q_0 > q_{Fund}$ (far-equilibrium region), where we choose $q_0 = 1.62$ (we choose the initial value q_0 to be lying near the value q_{Fund} of the fundamental state). Along the D-branch and when the values of the q-index become sufficiently close to antiequilibrium, let this be the n_{aEQ}-th iteration, i.e., $q_{n_{aEQ}} \cong \frac{5}{3}$, then, in the next ($n_{aEQ} + 1$)-th iteration, an isentropic switching shifts the value of the q-index out and near the edge of the cavity, namely, $q_{n_{aEQ}} + 1 \cong q_{ESC}$. Of course, isentropic switching back and forth is possible everywhere within the cavity, but only leads in a unique direction as particles switch from near the antiequilibrium state to where they can escape out of the cavity. Here we have $n_{aEQ} = 3$, and thus, from the fourth iteration and beyond, the dynamical behavior is similar to that of the case $q_0 < q_{Fund}$.

2.6.3 Numerical Application of the Discrete Transitions of Stationary States

In the following, we apply the map, Eq. (2.70b), to demonstrate the transition of stationary states within the framework of the discrete approach (as explained in Section 6.1), and their accumulation to the extreme states of equilibrium and antiequilibrium.

In this numerical calculation, the initial values of the map Eq. (2.70b) are randomly equidistributed in the interval $[q_{Fund} - \delta q, q_{Fund} + \delta q]$, with $\delta q \equiv 0.1 \cdot \left(\frac{5}{3} - q_{Esc}\right) \cong 0.0258$. Each initial value is iterated N_{ss} times through the map (2.70b) with the entropic step $\delta S = 10^{-4}$. Hence, N_{ss} transient stationary states are reached, and the relevant N_{ss} different values of the q-index are recorded.

We also allow isentropic switching [according to interpretation (2)] to take place after each transition of the system between two sequential stationary states. The choice of isentropic switching is randomly returned. Namely, after each transition $q_n \rightarrow q_{n+1} = f(q_n)$, the isentropic switching is applied $q_{n+1} \rightarrow D(q_{n+1})$ if a binary random variable, called "iso-S," returns iso-$S = 1$, while it is not applied when iso-$S = 0$.

Furthermore, we assume that the system is affected by the exterior factor that minimizes its entropy after the passage of N_{ss} steps. Therefore, the entropy is gradually increasing within $(N_{ss} - 1)$ steps, followed by an abrupt decrease in the N_{ss}-th step. The rejuvenation of the system's entropy leads to a value that is chosen to be randomly equidistributed close to the entropy minimum value (close to the fundamental state). Each of these rejuvenated entopic values is treated as a different initial value in the map, Eq. (2.70b). This cycle that involves certain $(N_{ss} - 1)$ entropic gradual increments and one abrupt decrement is repeated for $2 \times 10^5 / N_{ss}$ times, that is, a number of $2 \times 10^5 / N_{ss}$ different initial values. For each cycle, we have a number of N_{ss} iterations (discrete transitions), and thus N_{ss} values of q-indices. Therefore, the whole number of the values of q-indices for all the cycles is 2×10^5 for all the chosen values of N_{ss}.

We construct the normalized histogram $\mathcal{P}(q)$ for all 2×10^5 q-indices and plot it in Fig. 2.5. In panels (A)–(D), we plot $\mathcal{P}(q)$ for different number of iterations, i.e., (A) $N_{ss} = 2 \times 10$ (10^4 initial values), (B) $N_{ss} = 2 \times 10^2$ (10^3 initial values), (C) $N_{ss} = 2 \times 10^3$ (10^2 initial values), and (D) $N_{ss} = 2 \times 10^4$ (10 initial values). The more the iterations are, the more widely distributed the values are. For iterations less than 2×10^3 (and keeping the same entropic step $\delta S = 10^{-4}$), the stationary states remain within the cavity, $q \in [q_{Esc}, \frac{5}{3})$, and do not escape to the near-equilibrium region of $q \in [1, q_{Esc})$, e.g., see Fig. 2.5A–C. However, they do escape for larger number of iterations, i.e., for $N_{ss} = 2 \times 10^4$, e.g., see Fig. 2.5D. In addition, we have found the asymptotic behavior of $\mathcal{P}(q)$ near equilibrium, $\mathcal{P}(q) \sim e^{-2.45 M_q}$, and near antiequilibrium, $\mathcal{P}(q) = \frac{1}{1 - M_q}$, respectively. The near-equilibrium region collects the stationary states that came from both the A- and D-branches inside the cavity, and this is the reason that in the plot of Fig. 2.5D, it appears as a small abrupt step at $q \cong q_{Esc}$. Eventually, all the stationary states will move from the far-equilibrium $q \in [q_{Esc}, \frac{5}{3})$ to the near-equilibrium region $q \in [1, q_{Esc})$.

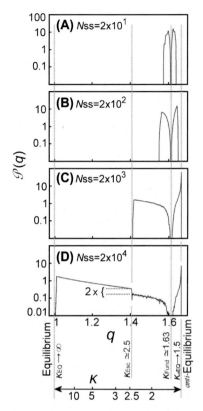

FIGURE 2.5
The probability distribution $\mathcal{P}(q)$ is depicted for a different number of iterations in our numerical simulation, i.e., (A) $N_{ss} = 2 \times 10$ (10^4 initial values), (B) $N_{ss} = 2 \times 10^2$ (10^3 initial values), (C) $N_{ss} = 2 \times 10^3$ (10^2 initial values), and (D) $N_{ss} = 2 \times 10^4$ (10 initial values). The extreme states of equilibrium and antiequilibrium states, the A- and D-branches, as well as the separator point q_{Esc} are indicated. For small numbers of iterations (A)–(C) where the stationary states are distributed only in the far-equilibrium region, then antiequilibrium is an "attractor," i.e., the states are accumulated close antiequilibrium. However, for a sufficiently large number of iterations, the antiequilibrium state drains, as the nearby states undergo isentropic transitions and escape toward equilibrium. The factor of two at $\kappa = \kappa_{Esc}$ comes from our assumption of switching, on average, every other time step. Adapted from Livadiotis and McComas (2010a).

For the assumptions used in this simple numerical experiment, it takes about $n_{aEQ} \cong 2 \times (1.5 - S_{Fund})/\delta S \cong 0.43/\delta S$ iterations for a stationary state to depart from the far-equilibrium to the near-equilibrium region. Here we assume that in the isentropic switching, states are equally likely to switch in both directions across the cavity and that a switch occurs on average every two time steps, giving the multiplier two in this expression. For $\delta S = 10^{-4}$, we have approximately a number of $n_{aEQ} \cong 4300$ iterations being spread inside the cavity, i.e., into the far-equilibrium region, out of N_{ss} total iterations. On the other hand, around $n_{EQ} \cong [(S_{EQ} - 1.5) + 2 \times (1.5 - S_{Fund})]/\delta S \cong 2.15/\delta S$ iterations are needed for a stationary state to reach equilibrium. For $\delta S = 10^{-4}$, we have $n_{EQ} \cong 21{,}500$ out of N_{ss} total iterations. In fact, having $N_{ss} = 2 \times 10^4$ (Fig. 2.5D), namely,

$N_{ss} \cong n_{EQ}$, equilibrium ($q = 1$) is barely reached. Actually, our choice of $N_{ss} = 2 \times 10^4$ was based exactly on this fact, since Maxwell distributions have been observed in solar wind, but not that frequently.

We find that around $(n_{EQ} - n_{aEQ})/n_{EQ} \cong 80\%$ of the stationary states depart from the cavity to the near-equilibrium region, accumulated close to the extreme state at equilibrium. On the other hand, around $n_{aEQ}/n_{EQ} \cong 20\%$ of the stationary states stay in the cavity, while around half of them, $\sim 10\%$ (because of the isentropic mechanism) are accumulated close to the extreme state of antiequilibrium. Therefore, the number of stationary states that are accumulated near equilibrium compared to those near antiequilibrium is $\sim 8:1$, but the former is spread over an eight times larger interval,

$$(q_{Esc} - 1)\Big/\Big(\tfrac{5}{3} - q_{Fund}\Big) \cong 8 : 1.$$

It is apparent now that both branches ultimately lead all the stationary states (under different dynamical procedures) to thermal equilibrium. First, and after a sufficient time, one should observe the accumulated stationary states near equilibrium and antiequilibrium. However, the respective dominant peaks, that is, the sharp peak at antiequilibrium and the broad peak at equilibrium (e.g., Fig. 2.5D) do not constitute a permanent configuration. On the contrary, the former is getting weaker, while the latter is getting stronger over the passage of time because stationary states transit from the accumulation at antiequilibrium and escape states via the isentropic mechanism. Therefore, the transition of stationary states, under small constant augments of entropy (e.g., $\delta S = 10^{-4}$), affirms the gradual diversion of stationary states from antiequilibrium to equilibrium, and the relevant shifting of the probability distribution of particles to larger values of velocities (or energies).

2.6.4 The Five Stages of Stationary State Transitions

Here we present five possible stages of stationary state transitions, corresponding to variations of the kappa index (independently of the temperature), in regards to the entropic measure given in Eq. (2.68b) and shown in Fig. 2.6.

2.6.4.1 The Intermediate Transitions Toward Antiequilibrium

Starting from stationary states near the fundamental state, interior (irreversible) procedures that increase the entropy of stationary states move the system either toward equilibrium (entropy increases along the A-branch) or toward antiequilibrium (entropy increases along the D-branch).

2.6.4.2 The Isentropic Switching and Its Importance

The stationary states inside the cavity can be affected by an isentropic switching between the two branches. In this way, the transitions of stationary states toward antiequilibrium along the D-branch can alter to the transitions toward equilibrium along the A-branch, and vice versa. This mostly affects the stationary states that are near antiequilibrium for which an increase of the entropy leads to stationary states out and near the edge of the cavity ($\kappa > \kappa_{Esc}$).

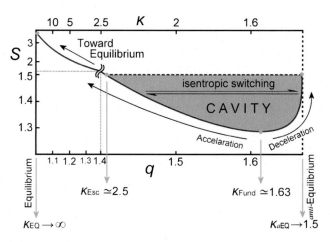

FIGURE 2.6

The entropy S is depicted with respect to the q- and κ-indices. We indicate the acceleration $\infty \geq \kappa \geq \kappa_{Fund}$ and deceleration $\kappa_{Fund} \geq \kappa > \frac{3}{2}$ branches, the isentropic switching that can take place between the two branches inside the cavity $\kappa_{Esc} \geq \kappa > \frac{3}{2}$, and the critical kappa indices corresponding to the extreme states of equilibrium $\kappa_{EQ} \to \infty$ and antiequilibrium $\kappa_{aEQ} \to \frac{3}{2}$, the escape state, $\kappa_{Esc} \cong 2.5$ (separates the near-equilibrium $\infty \geq \kappa > \kappa_{Esc}$ from the far-equilibrium $\kappa_{Esc} \geq \kappa > \frac{3}{2}$ regions), and the fundamental state, $\kappa_{Fund} \cong 1.63$ (minimum entropy). Adapted and modified from Livadiotis and McComas (2010a).

2.6.4.3 The Double Role of Antiequilibrium

In intermediate times, the stationary states do not escape yet from the cavity, but instead accumulate near antiequilibrium. This is mostly the case when the interior (irreversible) procedures that increase the entropy of stationary states are rare, e.g., in the quiet times of solar wind. However, after a sufficient number of transitions, the stationary states finally escape from the cavity because of the isentropic mechanism. Then the particles depart from antiequilibrium and transit toward equilibrium.

2.6.4.4 The Final Transitions Toward Equilibrium

The stationary states ultimately transit toward equilibrium, either directly, as entropy increases along the A-branch, or indirectly, as entropy increases along the D-branch and after an isentropic switching that leads the stationary states to escape from the cavity. In other words, the particles depart from antiequilibrium and transit toward equilibrium.

2.6.4.5 Back to Far-Equilibrium Region

External factors, which can decrease the entropy of the system, move it into a different stationary state, closer to the fundamental state (within the far-equilibrium region). In the case of solar wind, for example, newly formed PUIs may play just such a role because their motion is highly ordered, and thus the entropy of stationary states decreases.

2.7 Concluding Remarks

In this chapter, we presented the derivation and properties of the nonextensive entropic formulation for particle populations described by kappa distributions of their velocities (or Hamiltonians):

- First, we presented the paradox of units, which is connected with the entropic formulation in the continuous description, and how this is resolved by using the length and speed scale parameters, naturally emerged from the statistics of the number of microstates in an infinitesimal phase—space volume.
- Once the formulation is adjusted according to the length and speed scale parameters, the argument ϕ_q is derived and its properties are examined; this is related with the entropic formula, but also with the ratio of the actual temperature and the Lagrangian temperature-like parameter of the system.
- Then we derived the entropic formula for kappa distributions of velocities. We studied the properties of the entropy and its formula deduction in two thermodynamic limits, (1) $N_C < \kappa_0 \to \infty$, and (2) $\kappa_0 < N_C \to \infty$; these coincide also with the formulations in equilibrium and antiequilibrium, respectively.
- A special entropic measure, which depends only on the kappa index and characterizes the transition of the system among stationary states, is analytically derived and studied. We provided a survey of the relevant literature regarding the observed kappa indices and compared those to the characteristics of the presented entropic measure.
- A discrete dynamics analysis of transitions between stationary states was performed. We analytically derive the characteristic discrete dynamical map of stationary states transitions, develop numerical applications of this map, and examine the five stages of stationary state transitions.

These developments will help the study of entropy, the information exchange in plasma processes and transitions, and improve the understanding of thermodynamics, in general, for particle populations described by kappa distributions or combinations thereof.

2.8 Science Questions for Future Research

Future analyses and observations need to address the following questions:

1. How does the presence of a potential energy affect the entropy?
2. What is the entropy formulae for multispecies/pickup ions/electrons?
3. What is the entropy variation across a shock in plasmas?

Acknowledgment

The work in this chapter was supported in part by the project NNX17AB74G of NASA's HGI Program.

CHAPTER 3

Phase Space Kappa Distributions With Potential Energy

G. Livadiotis

Southwest Research Institute, San Antonio, TX,
United States

Chapter Outline

3.1 **Summary** 106
3.2 **Introduction** 106
3.3 **The Hamiltonian Distribution** 110
3.4 **Normalization of the Phase Space Kappa Distribution** 113
3.5 **Marginal Distributions** 116
3.6 **Mean Kinetic Energy in the Presence of a Potential Energy** 120
3.7 **Degeneration of the Kappa Index in the Presence of a Potential Energy** 123
   3.7.1 Rationale 123
   3.7.2 1-D Linear Gravitational Potential 124
   3.7.3 Positive Attractive Power Law Potential (Oscillator Type) 127
3.8 **Local Kappa Distribution** 134
3.9 **Negative Potentials** 142
   3.9.1 In General 142

3.9.2 The "Negative" Kappa Distribution 143
3.9.3 Formulation of Phase Space Distributions 145
   3.9.3.1 Positive Potential $\Phi(\vec{r}) > 0$ 145
   3.9.3.2 Negative Potential, $\Phi(\vec{r}) < 0$ 149
3.9.4 Negative Attractive Power Law Potential (Gravitational Type) 153
3.9.5 Potentials With Positive and Negative Values Acquiring Stable/Unstable Equilibrium Points 163
3.10 **Gravitational Potentials** 164
   3.10.1 Linear Gravitational Potential: The Barometric Formula 164
   3.10.2 Spherical Gravitational Potential 166

Kappa Distributions. http://dx.doi.org/10.1016/B978-0-12-804638-8.00003-6

3.10.3 Virial Theorem and Jeans 'Radius' 170

3.11 **Potentials with Angular Dependence** 171

3.12 **Potentials Forming Anisotropic Distribution of Velocity** 174

3.13 **Concluding Remarks** 175

3.14 **Science Questions for Future Research** 176

3.1 Summary

This chapter presents the theory, formulations, and properties of the kappa distributions that describe particle systems characterized by a nonzero potential energy. Among others, we investigate: (1) the phase space kappa distribution of the Hamiltonian given by the sum of the velocity-dependent kinetic energy and the position-dependent potential energy; (2) the derived two marginal distributions of the phase space distribution, that is, the distribution of the velocity (or kinetic energy) after integrating over the position, and the distribution of the position after integrating over the velocity; (3) the mean kinetic energy, that is, the kinetic definition of temperature, in the presence of a positional potential energy, showing that is the same as in the absence of the potential energy; (4) the degeneration of the kappa index in the presence of a potential energy; (5) the generic formulation of the phase space and positional kappa distributions for positive and negative potentials; (6) the specific two main attractive power law potentials, i.e., the positive potential with positive power law exponent (oscillator type) and the negative potential with negative power law exponent (gravitational type); (7) distributions for potentials with both positive and negative values; (8) gravitational potentials, e.g., the linear gravitational potential and the barometric formula, the spherical gravitational potential, the virial theorem, and the Jeans radius; (9) the local distributions, namely, the kappa distribution of velocity, the density, temperature, and thermal pressure, all expressed as functions of the position; and (10) potentials, depending on the velocity, may cause anisotropic velocity kappa distributions. These developments allow the researcher to describe any particle population of space, geophysical, laboratory, or other plasmas that are subject to a nonnegligible potential energy.

Science Question: How do kappa distributions vary with position in the presence of a potential?

Keywords: Degeneration; Gravitational pull; Hamiltonian; Oscillator; Phase space distribution; Polytrope.

3.2 Introduction

Numerous studies have shown that kappa distributions (or combinations thereof) are frequently observed in several space, geophysical, and other plasmas, where the deviation from Maxwell is more evident in the high-energy tails of the observed distributions. However, the vast majority of these studies were used for the statistical description of the velocity or kinetic energy of particles, but not of the potential energy. In particular, the kappa distributions of particle velocities and their characteristics in energy flux spectra were employed to describe space plasma populations: in (1) the *inner heliosphere*, including solar wind (e.g., Collier et al., 1996; Maksimovic et al. 1997a, 1997b, 2005; Pierrard et al., 1999; Mann et al., 2002; Marsch, 2006; Zouganelis 2008; Štverák et al., 2009; Livadiotis and McComas, 2010a, 2011a, 2013a, 2013c; Yoon, 2014; Pierrard and Pieters, 2015; Pavlos et al., 2016), solar spectra (e.g., Chapter 13; Dzifčáková and Dudík, 2013; Dzifčáková et al., 2015), solar corona (e.g., Owocki and Scudder, 1983; Vocks et al., 2008; Lee et al., 2013; Cranmer, 2014), solar energetic particles (e.g., Xiao et al., 2008; Laming et al., 2013), corotating interaction regions (e.g., Chotoo et al., 2000), and solar flares–related (e.g., Mann et al., 2009; Livadiotis and McComas, 2013b; Bian et al., 2014; Jeffrey et al., 2016); (2) the *planetary magnetospheres*, including magnetosheath (e.g., Binsack, 1966; Olbert, 1968; Vasyliūnas, 1968; Formisano et al., 1973; Ogasawara et al., 2013), near magnetopause (e.g., Ogasawara et al., 2015), magnetotail (e.g., Grabbe, 2000), ring current (e.g., Pisarenko et al., 2002), plasma sheet (e.g., Christon, 1987; Wang et al., 2003; Kletzing et al., 2003), magnetospheric substorms (e.g., Hapgood et al., 2011), magnetospheres of giant planets (Chapter 12), such as Jovian (e.g., Collier and Hamilton, 1995; Mauk et al., 2004), Saturnian (e.g., Schippers et al., 2008; Dialynas et al., 2009; Livi et al., 2014; Carbary et al., 2014), Uranian (e.g., Mauk et al., 1987), Neptunian (Krimigis et al., 1989), magnetospheres of planetary moons, such as Io (e.g., Moncuquet et al., 2002) and Enceladus (e.g., Jurac et al., 2002), or cometary magnetospheres (e.g., Broiles et al., 2016a, 2016b); (3) the *outer heliosphere and the inner heliosheath* (e.g., Decker and Krimigis, 2003; Decker et al., 2005; Heerikhuisen et al., 2008, 2010, 2014, 2015; Zank et al., 2010; Livadiotis et al., 2011, 2012, 2013; Livadiotis and McComas, 2011a, 2011b, 2012, 2013a, 2013c, 2013d; Livadiotis, 2014; Fuselier et al., 2014; Zirnstein and McComas, 2015); (4) *beyond the heliosphere*, including HII regions (e.g., Nicholls et al., 2012), planetary nebula (e.g., Nicholls et al., 2012; Zhang et al., 2014), and supernova magnetospheres (e.g., Raymond et al., 2010); or (5) *other space plasma–related analyses* (e.g., Milovanov and Zelenyi, 2000; Saito et al., 2000; Du 2004; Yoon et al., 2006; Raadu and Shafiq, 2007; Livadiotis, 2009, 2014a, 2015a, 2015b, 2015c, 2015e, 2016a, 2016b, 2016d; Tribeche et al., 2009; Hellberg et al., 2009; Livadiotis and McComas, 2009, 2010a, 2010b, 2010c, 2011b, 2014a; Baluku et al., 2010; Le Roux et al., 2010; Eslami et al., 2011; Kourakis et al., 2012; Randol and Christian, 2014; 2016; Varotsos et al., 2014; Fisk and Gloeckler, 2014; Liu et al., 2015; Viñas et al., 2015; Ourabah et al., 2015; Dos Santos et al., 2016; Nicolaou and Livadiotis, 2016).

Several physical mechanisms that can be applied in different plasma environments successfully explain the observed kappa distributions, where some of them are detailed and described in this book (e.g., Chapters 6, 8, 15, 16). Besides their empirical successful usage and their extraction from different mechanisms, kappa distributions are naturally exported (Chapter 1; Milovanov and Zelenyi, 2000; Leubner, 2002; Livadiotis and McComas, 2009; Livadiotis, 2015a) from the foundations of nonextensive statistical mechanics (Chapter 1; Tsallis, 1988, 2009; Tsallis et al., 1999).

Maxwell distributions have also been used in space science (e.g., Hammond et al., 1996), especially due to their simplicity; for example, they are often used to fit the "core" of the observed distributions, that is, the part of the distribution around its maximum. Another example is the convenient factorization of the exponential function that allows the usage of the Maxwellian distribution of velocities even when a nonzero potential energy exists. Indeed, the distribution of the Hamiltonian, the sum of the velocity-dependent kinetic energy and the position-dependent potential energy $H(\vec{r}, \vec{u}) = \varepsilon_K(\vec{u}) + \Phi(\vec{r})$ leads to the Maxwell distribution of velocities,

$$P(\vec{r}, \vec{u}) \propto \exp[H(\vec{r}, \vec{u})] \propto \exp[\varepsilon_K(\vec{u}) + \Phi(\vec{r})]$$
$$= \exp[\varepsilon_K(\vec{u})] \cdot \exp[\Phi(\vec{r})] \propto P(\vec{u}) \cdot P(\vec{r}), \tag{3.1a}$$

after the normalizations

$$P(\vec{u}) = \int_{-\infty}^{\infty} P(\vec{r}, \vec{u}) d\vec{r} \propto \left\{ \int_{-\infty}^{\infty} \exp[\Phi(\vec{r})] d\vec{r} \right\} \cdot \exp[\varepsilon_K(\vec{u})] \propto \exp[\varepsilon_K(\vec{u})]. \tag{3.1b}$$

$$P(\vec{r}) = \int_{-\infty}^{\infty} P(\vec{r}, \vec{u}) d\vec{u} \propto \left[\int_{-\infty}^{\infty} \exp[\varepsilon_K(\vec{u})] d\vec{u} \right] \cdot \exp[\Phi(\vec{r})] \propto \exp[\Phi(\vec{r})]. \tag{3.1c}$$

In the case of kappa distributions, the separation of the kinetic and potential energies is not possible; thus the integration of the Hamiltonian kappa distribution over the positions, as well as the connection between Hamiltonian and velocity distributions, is surely nontrivial. This explains the inconvenient role of kappa distributions compared to the Maxwell distribution when it comes to the consideration of potential energies. However, kappa distributions, even though nonexponentials, obey to the same condition as the Maxwell–Boltzmann exponential formulation:

The integration of the distribution of the Hamiltonian over the positions leads to the kappa distribution of velocities, that is, to the same formula as in the absence of a potential energy.

The kappa distribution of velocities characterizes the system in the absence of a potential energy,

$$P(\vec{u}; \kappa_0, T) = \left(\pi \kappa_0 \theta^2 \right)^{-\frac{1}{2} d_K} \cdot \frac{\Gamma\left(\kappa_0 + 1 + \frac{1}{2} d_K \right)}{\Gamma(\kappa_0 + 1)} \cdot \left[1 + \frac{1}{\kappa_0} \cdot \frac{(\vec{u} - \vec{u}_b)^2}{\theta^2} \right]^{-\kappa_0 - 1 - \frac{1}{2} d_K}, \tag{3.2a}$$

where $\theta \equiv \sqrt{2 k_B T / m}$ is the thermal speed of particle of mass m, that is, the particle temperature T expressed in speed units; both the particle $\vec{u}$ and bulk $\vec{u}_b$ velocities are in an inertial reference frame. The kappa distribution of the kinetic energy $\varepsilon_K(\vec{u}) = \frac{1}{2} m(\vec{u} - \vec{u}_b)^2$ is given by

$$P(\varepsilon_K; \kappa_0, T) d\varepsilon_K = \frac{(\kappa_0 k_B T)^{-\frac{1}{2} d_K}}{B\left(\frac{1}{2} d_K, \kappa_0 + 1 \right)} \cdot \left(1 + \frac{1}{\kappa_0} \cdot \frac{\varepsilon_K}{k_B T} \right)^{-\kappa_0 - 1 - \frac{1}{2} d_K} \varepsilon_K^{\frac{1}{2} d_K - 1} d\varepsilon_K, \tag{3.2b}$$

where $B(x,y) \equiv \Gamma(x)\Gamma(y)/\Gamma(x+y)$ is the beta function. The kinetic energy is expressed in the comoving system of the particle flow with bulk velocity $\vec{u}_b$. The kinetic degrees of freedom for the one-particle's phase space are given by $d_K \equiv 2<\varepsilon_K>/(k_BT)$. The quantity κ_0 indicates the kappa index of zero degrees of freedom, that is, a kappa index modified to be invariant of the kinetic degrees of freedom of the system (Livadiotis and McComas, 2011b; Livadiotis, 2015c). In contrast to κ_0, the well-known kappa index κ depends on the particle kinetic degrees of freedom d_K and the number of correlated particles N, according to

$$\kappa_f = \kappa_0 + \frac{1}{2}f, \quad f = N \cdot d_K. \tag{3.2c}$$

Eqs. (3.2a) and (3.2b) are special cases of the kappa distribution of free particles, each of which has the Hamiltonian equal to its kinetic energy $H(\vec{u}) = \varepsilon_K(\vec{u})$,

$$P(\vec{u};\kappa_0,T) \propto \left[1 + \frac{1}{\kappa_0} \cdot \frac{H(\vec{u})}{k_BT}\right]^{-\kappa_0-1-\frac{1}{2}d_K} \propto \left[1 + \frac{1}{\kappa_0} \cdot \frac{(\vec{u}-\vec{u}_b)^2}{\theta^2}\right]^{-\kappa_0-1-\frac{1}{2}d_K}. \tag{3.3}$$

Therefore, both Maxwell−Boltzmann (exponential) and kappa distributions are functions that obey to the previously mentioned condition:

$$\int_{-\infty}^{\infty} f[\varepsilon_K(\vec{u}) + \Phi(\vec{r})]d\vec{r} = f[\varepsilon_K(\vec{u})], \quad f(x) = \frac{1}{Z} \cdot \left(1 + \frac{1}{\kappa_0} \cdot x\right)^{-\kappa_0-1-\langle x \rangle} \tag{3.4a}$$

where

$$Z \equiv \int_{-\infty}^{\infty} \int_{-\infty}^{\infty} \left[1 + \frac{1}{\kappa_0} \cdot x(\vec{r},\vec{u})\right]^{-\kappa_0-1-\langle x \rangle} d\vec{r}d\vec{u},$$

$$\langle x \rangle = \frac{1}{Z} \cdot \int_{-\infty}^{\infty} \int_{-\infty}^{\infty} \left[1 + \frac{1}{\kappa_0} \cdot x(\vec{r},\vec{u})\right]^{-\kappa_0-1-\langle x \rangle} x(\vec{r},\vec{u})d\vec{r}d\vec{u}. \tag{3.4b}$$

This chapter presents the consistent development of the phase space kappa distribution, that is, the kappa distribution of the Hamiltonian function. Starting from the kappa distribution of the Hamiltonian function, we develop the distributions that describe either the complete phase space or the marginal spaces of positions and velocities. With the results provided here, it is now straightforward to apply these developments to describe any particle population of space, geophysical, laboratory, or other plasmas subject to a nonnegligible potential energy.

The chapter is structured as follows: in Section 3.3, we develop the Hamiltonian kappa distribution, that is, the kappa distribution of the Hamiltonian given by the sum of the velocity-dependent kinetic energy and the position-dependent potential energy. We start considering positive potentials. In Section 3.4, we develop the normalization of the phase space kappa distribution, while in Section 3.5, we derive the marginal distributions of the phase space distribution, that is, the distribution of the velocity (or kinetic energy) after integrating over the position, and the distribution of the position after integrating over the velocity. In Section 3.6, we show that the mean kinetic energy, that is, the kinetic definition of temperature (Chapter 1) in the presence of a positional potential energy is the same as in the absence of the potential energy. The degeneration of the kappa index in the presence of a potential energy is presented in detail in Section 3.7, using as main examples the 1-D linear gravitational potential, and in general, the positive attractive power law potential (oscillator type). Furthermore, in Section 3.8, we develop the local kappa distribution, namely, the kappa distribution of velocity, the density, the temperature, and the thermal pressure, all as functions of the position. In Section 3.9, we investigate negative potentials. We present the "negative" kappa distribution for kinetic energy and then formulate the phase space kappa distributions. We specifically study the negative attractive power law potential (gravitational type). We also study potentials with positive and negative values acquiring stable/unstable equilibrium points. Then in Section 3.10, we focus on gravitational potentials, e.g., the linear gravitational potential and the barometric formula, the spherical gravitational potential, the virial theorem, and the Jeans radius. Section 3.11 examines the case of potentials with angular dependence. Section 3.12 briefly examines the case of potentials forming anisotropic distribution of velocity. Finally, the concluding remarks are given in Section 3.13, while three general science questions for future analyses are posed in Section 3.14.

3.3 The Hamiltonian Distribution

The phase space distribution is derived from the distribution of the Hamiltonian function $H(\overrightarrow{r}, \overrightarrow{u}) = \varepsilon_K(\overrightarrow{u}) + \Phi(\overrightarrow{r})$, where $\varepsilon_K(\overrightarrow{u}) = \frac{1}{2}m(\overrightarrow{u} - \overrightarrow{u}_b)^2$ is the (velocity-dependent) kinetic energy (in the comoving system), and $\Phi(\overrightarrow{r})$ is the (position-dependent) potential energy.

Out of thermal equilibrium, the kappa distribution of the Hamiltonian is given by

$$P(\overrightarrow{r}, \overrightarrow{u}; \kappa, T) \propto \left[1 + \frac{1}{\kappa} \cdot \frac{H(\overrightarrow{r}, \overrightarrow{u}) - \langle H \rangle}{k_B T}\right]^{-\kappa - 1}, \tag{3.5}$$

where $\langle H \rangle$ is the ensemble phase space average of the Hamiltonian function, given by

$$\langle H \rangle = \frac{\int_{-\infty}^{\infty} \int_{-\infty}^{\infty} \left[1 + \frac{1}{\kappa} \cdot \frac{H(\overrightarrow{r}, \overrightarrow{u}) - \langle H \rangle}{k_B T}\right]^{-\kappa - 1} H(\overrightarrow{r}, \overrightarrow{u}) d\overrightarrow{r} d\overrightarrow{u}}{\int_{-\infty}^{\infty} \int_{-\infty}^{\infty} \left[1 + \frac{1}{\kappa} \cdot \frac{H(\overrightarrow{r}, \overrightarrow{u}) - \langle H \rangle}{k_B T}\right]^{-\kappa - 1} d\overrightarrow{r} d\overrightarrow{u}}. \tag{3.6}$$

Some interesting relations may be derived from Eq. (3.6), e.g.,

$$\int_{-\infty}^{\infty}\int_{-\infty}^{\infty}\left[1+\frac{1}{\kappa}\cdot\frac{H(\vec{r},\vec{u})-\langle H\rangle}{k_B T}\right]^{-\kappa-1}[H(\vec{r},\vec{u})-\langle H\rangle]d\vec{r}\,d\vec{u} = 0, \text{ and}$$

(3.7a)

$$\int_{-\infty}^{\infty}\int_{-\infty}^{\infty}\left[1+\frac{1}{\kappa}\cdot\frac{H(\vec{r},\vec{u})-\langle H\rangle}{k_B T}\right]^{-\kappa}d\vec{r}\,d\vec{u}$$

$$=\int_{-\infty}^{\infty}\int_{-\infty}^{\infty}\left[1+\frac{1}{\kappa}\cdot\frac{H(\vec{r},\vec{u})-\langle H\rangle}{k_B T}\right]^{-\kappa-1}d\vec{r}\,d\vec{u}.$$

(3.7b)

The Hamiltonian average $\langle H\rangle$ represents the internal energy of the whole system. While it appears naturally in the formula of the distribution of the canonical ensemble, it solves one of the major problems of the early theory of nonextensive statistical mechanics, that is, the dependence of the potential energy on the zero energy level (Chapter 1; Tsallis et al., 1998; Livadiotis and McComas, 2009).

In terms of the invariant kappa index κ_0, the phase space distribution, Eq. (3.5), is rewritten as

$$P(\vec{r},\vec{u};\kappa,T)\propto\left[1+\frac{1}{\kappa-\frac{\langle H\rangle}{k_B T}}\cdot\frac{H(\vec{r},\vec{u})}{k_B T}\right]^{-\kappa-1} = \left[1+\frac{1}{\kappa_0}\cdot\frac{H(\vec{r},\vec{u})}{k_B T}\right]^{-\kappa_0-1-\frac{\langle H\rangle}{k_B T}},$$

(3.8)

where the kappa index is degenerated to both the kinetic and potential energy,

$$\kappa_0 \equiv \kappa-\frac{\langle H\rangle}{k_B T} = \kappa-\frac{\langle \varepsilon_K\rangle}{k_B T}-\frac{\langle \Phi\rangle}{k_B T} = \kappa-\frac{1}{2}d_K-\frac{1}{2}d_\Phi,$$

(3.9)

where the positional degrees of freedom can be defined similar to the kinetic ones,

$$\frac{1}{2}d_K = \frac{\langle \varepsilon_K\rangle}{k_B T}, \quad \frac{1}{2}d_\Phi \equiv \frac{\langle \Phi\rangle}{k_B T}.$$

(3.10)

Hence, the phase space distribution becomes

$$P(\vec{r},\vec{u};\kappa_0,T)\propto\left[1+\frac{1}{\kappa_0}\cdot\frac{\varepsilon_K(\vec{u})+\Phi(\vec{r})}{k_B T}\right]^{-\kappa_0-1-\frac{1}{2}d_K-\frac{1}{2}d_\Phi} \quad \text{in terms of velocity } \vec{u},$$

(3.11a)

$$P(\vec{r},\varepsilon_K;\kappa_0,T)\propto\left[1+\frac{1}{\kappa_0}\cdot\frac{\varepsilon_K+\Phi(\vec{r})}{k_B T}\right]^{-\kappa_0-1-\frac{1}{2}d_K-\frac{1}{2}d_\Phi}\varepsilon_K^{\frac{1}{2}d_K-1} \quad \text{in terms of kinetic}$$

energy ε_K.

(3.11b)

The positional degrees of freedom are given by

$$
\frac{1}{2}d_\Phi \equiv \frac{\langle\Phi\rangle}{k_BT} = \frac{\int_{-\infty}^{\infty}\left[1+\frac{1}{\kappa_0}\cdot\frac{\Phi(\vec{r})}{k_BT}\right]^{-\kappa_0-1-\frac{1}{2}d_\Phi}\frac{\Phi(\vec{r})}{k_BT}d\vec{r}}{\int_{-\infty}^{\infty}\left[1+\frac{1}{\kappa_0}\cdot\frac{\Phi(\vec{r})}{k_BT}\right]^{-\kappa_0-1-\frac{1}{2}d_\Phi}d\vec{r}}, \quad \text{or}
$$

$$
1+\frac{\frac{1}{2}d_\Phi}{\kappa_0} = \frac{\int_{-\infty}^{\infty}\left[1+\frac{1}{\kappa_0}\cdot\frac{\Phi(\vec{r})}{k_BT}\right]^{-\kappa_0-\frac{1}{2}d_\Phi}d\vec{r}}{\int_{-\infty}^{\infty}\left[1+\frac{1}{\kappa_0}\cdot\frac{\Phi(\vec{r})}{k_BT}\right]^{-\kappa_0-1-\frac{1}{2}d_\Phi}d\vec{r}}
$$

(3.12)

As an example, let the case of the central attractive potential $\Phi(\vec{r}) = \Phi(r) \propto r^b$, i.e., $F_r = -(\partial/\partial r)\Phi(r) < 0$. Then, the infinitesimal volume is $d\vec{r} \propto r^{d_r-1}dr \propto \Phi^{\frac{d_r}{b}-1}d\Phi$, thus

$$
\frac{1}{2}d_\Phi = \frac{\int_{-\infty}^{\infty}\left(1+\frac{1}{\kappa_0}\cdot\frac{\Phi}{k_BT}\right)^{-\kappa_0-1-\frac{1}{2}d_\Phi}\frac{\Phi}{k_BT}\Phi^{\frac{d_r}{b}-1}d\Phi}{\int_{-\infty}^{\infty}\left(1+\frac{1}{\kappa_0}\cdot\frac{\Phi}{k_BT}\right)^{-\kappa_0-1-\frac{1}{2}d_\Phi}\Phi^{\frac{d_r}{b}-1}d\Phi}
$$

$$
= \frac{\int_{-\infty}^{\infty}\left(1+\frac{1}{\kappa_0}\cdot X\right)^{-\kappa_0-1-\frac{1}{2}d_\Phi}X^{\frac{d_r}{b}}dX}{\int_{-\infty}^{\infty}\left(1+\frac{1}{\kappa_0}\cdot X\right)^{-\kappa_0-1-\frac{1}{2}d_\Phi}X^{\frac{d_r}{b}-1}dX}
$$

$$
= \frac{\Gamma_{\kappa_0+\frac{1}{2}d_\Phi}\left(\frac{d_r}{b}+1\right)}{\Gamma_{\kappa_0+\frac{1}{2}d_\Phi}\left(\frac{d_r}{b}\right)}\cdot\frac{\kappa_0}{\kappa_0+\frac{d_\Phi}{2}} = \frac{d_r}{b}\cdot\frac{\kappa_0}{\kappa_0+\frac{d_\Phi}{2}-\frac{d_r}{b}},
$$

or

$$
\frac{d_\Phi}{2}-\frac{d_r}{b} = \frac{d_r}{b}\cdot\frac{\frac{d_\Phi}{2}-\frac{d_r}{b}}{\kappa_0+\frac{d_\Phi}{2}-\frac{d_r}{b}} \Rightarrow \text{two solutions: } \kappa_0+\frac{d_\Phi}{2}=0, \quad \text{or}
$$

$$
\frac{d_\Phi}{2}-\frac{d_r}{b}=0, \quad \text{hence,}
$$

$$
\frac{1}{2}d_\Phi = \frac{1}{b}d_r.
$$

(3.13)

We have set $X \equiv \Phi/(k_B T)$ and used the properties of the q-gamma functions (Chapter 1; see also Livadiotis and McComas, 2009; Appendix A):

$$\frac{\Gamma_{\kappa_0+\frac{1}{2}d_\Phi}\left(\frac{d_r}{b}+1\right)}{\Gamma_{\kappa_0+\frac{1}{2}d_\Phi}\left(\frac{d_r}{b}\right)} = \frac{d_r}{b}\cdot\left(\kappa_0+\frac{d_\Phi}{2}\right)\cdot\frac{\Gamma\left(\kappa_0+\frac{d_\Phi}{2}-\frac{d_r}{b}\right)}{\Gamma\left(\kappa_0+1+\frac{d_\Phi}{2}-\frac{d_r}{b}\right)}$$

$$= \frac{d_r}{b}\cdot\left(\kappa_0+\frac{d_\Phi}{2}\right)\cdot\frac{\kappa_0+\frac{d_\Phi}{2}}{\kappa_0+\frac{d_\Phi}{2}-\frac{d_r}{b}}. \tag{3.14}$$

Therefore, we may obtain

$$\frac{1}{2}d_\Phi \equiv \frac{\langle\Phi\rangle}{k_B T} = \frac{d_r}{b}, \quad b \equiv \left(\frac{\partial\log\Phi}{\partial\log r}\right)_{\log r\to 0}, \tag{3.15}$$

where $d_r = \dim(\overrightarrow{r})$ (dimensions of the position vector, i.e., the physical space), and b is the exponent in $\Phi = A\cdot r^{\pm b}+O(r)$; if positive integer, it gives the order of the smallest nonzero derivative.

3.4 Normalization of the Phase Space Kappa Distribution

First, we derive the normalized kappa distribution that includes both the kinetic and potential energy. We rewrite Eq. (3.7a) as

$$P(\overrightarrow{r},\varepsilon_K;\kappa_0,T) = A\cdot\left[1+\frac{1}{\kappa_0}\cdot\frac{\varepsilon_K+\Phi(\overrightarrow{r})}{k_B T}\right]^{-\kappa_0-1-\frac{1}{2}d_K-\frac{1}{2}d_\Phi} \varepsilon_K^{\frac{1}{2}d_K-1}, \tag{3.16a}$$

with normalization constant

$$A = \left\{\int_{-\infty}^{\infty}\int_0^{\infty}\left[1+\frac{1}{\kappa_0}\cdot\frac{\varepsilon_K+\Phi(\overrightarrow{r})}{k_B T}\right]^{-\kappa_0-1-\frac{1}{2}d_K-\frac{1}{2}d_\Phi} \varepsilon_K^{\frac{1}{2}d_K-1} d\varepsilon_K d\overrightarrow{r}\right\}^{-1}. \tag{3.16b}$$

Here we assume that the kinetic energy may take any possible value, $0 \le \varepsilon_K < \infty$; later, we will consider the positional dependence of the lowest limit in cases where $\Phi(\overrightarrow{r}) < 0$, i.e., $|\Phi(\overrightarrow{r})| - \kappa_0 k_B T \le \varepsilon_K$ if $|\Phi(\overrightarrow{r})| > \kappa_0 k_B T$. Note that we consider the kinetic upper limit at infinity, though the most mobile particle has finite speed and kinetic energy. The reason is that while the highest speed or kinetic energy is observationally finite, it is theoretically unbounded, i.e., may increase unrestrictedly and without regard to any characteristic scale.

Now let's discuss the limits of the physical space, that of the position vector. The positional vector covers the whole volume of the physical space of the examined particle system. However, the application of the potential may extend even beyond of the outer exteriors of this system. Obviously, different particle systems may have different volumes, even when they are under the same dynamics and include the same number of particles; in general, no characteristic scale exists as a universal upper limit of the physical volume. The upper limit of the volume that a particle system may cover, theoretically, extends to infinity. In the same way, we consider the kinetic upper limit at infinity, though the observational highest speed or kinetic energy is always finite; we consider the positional upper limit at infinity, though the observational further exterior of the system's volume is always finite. The notation $-\infty < \vec{r} < \infty$ means that all the components of the position vector may take any possible value.

Therefore, the physical volume of a system may appear with a loose meaning. The distribution density $P(\vec{r})$ definitely decreases as $\vec{r}$ increases to infinity, tending to zero, but theoretically never becoming exactly zero in the finite observational universe. In the context of a probability, the probability distribution density is normalized to unity, $\int_{-\infty}^{\infty} P(\vec{r}) d\vec{r} = 1$. When multiplied by the number of particles N, the distribution density $f(\vec{r}) = N \cdot P(\vec{r})$ is normalized to this number, while a portion of this is given by the cumulative distribution function into a certain volume V, enclosed in the space between $\vec{r} = 0$ and the positional vector $\vec{R}(\Omega)$ of angular dependence $\Omega = (\vartheta, \varphi)$. Any portion of N is still a number of particles, namely, an integer number larger than unity. If V is the volume for which the cumulative distribution of $f(\vec{r})$ is $N - 1$,

$\int_{\vec{r}=0}^{\vec{R}} f(\vec{r}) d\vec{r} = N - 1$, then the integration of the distribution density over the

complementary volume V' is $\int_{\vec{R}}^{\vec{r} \to \infty} f(\vec{r}) d\vec{r} = 1$ because $\int_{\vec{r} \in V'} P(\vec{r}) d\vec{r} +$

$\int_{\vec{r} \in V} P(\vec{r}) d\vec{r} = 1$ or $\int_{\vec{r} \in V'} f(\vec{r}) d\vec{r} + \int_{\vec{r} \in V} f(\vec{r}) d\vec{r} = N$. The exterior of

the system can be considered that physical space that corresponds to less than one particle, $\int_{\vec{R}}^{\vec{r} \to \infty} f(\vec{r}) d\vec{r} \leq 1$.

Next we calculate the normalization constant, given in Eq. (3.16b).

$$A^{-1} = \int_{-\infty}^{\infty} \left[1 + \frac{1}{\kappa_0} \cdot \frac{\Phi(\vec{r})}{k_B T} \right]^{-\kappa_0 - 1 - \frac{1}{2} d_K - \frac{1}{2} d_\Phi}$$

$$\times \int_0^\infty \left[1 + \frac{1}{\kappa_0 + \frac{\Phi(\vec{r})}{k_B T}} \cdot \frac{\varepsilon_K}{k_B T} \right]^{-\kappa_0 - 1 - \frac{1}{2} d_K - \frac{1}{2} d_\Phi} \varepsilon_K^{\frac{1}{2} d_K - 1} d\varepsilon_K d\vec{r}. \qquad (3.17)$$

The kinetic integral becomes

$$
\int_0^\infty \left[1 + \cfrac{1}{\kappa_0 + \cfrac{\Phi(\vec{r})}{k_B T}} \cdot \cfrac{\varepsilon_K}{k_B T} \right]^{-\kappa_0 - 1 - \frac{1}{2}d_K - \frac{1}{2}d_\Phi} \varepsilon_K^{\frac{1}{2}d_K - 1} d\varepsilon_K
$$

$$
= (k_B T)^{\frac{1}{2}d_K} \left[\cfrac{\kappa_0 + \cfrac{d_K}{2} + \cfrac{d_\Phi}{2}}{\kappa_0 + \cfrac{\Phi(\vec{r})}{k_B T}} \right]^{-\frac{1}{2}d_K} \cdot \left[\int_0^\infty \left(1 + \cfrac{1}{\kappa_0 + \cfrac{d_K}{2} + \cfrac{d_\Phi}{2}} \cdot X \right)^{-\kappa_0 - 1 - \frac{1}{2}d_K - \frac{1}{2}d_\Phi} X^{\frac{1}{2}d_K - 1} dX \right]
$$

$$
= (k_B T)^{\frac{1}{2}d_K} \left[\cfrac{\kappa_0 + \cfrac{d_K}{2} + \cfrac{d_\Phi}{2}}{\kappa_0 + \cfrac{\Phi(\vec{r})}{k_B T}} \right]^{-\frac{1}{2}d_K} \cdot \Gamma_{\kappa_0 + \frac{1}{2}d_K + \frac{1}{2}d_\Phi}\left(\cfrac{d_K}{2} \right)
$$

$$
= (k_B T)^{\frac{1}{2}d_K} \left[1 + \cfrac{1}{\kappa_0} \cfrac{\Phi(\vec{r})}{k_B T} \right]^{\frac{1}{2}d_K} \cdot \kappa_0^{\frac{1}{2}d_K} \cfrac{\Gamma\left(\cfrac{d_K}{2} \right) \Gamma\left(\kappa_0 + 1 + \cfrac{d_\Phi}{2} \right)}{\Gamma\left(\kappa_0 + 1 + \cfrac{d_K}{2} + \cfrac{d_\Phi}{2} \right)},
$$

where we set $X \equiv \varepsilon_K / (k_B T)$, while the q-gamma gives

$$
\Gamma_{\kappa_0 + \frac{1}{2}d_K + \frac{1}{2}d_\Phi}\left(\cfrac{d_K}{2} \right) = \Gamma\left(\cfrac{d_K}{2} \right) \cdot \left(\kappa_0 + \cfrac{d_K}{2} + \cfrac{d_\Phi}{2} \right)^{\frac{1}{2}d_K - 1} \cdot \cfrac{\Gamma\left(\kappa_0 + 1 + \cfrac{d_\Phi}{2} \right)}{\Gamma\left(\kappa_0 + \cfrac{d_K}{2} + \cfrac{d_\Phi}{2} \right)}. \quad (3.18)
$$

Therefore, the normalization constant in Eq. (3.18) becomes

$$
A^{-1} = (k_B T)^{\frac{1}{2}d_K} \cdot \kappa_0^{\frac{1}{2}d_K} \cfrac{\Gamma\left(\cfrac{d_K}{2} \right) \Gamma\left(\kappa_0 + 1 + \cfrac{d_\Phi}{2} \right)}{\Gamma\left(\kappa_0 + 1 + \cfrac{d_K}{2} + \cfrac{d_\Phi}{2} \right)} \int_{-\infty}^\infty \left[1 + \cfrac{1}{\kappa_0} \cdot \cfrac{\Phi(\vec{r})}{k_B T} \right]^{-\kappa_0 - 1 - \frac{1}{2}d_\Phi} d\vec{r},
$$

$$(3.19)$$

and the phase space kappa (probability) distribution in terms of the kinetic energy is

$$
P(\vec{r}, \varepsilon_K; \kappa_0, T) = \kappa_0^{-\frac{1}{2}d_K} \cdot \cfrac{\Gamma\left(\kappa_0 + 1 + \cfrac{d_K}{2} + \cfrac{d_\Phi}{2} \right)}{\Gamma\left(\kappa_0 + \cfrac{d_\Phi}{2} + 1 \right) \cdot \Gamma\left(\cfrac{d_K}{2} \right)} \cdot (k_B T)^{-\frac{1}{2}d_K}
$$

$$
\times \left\{ \int_{-\infty}^\infty \left[1 + \cfrac{1}{\kappa_0} \cdot \cfrac{\Phi(\vec{r})}{k_B T} \right]^{-\kappa_0 - 1 - \frac{1}{2}d_\Phi} d\vec{r} \right\}^{-1} \quad (3.20a)
$$

$$
\times \left[1 + \cfrac{1}{\kappa_0} \cdot \cfrac{\varepsilon_K + \Phi(\vec{r})}{k_B T} \right]^{-\kappa_0 - 1 - \frac{1}{2}d_K - \frac{1}{2}d_\Phi} \varepsilon_K^{\frac{1}{2}d_K - 1}.
$$

If instead of the kinetic energy, we use the velocity, Eq. (3.20a) becomes

$$P(\vec{r}, \vec{u}; \kappa_0, T) = \pi^{-\frac{1}{2}d_K} \kappa_0^{-\frac{1}{2}d_K} \cdot \frac{\Gamma\left(\kappa_0 + 1 + \dfrac{d_K}{2} + \dfrac{d_\Phi}{2}\right)}{\Gamma\left(\kappa_0 + \dfrac{d_\Phi}{2} + 1\right)} \cdot \theta^{-d_K}$$

$$\times \left\{ \int_{-\infty}^{\infty} \left[1 + \frac{1}{\kappa_0} \cdot \frac{\Phi(\vec{r})}{k_B T} \right]^{-\kappa_0 - 1 - \frac{1}{2}d_\Phi} d\vec{r} \right\}^{-1}$$

$$\times \left\{ 1 + \frac{1}{\kappa_0} \cdot \left[\frac{(\vec{u} - \vec{u}_b)^2}{\theta^2} + \frac{\Phi(\vec{r})}{k_B T} \right] \right\}^{-\kappa_0 - 1 - \frac{1}{2}d_K - \frac{1}{2}d_\Phi} . \qquad (3.20b)$$

3.5 Marginal Distributions

The marginal distributions provide the positional and velocity distributions, which are derived from integrating the phase space distribution of the Hamiltonian over the velocities and over the positions, respectively (Fig. 3.1),

$$P(\vec{r}) = \int_{-\infty}^{\infty} P(\vec{r}, \vec{u}) \, d\vec{u} \quad \text{and} \quad P(\vec{u}) = \int_{-\infty}^{\infty} P(\vec{r}, \vec{u}) \, d\vec{r}. \qquad (3.21)$$

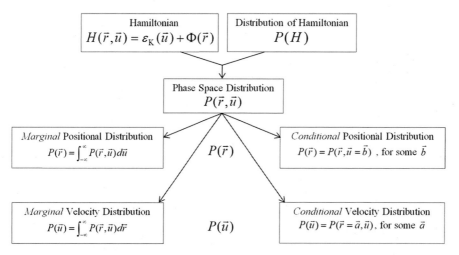

FIGURE 3.1
The phase space distribution is constructed as the distribution of the Hamiltonian, the sum of the kinetic and potential energy. The kinetic energy $\varepsilon_K(\vec{u})$ depends only on the velocity, while the potential energy $\Phi(\vec{r})$ can be dependent on both the position and the velocity in general. The distributions of the position (or the velocity) alone can be derived either as marginal distributions by integrating the velocity (or the position) or as conditional distributions, that is, by setting some value of the velocity $\vec{b}$ (or of the position $\vec{a}$). Adopted from Livadiotis (2015b).

If we integrate over one degree of freedom, let this be u_1, we have

$$P(\vec{r},\vec{U};\kappa_0,T) = \pi^{-\frac{1}{2}d_K}\kappa_0^{-\frac{1}{2}d_K}\cdot\frac{\Gamma\left(\kappa_0+1+\dfrac{d_K}{2}+\dfrac{d_\Phi}{2}\right)}{\Gamma\left(\kappa_0+\dfrac{d_\Phi}{2}+1\right)}\cdot\theta^{-d_K}$$

$$\times\left\{\int_{-\infty}^{\infty}\left[1+\frac{1}{\kappa_0}\cdot\frac{\Phi(\vec{r})}{k_BT}\right]^{-\kappa_0-1-\frac{1}{2}d_\Phi}d\vec{r}\right\}^{-1}$$

$$\times\left[1+\frac{1}{\kappa_0}\cdot\frac{\dfrac{m}{2}U^2+\Phi(\vec{r})}{k_BT}\right]^{-\kappa_0-1-\frac{1}{2}d_K-\frac{1}{2}d_\Phi}$$

$$\times\int_{-\infty}^{\infty}\left[1+\frac{1}{\kappa_0+\dfrac{\dfrac{m}{2}U^2+\Phi(\vec{r})}{k_BT}}\cdot\frac{(u_1-u_{b1})^2}{\theta^2}\right]^{-\kappa_0-1-\frac{1}{2}d_K-\frac{1}{2}d_\Phi}du_1\,d\vec{U}\,d\vec{r}, \quad (3.22)$$

where we set $\vec{u}\equiv(u_1,...,u_{d_K})$; $\vec{U}\equiv(u_2-u_{b2},...,u_{d_K}-u_{bd_K})$ and $U^2\equiv\sum_{i=2}^{d_K}(u_i-u_{b1})^2$; $\vec{u}_b$ are the flow bulk velocity; hence,

$$P(\vec{r},\vec{U};\kappa_0,T) = \pi^{-\frac{1}{2}d_K}\kappa_0^{-\frac{1}{2}d_K}\cdot\frac{\Gamma\left(\kappa_0+1+\dfrac{d_K}{2}+\dfrac{d_\Phi}{2}\right)}{\Gamma\left(\kappa_0+\dfrac{d_\Phi}{2}+1\right)}\cdot\theta^{-d_K+1}$$

$$\times\left\{\int_{-\infty}^{\infty}\left[1+\frac{1}{\kappa_0}\cdot\frac{\Phi(\vec{r})}{k_BT}\right]^{-\kappa_0-1-\frac{1}{2}d_\Phi}d\vec{r}\right\}^{-1} \qquad (3.23)$$

$$\times\left[1+\frac{1}{\kappa_0}\cdot\frac{\dfrac{m}{2}U^2+\Phi(\vec{r})}{k_BT}\right]^{-\kappa_0-1-\frac{1}{2}d_K-\frac{1}{2}d_\Phi}$$

$$\times\left[\frac{\kappa_0+\dfrac{d_K}{2}+\dfrac{d_\Phi}{2}}{\kappa_0+\dfrac{\dfrac{m}{2}U^2+\Phi(\vec{r})}{k_BT}}\right]^{-\frac{1}{2}}\Gamma_{\kappa_0+\frac{1}{2}d_K+\frac{1}{2}d_\Phi}\left(\frac{1}{2}\right),$$

and substituting $\Gamma_{\kappa_0 + \frac{1}{2}d_K + \frac{1}{2}d_\Phi}\left(\frac{1}{2}\right)$ (see Eq. 3.18), we find

$$P(\vec{r}, \vec{U}; \kappa_0, T) = \pi^{-\frac{1}{2}(d_K - 1)} \kappa_0^{-\frac{1}{2}(d_K - 1)}$$

$$\times \frac{\Gamma\left(\kappa_0 + 1 + \dfrac{d_K - 1}{2} + \dfrac{d_\Phi}{2}\right)}{\Gamma\left(\kappa_0 + \dfrac{d_\Phi}{2} + 1\right)} \theta^{-(d_K - 1)}$$

$$\times \left\{ \int_{-\infty}^{\infty} \left[1 + \frac{1}{\kappa_0} \cdot \frac{\Phi(\vec{r})}{k_B T}\right]^{-\kappa_0 - 1 - \frac{1}{2}d_\Phi} d\vec{r} \right\}^{-1}$$

$$\times \left[1 + \frac{1}{\kappa_0} \cdot \frac{\dfrac{m}{2}U^2 + \Phi(\vec{r})}{k_B T}\right]^{-\kappa_0 - 1 - \frac{1}{2}(d_K - 1) - \frac{1}{2}d_\Phi} . \tag{3.24}$$

Therefore, we have

$$\int_{-\infty}^{\infty} P(\vec{r}, \vec{u}; \kappa_0, T; d_k) du_1 = P(\vec{r}, \vec{u}; \kappa_0, T; d_K - 1), \tag{3.25}$$

and by induction, we end up with

$$\int_{-\infty}^{\infty} P(\vec{r}, \vec{u}; \kappa_0, T; d_K) du_1 \dots du_{d_K} = P(\vec{r}, \vec{u}; \kappa_0, T; 0) \equiv P(\vec{r}; \kappa_0, T), \tag{3.26}$$

that is, the positional distribution

$$P(\vec{r}; \kappa_0, T) = A_r \cdot \left[1 + \frac{1}{\kappa_0} \cdot \frac{\Phi(\vec{r})}{k_B T}\right]^{-\kappa_0 - 1 - \frac{1}{2}d_\Phi}, \tag{3.27a}$$

with normalization constant

$$A_r \equiv \left\{ \int_{-\infty}^{\infty} \left[1 + \frac{1}{\kappa_0} \cdot \frac{\Phi(\vec{r})}{k_B T}\right]^{-\kappa_0 - 1 - \frac{1}{2}d_\Phi} d\vec{r} \right\}^{-1}. \tag{3.27b}$$

Instead of integrating one by one the degrees of freedom, we may consider spherical coordinates and integrate simply on the velocity magnitude $(\vec{u} - \vec{u}_b)^2$. Then the positional distribution can be derived in only one step by integrating Eq. (3.20a) over the kinetic energy (Fig. 3.1),

$$\int_0^{\infty} P(\vec{r}, \varepsilon_K; \kappa_0, T) d\varepsilon_K = P(\vec{r}; \kappa_0, T). \tag{3.28}$$

Next, we integrate Eq. (3.20b) over the potential energy,

$$
\int_{-\infty}^{\infty} P(\vec{r}, \vec{u}; \kappa_0, T) d\vec{r} = \pi^{-\frac{1}{2}d_K} \kappa_0^{-\frac{1}{2}d_K} \cdot \theta^{-d_K} \cdot \left[1 + \frac{1}{\kappa_0} \cdot \frac{\varepsilon_K(\vec{u})}{k_B T} \right]^{-\kappa_0 - 1 - \frac{1}{2}d_K - \frac{1}{2}d_\Phi}
$$

$$
\times \frac{\Gamma\left(\kappa_0 + 1 + \dfrac{d_K}{2} + \dfrac{d_\Phi}{2} \right)}{\Gamma\left(\kappa_0 + \dfrac{d_\Phi}{2} + 1 \right)} \cdot \frac{\displaystyle\int_{-\infty}^{\infty} \left[1 + \frac{1}{\kappa_0 + \dfrac{\varepsilon_K(\vec{u})}{k_B T}} \cdot \frac{\Phi(\vec{r})}{k_B T} \right]^{-\kappa_0 - 1 - \frac{1}{2}d_K - \frac{1}{2}d_\Phi} d\vec{r}}{\displaystyle\int_{-\infty}^{\infty} \left[1 + \frac{1}{\kappa_0} \cdot \frac{\Phi(\vec{r})}{k_B T} \right]^{-\kappa_0 - 1 - \frac{1}{2}d_\Phi} d\vec{r}} \cdot
$$

$$
\tag{3.29}
$$

In our previous example of the central attractive potential $\Phi(\vec{r}) = \Phi(r) \propto r^b$, the last line of Eq. (3.29) gives

$$
\frac{\Gamma\left(\kappa_0 + 1 + \dfrac{d_K}{2} + \dfrac{d_\Phi}{2} \right)}{\Gamma\left(\kappa_0 + 1 + \dfrac{d_\Phi}{2} \right)} \cdot \frac{\displaystyle\int_{-\infty}^{\infty} \left[1 + \frac{1}{\kappa_0 + \dfrac{\varepsilon_K(\vec{u})}{k_B T}} \cdot \frac{\Phi(\vec{r})}{k_B T} \right]^{-\kappa_0 - 1 - \frac{1}{2}d_K - \frac{1}{2}d_\Phi} d\vec{r}}{\displaystyle\int_{-\infty}^{\infty} \left[1 + \frac{1}{\kappa_0} \cdot \frac{\Phi(\vec{r})}{k_B T} \right]^{-\kappa_0 - 1 - \frac{1}{2}d_\Phi} d\vec{r}}
$$

$$
= \frac{\Gamma\left(\kappa_0 + 1 + \dfrac{d_K}{2} + \dfrac{d_\Phi}{2} \right)}{\Gamma\left(\kappa_0 + 1 + \dfrac{d_\Phi}{2} \right)} \cdot \frac{\left[\dfrac{\kappa_0 + \dfrac{d_K}{2} + \dfrac{d_\Phi}{2}}{\kappa_0 + \dfrac{\varepsilon_K(\vec{u})}{k_B T}} \right]^{-\frac{d_r}{b}} \Gamma_{\kappa_0 + \frac{d_K}{2} + \frac{d_\Phi}{2}}\left(\dfrac{d_r}{b} \right)}{\left(\dfrac{\kappa_0 + \dfrac{d_\Phi}{2}}{\kappa_0} \right)^{-\frac{d_r}{b}} \Gamma_{\kappa_0 + \frac{d_\Phi}{2}}\left(\dfrac{d_r}{b} \right)}
$$

$$
= \left[1 + \frac{1}{\kappa_0} \cdot \frac{\varepsilon_K(\vec{u})}{k_B T} \right]^{\frac{d_r}{b}} \cdot \frac{\Gamma\left(\kappa_0 + 1 + \dfrac{d_K}{2} + \dfrac{d_\Phi}{2} - \dfrac{d_r}{b} \right)}{\Gamma\left(\kappa_0 + 1 + \dfrac{d_\Phi}{2} - \dfrac{d_r}{b} \right)} = \left[1 + \frac{1}{\kappa_0} \cdot \frac{\varepsilon_K(\vec{u})}{k_B T} \right]^{\frac{d_\Phi}{2}} \cdot \frac{\Gamma\left(\kappa_0 + 1 + \dfrac{d_K}{2} \right)}{\Gamma(\kappa_0 + 1)} \cdot
$$

given that $\dfrac{d_\Phi}{2} = \dfrac{d_r}{b}$, as shown in Eq. (3.15). Hence, Eq. (3.29) becomes

$$
\int_{-\infty}^{\infty} P(\vec{r}, \vec{u}; \kappa_0, T) d\vec{r} = \pi^{-\frac{1}{2}d_K} \kappa_0^{-\frac{1}{2}d_K} \cdot \frac{\Gamma\left(\kappa_0 + 1 + \dfrac{d_K}{2} \right)}{\Gamma(\kappa_0 + 1)} \cdot \theta^{-d_K} \cdot \left[1 + \frac{1}{\kappa_0} \cdot \frac{\varepsilon_K(\vec{u})}{k_B T} \right]^{-\kappa_0 - 1 - \frac{1}{2}d_K},
$$

$$
\tag{3.30}
$$

which coincides with the kappa distribution of velocities in the absence of potential energy

$$
\int_{-\infty}^{\infty} P(\vec{r}, \vec{u}; \kappa_0, T) d\vec{r} = P(\vec{u}; \kappa_0, T).
$$

$$
\tag{3.31}
$$

3.6 Mean Kinetic Energy in the Presence of a Potential Energy

The mean kinetic energy $\langle \varepsilon_K \rangle$ or the variance of the velocities $(\vec{u} - \vec{u}_b)^2$ is an important statistical moment since it constitutes the kinetic definition of temperature (Chapter 1; Livadiotis and McComas, 2009, 2010a; Livadiotis, 2009, 2014a, 2015a, 2015b, 2015c). It brings out the equipartition of energy at all the kinetic degrees of freedom $\langle \varepsilon_K \rangle = \frac{1}{2} d_K \cdot k_B T$. One characteristic property of this definition is that it is independent of the potential, given that the latter depends only on the position and not the velocity. At thermal equilibrium where the particles are described by a Maxwell–Boltzmann distribution, this appears to be a trivial feature because of the factorization of the exponential function, $\exp\{[\varepsilon_K(\vec{u}) + \Phi(\vec{r})]/(k_B T)\} = \exp[\varepsilon_K(\vec{u})/(k_B T)] \cdot \exp[\Phi(\vec{r})/(k_B T)]$. However, this is true also for kappa distributions. This is shown as follows, assuming a positive potential,

$$\langle \varepsilon_K \rangle = \int_{-\infty}^{\infty} \int_0^{\infty} \varepsilon_K P(\vec{r}, \varepsilon_K; \kappa_0, T) d\varepsilon_K d\vec{r}$$

$$= \kappa_0^{-\frac{1}{2}d_K} \cdot \frac{\Gamma\left(\kappa_0 + 1 + \frac{d_K}{2} + \frac{d_\Phi}{2}\right)}{\Gamma\left(\kappa_0 + \frac{d_\Phi}{2} + 1\right) \cdot \Gamma\left(\frac{d_K}{2}\right)} \cdot (k_B T)^{-\frac{1}{2}d_K}$$

$$\times \left\{ \int_{-\infty}^{\infty} \left[1 + \frac{1}{\kappa_0} \cdot \frac{\Phi(\vec{r})}{k_B T}\right]^{-\kappa_0 - 1 - \frac{1}{2}d_\Phi} d\vec{r} \right\}^{-1}$$

$$\times \int_{-\infty}^{\infty} \int_0^{\infty} \left[1 + \frac{1}{\kappa_0} \cdot \frac{\varepsilon_K + \Phi(\vec{r})}{k_B T}\right]^{-\kappa_0 - 1 - \frac{1}{2}d_K - \frac{1}{2}d_\Phi} \varepsilon_K^{\frac{1}{2}d_K} d\varepsilon_K d\vec{r}$$

$$\Rightarrow \langle \varepsilon_K \rangle = \kappa_0^{-\frac{1}{2}d_K} \cdot \frac{\Gamma\left(\kappa_0 + 1 + \frac{d_K}{2} + \frac{d_\Phi}{2}\right)}{\Gamma\left(\kappa_0 + \frac{d_\Phi}{2} + 1\right) \cdot \Gamma\left(\frac{d_K}{2}\right)} \cdot k_B T$$

$$\times \left\{ \int_{-\infty}^{\infty} \left[1 + \frac{1}{\kappa_0} \cdot \frac{\Phi(\vec{r})}{k_B T}\right]^{-\kappa_0 - 1 - \frac{1}{2}d_\Phi} d\vec{r} \right\}^{-1} \cdot \int_{-\infty}^{\infty} \left[1 + \frac{1}{\kappa_0} \cdot \frac{\Phi(\vec{r})}{k_B T}\right]^{-\kappa_0 - 1 - \frac{1}{2}d_K - \frac{1}{2}d_\Phi}$$

$$\times \int_0^{\infty} \left[1 + \frac{1}{\kappa_0 + \frac{\Phi(\vec{r})}{k_B T}} \cdot X\right]^{-\kappa_0 - 1 - \frac{1}{2}d_K - \frac{1}{2}d_\Phi} X^{\frac{1}{2}d_K} dX d\vec{r}$$

$$= k_B T \cdot \kappa_0 \cdot \frac{\Gamma\left(\kappa_0 + \frac{d_K}{2} + \frac{d_\Phi}{2}\right)}{\Gamma\left(\kappa_0 + \frac{d_\Phi}{2} + 1\right)} \cdot \left(\kappa_0 + \frac{d_K}{2} + \frac{d_\Phi}{2}\right)^{-\frac{1}{2}d_K} \frac{\Gamma_{\kappa_0 + \frac{d_K}{2} + \frac{d_\Phi}{2}}\left(\frac{d_K}{2} + 1\right)}{\Gamma\left(\frac{d_K}{2}\right)}$$

$$\times \frac{\displaystyle\int_{-\infty}^{\infty} \left[1 + \frac{1}{\kappa_0} \cdot \frac{\Phi(\vec{r})}{k_B T}\right]^{-\kappa_0 - \frac{1}{2}d_\Phi} d\vec{r}}{\displaystyle\int_{-\infty}^{\infty} \left[1 + \frac{1}{\kappa_0} \cdot \frac{\Phi(\vec{r})}{k_B T}\right]^{-\kappa_0 - 1 - \frac{1}{2}d_\Phi} d\vec{r}}$$

$$= \frac{d_K}{2} k_B T \frac{\kappa_0}{\kappa_0 + \frac{d_\Phi}{2}} \cdot \frac{\displaystyle\int_{-\infty}^{\infty} \left[1 + \frac{1}{\kappa_0} \cdot \frac{\Phi(\vec{r})}{k_B T}\right]^{-\kappa_0 - \frac{1}{2}d_\Phi} d\vec{r}}{\displaystyle\int_{-\infty}^{\infty} \left[1 + \frac{1}{\kappa_0} \cdot \frac{\Phi(\vec{r})}{k_B T}\right]^{-\kappa_0 - 1 - \frac{1}{2}d_\Phi} d\vec{r}}$$

$$= \frac{d_K}{2} k_B T \frac{\kappa_0}{\kappa_0 + \frac{d_\Phi}{2}} \cdot \left\langle 1 + \frac{1}{\kappa_0} \cdot \frac{\Phi(\vec{r})}{k_B T}\right\rangle,$$

where we have

$$\frac{\displaystyle\int_{-\infty}^{\infty} \left[1 + \frac{1}{\kappa_0} \cdot \frac{\Phi(\vec{r})}{k_B T}\right]^{-\kappa_0 - \frac{1}{2}d_\Phi} d\vec{r}}{\displaystyle\int_{-\infty}^{\infty} \left[1 + \frac{1}{\kappa_0} \cdot \frac{\Phi(\vec{r})}{k_B T}\right]^{-\kappa_0 - 1 - \frac{1}{2}d_\Phi} d\vec{r}}$$

$$= \frac{\displaystyle\int_{-\infty}^{\infty} \left[1 + \frac{1}{\kappa_0} \cdot \frac{\Phi(\vec{r})}{k_B T}\right]^{-\kappa_0 - 1\frac{1}{2}d_\Phi} \left[1 + \frac{1}{\kappa_0} \cdot \frac{\Phi(\vec{r})}{k_B T}\right] d\vec{r}}{\displaystyle\int_{-\infty}^{\infty} \left[1 + \frac{1}{\kappa_0} \cdot \frac{\Phi(\vec{r})}{k_B T}\right]^{-\kappa_0 - 1 - \frac{1}{2}d_\Phi} d\vec{r}} = \left\langle 1 + \frac{1}{\kappa_0} \cdot \frac{\Phi(\vec{r})}{k_B T}\right\rangle = 1 + \frac{\frac{d_\Phi}{2}}{\kappa_0},$$

(3.32)

that is, Eq. (3.12). Hence, we end up with

$$\langle \varepsilon_K \rangle = \int_{-\infty}^{\infty} \int_0^{\infty} \varepsilon_K P(\vec{r}, \varepsilon_K; \kappa_0, T) d\varepsilon_K d\vec{r} = \frac{d_K}{2} k_B T. \tag{3.33}$$

Therefore, the mean kinetic energy is the same as in the absence of a potential energy. We now investigate the higher statistical moments. We find

$$\left\langle \varepsilon_K^{\frac{1}{2}\alpha} \right\rangle = \int_{-\infty}^{\infty} \int_0^{\infty} \varepsilon_K^{\frac{1}{2}\alpha} P(\vec{r}, \varepsilon_K; \kappa_0, T) d\varepsilon_K d\vec{r}$$

$$= (k_B T)^{\frac{1}{2}\alpha} \kappa_0^{\frac{1}{2}\alpha} \cdot \frac{\Gamma\left(\frac{d_K}{2} + \frac{\alpha}{2}\right)}{\Gamma\left(\frac{d_K}{2}\right)} \cdot \frac{\Gamma\left(\kappa_0 + \frac{d_\Phi}{2} + 1 - \frac{\alpha}{2}\right)}{\Gamma\left(\kappa_0 + \frac{d_\Phi}{2} + 1\right)} \cdot \left\langle \left[1 + \frac{1}{\kappa_0} \cdot \frac{\Phi(\vec{r})}{k_B T}\right]^{\frac{1}{2}\alpha}\right\rangle.$$

(3.34)

where

$$\left\langle \left[1 + \frac{1}{\kappa_0} \cdot \frac{\Phi(\overrightarrow{r})}{k_B T}\right]^{\frac{1}{2}\alpha}\right\rangle = \frac{\displaystyle\int_{-\infty}^{\infty}\left[1 + \frac{1}{\kappa_0} \cdot \frac{\Phi(\overrightarrow{r})}{k_B T}\right]^{-\kappa_0 - 1 + \frac{1}{2}\alpha - \frac{1}{2}d_\Phi} d\overrightarrow{r}}{\displaystyle\int_{-\infty}^{\infty}\left[1 + \frac{1}{\kappa_0} \cdot \frac{\Phi(\overrightarrow{r})}{k_B T}\right]^{-\kappa_0 - 1 - \frac{1}{2}d_\Phi} d\overrightarrow{r}}.$$ (3.35)

As an example, we use again the case of the central attractive potential $\Phi(\overrightarrow{r}) = \Phi(r) \propto r^b$ and $d\overrightarrow{r} \propto r^{d_r - 1} dr \propto \Phi^{\frac{d_r}{b} - 1} d\Phi$, where we have

$$\frac{\displaystyle\int_{-\infty}^{\infty}\left(1 + \frac{1}{\kappa_0} \cdot \frac{\Phi}{k_B T}\right)^{-\kappa_0 - 1 + \frac{1}{2}\alpha - \frac{1}{2}d_\Phi} \Phi^{\frac{d_r}{b} - 1} d\Phi}{\displaystyle\int_{-\infty}^{\infty}\left(1 + \frac{1}{\kappa_0} \cdot \frac{\Phi}{k_B T}\right)^{-\kappa_0 - 1 - \frac{1}{2}d_\Phi} \Phi^{\frac{d_r}{b} - 1} d\Phi} = \frac{\displaystyle\int_{-\infty}^{\infty}\left(1 + \frac{1}{\kappa_0} \cdot X\right)^{-\kappa_0 - 1 + \frac{1}{2}\alpha - \frac{1}{2}d_\Phi} X^{\frac{d_r}{b} - 1} dX}{\displaystyle\int_{-\infty}^{\infty}\left(1 + \frac{1}{\kappa_0} \cdot X\right)^{-\kappa_0 - 1 - \frac{1}{2}d_\Phi} X^{\frac{d_r}{b} - 1} dX}$$

$$= \frac{\left(\dfrac{\kappa_0 - \frac{1}{2}\alpha + \frac{1}{2}d_\Phi}{\kappa_0}\right)^{-\frac{d_r}{b}} \Gamma_{\kappa_0 - \frac{1}{2}\alpha + \frac{1}{2}d_\Phi}\left(\dfrac{d_r}{b}\right)}{\left(\dfrac{\kappa_0 + \frac{1}{2}d_\Phi}{\kappa_0}\right)^{-\frac{d_r}{b}} \Gamma_{\kappa_0 + \frac{1}{2}d_\Phi}\left(\dfrac{d_r}{b}\right)} = \left(\dfrac{\kappa_0 - \frac{\alpha}{2} + \frac{d_\Phi}{2}}{\kappa_0 + \frac{d_\Phi}{2}}\right)^{-\frac{d_r}{b}}\left(\dfrac{\kappa_0 - \frac{\alpha}{2} + \frac{d_\Phi}{2}}{\kappa_0 + \frac{d_\Phi}{2}}\right)^{\frac{d_r}{b} - 1}$$

$$\times \frac{\dfrac{\Gamma\left(\kappa_0 + 1 - \dfrac{\alpha}{2} + \dfrac{d_\Phi}{2} - \dfrac{d_r}{b}\right)}{\Gamma\left(\kappa_0 - \dfrac{\alpha}{2} + \dfrac{d_\Phi}{2}\right)}}{\dfrac{\Gamma\left(\kappa_0 + 1 + \dfrac{d_\Phi}{2} - \dfrac{d_r}{b}\right)}{\Gamma\left(\kappa_0 + \dfrac{d_\Phi}{2}\right)}},$$

with $X \equiv \Phi/(k_B T)$; hence,

$$\left\langle \left[1 + \frac{1}{\kappa_0} \cdot \frac{\Phi(\overrightarrow{r})}{k_B T}\right]^{\frac{1}{2}\alpha}\right\rangle = \frac{\Gamma\left(\kappa_0 + 1 - \dfrac{\alpha}{2}\right)}{\Gamma(\kappa_0 + 1)} \cdot \frac{\Gamma\left(\kappa_0 + 1 + \dfrac{d_\Phi}{2}\right)}{\Gamma\left(\kappa_0 + 1 - \dfrac{\alpha}{2} + \dfrac{d_\Phi}{2}\right)}.$$ (3.36)

Substituting in Eq. (3.34), we derive

$$\left\langle \varepsilon_K^{\frac{1}{2}\alpha}\right\rangle = \int_{-\infty}^{\infty}\int_0^{\infty} \varepsilon_K^{\frac{1}{2}\alpha} P(\overrightarrow{r}, \varepsilon_K; \kappa_0, T) d\varepsilon_K d\overrightarrow{r}$$

$$= (k_B T)^{\frac{1}{2}\alpha} \kappa_0^{\frac{1}{2}\alpha} \cdot \frac{\Gamma\left(\dfrac{d_K}{2} + \dfrac{\alpha}{2}\right)}{\Gamma\left(\dfrac{d_K}{2}\right)} \cdot \frac{\Gamma\left(\kappa_0 + 1 - \dfrac{\alpha}{2}\right)}{\Gamma(\kappa_0 + 1)},$$ (3.37)

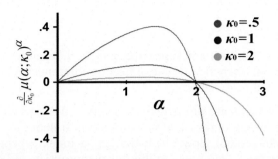

FIGURE 3.2
The derivative of the αth statistical moment with respect to the invariant kappa index κ_0, i.e., $\frac{\partial}{\partial \kappa_0}\mu(\alpha;\kappa_0)^{\alpha}$. The derivative becomes zero only for $\alpha = 0$ and 2, independently of the kappa index. Therefore, only these two statistical moments are independent of the kappa index. Adopted from Livadiotis (2014a).

which coincides with the kinetic moments in the absence of a potential energy (compared with Eq. 5.52a, where $m = \frac{1}{2}\alpha$).

According to the Maxwell's kinetic theory (1866), the mean kinetic energy provides the kinetic definition of temperature (Eq. 3.33; see also Chapter 1). This constitutes the equipartition theorem applied to each of the (d_K) kinetic degrees of freedom of a particle system at thermal equilibrium, where the mean kinetic energy per (half) degrees of freedom defines the kinetic energy $k_B T$, or, the temperature (in units of J/k_B). The temperature must be independent of other thermodynamic parameters, e.g., the kappa index. Indeed, the kinetic definition of temperature is applicable for both systems at or out of thermal equilibrium, and the equipartition theorem is identical for any kappa index. The second statistical moment of velocities, or the first statistical moment of kinetic energy, have this fundamental physical meaning of defining temperature, and thus no other moment is independent of the kappa index. Fig. 3.2 demonstrates the derivative of the αth statistical moment $\mu \equiv <X^{\frac{\alpha}{2}}>^{\frac{1}{\alpha}}$ (with $X \equiv \varepsilon/(k_B T) = (\vec{u} - \vec{u}_b)^2/\theta^2$), with respect to the invariant kappa index κ_0, showing that only the moments for $\alpha = 0$ and $\alpha = 2$ are independent of the kappa index.

3.7 Degeneration of the Kappa Index in the Presence of a Potential Energy

3.7.1 Rationale

In the absence of a potential energy, the formulation of the kappa distribution of the kinetic energy $\varepsilon_K = \frac{1}{2}m(\vec{u} - \vec{u}_b)^2$ may be written in terms of the invariant kappa index κ_0,

$$P(\varepsilon_K) \propto \left(1 + \frac{1}{\kappa - \frac{1}{2}d_K}\cdot\frac{\varepsilon_K}{k_B T}\right)^{-\kappa-1}\varepsilon_K^{\frac{1}{2}d_K-1} \Rightarrow P(\varepsilon_K) \propto \left(1 + \frac{1}{\kappa_0}\cdot\frac{\varepsilon_K}{k_B T}\right)^{-\kappa_0-1-\frac{1}{2}d_K}\varepsilon_K^{\frac{1}{2}d_K-1}.$$

$$(3.38)$$

The kappa index κ depends only on the kinetic degrees of freedom, d_K, namely,

$$\kappa(d_K) = \kappa_0 + \frac{1}{2}d_K, \tag{3.39}$$

where the invariant kappa index κ_0 is independent of the kinetic degrees of freedom. Every new degree of freedom "degenerates" the kappa index by adding one-half to its value.

The dependence property of the kappa index is problematic; the kappa index is an indicator of the certain nonequilibrium stationary state in which the particle system resides. However, if the kappa index depends on the dimensionality being used, then how can this be an appropriate indicator of the system's state? As it is been shown (Livadiotis and McComas, 2011b; Livadiotis, 2015c), it is the invariant index that can consistently play this role. For example, the most frequent kappa index in the solar wind is typically found to be around $\kappa \sim 4$ using 3-D distribution ($d_K = 3$), but we should not expect to find this value when dealing with other dimensionalities, for instant, a 1-D distribution ($d_K = 1$); indeed, $\kappa(d_K = 3) \sim 4$ means $\kappa(d_K = 1) \sim 3$, and for all cases, the invariant kappa index is $\kappa_0 \sim 2.5$ (note that Eq. 3.39 is similar to Eq. 3.2c, but here we consider one-particle distributions, $N = 1$).

Next, we will examine how the presence of a potential energy modifies this relation and further degenerates the kappa index due to the presence of the potential degrees of freedom d_Φ, defined in Eqs. (3.12) and (3.15). In order to show this, we assume that the kappa index is degenerated only by the kinetic degrees of freedom d_K and that it is not further degenerated by the potential degrees of freedom d_Φ so that the Hamiltonian probability distribution is given as the first part of Eq. (3.3), i.e.,

$$P(\vec{r}, \vec{u}; \kappa, T) \propto \left[1 + \frac{1}{\kappa - \frac{\langle H \rangle}{k_B T}} \cdot \frac{H(\vec{r}, \vec{u})}{k_B T}\right]^{-\kappa - 1} = \left[1 + \frac{1}{\kappa_0 - \frac{d_\Phi}{2}} \cdot \frac{H(\vec{r}, \vec{u})}{k_B T}\right]^{-\kappa_0 - 1 - \frac{1}{2}d_K}. \tag{3.40a}$$

At the end, we will show that this is transformed to the second part of Eq. (3.3), i.e.,

$$P(\vec{r}, \vec{u}; \kappa, T) \propto \left[1 + \frac{1}{\kappa_0} \cdot \frac{H(\vec{r}, \vec{u})}{k_B T}\right]^{-\kappa_0 - 1 - \frac{\langle H \rangle}{k_B T}} = \left[1 + \frac{1}{\kappa_0} \cdot \frac{H(\vec{r}, \vec{u})}{k_B T}\right]^{-\kappa_0 - 1 - \frac{1}{2}d_K - \frac{1}{2}d_\Phi}. \tag{3.40b}$$

3.7.2 1-D Linear Gravitational Potential

We start from a special case of our typical example of the central attractive potential $\Phi(\vec{r}) = \Phi(r) \propto r^b$; particularly, we examine the 1-D linear gravitational potential $\Phi(z) = mgz$, which is dependent on the altitude z. The phase space distribution of the Hamiltonian $H(z, \vec{u}) = \varepsilon_K(\vec{u}) + mgz$ is given by

$$P(z, \varepsilon_K; \kappa_0, T) \propto \left[1 + \frac{1}{\kappa_0 - \frac{1}{2}d_\Phi} \cdot \left(\frac{\varepsilon_K}{k_B T} + \frac{1}{2}d_\Phi \frac{z}{z_0}\right)\right]^{-\kappa_0 - 1 - \frac{1}{2}d_K} \varepsilon_K^{\frac{1}{2}d_K - 1}, \tag{3.41}$$

while the distribution of the potential energy becomes

$$P(z; \kappa_0, T) \propto \left(1 + \frac{1}{\kappa_0 - \frac{1}{2}d_\Phi} \cdot \frac{z}{z_0}\right)^{-\kappa_0 - 1}, \tag{3.42}$$

where $z_0 = \langle z \rangle$ is the average potential distance, defined from the effective degrees of freedom d_Φ,

$$\frac{1}{2}d_\Phi = \frac{\langle \Phi \rangle}{k_B T} = \frac{mg}{k_B T}z_0, \tag{3.43}$$

Hence, the positional distribution, Eq. (3.42), is written as

$$P(z; \kappa_0, T) \propto (1 + z/\alpha)^{-\kappa_0 - 1}, \tag{3.44}$$

with $\alpha \equiv z_0\left(\kappa_0 - \frac{1}{2}d_\Phi\right) \Big/ \frac{1}{2}d_\Phi$. The average distance z_0 and the normalization constant μ_2 are given by

$$z_0 = \langle z \rangle = \mu_1/\mu_0, \text{ with } \mu_m \equiv \int_0^\infty (1 + z/\alpha)^{-\kappa_0 - 1} z^m dz. \tag{3.45}$$

Setting $x \equiv z/\alpha$, we have

$$\mu_m = \alpha^{m+1} \int_0^\infty (1 + x)^{-\kappa_0 - 1} x^m dx = (\alpha/\kappa_0)^{m+1} \cdot \Gamma_{\kappa_0}(m + 1), \text{ or} \tag{3.46}$$

$$\mu_m = \alpha^{m+1} \cdot \frac{\Gamma(m + 1)\Gamma(\kappa_0 - m)}{\Gamma(\kappa_0 + 1)}. \tag{3.47}$$

Hence, the average radius is given by

$$z_0 = \langle z \rangle = \mu_1/\mu_0 = \alpha/(\kappa_0 - 1) = z_0 \frac{\kappa_0 - \frac{1}{2}d_\Phi}{\frac{1}{2}d_\Phi(\kappa_0 - 1)} \Rightarrow \frac{1}{2}d_\Phi = 1. \tag{3.48}$$

The normalization constant and average distance become

$$\mu_0 = \frac{\alpha}{\kappa_0} = \frac{\kappa_0 - 1}{\kappa_0} \cdot z_0, \quad z_0 = \frac{k_B T}{mg}, \tag{3.49}$$

where we substituted $\alpha \equiv z_0(\kappa_0 - 1)$.

Finally, the normalized phase space kappa distribution is given by

$$P(z, \varepsilon_K; \kappa_0, T) = \frac{\Gamma\left(\kappa_0 + 1 + \frac{d_K}{2}\right)}{(\kappa_0 - 1)^{\frac{1}{2}d_K + 1}\Gamma\left(\frac{d_K}{2}\right)\Gamma(\kappa_0)} \cdot z_0^{-1}(k_B T)^{-\frac{1}{2}d_K} \tag{3.50}$$

$$\times \left[1 + \frac{1}{\kappa_0 - 1} \cdot \left(\frac{\varepsilon_K}{k_B T} + \frac{z}{z_0}\right)\right]^{-\kappa_0 - 1 - \frac{1}{2}d_K} \varepsilon_K^{\frac{1}{2}d_K - 1}.$$

The positional kappa distribution is the marginal distribution derived from the integration of Eq. (3.50) over the velocity. This is the distribution shown in Eqs. (3.42) and (3.44), whose normalized function is given by

$$P(z; \kappa_0, z_0) = \frac{\kappa_0}{\kappa_0 - 1} \cdot z_0^{-1} \cdot \left(1 + \frac{1}{\kappa_0 - 1} \cdot \frac{z}{z_0} \right)^{-\kappa_0 - 1}. \tag{3.51}$$

The marginal kappa distribution of velocities (kinetic energy) is derived from the integration of Eq. (3.50) over the position, namely,

$$P(\varepsilon_K; \kappa_0, T) = \int_0^\infty P(\varepsilon_K, z; \kappa_0, T) dz = \frac{\Gamma\left(\kappa_0 + 1 + \frac{1}{2} d_K \right)}{(\kappa_0 - 1)^{\frac{1}{2} d_K + 1} \Gamma\left(\frac{1}{2} d_K \right) \Gamma(\kappa_0)}$$

$$\times z_0^{-1} (k_B T)^{-\frac{1}{2} d_K} \cdot \varepsilon_K^{\frac{1}{2} d_K - 1} \cdot \int_0^\infty \left[1 + \frac{1}{\kappa_0 - 1} \cdot \left(\frac{\varepsilon_K}{k_B T} + \frac{z}{z_0} \right) \right]^{-\kappa_0 - 1 - \frac{1}{2} d_K} dz, \tag{3.52}$$

or

$$P(\varepsilon_K; \kappa_0, T) = \frac{[(\kappa_0 - 1) k_B T]^{-\frac{1}{2} d_K}}{B\left[\frac{1}{2} d_K, (\kappa_0 - 1) + 1 \right]} \cdot \left[1 + \frac{1}{\kappa_0 - 1} \cdot \frac{\varepsilon_K}{k_B T} \right]^{-(\kappa_0 - 1) - 1 - \frac{1}{2} d_K} \varepsilon_K^{\frac{1}{2} d_K - 1}. \tag{3.53a}$$

We observe that this kinetic kappa distribution has one significant difference with the kinetic kappa distribution in the absence of any potential energy, that is,

$$P(\varepsilon_K; \kappa_0, T) = \frac{(\kappa_0 k_B T)^{-\frac{1}{2} d_K}}{B\left(\frac{1}{2} d_K, \kappa_0 + 1 \right)} \cdot \left(1 + \frac{1}{\kappa_0} \cdot \frac{\varepsilon_K}{k_B T} \right)^{-\kappa_0 - 1 - \frac{1}{2} d_K} \varepsilon_K^{\frac{1}{2} d_K - 1}. \tag{3.53b}$$

Obviously, the kappa index has been degenerated in the presence of the potential energy:

$$\kappa_0 \text{ (no potential)} \Rightarrow \kappa_0 + 1 \text{ (with potential } \Phi = mgz) \tag{3.54a}$$

or

$$\kappa(d_K) = \kappa_0 + \tfrac{1}{2} d_K \text{ (no potential)} \Rightarrow$$
$$\kappa(d_K) = \kappa_0 + \tfrac{1}{2} d_K + 1 \text{ (with potential } \Phi = mgz) \tag{3.54b}$$

Hence, applying the transformation $\kappa_0 - 1 \to \kappa_0$, the two distributions, Eqs. (3.53a) and (3.53b), become equal. It appears that the presence of this potential energy leads to an additional number of effective degrees of freedom. Still, we do not know the exact way this related the positional degrees of freedom in the kappa index. For example, if d_r is the dimensionality of the position vector, then what is the function $K(d_r)$ in $\kappa(d_K, d_r) = \kappa_0 + \tfrac{1}{2} d_K + K(d_r)$?

Before we answer this, let's rewrite correctly the phase space kappa distribution, given by Eq. (3.50), after the degeneration of the kappa index is considered,

$$P(z, \varepsilon_K; \kappa_0, T)dzd\varepsilon_K = \frac{\Gamma\left(\kappa_0 + 2 + \frac{1}{2}d_K\right)}{\kappa_0^{\frac{1}{2}d_K+1}\Gamma\left(\frac{1}{2}d_K\right)\Gamma(\kappa_0 + 1)} \cdot z_0^{-1}(k_B T)^{-\frac{1}{2}d_K} \cdot$$

$$\times \left[1 + \frac{1}{\kappa_0}\cdot\left(\frac{\varepsilon_K}{k_B T} + \frac{z}{z_0}\right)\right]^{-\kappa_0-2-\frac{1}{2}d_K} \varepsilon_K^{\frac{1}{2}d_K-1} dzd\varepsilon_K, \qquad (3.55a)$$

while the positional kappa distributions is written as

$$P(z; \kappa_0, T)dz = \frac{\kappa_0 + 1}{\kappa_0}\cdot z_0^{-1}\cdot\left(1 + \frac{1}{\kappa_0}\cdot\frac{z}{z_0}\right)^{-\kappa_0-2} dz, \text{ where } z_0(T) = \frac{k_B T}{mg},$$

$$(3.55b)$$

(the other marginal distribution is the standard kappa distribution of velocities shown in Eq. 3.53b).

3.7.3 Positive Attractive Power Law Potential (Oscillator Type)

Here we generalize the degeneration of the kappa index, using a potential that depends on a d_r-dimensional positional space; we examine the positive attractive power law potential $\Phi(r) = \frac{1}{b}k\cdot r^b$, i.e., $F_r = -(\partial/\partial r)\Phi(r) = -k\cdot r^{b-1} < 0$, $r \equiv |\vec{r}| \in \mathfrak{R}^{d_r}$. In order to avoid any confusion, we keep symbolizing the dimensionality of the velocity vector with d_K, i.e., $u \equiv |\vec{u}| \in \mathfrak{R}^{d_K}$. The phase space distribution of the Hamiltonian $H(\vec{r}, \vec{u}) = \varepsilon_K(\vec{u}) + \frac{1}{b}k\cdot|\vec{r}|^b$ is given by

$$P(r, \varepsilon_K; \kappa_0, T) \propto \left\{1 + \frac{1}{\kappa_0 - \frac{1}{2}d_\Phi}\cdot\left[\frac{\varepsilon_K}{k_B T} + \frac{1}{2}d_\Phi\left(\frac{r}{r_0}\right)^b\right]\right\}^{-\kappa_0-1-\frac{1}{2}d_K} \varepsilon_K^{\frac{1}{2}d_K-1}, \qquad (3.56a)$$

while the distribution of the potential energy becomes

$$P(r; \kappa_0, T) \sim \left[1 + \frac{\frac{1}{2}d_\Phi}{\kappa_0 - \frac{1}{2}d_\Phi}\cdot\left(\frac{r}{r_0}\right)^b\right]^{-\kappa_0-1}, \qquad (3.56b)$$

where the average potential radius $r_0 = \langle r^b \rangle^{\frac{1}{b}}$ is defined by the degrees of freedom, given by

$$\frac{1}{2}d_\Phi \equiv \frac{\langle\Phi\rangle}{k_B T} = \frac{\frac{1}{b}k}{k_B T}r_0^b. \qquad (3.57)$$

We have $r_0 = \langle r^b \rangle^{\frac{1}{b}} = \left(\mu_{b+d_r-1}/\mu_{d_r-1}\right)^{\frac{1}{b}}$, where

$$\mu_m \equiv \int_0^\infty \left[1 + \left(\frac{r}{\alpha}\right)^b\right]^{-\kappa_0-1} r^m dr, \qquad (3.58a)$$

with the involved constant now given by $\alpha \equiv r_0 \cdot \left[\left(\kappa_0 - \frac{1}{2}d_\Phi \right) \Big/ \frac{1}{2}d_\Phi \right]^{\frac{1}{b}}$. Thus

$$
\mu_m = \kappa_0^{-\frac{m+1}{b}} \alpha^{m+1} \int_0^\infty \left(1 + \frac{1}{\kappa_0} \cdot x \right)^{-\kappa_0 - 1} x^{\frac{m+1}{b} - 1} dx = \frac{1}{b} \kappa_0^{-\frac{m+1}{b}} \alpha^{m+1} \Gamma_{\kappa_0} \left(\frac{m+1}{b} \right)
$$

$$
= \frac{1}{b} \alpha^{m+1} B \left(\frac{m+1}{b}, \kappa_0 + 1 - \frac{m+1}{b} \right).
$$

(3.58b)

Hence,

$$
\mu_{b+d_r-1} = \frac{1}{b} \alpha^{b+d_r} B \left(1 + \frac{d_r}{b}, \kappa_0 - \frac{d_r}{b} \right), \mu_{d_r-1} = \frac{1}{b} \alpha^{d_r} B \left(\frac{d_r}{b}, \kappa_0 + 1 - \frac{d_r}{b} \right),
$$

(3.58c)

and,

$$
r_0 = \alpha \cdot \left[\frac{B \left(1 + \frac{d_r}{b}, \kappa_0 - \frac{d_r}{b} \right)}{B \left(\frac{d_r}{b}, \kappa_0 + 1 - \frac{d_r}{b} \right)} \right]^{\frac{1}{b}} = \alpha \cdot \left[\frac{\Gamma \left(1 + \frac{d_r}{b} \right) \Gamma \left(\kappa_0 - \frac{d_r}{b} \right)}{\Gamma \left(\frac{d_r}{b} \right) \Gamma \left(\kappa_0 + 1 - \frac{d_r}{b} \right)} \right]^{\frac{1}{b}}
$$

$$
= \alpha \cdot \left(\frac{\frac{d_r}{b}}{\kappa_0 - \frac{d_r}{b}} \right)^{\frac{1}{b}} = r_0 \cdot \left[\frac{\kappa_0 / \frac{1}{2}d_\Phi - 1}{\kappa_0 / \frac{d_r}{b} - 1} \right]^{\frac{1}{b}}, \quad \text{i.e.,}
$$

$$
\frac{1}{2}d_\Phi = \frac{d_r}{b}, r_0 = \left(\frac{\frac{d_r}{b}k_B T}{\frac{1}{b}k} \right)^{\frac{1}{b}}, \alpha \equiv r_0 \cdot \left[\left(\kappa_0 - \frac{d_r}{b} \right) \Big/ \frac{d_r}{b} \right]^{\frac{1}{b}}.
$$

(3.59)

Hence, the positional probability distribution becomes

$$
P(\overrightarrow{r}; \kappa_0, T) = A_r \cdot \left[1 + \frac{\frac{d_r}{b}}{\kappa_0 - \frac{d_r}{b}} \cdot \left(\frac{|\overrightarrow{r}|}{r_0} \right)^b \right]^{-\kappa_0 - 1},
$$

(3.60a)

where A_r is the normalization constant of the positional distribution

$$
A_r = \frac{1}{2} \left(\pi \kappa_0 r_0^2 \right)^{-\frac{d_r}{2}} \Gamma \left(\frac{d_r}{2} \right) \mu_{d_r-1} = \frac{b}{2} \pi^{-\frac{d_r}{2}} \left(\frac{d_r}{b} \right)^{\frac{d_r}{b}} \frac{\Gamma \left(\frac{d_r}{2} \right) \left(\kappa_0 - \frac{d_r}{b} \right)^{-\frac{d_r}{b}}}{B \left(\frac{d_r}{b}, \kappa_0 - \frac{d_r}{b} + 1 \right)} \cdot r_0^{-d_r}.
$$

(3.60b)

Then the whole phase space distribution is

$$
P(\overrightarrow{r}, \varepsilon_K; \kappa_0, T) d\overrightarrow{r} d\varepsilon_K = A_r A_K \cdot \left\{ 1 + \frac{1}{\kappa_0 - \frac{d_r}{b}} \cdot \left[\frac{\varepsilon_K}{k_B T} + \frac{d_r}{b} \left(\frac{|\overrightarrow{r}|}{r_0} \right)^b \right] \right\}^{-\kappa_0 - 1 - \frac{1}{2}d_K} \varepsilon_K^{\frac{1}{2}d_K - 1} d\overrightarrow{r} d\varepsilon_K,
$$

(3.61a)

where A_K is the normalization constant of the kinetic distribution

$$A_K = \frac{\left(\kappa_0 - \dfrac{d_r}{b}\right)^{-\frac{1}{2}d_K}}{B\left(\dfrac{d_K}{2}, \kappa_0 + 1\right)} \cdot (k_B T)^{-\frac{1}{2}d_K}. \tag{3.61b}$$

Namely,

$$P(\vec{r}, \varepsilon_K; \kappa_0, T)d\vec{r}d\varepsilon_K = \frac{\Gamma\left(\dfrac{d_r}{2}\right)}{2\pi^{\frac{d_r}{2}}} \cdot \frac{b\left(\dfrac{d_r}{b}\right)^{\frac{d_r}{b}}\left(\kappa_0 - \dfrac{d_r}{b}\right)^{-\frac{1}{b}d_r - \frac{1}{2}d_K}}{B\left(\dfrac{d_r}{b}, \kappa_0 - \dfrac{d_r}{b} + 1\right)B\left(\dfrac{d_K}{2}, \kappa_0 + 1\right)} \cdot (k_B T)^{-\frac{1}{2}d_K} r_0^{-d_r}$$

$$\times \left\{1 + \frac{1}{\kappa_0 - \dfrac{d_r}{b}} \cdot \left[\frac{\varepsilon_K}{k_B T} + \frac{d_r}{b}\left(\frac{|\vec{r}|}{r_0}\right)^b\right]\right\}^{-\kappa_0 - 1 - \frac{1}{2}d_K} \varepsilon_K^{\frac{1}{2}d_K - 1} d\vec{r}d\varepsilon_K, \tag{3.62a}$$

or

$$P(r, \varepsilon_K; \kappa_0, T)drd\varepsilon_K = \frac{b\left(\dfrac{d_r}{b}\right)^{\frac{d_r}{b}}\left(\kappa_0 - \dfrac{d_r}{b}\right)^{-\frac{1}{b}d_r - \frac{1}{2}d_K}}{B\left(\dfrac{d_r}{b}, \kappa_0 - \dfrac{d_r}{b} + 1\right)B\left(\dfrac{d_K}{2}, \kappa_0 + 1\right)} \cdot (k_B T)^{-\frac{1}{2}d_K} r_0^{-d_r}$$

$$\times \left\{1 + \frac{1}{\kappa_0 - \dfrac{d_r}{b}} \cdot \left[\frac{\varepsilon_K}{k_B T} + \frac{d_r}{b}\left(\frac{r}{r_0}\right)^b\right]\right\}^{-\kappa_0 - 1 - \frac{1}{2}d_K} \varepsilon_K^{\frac{1}{2}d_K - 1} r^{d_r - 1} drd\varepsilon_K. \tag{3.62b}$$

The marginal kappa distributions of position (radius) and velocity (kinetic energy) are respectively given by

$$P(\vec{r}; \kappa_0, T)d\vec{r} = \frac{\Gamma\left(\dfrac{d_r}{2}\right)}{2\pi^{\frac{d_r}{2}}} \cdot \frac{b\left(\dfrac{d_r}{b}\right)^{\frac{d_r}{b}}\left(\kappa_0 - \dfrac{d_r}{b}\right)^{-\frac{d_r}{b}}}{B\left(\dfrac{d_r}{b}, \kappa_0 - \dfrac{d_r}{b} + 1\right)} \cdot r_0^{-d_r} \cdot \left[1 + \frac{\dfrac{d_r}{b}}{\kappa_0 - \dfrac{d_r}{b}}\cdot\left(\frac{\vec{r}}{r_0}\right)^b\right]^{-\kappa_0 - 1} d\vec{r}, \text{ or} \tag{3.63a}$$

$$P(r; \kappa_0, T)dr = \frac{b\left(\dfrac{d_r}{b}\right)^{\frac{d_r}{b}}\left(\kappa_0 - \dfrac{d_r}{b}\right)^{-\frac{d_r}{b}}}{B\left(\dfrac{d_r}{b}, \kappa_0 - \dfrac{d_r}{b} + 1\right)} \cdot r_0^{-d_r} \cdot \left[1 + \frac{\dfrac{d_r}{b}}{\kappa_0 - \dfrac{d_r}{b}}\cdot\left(\frac{r}{r_0}\right)^b\right]^{-\kappa_0 - 1} r^{d_r - 1} dr, \tag{3.63b}$$

and

$$P(\varepsilon_K; \kappa_0, T)d\varepsilon_K = \frac{\left(\kappa_0 - \dfrac{d_r}{b}\right)^{-\frac{1}{2}d_K}}{B\left(\dfrac{d_K}{2}, \kappa_0 - \dfrac{d_r}{b} + 1\right)} \cdot (k_B T)^{-\frac{1}{2}d_K}$$

$$\times \left(1 + \frac{1}{\kappa_0 - \dfrac{d_r}{b}} \cdot \frac{\varepsilon_K}{k_B T}\right)^{-\left(\kappa_0 - \frac{1}{b}d_r\right) - 1 - \frac{1}{2}d_K} \varepsilon_K^{\frac{1}{2}d_u - 1} d\varepsilon_K. \quad (3.64)$$

Again, we observe that the kinetic kappa distribution in the presence of the potential energy (3.64) differs from the standard kinetic kappa distribution derived in the absence of a potential energy (3.53b). Their difference is in the kappa index, which has been degenerated because of the potential energy degrees of freedom, that is, the positional dimensionality d_r:

$$\kappa_0(\text{no potential}) \Rightarrow \kappa_0 + \frac{1}{b}d_r \ (\text{with potential } \Phi \propto r^b) \quad (3.65a)$$

or

$$\kappa(d_K) = \kappa_0 + \tfrac{1}{2}d_K \ (\text{no potential}) \Rightarrow$$
$$\kappa(d_K) = \kappa_0 + \frac{1}{2}d_K + \frac{1}{b}d_r \ (\text{with potential } \Phi \propto r^b). \quad (3.65b)$$

Hence, the function $K(d_r)$ in $\kappa(d_K, d_r) = \kappa_0 + \frac{1}{2}d_K + K(d_r)$ is $K(d_r) = \frac{1}{b} \cdot d_r$. Applying the transformation $\kappa_0 - \frac{1}{b}d_r \to \kappa_0$, the phase space distribution becomes

$$P(\vec{r}, \varepsilon_K; \kappa_0, T)d\vec{r}d\varepsilon_K = \frac{\Gamma\left(\dfrac{d_r}{2}\right)}{2\pi^{\frac{d_r}{2}}} \cdot \frac{b\left(\dfrac{d_r}{b}\right)^{\frac{d_r}{b}} \kappa_0^{-\frac{1}{b}d_r - \frac{1}{2}d_K}}{B\left(\dfrac{d_K}{2}, \dfrac{d_r}{b}\right)B\left(\dfrac{d_K}{2} + \dfrac{d_r}{b}, \kappa_0 + 1\right)} \cdot (k_B T)^{-\frac{1}{2}d_K} r_0^{-d_r}$$

$$\times \left\{1 + \frac{1}{\kappa_0} \cdot \left[\frac{\varepsilon_K}{k_B T} + \frac{d_r}{b}\left(\frac{|\vec{r}|}{r_0}\right)^b\right]\right\}^{-\kappa_0 - 1 - \frac{1}{2}d_K - \frac{1}{b}d_r}$$

$$\times \varepsilon_K^{\frac{1}{2}d_K - 1} d\vec{r}d\varepsilon_K, \quad (3.66a)$$

or

$$P(r, \varepsilon_K; \kappa_0, T)drd\varepsilon_K = \frac{b\left(\dfrac{d_r}{b}\right)^{\frac{d_r}{b}} \kappa_0^{-\frac{1}{b}d_r - \frac{1}{2}d_K} \cdot (k_B T)^{-\frac{1}{2}d_K} r_0^{-d_r}}{B\left(\dfrac{d_K}{2}, \dfrac{d_r}{b}\right)B\left(\dfrac{d_K}{2} + \dfrac{d_r}{b}, \kappa_0 + 1\right)}$$

$$\times \left\{1 + \frac{1}{\kappa_0} \cdot \left[\frac{\varepsilon_K}{k_B T} + \frac{d_r}{b}\left(\frac{r}{r_0}\right)^b\right]\right\}^{-\kappa_0 - 1 - \frac{1}{2}d_K - \frac{1}{b}d_r} \varepsilon_K^{\frac{1}{2}d_K - 1} r^{d_r - 1} drd\varepsilon_K.$$

$$(3.66b)$$

where we used

$$B\left(\frac{d_{\mathrm{r}}}{b}, \kappa_0 + 1\right) B\left(\frac{d_{\mathrm{K}}}{2}, \kappa_0 + \frac{d_{\mathrm{r}}}{b} + 1\right) = \frac{\Gamma\left(\frac{d_{\mathrm{r}}}{b}\right)\Gamma(\kappa_0 + 1)\Gamma\left(\frac{d_{\mathrm{K}}}{2}\right)}{\Gamma\left(\kappa_0 + 1 + \frac{d_{\mathrm{K}}}{2} + \frac{d_{\mathrm{r}}}{b}\right)}$$

$$= B\left(\frac{d_{\mathrm{K}}}{2}, \frac{d_{\mathrm{r}}}{b}\right) B\left(\frac{d_{\mathrm{K}}}{2} + \frac{d_{\mathrm{r}}}{b}, \kappa_0 + 1\right). \qquad (3.67)$$

The reader may exercise to derive Eq. (3.66a) starting from the generic phase space distribution function Eq. (3.20a). Further, the positional distribution Eqs. (3.63a) and (3.63b) is rewritten

$$P(\overrightarrow{r}; \kappa_0, T)d\overrightarrow{r} = \frac{\Gamma\left(\frac{d_{\mathrm{r}}}{2}\right)}{2\pi^{\frac{d_{\mathrm{r}}}{2}}} \cdot \frac{b\left(\frac{d_{\mathrm{r}}}{b}\right)^{\frac{d_{\mathrm{r}}}{b}}\kappa_0^{-\frac{d_{\mathrm{r}}}{b}}}{B\left(\frac{d_{\mathrm{r}}}{b}, \kappa_0 + 1\right)} \cdot r_0^{-d_{\mathrm{r}}} \cdot \left[1 + \frac{d_{\mathrm{r}}}{b} \cdot \left(\frac{|\overrightarrow{r}|}{r_0}\right)^b\right]^{-\kappa_0 - 1 - \frac{1}{b}d_{\mathrm{r}}} d\overrightarrow{r}, \text{ or}$$

$$(3.68a)$$

$$P(r; \kappa_0, T)dr = \frac{b\left(\frac{d_{\mathrm{r}}}{b}\right)^{\frac{d_{\mathrm{r}}}{b}}\kappa_0^{-\frac{d_{\mathrm{r}}}{b}}}{B\left(\frac{d_{\mathrm{r}}}{b}, \kappa_0 + 1\right)} \cdot r_0^{-d_{\mathrm{r}}} \cdot \left[1 + \frac{d_{\mathrm{r}}}{b} \cdot \left(\frac{r}{r_0}\right)^b\right]^{-\kappa_0 - 1 - \frac{1}{b}d_{\mathrm{r}}} r^{d_{\mathrm{r}} - 1} dr, \qquad (3.68b)$$

Fig. 3.3 plots the positional distribution function in terms of r^b:

$$P\left(r^b; \kappa_0\right)d\left(r^b\right) = \frac{\kappa_0^{-\frac{1}{b}d_{\mathrm{r}}}}{B\left(\frac{d_{\mathrm{r}}}{b}, \kappa_0 + 1\right)} \cdot \left(1 + \frac{1}{\kappa_0} \cdot r^b\right)^{-\kappa_0 - 1 - \frac{1}{b}d_{\mathrm{r}}} \left(r^b\right)^{d_{\mathrm{r}}/b - 1} d\left(r^b\right),$$

$$(3.68c)$$

where the radial distance is in units of $r_0\left(\frac{d_{\mathrm{r}}}{b}\right)^{-\frac{1}{b}} = \left[k_{\mathrm{B}}T/(\frac{1}{b}k)\right]^{\frac{1}{b}}$.

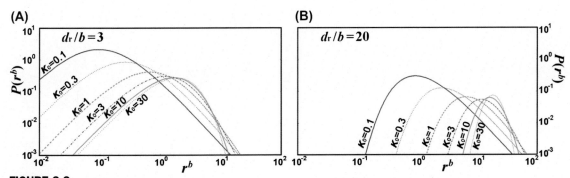

FIGURE 3.3
Positional distribution for the potential $\Phi(r) = \frac{1}{b}k \cdot r^b$, as shown in Eq. (3.68c), plotted for various values of the kappa index κ_0 and dimensionalities (A) $d_{\mathrm{r}}/b = 3$ and (B) $d_{\mathrm{r}}/b = 20$. Adapted from Livadiotis (2015c).

Therefore, the kappa index depends not only on the dimensionality of the velocity space, but also on the dimensionality of the positional space. The dependence is similar for both spaces, namely,

$$\kappa = \kappa_0 + \frac{1}{2}d_K + \frac{1}{b}d_r = \kappa_0 + \frac{\langle \varepsilon_K \rangle}{k_B T} + \frac{\langle \Phi \rangle}{k_B T} = \kappa_0 + \frac{\langle H \rangle}{k_B T}. \tag{3.69a}$$

The invariant kappa index κ_0 is independent of the positional and kinetic degrees of freedom, with a range of values $\kappa_0 \in (0, \infty)$. This is similar to the definition of the invariant kappa index κ_0 in Eq. (3.2c) (Chapter 1), but now the effect of the potential energy is also considered,

$$\kappa = \kappa_0 + \left(\frac{1}{2}d_K + \frac{1}{b}d_r \right) N. \tag{3.69b}$$

One last point remains to be referred to the kappa index degeneration: the zero-th level of the potential energy must be arbitrary. The theory of kappa distributions ensures for this condition because it involves the difference of the potential energy with its average value, e.g., see Eq. (3.5). However, the derived form in Eq. (3.40b) does not include this difference of potential energies; thus the invariance on the zero potential energy may be questioned. In order to show that this invariance is still valid, we transform the generic phase space distribution function, Eq. (3.20a), according to an arbitrary constant potential energy Φ_0, i.e., $\Phi \to \Phi + \Phi_0$,

$$P(\vec{r}, \varepsilon_K; \kappa_0, T) = \kappa_0^{-\frac{1}{2}d_K} \cdot \frac{\Gamma\left(\kappa_0 + 1 + \frac{d_K}{2} + \frac{d_\Phi}{2} + \frac{\Phi_0}{k_B T} \right)}{\Gamma\left(\kappa_0 + \frac{d_\Phi}{2} + \frac{\Phi_0}{k_B T} + 1 \right) \cdot \Gamma\left(\frac{d_K}{2} \right)} \cdot (k_B T)^{-\frac{1}{2}d_K}$$

$$\times \left\{ \int_{-\infty}^{\infty} \left[1 + \frac{1}{\kappa_0} \cdot \frac{\Phi_0 + \Phi(\vec{r})}{k_B T} \right]^{-\kappa_0 - 1 - \frac{1}{2}d_\Phi - \frac{\Phi_0}{k_B T}} d\vec{r} \right\}^{-1} \tag{3.70}$$

$$\times \left[1 + \frac{1}{\kappa_0} \cdot \frac{\varepsilon_K + \Phi_0 + \Phi(\vec{r})}{k_B T} \right]^{-\kappa_0 - 1 - \frac{1}{2}d_K - \frac{1}{2}d_\Phi - \frac{\Phi_0}{k_B T}} \varepsilon_K^{\frac{1}{2}d_K - 1}.$$

After calculations,

$$P(\vec{r}, \varepsilon_K; \kappa_0, T) = \left(\kappa_0 + \frac{\Phi_0}{k_B T} \right)^{-\frac{1}{2}d_K} \cdot \frac{\Gamma\left[\left(\kappa_0 + \frac{\Phi_0}{k_B T} \right) + 1 + \frac{d_K}{2} + \frac{d_\Phi}{2} \right]}{\Gamma\left[\left(\kappa_0 + \frac{\Phi_0}{k_B T} \right) + \frac{d_\Phi}{2} + 1 \right] \cdot \Gamma\left(\frac{d_K}{2} \right)} \cdot (k_B T)^{-\frac{1}{2}d_K}$$

$$\times \left\{ \int_{-\infty}^{\infty} \left[1 + \frac{1}{\kappa_0 + \frac{\Phi_0}{k_B T}} \cdot \frac{\Phi(\vec{r})}{k_B T} \right]^{-\left(\kappa_0 + \frac{\Phi_0}{k_B T} \right) - 1 - \frac{1}{2}d_\Phi} d\vec{r} \right\}^{-1}$$

$$\times \left[1 + \frac{1}{\kappa_0 + \frac{\Phi_0}{k_B T}} \cdot \frac{\varepsilon_K + \Phi(\vec{r})}{k_B T} \right]^{-\left(\kappa_0 + \frac{\Phi_0}{k_B T} \right) - 1 - \frac{1}{2}d_K - \frac{1}{2}d_\Phi} \varepsilon_K^{\frac{1}{2}d_K - 1},$$

$$\tag{3.71a}$$

but we recall that the kappa index degeneration is also affected by the new zero potential energy:

$$\kappa = \left(\kappa_0 + \frac{\Phi_0}{k_B T}\right) + \frac{\langle \varepsilon_K \rangle}{k_B T} + \frac{\langle \Phi \rangle}{k_B T}. \tag{3.71b}$$

We observe that Eqs. (3.71a) and (3.71b) depend on the quantity $\kappa_0 + \frac{\Phi_0}{k_B T}$, indicating that the invariant kappa index is now given by this quantity. Indeed, if all the degrees of freedom were zero, then the kappa index κ would coincide with its invariant value κ_0. The smallest value of the kappa index is zero, indicating the case of zero (kinetic and potential) degrees of freedom $d_K = d_\Phi = 0$, and zero invariant kappa index $\kappa_0 \rightarrow 0$. If κ_0 was the invariant kappa index, then its smallest value is $\kappa_0 \rightarrow 0$, and the smallest kappa index should be $\kappa \rightarrow \frac{\Phi_0}{k_B T}$. However, Eq. (3.71b) clearly shows that the invariant kappa index is given by $\kappa_0 + \frac{\Phi_0}{k_B T}$. Then after the transformation of the kappa index,

$$\kappa_0 \text{ (potential } \Phi) \Rightarrow \kappa_0 - \frac{\Phi_0}{k_B T} \text{ (potential } \Phi + \Phi_0), \tag{3.71c}$$

substituted in Eqs. (3.71a) and (3.71b), we obtain again the generic form of Eq. (3.20a),

$$P(\vec{r}, \varepsilon_K; \kappa_0, T) = \kappa_0^{-\frac{1}{2} d_K} \cdot \frac{\Gamma\left(\kappa_0 + 1 + \frac{d_K}{2} + \frac{d_\Phi}{2}\right)}{\Gamma\left(\kappa_0 + \frac{d_\Phi}{2} + 1\right) \cdot \Gamma\left(\frac{d_K}{2}\right)} \cdot (k_B T)^{-\frac{1}{2} d_K}$$

$$\times \left\{ \int_{-\infty}^{\infty} \left[1 + \frac{1}{\kappa_0} \cdot \frac{\Phi(\vec{r})}{k_B T}\right]^{-\kappa_0 - 1 - \frac{1}{2} d_\Phi} d\vec{r} \right\}^{-1} \tag{3.72a}$$

$$\times \left[1 + \frac{1}{\kappa_0} \cdot \frac{\varepsilon_K + \Phi(\vec{r})}{k_B T}\right]^{-\kappa_0 - 1 - \frac{1}{2} d_K - \frac{1}{2} d_\Phi} \varepsilon_K^{\frac{1}{2} d_K - 1},$$

and the degeneration of the kappa index

$$\kappa = \kappa_0 + \frac{\langle \varepsilon_K \rangle}{k_B T} + \frac{\langle \Phi \rangle}{k_B T} \quad \text{or} \quad \kappa(d_K, d_\Phi) = \kappa_0 + \frac{1}{2} d_K + \frac{1}{2} d_\Phi. \tag{3.72b}$$

Therefore, we end up with the same generic relations independently of the zero-th level of the potential energy given by the arbitrary constant Φ_0. The key is the re-degeneration of the kappa index so that the smallest possible kappa index to be equal to zero, which occurs when all the degrees of freedom are zero.

3.8 Local Kappa Distribution

The phase space distribution can be seen as the local velocity distribution, that is, the distribution of velocities that varies under different positions. The position dependence is hidden in the presence of the potential $\Phi(\overrightarrow{r})$. Finally, we will see how the positional dependence can shift from the potential to the temperature and density.

Eqs. (3.20a) and (3.20b) express the normalized phase space kappa distribution, e.g., $P(\overrightarrow{r}, \varepsilon_K)$ or $P(\overrightarrow{r}, \overrightarrow{u})$. This may be interpreted as the kinetic kappa distributions, but locally, in the position $\overrightarrow{r}$, i.e., $P(\overrightarrow{r}, \varepsilon_K)$ or $P(\overrightarrow{r}, \overrightarrow{u})$. However, a renormalization is needed so that their integration over velocities to lead to 1:

$$P(\overrightarrow{r}, \overrightarrow{u}; \kappa_0, T) = \frac{P(\overrightarrow{r}, \overrightarrow{u}; \kappa_0, T)}{\int_{-\infty}^{\infty} P(\overrightarrow{r}, \overrightarrow{u}; \kappa_0, T) d\overrightarrow{u}},$$

$$P(\varepsilon_K, \overrightarrow{r}; \kappa_0, T) = \frac{P(\overrightarrow{r}, \varepsilon_K; \kappa_0, T)}{\int_{0}^{\infty} P(\overrightarrow{r}, \varepsilon_K; \kappa_0, T) d\varepsilon_K},$$

(3.73a)

where the denominator is the positional distribution

$$\int_{0}^{\infty} P(\overrightarrow{r}, \varepsilon_K; \kappa_0, T) d\varepsilon_K = \int_{-\infty}^{\infty} P(\overrightarrow{r}, \overrightarrow{u}; \kappa_0, T) d\overrightarrow{u} = P(\overrightarrow{r}; \kappa_0, T), \quad (3.73b)$$

(see similar renormalization in Livadiotis and Desai, 2016).

Distributions, Eqs. (3.73a) and (3.73b), read the local kappa distributions of velocity and of kinetic energy. Similarly, we derive the local density and temperature (also see the detail derivations for the equilibrium and nonequilibrium cases in Section 5.3). First, from the marginal distribution of the potential energy, we derive the local density $n(\overrightarrow{r})$ using the normalization

$$n(\overrightarrow{r})/n(\overrightarrow{r}*) = P(\overrightarrow{r}; \kappa_0; T)/P(\overrightarrow{r}*; \kappa_0; T), \quad (3.74a)$$

where $\overrightarrow{r}*$ indicates the position of zero potential $\Phi(\overrightarrow{r}*) = 0$, and $n(\overrightarrow{r}*)$ is the local density at this position. Given Eqs. (3.27a) and (3.27b), we obtain

$$n(\overrightarrow{r}) = n(\overrightarrow{r}*) \cdot \left[1 + \frac{1}{\kappa_0} \cdot \frac{\Phi(\overrightarrow{r})}{k_B T} \right]^{-\kappa_0 - 1 - \frac{1}{2}d_\Phi}. \quad (3.74b)$$

Then the local temperature is derived from the local mean kinetic energy $\langle \varepsilon_K \rangle(\overrightarrow{r}) \equiv \frac{1}{2} d_K \cdot k_B T(\overrightarrow{r})$, that is,

$$\frac{1}{2}d_{\mathrm{K}}k_{\mathrm{B}}T(\vec{r}) = \frac{\displaystyle\int_{-\infty}^{\infty} P(\vec{r},\vec{u};T)\varepsilon_{\mathrm{K}}(\vec{u})d\vec{u}}{\displaystyle\int_{-\infty}^{\infty} P(\vec{r},\vec{u};T)d\vec{u}}$$

$$= \frac{\displaystyle\int_{0}^{\infty}\left[1+\frac{1}{\kappa_0}\cdot\frac{\varepsilon_{\mathrm{K}}+\Phi(\vec{r})}{k_{\mathrm{B}}T}\right]^{-\kappa_0-1-\frac{1}{2}d_{\mathrm{K}}-\frac{1}{2}d_{\Phi}}\varepsilon_{\mathrm{K}}^{\frac{1}{2}d_{\mathrm{K}}}d\varepsilon_{\mathrm{K}}}{\displaystyle\int_{0}^{\infty}\left[1+\frac{1}{\kappa_0}\cdot\frac{\varepsilon_{\mathrm{K}}+\Phi(\vec{r})}{k_{\mathrm{B}}T}\right]^{-\kappa_0-1-\frac{1}{2}d_{\mathrm{K}}-\frac{1}{2}d_{\Phi}}\varepsilon_{\mathrm{K}}^{\frac{1}{2}d_{\mathrm{K}}-1}d\varepsilon_{\mathrm{K}}}$$

$$= \frac{\displaystyle\int_{0}^{\infty}\left[1+\frac{\mathrm{X}}{\kappa_0+\dfrac{\Phi(\vec{r})}{k_{\mathrm{B}}T}}\right]^{-\kappa_0-1-\frac{1}{2}d_{\mathrm{K}}-\frac{1}{2}d_{\Phi}}\mathrm{X}^{\frac{1}{2}d_{\mathrm{K}}}d\mathrm{X}}{\displaystyle\int_{0}^{\infty}\left[1+\frac{\mathrm{X}}{\kappa_0+\dfrac{\Phi(\vec{r})}{k_{\mathrm{B}}T}}\right]^{-\kappa_0-1-\frac{1}{2}d_{\mathrm{K}}-\frac{1}{2}d_{\Phi}}\mathrm{X}^{\frac{1}{2}d_{\mathrm{K}}-1}d\mathrm{X}}\cdot k_{\mathrm{B}}T \qquad (3.75)$$

$$= \frac{\kappa_0+\dfrac{\Phi(\vec{r})}{k_{\mathrm{B}}T}}{\kappa_0+\dfrac{d_{\mathrm{K}}}{2}+\dfrac{d_{\Phi}}{2}}\cdot\frac{\Gamma_{\kappa_0+\frac{1}{2}d_{\mathrm{K}}+\frac{1}{2}d_{\Phi}}\left(\dfrac{d_{\mathrm{K}}}{2}+1\right)}{\Gamma_{\kappa_0+\frac{1}{2}d_{\mathrm{K}}+\frac{1}{2}d_{\Phi}}\left(\dfrac{d_{\mathrm{K}}}{2}\right)}\cdot k_{\mathrm{B}}T$$

$$= \frac{\kappa_0+\dfrac{\Phi(\vec{r})}{k_{\mathrm{B}}T}}{\kappa_0+\dfrac{d_{\Phi}}{2}}\cdot\frac{1}{2}d_{\mathrm{K}}\cdot k_{\mathrm{B}}T,$$

where we set $\mathrm{X}\equiv\varepsilon_{\mathrm{K}}/(k_{\mathrm{B}}T)$ and used the q-gamma functions

$$\frac{\Gamma_{\kappa_0+\frac{1}{2}d_{\mathrm{K}}+\frac{1}{2}d_{\Phi}}\left(\dfrac{1}{2}d_{\mathrm{K}}+1\right)}{\Gamma_{\kappa_0+\frac{1}{2}d_{\mathrm{K}}+\frac{1}{2}d_{\Phi}}\left(\dfrac{1}{2}d_{\mathrm{K}}\right)} = \frac{1}{2}d_{\mathrm{K}}\cdot\left(\kappa_0+\frac{1}{2}d_{\mathrm{K}}-\frac{1}{2}d_{\Phi}\right)\cdot\frac{\Gamma\left(\kappa_0+\dfrac{1}{2}d_{\Phi}\right)}{\Gamma\left(\kappa_0+1+\dfrac{1}{2}d_{\Phi}\right)}$$

$$= \frac{1}{2}d_{\mathrm{K}}\cdot\frac{\kappa_0+\dfrac{1}{2}d_{\mathrm{K}}-\dfrac{1}{2}d_{\Phi}}{\kappa_0+\dfrac{1}{2}d_{\Phi}}. \qquad (3.76)$$

Hence, we have

$$T(\vec{r}) = T\cdot\frac{\kappa_0}{\kappa_0+\dfrac{1}{2}d_{\Phi}}\cdot\left[1+\frac{1}{\kappa_0}\cdot\frac{\Phi(\vec{r})}{k_{\mathrm{B}}T}\right], \text{ or} \qquad (3.77a)$$

$$T(\overrightarrow{r}) = T(\overrightarrow{r}_*)\cdot\left[1 + \frac{1}{\kappa_0}\cdot\frac{\Phi(\overrightarrow{r})}{k_B T}\right], \text{ with } T(\overrightarrow{r}_*) \equiv \frac{\kappa_0}{\kappa_0 + \frac{1}{2}d_\Phi}\cdot T, \tag{3.77b}$$

where $T(\overrightarrow{r}_*)$ is the local temperature at the position of zero potential $\overrightarrow{r} = \overrightarrow{r}_*$.

Note that another characteristic position is when the potential becomes equal to its average value $\Phi(\overrightarrow{r}) = \langle\Phi\rangle$. Let this noted by $\overrightarrow{r} = \overrightarrow{r}_c$; hence, from Eq. (3.78a), we have

$$n(\overrightarrow{r}_c) = n(\overrightarrow{r}_*)\cdot\left(\frac{\kappa_0 + \frac{d_\Phi}{2}}{\kappa_0}\right)^{-\kappa_0 - 1 - \frac{1}{2}d_\Phi} , T(\overrightarrow{r}_c) = T(\overrightarrow{r}_*)\cdot\frac{\kappa_0 + \frac{d_\Phi}{2}}{\kappa_0} = T,$$

$$\left[\frac{n(\overrightarrow{r})}{n(\overrightarrow{r}_c)}\right] = \left[\frac{T(\overrightarrow{r})}{T(\overrightarrow{r}_c)}\right]^{-\kappa_0 - 1 - \frac{1}{2}d_\Phi} .$$

$$\tag{3.77c}$$

Therefore, we have

$$n(\overrightarrow{r}) = n(\overrightarrow{r}_*)\cdot\left[1 + \frac{1}{\kappa_0}\cdot\frac{\Phi(\overrightarrow{r})}{k_B T}\right]^{-\kappa_0 - 1 - \frac{1}{2}d_\Phi} ,$$

$$T(\overrightarrow{r}) = T(\overrightarrow{r}_*)\cdot\left[1 + \frac{1}{\kappa_0}\cdot\frac{\Phi(\overrightarrow{r})}{k_B T}\right],$$

$$p(\overrightarrow{r}) = p(\overrightarrow{r}_*)\cdot\left[1 + \frac{1}{\kappa_0}\cdot\frac{\Phi(\overrightarrow{r})}{k_B T}\right]^{-\kappa_0 - \frac{1}{2}d_\Phi} , \tag{3.78a}$$

where the thermal pressure is also given, with $p(\overrightarrow{r}) = n(\overrightarrow{r})k_B T(\overrightarrow{r})$ and $p(\overrightarrow{r}_*) = n(\overrightarrow{r}_*)k_B T(\overrightarrow{r}_*)$.

Combining the previously discussed relations, we find the polytropic relation (Chapter 5; Livadiotis, 2016c),

$$\left[\frac{n(\overrightarrow{r})}{n(\overrightarrow{r}_*)}\right] = \left[\frac{T(\overrightarrow{r})}{T(\overrightarrow{r}_*)}\right]^{-\kappa_0 - 1 - \frac{1}{2}d_\Phi} \equiv \left[\frac{T(\overrightarrow{r})}{T(\overrightarrow{r}_*)}\right]^\nu ,$$

with $\nu \equiv (a-1)^{-1} = -\kappa_0 - 1 - \frac{1}{2}d_\Phi.$ \tag{3.78b}

For small values of $\frac{d_\Phi}{2}$, we have $\nu \equiv (a-1)^{-1} \cong -\kappa_0 - 1 \cong -\left(\kappa - \frac{1}{2}\right)$ (see Chapters 5 and 11).

Specifically for the positive attractive power law potential $\Phi(r) = \frac{1}{b}k\cdot r^b$ and using the positional distribution, Eqs. (3.68a), (3.78a) and (3.78b) become

$$n(r) = n_0 \cdot \left[\frac{1 + \dfrac{\frac{d_r}{b}}{\kappa_0} \cdot \left(\dfrac{r}{r_0}\right)^b}{1 + \dfrac{\frac{d_r}{b}}{\kappa_0}} \right]^{-\kappa_0 - 1 - \frac{1}{b}d_r} \quad , T(r) = T_0 \cdot \left[\frac{1 + \dfrac{\frac{d_r}{b}}{\kappa_0} \cdot \left(\dfrac{r}{r_0}\right)^b}{1 + \dfrac{d_r}{\kappa_0}} \right],$$

$$p(r) = p_0 \cdot \left[\frac{1 + \dfrac{\frac{d_r}{b}}{\kappa_0} \cdot \left(\dfrac{r}{r_0}\right)^b}{1 + \dfrac{d_r}{\kappa_0}} \right]^{-\kappa_0 - \frac{1}{b}d_r} . \tag{3.79a}$$

where (n_0, T_0, p_0) are the density, temperature, and thermal pressure at radius $r = r_0$. The polytropic relation is

$$\frac{n(r)}{n_0} = \cdot \left[\frac{T(r)}{T_0}\right]^{-\kappa_0 - 1 - \frac{1}{b}d_r} , \nu = -\frac{1}{b}d_r - 1 - \kappa_0. \tag{3.79b}$$

Having derived the local density $n(\vec{r})$ and temperature $T(\vec{r})$, it is straightforward now to ask: what are the mean density and temperature? The mean density is defined by the number of particles N in the volume system V (as defined for open systems in the beginning of Section 3.4),

$$\langle n \rangle \equiv \frac{N}{V} = \frac{\displaystyle\int_{-\infty}^{\infty} n(\vec{r}) d\vec{r}}{\displaystyle\int_{-\infty}^{\infty} d\vec{r}} = n(\vec{r}_*) \cdot \frac{1}{V} \int_{-\infty}^{\infty} \left[1 + \frac{1}{\kappa_0} \frac{\Phi(\vec{r})}{k_B T}\right]^{-\kappa_0 - 1 - \frac{1}{2}d_\Phi} d\vec{r}$$

$$= n(\vec{r}_*) \cdot \frac{1}{A_r V}, \tag{3.80a}$$

where

$$N = \int_{-\infty}^{\infty} n(\vec{r}) d\vec{r} = n(\vec{r}_*) \cdot A_r^{-1}, \text{ or } n(\vec{r}_*) = N A_r, \tag{3.80b}$$

and we used the normalization constant of the positional distribution, A_r (see Eq. 3.27b). The mean temperature is given by averaging the local temperature,

$$\langle T \rangle = \frac{\displaystyle\int_{-\infty}^{\infty} n(\vec{r}) T(\vec{r}) d\vec{r}}{\displaystyle\int_{-\infty}^{\infty} n(\vec{r}) d\vec{r}} = T \cdot \frac{\kappa_0}{\kappa_0 + \frac{1}{2}d_\Phi} \frac{\displaystyle\int_{-\infty}^{\infty} \left[1 + \frac{1}{\kappa_0} \cdot \frac{\Phi(\vec{r})}{k_B T}\right]^{-\kappa_0 - \frac{1}{2}d_\Phi} d\vec{r}}{\displaystyle\int_{-\infty}^{\infty} \left[1 + \frac{1}{\kappa_0} \frac{\Phi(\vec{r})}{k_B T}\right]^{-\kappa_0 - 1 - \frac{1}{2}d_\Phi} d\vec{r}}$$

$$= T \cdot \frac{\kappa_0}{\kappa_0 + \frac{1}{2}d_\Phi} \cdot \left\langle 1 + \frac{1}{\kappa_0} \cdot \frac{\Phi(\vec{r})}{k_B T}\right\rangle = T. \tag{3.80c}$$

Therefore, the global temperature T is actually the average of the local temperature. The mean density and temperature are global characteristics of plasma in contrast to their local values. These global values will be used in the following derivations.

Next, we show the equivalence of two forms of the phase space distribution function. The two distribution forms are in the presence of a potential energy using global density and temperature and in the absence of a potential energy using local density and temperature.

We note the phase space distribution function as $f(\vec{r}, \varepsilon_K) \equiv N \cdot P(\vec{r}, \varepsilon_K)$, where f is normalized to the number of particles N; hence, from Eq. (3.20a), we obtain

$$
f(\vec{r}, \varepsilon_K) = N \cdot \kappa_0^{-\frac{1}{2}d_K} \cdot \frac{\Gamma\left(\kappa_0 + 1 + \dfrac{d_K}{2} + \dfrac{d_\Phi}{2}\right)}{\Gamma\left(\kappa_0 + \dfrac{d_\Phi}{2} + 1\right) \cdot \Gamma\left(\dfrac{d_K}{2}\right)} \cdot (k_B T)^{-\frac{1}{2}d_K}
$$

$$
\times \left\{ \int_{-\infty}^{\infty} \left[1 + \frac{1}{\kappa_0} \cdot \frac{\Phi(\vec{r})}{k_B T}\right]^{-\kappa_0 - 1 - \frac{1}{2}d_\Phi} d\vec{r} \right\}^{-1}
$$

$$
\times \left[1 + \frac{1}{\kappa_0} \cdot \frac{\varepsilon_K + \Phi(\vec{r})}{k_B T}\right]^{-\kappa_0 - 1 - \frac{1}{2}d_K - \frac{1}{2}d_\Phi} \varepsilon_K^{\frac{1}{2}d_K - 1}
$$

$$
= N \cdot \kappa_0^{-\frac{1}{2}d_K} \cdot \frac{\Gamma\left(\kappa_0 + 1 + \dfrac{d_K}{2} + \dfrac{d_\Phi}{2}\right)}{\Gamma\left(\kappa_0 + \dfrac{d_\Phi}{2} + 1\right) \cdot \Gamma\left(\dfrac{d_K}{2}\right)} \cdot (k_B T)^{-\frac{1}{2}d_K}
$$

$$
\times \left\{ \int_{-\infty}^{\infty} \left[1 + \frac{1}{\kappa_0} \cdot \frac{\Phi(\vec{r})}{k_B T}\right]^{-\kappa_0 - 1 - \frac{1}{2}d_\Phi} d\vec{r} \right\}^{-1}
$$

$$
\times \left[1 + \frac{1}{\kappa_0} \cdot \frac{\Phi(\vec{r})}{k_B T}\right]^{-\kappa_0 - 1 - \frac{1}{2}d_\Phi} \cdot \left[1 + \frac{1}{\kappa_0} \cdot \frac{\Phi(\vec{r})}{k_B T}\right]^{-\frac{1}{2}d_K}
$$

$$
\times \left[1 + \frac{1}{\kappa_0 + \dfrac{\Phi(\vec{r})}{k_B T}} \cdot \frac{\varepsilon_K}{k_B T}\right]^{-\kappa_0 - 1 - \frac{1}{2}d_K - \frac{1}{2}d_\Phi} \varepsilon_K^{\frac{1}{2}d_K - 1}
$$

$$
= N \cdot \kappa_0^{-\frac{1}{2}d_{\mathrm{K}}} \cdot \frac{\Gamma\left(\kappa_0 + 1 + \frac{d_{\mathrm{K}}}{2} + \frac{d_{\Phi}}{2}\right)}{\Gamma\left(\kappa_0 + \frac{d_{\Phi}}{2} + 1\right) \cdot \Gamma\left(\frac{d_{\mathrm{K}}}{2}\right)} \cdot \left\{ (k_{\mathrm{B}}T) \cdot \left[1 + \frac{1}{\kappa_0} \cdot \frac{\Phi(\overrightarrow{r})}{k_{\mathrm{B}}T} \right] \right\}^{-\frac{1}{2}d_{\mathrm{K}}}
$$

$$
\times \left\{ \int_{-\infty}^{\infty} \left[1 + \frac{1}{\kappa_0} \cdot \frac{\Phi(\overrightarrow{r})}{k_{\mathrm{B}}T} \right]^{-\kappa_0 - 1 - \frac{1}{2}d_{\Phi}} d\overrightarrow{r} \right\}^{-1} \cdot \left[1 + \frac{1}{\kappa_0} \cdot \frac{\Phi(\overrightarrow{r})}{k_{\mathrm{B}}T} \right]^{-\kappa_0 - 1 - \frac{1}{2}d_{\Phi}}
$$

$$
\times \left\{ 1 + \frac{1}{\kappa_0} \cdot \frac{\varepsilon_{\mathrm{K}}}{k_{\mathrm{B}}T \cdot \left[1 + \frac{1}{\kappa_0} \cdot \frac{\Phi(\overrightarrow{r})}{k_{\mathrm{B}}T} \right]} \right\}^{-\kappa_0 - 1 - \frac{1}{2}d_{\mathrm{K}} - \frac{1}{2}d_{\Phi}} \varepsilon_{\mathrm{K}}^{\frac{1}{2}d_{\mathrm{K}} - 1}
$$

$$
= NA_{\mathrm{r}} \left[1 + \frac{1}{\kappa_0} \cdot \frac{\Phi(\overrightarrow{r})}{k_{\mathrm{B}}T} \right]^{-\kappa_0 - 1 - \frac{1}{2}d_{\Phi}} \cdot \kappa_0^{-\frac{1}{2}d_{\mathrm{K}}} \cdot \frac{\Gamma\left(\kappa_0 + 1 + \frac{d_{\mathrm{K}}}{2} + \frac{d_{\Phi}}{2}\right)}{\Gamma\left(\kappa_0 + \frac{d_{\Phi}}{2} + 1\right) \cdot \Gamma\left(\frac{d_{\mathrm{K}}}{2}\right)}
$$

$$
\times \left\{ (k_{\mathrm{B}}T) \cdot \left[1 + \frac{1}{\kappa_0} \cdot \frac{\Phi(\overrightarrow{r})}{k_{\mathrm{B}}T} \right] \right\}^{-\frac{1}{2}d_{\mathrm{K}}}
$$

$$
\times \left\{ 1 + \frac{1}{\kappa_0} \cdot \frac{\varepsilon_{\mathrm{K}}}{k_{\mathrm{B}}T \cdot \left[1 + \frac{1}{\kappa_0} \cdot \frac{\Phi(\overrightarrow{r})}{k_{\mathrm{B}}T} \right]} \right\}^{-\kappa_0 - 1 - \frac{1}{2}d_{\mathrm{K}} - \frac{1}{2}d_{\Phi}} \varepsilon_{\mathrm{K}}^{\frac{1}{2}d_{\mathrm{K}} - 1}
$$

$$
= n(\overrightarrow{r}_*) \left[1 + \frac{1}{\kappa_0} \cdot \frac{\Phi(\overrightarrow{r})}{k_{\mathrm{B}}T} \right]^{-\kappa_0 - 1 - \frac{1}{2}d_{\Phi}} \cdot \left(\kappa_0 + \frac{d_{\Phi}}{2} \right)^{-\frac{1}{2}d_{\mathrm{K}}} \cdot \frac{\Gamma\left[\left(\kappa_0 + \frac{d_{\Phi}}{2} \right) + 1 + \frac{d_{\mathrm{K}}}{2} \right]}{\Gamma\left[\left(\kappa_0 + \frac{d_{\Phi}}{2} \right) + 1 \right] \cdot \Gamma\left(\frac{d_{\mathrm{K}}}{2} \right)}
$$

$$
\times \left\{ k_{\mathrm{B}}T(\overrightarrow{r}_*) \cdot \left[1 + \frac{1}{\kappa_0} \cdot \frac{\Phi(\overrightarrow{r})}{k_{\mathrm{B}}T} \right] \right\}^{-\frac{1}{2}d_{\mathrm{K}}}
$$

$$
\times \left\{ 1 + \frac{1}{\kappa_0 + \frac{d_{\Phi}}{2}} \cdot \frac{\varepsilon_{\mathrm{K}}}{k_{\mathrm{B}}T(\overrightarrow{r}_*) \cdot \left[1 + \frac{1}{\kappa_0} \cdot \frac{\Phi(\overrightarrow{r})}{k_{\mathrm{B}}T} \right]} \right\}^{-\left(\kappa_0 + \frac{1}{2}d_{\Phi} \right) - 1 - \frac{1}{2}d_{\mathrm{K}}} \varepsilon_{\mathrm{K}}^{\frac{1}{2}d_{\mathrm{K}} - 1}
$$

$$= n(\overrightarrow{r}) \cdot \left(\kappa_0 + \frac{d_\Phi}{2}\right)^{-\frac{1}{2}d_K} \cdot \frac{\Gamma\left[\left(\kappa_0 + \frac{d_\Phi}{2}\right) + 1 + \frac{d_K}{2}\right]}{\Gamma\left[\left(\kappa_0 + \frac{d_\Phi}{2}\right) + 1\right] \cdot \Gamma\left(\frac{d_K}{2}\right)} \cdot [k_B T(\overrightarrow{r})]^{-\frac{1}{2}d_K}$$

$$\times \left[1 + \frac{1}{\kappa_0 + \frac{d_\Phi}{2}} \cdot \frac{\varepsilon_K}{k_B T(\overrightarrow{r})}\right]^{-\left(\kappa_0 + \frac{1}{2}d_\Phi\right) - 1 - \frac{1}{2}d_K} \varepsilon_K^{\frac{1}{2}d_K - 1},$$

where we substituted the local density and temperature, shown in Eq. (3.78a). Therefore, we have transformed the phase space distribution density function from $f\left[\varepsilon_K, \Phi(\overrightarrow{r}); \kappa_0, T, n\right]$ to $f\left[\varepsilon_K, \Phi = 0; \kappa_0 + \frac{d_\Phi}{2}, T(\overrightarrow{r}), n(\overrightarrow{r})\right]$:

$$f[\varepsilon_K, \Phi(\overrightarrow{r}); \kappa_0, T, n] = n \cdot P[\varepsilon_K, \Phi(\overrightarrow{r}); \kappa_0, T]$$

$$= n \cdot \kappa_0^{-\frac{1}{2}d_K} \cdot \frac{\Gamma\left(\kappa_0 + 1 + \frac{d_K}{2} + \frac{d_\Phi}{2}\right)}{\Gamma\left(\kappa_0 + \frac{d_\Phi}{2} + 1\right) \cdot \Gamma\left(\frac{d_K}{2}\right)} \cdot (k_B T)^{-\frac{1}{2}d_K}$$

$$\times \left\{\int_{-\infty}^{\infty} \left[1 + \frac{1}{\kappa_0} \cdot \frac{\Phi(\overrightarrow{r})}{k_B T}\right]^{-\kappa_0 - 1 - \frac{1}{2}d_\Phi} \frac{d\overrightarrow{r}}{V}\right\}^{-1}$$

$$\times \left[1 + \frac{1}{\kappa_0} \cdot \frac{\varepsilon_K + \Phi(\overrightarrow{r})}{k_B T}\right]^{-\kappa_0 - 1 - \frac{1}{2}d_K - \frac{1}{2}d_\Phi} \varepsilon_K^{\frac{1}{2}d_K - 1} \qquad (3.81a)$$

$$\Rightarrow f\left[\varepsilon_K, \Phi = 0; \kappa_0 + \frac{d_\Phi}{2}, T(\overrightarrow{r}), n(\overrightarrow{r})\right] = n(\overrightarrow{r}) \cdot P\left[\varepsilon_K, \Phi = 0; \kappa_0 + \frac{d_\Phi}{2}, T(\overrightarrow{r})\right]$$

$$= n(\overrightarrow{r}) \cdot \left(\kappa_0 + \frac{d_\Phi}{2}\right)^{-\frac{1}{2}d_K} \cdot \frac{\Gamma\left[\left(\kappa_0 + \frac{d_\Phi}{2}\right) + 1 + \frac{d_K}{2}\right]}{\Gamma\left[\left(\kappa_0 + \frac{d_\Phi}{2}\right) + 1\right] \cdot \Gamma\left(\frac{d_K}{2}\right)} \cdot [k_B T(\overrightarrow{r})]^{-\frac{1}{2}d_K}$$

$$\times \left[1 + \frac{1}{\kappa_0 + \frac{d_\Phi}{2}} \cdot \frac{\varepsilon_K}{k_B T(\overrightarrow{r})}\right]^{-\left(\kappa_0 + \frac{1}{2}d_\Phi\right) - 1 - \frac{1}{2}d_K} \varepsilon_K^{\frac{1}{2}d_K - 1}.$$

$$(3.81b)$$

If we replace the kappa index so that $\kappa_0 + \frac{d_\Phi}{2} \to \widetilde{\kappa}_0$, then we may write

$$f[\varepsilon_K, \Phi = 0; \widetilde{\kappa}_0, T(\overrightarrow{r}), n(\overrightarrow{r})] = n(\overrightarrow{r}) \cdot P[\varepsilon_K, \Phi = 0; \widetilde{\kappa}_0, T(\overrightarrow{r})]$$

$$= n(\overrightarrow{r}) \cdot \widetilde{\kappa}_0^{-\frac{1}{2}d_K} \cdot \frac{\Gamma\left(\widetilde{\kappa}_0 + 1 + \dfrac{d_K}{2}\right)}{\Gamma(\widetilde{\kappa}_0 + 1) \cdot \Gamma\left(\dfrac{d_K}{2}\right)} \cdot [k_B T(\overrightarrow{r})]^{-\frac{1}{2}d_K} \tag{3.82a}$$

$$\times \left[1 + \frac{1}{\widetilde{\kappa}_0} \cdot \frac{\varepsilon_K}{k_B T(\overrightarrow{r})}\right]^{-\widetilde{\kappa}_0 - 1 - \frac{1}{2}d_K} \varepsilon_K^{\frac{1}{2}d_K - 1}.$$

The latter coincides with the distribution function in the absence of a potential energy, that is, the standard and frequently used kinetic kappa distribution in terms of the velocity or the kinetic energy,

$$f[\varepsilon_K, \Phi = 0; \kappa_0, T, n] = n \cdot P(\varepsilon_K, \Phi = 0; \kappa_0, T)$$

$$= n \cdot \kappa_0^{-\frac{1}{2}d_K} \cdot \frac{\Gamma\left(\kappa_0 + 1 + \dfrac{d_K}{2}\right)}{\Gamma(\kappa_0 + 1) \cdot \Gamma\left(\dfrac{d_K}{2}\right)} \cdot (k_B T)^{-\frac{1}{2}d_K} \tag{3.82b}$$

$$\times \left(1 + \frac{1}{\kappa_0} \cdot \frac{\varepsilon_K}{k_B T}\right)^{-\kappa_0 - 1 - \frac{1}{2}d_K} \varepsilon_K^{\frac{1}{2}d_K - 1}.$$

Of course, the transformation of the distribution function, Eq. (3.81a), into Eq. (3.81b) does not eliminate the presence of the potential energy. The latter forms the positional dependence of the density and temperature. It also modifies the kappa index. As explained in Section 3.7, the kappa index is further degenerated in the presence of the potential energy according to Eq. (3.72b),

$$\kappa_0 = \kappa - \frac{1}{2}d_K \Rightarrow \kappa_0 = \kappa - \frac{1}{2}d_K - \frac{1}{2}d_\Phi. \tag{3.83}$$

The kappa index κ_0 is physically meaningful and invariant of the dimensionality. Comparing the two equations, Eqs. (3.82a) and (3.82b), we recall that the invariant kappa index κ_0 ranges from zero to infinity, while the modified index $\widetilde{\kappa}_0$ in Eq. (3.82a) ranges from $\frac{d_\Phi}{2}$ to infinity.

In applications, we may ignore the existence of a potential energy, and thus use Eq. (3.82a) instead of Eq. (3.81a). One possible mistake is to consider the extracted values of $\widetilde{\kappa}_0$ as kappa indices (κ_0), while they should be derived from $\kappa_0 = \widetilde{\kappa}_0 - \frac{d_\Phi}{2}$. Such an analysis would have limited errors in cases where the potential energy is smaller than the thermal energy (e.g., Livadiotis et al., 2011, 2012, 2013).

3.9 Negative Potentials
3.9.1 In General

Let's recall our steps for deriving the phase space distribution for positive potentials. We write the Hamiltonian probability distribution is given by

$$
P(\vec{r}, \vec{u}; \kappa, T) \propto \left[1 + \frac{1}{\kappa} \cdot \frac{H(\vec{r}, \vec{u}) - \langle H \rangle}{k_B T}\right]^{-\kappa - 1} \propto \left[1 + \frac{1}{\kappa - \frac{\langle H \rangle}{k_B T}} \cdot \frac{H(\vec{r}, \vec{u})}{k_B T}\right]^{-\kappa - 1},
$$

(3.84)

which becomes the kinetic kappa distribution in the absence of a potential energy,

$$
P(\vec{u}; \kappa, T) \propto \left[1 + \frac{1}{\kappa - \frac{d_K}{2}} \cdot \frac{\varepsilon_K(\vec{u})}{k_B T}\right]^{-\kappa - 1} = \left[1 + \frac{1}{\kappa_0} \cdot \frac{\varepsilon_K(\vec{u})}{k_B T}\right]^{-\kappa_0 - 1 - \frac{1}{2}d_K},
$$

(3.85)

or, when a potential is applied

$$
P(\vec{r}, \vec{u}; \kappa, T) \propto \left[1 + \frac{1}{\kappa - \frac{d_K}{2} - \frac{d_\Phi}{2}} \cdot \frac{\varepsilon_K(\vec{u}) + \Phi(\vec{r})}{k_B T}\right]^{-\kappa - 1}
$$

$$
= \left[1 + \frac{1}{\kappa_0 - \frac{d_\Phi}{2}} \cdot \frac{\varepsilon_K(\vec{u}) + \Phi(\vec{r})}{k_B T}\right]^{-\kappa_0 - 1 - \frac{1}{2}d_K}.
$$

(3.86)

In the previous equation, the kappa index is further generated due to the potential degrees of freedom, i.e., the invariant kappa index given by $\kappa_0 = \kappa - \frac{d_K}{2}$ in the absence of the potential energy, becomes $\kappa_0 = \kappa - \frac{d_K}{2} - \frac{d_\Phi}{2}$, and

$$
P(\vec{r}, \vec{u}; \kappa, T) \propto \left[1 + \frac{1}{\kappa_0 - \frac{d_\Phi}{2}} \cdot \frac{\varepsilon_K(\vec{u}) + \Phi(\vec{r})}{k_B T}\right]^{-\kappa_0 - 1 - \frac{1}{2}d_K}
$$

$$
\Rightarrow \left[1 + \frac{1}{\kappa_0} \cdot \frac{\varepsilon_K(\vec{u}) + \Phi(\vec{r})}{k_B T}\right]^{-\kappa_0 - 1 - \frac{1}{2}d_K - \frac{1}{2}d_\Phi}.
$$

(3.87)

In the case of a negative potential, however, the probability distribution is not so trivially defined, and the degeneration of the kappa index is not so obvious. Some inconsistencies of the phase space distribution when $\Phi < 0$ are the following:

1. The quantity in the base $1 + \frac{1}{\kappa_0} \cdot H(\vec{r}, \vec{u})/(k_BT)$ may be negative and thus cannot be trivially defined.
2. Attractive potentials lead to stable systems, but negative potentials, $\Phi < 0$, require to be monotonically decreasing, e.g., $\Phi(r) = -\frac{1}{b}k \cdot r^{-b}$, $b > 0$, in order to be attractive; however, these potentials suffer from the long-standing problem of the divergence of the distribution for bounded radial potentials.
3. The invariant kappa index degeneration is complicated.

We will see how these difficulties can be resolved by employing the "negative" kappa distribution.

3.9.2 The "Negative" Kappa Distribution

The kappa distribution for negative kappa indices, called the "negative kappa distribution," is produced by setting $\kappa_0 \to -\kappa_0$ in the formalism of the standard kappa distribution; hence, the new kappa index will be still positive (Chapter 1; Livadiotis, 2015b). In terms of velocity

$$P(\vec{u}; \kappa_0, T) = \frac{\left(\pi \kappa_0 \theta^2\right)^{-\frac{1}{2}d_K} \Gamma(\kappa_0)}{\Gamma\left(\kappa_0 - \frac{d_K}{2}\right)} \cdot \left(1 - \frac{1}{\kappa_0} \frac{\vec{u}^2}{\theta^2}\right)^{\kappa_0 - 1 - \frac{d_K}{2}}_{+}, \tag{3.88a}$$

or in terms of kinetic energy,

$$P(\varepsilon_K; \kappa_0, T) = \frac{\left(\kappa_0 k_BT\right)^{-\frac{1}{2}d_K}}{B\left(\kappa_0 - \frac{1}{2}d_K, \frac{1}{2}d_K\right)} \cdot \left(1 - \frac{1}{\kappa_0} \cdot \frac{\varepsilon_K}{k_BT}\right)^{\kappa_0 - 1 - \frac{1}{2}d_K}_{+} \varepsilon_K^{\frac{1}{2}d_K - 1}. \tag{3.88b}$$

Eqs. (3.88a) and (3.88b) were written using the cut-off condition noted by the subscript "+," namely, $[y]_+ = y$, if $y \geq 0$ and $[y]_+ = 0$, if $y \leq 0$ (Eq. (1.13) in Chapter 1). Thus if the base of the kappa distribution is negative, then the distribution vanishes. In Eq. (3.88b), if $\varepsilon_K < \kappa_0 k_BT$, then $P(\varepsilon_K; \kappa_0, T) = 0$, i.e., the kinetic energy has a maximum value determined by the value of the kappa index and temperature.

The kappa index is still invariant of the dimensionality. In order to show this, we write

$$P(\vec{u}; \kappa_0, T) = \frac{\left(\pi \kappa_0 \theta^2\right)^{-\frac{1}{2}d_K} \Gamma(\kappa_0)}{\Gamma\left(\kappa_0 - \frac{d_K}{2}\right)} \cdot \left(1 - \frac{1}{\kappa_0} \cdot \frac{u_1^2 + \sum\limits_{i=2}^{d_K} u_i^2}{\theta^2}\right)^{\kappa_0 - 1 - \frac{d_K}{2}}_{+}. \tag{3.89}$$

Hence,

$$P(u_2, ..., u_{d_K}; \kappa_0, T) = \frac{\pi^{-\frac{1}{2}d_K} \left(\kappa_0 \theta^2\right)^{1-\kappa_0-\frac{1}{2}d_K} \Gamma(\kappa_0)}{\Gamma\left(\kappa_0 - \frac{d_K}{2}\right)} \cdot \int_{-\infty}^{\infty} \left(\kappa_0 \theta^2 - \sum_{i=2}^{d_K} u_i^2 - u_1^2\right)_+^{\kappa_0-1-\frac{d_K}{2}} du_1.$$

(3.90)

Setting $E \equiv \kappa_0 k_B T - \sum_{i=2}^{d_K} \varepsilon_i$, $\varepsilon_i \equiv \frac{1}{2} m u_i^2$, and $t \equiv E/(E - \varepsilon_1)$, we calculate the integral

$$2 \int_0^{\infty} \left(\kappa_0 \theta^2 - \sum_{i=2}^{d_K} u_i^2 - u_1^2\right)_+^{\kappa_0-1-\frac{1}{2}d_K} du_1$$

$$= \left(\frac{1}{2} m\right)^{-\kappa_0+\frac{1}{2}+\frac{1}{2}d_K} \int_0^E (E - \varepsilon_1)^{\kappa_0-1-\frac{1}{2}d_K} \varepsilon_1^{-\frac{1}{2}} d\varepsilon_1$$

$$= \left(\frac{1}{2} m\right)^{-\kappa_0+\frac{1}{2}+\frac{1}{2}d_K} E^{\kappa_0-\frac{1}{2}-\frac{1}{2}d_K} \int_0^{\infty} (1 + t)^{-\kappa_0-\frac{1}{2}+\frac{1}{2}d_K} t^{-\frac{1}{2}} dt$$

$$= \left(\frac{1}{2} m\right)^{-\kappa_0+\frac{1}{2}+\frac{1}{2}d_K} E^{\kappa_0-\frac{1}{2}-\frac{1}{2}d_K} \left(\kappa_0 - \frac{d_K}{2} - \frac{1}{2}\right)^{\frac{1}{2}} \Gamma_{\kappa_0-\frac{1}{2}d_K-\frac{1}{2}}\left(\frac{1}{2}\right)$$

$$= \frac{\pi^{\frac{1}{2}} \Gamma\left(\kappa_0 - \frac{d_K}{2}\right)}{\Gamma\left(\kappa_0 - \frac{d_K}{2} - \frac{1}{2}\right)} \left(\frac{1}{2} m\right)^{-\kappa_0+\frac{1}{2}d_K+\frac{1}{2}} E^{\kappa_0-\frac{1}{2}d_K-\frac{1}{2}}$$

$$= \frac{\pi^{\frac{1}{2}} \Gamma\left(\kappa_0 - \frac{d_K}{2}\right)}{\Gamma\left(\kappa_0 - \frac{d_K}{2} - \frac{1}{2}\right)} \left(\kappa_0 \theta^2 - \sum_{i=2}^{d_K} u_i^2\right)^{\kappa_0-\frac{1}{2}-\frac{1}{2}d_K},$$

where we substituted $E - \varepsilon_1 = E/(1 + t)$, $\varepsilon_1 = E \cdot t/(1 + t)$, and $d\varepsilon_1 = E/(1 + t)^2 dt$.

Therefore, by comparing the two distributions with dimensionalities d_K and $d_K - 1$:

$$P(u_1, ..., u_{d_K}; \kappa_0, T; d_K) = \frac{\left(\pi \kappa_0 \theta^2\right)^{-\frac{1}{2}d_K} \Gamma(\kappa_0)}{\Gamma\left(\kappa_0 - \frac{d_K}{2}\right)} \cdot \left(1 - \frac{1}{\kappa_0} \cdot \frac{\sum_{i=1}^{d_K} u_i^2}{\theta^2}\right)_+^{\kappa_0-1-\frac{1}{2}d_K}, \quad (3.91)$$

$$P(u_2, \ldots, u_{d_K}; \kappa_0, T; d_K - 1) = \frac{(\pi\kappa_0\theta^2)^{-\frac{1}{2}(d_K-1)}\Gamma(\kappa_0)}{\Gamma\left(\kappa_0 - \dfrac{d_K - 1}{2}\right)} \cdot \left(1 - \frac{1}{\kappa_0} \cdot \frac{\displaystyle\sum_{i=2}^{d_K} u_i^2}{\theta^2}\right)_+^{\kappa_0 - 1 - \frac{1}{2}(d_K - 1)},$$

$$(3.92)$$

we conclude that the κ_0 remains invariant under varying the dimensionality (similar property was shown for the positive kappa distribution in the Subsection 1.9.2. of Chapter 1). Notice that the minimum value of the invariant kappa index is not 0, since $\kappa_0 > \frac{1}{2}d_K$. Finally, note that the relation of the kappa index degeneration in the standard kinetic kappa distribution with positive kappa index is $\kappa(d_K) = \kappa_0 + \frac{1}{2}d_K$, while now is $\kappa(d_K) = -\kappa_0 + \frac{1}{2}d_K$.

3.9.3 Formulation of Phase Space Distributions

We develop the phase space kappa distributions for both positive and negative potentials, using both the formulations of positive and negative kappa indices. For now, we consider only the degeneration caused by the kinetic degrees of freedom $\kappa = \kappa_0 + \frac{d_K}{2}$.

3.9.3.1 Positive Potential $\Phi(\vec{r}) > 0$

3.9.3.1.1 Positive Potential: Positive Kappa Index, $\kappa_0 > 0$

The phase space distribution in position–kinetic energy coordinates $(\vec{r}, \varepsilon_K)$ is

$$P(\vec{r}, \varepsilon_K; \kappa_0, T) \propto \left\{1 + \frac{1}{\kappa_0} \cdot \frac{\varepsilon_K + \Phi(\vec{r}) - \langle\Phi\rangle}{k_B T}\right\}_+^{-\kappa_0 - 1 - \frac{1}{2}d_K} \varepsilon_K^{\frac{1}{2}d_K - 1}, \text{ or,} \qquad (3.93a)$$

$$P(\vec{r}, \varepsilon_K; \kappa_0, T) \propto \left[\kappa_0 - \frac{d_\Phi}{2} + \frac{\varepsilon_K + \Phi(\vec{r})}{k_B T}\right]_+^{-\kappa_0 - 1 - \frac{1}{2}d_K} \varepsilon_K^{\frac{1}{2}d_K - 1}. \qquad (3.93b)$$

The base of the distribution, Eq. (3.93b), is always positive if $\kappa_0 > \frac{d_\Phi}{2}$. Then the phase space distribution is

$$P(\vec{r}, \varepsilon_K; \kappa_0, T) \propto \left[1 + \frac{1}{\kappa_0 - \dfrac{d_\Phi}{2}} \cdot \frac{\varepsilon_K + \Phi(\vec{r})}{k_B T}\right]^{-\kappa_0 - 1 - \frac{1}{2}d_K} \varepsilon_K^{\frac{1}{2}d_K - 1}. \qquad (3.94)$$

Then, the positional distribution is given after the integration over the kinetic energy,

$$P(\vec{r}; \kappa_0, T) \propto \left[1 + \frac{1}{\kappa_0 - \dfrac{d_\Phi}{2}} \cdot \frac{\Phi(\vec{r})}{k_B T}\right]^{-\kappa_0 - 1}. \qquad (3.95)$$

The kappa index is subject to further degeneration caused by the potential (or positional) degrees of freedom: $\kappa_0 \Rightarrow \kappa_0 + \frac{d_\Phi}{2}$ so that the condition $\kappa_0 > \frac{d_\Phi}{2}$ becomes $\kappa_0 > 0$. The degeneration caused by kinetic and potential degrees of freedom is $\kappa = \kappa_0 + \frac{d_K}{2} + \frac{d_\Phi}{2}$.

The base in Eq. (3.93b) may also be negative if $\kappa_0 < \frac{d_\Phi}{2}$. We examine this case by considering either positive or negative kappa index. The phase space distribution, Eq. (3.93b), becomes

$$P(\overrightarrow{r}, \varepsilon_K; \kappa_0, T) \propto \left\{ \varepsilon_K - \left[\left(\frac{d_\Phi}{2} - \kappa_0 \right) k_B T - \Phi(\overrightarrow{r}) \right] \right\}_+^{-\kappa_0 - 1 - \frac{1}{2} d_K} \varepsilon_K^{\frac{1}{2} d_K - 1}. \qquad (3.96)$$

We examine separately the cases of large potential energies with a lower bound $\left(\frac{d_\Phi}{2} - \kappa_0 \right) k_B T \leq \Phi(\overrightarrow{r})$ and of small potential energies with an upper bound $\left(\frac{d_\Phi}{2} - \kappa_0 \right) k_B T \geq \Phi(\overrightarrow{r})$:

- Potentials with a lower bound: $\left(\frac{d_\Phi}{2} - \kappa_0 \right) k_B T \leq \Phi(\overrightarrow{r})$.

In this case, the base in Eq. (3.96) is always positive, i.e.,

$$P(\overrightarrow{r}, \varepsilon_K; \kappa_0, T) \propto \left[\varepsilon_K + \Phi(\overrightarrow{r}) - \left(\frac{d_\Phi}{2} - \kappa_0 \right) k_B T \right]^{-\kappa_0 - 1 - \frac{1}{2} d_K} \varepsilon_K^{\frac{1}{2} d_K - 1}. \qquad (3.97)$$

The positional distribution is given by integrating the energy,

$$P(\overrightarrow{r}; \kappa_0, T) \propto \left[\Phi(\overrightarrow{r}) - \left(\frac{d_\Phi}{2} - \kappa_0 \right) k_B T \right]^{-\kappa_0 - 1}. \qquad (3.98)$$

The integral of the distribution, Eq. (3.97), over the kinetic energy converges for kappa index $0 < \kappa_0$ for the mean energy (i.e., temperature) to exist. Hence, the range of the values of the kappa index is $0 < \kappa_0 < \frac{d_\Phi}{2}$. The inequality $\left(\frac{d_\Phi}{2} - \kappa_0 \right) k_B T \leq \Phi(\overrightarrow{r})$ gives a lower bound of the potential energy. Therefore, the positional distribution, Eq. (3.98), is suitable for the statistical description of repulsive potentials that decrease with distance, i.e., $\Phi(\overrightarrow{r}) = \frac{1}{b} k \cdot r^{-b}$. Due to the lower bound of the potential, a maximum distance can be defined by $r \leq \left\{ \frac{1}{b} k \Big/ \left[\left(\frac{d_\Phi}{2} - \kappa_0 \right) k_B T \right] \right\}^{\frac{1}{b}}$. Then the divergence of the distribution normalization for $r \to \infty$ is avoided. The distribution, Eq. (3.98), can also describe attractive potentials that increase with distance, i.e., $\Phi(\overrightarrow{r}) = \frac{1}{b} k \cdot r^b$; then extra limitations on the values of κ_0 are applied for the integral in Eq. (3.98) to converge, $\frac{3}{b} - 1 < \kappa_0$.

- Potentials with an upper bound: $\left(\frac{d_\Phi}{2} - \kappa_0 \right) k_B T \geq \Phi(\overrightarrow{r})$.

In this case, the base of Eq. (3.96) can be negative, i.e.,

$$
P(\vec{r}, \varepsilon_K; \kappa_0, T) \propto
\begin{cases}
\left\{ \varepsilon_K - \left[\left(\dfrac{d_\Phi}{2} - \kappa_0 \right) k_B T - \Phi(\vec{r}) \right] \right\}^{-\kappa_0 - 1 - \frac{1}{2}d_K} \varepsilon_K^{\frac{1}{2}d_K - 1}, & \text{if } \varepsilon_K \geq \left(\dfrac{d_\Phi}{2} - \kappa_0 \right) k_B T - \Phi(\vec{r}), \\[4mm]
0, & \text{if } \varepsilon_K \leq \left(\dfrac{d_\Phi}{2} - \kappa_0 \right) k_B T - \Phi(\vec{r}).
\end{cases}
$$

$$(3.99)$$

Thus the positional distribution is given by

$$
P(\vec{r}; \kappa_0, T) \propto \int_{\left(\frac{1}{2} d_\Phi - \kappa_0 \right) k_B T - \Phi(\vec{r})}^{\infty} \left\{ \varepsilon_K - \left[\left(\frac{d_\Phi}{2} - \kappa_0 \right) k_B T - \Phi(\vec{r}) \right] \right\}^{-\kappa_0 - 1 - \frac{1}{2}d_K} \varepsilon_K^{\frac{1}{2}d_K - 1} d\varepsilon_K.
$$

$$(3.100)$$

Applying the transformation $X \equiv \left[\left(\dfrac{d_\Phi}{2} - \kappa_0 \right) k_B T - \Phi(\vec{r}) \right] \cdot \left\{ \varepsilon_K - \left[\left(\dfrac{d_\Phi}{2} - \kappa_0 \right) \right. \right.$
$\left. \left. \cdot k_B T - \Phi(\vec{r}) \right] \right\}^{-1}$, Eq. (3.100) becomes

$$
P(\vec{r}; \kappa_0, T) \propto \left[\left(\frac{d_\Phi}{2} - \kappa_0 \right) k_B T - \Phi(\vec{r}) \right]^{-\kappa_0 - 1} \cdot \int_0^{\infty} X^{\kappa_0} (1 + X)^{\frac{1}{2}d_K - 1} dX. \quad (3.101)
$$

The integral converges only for $\kappa_0 + \frac{1}{2} d_K < 0$, a condition that cannot be satisfied since we considered $\kappa_0 > 0$. Therefore, this statistical description fails.

3.9.3.1.2 Positive Potential: Negative Kappa Index, $\kappa_0 < 0$

Next we examine the case of negative kappa index, $\kappa_0 < 0$. We substitute the kappa index $\kappa_0 \to -\kappa_0$ so that $-\kappa_0 < 0$ or $\kappa_0 > 0$, similar to the negative kappa distribution in Eqs. (3.88a) and (3.88b). Then the phase space distribution is

$$
P(\vec{r}, \varepsilon_K; \kappa_0, T) \propto \left[\kappa_0 + \frac{d_\Phi}{2} - \frac{\varepsilon_K + \Phi(\vec{r})}{k_B T} \right]_+^{\kappa_0 - 1 - \frac{1}{2}d_K} \varepsilon_K^{\frac{1}{2}d_K - 1}. \quad (3.102)
$$

Potentials with a lower bound $\left(\frac{d_\Phi}{2} + \kappa_0 \right) k_B T \leq \Phi(\vec{r})$ cannot be described because the base becomes always negative, $\left(\frac{d_\Phi}{2} + \kappa_0 \right) k_B T - \Phi(\vec{r}) - \varepsilon_K \leq 0$. Thus we examine the case of potentials with an upper bound, $\left(\frac{d_\Phi}{2} + \kappa_0 \right) k_B T \geq \Phi(\vec{r})$,

$$
P(\vec{r}, \varepsilon_K; \kappa_0, T) \propto \left[\left(\frac{d_\Phi}{2} + \kappa_0 \right) k_B T - \Phi(\vec{r}) - \varepsilon_K \right]_+^{\kappa_0 - 1 - \frac{1}{2}d_K} \varepsilon_K^{\frac{1}{2}d_K - 1}. \quad (3.103)
$$

The base in Eq. (3.103) can be negative, i.e.,

$$
P(\vec{r}, \varepsilon_K; \kappa_0, T) \propto
\begin{cases}
\left\{ \left[\left(\frac{d_\Phi}{2} + \kappa_0 \right) k_B T - \Phi(\vec{r}) \right] - \varepsilon_K \right\}^{-\kappa_0 - 1 - \frac{1}{2} d_K} \varepsilon_K^{\frac{1}{2} d_K - 1}, & \text{if } \varepsilon_K \leq \left(\frac{d_\Phi}{2} + \kappa_0 \right) k_B T - \Phi(\vec{r}), \\
0, & \text{if } \varepsilon_K \geq \left(\frac{d_\Phi}{2} + \kappa_0 \right) k_B T - \Phi(\vec{r}).
\end{cases}
$$

$$(3.104)$$

Thus the positional distribution is given by

$$
P(\vec{r}; \kappa_0, T) \propto \int_0^{\left(\frac{1}{2} d_\Phi + \kappa_0 \right) k_B T - \Phi(\vec{r})} \left\{ \left[\left(\frac{d_\Phi}{2} + \kappa_0 \right) k_B T - \Phi(\vec{r}) \right] - \varepsilon_K \right\}^{\kappa_0 - 1 - \frac{1}{2} d_K} \varepsilon_K^{\frac{1}{2} d_K - 1} d\varepsilon_K.
$$

$$(3.105)$$

Applying the transformation $X + 1 \equiv \left[\left(\frac{d_\Phi}{2} + \kappa_0 \right) k_B T - \Phi(\vec{r}) \right] \cdot \left\{ \left[\left(\frac{d_\Phi}{2} + \kappa_0 \right) \cdot k_B T - \Phi(\vec{r}) \right] - \varepsilon_K \right\}^{-1}$, Eq. (3.105) gives

$$
P(\vec{r}; \kappa_0, T) \propto \left[\left(\frac{d_\Phi}{2} + \kappa_0 \right) k_B T - \Phi(\vec{r}) \right]^{\kappa_0 - 1} \cdot \int_0^\infty (X + 1)^{-\kappa_0} X^{\frac{1}{2} d_K - 1} dX, \quad (3.106)
$$

in which the integral over X converges for kappa index $\frac{d_K}{2} < \kappa_0$ (note that for the mean energy, i.e., temperature, to exist, the integral converges under the same condition, $\frac{d_K}{2} < \kappa_0$). The positional distribution is

$$
P(\vec{r}; \kappa_0, T) \propto \left[\left(\frac{d_\Phi}{2} + \kappa_0 \right) k_B T - \Phi(\vec{r}) \right]^{\kappa_0 - 1}.
$$

$$(3.107)$$

The inequality $\left(\frac{d_\Phi}{2} + \kappa_0 \right) k_B T \geq \Phi(\vec{r})$ gives the upper bound of the potential energy. Therefore, the positional distribution, Eq. (3.107), is suitable for the statistical description of attractive potentials that increase with distance, i.e., $\Phi(\vec{r}) = kr^b$. Due to the upper bound of the potential, a maximum distance can be defined by $r \leq \left\{ \left[\left(\frac{d_\Phi}{2} + \kappa_0 \right) k_B T \right] / \left(\frac{1}{b} k \right) \right\}^{\frac{1}{b}}$. Therefore, there is no divergence of the distribution normalization for $r \to \infty$.

The infogram in Fig. 3.4 summarizes the results for positive potential.

$$\Phi(\vec{r}) > 0$$

$$\kappa_0 > 0$$

$$\kappa_0 > \tfrac{d_\Phi}{2}$$

Results are independent of bounds. Standard kappa distribution applies:

$$P(\vec{r}, \vec{u}; \kappa_0, T) \propto \left[(\kappa_0 - \tfrac{d_\Phi}{2}) k_B T + \Phi(\vec{r}) + \varepsilon_K(\vec{u}) \right]^{\kappa_0 - 1 - \frac{1}{2} d_K}$$

$$P(\vec{r}; \kappa_0, T) \propto \left[(\kappa_0 - \tfrac{d_\Phi}{2}) k_B T + \Phi(\vec{r}) \right]^{\kappa_0 - 1}$$

$$\kappa_0 < \tfrac{d_\Phi}{2}$$

Lower bound Potential: $(\tfrac{d_\Phi}{2} - \kappa_0) k_B T \leq \Phi(\vec{r})$

$$P(\vec{r}, \vec{u}; \kappa_0, T) \propto \left[\varepsilon_K(\vec{u}) + \Phi(\vec{r}) - (\tfrac{d_\Phi}{2} - \kappa_0) k_B T \right]^{-\kappa_0 - 1 - \frac{1}{2} d_K}$$

$$P(\vec{r}; \kappa_0, T) \propto \left[\Phi(\vec{r}) - (\tfrac{d_\Phi}{2} - \kappa_0) k_B T \right]^{-\kappa_0 - 1}$$

Upper bound Potential: $(\tfrac{d_\Phi}{2} - \kappa_0) k_B T \geq \Phi(\vec{r})$

Divergent distribution

$$\kappa_0 < 0 \; (\kappa_0 \to -\kappa_0)$$

Results are independent of $\tfrac{d_\Phi}{2}$.

Lower bound Potential: $(\tfrac{d_\Phi}{2} + \kappa_0) k_B T \leq \Phi(\vec{r})$

Undefined distribution

Upper bound Potential: $(\tfrac{d_\Phi}{2} + \kappa_0) k_B T \geq \Phi(\vec{r})$

$$P(\vec{r}, \vec{u}; \kappa_0, T) \propto \left[(\tfrac{d_\Phi}{2} + \kappa_0) k_B T - \Phi(\vec{r}) - \varepsilon_K(\vec{u}) \right]_+^{\kappa_0 - 1 - \frac{1}{2} d_K}$$

$$P(\vec{r}; \kappa_0, T) \propto \left[(\tfrac{d_\Phi}{2} + \kappa_0) k_B T - \Phi(\vec{r}) \right]_+^{\kappa_0 - 1}$$

FIGURE 3.4
The infogram explains the derivation of all the types of phase space and positional distributions for a positive potential (the degeneration due to positional degrees of freedom is not considered).

3.9.3.2 Negative Potential, $\Phi(\vec{r}) < 0$

In this case, the effective degrees of freedom of the potential are $d_\Phi \equiv -2 \langle \Phi \rangle / (k_B T) = 2 \langle |\Phi| \rangle / (k_B T)$; thus the phase space distribution $(\vec{r}, \varepsilon_K)$ is given by

$$P(\vec{r}, \varepsilon_K; \kappa_0, T) \propto \left\{ 1 + \frac{1}{\kappa_0} \cdot \frac{\varepsilon_K - [|\Phi(\vec{r})| - \langle |\Phi| \rangle]}{k_B T} \right\}_+^{-\kappa_0 - 1 - \frac{1}{2} d_K} \varepsilon_K^{\frac{1}{2} d_K - 1}, \qquad (3.108)$$

that is, for positive kappa index,

$$P(\vec{r}, \varepsilon_K; \kappa_0, T) \propto \left[\kappa_0 + \frac{d_\Phi}{2} + \frac{\varepsilon_K - |\Phi(\vec{r})|}{k_B T} \right]_+^{-\kappa_0 - 1 - \frac{1}{2} d_K} \varepsilon_K^{\frac{1}{2} d_K - 1}, \qquad (3.109a)$$

and for negative kappa index,

$$P(\overrightarrow{r}, \varepsilon_K; \kappa_0, T) \propto \left[\kappa_0 - \frac{d_\Phi}{2} + \frac{-\varepsilon_K + |\Phi(\overrightarrow{r})|}{k_B T} \right]_+^{\kappa_0 - 1 - \frac{1}{2}d_K} \varepsilon_K^{\frac{1}{2}d_K - 1}. \tag{3.109b}$$

3.9.3.2.1 Negative Potential: Positive Kappa Index, $\kappa_0 > 0$

The phase space distribution, Eq. (3.109a), becomes

$$P(\overrightarrow{r}, \varepsilon_K; \kappa_0, T) \propto \left\{ \varepsilon_K + \left[\left(\frac{d_\Phi}{2} + \kappa_0 \right) k_B T - |\Phi(\overrightarrow{r})| \right] \right\}_+^{-\kappa_0 - 1 - \frac{1}{2}d_K} \varepsilon_K^{\frac{1}{2}d_K - 1}. \tag{3.110}$$

- Potentials with an upper bound: $\left(\frac{d_\Phi}{2} + \kappa_0 \right) k_B T \geq |\Phi(\overrightarrow{r})|$.

Then the base of the distribution, Eq. (3.110), is always positive, and the phase space distribution is

$$P(\overrightarrow{r}, \varepsilon_K; \kappa_0, T) \propto \left\{ \varepsilon_K + \left[\left(\frac{d_\Phi}{2} + \kappa_0 \right) k_B T - |\Phi(\overrightarrow{r})| \right] \right\}^{-\kappa_0 - 1 - \frac{1}{2}d_K} \varepsilon_K^{\frac{1}{2}d_K - 1}. \tag{3.111}$$

After the integration over the kinetic energy, we derive the positional distribution

$$P(\overrightarrow{r}; \kappa_0, T) \propto \left[\left(\frac{d_\Phi}{2} + \kappa_0 \right) k_B T - |\Phi(\overrightarrow{r})| \right]^{-\kappa_0 - 1 - \frac{1}{2}d_K}. \tag{3.112}$$

- Potentials with a lower bound: $\left(\frac{d_\Phi}{2} + \kappa_0 \right) k_B T \leq |\Phi(\overrightarrow{r})|$.

The base of the distribution, Eq. (3.110), can be negative, and the phase space distribution is

$$P(\overrightarrow{r}, \varepsilon_K; \kappa_0, T) \propto \left\{ \varepsilon_K - \left[|\Phi(\overrightarrow{r})| - \left(\frac{d_\Phi}{2} + \kappa_0 \right) k_B T \right] \right\}_+^{-\kappa_0 - 1 - \frac{1}{2}d_K} \varepsilon_K^{\frac{1}{2}d_K - 1}, \tag{3.113}$$

and the positional distribution is given by

$$P(\overrightarrow{r}; \kappa_0, T) \propto \int_{|\Phi(\overrightarrow{r})| - \left(\frac{d_\Phi}{2} + \kappa_0 \right) k_B T}^{\infty} \left\{ \varepsilon_K - \left[|\Phi(\overrightarrow{r})| \right. \right.$$

$$\left. \left. - \left(\frac{d_\Phi}{2} + \kappa_0 \right) k_B T \right] \right\}^{-\kappa_0 - 1 - \frac{1}{2}d_K} \varepsilon_K^{\frac{1}{2}d_K - 1} d\varepsilon_K. \tag{3.114}$$

Applying the transformation $X \equiv \left[|\Phi(\vec{r})| - \left(\frac{d_\Phi}{2} + \kappa_0 \right) k_B T \right] \cdot \left\{ \varepsilon_K - \left[|\Phi(\vec{r})| - \right. \right.$

$\left. \left. \left(\frac{d_\Phi}{2} + \kappa_0 \right) k_B T \right] \right\}^{-1}$, Eq. (3.114) gives

$$P(\vec{r}; \kappa_0, T) \propto \left[|\Phi(\vec{r})| - \left(\frac{d_\Phi}{2} + \kappa_0 \right) k_B T \right]^{-\kappa_0 - 1} \cdot \int_0^\infty X^{\kappa_0} (1 + X)^{\frac{1}{2} d_K - 1} dX,$$

(3.115)

which is similar to Eq. (3.101), but the involved integral does not converge.

3.9.3.2.2 Negative Potential: Negative Kappa Index, $\kappa_0 < 0$

In the case of negative kappa index $\kappa_0 \rightarrow -\kappa_0$, the phase space distribution, Eq. (3.109b), becomes

$$P(\vec{r}, \varepsilon_K; \kappa_0, T) \propto \left\{ \left[\left(\kappa_0 - \frac{d_\Phi}{2} \right) k_B T + |\Phi(\vec{r})| \right] - \varepsilon_K \right\}_+^{\kappa_0 - 1 - \frac{1}{2} d_K} \varepsilon_K^{\frac{1}{2} d_K - 1}.$$

(3.116)

Let the case of $\kappa_0 > \frac{d_\Phi}{2}$. Then, the positional distribution is given by

$$P(\vec{r}; \kappa_0, T) \propto \int_0^{\left(\kappa_0 - \frac{1}{2} d_\Phi \right) k_B T + |\Phi(\vec{r})|} \left[\left(\kappa_0 - \frac{d_\Phi}{2} \right) k_B T + |\Phi(\vec{r})| - \varepsilon_K \right]^{\kappa_0 - 1 - \frac{1}{2} d_K} \varepsilon_K^{\frac{1}{2} d_K - 1} d\varepsilon_K.$$

(3.117)

Applying the transformation $X + 1 \equiv \left[\left(\kappa_0 - \frac{d_\Phi}{2} \right) k_B T + |\Phi(\vec{r})| \right] \cdot$

$\left\{ \left(\kappa_0 - \frac{d_\Phi}{2} \right) k_B T + |\Phi(\vec{r})| - \varepsilon_K \right\}^{-1}$, Eq. (3.117) gives

$$P(\vec{r}; \kappa_0, T) \propto \left[\left(\kappa_0 - \frac{d_\Phi}{2} \right) k_B T + |\Phi(\vec{r})| \right]^{\kappa_0 - 1} \cdot \int_0^\infty (X + 1)^{-\kappa_0} X^{\frac{1}{2} d_K - 1} dX,$$

(3.118)

where for the integral to exist must $\frac{d_k}{2} < \kappa_0$, which must be taken into account together with $\frac{d_\Phi}{2} < \kappa_0$. Hence,

$$P(\vec{r}; \kappa_0, T) \propto \left[\left(\kappa_0 - \frac{d_\Phi}{2} \right) k_B T + |\Phi(\vec{r})| \right]^{\kappa_0 - 1}.$$

(3.119)

Note that the distribution, Eq. (3.119), cannot describe repulsive potentials that increase with distance, i.e., $\Phi(\overrightarrow{r}) \propto -\frac{1}{b}kr^b$. The limitations on the values of κ_0 for the normalization of the distribution, Eq. (3.119), to converge are $\kappa_0 < 1 - \frac{d_r}{b}$, where d_r is the positional dimensionality. However, the two inequalities $\frac{d_K}{2} < \kappa_0 < 1 - \frac{d_r}{b}$ cannot be both true, unless $d_K = 1$ and $\frac{d_r}{b} < \frac{1}{2}$. Nevertheless, the distribution, Eq. (3.119), may also describe special types of bounded potentials, e.g., $\Phi(\overrightarrow{r}) \propto -(R^2 - r^2)^b$, which is restricted in $0 \le r \le R$.

Now let the case of $\kappa_0 < \frac{d_\Phi}{2}$. Hence, the phase space distribution is given by

$$P(\overrightarrow{r}, \varepsilon_K; \kappa_0, T) \propto \left\{ \left[|\Phi(\overrightarrow{r})| - \left(\frac{d_\Phi}{2} - \kappa_0 \right) k_B T \right] - \varepsilon_K \right\}_+^{\kappa_0 - 1 - \frac{1}{2}d_K} \varepsilon_K^{\frac{1}{2}d_K - 1}. \qquad (3.120)$$

Potentials with an upper bound $\left(\frac{d_\Phi}{2} - \kappa_0 \right) k_B T \ge \Phi(\overrightarrow{r})$ cannot be described because the base becomes negative, $-\left[\left(\frac{d_\Phi}{2} - \kappa_0 \right) k_B T - \Phi(\overrightarrow{r}) \right] - \varepsilon_K \le 0$. Thus we examine the case of potentials with a lower bound $\left(\frac{d_\Phi}{2} - \kappa_0 \right) k_B T \le |\Phi(\overrightarrow{r})|$.

The base of the distribution, Eq. (3.120), can be negative; thus the positional distribution is

$$P(\overrightarrow{r}; \kappa_0, T) \propto \int_0^{|\Phi(\overrightarrow{r})| - \left(\frac{1}{2}d_\Phi - \kappa_0 \right) k_B T} \left\{ \left[|\Phi(\overrightarrow{r})| - \left(\frac{d_\Phi}{2} - \kappa_0 \right) k_B T \right] \right.$$

$$\left. - \varepsilon_K \right\}^{\kappa_0 - 1 - \frac{1}{2}d_K} \varepsilon_K^{\frac{1}{2}d_K - 1} d\varepsilon_K. \qquad (3.121)$$

Applying the transformation $X + 1 \equiv \left[|\Phi(\overrightarrow{r})| - \left(\frac{d_\Phi}{2} - \kappa_0 \right) k_B T \right] \cdot \left\{ \left[|\Phi(\overrightarrow{r})| - \left(\frac{d_\Phi}{2} - \kappa_0 \right) k_B T \right] - \varepsilon_K \right\}^{-1}$, Eq. (3.121) gives

$$P(\overrightarrow{r}; \kappa_0, T) \propto \left[|\Phi(\overrightarrow{r})| - \left(\frac{d_\Phi}{2} - \kappa_0 \right) k_B T \right]^{\kappa_0 - 1} \cdot \int_0^\infty (X + 1)^{-\kappa_0} X^{\frac{1}{2}d_K - 1} dX, \text{ or}$$

$$\qquad (3.122a)$$

$$P(\overrightarrow{r}; \kappa_0, T) \propto \left[|\Phi(\overrightarrow{r})| - \left(\frac{d_\Phi}{2} - \kappa_0 \right) k_B T \right]^{\kappa_0 - 1}. \qquad (3.122b)$$

Similar to previous cases, the integral in Eq. (3.122a) converges for kappa index $\frac{d_K}{2} < \kappa_0$. Hence, the range of the values of the kappa index is $\frac{d_K}{2} < \kappa_0 < \frac{d_\Phi}{2}$. Inequality $\left(\frac{d_\Phi}{2} - \kappa_0\right)k_BT \leq \Phi(\vec{r})$ gives a lower bound of the potential energy.

Therefore, the positional distribution, Eq. (3.122b), is suitable for the statistical description of attractive potentials that decrease with distance, i.e., $\Phi(\vec{r}) = -\frac{1}{b}k \cdot r^{-b}$, for example, the gravitational potential, $\Phi(\vec{r}) = -GMm/r$ (G is the gravitational constant and M is the mass of the gravitational pull). Due to the lower bound of the potential, a maximum distance can be defined by $r \leq \left\{\frac{1}{b}k \Big/ \left[\left(\frac{d_\Phi}{2} - \kappa_0\right)k_BT\right]\right\}^{\frac{1}{b}}$, or $r \leq GMm \Big/ \left[\left(\frac{d_\Phi}{2} - \kappa_0\right)k_BT\right]$ for the gravitation potential. In this way, the known divergence of the distribution normalization for $r \to \infty$ (e.g., Livadiotis, 2009) is avoided (see also Subsections 4.1, 4.3). The distribution, Eq. (3.122b), can also describe repulsive potentials that increase with distance, but extra limitations on the values of κ_0 may be applied. For instance, the potential $\Phi(\vec{r}) \sim -\frac{1}{b}k \cdot r^b$ cannot be described with Eq. (3.122b) because the values of the kappa index are restricted by the double inequality $\frac{d_K}{2} < \kappa_0 < 1 - \frac{d_r}{b}$, which, again, cannot be true unless $d_K = 1$ and $\frac{d_r}{b} < \frac{1}{2}$.

The infogram in Fig. 3.5 summarizes the results for negative potential. Table shows the derived phase space and positional kappa distributions, and the range of values for kappa indices for positive and negative potentials. Note that only the degeneration of the kappa index caused by the kinetic degrees of freedom was considered, i.e., $\kappa = \kappa_0 + \frac{d_K}{2}$; for now, no further degeneration caused by the positional degrees of freedom was derived since this depends from the specific type of the potential.

3.9.4 Negative Attractive Power Law Potential (Gravitational Type)

Here we examine the negative attractive power law potential $\Phi(r) = -\frac{1}{b}k \cdot r^{-b}$, i.e., $F_r = -(\partial/\partial r)\Phi(r) = -k \cdot r^{-(b+1)} < 0$, $r \in \Re^{d_r}$ (compare with the positive analogue in Section 3.7.3). From Table 3.1, we obtain

$$P(\vec{r};\kappa_0,T) \propto \left[\frac{\frac{1}{b}k}{\left(\frac{d_\Phi}{2} - \kappa_0\right)k_BT} \cdot \frac{1}{r^b} - 1\right]^{\kappa_0-1}, r \leq \left\{\frac{1}{b}k \Big/ \left[\left(\frac{d_\Phi}{2} - \kappa_0\right)k_BT\right]\right\}^{\frac{1}{b}}.$$

$$(3.123)$$

The effective degrees of freedom of the interaction are

$$\frac{1}{2}d_\Phi \equiv -\frac{\langle\Phi\rangle}{k_BT} = \frac{\frac{1}{b}k}{k_BT}r_0^{-b},$$

$$(3.124)$$

$$\Phi(\vec{r}) < 0$$

$$\kappa_0 > 0$$

Results are independent of $\frac{d_\Phi}{2}$.

Lower bound Potential: $(\frac{d_\Phi}{2} + \kappa_0)k_B T \leq |\Phi(\vec{r})|$.

$\boxed{\text{Divergent distribution}}$

Upper bound Potential: $(\frac{d_\Phi}{2} + \kappa_0)k_B T \geq |\Phi(\vec{r})|$.

$$P(\vec{r}, \varepsilon_K; \kappa_0, T) \propto \left\{ [(\frac{d_\Phi}{2} + \kappa_0)k_B T - |\Phi(\vec{r})| + \varepsilon_K] \right\}^{-\kappa_0 - 1 - \frac{1}{2}d_K} \varepsilon_K^{\frac{1}{2}d_K - 1}$$

$$P(\vec{r}; \kappa_0, T) \propto \left[(\frac{d_\Phi}{2} + \kappa_0)k_B T - |\Phi(\vec{r})| \right]^{-\kappa_0 - 1 - \frac{1}{2}d_K}$$

$$\kappa_0 < 0 \ (\kappa_0 \to -\kappa_0)$$

$$\kappa_0 > \frac{d_\Phi}{2}$$

Results are independent of bounds of the potential.

$$P(\vec{r}, \varepsilon_K; \kappa_0, T) \propto \left\{ [(\kappa_0 - \frac{1}{2}d_\Phi)k_B T + |\Phi(\vec{r})|] - \varepsilon_K \right\}_+^{\kappa_0 - 1 - \frac{1}{2}d_K} \varepsilon_K^{\frac{1}{2}d_K - 1}$$

$$P(\vec{r}; \kappa_0, T) \propto \left[(\kappa_0 - \frac{1}{2}d_\Phi)k_B T + |\Phi(\vec{r})| \right]^{\kappa_0 - 1}$$

$$\kappa_0 < \frac{d_\Phi}{2}$$

Results are independent of bounds of the potential.

$$P(\vec{r}, \varepsilon_K; \kappa_0, T) \propto \left\{ [|\Phi(\vec{r})| - (\frac{d_\Phi}{2} - \kappa_0)k_B T] - \varepsilon_K \right\}_+^{\kappa_0 - 1 - \frac{1}{2}d_K} \varepsilon_K^{\frac{1}{2}d_K - 1}$$

$$P(\vec{r}; \kappa_0, T) \propto \left[|\Phi(\vec{r})| - (\frac{d_\Phi}{2} - \kappa_0)k_B T \right]^{\kappa_0 - 1}$$

FIGURE 3.5
The infogram explains the derivation of all the types of phase space and positional distributions for a negative potential (the degeneration due to positional degrees of freedom is not considered).

where $r_0 = \langle r^{-b} \rangle^{-\frac{1}{b}}$ gives the average potential radius. Hence, the distribution is written as

$$P(\vec{r}; \kappa_0, T) \propto \left[\frac{\frac{d_\Phi}{2}}{\frac{d_\Phi}{2} - \kappa_0} \cdot \left(\frac{r_0}{r} \right)^b - 1 \right]^{\kappa_0 - 1} = \left[\left(\frac{\alpha}{r} \right)^b - 1 \right]^{\kappa_0 - 1}, \ 0 \leq r \leq \alpha,$$

(3.125)

with the temperature hidden in the notion of r_0. The involved constant is given by $\alpha \equiv \left[\frac{d_\Phi}{2} / \left(\frac{d_\Phi}{2} - \kappa_0 \right) \right]^{\frac{1}{b}} \cdot r_0$. The particles must be distributed within the finite distance $0 \leq r \leq \alpha$ for the normalization integral to converge.

The average radius r_0 and the normalization constant μ_2 are given by

$$r_0 = \langle r^{-b} \rangle^{-\frac{1}{b}} = (\mu_{2-b}/\mu_2)^{-\frac{1}{b}}, \text{ with } \mu_m \equiv \int_0^\alpha \left[\left(\frac{\alpha}{r} \right)^b - 1 \right]^{\kappa_0 - 1} r^m dr.$$

(3.126)

Table 3.1 Phase Space and Positional Kappa Distributions for Positive and Negative Potentials

$\Phi(\vec{r})$	κ_0	$P(\vec{r},\varepsilon_K) \propto$	$P(\vec{r}) \propto$	$\tilde{\kappa}_0$	$\Phi_\infty \propto$
$0 \leq \Phi(\vec{r}) < \infty$	$\frac{d_\Phi}{2} < \kappa_0 < \infty$	$\left[1 + \dfrac{\varepsilon_K + \Phi(\vec{r})}{\left(\kappa_0 - \frac{d_\Phi}{2}\right)k_B T}\right]^{-\kappa_0 - 1 - \frac{1}{2}d_K} \varepsilon_K^{\frac{1}{2}d_K - 1}$	$\left[1 + \dfrac{\Phi(\vec{r})}{\left(\kappa_0 - \frac{d_\Phi}{2}\right)k_B T}\right]^{-\kappa_0 - 1}$	$\tilde{\kappa}_0 \equiv \kappa_0 - \frac{d_\Phi}{2}$ $0 < \tilde{\kappa}_0 < \infty$	r^b
$0 \leq \Phi(\vec{r}) \leq \left(\frac{d_\Phi}{2} + \kappa_0\right)k_B T$	$\frac{d_K}{2} < \kappa_0 < \infty$	$\left[1 - \dfrac{\varepsilon_K + \Phi(\vec{r})}{\left(\frac{d_\Phi}{2} + \kappa_0\right)k_B T}\right]^{\kappa_0 - 1 - \frac{1}{2}d_K} \varepsilon_K^{\frac{1}{2}d_K - 1}$	$\left[1 - \dfrac{\Phi(\vec{r})}{\left(\kappa_0 + \frac{d_\Phi}{2}\right)k_B T}\right]^{\kappa_0 - 1}$	$\tilde{\kappa}_0 \equiv \kappa_0 + \frac{d_\Phi}{2}$	r^b
$\left(\frac{d_\Phi}{2} - \kappa_0\right)k_B T \leq \Phi(\vec{r}) < \infty$	$\frac{d_K}{2} < \kappa_0 < \frac{d_\Phi}{2}$	$\left[\dfrac{\varepsilon_K + \Phi(\vec{r})}{\left(\frac{d_\Phi}{2} - \kappa_0\right)k_B T} - 1\right]^{-\kappa_0 - 1 - \frac{1}{2}d_K} \varepsilon_K^{\frac{1}{2}d_K - 1}$	$\left[\dfrac{\Phi(\vec{r})}{\left(\frac{d_\Phi}{2} - \kappa_0\right)k_B T} - 1\right]^{-\kappa_0 - 1}$	$\tilde{\kappa}_0 \equiv \frac{d_\Phi}{2} - \kappa_0$ $0 < \tilde{\kappa}_0 < \frac{d_\Phi}{2} - \frac{d_K}{2}$	r^{-b}
$-\infty < -\Phi(\vec{r}) \leq 0$	$\frac{d_K}{2} < \kappa_0, \frac{d_\Phi}{2} < \kappa_0$	$\left[1 + \dfrac{|\Phi(\vec{r})| - \varepsilon_K}{\left(\kappa_0 - \frac{d_\Phi}{2}\right)k_B T}\right]^{-\kappa_0 - 1 - \frac{1}{2}d_K} \varepsilon_K^{\frac{1}{2}d_K - 1}$	$\left[1 + \dfrac{|\Phi(\vec{r})|}{\left(\kappa_0 - \frac{d_\Phi}{2}\right)k_B T}\right]^{-\kappa_0 - 1}$	$\tilde{\kappa}_0 \equiv \kappa_0 - \frac{d_\Phi}{2}$	0
$-\left(\frac{1}{2}d_\Phi + \kappa_0\right)k_B T \leq \Phi(\vec{r}) \leq 0$	$0 < \kappa_0 < \infty$	$\left[1 + \dfrac{\varepsilon_K - |\Phi(\vec{r})|}{\left(\frac{d_\Phi}{2} + \kappa_0\right)k_B T}\right]^{-\kappa_0 - 1 - \frac{1}{2}d_K} \varepsilon_K^{\frac{1}{2}d_K - 1}$	$\left[1 - \dfrac{|\Phi(\vec{r})|}{\left(\kappa_0 + \frac{d_\Phi}{2}\right)k_B T}\right]^{-\kappa_0 - 1}$	$\tilde{\kappa}_0 \equiv \kappa_0 - \frac{d_\Phi}{2}$	$-r^b$
$-\infty < \Phi(\vec{r}) \leq \left(\kappa_0 - \frac{d_\Phi}{2}\right)k_B T$	$\frac{d_K}{2} < \kappa_0 < \frac{d_\Phi}{2}$	$\left[\dfrac{|\Phi(\vec{r})| - \varepsilon_K}{\left(\frac{d_\Phi}{2} - \kappa_0\right)k_B T} - 1\right]^{-\kappa_0 - 1 - \frac{1}{2}d_K} \varepsilon_K^{\frac{1}{2}d_K - 1}$	$\left[\dfrac{|\Phi(\vec{r})|}{\left(\frac{d_\Phi}{2} - \kappa_0\right)k_B T} - 1\right]^{-\kappa_0 - 1}$	$\tilde{\kappa}_0 \equiv \frac{d_\Phi}{2} - \kappa_0$ $0 < \tilde{\kappa}_0 < \frac{d_\Phi}{2} - \frac{d_K}{2}$	$-r^{-b}$

κ_0 is the kappa index degenerated only by kinetic degrees of freedom; $\tilde{\kappa}_0$ symbolizes the kappa index, which is further degenerated by the potential degrees of freedom (explained in Section 3.9.4).

Setting $X \equiv (r/\alpha)^b$, we obtain

$$
\begin{aligned}
\mu_m &= \frac{1}{b}\alpha^{m+1}\int_0^1 (1-X)^{\kappa_0-1}X^{\frac{m+1}{b}-\kappa_0}dX \\
&= \frac{1}{b}\alpha^{m+1}\cdot B\left[\kappa_0, \frac{m+1}{b}-\kappa_0+1\right], \quad \text{or} \\
\mu_m &= \alpha^{m+1}\frac{\Gamma(\kappa_0)\Gamma\left(\frac{m+1}{b}-\kappa_0+1\right)}{b\Gamma\left(\frac{m+1}{b}+1\right)}.
\end{aligned}
$$
(3.127)

Then the average radius is given by

$$
\begin{aligned}
r_0 &= \left(\mu_{d_r-1-b}/\mu_{d_r-1}\right)^{-\frac{1}{b}} = \left[\alpha^{-b}\cdot\frac{\Gamma\left(\frac{d_r}{b}-\kappa_0\right)}{\Gamma\left(\frac{d_r}{b}\right)}\cdot\frac{\Gamma\left(\frac{d_r}{b}+1\right)}{\Gamma\left(\frac{d_r}{b}-\kappa_0+1\right)}\right]^{-\frac{1}{b}} \\
&= \alpha\cdot\left(\frac{\frac{d_r}{b}}{\frac{d_r}{b}-\kappa_0}\right)^{-\frac{1}{b}} = r_0\cdot\left(\frac{\frac{d_\Phi}{2}}{\frac{d_\Phi}{2}-\kappa_0}\right)^{\frac{1}{b}}\cdot\left(\frac{\frac{d_r}{b}-\kappa_0}{\frac{d_r}{b}}\right)^{\frac{1}{b}},
\end{aligned}
$$

that leads to $\frac{d_\Phi}{2}\Big/\left(\frac{d_\Phi}{2}-\kappa_0\right) = \frac{d_r}{b}\Big/\left(\frac{d_r}{b}-\kappa_0\right)$, or $\frac{d_\Phi}{2} = \frac{d_r}{b}$. Hence, we derive

$$
\frac{1}{2}d_\Phi = \frac{d_r}{b}, r_0 = \left(\frac{\frac{1}{b}k}{\frac{d_r}{b}k_BT}\right)^{\frac{1}{b}}, \alpha \equiv \left(\frac{\frac{d_r}{b}}{\frac{d_r}{b}-\kappa_0}\right)^{\frac{1}{b}}\cdot r_0.
$$
(3.128)

The normalization constant is

$$
\mu_{d_r-1} = \frac{\left(\frac{d_r}{b}\right)^{\frac{d_r}{b}}}{d_r\Gamma\left(\frac{d_r}{b}\right)}\cdot\frac{\Gamma(\kappa_0)\Gamma\left(\frac{d_r}{b}-\kappa_0\right)}{\left(\frac{d_r}{b}-\kappa_0\right)^{\frac{d_r}{b}-1}}\cdot r_0^{d_r} = \frac{\left(\frac{d_r}{b}\right)^{\frac{d_r}{b}}B\left(\kappa_0,\frac{d_r}{b}-\kappa_0\right)}{d_r\left(\frac{d_r}{b}-\kappa_0\right)^{\frac{d_r}{b}-1}}\cdot r_0^{d_r}.
$$
(3.129)

Therefore, the positional kappa distribution is written as

$$
P(\vec{r};\kappa_0,T)d\vec{r} = \frac{\Gamma\left(\frac{d_r}{2}\right)}{2\pi^{\frac{d_r}{2}}}\cdot\frac{b\left(\frac{d_r}{b}\right)^{-\frac{d_r}{b}}\left(\frac{d_r}{b}-\kappa_0\right)^{\frac{d_r}{b}}}{B\left(\kappa_0,\frac{d_r}{b}-\kappa_0\right)}\cdot r_0^{-d_r}\cdot\left[\frac{\frac{d_r}{b}}{\frac{d_r}{b}-\kappa_0}\cdot\left(\frac{r_0}{r}\right)^b-1\right]^{\kappa_0-1}d\vec{r}.
$$
(3.130a)

Setting $P(\vec{r}; \kappa_0, T) = P(r; \kappa_0, r_0) \cdot B_{d_r} \cdot r^{d_r - 1}$,

$$P(r;\kappa_0,T)dr = \frac{b\left(\dfrac{d_r}{b}\right)^{-\frac{d_r}{b}}\left(\dfrac{d_r}{b}-\kappa_0\right)^{\frac{d_r}{b}}}{B\left(\kappa_0, \dfrac{d_r}{b}-\kappa_0+1\right)}\cdot r_0^{-d_r}\cdot\left[\frac{d_r}{\dfrac{d_r}{b}-\kappa_0}\cdot\left(\frac{r_0}{r}\right)^b - 1\right]^{\kappa_0-1} r^{d_r-1}dr,$$

(3.130b)

which is valid for $\frac{d_K}{2} < \kappa_0 < \frac{d_r}{b}$, i.e., this statistical description can be used only for exponents $b \leq 2d_r/d_K$.

The phase space kappa distribution is

$$P(\vec{r}, \varepsilon_K; \kappa_0, T)d\vec{r}d\varepsilon_K = \frac{\Gamma\left(\dfrac{d_r}{2}\right)}{2\pi^{\frac{d_r}{2}}}\cdot\frac{b\left(\dfrac{d_r}{b}\right)^{-\frac{d_r}{b}}\left(\dfrac{d_r}{b}-\kappa_0\right)^{\frac{1}{b}d_r-\frac{1}{2}d_K}}{B\left(\kappa_0, \dfrac{d_r}{b}-\kappa_0+1\right)B\left(\dfrac{d_K}{2}, \kappa_0-\dfrac{d_K}{2}\right)}\cdot(k_BT)^{-\frac{1}{2}d_K}r_0^{-d_r}$$

$$\times\left\{\frac{1}{\dfrac{d_r}{b}-\kappa_0}\cdot\left[\frac{d_r}{b}\left(\frac{r_0}{|\vec{r}|}\right)^b - \frac{\varepsilon_K}{k_BT}\right] - 1\right\}_+^{\kappa_0-1-\frac{1}{2}d_K} \varepsilon_K^{\frac{1}{2}d_K-1}d\vec{r}d\varepsilon_K,$$

(3.131a)

or

$$P(r, \varepsilon_K; \kappa_0, T)drd\varepsilon_K = \frac{b\left(\dfrac{d_r}{b}\right)^{-\frac{d_r}{b}}\left(\dfrac{d_r}{b}-\kappa_0\right)^{\frac{1}{b}d_r-\frac{1}{2}d_K}}{B\left(\kappa_0, \dfrac{d_r}{b}-\kappa_0+1\right)B\left(\dfrac{d_K}{2}, \kappa_0-\dfrac{d_K}{2}\right)}\cdot(k_BT)^{-\frac{1}{2}d_K}r_0^{-d_r}$$

$$\times\left\{\frac{1}{\dfrac{d_r}{b}-\kappa_0}\cdot\left[\frac{d_r}{b}\left(\frac{r_0}{r}\right)^b - \frac{\varepsilon_K}{k_BT}\right] - 1\right\}_+^{\kappa_0-1-\frac{1}{2}d_K} \varepsilon_K^{\frac{1}{2}d_K-1}r^{d_r-1}drd\varepsilon_K.$$

(3.131b)

The distribution of the kinetic energy is derived as the marginal of the phase space distribution, integrated over the positions

$$P(\varepsilon_K; \kappa_0, T)d\varepsilon_K = \frac{b\left(\dfrac{d_r}{b}\right)^{-\frac{d_r}{b}}\left(\dfrac{d_r}{b}-\kappa_0\right)^{\frac{1}{b}d_r-\frac{1}{2}d_K}}{B\left(\kappa_0, \dfrac{d_r}{b}-\kappa_0+1\right)B\left(\dfrac{d_K}{2}, \kappa_0-\dfrac{d_K}{2}\right)}\cdot(k_BT)^{-\frac{1}{2}d_K}r_0^{-d_r}$$

$$\times\int_0^{r_{max}}\left\{\frac{1}{\dfrac{d_r}{b}-\kappa_0}\cdot\left[\frac{d_r}{b}\left(\frac{r_0}{r}\right)^b - \frac{\varepsilon_K}{k_BT}\right] - 1\right\}_+^{\kappa_0-1-\frac{1}{2}d_K} r^{d_r-1}dr\varepsilon_K^{\frac{1}{2}d_K-1}d\varepsilon_K,$$

(3.132a)

where

$$r < r_{\max} \text{ with } r_{\max} = r_0 \left[\frac{\dfrac{d_r}{b}}{\left(\dfrac{d_r}{b} - \kappa_0\right) + \varepsilon_K/(k_B T)} \right]^{\frac{1}{b}}. \tag{3.132b}$$

Hence,

$$P(\varepsilon_K; \kappa_0, T) d\varepsilon_K = \frac{\left(\dfrac{d_r}{b} - \kappa_0\right)^{\frac{1}{b}d_r - \kappa_0 + 1}}{B\left(\kappa_0, \dfrac{d_r}{b} - \kappa_0 + 1\right) B\left(\dfrac{d_K}{2}, \kappa_0 - \dfrac{d_K}{2}\right)} \cdot (k_B T)^{-\frac{1}{2}d_K}$$

$$\times \left\{ \int_{\left(\frac{d_r}{b} - \kappa_0\right) + \frac{\varepsilon_K}{k_B T}}^{\infty} \left\{ X - \left[\left(\dfrac{d_r}{b} - \kappa_0\right) + \dfrac{\varepsilon_K}{k_B T}\right] \right\}^{\kappa_0 - 1 - \frac{1}{2}d_K} \right.$$

$$\left. \times X^{-d_r/b - 1} dX \right\} \varepsilon_K^{\frac{1}{2}d_K - 1} d\varepsilon_K, \tag{3.133a}$$

where $X \equiv \frac{d_r}{b}(r/r_0)^{-b}$, or $Y \equiv X - \left[\left(\dfrac{d_r}{b} - \kappa_0\right) + \varepsilon_K \middle/ (k_B T)\right]$,
$Z \equiv Y \middle/ \left[\left(\dfrac{d_r}{b} - \kappa_0\right) + \varepsilon_K \middle/ (k_B T)\right]$:

$$P(\varepsilon_K; \kappa_0, T) d\varepsilon_K = \frac{\left(\dfrac{d_r}{b} - \kappa_0\right)^{\frac{1}{b}d_r - \kappa_0 + 1}}{B\left(\kappa_0, \dfrac{d_r}{b} - \kappa_0 + 1\right) B\left(\dfrac{d_K}{2}, \kappa_0 - \dfrac{d_K}{2}\right)} \cdot (k_B T)^{-\frac{1}{2}d_K}$$

$$\times \left\{ \int_0^{\infty} \left[Y + \left(\dfrac{d_r}{b} - \kappa_0\right) + \dfrac{\varepsilon_K}{k_B T}\right]^{-\frac{1}{b}d_r - 1} Y^{\kappa_0 - 1 - \frac{1}{2}d_K} dY \right\}$$

$$\times \varepsilon_K^{\frac{1}{2}d_K - 1} d\varepsilon_K, \tag{3.133b}$$

and

$$P(\varepsilon_K; \kappa_0, T)d\varepsilon_K = \frac{\left(\dfrac{d_r}{b} - \kappa_0\right)^{-\frac{1}{2}d_K}}{B\left(\kappa_0, \dfrac{d_r}{b} - \kappa_0 + 1\right)B\left(\dfrac{d_K}{2}, \kappa_0 - \dfrac{d_K}{2}\right)} \cdot (k_B T)^{-\frac{1}{2}d_K}$$

$$\times \left[\int_0^\infty (1+Z)^{-\frac{1}{b}d_r - 1}Z^{\kappa_0 - 1 - \frac{1}{2}d_K}dZ\right] \qquad (3.133c)$$

$$\times \left[1 + \frac{\varepsilon_K}{\left(\dfrac{d_r}{b} - \kappa_0\right)k_B T}\right]^{\kappa_0 - 1 - \frac{1}{2}d_K - \frac{1}{b}d_r} \varepsilon_K^{\frac{1}{2}d_K - 1}d\varepsilon_K.$$

We have

$$\int_0^\infty (1+Z)^{-\frac{1}{b}d_r - 1}Z^{\kappa_0 - 1 - \frac{1}{2}d_K}dZ = \left(\frac{d_r}{b}\right)^{-\kappa_0 + \frac{1}{2}d_K}\Gamma_{\frac{d_r}{b}}\left(\kappa_0 - \frac{d_K}{2}\right)$$

$$= \frac{\Gamma\left(\kappa_0 - \dfrac{d_K}{2}\right)\Gamma\left(\dfrac{d_K}{2} + \dfrac{d_r}{b} + 1 - \kappa_0\right)}{\Gamma\left(\dfrac{d_r}{b} + 1\right)}, \qquad (3.134a)$$

and

$$B\left(\kappa_0, \frac{d_r}{b} - \kappa_0 + 1\right)B\left(\frac{d_K}{2}, \kappa_0 - \frac{d_K}{2}\right) = \Gamma\left(\frac{d_r}{b} - \kappa_0 + 1\right)\Gamma\left(\frac{d_K}{2}\right)\Gamma\left(\kappa_0 - \frac{d_K}{2}\right)$$

$$\times 1 \Big/ \Gamma\left(\frac{d_r}{b} + 1\right), \qquad (3.134b)$$

leading to the distribution of the kinetic energy

$$P(\varepsilon_K; \kappa_0, T) = \frac{\left(\dfrac{d_r}{b} - \kappa_0\right)^{-\frac{1}{2}d_K}\Gamma\left[\left(\dfrac{d_r}{b} - \kappa_0\right) + 1 + \dfrac{d_K}{2}\right]}{\Gamma\left[\left(\dfrac{d_r}{b} - \kappa_0\right) + 1\right]\Gamma\left(\dfrac{d_K}{2}\right)} \cdot (k_B T)^{-\frac{1}{2}d_K}$$

$$\qquad (3.135)$$

$$\times \left[1 + \frac{\varepsilon_K}{\left(\dfrac{d_r}{b} - \kappa_0\right)k_B T}\right]^{-\left(\frac{1}{b}d_r - \kappa_0\right) - 1 - \frac{1}{2}d_K} \varepsilon_K^{\frac{1}{2}d_K - 1}.$$

This is exactly the standard kappa distribution of the kinetic energy in the absence of the potential energy. The positional degrees of freedom degenerates the kappa index (similar to Eqs. 3.54a and 3.54b), i.e., $\left(\frac{d_r}{b} - \kappa_0\right) \Rightarrow \kappa_0$, that is,

$$-\kappa_0 \text{ (no potential)} \Rightarrow \kappa_0 - \frac{1}{b}d_r \text{ (with potential } \Phi \propto -r^{-b}) \tag{3.136a}$$

or

$$\kappa(d_K) = -\kappa_0 + \frac{1}{2}d_K \text{ (no potential)} \Rightarrow$$

$$\kappa(d_K, d_r) = \kappa_0 + \frac{1}{2}d_K - \frac{1}{b}d_r \text{ (with potential } \Phi \propto -r^{-b}). \tag{3.136b}$$

Finally, the phase-space kappa distribution becomes

$$P(\vec{r}, \varepsilon_K; \kappa_0, T)d\vec{r}\,d\varepsilon_K = \frac{\Gamma\left(\frac{d_r}{2}\right)}{2\pi^{\frac{d_r}{2}}} \cdot \frac{b\left(\frac{d_r}{b}\right)^{-\frac{d_r}{b}} \kappa_0^{\frac{1}{b}d_r - \frac{1}{2}d_K}}{B\left(\frac{d_r}{b} - \kappa_0, \kappa_0 + 1\right)B\left(\frac{d_K}{2}, -\kappa_0 - \frac{d_K}{2} + \frac{d_r}{b}\right)}$$

$$\times (k_B T)^{-\frac{1}{2}d_K} r_0^{-d_r} \cdot \left\{\frac{1}{\kappa_0}\left[\frac{d_r}{b}\left(\frac{r_0}{|\vec{r}|}\right)^b - \frac{\varepsilon_K}{k_B T}\right] - 1\right\}_+^{-\kappa_0 - 1 - \frac{1}{2}d_K + \frac{1}{b}d_r}$$

$$\times \varepsilon_K^{\frac{1}{2}d_K - 1} d\vec{r}\,d\varepsilon_K, \tag{3.137a}$$

or

$$P(r, \varepsilon_K; \kappa_0, T)dr\,d\varepsilon_K = \frac{b\left(\frac{d_r}{b}\right)^{-\frac{d_r}{b}} \kappa_0^{\frac{1}{b}d_r - \frac{1}{2}d_K}}{B\left(\frac{d_r}{b} - \kappa_0, \kappa_0 + 1\right)B\left(\frac{d_K}{2}, \frac{d_r}{b} - \frac{d_K}{2} - \kappa_0\right)} \cdot (k_B T)^{-\frac{1}{2}d_K} r_0^{-d_r}$$

$$\times \left\{\frac{1}{\kappa_0}\left[\frac{d_r}{b}\left(\frac{r_0}{r}\right)^b - \frac{\varepsilon_K}{k_B T}\right] - 1\right\}_+^{-\kappa_0 - 1 - \frac{1}{2}d_K + \frac{1}{b}d_r} \varepsilon_K^{\frac{1}{2}d_K - 1} r^{d_r - 1} dr\,d\varepsilon_K, \tag{3.137b}$$

which is valid for $0 < \kappa_0 < \frac{d_r}{b} - \frac{d_K}{2}$ (with exponents $b \leq 2d_r/d_K$). The positional kappa distribution, Eq. (3.130), becomes

$$P(\vec{r}; \kappa_0, T)d\vec{r} = \frac{\Gamma\left(\frac{d_r}{2}\right)}{2\pi^{\frac{d_r}{2}}} \cdot \frac{b\left(\frac{d_r}{b}\right)^{-\frac{d_r}{b}} \kappa_0^{\frac{d_r}{b}}}{B\left(\frac{d_r}{b} - \kappa_0, \kappa_0 + 1\right)} \cdot r_0^{-d_r} \cdot \left[\frac{\frac{d_r}{b}}{\kappa_0}\left(\frac{r_0}{|\vec{r}|}\right)^b - 1\right]^{-\kappa_0 - 1 + \frac{1}{b}d_r} d\vec{r}, \tag{3.138a}$$

or

$$P(r; \kappa_0, T)dr = \frac{b\left(\dfrac{d_r}{b}\right)^{-\frac{d_r}{b}} \kappa_0^{\frac{d_r}{b}}}{B\left(\dfrac{d_r}{b} - \kappa_0, \kappa_0 + 1\right)} \cdot r_0^{-d_r} \cdot \left[\dfrac{\dfrac{d_r}{b}}{\kappa_0} \cdot \left(\dfrac{r_0}{r}\right)^b - 1\right]^{-\kappa_0 - 1 + \frac{1}{b}d_r} r^{d_r - 1} dr,$$

(3.138b)

where the double beta function may be written as

$$B\left(\frac{d_r}{b} - \kappa_0, \kappa_0 + 1\right) B\left(\frac{d_K}{2}, \frac{d_r}{b} - \frac{d_K}{2} - \kappa_0\right)$$

$$= \frac{\Gamma(\kappa_0 + 1)\Gamma\left(\dfrac{d_K}{2}\right)\Gamma\left(\dfrac{d_r}{b} - \dfrac{d_K}{2} - \kappa_0\right)}{\Gamma\left(\dfrac{d_r}{b} + 1\right)}$$

(3.139)

$$= \frac{\Gamma(\kappa_0 + 1)\Gamma\left(\dfrac{d_K}{2}\right)}{\Gamma\left(\kappa_0 + 1 + \dfrac{d_K}{2}\right)} \frac{\Gamma\left(\kappa_0 + 1 + \dfrac{d_K}{2}\right)\Gamma\left(\dfrac{d_r}{b} - \dfrac{d_K}{2} - \kappa_0\right)}{\Gamma\left(\dfrac{d_r}{b} + 1\right)}$$

$$= B\left(\frac{d_K}{2}, \kappa_0 + 1\right) B\left(\kappa_0 + 1 + \frac{d_K}{2}, \frac{d_r}{b} - \frac{d_K}{2} - \kappa_0\right).$$

Similar to Eqs. (3.78a) and (3.78b), the density, temperature, and thermal pressure are given by

$$n(r) = n_0 \cdot \left[\frac{\dfrac{d_r}{b}}{\kappa_0} \cdot \left(\dfrac{r_0}{r}\right)^b - 1}{\dfrac{\dfrac{d_r}{b}}{\kappa_0} - 1}\right]^{-\kappa_0 - 1 + \frac{1}{b}d_r}, \quad T(r) = T_0 \cdot \left[\frac{\dfrac{d_r}{b}}{\kappa_0} \cdot \left(\dfrac{r_0}{r}\right)^b - 1}{\dfrac{\dfrac{d_r}{b}}{\kappa_0} - 1}\right],$$

(3.140a)

$$p(r) = p_0 \cdot \left[\frac{\dfrac{d_r}{b}}{\kappa_0} \cdot \left(\dfrac{r_0}{r}\right)^b - 1}{\dfrac{\dfrac{d_r}{b}}{\kappa_0} - 1}\right]^{-\kappa_0 + \frac{1}{b}d_r}.$$

where, again, (n_0, T_0, p_0) are the density, temperature, and thermal pressure at radius $r = r_0$. The polytropic relation is

$$\frac{n(r)}{n_0} = \cdot \left[\frac{T(r)}{T_0}\right]^{-\kappa_0 - 1 + \frac{1}{b}d_r}, \quad \nu = \frac{1}{b}d_r - 1 - \kappa_0. \tag{3.140b}$$

Compare with Eqs. (3.79a) and (3.79b) for the positive potential $\Phi(r) = \frac{1}{b}k \cdot r^b$ (see also Chapter 5, Section 5.3 on the polytropes). The two sets of equations can be written more compactly as follows:

For the potential $\Phi(r) = \pm\frac{1}{b}k \cdot r^{\pm b}$, we have

$$n(r) = n_0 \cdot \left[\frac{\frac{\mp b}{\kappa_0} \cdot \left(\frac{r}{r_0}\right)^{\pm b} - 1}{\frac{\mp b}{\kappa_0} - 1}\right]^{-\kappa_0 - 1 \mp \frac{1}{b}d_r},$$

$$T(r) = T_0 \cdot \left[\frac{\frac{\mp b}{\kappa_0} \cdot \left(\frac{r}{r_0}\right)^{\pm b} - 1}{\frac{\mp b}{\kappa_0} - 1}\right], \tag{3.141a}$$

$$p(r) = p_0 \cdot \left[\frac{\frac{\mp b}{\kappa_0} \cdot \left(\frac{r}{r_0}\right)^{\pm b} - 1}{\frac{\mp b}{\kappa_0} - 1}\right]^{-\kappa_0 \mp \frac{1}{b}d_r}.$$

and

$$\frac{n(r)}{n_0} = \left[\frac{T(r)}{T_0}\right]^{\nu}, \quad \nu = \mp\frac{1}{b}d_r - 1 - \kappa_0. \tag{3.141b}$$

In Fig. 3.6, we plot the positive potential $\Phi(r) \propto +r^2$ (oscillator) and the negative potential $\Phi(r) \propto -r^{-1}$ (gravitational pull) for $d_r = 3$ and kappa indices $\kappa_0 = 0.5$, 1, and 1.5. We observe the monotonically decreasing density and thermal pressure with increasing the central distance for both the positive and negative potentials. However, the temperature decreases due to the gravitational pull, but increases due to the oscillator potential. It will be interesting to examine the combination of both the potentials (e.g., see Section 3.9.5). Such a combination may describe the observed behavior of proton temperature in the heliosphere (e.g., Richardson and Smith, 2003).

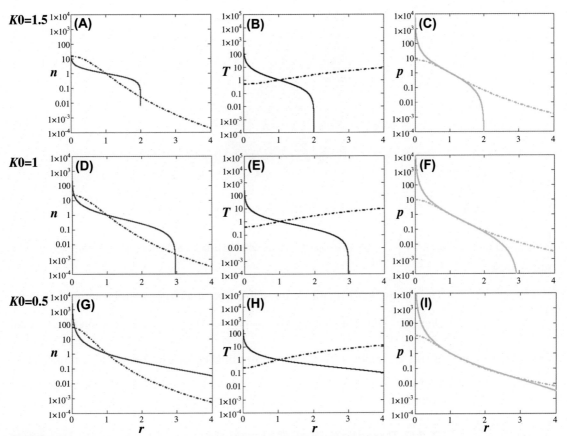

FIGURE 3.6

The positional profiles of density n, temperature T, and thermal pressure p, for the positive potential $\Phi(r) \propto +r^{+b}$ with exponent $+b = 2$ (*dash-dot line*) and the negative potential $\Phi(r) \propto -r^{-b}$ with exponent $-b = -1$ (*solid line*). The panels in the three rows correspond to different kappa indices, $\kappa_0 = 1.5$ (upper row), $\kappa_0 = 1$ (middle row), $\kappa_0 = 0.5$ (lower row).

3.9.5 Potentials With Positive and Negative Values Acquiring Stable/Unstable Equilibrium Points

Here we consider central potentials with a stable or an unstable point. The existence of these equilibria is caused by the two competing terms of the potential, the attractive and the repulsive terms. The used attractive term is the gravitational potential for both cases $\sim -GMm/r$, while the used repulsive terms are $\sim \left(\frac{1}{2}L^2/m\right)/r^2$ in the case of a stable equilibrium and $\sim -\frac{1}{2}m\omega^2 r^2$ in the case of an unstable equilibrium. The total potential is the sum of the opposing centrifugal potential energy with the potential energy of the gravitational field. In the case of the stable potential $\Phi_1(r) = \left(\frac{1}{2}L^2/m\right)/r^2 - GMm/r$, all the particles have the same orbital angular momentum (L). The distribution density vanishes at two boundary radii (a minimum and a maximum radius), while at the stable radius, the density is maximized. In the case of the unstable potential $\Phi_2(r) = -\frac{1}{2}m\omega^2 r^2 - GMm/r$, all the particles are subject to the same orbital angular frequency (ω). The distribution density vanishes between the two boundary radii.

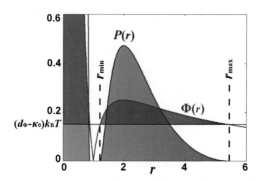

FIGURE 3.7

The stable potential $\Phi_1(r) = \left(\frac{1}{2}L^2/m\right)/r^2 - GMm/r$ and the respective kappa distribution density $P_1(r) \propto$

$\left\{ |\Phi_1(r)|/\left[\left(\frac{1}{2}d_\Phi - \kappa_0\right)k_B T\right] - 1 \right\}_+^{\kappa_0 - 1}$ (the degeneration due to positional degrees of freedom is not considered). Adopted from

Livadiotis (2015b).

Fig. 3.7 shows the graph of the stable potential $\Phi_1(r)$ and the kappa distribution

density $P_1(r) \propto \left\{ |\Phi_1(r)|/\left[\left(\frac{d_\Phi}{2} - \kappa_0\right)k_B T\right] - 1 \right\}_+^{\kappa_0 - 1}$. When the potential is smaller

than $\left(\frac{d_\Phi}{2} - \kappa_0\right)k_B T$, the distribution vanishes (due to the cut-off condition, noted

by "+"). Note that the degeneration due to positional degrees of freedom is not
considered.

3.10 Gravitational Potentials
3.10.1 Linear Gravitational Potential: The Barometric Formula

In Section 3.7.2, we derived the kappa distributions in the case of the linear
gravitational field. The positional kappa distribution is

$$P(z; \kappa_0, T)dz = \frac{\kappa_0 + 1}{\kappa_0} \cdot z_0^{-1} \cdot \left(1 + \frac{1}{\kappa_0}\cdot\frac{z}{z_0}\right)^{-\kappa_0 - 2} dz, \quad \text{where } z_0(T) = \frac{k_B T}{mg}.$$

(3.142)

The barometric formula is the atmospheric particle density profile, and it can be
obtained by $n(\overrightarrow{r}) = N \cdot P(\overrightarrow{r})$, where $\int_{-\infty}^{\infty} n(\overrightarrow{r})d\overrightarrow{r} = N$ is the total number of
particles. Here we have $n(z) = (N/S) \cdot P(z)$, where S is the total cross-sectional
area of the whole system; N and S are constants, so the local density is written as

$$n(z; \kappa_0, T) = n_0 \cdot \left(1 + \frac{1}{\kappa_0}\cdot\frac{mg}{k_B T}\cdot z\right)^{-\kappa_0 - 2},$$

$$n(z; \kappa_0, T) = n_0 \cdot \exp\left(-\frac{mg}{k_B T}\cdot z\right) \text{ for } \kappa_0 \to \infty,$$

(3.143a)

where n_0 is the density at the zero potential level $(z = 0)$, equal to $n_0 = n_{av} \cdot (1 + 1/\kappa_0) \cdot [mgL/(k_BT)]$ with $n_{av} = N/(SL)$ representing the average density over an arbitrary length L.

At thermal equilibrium $(\kappa_0 \rightarrow \infty)$, the density is given by the standard altitude-dependent formula (exponential in Eq. 3.143a). This characterizes the density and pressure profiles of the low atmosphere. However, a more complicated formula is found to model the higher atmospheric layers that matches with the kappa distribution in Eq. (3.143a) (for further details, see Champion et al., 1961).

The local temperature is given by Eqs. (3.77a) and (3.77b), i.e.,

$$T(z; \kappa_0, T) = T_0 \cdot \left(1 + \frac{1}{\kappa_0} \cdot \frac{mg}{k_BT} \cdot z\right), \tag{3.143b}$$

and the local thermal pressure

$$p(z; \kappa_0, T) = n(z; \kappa_0, T) k_BT(z; \kappa_0, T) = p_0 \cdot \left(1 + \frac{1}{\kappa_0} \cdot \frac{mg}{k_BT} \cdot z\right)^{-\kappa_0 - 1}, \tag{3.143c}$$

with $p_0 = n_0 k_B T_0$. The relevant polytropic index relation is

$$\frac{n(z)}{n_0} = \left[\frac{T(z)}{T_0}\right]^{-\kappa_0 - 2}, \quad \nu = -2 - \kappa_0. \tag{3.143d}$$

Similar to the standard kinetic kappa distribution, e.g., Eqs. (3.2a) and (3.2b), the positional distribution, Eq. (3.143a), is characterized by a power law behavior. Indeed, for sufficiently large altitude, $z \gg \kappa_0 k_BT/(mg)$, the density profile behaves as a power law,

$$n(z) \cong A \cdot (z/L)^{-\gamma}, \quad \text{for } z \gg \kappa_0 k_BT/(mg), \tag{3.144}$$

where $A \equiv n_0 \cdot [\kappa_0 k_BT/(mgL)]^{\kappa_0 + 2} = n_{av} \cdot (\kappa_0 + 1)[\kappa_0 k_BT/(mgL)]^{\kappa_0 + 1}$, while the negative power law exponent is equal to $\gamma \equiv \kappa_0 + 2$; L denotes the units of the altitude z.

The barometric power law has been observed even in other planetary magnetospheres. For example, the plasma density in the Saturnian magnetosphere exhibits a power law dependence on the radial distance from the planet with $\gamma \sim 4$ for protons (Gurnett et al., 2007; Thomsen et al., 2010). Using Eq. (3.143a), we find that $\gamma \sim 4$ corresponds to $\kappa_0 \sim 2$, or in terms of the standard kappa index for three degrees of freedom, $\kappa = 2 + \frac{1}{2}3 \sim 3.5$ (see also the end of Section 3.11). This is in remarkable agreement with the most frequent kappa indices of proton populations in the Saturnian magnetosphere. Indeed, Dialynas et al. (2009) fitted kappa distributions on Saturnian magnetospheric energetic ion spectral measurements and derived, among others, the kappa indices, further studied in Livadiotis and McComas (2010a, Fig. 4); as it was shown, the most frequent value of the proton kappa indices is $\kappa_{H+} \approx 3.5$. Finally, it is worth noting that the vertical density distribution of the OH cloud near Enceladus is fitted by a positional kappa distribution (c.f., Jurac et al., 2002, Fig. 12.9).

3.10.2 Spherical Gravitational Potential

Here we study the statistical description of the spherical gravitational potential (note that the same analysis holds for any other attractive central potential that drops with $\sim 1/r$, e.g., the electrostatic potential). We are interested to find the kappa distribution of a negative potential energy $\Phi(\vec{r}) < 0$; thus we focus on the following two cases of Table 3.1,

$$
P(\vec{r}; K, T) \propto \left[\frac{|\Phi(\vec{r})|}{\left(\frac{d_\Phi}{2} + K\right) k_B T} - 1 \right]^{-K-1} , |\Phi(\vec{r})| \geq \left(\frac{d_\Phi}{2} + K\right) k_B T, \quad K = -\kappa_0,
$$

(3.145a)

$$
P(\vec{r}; K, T) \propto \left[1 - \frac{|\Phi(\vec{r})|}{\left(\frac{d_\Phi}{2} + K\right) k_B T} \right]^{-K-1} , |\Phi(\vec{r})| \leq \left(\frac{1}{2} d_\Phi + K\right) k_B T, \quad K = \kappa_0,
$$

(3.145b)

In the case of the gravitational potential $\Phi(\vec{r}) = -GMm/r$, we obtain

$$
P(\vec{r}; K, T) \propto \left[\frac{GMm}{\left(\frac{1}{2} d_\Phi + K\right) k_B T} \cdot \frac{1}{r} - 1 \right]^{-K-1} ,
$$

$$
GMm/r \geq \left(\frac{d_\Phi}{2} + K\right) k_B T, K = -\kappa_0,
$$

(3.146a)

$$
P(\vec{r}; K, T) \propto \left[1 - \frac{GMm}{\left(\frac{d_\Phi}{2} + K\right) k_B T} \cdot \frac{1}{r} \right]^{-K-1} , GMm/r \leq \left(\frac{d_\Phi}{2} + K\right) k_B T, \quad K = \kappa_0.
$$

(3.146b)

The effective degrees of freedom d_Φ of the gravitational interaction are defined by

$$
\frac{1}{2} d_\Phi \equiv \frac{|\langle\Phi\rangle|}{k_B T} = \frac{GMm}{k_B T r_0},
$$

(3.147)

where $r_0 = \langle r^{-1}\rangle^{-1}$ is the average potential radius (hereafter, it will be called simply average radius). Hence, the distribution is written as

$$
P(\vec{r}; K, T) \propto \left(\frac{\alpha}{r} - 1\right)^{-K-1}, \quad r \leq \alpha, \quad \alpha \equiv GMm \Big/ \left[\left(\frac{d_\Phi}{2} + K\right) k_B T\right], \quad K = -\kappa_0,
$$

(3.148a)

$$P(\overrightarrow{r}; K, T) \propto \left(1 - \frac{\alpha}{r}\right)^{-K-1}, \quad r \geq \alpha, \quad \alpha \equiv GMm \left/ \left[\left(\frac{d_\Phi}{2} + K\right) k_B T\right], \quad K = \kappa_0.\right.$$

$$(3.148b)$$

The distribution, Eq. (3.148b), cannot be normalized because it does not converge for $r \to \infty$ (i.e., the partition function diverges). Indeed, for $r \gg \alpha$, the distribution becomes of the order of ~ 1, and the integrand diverges, i.e., $\lim_{r \to \infty} (1 - \alpha/r)^{-\kappa_0 - 1} r^2 = \lim_{r \to \infty} r^2 \to \infty$.

The previously mentioned type of divergence of the gravitational (or electrostatic) potential is a long-standing problem in classical statistical physics (e.g., see Dauxois et al., 2002; Livadiotis, 2009). One way to solve the "statistical paradox" is within the framework of kappa distributions. Eq. (3.148a) shows that the distribution becomes zero at $r = \alpha$ (radial node). The local density of the particle population $n(\overrightarrow{r})$ is proportional to the distribution; thus no particles can flow across this radial node. On the other hand, for particles existing beyond the node, their density would diverge unless they were influenced by other interactions. In the absence of any other interaction contributing to the total potential, the only physically meaningful configuration for the system to stabilize is that of particles being distributed within the distance $0 < r < \alpha$. Then the positional kappa distribution, Eq. (3.148a), becomes

$$P(\overrightarrow{r}; \kappa_0, T) \propto \left(\frac{\alpha}{r} - 1\right)^{\kappa_0 - 1}, \quad r \leq \alpha, \quad \text{with } \alpha \equiv \frac{GMm}{\left(\frac{d_\Phi}{2} - \kappa_0\right) k_B T} = \frac{\frac{d_\Phi}{2}}{\frac{d_\Phi}{2} - \kappa_0} \cdot r_0.$$

$$(3.149)$$

The average radius r_0 and the normalization constant μ_2 are given by

$$r_0 = \langle r^{-1} \rangle^{-1} = \mu_2/\mu_1, \quad \text{with, } \mu_m \equiv \int_0^\alpha \left(\frac{\alpha}{r} - 1\right)^{\kappa_0 - 1} r^m dr. \quad (3.150)$$

Setting $X \equiv r/\alpha$, we have, or

$$\mu_m = \alpha^{m+1} \int_0^1 (1 - X)^{\kappa_0 - 1} X^{m - \kappa_0 + 1} dX = \alpha^{m+1} B(\kappa_0, m - \kappa_0 + 2), \quad \text{or}$$

$$\mu_m = \alpha^{m+1} \frac{\Gamma(\kappa_0) \Gamma(m - \kappa_0 + 2)}{\Gamma(m + 2)}. \quad (3.151)$$

Hence, the average radius is given by

$$r_0 = \mu_2/\mu_1 = \alpha \cdot \frac{\Gamma(4 - \kappa_0)/\Gamma(4)}{\Gamma(3 - \kappa_0)/\Gamma(3)} = \alpha \cdot \frac{3 - \kappa_0}{3} = \frac{\frac{d_\Phi}{2}}{\frac{d_\Phi}{2} - \kappa_0} r_0 \cdot \frac{3 - \kappa_0}{3}, \quad \text{or}$$

$$\frac{1}{2} d_\Phi = 3, \quad \alpha = \frac{GMm}{(3 - \kappa_0) k_B T} = \frac{3}{3 - \kappa_0} \cdot r_0, r_0 = \frac{GMm}{3 k_B T}.$$

$$(3.152)$$

The normalization constant μ_2 converges for $2.5 < \kappa_0 < 3$ and equals

$$\mu_2 = \alpha^3 \frac{\Gamma(\kappa_0)\Gamma(4-\kappa_0)}{\Gamma(4)} = \frac{9}{2}r_0^3 \frac{(1-\kappa_0)(2-\kappa_0)}{(3-\kappa_0)^2}\Gamma(\kappa_0)\Gamma(1-\kappa_0)$$

$$= \frac{9\pi}{2}r_0^3 \frac{(1-\kappa_0)(2-\kappa_0)}{(3-\kappa_0)^2 \sin(\pi\kappa_0)}, \tag{3.153}$$

where we substituted α from Eq. (3.152) and used the Euler's reflection formula $\Gamma(1-x)\Gamma(x) = \pi/\sin(\pi x)$.

Therefore, the distribution is written as

$$P(\vec{r};\kappa_0,T)d\vec{r} = \frac{1}{18\pi^2}\cdot\frac{(3-\kappa_0)^2 \sin(\pi\kappa_0)}{(2-\kappa_0)(1-\kappa_0)}\cdot r_0^{-3}\cdot\left(\frac{3}{3-\kappa_0}\cdot\frac{r_0}{r}-1\right)^{\kappa_0-1} d\vec{r}, \tag{3.154a}$$

or, setting $P(r;\kappa_0,T)\equiv P(\vec{r};\kappa_0,T)\cdot 4\pi r^2$,

$$P(r;\kappa_0,T)dr = \frac{2}{9\pi}\cdot\frac{(3-\kappa_0)^2 \sin(\pi\kappa_0)}{(2-\kappa_0)(1-\kappa_0)}\cdot r_0^{-3}\cdot\left(\frac{3}{3-\kappa_0}\cdot\frac{r_0}{r}-1\right)^{\kappa_0-1} r^2 dr. \tag{3.154b}$$

Finally, we derive the phase space kappa distribution

$$P(\vec{r},\varepsilon_K;\kappa_0,T)d\vec{r}d\varepsilon_K = \frac{(3-\kappa_0)^{2-\frac{1}{2}d_K}\Gamma(\kappa_0-2)\sin(\pi\kappa_0)}{18\pi^2\Gamma\left(\frac{d_K}{2}\right)(\kappa_0-1)\Gamma\left(\kappa_0-\frac{d_K}{2}\right)}\cdot(k_BT)^{-\frac{1}{2}d_K}r_0^{-3}$$

$$\times\left[\frac{1}{3-\kappa_0}\cdot\left(\frac{3r_0}{r}-\frac{\varepsilon_K}{k_BT}\right)-1\right]_+^{\kappa_0-1-\frac{1}{2}d_K}\varepsilon_K^{\frac{1}{2}d_K-1}d\vec{r}d\varepsilon_K, \tag{3.155a}$$

or

$$P(r,\varepsilon_K;\kappa_0,T)dr d\varepsilon_K = \frac{2(3-\kappa_0)^{2-\frac{1}{2}d_K}\Gamma(\kappa_0-2)\sin(\pi\kappa_0)}{9\pi\Gamma\left(\frac{d_K}{2}\right)(\kappa_0-1)\Gamma\left(\kappa_0-\frac{d_K}{2}\right)}\cdot(k_BT)^{-\frac{1}{2}d_K}r_0^{-3}$$

$$\times\left[\frac{1}{3-\kappa_0}\cdot\left(\frac{3r_0}{r}-\frac{\varepsilon_K}{k_BT}\right)-1\right]_+^{\kappa_0-1-\frac{1}{2}d_K}r^2\varepsilon_K^{\frac{1}{2}d_K-1}dr d\varepsilon_K. \tag{3.155b}$$

After including the degeneration of the kappa index due to the potential energy, shown in Eqs. (3.136a) and (3.136b), the phase space and positional distributions become (Fig. 3.8)

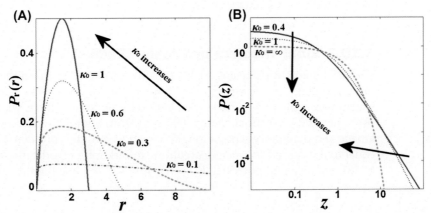

FIGURE 3.8
Positional kappa distributions of (A) spherical, and (B) linear gravitational potential, according to Eqs. (3.157) and (3.142), respectively. Adopted and modified from Livadiotis (2015b).

$$P(r, \varepsilon_K; \kappa_0, T) dr d\varepsilon_K = \frac{2}{9\pi} \cdot \frac{\kappa_0^{2-\frac{1}{2}d_K} \Gamma(1 - \kappa_0) \sin(\pi \kappa_0)}{\Gamma\left(\frac{d_K}{2}\right)(2 - \kappa_0)\Gamma\left(3 - \frac{d_K}{2} - \kappa_0\right)} \cdot (k_B T)^{-\frac{1}{2}d_K} r_0^{-3}$$

$$\times \left[\frac{1}{\kappa_0} \cdot \left(\frac{3r_0}{r} - \frac{\varepsilon_K}{k_B T}\right) - 1\right]_+^{-\kappa_0 - \frac{1}{2}d_K + 2} r^2 \varepsilon_K^{\frac{1}{2}d_K - 1} dr d\varepsilon_K,$$

(3.156)

and

$$P(r; \kappa_0, T) dr = \frac{2}{9\pi} \cdot \frac{\kappa_0^2 \sin(\pi \kappa_0)}{(\kappa_0 - 1)(\kappa_0 - 2)} \cdot r_0^{-3} \cdot \left(\frac{3}{\kappa_0} \cdot \frac{r_0}{r} - 1\right)^{-\kappa_0 + 2} r^2 dr. \quad (3.157)$$

Similar to Eq. (3.140a), the density, temperature, and thermal pressure are given by

$$n(r) = n_0 \cdot \left(\frac{\frac{3}{\kappa_0} \cdot \frac{r_0}{r} - 1}{\frac{3}{\kappa_0} - 1}\right)^{-\kappa_0 + 2}, \quad T(r) = T_0 \cdot \left(\frac{\frac{3}{\kappa_0} \cdot \frac{r_0}{r} - 1}{\frac{3}{\kappa_0} - 1}\right),$$

$$p(r) = p_0 \cdot \left(\frac{\frac{3}{\kappa_0} \cdot \frac{r_0}{r} - 1}{\frac{3}{\kappa_0} - 1}\right)^{-\kappa_0 + 3}.$$

(3.158a)

where (n_0, T_0, p_0) are the density, temperature, and thermal pressure at radius $r = r_0$.

Finally, the polytropic relation becomes

$$\frac{n(r)}{n_0} = \left[\frac{T(r)}{T_0}\right]^{-\kappa_0 + 2}, \quad \nu = 2 - \kappa_0. \quad (3.158b)$$

(compare with Eq. 3.143d for linear gravitational field. See also Chapter 5, Section 5.3 on the polytropes). For the isobaric polytropic index observed in the inner heliosheath and planetary magnetosheaths (Table 5.1), $\nu \approx -1$, corresponds to the kappa index $\kappa_0 \approx 3$ or to the standard 3-D kappa index $\kappa = 3 + \frac{1}{2}3 \sim 4.5$.

3.10.3 Virial Theorem and Jeans 'Radius'

The virial theorem (Clausius, 1870) provides a general equation that relates in a proportionality of the average over time of the kinetic energy $\langle \varepsilon_K \rangle$ of a stable system consisting of N particles and bound by a potential force, with the average over time of the potential energy $\langle \Phi \rangle$. In order to obtain the time averages of the kinetic and potential energy, a long time period must be considered, but they are equal to the ensemble averages according to Birkhoff's ergodic theorem (Collins, 1978). For a central potential $\Phi(\overrightarrow{r}) = \Phi(r) \propto \pm r^{\pm b}$, the theorem states

$$\langle \varepsilon_K \rangle = \pm \frac{b}{2} \cdot \langle \Phi \rangle, \tag{3.159}$$

for either positive or negative exponents b. For linear oscillators, $b = 2$, we obtain $\langle \varepsilon_K \rangle = \langle \Phi \rangle$, while for the gravitational potential, the exponent is $b = -1$, thus $\langle \varepsilon_K \rangle = -\frac{1}{2} \cdot \langle \Phi \rangle$.

Indeed, for the generalized oscillator potential $\Phi(r) \propto r^b$, we derived in Section 3.7.3 that

$$\langle \Phi \rangle = \frac{d_r}{b} k_B T = \frac{2}{b} \frac{d_r}{d_K} \frac{d_K}{2} k_B T = \frac{2}{b} \frac{d_r}{d_K} \langle \varepsilon_K \rangle, \tag{3.160}$$

while for the generalized gravitational potential $\Phi(r) \propto -r^{-b}$, we derived in Section 3.9.4 that

$$-\langle \Phi \rangle = \frac{2}{b} \frac{d_r}{d_K} \langle \varepsilon_K \rangle. \tag{3.161}$$

Hence, we end up with Eq. (3.159), given that $d_r = d_K$, i.e., positional and velocity spaces have the same dimensionality. For example, in the previous Section 3.10.2, we observe that

$$-\langle \Phi \rangle = \langle GMm/r \rangle = GMm \cdot r_0 = 3k_B T = 2\langle \varepsilon_K \rangle, \text{or,} \langle \varepsilon_K \rangle = -\frac{1}{2}\langle \Phi \rangle. \tag{3.162}$$

This remarkable equation constitutes a verification of the virial theorem for the gravitational potential, i.e., for the exponent -1.

The average radius r_0 is related to the *Jeans* radius r_J (Jeans, 1902), which is a measure of the size of a cloud of gas and dust particles where gravity balances gas pressure. For larger sizes, gravity overwhelms gas pressure and the cloud will collapse. The Jeans radius is given by $r_J = \sqrt{\frac{5}{3} R^{\frac{3}{2}} r_0^{-\frac{1}{2}}}$, where R is the cloud radius. The latter can be interpreted as the maximum radius α, given in Eq. (3.152), i.e., $R = (3/\kappa_0) r_0$. Then we obtain $r_J = 3\sqrt{5}\kappa_0^{-\frac{3}{2}} r_0$. On the other hand, the most frequent radius $r_* = \frac{3}{2} r_0$ is given by the maximum of the radial probability distribution, Eq. (3.157), and is independent of the kappa index. Hence, the arrangement of three radii is

$$r_* = \frac{3}{2} r_0 < R = (3/\kappa_0) r_0 = (2/\kappa_0) r_* < r_J = 3\sqrt{5}\kappa_0^{-\frac{3}{2}} r_0 = \left(\sqrt{5} \big/ \kappa_0\right) R. \tag{3.163}$$

(The inequalities can be easily shown given the maximum value of the invariant kappa index, that is, $\kappa_0 < \frac{d_r}{b} - \frac{d_K}{2} = \frac{3}{2}$).

3.11 Potentials with Angular Dependence

In the case of an angular potential, which depends only on the spherical angles, $\Phi(\Omega)$, i.e., for a 3-dimensional sphere, $\Omega = (\vartheta, \varphi)$, the phase-space distribution for positive potentials is given by

$$P(\vec{r}, \vec{u}; \kappa_0, T) = \pi^{-\frac{1}{2}d_K}\kappa_0^{-\frac{1}{2}d_K} \cdot \frac{\Gamma\left(\kappa_0 + 1 + \frac{d_K}{2} + \frac{d_\Phi}{2}\right)}{\Gamma\left(\kappa_0 + \frac{d_\Phi}{2} + 1\right)} \cdot \theta^{-d_K}$$

$$\times \left\{ \int_0^{2\pi} \int_{-1}^{1} \left[1 + \frac{1}{\kappa_0} \cdot \frac{\Phi(\vartheta, \varphi)}{k_B T} \right]^{-\kappa_0 - 1 - \frac{1}{2}d_\Phi} d\cos\vartheta \, d\varphi \right\}^{-1}$$

$$\times \left\{ 1 + \frac{1}{\kappa_0} \cdot \left[\frac{(\vec{u} - \vec{u}_b)^2}{\theta^2} + \frac{\Phi(\vartheta, \varphi)}{k_B T} \right] \right\}^{-\kappa_0 - 1 - \frac{1}{2}d_K - \frac{1}{2}d_\Phi},$$

(3.164a)

or for negative potentials, assuming that $|\Phi(\vartheta, \varphi)| < \kappa_0 k_B T$,

$$P(\vec{r}, \vec{u}; \kappa_0, T) = \pi^{-\frac{1}{2}d_K}\kappa_0^{-\frac{1}{2}d_K} \cdot \frac{\Gamma\left(\kappa_0 + 1 + \frac{d_K}{2} - \frac{d_\Phi}{2}\right)}{\Gamma\left(\kappa_0 - \frac{d_\Phi}{2} + 1\right)} \cdot \theta^{-d_K}$$

$$\times \left\{ \int_0^{2\pi} \int_{-1}^{1} \left[1 - \frac{1}{\kappa_0} \cdot \frac{|\Phi(\vartheta, \varphi)|}{k_B T} \right]^{-\kappa_0 - 1 + \frac{1}{2}d_\Phi} d\cos\vartheta \, d\varphi \right\}^{-1}$$

$$\times \left\{ 1 + \frac{1}{\kappa_0} \cdot \left[\frac{(\vec{u} - \vec{u}_b)^2}{\theta^2} - \frac{|\Phi(\vartheta, \varphi)|}{k_B T} \right] \right\}^{-\kappa_0 - 1 - \frac{1}{2}d_K + \frac{1}{2}d_\Phi},$$

(3.164b)

with potential degrees of freedom

$$\frac{1}{2}d_\Phi = \frac{|\Phi(\vartheta, \varphi)|}{k_B T},$$

(3.164c)

and kappa index degeneration

$$\kappa = \kappa_0 + \frac{1}{2}d_K \pm \frac{1}{2}d_\Phi.$$

(3.164d)

As an example of an angular potential, which is dependent of the polar angle ϑ but with azimuthal symmetry, is given by a system of paramagnetic particles with freely-rotating magnetic moments $\vec{\mu}$ in the presence of an external magnetic field $\vec{B}$. The particle potential energy is given by the polar-angle dependent potential $\Phi(\vartheta) = -\vec{\mu} \cdot \vec{B} = -\mu B \cos \vartheta$, or $\Phi(\vartheta)/(k_B T) = -\beta \cos \vartheta$ with $\beta \equiv \mu B/(k_B T)$, which is examined in Chapter 5, Section 8 (see also Livadiotis 2015b; 2016a). In this case, the phase-space distribution is given by

$$P(\vartheta, \varepsilon_K ; \kappa_0, \beta) \, d\varepsilon_K d\cos\vartheta = \frac{\Gamma\left(\kappa_0 + \frac{1}{2}d_K + 1\right)(\kappa_0 + \beta\langle\cos\vartheta\rangle + \beta)^{\kappa_0}(1 + \beta\langle\cos\vartheta\rangle - \beta)}{2\Gamma(\kappa_0)\Gamma\left(\frac{1}{2}d_K\right)(\kappa_0 + \beta\langle\cos\vartheta\rangle)^{\kappa_0+1+\frac{1}{2}d_K}}$$

$$\times (k_B T)^{-\frac{1}{2}d_K} \cdot \left[1 + \frac{1}{\kappa_0 + \beta\langle\cos\vartheta\rangle} \cdot \left(\frac{\varepsilon_K}{k_B T} - \beta\cos\vartheta\right)\right]^{-\kappa_0 - 1 - \frac{1}{2}d_K}$$

$$\times \varepsilon_K^{\frac{1}{2}d_K - 1} d\varepsilon_K d\cos\vartheta.$$

(3.165)

where the average $\langle\cos\vartheta\rangle$ is given by

$$\frac{1 + \beta\langle\cos\vartheta\rangle + \beta}{1 + \beta\langle\cos\vartheta\rangle - \beta} = \left(\frac{\kappa_0 + \beta\langle\cos\vartheta\rangle + \beta}{\kappa_0 + \beta\langle\cos\vartheta\rangle - \beta}\right)^{\kappa_0}.$$

(3.166)

The marginal kinetic energy distribution, derived from the integration of polar angle is given by

$$P(\varepsilon_K ; \kappa_0, \beta) \, d\varepsilon_K = \frac{\Gamma\left(\kappa_0 + \frac{1}{2}d_K\right)(\kappa_0 + \beta\langle\cos\vartheta\rangle + \beta)^{\kappa_0}(1 + \beta\langle\cos\vartheta\rangle - \beta)}{2\beta\,\Gamma(\kappa_0)\Gamma\left(\frac{1}{2}d_K\right)(\kappa_0 + \beta\langle\cos\vartheta\rangle)^{\kappa_0+\frac{1}{2}d_K}} \cdot (k_B T)^{-\frac{1}{2}d_K}$$

$$\times \left\{\left[1 + \frac{\varepsilon_K - \beta k_B T}{(\kappa_0 + \beta\langle\cos\vartheta\rangle)k_B T}\right]^{-\kappa_0 - \frac{1}{2}d_K}\right.$$

$$\left. - \left[1 + \frac{\varepsilon_K + \beta k_B T}{(\kappa_0 + \beta\langle\cos\vartheta\rangle)k_B T}\right]^{-\kappa_0 - \frac{1}{2}d_K}\right\} \cdot \varepsilon_K^{\frac{1}{2}d_K - 1} d\varepsilon_K.$$

(3.167)

The potential degrees of freedom are

$$\frac{1}{2}d_\Phi = \frac{\langle|\Phi(\vartheta)|\rangle}{k_B T} = \beta\langle\cos\vartheta\rangle.$$

(3.168)

The degeneration of the kappa index caused by the potential degrees of freedom is

$$\kappa(d_K) = \kappa_0 + \frac{1}{2}d_K(\text{no potential}) \Rightarrow \kappa(d_K, d_\Phi) = \kappa_0 + \frac{1}{2}d_K - \frac{1}{2}d_\Phi(\text{with potential}),$$

(3.169a)

namely,

$$\kappa(d_K) = \kappa_0 + \frac{1}{2}d_K(\Phi = 0) \Rightarrow \kappa(d_K, d_\Phi) = \kappa_0 + \frac{1}{2}d_K - \beta\langle\cos\vartheta\rangle(\Phi \propto - \cos\vartheta).$$

(3.169b)

Hence, the distribution of kinetic energy is actually given by

$$P(\varepsilon_K ; \kappa_0, \beta)\, d\varepsilon_K = \frac{\kappa_0^{-\frac{1}{2}d_K} \cdot (k_B T)^{-\frac{1}{2}d_K}}{B\left(\kappa_0 - \beta\langle\cos\vartheta\rangle, \frac{1}{2}d_K\right)} \cdot \left(1 + \frac{\beta}{\kappa_0}\right)^{\kappa_0 - \beta\langle\cos\vartheta\rangle} \cdot \left(\beta^{-1} + \langle\cos\vartheta\rangle - 1\right)$$

$$\times \frac{1}{2}\left\{\left[1 + \frac{\varepsilon_K - \beta\, k_B T}{\kappa_0 k_B T}\right]^{-\kappa_0 - \frac{1}{2}d_K + \beta\langle\cos\vartheta\rangle}\right.$$

$$\left. - \left[1 + \frac{\varepsilon_K + \beta\, k_B T}{\kappa_0 k_B T}\right]^{-\kappa_0 - \frac{1}{2}d_K + \beta\langle\cos\vartheta\rangle}\right\} \cdot \varepsilon_K^{\frac{1}{2}d_K - 1}\, d\varepsilon_K.$$

$$(3.170a)$$

This is a linear superposition of two kappa distributions: The two extreme values of the potential energy are given for polar angles $\cos\vartheta = 1$ (parallel) and $\cos\vartheta = -1$ (antiparallel). The total energy is respectively given by $\varepsilon_K - \beta\, k_B T$ and $\varepsilon_K + \beta\, k_B T$. Equation (3.170) involves the probability distribution to have these two energy values; it is actually, the distribution of having parallel but not antiparallel, magnetic moment.

The whole phase-space distribution is now given by

$$P(\vartheta, \varepsilon_K ; \kappa_0, \beta)\, d\varepsilon_K d\cos\vartheta$$

$$= \frac{\Gamma\left(\kappa_0 - \beta\langle\cos\vartheta\rangle + \frac{1}{2}d_K + 1\right)(\kappa_0 + \beta)^{\kappa_0 - \beta\langle\cos\vartheta\rangle}(1 + \beta\langle\cos\vartheta\rangle - \beta)}{2\Gamma(\kappa_0 - \beta\langle\cos\vartheta\rangle)\Gamma\left(\frac{1}{2}d_K\right)\kappa_0^{\kappa_0 + 1 + \frac{1}{2}d_K - \beta\langle\cos\vartheta\rangle}}$$

$$\times (k_B T)^{-\frac{1}{2}d_K} \cdot \left[1 + \frac{1}{\kappa_0} \cdot \left(\frac{\varepsilon_K}{k_B T} - \beta\cos\vartheta\right)\right]^{-\kappa_0 - 1 - \frac{1}{2}d_K + \beta\langle\cos\vartheta\rangle} \varepsilon_K^{\frac{1}{2}d_K - 1}\, d\varepsilon_K d\cos\vartheta .$$

$$(3.170b)$$

The marginal potential energy distribution is

$$P(\cos\vartheta ; \kappa_0, \beta)\, d\cos\vartheta = \frac{\kappa_0(\kappa_0 + \beta\langle\cos\vartheta\rangle + \beta)^{\kappa_0}(1 + \beta\langle\cos\vartheta\rangle - \beta)}{2(\kappa_0 + \beta\langle\cos\vartheta\rangle)^{\kappa_0 + 1}}$$

$$\times \left(1 - \frac{\beta\cos\vartheta}{\kappa_0 + \beta\langle\cos\vartheta\rangle}\right)^{-\kappa_0 - 1} d\cos\vartheta,$$

$$(3.171)$$

while after the degeneration of the kappa index due to the potential degrees of freedom as shown in (3.169b), the polar angle distribution becomes

$$P(\cos\vartheta ; \kappa_0, \beta)\, d\cos\vartheta = \frac{1}{2}(\kappa_0 - \beta\langle\cos\vartheta\rangle)(\kappa_0 + \beta)^{\kappa_0 - \beta\langle\cos\vartheta\rangle}(1 + \beta\langle\cos\vartheta\rangle - \beta)$$

$$\times (\kappa_0 - \beta\cos\vartheta)^{-\kappa_0 - 1 + \beta\langle\cos\vartheta\rangle} d\cos\vartheta,$$

$$(3.172)$$

which is valid for kappa indices $\kappa_0 > \beta$. Now the average $\langle \cos \vartheta \rangle$ is given by

$$\frac{1 + \beta \langle \cos \vartheta \rangle + \beta}{1 + \beta \langle \cos \vartheta \rangle - \beta} = \left(\frac{\kappa_0 + \beta}{\kappa_0 - \beta} \right)^{\kappa_0 - \beta \langle \cos \vartheta \rangle} . \tag{3.173}$$

3.12 Potentials Forming Anisotropic Distribution of Velocity

In the presence of a potential energy that depends on both the position and the velocity, the marginal kappa distribution of the velocity (obtained by integrating the phase space distribution over the position) can be different from the standard kappa distribution, Eqs. (3.2a) and (3.2b), that is, in the absence of a potential energy. One of the important conclusions is that the presence of a potential energy may cause anisotropy to the distribution of the velocity (kinetic energy).

Anisotropy in the comoving system cannot be caused by a positional dependence on the distribution parameters, i.e., $\vec{u}_b(\vec{r})$, $T(\vec{r})$, or by a positional dependence of the potential, $\Phi(\vec{r})$. Anisotropy can occur only if the potential has velocity dependence $\Phi(\vec{r}, \vec{u})$.

As an example, consider the potential $\Phi(\vec{r}, \vec{u}) = \Phi_1(\vec{r}) \cdot u_x^2 + \Phi_2(\vec{r}) \cdot u_y^2 + \Phi_3(\vec{r}) \cdot u_z^2$. The Hamiltonian will be $H(\vec{r}, \vec{u}) = \left[\frac{m}{2} + \Phi_1(\vec{r}) \right] \cdot u_x^2 + \left[\frac{m}{2} + \Phi_2(\vec{r}) \right] \cdot u_y^2 + \left[\frac{m}{2} + \Phi_3(\vec{r}) \right] \cdot u_z^2$, and the Hamiltonian factor $H(\vec{r}, \vec{u})/(k_B T)$ will be replaced by $H(\vec{r}, \vec{u}) = \sum_{i=x,y,z} \left[\frac{m}{2} u_i^2 / (k_B T_i) \right]$, where $T_i \equiv T / \left[1 + \frac{2}{m} \Phi_i(\vec{r}) \right]$, for $i = x, y, z$.

As another example, we examine again the particles with magnetic momentum μ in a magnetic field B, with potential energy $\Phi(\vartheta) = -\mu B \cos \vartheta$. The polar angle ϑ is a positional spherical coordinate, counting the angle from the magnetic field vector and the magnetic momentum vector of each particle. Namely, there is only positional dependence of the potential $\Phi(\vec{r})$, and thus no anisotropy is expected. Furthermore, we consider particles that due to their shape, their velocity tends to be aligned to their magnetic moment. In other words, the polar angle ϑ identifies both the positional and velocity coordinates, and thus the phase space distribution now reads the pitch angle distribution (c.f., Lin et al., 1997, Fig. 3). Hence, we have $\varepsilon_{K,\parallel} = \varepsilon_K \cos^2 \vartheta$ and $\varepsilon_{K,\perp} = \varepsilon_K \sin^2 \vartheta$ (parallel and perpendicular direction with respect to the magnetic field).

The mean kinetic energy can be easily calculated, $\langle \varepsilon_K \rangle = \frac{d_K}{2} k_B T$. This result is due to the energy equipartition theorem in general (Chapter 1). On the other hand, the mean kinetic energy in directions parallel and perpendicular to the field can be derived from Eq. (3.170b) as follows:

$$
\langle \varepsilon_{K,\text{II}}/(k_B T) \rangle = \frac{\Gamma\!\left(\kappa_0 + \frac{5}{2} - \beta\langle\cos\vartheta\rangle\right)(\kappa_0 + \beta\,)^{\kappa_0 - \beta\langle\cos\vartheta\rangle}(1 + \beta\langle\cos\vartheta\rangle - \beta)}{\Gamma\!\left(\frac{1}{2}\right)\Gamma(\kappa_0 - \beta\langle\cos\vartheta\rangle)}
$$

$$
\times \int_{-1}^{1} \cos^2\vartheta \cdot \left[\int_{0}^{\infty} (\kappa_0 - \beta\cos\vartheta + X)^{-\kappa_0 - \frac{5}{2} + \beta\langle\cos\vartheta\rangle} X^{\frac{3}{2}} dX\right] d\cos\vartheta,
$$

$$
\tag{3.174}
$$

where we set $X \equiv \varepsilon_K/(k_B T)$ and $d_K = 3$, leading to

$$
\langle \varepsilon_{K,\text{II}} \rangle = \frac{1}{2} k_B T \cdot I(\kappa_0; \beta) \quad \text{and} \quad \langle \varepsilon_{K,\perp} \rangle = \frac{3}{2} k_B T - \frac{1}{2} k_B T \cdot I(\kappa_0; \beta) \,, \quad \text{with} \tag{3.175a}
$$

$$
I(\kappa_0; \beta) = \frac{3(\kappa_0 + \beta\,)^{\kappa_0 - \beta\langle\cos\vartheta\rangle}(1 + \beta\langle\cos\vartheta\rangle - \beta)}{2\beta^3 \kappa_0^{\kappa_0 - 3 - \beta\langle\cos\vartheta\rangle}}
$$

$$
\cdot \left[F\!\left(\frac{\beta}{\kappa_0}; \kappa_0 - \beta\langle\cos\vartheta\rangle\right) - F\!\left(-\frac{\beta}{\kappa_0}; \kappa_0 - \beta\langle\cos\vartheta\rangle\right) \right] \tag{3.175b}
$$

with

$$
F(t; \kappa) = \frac{(\kappa - 1)(\kappa - 2)t^2 - 2(\kappa - 1)t + 2}{(\kappa - 1)(\kappa - 2)(\kappa - 3)} \cdot (1 - t)^{-\kappa + 1}. \tag{3.175c}
$$

Setting $\langle \varepsilon_{K,\text{II}} \rangle \equiv \frac{1}{2} k_B T_{\text{II}}$ and $\langle \varepsilon_{K,\perp} \rangle \equiv k_B T_{\perp}$, or $T_{\text{II}}/T = I(\kappa_0; \beta)$ and $T_{\perp}/T = [3 - I(\kappa_0; \beta)]/2$, we obtain the anisotropy $(T_{\text{II}} - T_{\perp})/T = \frac{3}{2}[I(\kappa_0; \beta) - 1] \cong \frac{1}{5}(1 + 1/\kappa_0) \cdot \beta^2 + O(\beta^3)$. We observe that the developed anisotropy faints as β decreases (i.e., the magnetic energy μB decreases or the thermal energy $k_B T$ increases), and as κ_0 increases (i.e., the system resides in stationary states closer to thermal equilibrium).

Therefore, the anisotropy that characterized observed distributions in various space plasmas may be caused by the presence of a potential energy, such as the magnetic energy due to the particle magnetic moment in a magnetic field. As an example, we apply our results to the electron plasma in the magnetotail. Observations from THEMIS spacecraft indicated a typical value of kappa index $\kappa \sim 3.5$ (Ogasawara et al., 2013), i.e., $\kappa_0 \sim 2$ ($\kappa_0 = \kappa - 1.5$).

Also, measurements of $B \sim 7.4$ nT and $k_B T \sim 1.9$ keV correspond to thermal anisotropy $T_{\text{II}}/T_{\perp} \sim 1.1$ (Artemyev et al., 2013). Then we find $\beta \sim 0.6$ that gives the electron magnetic moment $\mu \sim 160$ eV/nT. This coincides with the average electron magnetic moment found by Kaufmann et al. (2004).

3.13 Concluding Remarks

This chapter presented the Hamiltonian kappa distribution, that is, the phase space kappa distribution of the Hamiltonian as given by the sum of the velocity-dependent kinetic energy and the position-dependent potential energy. Specifically, we derived and investigated:

- the two marginal distributions of the phase space distribution, that is, the distribution of the velocity (or kinetic energy) after integrating over the position and the distribution of the position after integrating over the velocity;

- the mean kinetic energy, that is, the kinetic definition of temperature, in the presence of a positional potential energy, showing that is the same as in the absence of the potential energy;

- the degeneration of the kappa index in the presence of a potential energy;

- the generic formulation of the phase space and positional kappa distributions for positive and negative potentials;

- the two main attractive power law potentials: the positive potential with positive power law exponent (oscillator type), and the negative potential with negative power law exponent (gravitational type);

- distributions for potentials with both positive and negative values;

- gravitational potentials, e.g., the linear gravitational potential and the barometric formula, the spherical gravitational potential, the virial theorem, and the Jeans radius;

- the local distributions, namely, the kappa distribution of velocity, the density, temperature, and thermal pressure, all expressed as functions of the position;

- the long-standing problem in classical statistical physics of the divergence of the phase space distribution of the gravitational or other potential decreasing with distance can be resolved using the "negative kappa distribution";

- potentials that depend on the velocity and cause anisotropies in the kappa distributions of velocities.

These developments allow the researcher to describe any particle population of space, geophysical, laboratory, or other plasmas that are subject to a non-negligible potential energy.

3.14 Science Questions for Future Research

Future analyses and observations need to address the following questions:

1. How the multiparticle distribution of the Hamiltonian would be?
2. How can the degeneration of κ be systematically defined for any ϕ?
3. Can types of potentials explain the observed anisotropic distributions?

G. Livadiotis
Southwest Research Institute, San Antonio, TX, United States

Chapter Outline

4.1 **Summary** 185
4.2 **Introduction** 186
4.3 **Isotropic Distributions (Without Potential)** 187
    4.3.1 Standard (Positive) Multidimensional Kappa Distributions 187
        4.3.1.1 Distributions, for d_K Degrees of Freedom, Using the Invariant Kappa Index κ_0 187
        4.3.1.2 Distributions, for d_K Degrees of Freedom, Using the 3-D Kappa Index $\kappa_3 = \kappa_0 + \frac{3}{2}$ 188
        4.3.1.3 Distributions, for d_K Degrees of Freedom, Using the d_K-Dimensional Kappa Index $\kappa = \kappa_0 + \frac{1}{2}d_K$ 188
    4.3.2 Standard (Positive) Multidimensional Kappa Distributions in an Inertial Reference Frame 189
        4.3.2.1 Distributions, for d_K Degrees of Freedom, Using the Invariant Kappa Index κ_0 189
        4.3.2.2 Distributions, for $d_K = 3$ Degrees of Freedom, Using the Invariant Kappa Index κ_0 190
        4.3.2.3 Distributions, for $d_K = 3$ Degrees of Freedom, Using the 3-D Kappa Index $\kappa = \kappa_0 + \frac{3}{2}$ 192
    4.3.3 Negative, Multidimensional Kappa Distribution 193
        4.3.3.1 Distributions Using the

Kappa Distributions. http://dx.doi.org/10.1016/B978-0-12-804638-8.00004-8

Invariant Kappa
Index κ_0 193
4.3.3.2 Distributions
Using the
Kappa Index
$\kappa = \kappa_0 - \frac{1}{2}d_K$
193
4.3.4 Superposition of
Multidimensional Kappa
Distributions 194
4.3.4.1 Linear
Superposition
of Positive and
Negative
Distributions in
Terms of the
Kinetic Energy
ε_K 194
4.3.4.2 Linear
Superposition
Given a Density
of Kappa
Indices $D(\kappa_0)$
195
4.3.4.3 Linear
Superposition
in Terms of the
Kinetic Energy
ε_K with
Components of
Different
Temperature
196
4.3.4.4 Nonlinear
Superposition
of Positive and
Negative
Distributions in
Terms of the
Kinetic Energy
ε_K 196
4.4 **Anisotropic Distributions**
(Without Potential) 196

4.4.1 Correlated Degrees of
Freedom 196
4.4.1.1 Distributions,
for d_K Degrees
of Freedom,
Using the
Invariant Kappa
Index κ_0 196
4.4.1.2 Distribution in
Terms of the
Velocity $\vec{u}$ for
d_K Degrees of
Freedom, Using
the d_K-
Dimensional
Kappa Index
$\kappa = \kappa_0 + \frac{1}{2}d_K$
197
4.4.1.3 Distribution in
Terms of the
Velocity $\vec{u}$ for
d_K Degrees of
Freedom, Using
the 3-D Kappa
Index $\kappa_3 = \kappa_0$
$+ \frac{1}{2}(d_K - 3)$
198
4.4.1.4 Distributions,
for $d_K = 3$
Degrees of
Freedom, Using
the Invariant
Kappa Index
κ_0 198
4.4.1.5 Distribution, for
$d_K = 3$ Degrees
of Freedom,
Using the 3-D
Kappa Index
$\kappa = \kappa_0 + \frac{3}{2}$
199
4.4.2 Correlation Between the
Projection at a Certain

Direction and the Perpendicular Plane 200

4.4.2.1 Distributions, for $d_{\kappa} = 3$ Degrees of Freedom, Using the Invariant Kappa Index κ_0 200

4.4.2.2 Distribution in Terms of the Velocity $\vec{u} = \left(u_{\|}, \vec{u}_{\perp} \right)$ for $d_{\kappa} = 3$ Degrees of Freedom, Using the 3-D Kappa Index $\kappa = \kappa_0 + \frac{3}{2}$ 201

4.4.2.3 The Temperature is Given by 201

4.4.3 Self-Correlated Degrees of Freedom 201

4.4.3.1 Distributions, for d_{κ} Degrees of Freedom, Using the Invariant Kappa Index κ_0 201

4.4.3.2 Distribution in Terms of the Velocity, $\vec{u}$, for d_{κ} Degrees of Freedom, Using the d_{κ}-Dimensional Kappa Index $\kappa = \kappa_0 + \frac{1}{2} d_{\kappa}$ 202

4.4.3.3 Distribution in Terms of the Velocity, $\vec{u}$, for

d_{κ} Degrees of Freedom, Using the 3-D Kappa Index $\kappa = \kappa_0 + \frac{3}{2}$ 202

4.4.3.4 Distributions, for $d_{\kappa} = 3$ Degrees of Freedom, Using the Invariant Kappa Index κ_0 202

4.4.3.5 Distribution, for $d_{\kappa} = 3$ Degrees of Freedom, Using the 3-D Kappa Index $\kappa = \kappa_0 + \frac{3}{2}$ 203

4.4.4 Self-Correlated Projections at a Direction and Perpendicular Plane 204

4.4.4.1 Distributions, for $d_{\kappa} = 3$ Degrees of Freedom, Using the Invariant Kappa Index κ_0 204

4.4.4.2 Distribution in Terms of the Velocity $\vec{u} = \left(u_{\|}, \vec{u}_{\perp} \right)$ for $d_{\kappa} = 3$ Degrees of Freedom, Using the 3-D Kappa Index $\kappa = \kappa_0 + \frac{3}{2}$ 205

4.4.4.3 The Temperature is Given by 205

4.4.5 Self-Correlated Degrees of Freedom With Different Kappa 205

    4.4.5.1 Distributions, for d_κ Degrees of Freedom, Using the Invariant Kappa Index κ_0 205

    4.4.5.2 Distribution in Terms of the Velocity $\vec{u}$ for d_κ Degrees of Freedom, Using the 1-D Kappa Index $\kappa_i = \kappa_{0i} + \frac{1}{2}$ 206

    4.4.5.3 Distribution in Terms of the Velocity $\vec{u}$ for d_κ Degrees of Freedom, Using the 3-D Kappa Index $\kappa = \kappa_0 + \frac{3}{2}$ 206

4.4.6 Self-Correlated Projections at a Direction and Perpendicular Plane With Different Kappa 207

    4.4.6.1 Distributions, for $d_\kappa = 3$ Degrees of Freedom, Using the Invariant Kappa Index κ_0 207

    4.4.6.2 Distribution in Terms of the Velocity $\vec{u} = \left(u_\parallel, \vec{u}_\perp \right)$

    for $d_\kappa = 3$ Degrees of Freedom, Using the Kappa Indices 207

    4.4.6.3 The Temperature is Given by 208

4.4.7 Self-Correlated Projections of Different Dimensionality and Kappa 208

    4.4.7.1 Distributions, for d_κ Degrees of Freedom of Mixed Correlation, i.e., i: 1, …, M Uncorrelated Groups of Correlated Degrees of Freedom, With Each Group Having Different Degrees of Freedom f_i, Using Different (Invariant) Kappa Index, κ_{0i}, but Common Temperature T or θ 208

    4.4.7.2 Distributions, for M Correlated Groups, Each With f_i Degrees of Freedom, Self-Correlated With Kappa Index κ_{0i} and Temperature θ_i 209

4.4.8 Different Self-Correlation
and Intercorrelation
Between Degrees of
Freedom 210

    4.4.8.1 Distributions,
for M Correlated
Groups, Each
With f_i Degrees
of Freedom,
Self-Correlated
With Invariant
Kappa Index,
κ_{0i}, and
Intercorrelated
With Invariant
Kappa Index κ_{0i}^{int}
210

    4.4.8.2 Distributions,
for M Correlated
Groups, Each
With f_i Degrees
of Freedom,
Self-Correlated
With Kappa
Index, κ_i, and
Intercorrelated
With Kappa
Index κ_i^{int} 211

    4.4.8.3 Distributions,
for M Correlated
Groups, Each
With f_i Degrees
of Freedom,
Self-Correlated
With Invariant
Kappa Index,
$\kappa_{0i} = \kappa_0$, and
Intercorrelated
With Invariant
Kappa Index κ_0^{int}
213

    4.4.8.4 Distributions,
for $d_K = 3$
Correlated
Groups, Each

With $f_i = 1$
Degrees of
Freedom, Self-
Correlated With
Kappa Index,
κ_{0i}, and
Intercorrelated
With Kappa
Index κ_0^{int} 213

    4.4.8.5 Distributions,
for Two
Correlated
Groups, With
$f_1 = 1$ and
$f_2 = 2$ Degrees
of Freedom,
Self-Correlated
With Kappa
Index κ_{0i}, and
Intercorrelated
With Kappa
Index κ_0^{int} 216

**4.5 Distributions With
Potential 217**

    4.5.1 General Hamiltonian
Distribution 218

        4.5.1.1 Phase Space
Distribution
218

        4.5.1.2 Hamiltonian
Function 218

        4.5.1.3 Hamiltonian
Degrees of
Freedom
Summing Up
the Kinetic
and Potential
Degrees of
Freedom 218

    4.5.2 Positive Attractive
Potential $\Phi\left(\vec{r}\right) > 0$
218

        4.5.2.1 Phase Space
Distributions
218

4.5.2.2 Positional
Distribution
Function 219
4.5.2.3 Potential
Degrees of
Freedom 219
4.5.3 Negative Attractive
Potential $\Phi\left(\vec{r}\right) < 0$
220
4.5.3.1 Phase Space
Distributions
220
4.5.3.2 Positional
Distribution
Function 221
4.5.3.3 Potential
Degrees of
Freedom 221
4.5.4 Small Positive/Negative
Attractive/Repulsive
Potential
$|\Phi\left(\vec{r}\right)/(\kappa_0 k_B T)| \ll 1$
Defined in a Finite
Volume $\vec{r} \in V$ 221
4.5.4.1 Phase Space
Distribution
221
4.5.4.2 Positional
Distribution
Function 222
4.5.4.3 Potential
Degrees of
Freedom 222
4.5.5 Equivalent Local
Distribution 222
4.5.5.1 Phase Space
Distribution
With Potential
$\Phi\left(\vec{r}\right)$ 223
4.5.5.2 Equivalent Local
Distribution
With No
Potential Energy
223

4.5.6 Positive Power Law
Central Potential
(Oscillation Type)
$\Phi(r) = \frac{1}{b} k r^b$
224
4.5.6.1 Phase Space
Distributions
224
4.5.6.2 Potential
Degrees of
Freedom 224
4.5.6.3 Positional
Distribution
225
4.5.7 Negative Power Law
Central Potential
(Gravitational Type)
$\Phi(r) = -\frac{1}{b} k r^{-b}$ 226
4.5.7.1 Phase Space
Distributions
226
4.5.7.2 Potential
Degrees of
Freedom 226
4.5.7.3 Positional
Distribution
227
4.5.8 Properties for $\Phi(r) = \pm$
$\frac{1}{b} k r^{\pm b}$ 228
4.5.8.1 Parameters
228
4.5.8.2 Degeneration of
the Kappa
Index 228
4.5.8.3 Local
Parameters
228
4.5.8.4 Polytropic Index
229
4.5.8.5 Statistical
Moments
229
4.5.9 Marginal and Conditional
Distributions 230

4.5.9.1 Marginal
Distributions
230
4.5.9.2 Conditional
Distributions
230
4.5.10 Angular Potentials
$\Phi(\vartheta,\varphi)$ 231
4.5.10.1 Phase Space
Distribution
231
4.5.10.2 Potential
Degrees of
Freedom   232
4.5.10.3 Kappa Index
Degeneration
232
4.5.11 Magnetization Potential
$\Phi(\vartheta) \propto -cos\vartheta$ 232
4.5.11.1 Potential Energy
232
4.5.11.2 Phase Space
Distribution, in
Terms of the
Positional Polar
Angle ϑ and the
Kinetic Energy
ε_K, for 1
Positional and
d_K Kinetic
Degrees of
Freedom   232
4.5.11.3 Distribution of
the Kinetic
Energy ε_K
232
4.5.11.4 Distribution of
the Polar Angle
$cos\vartheta$ 233
4.5.11.5 Degeneration of
the Kappa
Index   233
4.6 Multiparticle
Distributions   234

4.6.1 Standard N-Particle ($N \cdot d$)
−Dimensional Kappa
Distributions   234
4.6.1.1 Distributions, in
Terms of the
Velocity, $\vec{u}$, for
N Particles, d_K
Degrees of
Freedom per
Particle, Using
the Invariant
Kappa Index
κ_0 234
4.6.1.2 Distributions, in
Terms of the
Kinetic Energy,
$\varepsilon_{K(n)} = \frac{1}{2}m \times$
$\left(\vec{u}_{(n)} - \vec{u}_b\right)^2$,
for N Particles,
d_K Degrees of
Freedom per
Particle, Using
the Invariant
Kappa Index
κ_0 234
4.6.2 Negative N-Particle ($N \cdot d$)
−Dimensional Kappa
Distribution   235
4.6.2.1 Distributions, in
Terms of the
Velocity, $\vec{u}$, for
N Particles, d_K
Degrees of
Freedom per
Particle, Using
the Invariant
Kappa Index
κ_0 235
4.6.2.2 Distributions, in
Terms of the
Kinetic Energy,
$\varepsilon_{K(n)}$, for N
Particles, d_K

Degrees of Freedom per Particle, Using the Invariant Kappa Index κ_0 235

4.6.3 N-Particle ($N \cdot d$) —Dimensional Kappa Distributions With Potential   235

   4.6.3.1 Phase Space Distribution, for N Particles, d_K Degrees of Freedom per Particle, Using the Invariant Kappa Index κ_0 235

   4.6.3.2 Hamiltonian Function   236

   4.6.3.3 Hamiltonian Degrees of Freedom Summing up the Kinetic and Potential Degrees of Freedom   236

4.6.4 Standard N-Particle ($N \cdot d$) —Dimensional Kappa Distributions   236

   4.6.4.1 Distributions, in Terms of the Velocity $\vec{u}$ With d_K Degrees of Freedom per Particle, Using the Invariant Kappa Index κ_0 236

   4.6.4.2 Distributions, in Terms of the Kinetic Energies $\{\varepsilon_{K(n)}\}_{n=1}^{N}$ With d_K Degrees of Freedom per Particle, Using the Invariant Kappa Index κ_0 237

   4.6.4.3 Phase Space Distributions $\{H_{(n)} = H(\vec{r}_{(n)}, \vec{u}_{(n)})\}_{n=1}^{N_c}$ With d Degrees of Freedom per Particle, Summing d_K Kinetic and d_Φ Potential Degrees of Freedom, Using the Invariant Kappa Index κ_0 238

4.6.5 Multispecies Distributions   240

   4.6.5.1 Distributions, in Terms of the Velocity, of N Different Particle Species, $\vec{u}_{(1)}, \vec{u}_{(2)}, \cdots, \vec{u}_{(N)}$, With d_K Degrees of Freedom per Particle, Using the Invariant Kappa Index, κ_0 240

   4.6.5.2 Distributions, in Terms of the Kinetic Energies, $\{\varepsilon_{K(n)}\}_{n=1}^{N}$, With d_K Degrees of Freedom per

Particle, Using
the Invariant
Kappa Index,
κ_O 241
4.6.5.3 Phase Space
Distributions of
N Species
$\{H_{(n)} = H(\vec{r}_{(n)},$
$\vec{u}_{(n)})\}_{n=1}^{N}$ With
d Degrees of
Freedom per
Particle,
Summing d_K
Kinetic and d_Φ
Potential
Degrees of
Freedom, Using
the Invariant
Kappa Index, κ_O
242

4.7 Non-Euclidean–
  Normed Distributions   243
    4.7.1 Standard
      d_K–Dimensional Kappa
      Distribution of Velocity,
      $\vec{u}$ 243
    4.7.2 Standard d_K–
      Dimensional Kappa
      Distribution of Kinetic
      Energy, ε_K 244
    4.7.3 Argument x_* 244
4.8 Discrete Distributions   245
    4.8.1 Distribution of Energy
      245
    4.8.2 Partition Function   245
    4.8.3 Internal Energy   245
4.9 Concluding Remarks   245
**4.10 Science Questions for Future
  Research   246**

4.1 Summary

The chapter presents all the formulae of kappa distributions necessary to describe particle populations out of thermal equilibrium in plasmas and beyond. In summary, we first provide the isotropic and anisotropic distributions in the absence of a potential energy, while then we continue with the formulae of kappa distributions in the presence of a potential energy; we present the formulation of multiparticle and multispecies kappa distributions, and the generalized kappa distributions based on non-Euclidean norms or under the discrete description of energy. The presented formulations and their guidelines constitute the most updated "toolbox" of useful and statistically well-grounded equations for future analyses that seek to apply kappa distributions in data analysis, simulations, modeling, theory, and other work in space, geophysical, laboratory, and other plasmas, or related particle systems.

Science Question: What are the various kappa distribution formulae that describe particle populations out of thermal equilibrium?

Keywords: Angular; Anisotropic; Correlation; Degrees of freedom; Distributions: phase space; Hamiltonian; Isotropic; Kinetic energy; Multiparticle; Multispecies; One-particle; Positional; Speed; Velocity.

4.2 Introduction

This chapter provides the kappa distribution formulae, which are the basic and frequently used tools for describing the statistics of systems out of thermal equilibrium. Useful information about the properties and the parameters of these formulae is also given. There are several categories and subcategories of distributions. For instance, multiparticle distributions can be reduced to one-particle distributions, which are more convenient to handle, but less accurate for describing the statistics of particle systems, such as plasmas. Similarly, the distributions of Hamiltonian are more general, but they can be reduced to distributions of velocities for cases where the potential energy can be ignored.

The provided toolbox should be used for future space and plasma physics analyses that seek to apply kappa distributions in data analyses, simulations, modeling, or other theoretical work. Using these equations guarantees results that remain firmly grounded on the foundation of nonextensive statistical mechanics. The chapter is structured as follows: first, we start with the isotropic distributions in the absence of a potential energy. Section 4.3 presents the standard (positive) multidimensional kappa distributions in the plasma flow (comoving) reference frame or any other inertial reference frame (in which the plasma flows with constant velocity). Next, the negative, multidimensional kappa distribution is presented; (as we will see, the nomenclature of positive and negative corresponds to the values of the kappa indices). Section 4.4 presents the anisotropic distributions, again, in the absence of a potential energy. The following important subcategories are described according to the degrees of freedom per particle: correlated degrees of freedom, correlation between the projection at a certain direction and the perpendicular plane, self-correlated degrees of freedom, self-correlated projections at a direction and perpendicular plane, self-correlated degrees of freedom with different kappa indices, self-correlated projections at a direction and perpendicular plane with different kappa, self-correlated projections of different dimensionality and kappa, and different self-correlation and intercorrelation between degrees of freedom. Section 4.5 presents the formulae of kappa distributions in the presence of a potential energy. In particular, we describe the subcategories: Hamiltonian kappa distribution, in general; in the presence of positive or negative attractive potentials, or of any small positive/negative attractive/repulsive potentials defined in a finite volume, $\vec{r} \in V$; reduction of the distribution of the kinetic and potential energy to the equivalent local distribution of solely the kinetic energy; kappa distributions for positive or negative power law central attractive potentials, that is, of oscillation or gravitational types, respectively; properties of the distributions with these potentials (local density, temperature, thermal pressure, polytropic index); marginal and conditional distributions; and distributions with angular potentials, e.g., magnetization potential. Section 4.6 provides the formulae of (positive or negative) multiparticle kappa distributions, and of multispecies kappa distributions, both in the presence or not of a potential energy. Section 4.7 provides the formulation of generalized L_p kappa distributions, which are based on non-Euclidean L_p-norms. Section 4.8 provides briefly the formulation of discrete kappa distributions (distribution of energy, partition function, internal energy).

Finally, the concluding remarks are given in Section 4.9, while three general science questions for future analyses are posed in Section 4.10.

4.3 Isotropic Distributions (Without Potential) (Livadiotis and McComas, 2009; 2011b)

4.3.1 Standard (Positive) Multidimensional Kappa Distributions

4.3.1.1 Distributions, for d_K Degrees of Freedom, Using the Invariant Kappa Index κ_0

4.3.1.1.1 In Terms of the Velocity $\vec{u}$

$$P(\vec{u};\kappa_0,\theta) = \frac{(\pi\kappa_0\theta^2)^{-\frac{1}{2}d_K}\Gamma\left(\kappa_0+1+\frac{1}{2}d_K\right)}{\Gamma(\kappa_0+1)}\cdot\left[1+\frac{1}{\kappa_0}\cdot\frac{(\vec{u}-\vec{u}_b)^2}{\theta^2}\right]^{-\kappa_0-1-\frac{1}{2}d_K}.$$

(4.1a)

Normalization

$$\int_{-\infty}^{\infty}P(\vec{u};\kappa_0,\theta)d\vec{u} = 1,$$

(4.1b)

where $\theta = \sqrt{2k_BT/m}$.

4.3.1.1.2 In Terms of the Kinetic Energy $\varepsilon_K = \frac{1}{2}m(\vec{u}-\vec{u}_b)^2$

$$P(\varepsilon_K;\kappa_0,T) = \frac{(\kappa_0 k_B T)^{-\frac{1}{2}d_K}}{B\left(\frac{d_K}{2},\kappa_0+1\right)}\cdot\left(1+\frac{1}{\kappa_0}\cdot\frac{\varepsilon_K}{k_B T}\right)^{-\kappa_0-1-\frac{1}{2}d_K}\varepsilon_K^{\frac{1}{2}d_K-1}.$$

(4.2a)

Normalization

$$\int_0^{\infty}P(\varepsilon_K;\kappa_0,T)d\varepsilon_K = 1,$$

(4.2b)

where the Beta function is defined by $B(x,y)\equiv\Gamma(x)\cdot\Gamma(y)/\Gamma(x+y)$.

4.3.1.1.3 In Terms of Normalized Kinetic Energy $\xi\equiv\varepsilon_K/(\kappa_0 k_B T)$

$$P(\xi;\kappa_0;T) = F(\xi;2\kappa_0+2;f).$$

(4.3a)

Normalization

$$\int_0^{\infty}P(\xi;\kappa_0,T)d\xi = 1,$$

(4.3b)

where the *F*-distribution is defined by

$$F(x; m, n) = \frac{1}{B\left(\dfrac{m}{2}, \dfrac{n}{2}\right)} \cdot x^{\frac{n}{2}-1} \cdot (1+x)^{-\frac{m+n}{2}}. \tag{4.3c}$$

4.3.1.2 Distributions, for d_K Degrees of Freedom, Using the 3-D Kappa Index $\kappa_3 = \kappa_0 + \frac{3}{2}$

4.3.1.2.1 Distribution in Terms of the Velocity $\vec{u}$

$$P(\vec{u}; \kappa_3, \theta) = \frac{\left[\left(\kappa_3 - \dfrac{3}{2}\right)\theta^2\right]^{-\frac{1}{2}d_K} \Gamma\left(\kappa_3 - \dfrac{1}{2} + \dfrac{d_K}{2}\right)}{\Gamma\left(\kappa_3 - \dfrac{1}{2}\right)}$$

$$\times \left[1 + \frac{1}{\kappa_3 - \dfrac{3}{2}} \cdot \frac{(\vec{u} - \vec{u}_b)^2}{\theta^2}\right]^{-\kappa_3 - 1 - \frac{1}{2}(d_K - 3)}. \tag{4.4}$$

4.3.1.2.2 In Terms of the Kinetic Energy ε_K

$$P(\varepsilon_K; \kappa_3, T) = \frac{\left[\left(\kappa_3 - \dfrac{3}{2}\right)k_B T\right]^{-\frac{1}{2}d_K}}{B\left(\dfrac{d_K}{2}, \kappa_3 - \dfrac{1}{2}\right)} \cdot \left(1 + \frac{1}{\kappa_3 - \dfrac{3}{2}} \cdot \frac{\varepsilon_K}{k_B T}\right)^{-\kappa_3 - 1 - \frac{1}{2}(d_K - 3)} \varepsilon_K^{\frac{1}{2}d_K - 1}. \tag{4.5}$$

4.3.1.3 Distributions, for d_K Degrees of Freedom, Using the d_K-Dimensional Kappa Index $\kappa = \kappa_0 + \frac{1}{2}d_K$

4.3.1.3.1 In Terms of the Velocity $\vec{u}$

$$P(\vec{u}; \kappa, \theta) = \frac{\left[\left(\kappa - \dfrac{d_K}{2}\right)\theta^2\right]^{-\frac{1}{2}d_K} \Gamma(\kappa + 1)}{\Gamma\left(\kappa + 1 - \dfrac{d_K}{2}\right)} \cdot \left[1 + \frac{1}{\kappa - \dfrac{d_K}{2}} \cdot \frac{(\vec{u} - \vec{u}_b)^2}{\theta^2}\right]^{-\kappa - 1}. \tag{4.6}$$

4.3.1.3.2 In Terms of the Kinetic Energy ε_K

$$P(\varepsilon_K; \kappa, T) = \frac{\left[\left(\kappa - \frac{d_K}{2}\right)k_B T\right]^{-\frac{1}{2}d_K}}{B\left(\frac{d_K}{2}, \kappa + 1 - \frac{d_K}{2}\right)} \cdot \left(1 + \frac{1}{\kappa - \frac{d_K}{2}} \cdot \frac{\varepsilon_K}{k_B T}\right)^{-\kappa - 1} \varepsilon_K^{\frac{1}{2}d_K - 1}. \tag{4.7}$$

Comment: For the $(d_K = 3)$-dimensional distribution, the two descriptions, using the 3-D kappa index: $\kappa_3 = \kappa_0 + \frac{3}{2}$, or the d_K-dimensional kappa index: $\kappa = \kappa_0 + \frac{1}{2}d_K$, obviously, coincide.

4.3.2 Standard (Positive) Multidimensional Kappa Distributions in an Inertial Reference Frame (Livadiotis and McComas, 2013a)

4.3.2.1 Distributions, for d_K Degrees of Freedom, Using the Invariant Kappa Index κ_0

4.3.2.1.1 In Terms of the Kinetic Energy $E_K = \frac{1}{2}m\vec{u}^2$ and the Polar Angle, ϑ, Set Between $\vec{u}$ and $\vec{u}_b$

$$P(E_K, \cos\vartheta; \kappa_0, T) = \frac{(\kappa_0 k_B T)^{-\frac{1}{2}d_K}}{B\left(\frac{1}{2}, \frac{d_K - 1}{2}\right)B\left(\frac{d_K}{2}, \kappa_0 + 1\right)}$$

$$\times \left(1 + \frac{1}{\kappa_0} \cdot \frac{E_K + E_{K,b} - 2\sqrt{E_{K,b}}\sqrt{E_K}\cos\vartheta}{k_B T}\right)^{-\kappa_0 - 1 - \frac{1}{2}d_K}$$

$$E_K^{\frac{1}{2}d_K - 1}\left(1 - \cos^2\vartheta\right)^{\frac{1}{2}(d_K - 3)}.$$

$$\tag{4.8a}$$

Normalization

$$\int_{-1}^{1} \int_{0}^{\infty} P(E_K, \cos\vartheta; \kappa_0, T) dE_K d\cos\vartheta = 1, \tag{4.8b}$$

where $E_{K,b} \equiv \frac{1}{2}m\vec{u}_b^2$.

4.3.2.1.2 In Terms of the Kinetic Energy E_K

$$P(E_K; \kappa_0, T) = \frac{(\kappa_0 k_B T)^{-\frac{1}{2}d_K}}{B\left(\frac{1}{2}, \frac{d_K - 1}{2}\right)B\left(\frac{d_K}{2}, \kappa_0 + 1\right)} \cdot \left(1 + \frac{1}{\kappa_0} \cdot \frac{E_K + E_{K,b}}{k_B T}\right)^{-\kappa_0 - 1 - \frac{1}{2}d_K} E_K^{\frac{1}{2}d_K - 1}$$

$$\times g\left(\xi = \frac{2\sqrt{E_{K,b}}\sqrt{E_K}}{\kappa_0 k_B T + E_K + E_{K,b}}; \kappa = \kappa_0 + \frac{d_K}{2}, \lambda = \frac{d_K - 3}{2}\right),$$

$$\tag{4.9a}$$

where

$$g(\xi; \kappa, \lambda) \equiv \int_{-1}^{1} (1 - \xi \cdot x)^{-\kappa-1} (1 - x^2)^{\lambda} dx. \tag{4.9b}$$

Normalization

$$\int_{0}^{\infty} P(E_K; \kappa_0, T) dE_K = 1. \tag{4.9c}$$

4.3.2.1.3 In Terms of the Polar Angle ϑ

$$P(\cos \vartheta; \kappa_0, \xi_b) = \frac{2(1 - \cos^2 \vartheta)^{\frac{1}{2}(d_K - 3)}}{B\left(\frac{1}{2}, \frac{d_K - 1}{2}\right) B\left(\frac{d_K}{2}, \kappa_0 + 1\right)} \cdot \left[1 + \frac{1}{\kappa_0} \cdot \xi_b (1 - \cos^2 \vartheta)\right]^{-\kappa_0 - 1}$$

$$\times f\left[h\left(\cos \vartheta; \kappa_0 \xi_b^{-1}\right); \kappa_0 + \frac{1}{2} d_K; d_K\right], \tag{4.10a}$$

$$f(x; \kappa; d) \equiv \int_{0}^{\infty} \left[1 + (y - x)^2\right]^{-\kappa-1} y^{d-1} dy, \quad h(x; y) \equiv x \cdot (y + 1 - x^2)^{-\frac{1}{2}}, \quad \xi_b \equiv \frac{E_{K,b}}{k_B T}. \tag{4.10b}$$

Normalization

$$\int_{-1}^{1} P(\cos \vartheta; \kappa_0, \xi_b) d \cos \vartheta = 1. \tag{4.10c}$$

4.3.2.2 Distributions, for $d_K = 3$ Degrees of Freedom, Using the Invariant Kappa Index κ_0

4.3.2.2.1 In Terms of the Kinetic Energy $E_K = \frac{1}{2} m \vec{u}^2$ and the Polar Angle ϑ

$$P(E_K, \cos \vartheta; \kappa_0, T) = (\kappa_0 k_B T)^{-\frac{3}{2}} \frac{\Gamma\left(\kappa_0 + \frac{5}{2}\right)}{\sqrt{\pi} \Gamma(\kappa_0 + 1)}$$

$$\times \left(1 + \frac{1}{\kappa_0} \cdot \frac{E_K + E_{K,b} - 2\sqrt{E_{K,b}}\sqrt{E_K} \cos \vartheta}{k_B T}\right)^{-\kappa_0 - \frac{5}{2}} E_K^{\frac{1}{2}}. \tag{4.11}$$

4.3.2.2.2 In Terms of the Kinetic Energy E_K

$$P(E_K; \kappa_0, T) = \left(\pi\kappa_0 k_B T E_{K,b}\right)^{-\frac{1}{2}} \frac{\Gamma\left(\kappa_0 + \frac{3}{2}\right)}{\Gamma(\kappa_0 + 1)} \cdot \frac{1}{2} \left\{\left[1 + \frac{\left(\sqrt{E_K} - \sqrt{E_{K,b}}\right)^2}{\kappa_0 k_B T}\right]^{-\kappa_0 - \frac{3}{2}}\right.$$

$$\left. - \left[1 + \frac{\left(\sqrt{E_K} + \sqrt{E_{K,b}}\right)^2}{\kappa_0 k_B T}\right]^{-\kappa_0 - \frac{3}{2}}\right\}.$$

(4.12)

4.3.2.2.3 In Terms of the Speed u

$$P(u; \kappa_0, \theta) = \left(\pi\kappa_0 \theta^2 u_b^2\right)^{-\frac{1}{2}} \frac{\Gamma\left(\kappa_0 + \frac{3}{2}\right)}{\Gamma(\kappa_0 + 1)}$$

$$\times \left\{\left[1 + \frac{(u - u_b)^2}{\kappa_0 \theta^2}\right]^{-\kappa_0 - \frac{3}{2}} - \left[1 + \frac{(u + u_b)^2}{\kappa_0 \theta^2}\right]^{-\kappa_0 - \frac{3}{2}}\right\} \cdot u.$$

(4.13a)

Normalization

$$\int_0^\infty P(u; \kappa_0, \theta) du = 1.$$

(4.13b)

4.3.2.2.4 In Terms of the Polar Angle ϑ (Compare With Eq. 4.10b)

$$P(\cos\vartheta; \kappa_0, \xi_b) = \frac{2\Gamma\left(\kappa_0 + \frac{5}{2}\right)}{\sqrt{\pi}\Gamma(\kappa_0 + 1)} \cdot \left[1 + \frac{1}{\kappa_0} \cdot \xi_b\left(1 - \cos^2\vartheta\right)\right]^{-\kappa_0 - 1}$$

$$\times f\left[h\left(\cos\vartheta; \kappa_0\xi_b^{-1}\right); \kappa_0 + \frac{3}{2}, 3\right].$$

(4.14)

4.3.2.3 Distributions, for $d_K = 3$ Degrees of Freedom, Using the 3-D Kappa Index $\kappa = \kappa_0 + \dfrac{3}{2}$

4.3.2.3.1 In Terms of the Kinetic Energy $E_K = \frac{1}{2}m\vec{u}^2$ and the Polar Angle ϑ

$$P(E_K, \cos\vartheta; \kappa, T) = \left[\left(\kappa - \frac{3}{2}\right)k_B T\right]^{-\frac{3}{2}} \frac{\Gamma(\kappa+1)}{\sqrt{\pi}\,\Gamma\left(\kappa - \frac{1}{2}\right)}$$

$$\times \left(1 + \frac{1}{\kappa - \frac{3}{2}} \cdot \frac{E_K + E_{K,b} - 2\sqrt{E_{K,b}}\sqrt{E_K}\cos\vartheta}{k_B T}\right)^{-\kappa-1} E_K^{\frac{1}{2}}.$$

$$(4.15)$$

4.3.2.3.2 In Terms of the Kinetic Energy E_K

$$P(E_K; \kappa, T) = \left[\pi\left(\kappa - \frac{3}{2}\right)k_B T\, E_{K,b}\right]^{-\frac{1}{2}} \frac{\Gamma(\kappa)}{\Gamma\left(\kappa - \frac{1}{2}\right)} \cdot \frac{1}{2}\left\{\left[1 + \frac{(\sqrt{E_K} - \sqrt{E_{K,b}})^2}{\left(\kappa - \frac{3}{2}\right)k_B T}\right]^{-\kappa}\right.$$

$$\left. - \left[1 + \frac{(\sqrt{E_K} + \sqrt{E_{K,b}})^2}{\left(\kappa - \frac{3}{2}\right)k_B T}\right]^{-\kappa}\right\}.$$

$$(4.16)$$

4.3.2.3.3 In Terms of the Speed u

$$P(u; \kappa, \theta) = \left[\pi\left(\kappa - \frac{3}{2}\right)\theta^2 u_b^2\right]^{-\frac{1}{2}} \frac{\Gamma(\kappa)}{\Gamma\left(\kappa - \frac{1}{2}\right)} \cdot \left\{\left[1 + \frac{(u - u_b)^2}{\left(\kappa - \frac{3}{2}\right)\theta^2}\right]^{-\kappa}\right.$$

$$\left. - \left[1 + \frac{(u + u_b)^2}{\left(\kappa - \frac{3}{2}\right)\theta^2}\right]^{-\kappa}\right\} \cdot u.$$

$$(4.17)$$

4.3.2.3.4 In Terms of the Polar Angle ϑ

$$P(\cos \vartheta; \kappa, \xi_b) = \frac{2\Gamma(\kappa + 1)}{\sqrt{\pi}\,\Gamma\left(\kappa - \frac{1}{2}\right)} \cdot \left[1 + \frac{1}{\kappa - \frac{3}{2}} \cdot \xi_b(1 - \cos^2 \vartheta)\right]^{-\kappa + \frac{1}{2}}$$

$$\times f\left\{h\left[\cos \vartheta; \left(\kappa - \frac{3}{2}\right)\xi_b^{-1}\right]; \kappa, 3\right\}. \tag{4.18}$$

4.3.3 Negative, Multidimensional Kappa Distribution (Livadiotis, 2015b)

4.3.3.1 Distributions Using the Invariant Kappa Index κ_0
4.3.3.1.1 In Terms of the Velocity $\vec{u}$

$$P(\vec{u}; \kappa_0, \theta) = \left(\pi \kappa_0 \theta^2\right)^{-\frac{1}{2}d_K} \cdot \frac{\Gamma(\kappa_0)}{\Gamma\left(\kappa_0 - \frac{d_K}{2}\right)} \cdot \left[1 - \frac{1}{\kappa_0} \cdot \frac{(\vec{u} - \vec{u}_b)^2}{\theta^2}\right]_+^{\kappa_0 - 1 - \frac{1}{2}d_K}.$$

$$\tag{4.19a}$$

where the symbol "$+$" denotes the *cut-off condition* (also see Eq. 1.13 in Chapter 1; Livadiotis and McComas, 2009; Tsallis, 2009b):

$$x_+ = x \text{ for } x \geq 0 \text{ and } x_+ = 0 \text{ for } x < 0. \tag{4.19b}$$

4.3.3.1.2 In Terms of the Kinetic Energy ε_K

$$P(\varepsilon_K; \kappa_0, T) = \frac{(\kappa_0 k_B T)^{-\frac{1}{2}d_K}}{B\left(\frac{d_K}{2}; \kappa_0 - \frac{d_K}{2}\right)} \cdot \left(1 - \frac{1}{\kappa_0} \cdot \frac{\varepsilon_K}{k_B T}\right)_+^{\kappa_0 - 1 - \frac{1}{2}d_K} \varepsilon_K^{\frac{1}{2}d_K - 1}. \tag{4.20}$$

4.3.3.2 Distributions Using the Kappa Index $\kappa = \kappa_0 - \frac{1}{2}d_K$
4.3.3.2.1 In Terms of the Velocity $\vec{u}$

$$P(\vec{u}; \kappa, \theta) = \left[\pi\left(\kappa + \frac{d_K}{2}\right)\theta^2\right]^{-\frac{1}{2}d_K} \cdot \frac{\Gamma\left(\kappa + \frac{d_K}{2}\right)}{\Gamma(\kappa)} \cdot \left[1 - \frac{1}{\kappa + \frac{d_K}{2}} \cdot \frac{(\vec{u} - \vec{u}_b)^2}{\theta^2}\right]_+^{\kappa - 1}.$$

$$\tag{4.21}$$

4.3.3.2.2 In Terms of the Kinetic Energy ε_K

$$P(\varepsilon_K; \kappa_0, T) = \frac{\left[\left(\kappa + \frac{d_K}{2}\right) k_B T\right]^{-\frac{1}{2} d_K}}{B\left(\frac{d_K}{2}; \kappa\right)} \cdot \left(1 - \frac{1}{\kappa + \frac{d_K}{2}} \cdot \frac{\varepsilon_K}{k_B T}\right)_+^{\kappa - 1} \varepsilon_K^{\frac{1}{2} d_K - 1}. \tag{4.22}$$

4.3.4 Superposition of Multidimensional Kappa Distributions

4.3.4.1 Linear Superposition of Positive and Negative Distributions in Terms of the Kinetic Energy ε_K

4.3.4.1.1 Positive to Negative Proportion is $c \div (1 - c)$

$$P(\varepsilon_K; \kappa_0^+, \kappa_0^-, T; c) = c \cdot P(\varepsilon_K; \kappa_0^+, T) + (1 - c) \cdot P(\varepsilon_K; \kappa_0^-, T), \quad 0 \le c \le 1, \tag{4.23a}$$

where $P(\varepsilon_K; \kappa_0^+, T)$ is the positive and $P(\varepsilon_K; \kappa_0^-, T)$ is the negative kappa distributions, i.e.,

$$P(\varepsilon_K; \kappa_0^+, \kappa_0^-, T; c) = \frac{(k_B T)^{-\frac{1}{2} d_K}}{\Gamma\left(\frac{d_K}{2}\right)} \cdot \varepsilon_K^{\frac{1}{2} d_K - 1}$$

$$\times \left[c \cdot \frac{\left(\kappa_0^+\right)^{-\frac{1}{2} d_K} \Gamma\left(\kappa_0^+ + 1 + \frac{d_K}{2}\right)}{\Gamma\left(\kappa_0^+ + 1\right)} \left(1 + \frac{1}{\kappa_0^+} \cdot \frac{\varepsilon_K}{k_B T}\right)^{-\kappa_0^+ - 1 - \frac{1}{2} d_K} \right.$$

$$\left. + (1 - c) \cdot \frac{\left(\kappa_0^-\right)^{-\frac{1}{2} d_K} \Gamma\left(\kappa_0^-\right)}{\Gamma\left(\kappa_0^- - \frac{d_K}{2}\right)} \left(1 - \frac{1}{\kappa_0^-} \cdot \frac{\varepsilon_K}{k_B T}\right)_+^{-\kappa_0^- - 1 - \frac{1}{2} d_K} \right]. \tag{4.23b}$$

4.3.4.1.2 Positive to Negative Proportion is 1:1

$$P(\varepsilon_K; \kappa_0^+, \kappa_0^-, T) = \frac{1}{2} \cdot \left[P(\varepsilon_K; \kappa_0^+, T) + P(\varepsilon_K; \kappa_0^-, T) \right], \quad \text{or} \tag{4.24a}$$

$$P\left(\varepsilon_{\mathrm{K}};\kappa_0^+,\kappa_0^-,T\right) = \frac{(k_{\mathrm{B}}T)^{-\frac{1}{2}d_{\mathrm{K}}}}{2\Gamma\left(\dfrac{d_{\mathrm{K}}}{2}\right)} \cdot \varepsilon_{\mathrm{K}}^{\frac{1}{2}d_{\mathrm{K}}-1}$$

$$\times \left[\frac{\left(\kappa_0^+\right)^{-\frac{1}{2}d_{\mathrm{K}}}\Gamma\left(\kappa_0^+ + 1 + \dfrac{d_{\mathrm{K}}}{2}\right)}{\Gamma(\kappa_0^+ + 1)} \left(1 + \frac{1}{\kappa_0^+}\cdot\frac{\varepsilon_{\mathrm{K}}}{k_{\mathrm{B}}T}\right)^{-\kappa_0^+ - 1 - \frac{1}{2}d_{\mathrm{K}}} \right.$$

$$\left. + \frac{\left(\kappa_0^-\right)^{-\frac{1}{2}d_{\mathrm{K}}}\Gamma\left(\kappa_0^-\right)}{\Gamma\left(\kappa_0^- - \dfrac{d_{\mathrm{K}}}{2}\right)} \left(1 - \frac{1}{\kappa_0^-}\cdot\frac{\varepsilon_{\mathrm{K}}}{k_{\mathrm{B}}T}\right)_+^{-\kappa_0^- - 1 - \frac{1}{2}d_{\mathrm{K}}} \right].$$

$$(4.24\mathrm{b})$$

4.3.4.1.3 Positive to Negative Proportion is 1:1, and Common Kappa Index (Leubner and Voros, 2005)

$$P(\varepsilon_{\mathrm{K}};\kappa_0,T) = \frac{1}{2}\cdot[P(\varepsilon_{\mathrm{K}};\kappa_0,T) + P(\varepsilon_{\mathrm{K}};\kappa_0,T)],\ \ \text{or} \qquad (4.25\mathrm{a})$$

$$P(\varepsilon_{\mathrm{K}};\kappa_0,T) = \frac{(\kappa_0 k_{\mathrm{B}}T)^{-\frac{1}{2}d_{\mathrm{K}}}}{2\Gamma\left(\dfrac{d_{\mathrm{K}}}{2}\right)} \cdot \varepsilon_{\mathrm{K}}^{\frac{1}{2}d_{\mathrm{K}}-1}$$

$$\times \left[\frac{\Gamma\left(\kappa_0 + 1 + \dfrac{d_{\mathrm{K}}}{2}\right)}{\Gamma(\kappa_0 + 1)} \left(1 + \frac{1}{\kappa_0}\cdot\frac{\varepsilon_{\mathrm{K}}}{k_{\mathrm{B}}T}\right)^{-\kappa_0 - 1 - \frac{1}{2}d_{\mathrm{K}}} \right. \qquad (4.25\mathrm{b})$$

$$\left. + \frac{\Gamma(\kappa_0)}{\Gamma\left(\kappa_0 - \dfrac{d_{\mathrm{K}}}{2}\right)} \left(1 - \frac{1}{\kappa_0}\cdot\frac{\varepsilon_{\mathrm{K}}}{k_{\mathrm{B}}T}\right)_+^{-\kappa_0 - 1 - \frac{1}{2}d_{\mathrm{K}}} \right].$$

4.3.4.2 Linear Superposition Given a Density of Kappa Indices D(κ₀)

$$P(\varepsilon_{\mathrm{K}};T) = c\sum_{\kappa_0^+}D\left(\kappa_0^+\right)P\left(\varepsilon_{\mathrm{K}};\kappa_0^+,T\right) + (1-c)\sum_{\kappa_0^-}D\left(\kappa_0^-\right)P\left(\varepsilon_{\mathrm{K}};\kappa_0^-,T\right). \qquad (4.26)$$

*4.3.4.3 Linear Superposition in Terms of the Kinetic Energy ε_K
with Components of Different Temperature*

$$P(\varepsilon_K; T_s) = c\sum_{\kappa_0^+} D(\kappa_0^+)P[\varepsilon_K; \kappa_0^+, T(\kappa_0^+)] + (1-c)\sum_{\kappa_0^-} D(\kappa_0^-)P[\varepsilon_K; \kappa_0^-, T(\kappa_0^-)],$$

$$(4.27a)$$

$$T_s = c\sum_{\kappa_0^+} D(\kappa_0^+)\cdot T(\kappa_0^+) + (1-c)\sum_{\kappa_0^-} D(\kappa_0^-)\cdot T(\kappa_0^-) = \langle T\rangle. \qquad (4.27b)$$

*4.3.4.4 Nonlinear Superposition of Positive and Negative Distributions
in Terms of the Kinetic Energy ε_K*

$$P(\varepsilon_K; T) = \phi^{-1}\left\{ c\sum_{\kappa_0^+} D(\kappa_0^+)\cdot\phi[P(\varepsilon_K; \kappa_0^+, T)] \right.$$
$$\left. + (1-c)\sum_{\kappa_0^-} D(\kappa_0^-)\cdot\phi[P(\varepsilon_K; \kappa_0^-, T)] \right\},$$

$$(4.28)$$

where ϕ is some monotonic function (Livadiotis and McComas, 2013a).

4.4 Anisotropic Distributions (Without Potential)

The reader may take a look to the references (Pierrard and Lazar, 2010; Lazar et al., 2012; Livadiotis, 2015a).

4.4.1 Correlated Degrees of Freedom

4.4.1.1 Distributions, for d_K Degrees of Freedom, Using the Invariant Kappa Index κ_0

4.4.1.1.1 In Terms of the Velocity $\vec{u}$

$$P\left(\vec{u}; \kappa_0, \{\theta_i\}_{i=1}^{d_K}\right) = (\pi\kappa_0)^{-\frac{1}{2}d_K}\cdot\frac{\Gamma\left(\kappa_0+1+\dfrac{d_K}{2}\right)}{\Gamma(\kappa_0+1)}$$
$$\times \prod_{i=1}^{d_K}\theta_i^{-1}\cdot\left[1+\frac{1}{\kappa_0}\cdot\sum_{i=1}^{d_K}\frac{(u_i-u_{bi})^2}{\theta_i^2}\right]^{-\kappa_0-1-\frac{1}{2}d_K}, \qquad (4.29)$$

where $\theta_i = \sqrt{2k_B T_i/m}$, for $i: 1, ..., d_K$.

4.4.1.1.2 In Terms of the Kinetic Energy per Degree of Freedom $\varepsilon_{Ki} = \frac{1}{2}m(u_i - u_{bi})^2$ for i: 1, ..., d_K

$$P\left(\{\varepsilon_{Ki}\}_{i=1}^{d_K}; \kappa_0, \{T_i\}_{i=1}^{d_K}\right) = (\pi\kappa_0)^{-\frac{1}{2}d_K} \cdot \frac{\Gamma\left(\kappa_0 + 1 + \frac{d_K}{2}\right)}{\Gamma(\kappa_0 + 1)}$$

$$\times \prod_{i=1}^{d_K} (k_B T_i)^{-\frac{1}{2}} \cdot \left(1 + \frac{1}{\kappa_0} \cdot \sum_{i=1}^{d_K} \frac{\varepsilon_{Ki}}{k_B T_i}\right)^{-\kappa_0 - 1 - \frac{1}{2}d_K} \cdot \prod_{i=1}^{d_K} \varepsilon_{Ki}^{-\frac{1}{2}}.$$

(4.30a)

Normalization

$$\int_0^\infty \cdots \int_0^\infty P\left(\{\varepsilon_{Ki}\}_{i=1}^{d_K}; \kappa_0, \{T_i\}_{i=1}^{d_K}\right) d\varepsilon_{K1} \cdots d\varepsilon_{Kd_K} = 1.$$

(4.30b)

4.4.1.1.3 The Temperature is Given by

$$T = \langle T_i \rangle = \frac{1}{d_K} \sum_{i=1}^{d_K} T_i.$$

(4.31)

4.4.1.2 Distribution in Terms of the Velocity $\vec{u}$ for d_K Degrees of Freedom, Using the d_K-Dimensional Kappa Index $\kappa = \kappa_0 + \frac{1}{2}d_K$

$$P\left(\vec{u}; \kappa, \{\theta_i\}_{i=1}^{d_K}\right) = \left[\pi\left(\kappa - \frac{d_K}{2}\right)\right]^{-\frac{1}{2}d_K} \cdot \frac{\Gamma(\kappa + 1)}{\Gamma\left(\kappa - \frac{d_K}{2} + 1\right)}$$

$$\times \prod_{i=1}^{d_K} \theta_i^{-1} \cdot \left[1 + \frac{1}{\kappa - \frac{d_K}{2}} \cdot \sum_{i=1}^{d_K} \frac{(u_i - u_{bi})^2}{\theta_i^2}\right]^{-\kappa - 1}.$$

(4.32)

4.4.1.3 Distribution in Terms of the Velocity $\vec{u}$ for d_K Degrees of Freedom, Using the 3-D Kappa Index $\kappa_3 = \kappa_0 + \frac{1}{2}(d_K - 3)$

$$P\left(\vec{u}; \kappa_3, \{\theta_i\}_{i=1}^{d_K}\right) = \left[\pi\left(\kappa_3 - \frac{3}{2}\right)\right]^{-\frac{1}{2}d_K} \cdot \frac{\Gamma\left(\kappa_3 + 1 + \frac{d_K - 3}{2}\right)}{\Gamma\left(\kappa_3 - \frac{3}{2} + 1\right)}$$

$$\times \prod_{i=1}^{d_K} \theta_i^{-1} \cdot \left[1 + \frac{1}{\kappa_3 - \frac{3}{2}} \cdot \sum_{i=1}^{d_K} \frac{(u_i - u_{bi})^2}{\theta_i^2}\right]^{-\kappa_3 - 1 - \frac{1}{2}(d_K - 3)}.$$

(4.33)

4.4.1.4 Distributions, for $d_K = 3$ Degrees of Freedom, Using the Invariant Kappa Index κ_0

4.4.1.4.1 In Terms of the Velocity $\vec{u}$

$$P\left(\vec{u}; \kappa_0, \theta_x, \theta_y, \theta_z\right) = (\pi\kappa_0)^{-\frac{3}{2}} \cdot \frac{\Gamma\left(\kappa_0 + \frac{5}{2}\right)}{\Gamma(\kappa_0 + 1)} \cdot \theta_x^{-1}\theta_y^{-1}\theta_z^{-1}$$

$$\times \left\{1 + \frac{1}{\kappa_0} \cdot \left[\frac{(u_x - u_{bx})^2}{\theta_x^2} + \frac{(u_y - u_{by})^2}{\theta_y^2} + \frac{(u_z - u_{bz})^2}{\theta_z^2}\right]\right\}^{-\kappa_0 - \frac{5}{2}}.$$

(4.34)

4.4.1.4.2 In Terms of the Kinetic Energy Per Degree of Freedom $\varepsilon_{Ki} = \frac{1}{2}m(u_i - u_{bi})^2$ for i: x, y, z

$$P\left(\varepsilon_{Kx}, \varepsilon_{Ky}, \varepsilon_{Kz}; \kappa_0, T_x, T_y, T_z\right) = (\pi\kappa_0)^{-\frac{3}{2}} \cdot \frac{\Gamma\left(\kappa_0 + \frac{5}{2}\right)}{\Gamma(\kappa_0 + 1)} \cdot \left[(k_B T_x)(k_B T_y)(k_B T_z)\right]^{-\frac{1}{2}}$$

$$\times \left[1 + \frac{1}{\kappa_0} \cdot \left(\frac{\varepsilon_{Kx}}{k_B T_x} + \frac{\varepsilon_{Ky}}{k_B T_y} + \frac{\varepsilon_{Kz}}{k_B T_z}\right)\right]^{-\kappa_0 - \frac{5}{2}} \varepsilon_{Kx}^{-\frac{1}{2}}\varepsilon_{Ky}^{-\frac{1}{2}}\varepsilon_{Kz}^{-\frac{1}{2}}.$$

(4.35a)

Normalization

$$\int_0^\infty \int_0^\infty \int_0^\infty P(\varepsilon_{Kx}, \varepsilon_{Ky}, \varepsilon_{Kz}; \kappa_0, T_x, T_y, T_z) d\varepsilon_{Kx} d\varepsilon_{Ky} d\varepsilon_{Kz} = 1. \tag{4.35b}$$

4.4.1.5 Distribution, for $d_K = 3$ Degrees of Freedom, Using the 3-D Kappa Index $\kappa = \kappa_0 + \frac{3}{2}$

4.4.1.5.1 In Terms of the Velocity $\vec{u}$

$$P(\vec{u}; \kappa, \theta_x, \theta_y, \theta_z) = \left[\pi\left(\kappa - \frac{3}{2}\right)\right]^{-\frac{3}{2}} \cdot \frac{\Gamma(\kappa+1)}{\Gamma\left(\kappa - \frac{1}{2}\right)} \cdot \theta_x^{-1} \theta_y^{-1} \theta_z^{-1}$$

$$\times \left\{1 + \frac{1}{\kappa - \frac{3}{2}} \cdot \left[\frac{(u_x - u_{bx})^2}{\theta_x^2} + \frac{(u_y - u_{by})^2}{\theta_y^2} + \frac{(u_z - u_{bz})^2}{\theta_z^2}\right]\right\}^{-\kappa-1}.$$

$$\tag{4.36}$$

4.4.1.5.2 In Terms of the Kinetic Energy per Degree of Freedom ε_{Ki} for i: x, y, z

$$P(\varepsilon_{Kx}, \varepsilon_{Ky}, \varepsilon_{Kz}; \kappa, T_x, T_y, T_z) = \left[\pi\left(\kappa - \frac{3}{2}\right)\right]^{-\frac{3}{2}} \cdot \frac{\Gamma(\kappa+1)}{\Gamma\left(\kappa - \frac{1}{2}\right)} \cdot [(k_B T_x)(k_B T_y)(k_B T_z)]^{-\frac{1}{2}}$$

$$\times \left[1 + \frac{1}{\kappa - \frac{3}{2}} \cdot \left(\frac{\varepsilon_{Kx}}{k_B T_x} + \frac{\varepsilon_{Ky}}{k_B T_y} + \frac{\varepsilon_{Kz}}{k_B T_z}\right)\right]^{-\kappa-1} \varepsilon_{Kx}^{-\frac{1}{2}} \varepsilon_{Ky}^{-\frac{1}{2}} \varepsilon_{Kz}^{-\frac{1}{2}}.$$

$$\tag{4.37}$$

4.4.1.5.3 The Temperature is Given by

$$T = (T_x + T_y + T_z)/3. \tag{4.38}$$

4.4.2 Correlation Between the Projection at a Certain Direction and the Perpendicular Plane

This formalism is most frequently used for describing plasma populations in the presence of an ambient magnetic field.

4.4.2.1 Distributions, for $d_K = 3$ Degrees of Freedom, Using the Invariant Kappa Index κ_0

4.4.2.1.1 In Terms of the Velocity $\vec{u} = \left(u_{\parallel}, \vec{u}_{\perp} \right)$

$$P\left(\vec{u}; \kappa_0, \theta_{\parallel}, \theta_{\perp}\right) = (\pi\kappa_0)^{-\frac{3}{2}} \cdot \frac{\Gamma\left(\kappa_0 + \frac{5}{2}\right)}{\Gamma(\kappa_0 + 1)} \cdot \theta_{\parallel}^{-1}\theta_{\perp}^{-2}$$

$$\times \left\{ 1 + \frac{1}{\kappa_0} \cdot \left[\frac{\left(u_{\parallel} - u_{b\parallel}\right)^2}{\theta_{\parallel}^2} + \frac{\left(\vec{u}_{\perp} - \vec{u}_{b\perp}\right)^2}{\theta_{\perp}^2} \right] \right\}^{-\kappa_0 - \frac{5}{2}}.$$

(4.39a)

Normalization

$$\int_{-\infty}^{\infty} P\left(\vec{u}; \kappa_0, \theta_{\parallel}, \theta_{\perp}\right) d\vec{u} = 1.$$

(4.39b)

4.4.2.1.2 In Terms of the Kinetic Energy, $\varepsilon_{K\parallel} = \frac{1}{2}m\left(u_{\parallel} - u_{b\parallel}\right)^2$ $\varepsilon_{K\perp} = \frac{1}{2}m(\vec{u}_{\perp} - \vec{u}_{b\perp})^2$

$$P\left(\varepsilon_{K\parallel}, \varepsilon_{K\perp}; \kappa_0, T_{\parallel}, T_{\perp}\right) = \frac{\frac{1}{2}\kappa_0^{-\frac{3}{2}}}{B\left(\frac{3}{2}, \kappa_0 + 1\right)} \cdot \left(k_B T_{\parallel}\right)^{-\frac{1}{2}}(k_B T_{\perp})^{-1}$$

(4.40a)

$$\times \left[1 + \frac{1}{\kappa_0} \cdot \left(\frac{\varepsilon_{K\parallel}}{k_B T_{\parallel}} + \frac{\varepsilon_{K\perp}}{k_B T_{\perp}} \right) \right]^{-\kappa_0 - \frac{5}{2}} \varepsilon_{K\parallel}^{-\frac{1}{2}}.$$

Normalization

$$\int_0^{\infty} \int_0^{\infty} P\left(\varepsilon_{K\parallel}, \varepsilon_{K\perp}; \kappa_0, T_{\parallel}, T_{\perp}\right) d\varepsilon_{K\parallel} d\varepsilon_{K\perp} = 1.$$

(4.40b)

4.4.2.2 Distribution in Terms of the Velocity $\vec{u} = \left(u_\parallel, \vec{u}_\perp\right)$
for $d_K = 3$ Degrees of Freedom, Using the 3-D Kappa Index $\kappa = \kappa_0 + \frac{3}{2}$

$$P\left(\vec{u}; \kappa, \theta_\parallel, \theta_\perp\right) = \left[\pi\left(\kappa - \frac{3}{2}\right)\right]^{-\frac{3}{2}} \cdot \frac{\Gamma(\kappa + 1)}{\Gamma\left(\kappa - \frac{1}{2}\right)} \cdot \theta_\parallel^{-1} \theta_\perp^{-2}$$

$$\times \left\{1 + \frac{1}{\kappa - \frac{3}{2}} \cdot \left[\frac{\left(u_\parallel - u_{b\parallel}\right)^2}{\theta_\parallel^2} + \frac{\left(\vec{u}_\perp - \vec{u}_{b\perp}\right)^2}{\theta_\perp^2}\right]\right\}^{-\kappa - 1}.$$

$$(4.41)$$

4.4.2.3 The Temperature is Given by

$$T = \left(T_\parallel + 2T_\perp\right)/3. \qquad (4.42)$$

4.4.3 Self-Correlated Degrees of Freedom

4.4.3.1 Distributions, for d_K Degrees of Freedom, Using the Invariant Kappa Index κ_0

4.4.3.1.1 In Terms of the Velocity $\vec{u}$

$$P\left(\vec{u}; \kappa_0, \{\theta_i\}_{i=1}^{d_K}\right) = \prod_{i=1}^{d_K} P(u_i; \kappa_0, \theta_i)$$

$$= (\pi\kappa_0)^{-\frac{1}{2}d_K} \cdot \left[\frac{\Gamma\left(\kappa_0 + \frac{3}{2}\right)}{\Gamma(\kappa_0 + 1)}\right]^{d_K} \qquad (4.43)$$

$$\times \prod_{i=1}^{d_K}\left\{\theta_i^{-1}\left[1 + \frac{1}{\kappa_0} \cdot \frac{(u_i - u_{bi})^2}{\theta_i^2}\right]^{-\kappa_0 - \frac{3}{2}}\right\}.$$

4.4.3.1.2 In Terms of the Kinetic Energy per Degree of Freedom $\varepsilon_{Ki} = \frac{1}{2}m(u_i - u_{bi})^2$ for i: 1, ..., d_K

$$P\left(\{\varepsilon_{Ki}\}_{i=1}^{d_K}; \kappa_0, \{T_i\}_{i=1}^{d_K}\right) = \prod_{i=1}^{d_K} P(\varepsilon_{Ki}; \kappa_0, T_i) = (\pi\kappa_0)^{-\frac{1}{2}d_K} \cdot \left[\frac{\Gamma\left(\kappa_0 + \frac{3}{2}\right)}{\Gamma(\kappa_0 + 1)}\right]^{d_K}$$

$$\times \prod_{i=1}^{d_K}(k_B T_i)^{-\frac{1}{2}} \prod_{i=1}^{d_K}\left[1 + \frac{1}{\kappa_0} \cdot \frac{\varepsilon_{Ki}}{k_B T_i}\right]^{-\kappa_0 - \frac{3}{2}} \cdot \varepsilon_{Ki}^{-\frac{1}{2}}.$$

$$(4.44)$$

4.4.3.1.3 The Temperature is Given by

$$T = \langle T_i \rangle = \frac{1}{d_K} \sum_{i=1}^{d_K} T_i. \tag{4.45}$$

4.4.3.2 Distribution in Terms of the Velocity, $\vec{u}$, for d_K Degrees of Freedom, Using the d_K-Dimensional Kappa Index $\kappa = \kappa_0 + \frac{1}{2}d_K$

$$P\left(\vec{u}; \kappa, \{\theta_i\}_{i=1}^{d_K}\right) = \left[\pi\left(\kappa - \frac{d_K}{2}\right)\right]^{-\frac{1}{2}d_K} \cdot \left[\frac{\Gamma\left(\kappa + 1 - \frac{d_K - 1}{2}\right)}{\Gamma\left(\kappa - \frac{d_K}{2} + 1\right)}\right]^{d_K}$$

$$\times \prod_{i=1}^{d_K}\left\{\theta_i^{-1}\left[1 + \frac{1}{\kappa - \frac{d_K}{2}} \cdot \frac{(u_i - u_{bi})^2}{\theta_i^2}\right]^{-\kappa + \frac{1}{2}(d_K - 3)}\right\}. \tag{4.46}$$

4.4.3.3 Distribution in Terms of the Velocity, $\vec{u}$, for d_K Degrees of Freedom, Using the 3-D Kappa Index $\kappa = \kappa_0 + \frac{3}{2}$

$$P\left(\vec{u}; \kappa_3, \{\theta_i\}_{i=1}^{d_K}\right) = \left[\pi\left(\kappa_3 - \frac{3}{2}\right)\right]^{-\frac{1}{2}d_K} \cdot \left[\frac{\Gamma(\kappa_3)}{\Gamma\left(\kappa_3 - \frac{1}{2}\right)}\right]^{d_K}$$

$$\times \prod_{i=1}^{d_K}\left\{\theta_i^{-1}\left[1 + \frac{1}{\kappa_3 - \frac{3}{2}} \cdot \frac{(u_i - u_{bi})^2}{\theta_i^2}\right]^{-\kappa_3}\right\}. \tag{4.47}$$

4.4.3.4 Distributions, for $d_K = 3$ Degrees of Freedom, Using the Invariant Kappa Index κ_0

4.4.3.4.1 In Terms of the Velocity $\vec{u}$

$$P\left(\vec{u}; \kappa_0, \theta_x, \theta_y, \theta_z\right) = (\pi\kappa_0)^{-\frac{3}{2}} \cdot \frac{\Gamma\left(\kappa_0 + \frac{3}{2}\right)^3}{\Gamma(\kappa_0 + 1)^3} \cdot \theta_x^{-1}\theta_y^{-1}\theta_z^{-1} \cdot \left[1 + \frac{1}{\kappa_0} \cdot \frac{(u_x - u_{b_x})^2}{\theta_x^2}\right]^{-\kappa_0 - \frac{3}{2}}$$

$$\times \left[1 + \frac{1}{\kappa_0} \cdot \frac{(u_y - u_{b_y})^2}{\theta_y^2}\right]^{-\kappa_0 - \frac{3}{2}} \cdot \left[1 + \frac{1}{\kappa_0} \cdot \frac{(u_z - u_{b_z})^2}{\theta_z^2}\right]^{-\kappa_0 - \frac{3}{2}}. \tag{4.48}$$

4.4.3.4.2 In Terms of the Kinetic Energy per Degree of Freedom
$\varepsilon_{Ki} = \frac{1}{2}m(u_i - u_{bi})^2$ for i: x, y, z

$$P(\varepsilon_{Kx}, \varepsilon_{Ky}, \varepsilon_{Kz}; \kappa_0, T_x, T_y, T_z) = (\pi\kappa_0)^{-\frac{3}{2}} \cdot \frac{\Gamma\left(\kappa_0 + \frac{3}{2}\right)^3}{\Gamma(\kappa_0 + 1)^3} \cdot [(k_B T_x)(k_B T_y)(k_B T_z)]^{-\frac{1}{2}}$$

$$\times \left(1 + \frac{1}{\kappa_0} \cdot \frac{\varepsilon_{Kx}}{k_B T_x}\right)^{-\kappa_0 - \frac{3}{2}} \left(1 + \frac{1}{\kappa_0} \cdot \frac{\varepsilon_{Ky}}{k_B T_y}\right)^{-\kappa_0 - \frac{3}{2}}$$

$$\times \left(1 + \frac{1}{\kappa_0} \cdot \frac{\varepsilon_{Kz}}{k_B T_z}\right)^{-\kappa_0 - \frac{3}{2}} \varepsilon_{Kx}^{-\frac{1}{2}} \varepsilon_{Ky}^{-\frac{1}{2}} \varepsilon_{Kz}^{-\frac{1}{2}}.$$

$$(4.49)$$

4.4.3.5 Distribution, for $d_K = 3$ Degrees of Freedom, Using the 3-D Kappa Index $\kappa = \kappa_0 + \frac{3}{2}$

4.4.3.5.1 In Terms of the Velocity $\vec{u}$

$$P(\vec{u}; \kappa, \theta_x, \theta_y, \theta_z) = \left[\pi\left(\kappa - \frac{3}{2}\right)\right]^{-\frac{3}{2}} \cdot \frac{\Gamma(\kappa)^3}{\Gamma\left(\kappa - \frac{1}{2}\right)^3} \cdot \theta_x^{-1} \theta_y^{-1} \theta_z^{-1}$$

$$\times \left[1 + \frac{1}{\kappa - \frac{3}{2}} \cdot \frac{(u_x - u_{b_x})^2}{\theta_x^2}\right]^{-\kappa} \cdot \left[1 + \frac{1}{\kappa - \frac{3}{2}} \cdot \frac{(u_y - u_{b_y})^2}{\theta_y^2}\right]^{-\kappa}$$

$$\times \left[1 + \frac{1}{\kappa - \frac{3}{2}} \cdot \frac{(u_z - u_{b_z})^2}{\theta_z^2}\right]^{-\kappa}.$$

$$(4.50)$$

4.4.3.5.2 In Terms of the Kinetic Energy per Degree of Freedom
ε_{Ki} for i: x, y, z

$$P(\varepsilon_{Kx}, \varepsilon_{Ky}, \varepsilon_{Kz}; \kappa, T_x, T_y, T_z) = \left[\pi\left(\kappa - \frac{3}{2}\right)\right]^{-\frac{3}{2}} \cdot \frac{\Gamma(\kappa)^3}{\Gamma\left(\kappa - \frac{1}{2}\right)^3} \cdot [(k_B T_x)(k_B T_y)(k_B T_z)]^{-\frac{1}{2}}$$

$$\times \left(1 + \frac{1}{\kappa - \frac{3}{2}} \cdot \frac{\varepsilon_{Kx}}{k_B T_x}\right)^{-\kappa} \left(1 + \frac{1}{\kappa - \frac{3}{2}} \cdot \frac{\varepsilon_{Ky}}{k_B T_y}\right)^{-\kappa}$$

$$\times \left(1 + \frac{1}{\kappa - \frac{3}{2}} \cdot \frac{\varepsilon_{Kz}}{k_B T_z}\right)^{-\kappa} \varepsilon_{Kx}^{-\frac{1}{2}} \varepsilon_{Ky}^{-\frac{1}{2}} \varepsilon_{Kz}^{-\frac{1}{2}}.$$

$$(4.51)$$

4.4.3.5.3 The Temperature is Given by

$$T = (T_x + T_y + T_z)/3. \tag{4.52}$$

Comment: The exponent of each component is $-\kappa$ instead of $-\kappa - 1$. This is because each of the direction is described by a 1-D kappa distribution. As it is mentioned, the kappa index depends on the degrees of freedom, i.e., $\kappa(d_K) = \kappa_0 + \frac{1}{2}d_K$; hence, $\kappa(d_K) = \kappa + \frac{1}{2}(d_K - 3)$, where we symbolize with κ the standard d_K-dimensional kappa index and with κ_3 the 3-D kappa index. Thus, the exponent of the d_K-dimensional kappa distribution is $-\kappa(d_K) - 1$, which, in terms of κ_3, becomes $-\kappa_3 - 1 - \frac{1}{2}(d_K - 3)$. For $d_K = 1$, the exponent becomes $-\kappa_3$, while for $d_K = 2$, it becomes $-\kappa_3 - \frac{1}{2}$.

4.4.4 Self-Correlated Projections at a Direction and Perpendicular Plane

This formalism may be used in cases of very strong ambient magnetic fields. It has been introduced in Lazar and Poedts (2014) and called "product-bi-kappa distributions" (e.g., Lazar and Poedts, 2014; dos Santos et al., 2015).

4.4.4.1 Distributions, for $d_K = 3$ Degrees of Freedom, Using the Invariant Kappa Index κ_0

4.4.4.1.1 In Terms of the Velocity $\vec{u} = \left(u_\parallel, \vec{u}_\perp \right)$

$$P\left(\vec{u}; \kappa_0, \theta_\parallel, \theta_\perp \right) = (\pi \kappa_0)^{-\frac{3}{2}} \cdot \frac{\Gamma\left(\kappa_0 + \frac{3}{2} \right)(\kappa_0 + 1)}{\Gamma(\kappa_0 + 1)} \cdot \theta_\parallel^{-1} \theta_\perp^{-2}$$

$$\times \left[1 + \frac{1}{\kappa_0} \cdot \frac{\left(u_\parallel - u_{b\parallel} \right)^2}{\theta_\parallel^2} \right]^{-\kappa_0 - \frac{3}{2}} \tag{4.53}$$

$$\times \left[1 + \frac{1}{\kappa_0} \cdot \frac{\left(\vec{u}_\perp - \vec{u}_{b\perp} \right)^2}{\theta_\perp^2} \right]^{-\kappa_0 - 2}.$$

4.4.4.1.2 In Terms of the Kinetic Energy, $\varepsilon_{K\parallel} = \frac{1}{2} m \left(u_\parallel - u_{b\parallel} \right)^2$

$\varepsilon_{K\perp} = \frac{1}{2} m (\vec{u}_\perp - \vec{u}_{b\perp})^2$

$$P\left(\varepsilon_{K\parallel}, \varepsilon_{K\perp}; \kappa_0, T_\parallel, T_\perp \right) = \frac{(\kappa_0 + 1) \cdot \kappa_0^{-\frac{3}{2}}}{B\left(\frac{1}{2}, \kappa_0 + 1 \right)} \cdot \left(k_B T_\parallel \right)^{-\frac{1}{2}} (k_B T_\perp)^{-1}$$

$$\times \left[1 + \frac{1}{\kappa_0} \cdot \frac{\varepsilon_{K\parallel}}{k_B T_\parallel} \right]^{-\kappa_0 - \frac{3}{2}} \cdot \left[1 + \frac{1}{\kappa_0} \cdot \frac{\varepsilon_{K\perp}}{k_B T_\perp} \right]^{-\kappa_0 - 2} \varepsilon_{K\parallel}^{-\frac{1}{2}}.$$

$$\tag{4.54}$$

4.4.4.2 Distribution in Terms of the Velocity $\vec{u} = \left(u_{\parallel}, \vec{u}_{\perp}\right)$ *for* $d_K = 3$
Degrees of Freedom, Using the 3-D Kappa Index $\kappa = \kappa_0 + \frac{3}{2}$

$$P\left(\vec{u}; \kappa, \theta_{\parallel}, \theta_{\perp}\right) = \left[\pi\left(\kappa - \frac{3}{2}\right)\right]^{-\frac{3}{2}} \cdot \frac{\Gamma(\kappa)\left(\kappa - \frac{1}{2}\right)}{\Gamma\left(\kappa - \frac{1}{2}\right)} \cdot \theta_{\parallel}^{-1}\theta_{\perp}^{-2}$$

$$\times \left[1 + \frac{1}{\kappa - \frac{3}{2}} \cdot \frac{\left(u_{\parallel} - u_{b\parallel}\right)^2}{\theta_{\parallel}^2}\right]^{-\kappa} \tag{4.55}$$

$$\times \left[1 + \frac{1}{\kappa - \frac{3}{2}} \cdot \frac{\left(\vec{u}_{\perp} - \vec{u}_{b\perp}\right)^2}{\theta_{\perp}^2}\right]^{-\kappa - \frac{1}{2}}.$$

4.4.4.3 The Temperature is Given by

$$T = \left(T_{\parallel} + 2T_{\perp}\right)\big/3. \tag{4.56}$$

Comment: The exponent of the 1-D kappa distribution (left-hand side) is $-\kappa$, and the exponent of the 2-D kappa distribution (right-hand side) is $-\kappa - \frac{1}{2}$.

4.4.5 Self-Correlated Degrees of Freedom With Different Kappa

4.4.5.1 Distributions, for d_K Degrees of Freedom, Using the Invariant Kappa Index κ_0

4.4.5.1.1 In Terms of the Velocity $\vec{u}$

$$P\left(\vec{u}; \{\kappa_{0i}\}_{i=1}^{d_K}, \{\theta_i\}_{i=1}^{d_K}\right) = \prod_{i=1}^{d_K} P(u_i; \kappa_{0i}, \theta_i)$$

$$= \prod_{i=1}^{d_K} \left\{ \left(\pi\kappa_{0i}\theta_i^2\right)^{-\frac{1}{2}} \cdot \frac{\Gamma\left(\kappa_{0i} + \frac{3}{2}\right)}{\Gamma(\kappa_{0i} + 1)} \right.$$

$$\left. \times \left[1 + \frac{1}{\kappa_{0i}} \cdot \frac{(u_i - u_{bi})^2}{\theta_i^2}\right]^{-\kappa_{0i} - \frac{3}{2}} \right\}. \tag{4.57}$$

4.4.5.1.2 In Terms of the Kinetic Energy per Degree of Freedom $\varepsilon_{Ki} = \frac{1}{2}m(u_i - u_{bi})^2$ for i: 1, ..., d_K

$$P\left(\{\varepsilon_{Ki}\}_{i=1}^{d_K}; \{\kappa_{0i}\}_{i=1}^{d_K}, \{T_i\}_{i=1}^{d_K}\right) = \prod_{i=1}^{d_K} P(\varepsilon_{Ki}; \kappa_{0i}, T_i)$$

$$= \prod_{i=1}^{d_K} \left\{ (\pi\kappa_{0i})^{-\frac{1}{2}d_K} \cdot \left[\frac{\Gamma\left(\kappa_{0i} + \frac{3}{2}\right)}{\Gamma(\kappa_{0i} + 1)} \right]^{d_K} \right.$$

$$\left. \times (k_B T_i)^{-\frac{1}{2}} \cdot \left(1 + \frac{1}{\kappa_{0i}} \cdot \frac{\varepsilon_{Ki}}{k_B T_i}\right)^{-\kappa_{0i} - \frac{3}{2}} \cdot \varepsilon_{Ki}^{-\frac{1}{2}} \right\}.$$

$$(4.58)$$

4.4.5.1.3 The Temperature is Given by

$$T = \langle T_i \rangle = \frac{1}{d_K} \sum_{i=1}^{d_K} T_i. \tag{4.59}$$

4.4.5.2 Distribution in Terms of the Velocity $\vec{u}$ for d_K Degrees of Freedom, Using the 1-D Kappa Index $\kappa_i = \kappa_{0i} + \frac{1}{2}$

$$P\left(\vec{u}; \{\kappa_i\}_{i=1}^{d_K}, \{\theta_i\}_{i=1}^{d_K}\right) = \prod_{i=1}^{d_K} \left\{ \left[\pi\left(\kappa_i - \frac{1}{2}\right)\theta_i^2 \right]^{-\frac{1}{2}} \cdot \frac{\Gamma(\kappa_i + 1)}{\Gamma\left(\kappa_i + \frac{1}{2}\right)} \right.$$

$$\left. \times \left[1 + \frac{1}{\kappa_i - \frac{1}{2}} \frac{(u_i - u_{bi})^2}{\theta_i^2} \right]^{-\kappa_i - 1} \right\}. \tag{4.60}$$

4.4.5.3 Distribution in Terms of the Velocity $\vec{u}$ for d_K Degrees of Freedom, Using the 3-D Kappa Index $\kappa = \kappa_0 + \frac{3}{2}$

$$P\left(\vec{u}; \{\kappa_{3i}\}_{i=1}^{d_K}, \{\theta_i\}_{i=1}^{d_K}\right) = \prod_{i=1}^{d_K} \left\{ \left[\pi\left(\kappa_{3i} - \frac{3}{2}\right)\theta_i^2 \right]^{-\frac{1}{2}} \cdot \frac{\Gamma(\kappa_{3i})}{\Gamma\left(\kappa_{3i} - \frac{1}{2}\right)} \right.$$

$$\left. \times \left[1 + \frac{1}{\kappa_{3i} - \frac{3}{2}} \frac{(u_i - u_{bi})^2}{\theta_i^2} \right]^{-\kappa_{3i}} \right\}. \tag{4.61}$$

4.4.6 Self-Correlated Projections at a Direction and Perpendicular Plane With Different Kappa

4.4.6.1 Distributions, for $d_K = 3$ Degrees of Freedom, Using the Invariant Kappa Index κ_0

4.4.6.1.1 In Terms of the Velocity $\vec{u} = \left(u_\|, \vec{u}_\perp \right)$

$$P\left(\vec{u}; \kappa_{\|0}, \kappa_{\perp 0}, \theta_\|, \theta_\perp\right) = \pi^{-\frac{3}{2}}\left(1 + \kappa_{\perp 0}^{-1}\right) \cdot \kappa_{\|0}^{-\frac{1}{2}} \frac{\Gamma\left(\kappa_{\|0} + \frac{3}{2}\right)}{\Gamma\left(\kappa_{\|0} + 1\right)} \cdot \theta_\|^{-1} \theta_\perp^{-2}$$

$$\times \left[1 + \frac{1}{\kappa_{\|0}} \cdot \frac{\left(u_\| - u_{b\|}\right)^2}{\theta_\|^2}\right]^{-\kappa_{\|0}-\frac{3}{2}}$$

$$\times \left[1 + \frac{1}{\kappa_{\perp 0}} \cdot \frac{\left(\vec{u}_\perp - \vec{u}_{b\perp}\right)^2}{\theta_\perp^2}\right]^{-\kappa_{\perp 0}-2}. \tag{4.62}$$

4.4.6.1.2 In Terms of the Kinetic Energy, $\varepsilon_{K\|} = \frac{1}{2}m\left(u_\| - u_{b\|}\right)^2$, $\varepsilon_{K\perp} = \frac{1}{2}m(\vec{u}_\perp - \vec{u}_{b\perp})^2$ (also see Eq. (4.54))

$$P\left(\varepsilon_{K\|}, \varepsilon_{K\perp}; \kappa_{\|0}, \kappa_{\perp 0}, T_\|, T_\perp\right) = \frac{\left(1 + \kappa_{\perp 0}^{-1}\right) \cdot \kappa_{\|0}^{-\frac{1}{2}}}{B\left(\frac{1}{2}, \kappa_{\|0} + 1\right)} \cdot \left(k_B T_\|\right)^{-\frac{1}{2}} (k_B T_\perp)^{-1}$$

$$\times \left[1 + \frac{1}{\kappa_{\|0}} \cdot \frac{\varepsilon_{K\|}}{k_B T_\|}\right]^{-\kappa_{\|0}-\frac{3}{2}} \cdot \left[1 + \frac{1}{\kappa_{\perp 0}} \frac{\varepsilon_{K\perp}}{k_B T_\perp}\right]^{-\kappa_{\perp 0}-2} \varepsilon_{K\|}^{-\frac{1}{2}}.$$

$$\tag{4.63}$$

4.4.6.2 Distribution in Terms of the Velocity $\vec{u} = \left(u_\|, \vec{u}_\perp \right)$ for $d_K = 3$ Degrees of Freedom, Using the Kappa Indices $\kappa_\| = \kappa_{\|0} + \frac{1}{2}$, $\kappa_\perp = \kappa_{\perp 0} + 1$

$$P\left(\vec{u}; \kappa_\|, \kappa_\perp, \theta_\|, \theta_\perp\right) = \pi^{-\frac{3}{2}} \frac{\kappa_\perp}{\kappa_\perp - 1} \cdot \left(\kappa_\| - \frac{1}{2}\right)^{-\frac{1}{2}} \frac{\Gamma\left(\kappa_\| + 1\right)}{\Gamma\left(\kappa_\| + \frac{1}{2}\right)} \cdot \theta_\|^{-1} \theta_\perp^{-2}$$

$$\times \left[1 + \frac{1}{\kappa_\| - \frac{1}{2}} \cdot \frac{\left(u_\| - u_{b\|}\right)^2}{\theta_\|^2}\right]^{-\kappa_\|-1}$$

$$\times \left[1 + \frac{1}{\kappa_\perp - 1} \cdot \frac{\left(\vec{u}_\perp - \vec{u}_{b\perp}\right)^2}{\theta_\perp^2}\right]^{-\kappa_\perp-1}. \tag{4.64}$$

4.4.6.3 The Temperature is Given by

$$T = \left(T_{\parallel} + 2T_{\perp}\right)\Big/3.$$

(4.65)

4.4.7 Self-Correlated Projections of Different Dimensionality and Kappa

4.4.7.1 Distributions, for d_K Degrees of Freedom of Mixed Correlation, i.e., i: 1, ..., M Uncorrelated Groups of Correlated Degrees of Freedom, With Each Group Having Different Degrees of Freedom f_i, Using Different (Invariant) Kappa Index, κ_{0i}, but Common Temperature T or θ

4.4.7.1.1 In Terms of the Velocity $\vec{u}$

$$P\left(\vec{u}; \{\kappa_{0i}\}_{i=1}^{M}, \theta\right) = \prod_{i=1}^{M}\left\{(\pi\kappa_{0i}\theta^2)^{-\frac{1}{2}f_i} \cdot \frac{\Gamma\left(\kappa_{0i} + 1 + \frac{f_i}{2}\right)}{\Gamma(\kappa_{0i} + 1)}\right.$$

$$\left. \times \left[1 + \frac{1}{\kappa_{0i}} \cdot \frac{(\vec{u} - \vec{u}_b)^2\big|_{(f_i)}}{\theta^2}\right]^{-\kappa_{0i} - 1 - \frac{1}{2}f_i}\right\},$$

(4.66a)

where we set the magnitude of the vector $\vec{u} - \vec{u}_b$ as

$$(\vec{u} - \vec{u}_b)^2 = \sum_{j=1}^{d_K}(u_j - u_{bj})^2 = \sum_{i=1}^{M}(\vec{u} - \vec{u}_b)^2\big|_{(f_i)}$$

(4.66b)

and

$$(\vec{u} - \vec{u}_b)^2\big|_{(f_i)} = \sum_{j=j_{bef}+1}^{j_{bef}+f_i}(u_j - u_{bj})^2, \quad j_{bef} = \begin{cases} \sum_{i'=1}^{i-1} f_{i'} & \text{if } i > 0, \\ 0 & \text{if } i = 0. \end{cases}$$

4.4.7.1.2 In Terms of the Kinetic Energy $\varepsilon_{Ki} = \frac{1}{2}m(\vec{u} - \vec{u}_b)^2\big|_{(f_i)}$ for i: 1, ..., M

$$P\left(\{\varepsilon_{Ki}\}_{i=1}^{M}; \{\kappa_{0i}\}_{i=1}^{M}, T\right) = \prod_{i=1}^{M}\left[\frac{(\kappa_{0i}k_BT)^{-\frac{1}{2}f_i}}{B\left(\frac{f_i}{2}, \kappa_{0i} + 1\right)} \cdot \left(1 + \frac{1}{\kappa_{0i}} \cdot \frac{\varepsilon_{Ki}}{k_BT}\right)^{-\kappa_{0i} - 1 - \frac{1}{2}f_i} \varepsilon_{Ki}^{\frac{1}{2}f_i - 1}\right].$$

(4.67)

4.4.7.1.3 The Total Degrees of Freedom d_K Are Given by

$$d_K = \sum_{i=1}^{M} f_i. \tag{4.68}$$

4.4.7.2 Distributions, for M Correlated Groups, Each With f_i Degrees of Freedom, Self-Correlated With Kappa Index κ_{0i} and Temperature θ_i

4.4.7.2.1 In Terms of the Velocity $\vec{u}$

$$P\left(\vec{u}; \{\kappa_{0i}\}_{i=1}^{M}, \{\theta_i\}_{i=1}^{M}\right) = \prod_{i=1}^{M} \left\{ \left(\pi\kappa_{0i}\theta_i^2\right)^{-\frac{1}{2}f_i} \cdot \frac{\Gamma\left(\kappa_{0i} + 1 + \frac{f_i}{2}\right)}{\Gamma(\kappa_{0i} + 1)} \right.$$

$$\left. \times \left[1 + \frac{1}{\kappa_{0i}} \cdot \frac{(\vec{u} - \vec{u}_b)^2 \big|_{(f_i)}}{\theta_i^2} \right]^{-\kappa_{0i} - 1 - \frac{1}{2}f_i} \right\}. \tag{4.69}$$

4.4.7.2.2 In Terms of the Kinetic Energy $\varepsilon_{Ki} = \frac{1}{2}m(\vec{u} - \vec{u}_b)^2 \big|_{(f_i)}$

$$P\left(\{\varepsilon_{Ki}\}_{i=1}^{M}; \{\kappa_{0i}\}_{i=1}^{M}, \{T_i\}_{i=1}^{M}\right) = \prod_{i=1}^{M} \left[\frac{(\kappa_{0i}k_B T_i)^{-\frac{1}{2}f_i}}{B\left(\frac{f_i}{2}, \kappa_{0i} + 1\right)} \cdot \left(1 + \frac{1}{\kappa_{0i}} \cdot \frac{\varepsilon_{Ki}}{k_B T_i}\right)^{-\kappa_{0i} - 1 - \frac{1}{2}f_i} \varepsilon_{Ki}^{\frac{1}{2}f_i - 1} \right]. \tag{4.70}$$

4.4.7.2.3 The Total Degrees of Freedom d_K Are Given by

$$d_K = \sum_{i=1}^{M} f_i. \tag{4.71}$$

4.4.7.2.4 The Temperature is Given by

$$T = \langle T_i \rangle = \frac{1}{d_K} \sum_{i=1}^{M} f_i T_i. \tag{4.72}$$

4.4.8 Different Self-Correlation and Intercorrelation Between Degrees of Freedom

4.4.8.1 Distributions, for M Correlated Groups, Each With f_i Degrees of Freedom, Self-Correlated With Invariant Kappa Index, κ_{0i}, and Intercorrelated With Invariant Kappa Index κ_{0i}^{int}

4.4.8.1.1 In Terms of the Velocity $\vec{u}$

$$P\left(\vec{u};\{\kappa_{0i}\}_{i=1}^{M},\{\kappa_{0i}^{int}\}_{i=1}^{M},\kappa_{0}^{int},\{\theta_i\}_{i=1}^{M}\right)=\frac{1}{Z}$$

$$\times\left\{1-\sum_{i=1}^{M}\frac{\kappa_{0i}^{int}+\frac{f_i}{2}}{\kappa_{0}^{int}+\frac{d_K}{2}}\cdot\left[1-\left(1+\frac{1}{\kappa_{0i}+\frac{f_i}{2}}\cdot\frac{(\vec{u}-\vec{u}_b)^2\big|_{(f_i)}-\frac{f_i}{2}\theta_i^2}{\theta_i^2}\right)^{\frac{\kappa_{0i}+1+\frac{f_i}{2}}{\kappa_{0i}^{int}+1+\frac{f_i}{2}}}\right]\right\}^{-\kappa_{0}^{int}-1-\frac{1}{2}d_K}$$

(4.73a)

The partition function is given by the normalization:

$$Z=\int_{-\infty}^{\infty}\left\{1-\sum_{i=1}^{M}\frac{\kappa_{0i}^{int}+\frac{f_i}{2}}{\kappa_{0}^{int}+\frac{d_K}{2}}\cdot\left[1-\left(1+\frac{1}{\kappa_{0i}+\frac{f_i}{2}}\cdot\frac{(\vec{u}-\vec{u}_b)^2\big|_{(f_i)}-\frac{f_i}{2}\theta_i^2}{\theta_i^2}\right)^{\frac{\kappa_{0i}+1+\frac{f_i}{2}}{\kappa_{0i}^{int}+1+\frac{f_i}{2}}}\right]\right\}^{-\kappa_{0}^{int}-1-\frac{1}{2}d_K}d\vec{u}.$$

(4.73b)

Comment: The kappa index κ_0^{int} characterizes overall the intercorrelation. While it may be used as an independent parameter, it must be aligned to the following conditions: (1) if $\kappa_{0i}^{int}=0$ for *every* i, then $\kappa_0^{int}=0$; (2) if $\kappa_{0i}^{int}\to\infty$ for *any* i, then $\kappa_0^{int}\to\infty$; (3) if $\kappa_{0i}^{int}=$ *constant* for any i, then $\kappa_{0i}^{int}=\kappa_0^{int}$.

4.4.8.1.2 In Terms of the Kinetic Energy $\varepsilon_{Ki} = \frac{1}{2}m(\vec{u} - \vec{u}_b)^2\big|_{(f_i)}$

$$P\left(\{\varepsilon_{Ki}\}_{i=1}^M; \{\kappa_{0i}\}_{i=1}^M, \{\kappa_{0i}^{\mathrm{int}}\}_{i=1}^M, \kappa_0^{\mathrm{int}}, \{T_i\}_{i=1}^M\right)$$

$$= \frac{1}{Z} \cdot \prod_{i=1}^M \varepsilon_{Ki}^{\frac{1}{2}f_i - 1} \cdot \left\{ 1 - \sum_{i=1}^M \frac{\kappa_{0i}^{\mathrm{int}} + \frac{f_i}{2}}{\kappa_0^{\mathrm{int}} + \frac{d_K}{2}} \cdot \left[1 - \left(1 + \frac{1}{\kappa_{0i} + \frac{f_i}{2}} \cdot \frac{\varepsilon_{Ki} - \frac{f_i}{2}k_B T_i}{k_B T_i} \right)^{\frac{\kappa_{0i} + 1 + \frac{f_i}{2}}{\kappa_{0i}^{\mathrm{int}} + 1 + \frac{f_i}{2}}} \right] \right\}^{-\kappa_0^{\mathrm{int}} - 1 - \frac{1}{2}d_K}$$

(4.74a)

The partition function is given by the normalization:

$$Z = \int_0^\infty \cdots \int_0^\infty \{\cdots\}^{-\kappa_0^{\mathrm{int}} - 1 - \frac{1}{2}d_K} \prod_{i=1}^M \varepsilon_{Ki}^{\frac{1}{2}f_i - 1} d\varepsilon_{K1} \cdots d\varepsilon_{KM}$$

$$\text{where } \{\cdots\} = \left\{ 1 - \sum_{i=1}^M \frac{\kappa_{0i}^{\mathrm{int}} + \frac{f_i}{2}}{\kappa_0^{\mathrm{int}} + \frac{d_K}{2}} \cdot \left[1 - \left(1 + \frac{1}{\kappa_{0i} + \frac{f_i}{2}} \cdot \frac{\varepsilon_{Ki} - \frac{f_i}{2}k_B T_i}{k_B T_i} \right)^{\frac{\kappa_{0i} + 1 + \frac{f_i}{2}}{\kappa_{0i}^{\mathrm{int}} + 1 + \frac{f_i}{2}}} \right] \right\}^{-\kappa_0^{\mathrm{int}} - 1 - \frac{1}{2}d_K}$$

(4.74b)

4.4.8.2 Distributions, for M Correlated Groups, Each With f_i Degrees of Freedom, Self-Correlated With Kappa Index, κ_i, and Intercorrelated With Kappa Index κ_i^{int}

4.4.8.2.1 In Terms of the Velocity $\vec{u}$

$$P\left(\vec{u}; \{\kappa_i\}_{i=1}^M, \{\kappa_i^{\mathrm{int}}\}_{i=1}^M, \kappa^{\mathrm{int}}, \{\theta_i\}_{i=1}^M\right)$$

$$= \frac{1}{Z} \cdot \left\{ 1 - \sum_{i=1}^M \frac{\kappa_i^{\mathrm{int}}}{\kappa^{\mathrm{int}}} \cdot \left\{ 1 - \left[1 + \frac{1}{\kappa_i} \cdot \frac{(\vec{u} - \vec{u}_b)^2\big|_{(f_i)} - \frac{f_i}{2}\theta_i^2}{\theta_i^2} \right]^{\frac{\kappa_i + 1}{\kappa_i^{\mathrm{int}} + 1}} \right\} \right\}^{-\kappa^{\mathrm{int}} - 1}$$

(4.75)

(Partition functions as above.)

4.4.8.2.2 In Terms of the Kinetic Energy $\varepsilon_{Ki} = \frac{1}{2}m(\vec{u} - \vec{u}_b)^2\Big|_{(f_i)}$

$$P\left(\{\varepsilon_{Ki}\}_{i=1}^M; \{\kappa_i\}_{i=1}^M, \{\kappa_i^{int}\}_{i=1}^M, \kappa^{int}, \{T_i\}_{i=1}^M\right)$$

$$= \frac{1}{Z} \cdot \prod_{i=1}^M \varepsilon_{Ki}^{\frac{1}{2}f_i - 1} \cdot \left\{ 1 - \sum_{i=1}^M \frac{\kappa_i^{int}}{\kappa^{int}} \cdot \left[1 - \left(1 + \frac{1}{\kappa_i} \cdot \frac{\varepsilon_{Ki} - \frac{f_i}{2}k_B T_i}{k_B T_i} \right)^{-\frac{\kappa_i + 1}{\kappa_i^{int} + 1}} \right] \right\}^{-\kappa^{int} - 1} .$$

(4.76)

4.4.8.2.3 Internal Energy

$$U_i = \frac{1}{2}f_i k_B T_i, \quad \frac{2}{m}U_i = \frac{1}{2}f_i \theta_i^2.$$

(4.77)

4.4.8.2.4 Degeneration of the Kappa Index

$$\kappa_i = \kappa_{0i} + \frac{1}{2}f_i, \quad \kappa_i^{int} = \kappa_{0i}^{int} + \frac{1}{2}f_i, \quad \kappa^{int} = \kappa_0^{int} + \frac{1}{2}d_K.$$

(4.78)

4.4.8.2.5 Nonlinear Superposition
Eq. (4.76) can be written as

$$\kappa^{int} \cdot \left\{ 1 - [ZP(\{\varepsilon_{Ki}\}_{i=1}^M)]^{-\frac{1}{\kappa^{int}+1}} \right\} = \sum_{i=1}^M \kappa_i^{int} \cdot \left\{ 1 - [Z_i P(\varepsilon_{Ki})]^{-\frac{1}{\kappa_i^{int}+1}} \right\}, \quad (4.79a)$$

or using the ordinary distributions and their relation to the escort distributions, $P^{-\frac{1}{\kappa+1}} \propto p^{-\frac{1}{\kappa}}$ (Chapter 1; Livadiotis and McComas, 2009)

$$\kappa^{int} \cdot \left\{ 1 - [\tilde{Z} p(\{\varepsilon_{Ki}\}_{i=1}^M)]^{-\frac{1}{\kappa^{int}}} \right\} = \sum_{i=1}^M \kappa_i^{int} \cdot \left\{ 1 - [\tilde{Z}_i p(\varepsilon_{Ki})]^{-\frac{1}{\kappa_i^{int}}} \right\}. \quad (4.79b)$$

Comment: This is a special case of Eq. (4.28), where the distribution on the kappa indices is $D(\kappa_0) = \sum_{i=1}^M \delta(\kappa_0 - \kappa_{0i}^{int})$, while the monotonic function ϕ is given by the deformed logarithm function (Chapter 1):

$$\phi(x) = \ln_\kappa(x) \equiv \kappa \cdot \left(1 - x^{-\frac{1}{\kappa}} \right), \quad \text{with } x = Z \cdot P(\{\varepsilon_{Ki}\}_{i=1}^M).$$

(4.79c)

4.4.8.3 Distributions, for M Correlated Groups, Each With f_i Degrees of Freedom, Self-Correlated With Invariant Kappa Index, $\kappa_{0i} = \kappa_0$, and Intercorrelated With Invariant Kappa Index κ_0^{int}

4.4.8.3.1 In Terms of the Velocity $\vec{u}$

$$P\left(\vec{u}; \kappa_0, \kappa_0^{int}, \{\theta_i\}_{i=1}^M\right)$$

$$= \frac{1}{Z} \cdot \left\{ 1 - \sum_{i=1}^M \frac{\kappa_0^{int} + \frac{f_i}{2}}{\kappa_0^{int} + \frac{d_K}{2}} \cdot \left[1 - \left(1 + \frac{1}{\kappa_0 + \frac{f_i}{2}} \cdot \frac{(\vec{u} - \vec{u_b})^2\big|_{(f_i)} - \frac{f_i}{2}\theta_i^2}{\theta_i^2} \right)^{\frac{\kappa_0 + 1 + \frac{f_i}{2}}{\kappa_0^{int} + 1 + \frac{f_i}{2}}} \right] \right\}^{-\kappa_0^{int} - 1 - \frac{1}{2}d_K}$$

(4.80)

4.4.8.3.2 In Terms of the Kinetic Energy $\varepsilon_{Ki} = \frac{1}{2}m(\vec{u} - \vec{u_b})^2\big|_{(f_i)}$

$$P\left(\{\varepsilon_{Ki}\}_{i=1}^M; \kappa_0, \kappa_0^{int}, \{T_i\}_{i=1}^M\right)$$

$$= \frac{1}{Z} \cdot \prod_{i=1}^M \varepsilon_{Ki}^{\frac{1}{2}f_i - 1} \cdot \left\{ 1 - \sum_{i=1}^M \frac{\kappa_0^{int} + \frac{f_i}{2}}{\kappa_0^{int} + \frac{d_K}{2}} \cdot \left[1 - \left(1 + \frac{1}{\kappa_0 + \frac{f_i}{2}} \cdot \frac{\varepsilon_{Ki} - \frac{f_i}{2}k_B T_i}{k_B T_i} \right)^{\frac{\kappa_0 + 1 + \frac{f_i}{2}}{\kappa_0^{int} + 1 + \frac{f_i}{2}}} \right] \right\}^{-\kappa_0^{int} - 1 - \frac{1}{2}d_K}.$$

(4.81)

4.4.8.4 Distributions, for $d_K = 3$ Correlated Groups, Each With $f_i = 1$ Degrees of Freedom, Self-Correlated With Kappa Index, κ_{0i}, and Intercorrelated With Kappa Index κ_0^{int}

4.4.8.4.1 In Terms of the Velocity $\vec{u}$

$$P\left(\vec{u}; \{\kappa_{0i}\}_{i=1}^3, \kappa_0^{int}, \{\theta_i\}_{i=1}^3\right)$$

$$= \frac{1}{Z} \cdot \left\{ 1 - \frac{\kappa_0^{int} + \frac{1}{2}}{\kappa_0^{int} + \frac{3}{2}} \sum_{i=1}^3 \left\{ 1 - \left[\left(\frac{\kappa_{0i}}{\kappa_{0i} + \frac{1}{2}} \right) \cdot \left[1 + \frac{1}{\kappa_{0i}} \cdot \frac{(u_i - u_{bi})^2}{\theta_i^2} \right]^{\frac{\kappa_{0i} + \frac{3}{2}}{\kappa_0^{int} + \frac{3}{2}}} \right] \right\} \right\}^{-\kappa_0^{int} - \frac{5}{2}}.$$

(4.82)

The partition function is given by the normalization:

$$
Z = \int_{-\infty}^{\infty} \left\{ 1 - \frac{\kappa_0^{int} + \frac{1}{2}}{\kappa_0^{int} + \frac{3}{2}} \cdot \sum_{i=1}^{3} \left\{ 1 - \left[\left(\frac{\kappa_{0i}}{\kappa_{0i} + \frac{1}{2}} \right) \cdot \left[1 + \frac{1}{\kappa_{0i}} \cdot \frac{(u_i - u_{bi})^2}{\theta_i^2} \right] \right]^{\frac{\kappa_{0i} + \frac{3}{2}}{\kappa_0^{int} + \frac{3}{2}}} \right\} \right\}^{-\kappa_0^{int} - \frac{5}{2}} d\vec{u}.
$$

(4.83)

4.4.8.4.2 In Terms of the Kinetic Energy $\varepsilon_{Ki} = \frac{1}{2} m (u_i - u_{bi})^2$ for i: x, y, z

$$
P\left(\{\varepsilon_{Ki}\}_{i=1}^{3}; \{\kappa_{0i}\}_{i=1}^{3}, \kappa_0^{int}, \{T_i\}_{i=1}^{3}\right)
$$

$$
= \frac{1}{Z} \cdot \prod_{i=1}^{3} \varepsilon_{Ki}^{-\frac{1}{2}} \cdot \left\{ 1 - \frac{\kappa_0^{int} + \frac{1}{2}}{\kappa_0^{int} + \frac{3}{2}} \cdot \sum_{i=1}^{3} \left\{ 1 - \left[\left(\frac{\kappa_{0i}}{\kappa_{0i} + \frac{1}{2}} \right) \cdot \left(1 + \frac{1}{\kappa_{0i}} \cdot \frac{\varepsilon_{Ki}}{k_B T_i} \right) \right]^{\frac{\kappa_{0i} + \frac{3}{2}}{\kappa_0^{int} + \frac{3}{2}}} \right\} \right\}^{-\kappa_0^{int} - \frac{5}{2}}.
$$

(4.84)

The partition function is given by the normalization:

$$
Z = \int_{0}^{\infty} \int_{0}^{\infty} \int_{0}^{\infty} \left\{ 1 - \frac{\kappa_0^{int} + \frac{1}{2}}{\kappa_0^{int} + \frac{3}{2}} \cdot \sum_{i=1}^{3} \left\{ 1 - \left[\left(\frac{\kappa_{0i}}{\kappa_{0i} + \frac{1}{2}} \right) \cdot \left(1 + \frac{1}{\kappa_{0i}} \cdot \frac{\varepsilon_{Ki}}{k_B T_i} \right) \right]^{\frac{\kappa_{0i} + \frac{3}{2}}{\kappa_0^{int} + \frac{3}{2}}} \right\} \right\}^{-\kappa_0^{int} - \frac{5}{2}}
$$

$$
\times \prod_{i=1}^{3} \varepsilon_{Ki}^{-\frac{1}{2}} d\varepsilon_{K1} d\varepsilon_{K2} d\varepsilon_{K3}.
$$

(4.85)

4.4.8.4.3 Correlated Degrees in Uncorrelated Groups $\kappa_0^{\text{int}} \to \infty$

Using the approximations:

$$\left\{ 1 - \frac{\kappa_0^{\text{int}} + \frac{1}{2}}{\kappa_0^{\text{int}} + \frac{3}{2}} \cdot \sum_{i=1}^{3} \left\{ 1 - \left[\left(\frac{\kappa_{0i}}{\kappa_{0i} + \frac{1}{2}} \right) \cdot \left(1 + \frac{1}{\kappa_{0i}} \cdot \frac{\varepsilon_{Ki}}{k_B T_i} \right) \right]^{\frac{\kappa_{0i} + \frac{3}{2}}{\kappa_0^{\text{int}} + \frac{3}{2}}} \right\} \right\}^{-\kappa_0^{\text{int}} - \frac{5}{2}}$$

$$\approx \left\{ 1 + \frac{1}{\kappa_0^{\text{int}} + \frac{3}{2}} \cdot \ln \left\{ \prod_{i=1}^{3} \left[\left(\frac{\kappa_{0i}}{\kappa_{0i} + \frac{1}{2}} \right) \cdot \left(1 + \frac{1}{\kappa_{0i}} \cdot \frac{\varepsilon_{Ki}}{k_B T_i} \right) \right]^{\kappa_{0i} + \frac{3}{2}} \right\} \right\}^{-\kappa_0^{\text{int}} - \frac{5}{2}}$$

$$\approx \exp \left\{ - \ln \left\{ \prod_{i=1}^{3} \left[\left(\frac{\kappa_{0i}}{\kappa_{0i} + \frac{1}{2}} \right) \cdot \left(1 + \frac{1}{\kappa_{0i}} \cdot \frac{\varepsilon_{Ki}}{k_B T_i} \right) \right]^{\kappa_{0i} + \frac{3}{2}} \right\} \right\},$$

we find

$$P\left(\{\varepsilon_{Ki}\}_{i=1}^{3} ; \{\kappa_{0i}\}_{i=1}^{3}, \kappa_0^{\text{int}} \to \infty, \{T_i\}_{i=1}^{3} \right) \propto \prod_{i=1}^{3} \left\{ \left(1 + \frac{1}{\kappa_{0i}} \cdot \frac{\varepsilon_{Ki}}{k_B T_i} \right)^{-\kappa_{0i} - \frac{3}{2}} \cdot \varepsilon_{Ki}^{-\frac{1}{2}} \right\}, \quad (4.86)$$

coinciding, thus, with Eq. (4.58) in 4.4.5.1.2.

4.4.8.4.4 Correlated Degrees in Equally Correlated Groups $\kappa_0^{\text{int}} \to \kappa_0$

$$P\left(\{\varepsilon_{Ki}\}_{i=1}^{3} ; \{\kappa_{0i} \to \kappa_0\}_{i=1}^{3}, \kappa_0^{\text{int}} \to \kappa_0, \{T_i\}_{i=1}^{3} \right) \propto \left(1 + \frac{1}{\kappa_0} \cdot \frac{\sum_{i=1}^{3} \varepsilon_{Ki}}{k_B T_i} \right)^{-\kappa_0 - \frac{5}{2}} \cdot \prod_{i=1}^{3} \varepsilon_{Ki}^{-\frac{1}{2}},$$

$$(4.87)$$

coinciding, thus, with Eq. (4.30a) in 4.4.1.1.2.

4.4.8.5 Distributions, for Two Correlated Groups, With $f_1 = 1$ and $f_2 = 2$ Degrees of Freedom, Self-Correlated With Kappa Index κ_{0i}, and Intercorrelated With Kappa Index κ_0^{int}

This is the generalization of the cases of "correlation between the projection at a certain direction and the perpendicular plane" and "self-correlated projections at a direction and perpendicular plane," in Subsections 4.4.2 and 4.4.4 (or 4.4.6), respectively.

4.4.8.5.1 Distribution in Terms of the Velocity $\vec{u} = \left(u_{\parallel}, \vec{u}_{\perp} \right)$

$$
P\left(\vec{u}; \kappa_0, \kappa_0^{int}, \theta_{\parallel}, \theta_{\perp}\right) = \frac{1}{Z} \cdot \left\{ \frac{\kappa_0^{int} + \frac{1}{2}}{\kappa_0^{int}} \cdot \left[1 + \frac{1}{\kappa_0 + \frac{1}{2}} \cdot \frac{\left(u_{\parallel} - u_{b\parallel}\right)^2 - \frac{1}{2}\theta_{\parallel}^2}{\theta_{\parallel}^2} \right]^{\frac{\kappa_0 + \frac{3}{2}}{\kappa_0^{int} + \frac{3}{2}}} \right.
$$

$$
\left. + \frac{\kappa_0^{int} + 1}{\kappa_0^{int}} \cdot \left[1 + \frac{1}{\kappa_0 + 1} \cdot \frac{\left(\vec{u}_{\perp} - \vec{u}_{b\perp}\right)^2 - \theta_{\perp}^2}{\theta_{\perp}^2} \right]^{\frac{\kappa_0 + 2}{\kappa_0^{int} + 2}} - 1 \right\}^{-\kappa_0^{int} - \frac{5}{2}} .
$$

(4.88a)

Normalization

$$
Z = \int_{-\infty}^{\infty} \left\{ \frac{\kappa_0^{int} + \frac{1}{2}}{\kappa_0^{int}} \cdot \left[1 + \frac{1}{\kappa_0 + \frac{1}{2}} \cdot \frac{\left(u_{\parallel} - u_{b\parallel}\right)^2 - \frac{1}{2}\theta_{\parallel}^2}{\theta_{\parallel}^2} \right]^{\frac{\kappa_0 + \frac{3}{2}}{\kappa_0^{int} + \frac{3}{2}}} \right.
$$

$$
\left. + \frac{\kappa_0^{int} + 1}{\kappa_0^{int}} \cdot \left[1 + \frac{1}{\kappa_0 + 1} \cdot \frac{\left(\vec{u}_{\perp} - \vec{u}_{b\perp}\right)^2 - \theta_{\perp}^2}{\theta_{\perp}^2} \right]^{\frac{\kappa_0 + 2}{\kappa_0^{int} + 2}} - 1 \right\}^{-\kappa_0^{int} - \frac{5}{2}} d\vec{u}.
$$

(4.88b)

4.4.8.5.2 Distribution in Terms of the Kinetic Energy,
$\varepsilon_{K\parallel} = \frac{1}{2}m\left(u_\parallel - u_{b\parallel}\right)^2$ and $\varepsilon_{K\perp} = \frac{1}{2}m\left(\vec{u}_\perp - \vec{u}_{b\perp}\right)^2$

$$P\left(\varepsilon_{K\parallel}, \varepsilon_{K\perp}; \kappa_0, \kappa_0^{\text{int}}, T_\parallel, T_\perp\right) = \frac{1}{Z} \cdot \left[\frac{\kappa_0^{\text{int}} + \frac{1}{2}}{\kappa_0^{\text{int}}} \cdot \left(1 + \frac{1}{\kappa_0 + \frac{1}{2}} \cdot \frac{\varepsilon_{K\parallel} - \frac{1}{2}k_B T_\parallel}{k_B T_\parallel}\right)^{\frac{\kappa_0 + \frac{3}{2}}{\kappa_0^{\text{int}} + \frac{3}{2}}} \right.$$
$$\left. + \frac{\kappa_0^{\text{int}} + 1}{\kappa_0^{\text{int}}} \cdot \left[1 + \frac{1}{\kappa_0 + 1} \cdot \frac{\varepsilon_{K\perp} - k_B T_\perp}{k_B T_\perp}\right]^{\frac{\kappa_0 + 2}{\kappa_0^{\text{int}} + 2}} - 1 \right]^{-\kappa_0^{\text{int}} - \frac{5}{2}} \varepsilon_{K\parallel}^{-\frac{1}{2}}.$$

(4.89a)

Normalization

$$Z = \int_0^\infty \int_0^\infty \left[\frac{\kappa_0^{\text{int}} + \frac{1}{2}}{\kappa_0^{\text{int}}} \cdot \left(1 + \frac{1}{\kappa_0 + \frac{1}{2}} \cdot \frac{\varepsilon_{K\parallel} - \frac{1}{2}k_B T_\parallel}{k_B T_\parallel}\right)^{\frac{\kappa_0 + \frac{3}{2}}{\kappa_0^{\text{int}} + \frac{3}{2}}} \right.$$
$$\left. + \frac{\kappa_0^{\text{int}} + 1}{\kappa_0^{\text{int}}} \cdot \left[1 + \frac{1}{\kappa_0 + 1} \cdot \frac{\varepsilon_{K\perp} - k_B T_\perp}{k_B T_\perp}\right]^{\frac{\kappa_0 + 2}{\kappa_0^{\text{int}} + 2}} - 1 \right]^{-\kappa_0^{\text{int}} - \frac{5}{2}} \varepsilon_{K\parallel}^{-\frac{1}{2}} d\varepsilon_{K\parallel} d\varepsilon_{K\perp}$$

(4.89b)

4.4.8.5.3 Correlated Degrees in Equally Correlated Groups
$\kappa_0^{\text{int}} \to \kappa_0$ (See 4.4.2.1.2)

$$P\left(\varepsilon_{K\parallel}, \varepsilon_{K\perp}; \kappa_0, T_\parallel, T_\perp\right) \propto \left[1 + \frac{1}{\kappa_0} \cdot \left(\frac{\varepsilon_{K\parallel}}{k_B T_\parallel} + \frac{\varepsilon_{K\perp}}{k_B T_\perp}\right)\right]^{-\kappa_0 - \frac{5}{2}} \varepsilon_{K\parallel}^{-\frac{1}{2}}. \qquad (4.89c)$$

4.4.8.5.4 Correlated Degrees in Uncorrelated Groups
$\kappa_0^{\text{int}} \to \infty$ (See 4.4.4.1.2)

$$P\left(\varepsilon_{K\parallel}, \varepsilon_{K\perp}; \kappa_0, T_\parallel, T_\perp\right) \propto \left[1 + \frac{1}{\kappa_0} \cdot \frac{\varepsilon_{K\parallel}}{k_B T_\parallel}\right]^{-\kappa_0 - \frac{3}{2}} \cdot \left[1 + \frac{1}{\kappa_0} \cdot \frac{\varepsilon_{K\perp}}{k_B T_\perp}\right]^{-\kappa_0 - 2} \varepsilon_{K\parallel}^{-\frac{1}{2}}. \qquad (4.89d)$$

4.5 Distributions With Potential

The kappa distributions in the presence of a potential energy have been first systematically studied in Livadiotis (2015b). Special cases can be found in Livadiotis

(2015c, 2016a). General Hamiltonian distributions formulae can be found also in Livadiotis et al. (2012) and Livadiotis and McComas (2013a, 2014a).

4.5.1 General Hamiltonian Distribution

4.5.1.1 Phase Space Distribution
(Livadiotis et al., 2012; Livadiotis and McComas, 2013a, 2014a)

$$P(\vec{r}, \vec{u}; \kappa_0, T) = \left\{ \int_{-\infty}^{\infty} \int_{-\infty}^{\infty} \left[1 + \frac{1}{\kappa_0} \cdot \frac{H(\vec{r}, \vec{u})}{k_B T} \right]^{-\kappa_0 - 1 - \frac{1}{2}d} d\vec{r} d\vec{u} \right\}^{-1}$$

$$\times \left[1 + \frac{1}{\kappa_0} \cdot \frac{H(\vec{r}, \vec{u})}{k_B T} \right]^{-\kappa_0 - 1 - \frac{1}{2}d} .$$

(4.90a)

Normalization

$$\int_{-\infty}^{\infty} \int_{-\infty}^{\infty} P(\vec{r}, \vec{u}; \kappa_0, T) d\vec{r} d\vec{u} = 1.$$

(4.90b)

4.5.1.2 Hamiltonian Function

$$H(\vec{r}, \vec{u}) = \varepsilon_K(\vec{u}) + \Phi(\vec{r}).$$

(4.90c)

4.5.1.3 Hamiltonian Degrees of Freedom Summing Up the Kinetic and Potential Degrees of Freedom

$$\frac{1}{2}d \equiv \frac{\langle H(\vec{r}, \vec{u}) \rangle}{k_B T} = \frac{\langle \varepsilon_K(\vec{u}) \rangle}{k_B T} + \frac{\langle \Phi(\vec{r}) \rangle}{k_B T} = \frac{1}{2}d_K + \frac{1}{2}d_\Phi.$$

(4.90d)

4.5.2 Positive Attractive Potential $\Phi(\vec{r}) > 0$

4.5.2.1 Phase Space Distributions
4.5.2.1.1 In Terms of the Velocity $\vec{u}$

$$P(\vec{r}, \vec{u}; \kappa_0, \theta) = \pi^{-\frac{1}{2}d_K} \kappa_0^{-\frac{1}{2}d_K} \cdot \frac{\Gamma\left(\kappa_0 + 1 + \frac{d_K}{2} + \frac{d_\Phi}{2} \right)}{\Gamma\left(\kappa_0 + 1 + \frac{d_\Phi}{2} \right)} \cdot \theta^{-d_K}$$

$$\times \left\{ \int_{-\infty}^{\infty} \left[1 + \frac{1}{\kappa_0} \cdot \frac{\Phi(\vec{r})}{k_B T} \right]^{-\kappa_0 - 1 - \frac{1}{2}d_\Phi} d\vec{r} \right\}^{-1}$$

$$\times \left\{ 1 + \frac{1}{\kappa_0} \cdot \left[\frac{(\vec{u} - \vec{u}_b)^2}{\theta^2} + \frac{\Phi(\vec{r})}{k_B T} \right] \right\}^{-\kappa_0 - 1 - \frac{1}{2}d_K - \frac{1}{2}d_\Phi} .$$

(4.91a)

Normalization

$$\int_{-\infty}^{\infty} \int_{-\infty}^{\infty} P(\vec{r}, \vec{u}; \kappa_0, \theta) d\vec{r} d\vec{u} = 1.$$

(4.91b)

4.5.2.1.2 In Terms of the Kinetic Energy ε_K (compare With Eq. (4.2a))

$$P(\vec{r}, \varepsilon_K; \kappa_0, T) = \frac{(\kappa_0 k_B T)^{-\frac{1}{2}d_K}}{B\left(\dfrac{d_K}{2}, \kappa_0 + 1 + \dfrac{d_\Phi}{2}\right)}$$

$$\times \left\{ \int_{-\infty}^{\infty} \left[1 + \frac{1}{\kappa_0} \cdot \frac{\Phi(\vec{r})}{k_B T} \right]^{-\kappa_0 - 1 - \frac{1}{2}d_\Phi} d\vec{r} \right\}^{-1}$$

$$\times \left[1 + \frac{1}{\kappa_0} \cdot \frac{\varepsilon_K + \Phi(\vec{r})}{k_B T} \right]^{-\kappa_0 - 1 - \frac{1}{2}d_K - \frac{1}{2}d_\Phi} \varepsilon_K^{\frac{1}{2}d_K - 1}.$$

(4.92a)

Normalization

$$\int_{0}^{\infty} \int_{-\infty}^{\infty} P(\vec{r}, \varepsilon_K; \kappa_0, T) d\vec{r} d\varepsilon_K = 1.$$

(4.92b)

4.5.2.2 Positional Distribution Function

$$P(\vec{r}; \kappa_0, T) = \left\{ \int_{-\infty}^{\infty} \left[1 + \frac{1}{\kappa_0} \cdot \frac{\Phi(\vec{r})}{k_B T} \right]^{-\kappa_0 - 1 - \frac{1}{2}d_\Phi} d\vec{r} \right\}^{-1} \cdot \left[1 + \frac{1}{\kappa_0} \cdot \frac{\Phi(\vec{r})}{k_B T} \right]^{-\kappa_0 - 1 - \frac{1}{2}d_\Phi}.$$

(4.93a)

Normalization

$$\int_{-\infty}^{\infty} P(\vec{r}; \kappa_0, T) d\vec{r} = 1.$$

(4.93b)

4.5.2.3 Potential Degrees of Freedom

$$\frac{1}{2} d_\Phi \equiv \frac{\langle \Phi(\vec{r}) \rangle}{k_B T}.$$

(4.94)

4.5.3 Negative Attractive Potential $\Phi(\vec{r}) < 0$

4.5.3.1 Phase Space Distributions

4.5.3.1.1 In Terms of the Velocity $\vec{u}$

$$P(\vec{r}, \vec{u}; \kappa_0, T) = \pi^{-\frac{1}{2}d_K} \kappa_0^{-\frac{1}{2}d_K} \cdot \frac{\Gamma\left(\kappa_0 + 1 + \dfrac{d_K}{2} - \dfrac{d_\Phi}{2}\right)}{\Gamma\left(\kappa_0 + 1 - \dfrac{d_\Phi}{2}\right)} \cdot \theta^{-d_K}$$

$$\times \left\{ \int_{-\infty}^{\infty} \left[\frac{\left|\Phi(\vec{r})\right|}{\kappa_0 k_B T} - 1 \right]_+^{-\kappa_0 - 1 + \frac{1}{2}d_\Phi} d\vec{r} \right\}^{-1}$$

$$\times \left\{ \frac{1}{\kappa_0} \cdot \left[\frac{\left|\Phi(\vec{r})\right|}{k_B T} - \frac{(\vec{u} - \vec{u}_b)^2}{\theta^2} \right] - 1 \right\}_+^{-\kappa_0 - 1 - \frac{1}{2}d_K + \frac{1}{2}d_\Phi} .$$

(4.95)

4.5.3.1.2 In Terms of the Kinetic Energy ε_K

$$P(\vec{r}, \varepsilon_K; \kappa_0, T) = \frac{(\kappa_0 k_B T)^{-\frac{1}{2}d_K}}{B\left(\dfrac{d_K}{2}, \kappa_0 + 1 - \dfrac{d_\Phi}{2}\right)}$$

$$\times \left\{ \int_{-\infty}^{\infty} \left[\frac{\left|\Phi(\vec{r})\right|}{\kappa_0 k_B T} - 1 \right]_+^{-\kappa_0 - 1 + \frac{1}{2}d_\Phi} d\vec{r} \right\}^{-1}$$

$$\times \left[\frac{\left|\Phi(\vec{r})\right| - \varepsilon_K}{\kappa_0 k_B T} - 1 \right]_+^{-\kappa_0 - 1 - \frac{1}{2}d_K + \frac{1}{2}d_\Phi} .$$

(4.96)

4.5.3.1.3 Restrictions

$$-\infty < \Phi(\vec{r}) \le -\kappa_0 k_B T \quad \text{or} \quad \kappa_0 k_B T \le \left|\Phi(\vec{r})\right| \le \infty, \quad 0 < \kappa_0 < \frac{d_\Phi}{2} - \frac{d_K}{2}. \quad (4.97)$$

4.5.3.2 Positional Distribution Function

$$
P(\overrightarrow{r}; \kappa_0, T) = \left\{ \int_{-\infty}^{\infty} \left[\frac{|\Phi(\overrightarrow{r})|}{\kappa_0 k_B T} - 1 \right]_{+}^{-\kappa_0 - 1 + \frac{d_\Phi}{2}} d\overrightarrow{r} \right\}^{-1} \cdot \left[\frac{|\Phi(\overrightarrow{r})|}{\kappa_0 k_B T} - 1 \right]_{+}^{-\kappa_0 - 1 + \frac{d_\Phi}{2}} .
$$

$$(4.98)$$

See: $x_+ = x$ for $x \geq 0$ and $x_+ = 0$ for $x < 0$ (See Eq. (1.13) in Chapter 1).

4.5.3.3 Potential Degrees of Freedom

$$
\frac{1}{2} d_\Phi = \frac{\langle |\Phi(\overrightarrow{r})| \rangle}{k_B T} .
$$

$$(4.99)$$

4.5.4 Small Positive/Negative Attractive/Repulsive Potential $\left| \Phi(\overrightarrow{r}) / (\kappa_0 k_B T) \right| \ll 1$ Defined in a Finite Volume $\overrightarrow{r} \in V$

4.5.4.1 Phase Space Distribution

4.5.4.1.1 In Terms of the Velocity $\overrightarrow{u}$

$$
P(\overrightarrow{r}, \overrightarrow{u}; \kappa_0, T) = \pi^{-\frac{1}{2} d_K} \kappa_0^{-\frac{1}{2} d_K} \cdot \frac{\Gamma\left(\kappa_0 + 1 + \frac{d_K}{2} \pm \frac{d_\Phi}{2} \right)}{\Gamma\left(\kappa_0 + 1 \pm \frac{d_\Phi}{2} \right)} \cdot \theta^{-d_K}
$$

$$
\times \left\{ \int_{\overrightarrow{r} \in V} \left[1 \pm \frac{1}{\kappa_0} \cdot \frac{|\Phi(\overrightarrow{r})|}{k_B T} \right]^{-\kappa_0 - 1 \mp \frac{1}{2} d_\Phi} d\overrightarrow{r} \right\}^{-1}
$$

$$
\times \left\{ 1 + \frac{1}{\kappa_0} \cdot \left[\frac{(\overrightarrow{u} - \overrightarrow{u}_b)^2}{\theta^2} \pm \frac{|\Phi(\overrightarrow{r})|}{k_B T} \right] \right\}^{-\kappa_0 - 1 - \frac{1}{2} d_K \mp \frac{1}{2} d_\Phi} .
$$

$$(4.100)$$

4.5.4.1.2 In Terms of the Kinetic Energy ε_K

$$P\left(\overrightarrow{r}, \varepsilon_K; \kappa_0, T\right) = \frac{\left(\kappa_0 k_B T\right)^{-\frac{1}{2}d_K}}{B\left(\dfrac{d_K}{2}, \kappa_0 + 1 \pm \dfrac{d_\Phi}{2}\right)} \cdot$$

$$\times \left\{ \int_{\overrightarrow{r} \in V} \left[1 \pm \frac{1}{\kappa_0} \cdot \frac{\left|\Phi(\overrightarrow{r})\right|}{k_B T}\right]^{-\kappa_0 - 1 \mp \frac{1}{2}d_\Phi} d\overrightarrow{r} \right\}^{-1}$$

$$\times \left[1 + \frac{1}{\kappa_0} \cdot \frac{\varepsilon_K \pm \left|\Phi(\overrightarrow{r})\right|}{k_B T}\right]^{-\kappa_0 - 1 - \frac{1}{2}d_K \mp \frac{1}{2}d_\Phi} \varepsilon_K^{\frac{1}{2}d_K - 1}. \qquad (4.101)$$

4.5.4.2 Positional Distribution Function

$$P\left(\overrightarrow{r}; \kappa_0, T\right) = \left\{ \int_{\overrightarrow{r} \in V} \left[1 \pm \frac{1}{\kappa_0} \cdot \frac{\left|\Phi(\overrightarrow{r})\right|}{k_B T}\right]^{-\kappa_0 - 1 \mp \frac{1}{2}d_\Phi} d\overrightarrow{r} \right\}^{-1}$$

$$\qquad (4.102)$$

$$\times \left[1 \pm \frac{1}{\kappa_0} \cdot \frac{\left|\Phi(\overrightarrow{r})\right|}{k_B T}\right]^{-\kappa_0 - 1 \mp \frac{1}{2}d_\Phi}.$$

Comment: This is similar to Eq. (4.93a) for $\Phi(\overrightarrow{r}) = \pm\left|\Phi(\overrightarrow{r})\right|$ (4.5.2.2), for positions spanning a finite volume $\overrightarrow{r} \in V$.

4.5.4.3 Potential Degrees of Freedom

$$\frac{1}{2}d_\Phi = \frac{\langle\left|\Phi(\overrightarrow{r})\right|\rangle}{k_B T}. \qquad (4.103)$$

4.5.5 Equivalent Local Distribution

Reduction of the distribution function of the kinetic and potential energies to the simpler but equivalent local distribution of solely the kinetic energy.

4.5.5.1 Phase Space Distribution With Potential $\Phi(\vec{r})$

$$f(\vec{r}, \varepsilon_K) \equiv N \cdot P(\vec{r}, \varepsilon_K) = N \cdot \frac{(\kappa_0 k_B T)^{-\frac{1}{2}d_K}}{B\left(\dfrac{d_K}{2}, \kappa_0 + 1 + \dfrac{d_\Phi}{2}\right)}$$

$$\times \left\{ \int_{-\infty}^{\infty} \left[1 + \frac{1}{\kappa_0} \cdot \frac{\Phi(\vec{r})}{k_B T} \right]^{-\kappa_0 - 1 - \frac{1}{2}d_\Phi} d\vec{r} \right\}^{-1}$$

$$\times \left[1 + \frac{1}{\kappa_0} \cdot \frac{\varepsilon_K + \Phi(\vec{r})}{k_B T} \right]^{-\kappa_0 - 1 - \frac{1}{2}d_K - \frac{1}{2}d_\Phi} \varepsilon_K^{\frac{1}{2}d_K - 1}.$$

$$(4.104a)$$

Normalization to N number of particles

$$\int_{-\infty}^{\infty} \int_{0}^{\infty} f(\vec{r}, \varepsilon_K) d\vec{r} d\varepsilon_K = N \cdot \int_{-\infty}^{\infty} \int_{0}^{\infty} P(\vec{r}, \varepsilon_K) d\vec{r} d\varepsilon_K = N. \qquad (4.104b)$$

4.5.5.2 Equivalent Local Distribution With No Potential Energy

$$f(\vec{r}, \varepsilon_K) = n(\vec{r}) \cdot P\left[\varepsilon_K, \Phi = 0; \tilde{\kappa}_0 = \kappa_0 + \frac{d_\Phi}{2}, T(\vec{r}) \right]$$

$$= n(\vec{r}) \cdot \frac{\left[\tilde{\kappa}_0 k_B T(\vec{r}) \right]^{-\frac{1}{2}d_K}}{B\left(\dfrac{d_K}{2}, \tilde{\kappa}_0 + 1\right)}$$

$$\times \left[1 + \frac{1}{\tilde{\kappa}_0} \cdot \frac{\varepsilon_K}{k_B T(\vec{r})} \right]^{-\kappa_0 - 1 - \frac{1}{2}d_K} \varepsilon_K^{\frac{1}{2}d_K - 1}. \qquad (4.104c)$$

4.5.6 Positive Power Law Central Potential (Oscillation Type) $\Phi(r) = \frac{1}{b}kr^b$ (Livadiotis, 2015c)

4.5.6.1 Phase Space Distributions

4.5.6.1.1 In Terms of the Position Vector $\vec{r}$ and the Kinetic Energy ε_K

$$
P(\vec{r}, \varepsilon_K; \kappa_0, T) = \frac{\Gamma\left(\frac{d_r}{2}\right)}{2\pi^{\frac{d_r}{2}}} \cdot \frac{b\left(\frac{d_r}{b}\right)^{\frac{d_r}{b}} \kappa_0^{-\frac{1}{b}d_r - \frac{1}{2}d_K}}{B\left(\frac{d_K}{2}, \frac{d_r}{b}\right) B\left(\frac{d_K}{2} + \frac{d_r}{b}, \kappa_0 + 1\right)} \cdot (k_B T)^{-\frac{1}{2}d_K} r_0^{-d_r}
$$

$$
\times \left\{ 1 + \frac{1}{\kappa_0} \cdot \left[\frac{\varepsilon_K}{k_B T} + \frac{d_r}{b} \left(\frac{|\vec{r}|}{r_0} \right)^b \right] \right\}^{-\kappa_0 - 1 - \frac{1}{2}d_K - \frac{1}{b}d_r} \varepsilon_K^{\frac{1}{2}d_K - 1}.
$$

(4.105)

4.5.6.1.2 In Terms of the Position Distance r and the Kinetic Energy ε_K

$$
P(r, \varepsilon_K; \kappa_0, T) = \frac{b\left(\frac{d_r}{b}\right)^{\frac{d_r}{b}} \kappa_0^{-\frac{1}{b}d_r - \frac{1}{2}d_K} \cdot (k_B T)^{-\frac{1}{2}d_K} r_0^{-d_r}}{B\left(\frac{d_K}{2}, \frac{d_r}{b}\right) B\left(\frac{d_K}{2} + \frac{d_r}{b}, \kappa_0 + 1\right)}
$$

$$
\times \left\{ 1 + \frac{1}{\kappa_0} \cdot \left[\frac{\varepsilon_K}{k_B T} + \frac{d_r}{b} \left(\frac{r}{r_0} \right)^b \right] \right\}^{-\kappa_0 - 1 - \frac{1}{2}d_K - \frac{1}{b}d_r} \varepsilon_K^{\frac{1}{2}d_K - 1} r^{d_r - 1}.
$$

(4.106a)

Normalization

$$
\int_0^\infty \int_0^\infty P(r, \varepsilon_K; \kappa_0, T) dr d\varepsilon_K = 1,
$$

(4.106b)

where d_r is the position vector dimensionality.

4.5.6.2 Potential Degrees of Freedom

$$
\frac{1}{2}d_\Phi \equiv \frac{\langle \Phi(\vec{r}) \rangle}{k_B T} = \frac{d_r}{b}.
$$

(4.107)

4.5.6.3 Positional Distribution
4.5.6.3.1 In Terms of the Position Vector $\vec{r}$

$$P(\vec{r};\kappa_0,r_0) = \frac{\Gamma\left(\dfrac{d_{\mathrm{r}}}{2}\right)}{2\pi^{\frac{d_{\mathrm{r}}}{2}}} \cdot \frac{b\left(\dfrac{d_{\mathrm{r}}}{b}\right)^{\frac{d_{\mathrm{r}}}{b}}\kappa_0^{-\frac{d_{\mathrm{r}}}{b}}}{\mathrm{B}\left(\dfrac{d_{\mathrm{r}}}{b},\kappa_0+1\right)} \cdot r_0^{-d_{\mathrm{r}}} \cdot \left[1+\frac{\dfrac{d_{\mathrm{r}}}{b}}{\kappa_0}\cdot\left(\frac{|\vec{r}|}{r_0}\right)^b\right]^{-\kappa_0-1-\frac{1}{b}d_{\mathrm{r}}}.$$

$$(4.108)$$

4.5.6.3.2 In Terms of the Position Distance r

$$P(r;\kappa_0,r_0) = \frac{b\left(\dfrac{d_{\mathrm{r}}}{b}\right)^{\frac{d_{\mathrm{r}}}{b}}\kappa_0^{-\frac{d_{\mathrm{r}}}{b}}}{\mathrm{B}\left(\dfrac{d_{\mathrm{r}}}{b},\kappa_0+1\right)} \cdot r_0^{-d_{\mathrm{r}}} \cdot \left[1+\frac{\dfrac{d_{\mathrm{r}}}{b}}{\kappa_0}\cdot\left(\frac{r}{r_0}\right)^b\right]^{-\kappa_0-1-\frac{1}{b}d_{\mathrm{r}}} r^{d_{\mathrm{r}}-1}. \qquad (4.109a)$$

Normalization

$$\int_0^\infty P(r;\kappa_0,r_0)dr = 1. \qquad (4.109b)$$

4.5.6.3.3 In Terms of the Potential Energy Φ

$$P(R;\kappa_0) = \frac{\kappa_0^{-\frac{1}{b}d_{\mathrm{r}}}}{\mathrm{B}\left(\dfrac{d_{\mathrm{r}}}{b},\kappa_0+1\right)} \cdot \left(1+\frac{1}{\kappa_0}\cdot R\right)^{-\kappa_0-1-\frac{1}{b}d_{\mathrm{r}}} R^{d_{\mathrm{r}}/b-1}, \quad R\equiv\frac{d_{\mathrm{r}}}{b}\cdot(r/r_0)^b \propto \Phi(r).$$

$$(4.110a)$$

Normalization

$$\int_0^\infty P(R;\kappa_0)dR = 1. \qquad (4.110b)$$

4.5.7 Negative Power Law Central Potential (Gravitational Type) $\Phi(r) = -\frac{1}{b}kr^{-b}$ (Livadiotis, 2015b)

4.5.7.1 Phase Space Distributions

4.5.7.1.1 In Terms of the Position Vector $\vec{r}$ and the Kinetic Energy ε_K

$$P(\vec{r}, \varepsilon_K; \kappa_0, T) = \frac{\Gamma\left(\frac{d_r}{2}\right)}{2\pi^{\frac{d_r}{2}}} \cdot \frac{b\left(\frac{d_r}{b}\right)^{-\frac{d_r}{b}} \kappa_0^{\frac{1}{b}d_r - \frac{1}{2}d_K}}{B\left(\frac{d_r}{b} - \kappa_0, \kappa_0 + 1\right) B\left(\frac{d_K}{2}, -\kappa_0 - \frac{d_K}{2} + \frac{d_r}{b}\right)} \cdot (k_B T)^{-\frac{1}{2}d_K} r_0^{-d_r}$$

$$\times \left\{ \frac{1}{\kappa_0} \cdot \left[\frac{d_r}{b} \left(\frac{r_0}{|\vec{r}|} \right)^b - \frac{\varepsilon_K}{k_B T} \right] - 1 \right\}_+^{-\kappa_0 - 1 - \frac{1}{2}d_K + \frac{1}{b}d_r} \varepsilon_K^{\frac{1}{2}d_K - 1}.$$

(4.111)

4.5.7.1.2 In Terms of the Position Distance r and the Kinetic Energy ε_K

$$P(r, \varepsilon_K; \kappa_0, T) = \frac{b\left(\frac{d_r}{b}\right)^{-\frac{d_r}{b}} \kappa_0^{\frac{1}{b}d_r - \frac{1}{2}d_K}}{B\left(\frac{d_r}{b} - \kappa_0, \kappa_0 + 1\right) B\left(\frac{d_K}{2}, \frac{d_r}{b} - \frac{d_K}{2} - \kappa_0\right)} \cdot (k_B T)^{-\frac{1}{2}d_K} r_0^{-d_r}$$

$$\times \left\{ \frac{1}{\kappa_0} \cdot \left[\frac{d_r}{b} \left(\frac{r_0}{r} \right)^b - \frac{\varepsilon_K}{k_B T} \right] - 1 \right\}_+^{-\kappa_0 - 1 - \frac{1}{2}d_K + \frac{1}{b}d_r} \varepsilon_K^{\frac{1}{2}d_K - 1} r^{d_r - 1}.$$

(4.112)

4.5.7.2 Potential Degrees of Freedom

$$\frac{1}{2}d_\Phi \equiv \frac{\langle |\Phi(\vec{r})| \rangle}{k_B T} = \frac{d_r}{b}.$$

(4.113)

4.5.7.3 Positional Distribution
4.5.7.3.1 In Terms of the Position Vector $\vec{r}$

$$P(\vec{r}; \kappa_0, r_0)d\vec{r} = \frac{\Gamma\left(\dfrac{d_r}{2}\right)}{2\pi^{\frac{d_r}{2}}} \cdot \frac{b\left(\dfrac{d_r}{b}\right)^{-\frac{d_r}{b}} \kappa_0^{\frac{d_r}{b}}}{B\left(\dfrac{d_r}{b} - \kappa_0, \kappa_0 + 1\right)} \cdot r_0^{-d_r} \cdot \left[\frac{d_r}{b} \cdot \left(\frac{r_0}{|\vec{r}|}\right)^b - 1\right]^{-\kappa_0 - 1 + \frac{1}{b}d_r} d\vec{r},$$

(4.114)

4.5.7.3.2 In Terms of the Position Distance r

$$P(r; \kappa_0, r_0)dr = \frac{b\left(\dfrac{d_r}{b}\right)^{-\frac{d_r}{b}} \kappa_0^{\frac{d_r}{b}}}{B\left(\dfrac{d_r}{b} - \kappa_0, \kappa_0 + 1\right)} \cdot r_0^{-d_r} \cdot \left[\frac{d_r}{b} \cdot \left(\frac{r_0}{r}\right)^b - 1\right]^{-\kappa_0 - 1 + \frac{1}{b}d_r} r^{d_r - 1}dr,$$

(4.115)

4.5.7.3.3 In Terms of the Potential Energy Φ

$$P(R; \kappa_0)dR = \frac{\kappa_0^{\frac{1}{b}d_r}}{B\left(\dfrac{d_r}{b} - \kappa_0, \kappa_0 + 1\right)} \cdot \left(\frac{1}{\kappa_0} \cdot R - 1\right)^{-\kappa_0 - 1 + \frac{1}{b}d_r} R^{-d_r/b - 1}dR,$$

(4.116)

$$R \equiv \frac{d_r}{b} \cdot (r/r_0)^{-b} \propto \Phi(r).$$

4.5.7.3.4 Restrictions

$$0 < \kappa_0 < \frac{d_r}{b} - \frac{d_K}{2}, \quad b \leq 2d_r/d_K.$$

(4.117)

4.5.8 Properties for $\Phi(r) = \pm\frac{1}{b}kr^{\pm b}$

4.5.8.1 Parameters

$$r_0 \equiv \left\langle r^{\pm b} \right\rangle^{\pm\frac{1}{b}} = \left(\frac{\frac{d_r}{b}k_B T}{\frac{1}{b}k} \right)^{\pm\frac{1}{b}} , \quad \frac{1}{2}d_\Phi \equiv \pm\frac{\langle\Phi\rangle}{k_B T} = \frac{\frac{1}{b}k}{k_B T}r_0^{\pm b} = \frac{1}{b}d_r. \tag{4.118}$$

4.5.8.2 Degeneration of the Kappa Index

$$\kappa(d_K) = \kappa_0 + \frac{1}{2}d_K \quad (\text{no potential}) \Rightarrow \kappa(d_K, d_r)$$

$$= \kappa_0 + \frac{1}{2}d_K \pm \frac{1}{b}d_r \quad \left(\text{with potential } \Phi(r) = \pm\frac{1}{b}kr^{\pm b}\right). \tag{4.119}$$

4.5.8.3 Local Parameters
4.5.8.3.1 Density

$$n(r) = n_0 \cdot \left[\frac{\frac{d_r}{\mp b} \cdot \left(\frac{r}{r_0}\right)^{\pm b} - 1}{\frac{d_r}{\mp b} - 1} \right]^{-\kappa_0 - 1 \mp \frac{1}{b}d_r} . \tag{4.120a}$$

4.5.8.3.2 Temperature

$$T(r) = T_0 \cdot \left[\frac{\frac{d_r}{\mp b} \cdot \left(\frac{r}{r_0}\right)^{\pm b} - 1}{\frac{d_r}{\mp b} - 1} \right] . \tag{4.120b}$$

4.5.8.3.3 Thermal Pressure

$$
p(r) = p_0 \cdot \left[\frac{\frac{d_r}{\mp b} \cdot \left(\frac{r}{r_0} \right)^{\pm b} - 1}{\frac{d_r}{\mp b} - 1} \right]^{-\kappa_0 \mp \frac{1}{b} d_r} .
\tag{4.120c}
$$

4.5.8.4 Polytropic Index

$$
\frac{n(r)}{n_0} = \left[\frac{T(r)}{T_0} \right]^{\nu}, \quad \nu = \mp \frac{1}{b} d_r - 1 - \kappa_0,
\tag{4.120d}
$$

where (n_0, T_0, p_0): density, temperature, and thermal pressure at radius $r = r_0$, given by Eq. (4.118). (See Meyer-Vernet et al., 1995; Livadiotis, 2015b, 2015c, 2016c.) The polytropic relation (Eq. 4.120d) can be modified in cases where (n_0, T_0, p_0) depend on the position (see Chapter 5, Section 5.3).

4.5.8.5 Statistical Moments
4.5.8.5.1 Kinetic Moments
(Livadiotis, 2014a)

$$
\left\langle |\vec{u}|^{\alpha} \right\rangle = \int_{-\infty}^{\infty} |\vec{u}|^{\alpha} \int_{-\infty}^{\infty} P(\vec{r}, \vec{u}) d\vec{r} d\vec{u}
$$

$$
= \theta^{\alpha} \kappa_0^{\frac{1}{2}\alpha} \cdot \frac{\Gamma\left(\frac{d_K + \alpha}{2} \right)}{\Gamma\left(\frac{d_K}{2} \right)} \cdot \frac{\Gamma\left(\kappa_0 + 1 - \frac{\alpha}{2} \right)}{\Gamma(\kappa_0 + 1)},
\tag{4.121a}
$$

$$
\left\langle \varepsilon_K^{\frac{1}{2}\alpha} \right\rangle = \int_0^{\infty} \varepsilon_K^{\frac{1}{2}\alpha} \int_{-\infty}^{\infty} P(\vec{r}, \varepsilon_K) d\vec{r} d\varepsilon_K
$$

$$
= (k_B T)^{\frac{1}{2}\alpha} \kappa_0^{\frac{1}{2}\alpha} \frac{\Gamma\left(\frac{d_K + \alpha}{2} \right)}{\Gamma\left(\frac{d_K}{2} \right)} \cdot \frac{\Gamma\left(\kappa_0 + 1 - \frac{\alpha}{2} \right)}{\Gamma(\kappa_0 + 1)} .
\tag{4.121b}
$$

4.5.8.5.2 Potential Moments (Based on Eqs. 4.106a, 4.111)

$$\langle |\vec{r}|^\alpha \rangle_{\Phi>0} = \int_{-\infty}^{\infty} |\vec{r}|^\alpha \int_{-\infty}^{\infty} P(\vec{r}, \vec{u}; \Phi > 0) d\vec{u} d\vec{r}$$

$$= r_0^\alpha \kappa_0^{\frac{1}{b}\alpha} \cdot \left(\frac{d_r}{b}\right)^{-\frac{\alpha}{b}} \frac{\Gamma\left(\dfrac{d_r + \alpha}{b}\right)}{\Gamma\left(\dfrac{d_r}{b}\right)} \cdot \frac{\Gamma\left(\kappa_0 + 1 - \dfrac{\alpha}{b}\right)}{\Gamma(\kappa_0 + 1)}, \tag{4.122a}$$

$$\langle |\vec{r}|^\alpha \rangle_{\Phi<0} = \int_{-\infty}^{\infty} |\vec{r}|^\alpha \int_{-\infty}^{\infty} P(\vec{r}, \vec{u}; \Phi < 0) d\vec{u} d\vec{r}$$

$$= r_0^\alpha \kappa_0^{-\frac{1}{b}\alpha} \cdot \left(\frac{d_r}{b}\right)^{\frac{\alpha}{b}} \frac{\Gamma\left(\dfrac{d_r}{b} + 1\right)}{\Gamma\left(\dfrac{d_r + \alpha}{b} + 1\right)} \cdot \frac{\Gamma\left(\kappa_0 + 1 + \dfrac{\alpha}{b}\right)}{\Gamma(\kappa_0 + 1)}. \tag{4.122b}$$

4.5.9 Marginal and Conditional Distributions (Livadiotis, 2015b)

4.5.9.1 Marginal Distributions

The marginal distributions provide the positional and velocity distributions, which are derived from integrating the phase space distribution of the Hamiltonian over the velocities and the positions, respectively,

$$P(\vec{r}) = \int_{-\infty}^{\infty} P(\vec{r}, \vec{u}) d\vec{u} \quad \text{and} \quad P(\vec{u}) = \int_{-\infty}^{\infty} P(\vec{r}, \vec{u}) d\vec{r}. \tag{4.123}$$

4.5.9.2 Conditional Distributions

The conditional distributions provide the positional and velocity distributions at a certain velocity $\vec{u} = \vec{b}$ or position $\vec{r} = \vec{a}$

$$P(\vec{r}) = P\left(\vec{r}, \vec{u} = \vec{b}\right) \quad \text{and} \quad P(\vec{u}) = P(\vec{r} = \vec{a}, \vec{u}), \tag{4.124a}$$

while a local probability distribution may be defined after the normalization:

$$P(\vec{r}) = \frac{P\left(\vec{r}, \vec{u} = \vec{b}\right)}{\int_{-\infty}^{\infty} P\left(\vec{r}, \vec{u} = \vec{b}\right) d\vec{r}} \quad \text{and} \quad P(\vec{u}) = \frac{P(\vec{r} = \vec{a}, \vec{u})}{\int_{-\infty}^{\infty} P(\vec{r} = \vec{a}, \vec{u}) d\vec{u}}. \tag{4.124b}$$

4.5.10 Angular Potentials $\Phi(\vartheta, \varphi)$

This may apply for positive or small negative potentials with $|\Phi(\vartheta, \varphi)| < \kappa_0 k_B T$.

4.5.10.1 Phase Space Distribution

4.5.10.1.1 In Terms of the Angular Dependence (ϑ, φ) of the $(d_r = 3)$-dimensional Position Vector $\vec{r}$ and the (d_K)-Dimensional Velocity $\vec{u}$

$$
P(\vartheta, \varphi, \vec{u}; \kappa_0, T) = \pi^{-\frac{1}{2}d_K} \kappa_0^{-\frac{1}{2}d_K} \cdot \frac{\Gamma\left(\kappa_0 + 1 + \frac{d_K}{2} \pm \frac{d_\Phi}{2}\right)}{\Gamma\left(\kappa_0 + 1 \pm \frac{d_\Phi}{2}\right)} \cdot \theta^{-d_K}
$$

$$
\times \left\{ \int_0^{2\pi} \int_{-1}^{1} \left[1 + \frac{1}{\kappa_0} \cdot \frac{\Phi(\vartheta, \varphi)}{k_B T} \right]^{-\kappa_0 - 1 \mp \frac{1}{2}d_\Phi} d\cos\vartheta d\varphi \right\}^{-1}
$$

$$
\times \left\{ 1 + \frac{1}{\kappa_0} \cdot \left[\frac{(\vec{u} - \vec{u}_b)^2}{\theta^2} + \frac{\Phi(\vartheta, \varphi)}{k_B T} \right] \right\}^{-\kappa_0 - 1 - \frac{1}{2}d_K \mp \frac{1}{2}d_\Phi} .
$$

(4.125a)

where the upper/lower sign corresponds to positive/negative potential.

Normalization

$$
\int_{-\infty}^{\infty} \int_0^{2\pi} \int_{-1}^{1} P(\vartheta, \varphi, \varepsilon_K; \kappa_0, T) d\cos\vartheta d\varphi \, d\vec{u} = 1.
$$

(4.125b)

4.5.10.1.2 In Terms of the Position Vector $\vec{r}$ and the Kinetic Energy ε_K

$$
P(\vartheta, \varphi, \varepsilon_K; \kappa_0, T) = \frac{(\kappa_0 k_B T)^{-\frac{1}{2}d_K}}{B\left(\frac{d_K}{2}, \kappa_0 + 1 \pm \frac{d_\Phi}{2}\right)}
$$

$$
\times \left\{ \int_0^{2\pi} \int_{-1}^{1} \left[1 + \frac{1}{\kappa_0} \cdot \frac{\Phi(\vartheta, \varphi)}{k_B T} \right]^{-\kappa_0 - 1 \mp \frac{1}{2}d_\Phi} d\cos\vartheta d\varphi \right\}^{-1}
$$

$$
\times \left[1 + \frac{1}{\kappa_0} \cdot \frac{\varepsilon_K + \Phi(\vartheta, \varphi)}{k_B T} \right]^{-\kappa_0 - 1 - \frac{1}{2}d_K \mp \frac{1}{2}d_\Phi} \varepsilon_K^{\frac{1}{2}d_K - 1}.
$$

(4.126a)

Normalization

$$
\int_0^{\infty} \int_0^{2\pi} \int_{-1}^{1} P(\vartheta, \varphi, \varepsilon_K; \kappa_0, T) \, d\cos\vartheta d\varphi \, d\varepsilon_K = 1.
$$

(4.126b)

4.5.10.2 Potential Degrees of Freedom

$$\frac{1}{2}d_\Phi = \frac{|\Phi(\vartheta,\varphi)|}{k_BT}.$$
(4.127)

4.5.10.3 Kappa Index Degeneration

$$\kappa = \kappa_0 + \frac{1}{2}d_K \pm \frac{1}{2}d_\Phi.$$
(4.128)

4.5.11 Magnetization Potential $\Phi(\vartheta) \propto -\cos\vartheta$ (Livadiotis, 2015b, 2015c, 2016a)

4.5.11.1 Potential Energy

$$\Phi(\vartheta) = -\vec{\mu}\cdot\vec{B} = -\mu B\cos\vartheta, \quad \beta \equiv \mu B/(k_BT).$$
(4.129)

4.5.11.2 Phase Space Distribution, in Terms of the Positional Polar Angle ϑ and the Kinetic Energy ε_K, for 1 Positional and d_K Kinetic Degrees of Freedom

$$P(\cos\vartheta,\varepsilon_K;\kappa_0,\beta) = \frac{\Gamma\left(\kappa_0 - \beta\langle\cos\vartheta\rangle + \frac{1}{2}d_K + 1\right)(\kappa_0+\beta)^{\kappa_0-\beta\langle\cos\vartheta\rangle}(1+\beta\langle\cos\vartheta\rangle-\beta)}{2\Gamma(\kappa_0-\beta\langle\cos\vartheta\rangle)\Gamma\left(\frac{1}{2}d_K\right)\kappa_0^{\kappa_0+1+\frac{1}{2}d_K-\beta\langle\cos\vartheta\rangle}}$$

$$\times (k_BT)^{-\frac{1}{2}d_K}\cdot\left[1+\frac{1}{\kappa_0}\cdot\left(\frac{\varepsilon_K}{k_BT}-\beta\cos\vartheta\right)\right]^{-\kappa_0-1-\frac{1}{2}d_K+\beta\langle\cos\vartheta\rangle}\varepsilon_K^{\frac{1}{2}d_K-1}.$$
(4.130a)

Normalization

$$\int_0^\infty\int_{-1}^1 P(\cos\vartheta,\varepsilon_K;\kappa_0,\beta)d\cos\vartheta\, d\varepsilon_K = 1.$$
(4.130b)

4.5.11.3 Distribution of the Kinetic Energy ε_K

$$P(\varepsilon_K;\kappa_0,\beta) = \frac{(\kappa_0 k_BT)^{-\frac{1}{2}d_K}}{B\left(\kappa_0-\beta\langle\cos\vartheta\rangle,\frac{1}{2}d_K\right)}\cdot\left(1+\frac{\beta}{\kappa_0}\right)^{\kappa_0-\beta\langle\cos\vartheta\rangle}$$

$$\times (\beta^{-1}+\langle\cos\vartheta\rangle-1)\cdot\frac{1}{2}\left\{\left[1+\frac{\varepsilon_K-\beta k_BT}{\kappa_0 k_BT}\right]^{-\kappa_0-\frac{1}{2}d_K+\beta\langle\cos\vartheta\rangle}\right.$$

$$\left.-\left[1+\frac{\varepsilon_K+\beta k_BT}{\kappa_0 k_BT}\right]^{-\kappa_0-\frac{1}{2}d_K+\beta\langle\cos\vartheta\rangle}\right\}\cdot\varepsilon_K^{\frac{1}{2}d_K-1}.$$
(4.131)

Comment: The distribution of the kinetic energy is a superposition of two kappa distributions, one with energy $\varepsilon_K - \beta k_B T$ and one with energy $\varepsilon_K + \beta k_B T$.

4.5.11.4 Distribution of the Polar Angle $\cos\vartheta$
4.5.11.4.1 Distribution

$$P(\cos\vartheta; \kappa_0, \beta) = \frac{1}{2}(\kappa_0 - \beta\langle\cos\vartheta\rangle)(\kappa_0 + \beta)^{\kappa_0 - \beta\langle\cos\vartheta\rangle}(1 + \beta\langle\cos\vartheta\rangle - \beta)$$
$$\times (\kappa_0 - \beta\cos\vartheta)^{-\kappa_0 - 1 + \beta\langle\cos\vartheta\rangle}.$$

(4.132a)

Normalization

$$\int_{-1}^{1} P(\cos\vartheta; \kappa_0, \beta)d\cos\vartheta = 1.$$

(4.132b)

4.5.11.4.2 Restrictions

$$\kappa_0 > \beta.$$

(4.133)

4.5.11.4.3 Average $\langle\cos\vartheta\rangle$ is Given by

$$\frac{1 + \beta\langle\cos\vartheta\rangle + \beta}{1 + \beta\langle\cos\vartheta\rangle - \beta} = \left(\frac{\kappa_0 + \beta}{\kappa_0 - \beta}\right)^{\kappa_0 - \beta\langle\cos\vartheta\rangle}.$$

(4.134)

4.5.11.5 Degeneration of the Kappa Index

$$\kappa(d_K) = \kappa_0 + \frac{1}{2}d_K \quad \text{(no potential)} \Rightarrow$$

$$\kappa(d_K, d_\Phi) = \kappa_0 + \frac{1}{2}d_K - \beta\langle\cos\vartheta\rangle \quad \text{(with potential)},$$

(4.135a)

where

$$\frac{1}{2}d_\Phi = \beta\langle\cos\vartheta\rangle.$$

(4.135b)

Comment: The application was first examined using the kappa index κ_0 given by $\kappa(d_K) = \kappa_0 + \frac{1}{2}d_K$ (Livadiotis, 2015b; 2016a), and this kappa index is used in Chapter 5, Section 5.8. For instance, the reader may compare Eq. (5.159a) with Eq. (4.134), where the degeneration caused by the potential degrees of freedom is also considered, using $\kappa_0 \Rightarrow \kappa_0 - \beta\langle\cos\vartheta\rangle$.

4.6 Multiparticle Distributions

4.6.1 Standard N-Particle (N·d)−Dimensional Kappa Distributions (Livadiotis and McComas, 2011b)

4.6.1.1 Distributions, in Terms of the Velocity, $\vec{u}$, for N Particles, d_K Degrees of Freedom per Particle, Using the Invariant Kappa Index κ_0

$$
P\left(\left\{\vec{u}_{(n)}\right\}_{n=1}^{N};\kappa_0;\theta\right) = \left(\pi\kappa_0\theta^2\right)^{-\frac{1}{2}d_K N}\cdot\frac{\Gamma\left(\kappa_0+1+\dfrac{d_K}{2}N\right)}{\Gamma(\kappa_0+1)}
$$
$$
\times\left[1+\frac{1}{\kappa_0}\cdot\frac{1}{\theta^2}\sum_{n=1}^{N}\left(\vec{u}_{(n)}-\vec{u}_b\right)^2\right]^{-\kappa_0-1-\frac{1}{2}d_K N}.
$$

(4.136a)

Normalization

$$
\int_{-\infty}^{\infty}\cdots\int_{-\infty}^{\infty}P\left(\left\{\vec{u}_{(n)}\right\}_{n=1}^{N};\kappa_0;\theta\right)d\vec{u}_{(1)}\cdots d\vec{u}_{(N)} = 1.
$$

(4.136b)

4.6.1.2 Distributions, in Terms of the Kinetic Energy, $\varepsilon_{K(n)} = \frac{1}{2}m\times$ $\left(\vec{u}_{(n)}-\vec{u}_b\right)^2$, for N Particles, d_K Degrees of Freedom per Particle, Using the Invariant Kappa Index κ_0

$$
P\left(\left\{\varepsilon_{K(n)}\right\}_{n=1}^{N};\kappa_0;T\right) = \left(\kappa_0 k_B T\right)^{-\frac{1}{2}d_K N}\cdot\frac{\Gamma\left(\kappa_0+1+\dfrac{d_K}{2}N\right)}{\Gamma(\kappa_0+1)\cdot\Gamma\left(\dfrac{d_K}{2}\right)^N}
$$
$$
\times\left(1+\frac{1}{\kappa_0}\cdot\frac{1}{k_B T}\sum_{n=1}^{N}\varepsilon_{K(n)}\right)^{-\kappa_0-1-\frac{1}{2}d_K N}\cdot\prod_{n=1}^{N}\varepsilon_{K(n)}^{\frac{1}{2}d_K-1}.
$$

(4.137a)

Normalization

$$
\int_{0}^{\infty}\cdots\int_{0}^{\infty}P\left(\left\{\varepsilon_{K(n)}\right\}_{n=1}^{N};\kappa_0;T\right)d\varepsilon_{K(1)}\cdots d\varepsilon_{K(N)} = 1.
$$

(4.137b)

4.6.2 Negative *N*-Particle (*N·d*)−Dimensional Kappa Distribution

4.6.2.1 Distributions, in Terms of the Velocity, $\vec{u}$, for N Particles, d_K Degrees of Freedom per Particle, Using the Invariant Kappa Index κ_0

$$
P\left(\left\{\vec{u}_{(n)}\right\}_{n=1}^{N};\kappa_0;\theta\right) = \left(\pi\kappa_0\theta^2\right)^{-\frac{1}{2}d_K N} \cdot \frac{\Gamma(\kappa_0)}{\Gamma\left(\kappa_0 - \dfrac{d_K}{2}N\right)} \cdot
$$

$$
\times \left[1 - \frac{1}{\kappa_0}\cdot\frac{1}{\theta^2}\sum_{n=1}^{N}\left(\vec{u}_{(n)} - \vec{u}_b\right)^2\right]_+^{\kappa_0 - 1 - \frac{1}{2}d_K N}.
$$

$$\tag{4.138}$$

4.6.2.2 Distributions, in Terms of the Kinetic Energy, $\varepsilon_{K(n)}$, for N Particles, d_K Degrees of Freedom per Particle, Using the Invariant Kappa Index κ_0

$$
P\left(\left\{\varepsilon_{K(n)}\right\}_{n=1}^{N};\kappa_0;T\right) = \left(\kappa_0 k_B T\right)^{-\frac{1}{2}d_K N} \cdot \frac{\Gamma(\kappa_0)}{\Gamma\left(\kappa_0 - \dfrac{d_K}{2}N\right)\cdot\Gamma\left(\dfrac{d_K}{2}\right)^N}
$$

$$
\times \left(1 - \frac{1}{\kappa_0}\cdot\frac{1}{k_B T}\sum_{n=1}^{N}\varepsilon_{K(n)}\right)^{\kappa_0 - 1 - \frac{1}{2}d_K N} \cdot \prod_{n=1}^{N}\varepsilon_{K(n)}^{\frac{1}{2}d_K - 1}.
$$

$$\tag{4.139}$$

4.6.3 *N*-Particle (*N·d*)−Dimensional Kappa Distributions With Potential (Livadiotis, 2015c)

4.6.3.1 Phase Space Distribution, for N Particles, d_K Degrees of Freedom per Particle, Using the Invariant Kappa Index κ_0

$$
P\left(\left\{\vec{r}_{(n)}, \vec{u}_{(n)}\right\}_{n=1}^{N};\kappa_0,T\right) = \left\{\int_{-\infty}^{\infty}\cdots\int_{-\infty}^{\infty}\left[1 + \frac{1}{\kappa_0}\cdot\frac{H\left(\left\{\vec{r}_{(n)}, \vec{u}_{(n)}\right\}_{n=1}^{N}\right)}{k_B T}\right]^{-\kappa_0 - 1 - \frac{1}{2}dN}\right.
$$

$$
\left. d\vec{r}_{(1)}d\vec{u}_{(1)}\cdots d\vec{r}_{(N)}d\vec{u}_{(N)}\right\}^{-1} \cdot \left[1 + \frac{1}{\kappa_0}\cdot\frac{H\left(\left\{\vec{r}_{(n)}, \vec{u}_{(n)}\right\}_{n=1}^{N}\right)}{k_B T}\right]^{-\kappa_0 - 1 - \frac{1}{2}dN}.
$$

$$\tag{4.140a}$$

Normalization

$$\int_{-\infty}^{\infty} \cdots \int_{-\infty}^{\infty} P\left(\left\{\vec{r}_{(n)}, \vec{u}_{(n)}\right\}_{n=1}^{N}; \kappa_0, T\right) d\vec{r}_{(1)} d\vec{u}_{(1)} \cdots d\vec{r}_{(N)} d\vec{u}_{(N)} = 1.$$

(4.140b)

4.6.3.2 Hamiltonian Function

$$H\left(\left\{\vec{r}_{(n)}, \vec{u}_{(n)}\right\}_{n=1}^{N}\right) = \frac{1}{2} m \sum_{n=1}^{N} \left(\vec{u}_{(n)} - \vec{u}_b\right)^2 + \Phi\left(\left\{\vec{r}_{(n)}\right\}_{n=1}^{N}\right).$$ (4.140c)

4.6.3.3 Hamiltonian Degrees of Freedom Summing up the Kinetic and Potential Degrees of Freedom

$$\frac{1}{2}d \equiv \frac{1}{N} \cdot \frac{\left\langle H\left(\left\{\vec{r}_{(n)}, \vec{u}_{(n)}\right\}_{n=1}^{N}\right)\right\rangle}{k_B T} = \frac{1}{N} \cdot \sum_{n=1}^{N} \frac{\left\langle \epsilon_K\left(\vec{u}_{(n)}\right)\right\rangle}{k_B T} + \frac{1}{N} \cdot \frac{\left\langle \Phi\left(\left\{\vec{r}_{(n)}\right\}_{n=1}^{N}\right)\right\rangle}{k_B T}$$

$$= \frac{1}{2}d_K + \frac{1}{2}d_\Phi, \quad \frac{1}{2}d_\Phi \equiv \frac{1}{N} \cdot \frac{\left\langle \Phi\left(\left\{\vec{r}_{(n)}\right\}_{n=1}^{N}\right)\right\rangle}{k_B T}.$$

(4.140d)

4.6.4 Standard N-Particle (N·d)—Dimensional Kappa Distributions

4.6.4.1 Distributions, in Terms of the Velocity $\vec{u}$ With d_K Degrees of Freedom per Particle, Using the Invariant Kappa Index κ_0

4.6.4.1.1 In Clusters of N_C Uncorrelated Particles

$$P\left(\left\{\vec{u}_{(n)}\right\}_{n=1}^{N_C}; \kappa_0; \theta\right) = \prod_{n=1}^{N_C} P\left(\left\{\vec{u}_{(n)}\right\}_{n=1}^{N_C}; \kappa_0; \theta\right)$$

$$= \left[\left(\pi \kappa_0 \theta^2\right)^{-\frac{1}{2}d_K} \cdot \frac{\Gamma\left(\kappa_0 + 1 + \frac{d_K}{2}\right)}{\Gamma(\kappa_0 + 1)}\right]^{N_C}$$ (4.141a)

$$\times \prod_{n=1}^{N_C} \left[1 + \frac{1}{\kappa_0} \cdot \frac{\left(\vec{u}_{(n)} - \vec{u}_b\right)^2}{\theta^2}\right]^{-\kappa_0 - 1 - \frac{1}{2}d_K}.$$

4.6.4.1.2 In Clusters of N_C Correlated Particles; See (4.6.1.1)

$$P\left(\left\{\vec{u}_{(n)}\right\}_{n=1}^{N_C}; \kappa_0; \theta\right) \propto \left[1 + \frac{1}{\kappa_0} \cdot \frac{1}{\theta^2} \sum_{n=1}^{N_C} \left(\vec{u}_{(n)} - \vec{u}_b\right)^2\right]^{-\kappa_0 - 1 - \frac{1}{2}d_K N_C}. \qquad (4.141b)$$

4.6.4.1.3 In Clusters of N_C Correlated Particles, But the Correlation Among the Clusters (κ_0^{int}) Differs From That Among the Particles (κ_0)

$$P\left(\left\{\vec{u}_{(n)}\right\}_{n=1}^{N}; \kappa_0, \kappa_0^{\text{int}}, \theta\right) = \frac{1}{Z}$$

$$\times \left\{1 - \frac{\kappa_0^{\text{int}} + \frac{d_K}{2}}{\kappa_0^{\text{int}} + \frac{d_K N_C}{2}} \cdot \sum_{n=1}^{N_C} \left[1 - \left(1 + \frac{1}{\kappa_0 + \frac{d_K}{2}} \cdot \frac{\left(\vec{u}_{(n)} - \vec{u}_b\right)^2 - \frac{d_K}{2}\theta^2}{\theta^2}\right)^{\frac{\kappa_0 + 1 + \frac{d_K}{2}}{\kappa_0^{\text{int}} + 1 + \frac{d_K}{2}}}\right]\right\}^{-\kappa_0^{\text{int}} - 1 - \frac{1}{2}d_K N_C}.$$

$$(4.141c)$$

4.6.4.2 Distributions, in Terms of the Kinetic Energies $\left\{\varepsilon_{K(n)}\right\}_{n=1}^{N}$ With d_K Degrees of Freedom per Particle, Using the Invariant Kappa Index κ_0

4.6.4.2.1 In Clusters of N_C Uncorrelated Particles

$$P\left(\left\{\varepsilon_{K(n)}\right\}_{n=1}^{N_C}; \kappa_0; T\right) = \prod_{n=1}^{N_C} P\left(\varepsilon_{K(n)}; \kappa_0; T\right)$$

$$= \left[\frac{(\kappa_0 k_B T)^{-\frac{1}{2}d_K}}{B\left(\frac{d_K}{2}, \kappa_0 + 1\right)}\right]^{N_C} \cdot \prod_{n=1}^{N_C} \left(1 + \frac{1}{\kappa_0} \cdot \frac{\varepsilon_{K(n)}}{k_B T}\right)^{-\kappa_0 - 1 - \frac{1}{2}d_K} \prod_{n=1}^{N_C} \varepsilon_{K(n)}^{\frac{1}{2}d_K - 1}$$

$$(4.142a)$$

4.6.4.2.2 In Clusters of N_C Correlated Particles; See (4.6.1.2)

$$P\left(\left\{\varepsilon_{K(n)}\right\}_{n=1}^{N_C};\kappa_0;T\right)\propto\left(1+\frac{1}{\kappa_0}\cdot\frac{1}{k_BT}\sum_{n=1}^{N_C}\varepsilon_{K(n)}\right)^{-\kappa_0-1-\frac{1}{2}d_KN_C}\prod_{n=1}^{N_C}\varepsilon_{K(n)}^{\frac{1}{2}d_K-1} \quad (4.142b)$$

4.6.4.2.3 In Clusters of N_C Correlated Particles, But the Correlation Among the Clusters (κ_0^{int}) Differs From That Among the Particles (κ_0)

$$P\left(\left\{\varepsilon_{K(n)}\right\}_{n=1}^{N_C};\kappa_0,\kappa_0^{int},T\right)=\frac{1}{Z}\cdot\left\{1-\frac{\kappa_0^{int}+\dfrac{d_K}{2}}{\kappa_0^{int}+\dfrac{d_KN_C}{2}}\right.$$

$$\left.\cdot\sum_{n=1}^{N_C}\left[1-\left(1+\frac{1}{\kappa_0+\dfrac{d_K}{2}}\cdot\frac{\varepsilon_{K(n)}-\dfrac{d_K}{2}k_BT}{k_BT}\right)^{\frac{\kappa_0+1+\frac{d_K}{2}}{\kappa_0^{int}+1+\frac{d_K}{2}}}\right]^{-\kappa_0^{int}-1-\frac{1}{2}d_KN_C}\right\}$$

$$\times\prod_{n=1}^{N_C}\varepsilon_{K(n)}^{\frac{1}{2}d_K-1} \quad (4.142c)$$

4.6.4.3 Phase Space Distributions $\left\{H_{(n)}=H\left(\vec{r}_{(n)},\vec{u}_{(n)}\right)\right\}_{n=1}^{N_C}$ With d Degrees of Freedom per Particle, Summing d_K Kinetic and d_Φ Potential Degrees of Freedom, Using the Invariant Kappa Index κ_0

4.6.4.3.1 In Clusters of N_C Uncorrelated Particles

$$P\left(\left\{H_{(n)}\right\}_{n=1}^{N_C};\kappa_0;T\right)=\prod_{n=1}^{N_C}P\left(H_{(n)};\kappa_0;T\right)$$

$$=\frac{1}{Z}\cdot\prod_{n=1}^{N_C}\left[1+\frac{1}{\kappa_0}\cdot\frac{H\left(\vec{r}_{(n)},\vec{u}_{(n)}\right)}{k_BT}\right]^{-\kappa_0-1-\frac{1}{2}d}. \quad (4.143a)$$

Normalization

$$Z = \prod_{n=1}^{N_C} \left\{ \int_{-\infty}^{\infty} \int_{-\infty}^{\infty} \left[1 + \frac{1}{\kappa_0} \cdot \frac{H\left(\overrightarrow{r}_{(n)}, \overrightarrow{u}_{(n)}\right)}{k_B T} \right]^{-\kappa_0 - 1 - \frac{1}{2}d} d\overrightarrow{r}_{(n)} d\overrightarrow{u}_{(n)} \right\}.$$

(4.143b)

4.6.4.3.2 In Clusters of N_C Correlated Particles

$$P\left(\left\{H_{(n)}\right\}_{n=1}^{N_C}; \kappa_0; T\right) = \frac{1}{Z} \cdot \left[1 + \frac{1}{\kappa_0} \cdot \frac{1}{k_B T} \sum_{n=1}^{N_C} H\left(\overrightarrow{r}_{(n)}, \overrightarrow{u}_{(n)}\right) \right]^{-\kappa_0 - 1 - \frac{1}{2}d \cdot N_C}.$$

(4.144a)

Normalization

$$Z = \int_{-\infty}^{\infty} \int_{-\infty}^{\infty} \left[1 + \frac{1}{\kappa_0} \cdot \frac{1}{k_B T} \sum_{n=1}^{N_C} H\left(\overrightarrow{r}_{(n)}, \overrightarrow{u}_{(n)}\right) \right]^{-\kappa_0 - 1 - \frac{1}{2}d \cdot N_C}$$

(4.144b)

$$d\overrightarrow{r}_{(1)} d\overrightarrow{u}_{(1)} \cdots d\overrightarrow{r}_{(N_C)} d\overrightarrow{u}_{(N_C)}.$$

4.6.4.3.3 In Clusters of N_C Correlated Particles, But the Correlation Among the Clusters $\left(\kappa_0^{int}\right)$ Differs From That Among the Particles (κ_0)

$$P\left(\left\{H_{(n)}\right\}_{n=1}^{N_C}; \kappa_0, \kappa_0^{int}, T\right) = \frac{1}{Z}$$

$$\times \left\{ 1 - \frac{\kappa_0^{int} + \frac{d}{2}}{\kappa_0^{int} + \frac{d \cdot N_C}{2}} \cdot \sum_{n=1}^{N_C} \left\{ 1 - \left[1 + \frac{1}{\kappa_0 + \frac{d}{2}} \cdot \frac{H\left(\overrightarrow{r}_{(n)}, \overrightarrow{u}_{(n)}\right) - \frac{d}{2}k_B T}{k_B T} \right]^{\frac{\kappa_0 + 1 + \frac{d}{2}}{\kappa_0^{int} + 1 + \frac{d}{2}}} \right\} \right\}^{-\kappa_0^{int} - 1 - \frac{1}{2}d \cdot N_C}.$$

(4.145a)

Normalization

$$Z = \int_{-\infty}^{\infty} \cdots \int_{-\infty}^{\infty} \{\cdots\} d\vec{r}_{(1)} d\vec{u}_{(1)} \cdots d\vec{r}_{(N_C)} d\vec{u}_{(N_C)}$$

$$\{\cdots\} = \left\{ 1 - \frac{\kappa_0^{\mathrm{int}} + \frac{d}{2}}{\kappa_0^{\mathrm{int}} + \frac{d \cdot N_C}{2}} \cdot \sum_{n=1}^{N_C} \left\{ 1 - \left[1 + \frac{1}{\kappa_0 + \frac{d}{2}} \cdot \frac{H\left(\vec{r}_{(n)}, \vec{u}_{(n)}\right) - \frac{d}{2}k_B T}{k_B T} \right]^{-\kappa_0^{\mathrm{int}}+1+\frac{d}{2}} \right\}^{\kappa_0^{\mathrm{int}}+1+\frac{d}{2}} \right\}^{-\kappa_0^{\mathrm{int}}-1-\frac{1}{2}d \cdot N_C}$$

$$(4.145b)$$

4.6.5 Multispecies Distributions (Livadiotis and McComas, 2014a)

4.6.5.1 Distributions, in Terms of the Velocity, of N Different Particle Species, $\vec{u}_{(1)}, \vec{u}_{(2)}, \cdots, \vec{u}_{(N)}$, With d_K Degrees of Freedom per Particle, Using the Invariant Kappa Index, κ_0

4.6.5.1.1 N Uncorrelated Species

$$P\left(\left\{\vec{u}_{(n)}\right\}_{n=1}^{N}; \left\{\kappa_{0(n)}\right\}_{n=1}^{N}; \left\{\theta_{(n)}\right\}_{n=1}^{N}\right) = \prod_{n=1}^{N} P\left(\vec{u}_{(n)}; \kappa_{0(n)}; \theta_{(n)}\right)$$

$$= \prod_{n=1}^{N} \left(\pi \kappa_{0(n)} \theta_{(n)}^2\right)^{-\frac{1}{2}d_K} \cdot \frac{\Gamma\left(\kappa_{0(n)} + 1 + \frac{d_K}{2}\right)}{\Gamma\left(\kappa_{0(n)} + 1\right)} \left[1 + \frac{1}{\kappa_{0(n)}} \cdot \frac{\left(\vec{u}_{(n)} - \vec{u}_b\right)^2}{\theta_{(n)}^2}\right]^{-\kappa_{0(n)}-1-\frac{1}{2}d_K}$$

$$(4.146a)$$

4.6.5.1.2 N Correlated Species (Compare With Eqs. 4.1a, 4.136a)

$$P\left(\left\{\vec{u}_{(n)}\right\}_{n=1}^{N}; \kappa_0; \left\{\theta_{(n)}\right\}_{n=1}^{N}\right) = (\pi \kappa_0)^{-\frac{1}{2}d_K N} \cdot \frac{\Gamma\left(\kappa_0 + 1 + \frac{d_K}{2}N\right)}{\Gamma(\kappa_0 + 1)}$$

$$\cdot \prod_{n=1}^{N} \theta_{(n)}^{-d_K} \cdot \left[1 + \frac{1}{\kappa_0} \cdot \sum_{n=1}^{N} \frac{\left(\vec{u}_{(n)} - \vec{u}_b\right)^2}{\theta_{(n)}^2}\right]^{-\kappa_0-1-\frac{1}{2}d_K N}$$

$$(4.146b)$$

Comment: The correlation among the species is the same with the correlation among the particles.

4.6.5.1.3 *N* Correlated Species, But the Correlation Among the Species (κ_0^{int}) Differs From That Among the Particles (κ_0)

$$
P\left(\left\{\vec{u}_{(n)}\right\}_{n=1}^{N}; \kappa_0, \kappa_0^{int}, \left\{\theta_{(n)}\right\}_{n=1}^{N}\right) = \frac{1}{Z} \cdot \left\{ 1 - \frac{\kappa_0^{int} + \dfrac{d_K}{2}}{\kappa_0^{int} + \dfrac{d_K N}{2}} \cdot \right.
$$

$$
\left. \sum_{n=1}^{N}\left[1 - \left(1 + \frac{1}{\kappa_0 + \dfrac{d_K}{2}} \cdot \frac{\left(\vec{u}_{(n)} - \vec{u}_b\right)^2 - \dfrac{d_K}{2}\theta_{(n)}^2}{\theta_{(n)}^2}\right)^{\frac{\kappa_0 + 1 + \frac{d_K}{2}}{\kappa_0^{int} + 1 + \frac{d_K}{2}}}\right] \right\}^{-\kappa_0^{int} - 1 - \frac{1}{2}d_K N}.
$$

$$(4.146c)$$

4.6.5.2 *Distributions, in Terms of the Kinetic Energies,* $\left\{\varepsilon_{K(n)}\right\}_{n=1}^{N}$, *With* d_K *Degrees of Freedom per Particle, Using the Invariant Kappa Index,* κ_0

4.6.5.2.1 *N* Uncorrelated Species

$$
P\left(\left\{\varepsilon_{K(n)}\right\}_{n=1}^{N}; \left\{\kappa_{0(n)}\right\}_{n=1}^{N}; \left\{T_{(n)}\right\}_{n=1}^{N}\right) = \prod_{n=1}^{N} P\left(\varepsilon_{K(n)}; \kappa_{0(n)}; T_{(n)}\right)
$$

$$
= \prod_{n=1}^{N} \frac{\left(\kappa_{0(n)} k_B T_{(n)}\right)^{-\frac{1}{2}d_K}}{B\left(\dfrac{d_K}{2}, \kappa_{0(n)} + 1\right)} \cdot \left(1 + \frac{1}{\kappa_{0(n)}} \cdot \frac{\varepsilon_{K(n)}}{k_B T_{(n)}}\right)^{-\kappa_{0(n)} - 1 - \frac{1}{2}d_K} \prod_{n=1}^{N} \varepsilon_{K(n)}^{\frac{1}{2}d_K - 1}.
$$

$$(4.147a)$$

4.6.5.2.2 *N* Correlated Species

$$
P\left(\left\{\varepsilon_{K(n)}\right\}_{n=1}^{N}; \kappa_0; \left\{T_{(n)}\right\}_{n=1}^{N}\right) \propto \left(1 + \frac{1}{\kappa_0} \cdot \sum_{n=1}^{N} \frac{\varepsilon_{K(n)}}{k_B T_{(n)}}\right)^{-\kappa_0 - 1 - \frac{1}{2}d_K N} \prod_{n=1}^{N} \varepsilon_{K(n)}^{\frac{1}{2}d_K - 1}.
$$

$$(4.147b)$$

4.6.5.2.3 N Correlated Species, But the Correlation Among the Species (κ_0^{int}) Differs From That Among the Particles (κ_0)

$$
P\left(\left\{\varepsilon_{K(n)}\right\}_{n=1}^{N}; \kappa_0, \kappa_0^{\text{int}}, \left\{T_{(n)}\right\}_{n=1}^{N}\right) = \frac{1}{Z} \cdot \left\{ 1 - \frac{\kappa_0^{\text{int}} + \frac{d_K}{2}}{\kappa_0^{\text{int}} + \frac{d_K N}{2}} \right.
$$

$$
\times \sum_{n=1}^{N} \left[1 - \left(1 + \frac{1}{\kappa_0 + \frac{d_K}{2}} \cdot \frac{\varepsilon_{K(n)} - \frac{d_K}{2} k_B T_{(n)}}{k_B T_{(n)}} \right)^{\frac{\kappa_0 + 1 + \frac{d_K}{2}}{\kappa_0^{\text{int}} + 1 + \frac{d_K}{2}}} \right]^{-\kappa_0^{\text{int}} - 1 - \frac{1}{2} d_K N} \left. \right\} \prod_{n=1}^{N} \varepsilon_{K(n)}^{\frac{1}{2} d_K - 1}.
$$

$$(4.147c)$$

4.6.5.3 Phase Space Distributions of N Species
$\left\{ H_{(n)} = H\left(\vec{r}_{(n)}, \vec{u}_{(n)} \right) \right\}_{n=1}^{N}$ With d Degrees of Freedom per Particle, Summing d_K Kinetic and d_Φ Potential Degrees of Freedom, Using the Invariant Kappa Index, κ_0

4.6.5.3.1 N Uncorrelated Species

$$
P\left(\left\{H_{(n)}\right\}_{n=1}^{N}; \left\{\kappa_{0(n)}\right\}_{n=1}^{N}; \left\{T_{(n)}\right\}_{n=1}^{N}\right)
$$

$$
= \prod_{n=1}^{N} P\left(H_{(n)}; \kappa_{0(n)}; T_{(n)}\right) = \frac{1}{Z} \cdot \prod_{n=1}^{N} \left[1 + \frac{1}{\kappa_{0(n)}} \cdot \frac{H\left(\vec{r}_{(n)}, \vec{u}_{(n)}\right)}{k_B T_{(n)}} \right]^{-\kappa_{0(n)} - 1 - \frac{1}{2} d}.
$$

$$(4.148a)$$

4.6.5.3.2 N Correlated Species

$$
P\left(\left\{H_{(n)}\right\}_{n=1}^{N}; \kappa_0; \left\{T_{(n)}\right\}_{n=1}^{N}\right) = \frac{1}{Z} \cdot \left[1 + \frac{1}{\kappa_0} \cdot \sum_{n=1}^{N} \frac{H\left(\vec{r}_{(n)}, \vec{u}_{(n)}\right)}{k_B T_{(n)}} \right]^{-\kappa_0 - 1 - \frac{1}{2} d \cdot N}.
$$

$$(4.148b)$$

4.6.5.3.3 N Correlated Species, But the Correlation Among the Species $\left(\kappa_0^{\text{int}}\right)$ Differs From That Among the Particles (κ_0)

$$
P\left(\left\{H_{(n)}\right\}_{n=1}^{N}; \kappa_0, \kappa_0^{\text{int}}, \left\{T_{(n)}\right\}_{n=1}^{N}\right) = \frac{1}{Z} \cdot \left\{ 1 - \frac{\kappa_0^{\text{int}} + \dfrac{d}{2}}{\kappa_0^{\text{int}} + \dfrac{d \cdot N}{2}} \right.
$$

$$
\times \sum_{n=1}^{N} \left\{ 1 - \left[1 + \frac{1}{\kappa_0 + \dfrac{d}{2}} \cdot \frac{H\left(\vec{r}_{(n)}, \vec{u}_{(n)}\right) - \dfrac{d}{2} k_B T_{(n)}}{k_B T_{(n)}} \right]^{\frac{\kappa_0 + 1 + \frac{d}{2}}{\kappa_0^{\text{int}} + 1 + \frac{d}{2}}} \right\}^{-\kappa_0^{\text{int}} - 1 - \frac{d}{2} d \cdot N} \right\}.
$$

(4.148c)

4.7 Non-Euclidean–Normed Distributions (Livadiotis, 2007, 2008, 2012, 2016b)

4.7.1 Standard d_K–Dimensional Kappa Distribution of Velocity, $\vec{u}$

$$
P(\vec{u}; \kappa_0, \theta, p) = \frac{1}{Z} \cdot \left\{ 1 + \frac{1}{\kappa_0 \cdot (p-1)} \cdot \left\{ \frac{d_K}{2} + x_*^{p-1} \cdot \left| \frac{(\vec{u} - \vec{u}_b)^2}{\theta^2} - \frac{d_K}{2} \right|^{p-2} \right. \right.
$$

$$
\times \left. \left. \left[\frac{(\vec{u} - \vec{u}_b)^2}{\theta^2} - \frac{d_K}{2} \right] \right\} \right\}^{-\kappa_0 - 1 - \frac{1}{2} d_K \cdot (p-1)^{-1}},
$$

(4.149a)

with the high-energy approximation (e.g., see also Qureshi et al., 2003, 2014; Randol and Christian, 2014, 2016),

$$
P(\vec{u}; \kappa_0, \theta, p) \cong \frac{1}{Z} \cdot \left[1 + \frac{x_*^{p-1}}{\kappa_0 \cdot (p-1)} \cdot \frac{\left| \vec{u} - \vec{u}_b \right|^{2(p-1)}}{\theta^{2(p-1)}} \right]^{-\kappa_0 - 1 - \frac{1}{2} d_K \cdot (p-1)^{-1}},
$$

(4.149b)

for $\left| \vec{u} - \vec{u}_b \right|^2 \gg \frac{1}{2} d_K \theta^2.$ (4.149b)

Partition function:

$$
Z = \int_{-\infty}^{\infty} \left\{ 1 + \frac{1}{\kappa_0 \cdot (p-1)} \cdot \left\{ \frac{d_K}{2} + x_*^{p-1} \cdot \left| \frac{(\vec{u} - \vec{u}_b)^2}{\theta^2} - \frac{d_K}{2} \right|^{p-2} \right. \right.
$$

$$
\times \left. \left. \left[\frac{(\vec{u} - \vec{u}_b)^2}{\theta^2} - \frac{d_K}{2} \right] \right\} \right\}^{-\kappa_0 - 1 - \frac{1}{2} d_K \cdot (p-1)^{-1}} d\vec{u}.
$$

(4.149c)

4.7.2 Standard d_K–Dimensional Kappa Distribution of Kinetic Energy, ε_K

$$P(\varepsilon_K; \kappa_0, T, p) = \frac{1}{Z}$$

$$\times \left\{ 1 + \frac{1}{\kappa_0 \cdot (p-1)} \cdot \left[\frac{d_K}{2} + x_*^{p-1} \cdot \left| \frac{\varepsilon_K}{k_B T} - \frac{d_K}{2} \right|^{p-2} \left(\frac{\varepsilon_K}{k_B T} - \frac{d_K}{2} \right) \right] \right\}^{-\kappa_0 - 1 - \frac{1}{2}d_K \cdot (p-1)^{-1}} \varepsilon_K^{\frac{1}{2}d_K - 1},$$

$$\text{(4.150a)}$$

with the high-energy approximation.

$$P(\varepsilon_K; \kappa_0, T, p) \cong \frac{1}{Z} \cdot \left[1 + \frac{x_*^{p-1}}{\kappa_0 \cdot (p-1)} \cdot \left(\frac{\varepsilon_K}{k_B T} \right)^{p-1} \right]^{-\kappa_0 - 1 - \frac{1}{2}d_K \cdot (p-1)^{-1}} \varepsilon_K^{\frac{1}{2}d_K - 1},$$

for $\varepsilon_K \gg \frac{1}{2}d_K k_B T$. $\qquad\qquad$ (4.150b)

Partition function:

$$Z = (k_B T/x_*)^{-\frac{1}{2}d_K} \cdot \int_0^\infty \left\{ 1 + \frac{1}{\kappa_0 \cdot (p-1)} \cdot \left[\frac{d_K}{2} + \left| x - x_* \frac{d_K}{2} \right|^{p-2} \right. \right.$$

$$\left. \left. \times \left(x - x_* \frac{d_K}{2} \right) \right] \right\}^{-\kappa_0 - 1 - \frac{1}{2}d_K \cdot (p-1)^{-1}} x^{\frac{1}{2}d_K - 1} dx. \qquad \text{(4.150c)}$$

4.7.3 Argument x_*

$$\int_0^\infty \left\{ 1 + \frac{1}{\kappa_0 \cdot (p-1)} \cdot \left[\frac{d_K}{2} + \left| x - x_* \frac{d_K}{2} \right|^{p-2} \left(x - x_* \frac{d_K}{2} \right) \right] \right\}^{-\kappa_0 - 1 - \frac{1}{2}d_K \cdot (p-1)^{-1}}$$

$$\times \left| x - x_* \frac{d_K}{2} \right|^{p-2} \left(x - x_* \frac{d_K}{2} \right) x^{\frac{1}{2}d_K - 1} dx = 0,$$

$$\text{(4.151)}$$

e.g., $x_* = 1$ for $p = 2$ and any κ_0.

4.8 Discrete Distributions (Tsallis et al., 1998)

4.8.1 Distribution of Energy

$$P_i(\varepsilon_i; \kappa, T) = \frac{1}{Z\left(\{\varepsilon_j\}_{j=1}^{W}\right)} \cdot \left[1 + \frac{1}{\kappa} \cdot \frac{\varepsilon_i - U\left(\{\varepsilon_j\}_{j=1}^{W}; \kappa, T\right)}{k_B T}\right]^{-\kappa-1} \qquad (4.152a)$$

4.8.2 Partition Function

$$Z\left(\{\varepsilon_j\}_{j=1}^{W}; \kappa, T\right) = \sum_{i=1}^{W} \left[1 + \frac{1}{\kappa} \cdot \frac{\varepsilon_i - U\left(\{\varepsilon_j\}_{j=1}^{W}; \kappa, T\right)}{k_B T}\right]^{-\kappa-1} \qquad (4.152b)$$

4.8.3 Internal Energy

The internal energy $U\left(\{\varepsilon_j\}_{j=1}^{W}; \kappa, T\right) \equiv \langle \varepsilon_i \rangle$ is implicitly given by

$$\sum_{i=1}^{W} \left(1 + \frac{1}{\kappa} \cdot \frac{\varepsilon_i - U}{k_B T}\right)^{-\kappa-1} \cdot (\varepsilon_i - U) = 0. \qquad (4.152c)$$

4.9 Concluding Remarks

The chapter provided all the formulae of kappa distributions necessary to describe particle populations out of thermal equilibrium in plasmas and beyond. In particular, it was shown:

- isotropic distributions, in the absence of a potential energy: the positive and negative multidimensional kappa distributions, in the plasma flow (comoving) or an inertial reference frame;
- anisotropic distributions, in the absence of a potential energy, provided for the cases: correlated degrees of freedom, correlation between the projection at a certain direction and the perpendicular plane, self-correlated degrees of freedom, self-correlated projections at a direction and perpendicular plane, self-correlated degrees of freedom with different kappa indices, self-correlated projections at a direction and perpendicular plane with different kappa, self-correlated projections of different dimensionality and kappa, different self-correlation, and intercorrelation between degrees of freedom;

- formulae of kappa distributions in the presence of a potential energy: general Hamiltonian kappa distribution; in the presence of positive or negative attractive potentials; or, of any small positive/negative attractive/repulsive potentials defined in a finite volume; reduction of the distribution of the kinetic and potential energy to the equivalent local distribution of solely the kinetic energy; kappa distributions for positive (oscillation type) or negative (gravitational type) power law central potentials; properties of power law central potentials (local density, temperature, thermal pressure, polytropic index); marginal and conditional distributions; distributions with angular potentials (e.g., magnetization potential);
- formulae of (positive or negative) multiparticle kappa distributions in the presence or not of a potential energy and of multispecies kappa distributions;
- formulation of generalized L_p kappa distributions, which are based on non-Euclidean L_p-norms; and
- formulation of discrete kappa distributions (distribution of energy, partition function, internal energy).

These developments allow the researcher to express and work with the distributions of any particle populations out of thermal equilibrium in space, geophysical, laboratory, or other plasmas.

4.10 Science Questions for Future Research

Future analyses and observations need to address the following questions:

1. What is the magnetization angular distribution in an inertial frame?
2. What is the temperature if particle and group correlations are different?
3. Can a superposition on κ form a complete/orthogonal basis?

PART 2
Plasma Physics

5. Basic Plasma Parameters Described by Kappa
   Distributions   249
   G. Livadiotis
6. Superstatistics: Superposition of Maxwell–Boltzmann
   Distributions   313
   C. Beck, E.G.D. Cohen
7. Linear Kinetic Waves in Plasmas Described by Kappa
   Distributions   329
   A.F. Viñas, R. Gaelzer, P.S. Moya, R. Mace, J.A. Araneda
8. Nonlinear Wave–Particle Interaction and Electron Kappa
   Distribution   363
   P.H. Yoon, G. Livadiotis
9. Solitary Waves in Plasmas Described by Kappa
   Distributions   399
   G.S. Lakhina, S. Singh

CHAPTER 5

Basic Plasma Parameters Described by Kappa Distributions

G. Livadiotis

Southwest Research Institute, San Antonio, TX, United States

Chapter Outline

5.1 **Summary** 250
5.2 **Introduction** 250
5.3 **Polytropes** 252
  5.3.1 Simple Polytropes and Their Characteristic Exponent: The Polytropic Index 252
  5.3.2 Generalized Polytropic Relations 254
  5.3.3 Connection With the Kappa Index 257
    5.3.3.1 *Thermal Equilibrium, $\kappa_O \to \infty$* 257
    5.3.3.2 *Out of Thermal Equilibrium, $\kappa_O < \infty$* 259
  5.3.4 Complicated Relations Between Polytropic and Kappa Indices 261
5.4 **Correlation Between Particle Energies** 267
5.5 **Debye Length in Equilibrium and Nonequilibrium Plasmas** 272

5.5.1 General Aspects 272
5.5.2 Poisson Equation for the Electrostatic Potential 272
5.5.3 Symmetric Poisson Equation and Solutions 276
    5.5.3.1 *1-D Potential: Planar Charge Perturbation* 276
    5.5.3.2 *2-D Potential: Linear Charge Perturbation* 277
    5.5.3.3 *3-D Potential: Point Charge Perturbation* 278
    5.5.3.4 *d−D Potential* 279
  5.5.4 The Case of Large Potential Energy 279
  5.5.5 Main Interpretations 281
5.6 **Electrical Conductivity** 286
5.7 **Collision Frequency and Mean Free Path** 292
5.8 **Magnetization: The Curie Constant** 295

Kappa Distributions. http://dx.doi.org/10.1016/B978-0-12-804638-8.00005-X

5.9 Large-Scale Quantization Constant 300
5.9.1 The Role of Correlations 300
5.9.2 The Smallest Phase Space Parcel in Plasmas 302

5.9.3 Estimation of $\hbar*$ for Space Plasmas 304
5.9.4 Application: Missing Plasma Parameters 308
5.10 Concluding Remarks 310
5.11 Science Questions for Future Research 311
Acknowledgments 311

5.1 Summary

This chapter presents the derivation of basic plasma parameters within the framework of kappa distribution theory. We examine the thermodynamic processes of polytropes and the physics behind the notion of polytropic index, the correlations between any two particle energies or other statistical moments, the Debye shielding and length scale, the electrical conductivity, the collision frequency for electrons colliding with ions and the respective mean free path, the plasma magnetization, the Pierre Curie's law and the Curie constant, and the topic of large-scale quantization constant. It is now straightforward to apply these developments to study space, geophysical, laboratory, or other plasmas out of thermal equilibrium.

Science Question: How are the kappa distributions involved in basic plasma parameters and concepts?

Keywords: Collision frequency; Correlation; Curie constant; Debye length; Electrical conductivity; Large-scale quantization; Mean free path; Polytrope.

5.2 Introduction

Particle populations in collisionless and weakly coupled plasmas are described by kappa distributions (or combinations thereof). Several mechanisms exist for generating these distributions, where some of them are detailed in this book. Phenomenologically, however, the presence of these distributions is interwoven with the existence of correlations between plasma particles; the stronger the correlation, the further the plasma resides from thermal equilibrium. In this sense, the observation of a kappa distribution, or equivalently, the existence of particle correlations, certifies for a collisionless plasma, otherwise collisions would have destroyed correlations.

The statistical behavior of plasmas out of thermal equilibrium is related to their nature as collisionless and weakly coupled plasmas, with large numbers of particles in their Debye spheres. Weakly coupled plasmas are not governed by interactions between individual particles, but instead the overall collective electrostatic forces of many particles. Strong interactions between individual particles are relatively rare and cause little if any sudden change in the particles' motion, which is largely governed by its kinetic energy. Therefore, these plasmas show strong collective behavior that characterizes the correlated particles within a Debye sphere without localized phenomena between individual particles due to interactions or collisions. This behavior leads these systems to exotic statistical states that cannot be understood by the classical statistical description of thermal equilibrium.

Kappa distributions are based on the solid statistical background of nonextensive statistical mechanics (Chapter 1) and describe systems that are in stationary states but out of thermal equilibrium. Stationary states refer to the distribution function of phase space of a system, meaning that must be, at least temporarily, invariant, even though is not given by the classical Boltzmann–Gibbs distribution of energy or the equivalent Maxwell distribution of velocities. Instead, the distribution function for stationary states out of thermal equilibrium is described via kappa distributions.

This chapter presents the effect of the kappa distribution theory in basic parameters and concepts involved in statistical plasma physics. We examine the thermodynamic processes called polytropes and the physics behind the polytropic index, the correlations between particle energies and moments, the Debye shielding and length scale, the electrical conductivity, the collision frequency and the mean free path, the plasma magnetization and the Curie constant, and the topic of large-scale quantization. The developed ideas and results will be useful and evaluated in the general plasma physics research, as well as in the context of the following chapters of Parts B and C.

The chapter is structured as follows: in the next Section 5.3, we study the thermodynamic processes called polytropes, the particle density and temperature in the presence of potential energy and how they involved producing a polytrope, and the relation between the polytropic and kappa indices. In Section 5.4, we show how to calculate these correlations via the Pearson's formula by deriving the covariance between the moments of two particles using the two-particle joint kappa distribution function. In Section 5.5, we investigate the Debye shielding in plasmas and derive the Debye length in several dimensionalities, symmetries, and coordinate systems, and discuss the main interpretations of this scale. In Section 5.6, we derive the electrical conductivity in plasmas described by kappa distributions using the Fokker–Planck equation. In Section 5.7, we use the notion of the electrical conductivity to derive the ion–electron collision frequency and the related plasma mean free path. In Section 5.8, we derive the magnetization of a system, the Pierre Curie's law, and the adaptation of the Curie constant for paramagnetic particles out of thermal equilibrium described by kappa distributions. The analysis uses the theory and formulation of the kappa distributions that describe particle systems with a nonzero potential energy. In Section 5.9, we present the concept of the large-scale quantization constant, where observations of space and other plasmas identify a new, larger phase space minimum, about 12 orders of magnitude larger than the Planck's constant. Finally, the concluding remarks are given in Section 5.10, while three general science questions for future analyses are posed in Section 5.11.

5.3 Polytropes

5.3.1 Simple Polytropes and Their Characteristic Exponent: The Polytropic Index

A polytropic relation connects the values of thermal variables (such as, density n, temperature T, thermal pressure P, the kappa index κ or κ_0, etc.) along a certain streamline of the plasma flow. A polytrope is a certain thermodynamic process characterized by such a relation. Typically, this is a power law between two thermodynamic variables, that is,

$$[P(\overrightarrow{r})/P(\overrightarrow{r}_*)] = [n(\overrightarrow{r})/n(\overrightarrow{r}_*)]^a, \text{ or}$$
$$[n(\overrightarrow{r})/n(\overrightarrow{r}_*)] = [T(\overrightarrow{r})/T(\overrightarrow{r}_*)]^\nu, \text{ with } \nu \equiv 1/(a-1), \tag{5.1}$$

where the exponent a denotes the typical polytropic index; ν is a secondary polytropic index, which corresponds to the effective degrees of freedom $\frac{1}{2}d_{eff}$ (Livadiotis, 2015e); $n(\overrightarrow{r})$, $T(\overrightarrow{r})$, and $P(\overrightarrow{r})$, are, respectively, the local density, temperature, and thermal pressure, along the streamline; and $n_* \equiv n(\overrightarrow{r}_*)$, $T_* \equiv T(\overrightarrow{r}_*)$, and $P_* \equiv P(\overrightarrow{r}_*)$ are the values at some location on the streamline.

Fig. 5.1 shows the arrangement of all the polytropic indices and the corresponding thermodynamic processes in plasmas (for fixed polytropic index a or

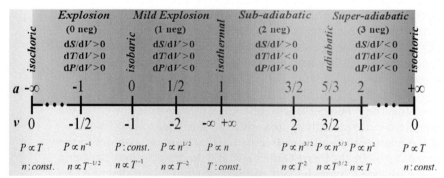

FIGURE 5.1
Polytropic spectrum: Arrangement of thermodynamic processes along the interval of the (constant) polytropic index a or $\nu \equiv 1/(a-1)$. Starting from $a \to -\infty$ (isochoric) and moving with increasing a (left to right), we find the four intervals of "Explosion," "Mild Explosion," "Subadiabatic," and "Superadiabatic," which are characterized by a number of 0, 1, 2, and 3 negative terms (noted in *blue*) out of the three {dS/dV, dT/dV, dP/dV}. The entropy S depends on the positional dimensions, which is taken to be $d = 3$. The adiabatic process corresponds to $a = 1 + 2/d$; thus $a = 5/3$ for $d = 3$. Adapted by Livadiotis (2016).

$v \equiv 1/(a-1))$. The dependence of thermal pressure P, temperature T, and entropy S on the volume V or density n is important for setting this arrangement:

$$P \propto n^a, \text{ or } P \propto V^{-a}, \text{ thus, } \frac{dP}{dV} \propto -a = -1/v - 1. \tag{5.2a}$$

$$n \propto T^v, \text{ or } T \propto V^{-1/v} \propto V^{1-a}, \text{ thus, } \frac{dT}{dV} \propto (1-a) = -1/v. \tag{5.2b}$$

$$S = \text{const.} + \left(\frac{d}{2} - v\right) \cdot \ln T, \text{ thus, } \frac{dS}{dV} \propto \left(\frac{d}{2} - v\right) \cdot \frac{dT}{dV} \propto \frac{d}{2} \Big/ v - 1$$

$$= \frac{d}{2}(a-1) - 1. \tag{5.2c}$$

Note that Eq. (5.2c) comes from the Sackur–Tetrode entropic formula $S = \text{const.} + \ln\left(T^{d/2}/n\right)$, e.g., Chapter 2; Livadiotis and McComas (2013b); Livadiotis (2014a); here, d denotes the positional dimension (that can be also denoted with d_r); it must not be confused with the effective dimensionality $d_{\text{eff}} \equiv 2v = 2/(a-1)$. Inequalities, Eqs. (5.2a–c), are demonstrated in Fig. 5.1.

During the last two decades, the polytropic index of solar wind has been estimated by analyzing various datasets of solar wind protons (e.g., Totten et al., 1995; Newbury et al., 1997; Kartalev et al., 2006; Nicolaou et al., 2014). Earlier et al. (1984) estimated the polytropic index at the terrestrial bow shock applying the Rankine–Hugoniot conditions upstream and downstream the shock. They estimated a polytropic index between 1.6 and 1.7, which is consistent with the adiabatic 5/3, while Zhuang and Russell (1981) found that a polytropic index $a \approx 2$ fits better the terrestrial bow shock. As it is shown in Table 5.1, space plasmas are mostly characterized by polytropic indices close to the adiabatic or the isobaric values. For example, interplanetary or bow shocks are characterized by polytropic indices close to its adiabatic value $a = 5/3$, while numerous analyses were performed considering this adiabatic polytrope (e.g., Zank, 1999).

On the other hand, using the method of correlation maximization to the data derived from Livadiotis et al. (2011), Livadiotis and McComas (2013b) found that the polytropic index in the inner heliosheath is near zero (also, see: Livadiotis and McComas, 2012, Section 4 of the present paper). These anticorrelations of $n-T$, consistent with constant or quasiconstant thermal pressure P (sometimes denoted with p), were also found in the low-latitude boundary layer at the terrestrial magnetosheath (Sckopke et al., 1981), in the terrestrial central plasma sheet (Pang et al., 2015a, 2015b), and in the Jovian magnetosheath (Nicolaou et al., 2015). It is not clear why thermodynamic processes in these space plasmas are isobaric (Fig. 5.1); a possible explanation may be the relation between the kappa and polytropic indices, as we will show further as follows.

Table 5.1 Studies for Determining the Solar Wind Polytropic Index

Study References	Resulting a	Datasets	Space Plasma
[1] Zhuang and Russell (1981)	2	ISEE-1	Bow shock
[2] Tatrallyay et al. (1984)	1.85	PVO	Venus bow shock
[3] Winterhalter et al. (1984)	1.6−1.7	ISEE-1	Bow shock
[4] Baumjohann and Paschmann (1989)	1.67 ± 0.5, 1.4	AMPTE/IRM	Plasma sheet
[5] Newbury et al. (1997)	5/3, 2	PVO	Solar wind
[6] Kartalev et al. (2006)	0.5−2.5	Wind	Solar wind
[7] Nicolaou et al. (2014)	1.8 ± 2.4	OMNI	Solar wind
[8] Totten et al. (1995)	1.46, 1.58	Helios-1	Solar wind
[9] Osherovich et al. (1993)	0.5, 0.6	IMP-8	Magnetic clouds
[10] Hammond et al. (1996)	0.73, 0.78	Ulysses/CMEs	CMEs
[11] Sckopke et al. (1981)	∼0	ISEE-1 and -2	Magnetosheath
[12] Livadiotis et al. (2011)	-0.04 ± 0.07	IBEX	Inner heliosheath
[13] Pang et al. (2015a, 2015b)	−0.15, 0−1	Cluster	Plasma sheet
[14] Nicolaou et al. (2015)	∼0	New Horizons	Jovian magnetosheath
[15] Livadiotis (2016c)	∼0	IBEX	Inner heliosheath
[16] Livadiotis and Desai (2016)	∼1.66	Wind	Solar wind

Notes: The first two cases [1,2] have polytropic indices $5/3 < a$, noted in the region of "Superadiabatic" processes. Cases [3−7,16] have polytropic indices close to $a \sim 5/3$, noted as a quasiadiabatic process. The case [8] direct to $1 < a < 5/3$, noted as "Subadiabatic" process. Magnetic clouds [9] and CMEs [10] studies found polytropic indices $0 < a < 1$, noted in the region of "Mild Explosion." The cases [11−15] indicate isobaric processes (in [12], see also: Livadiotis and McComas, 2012, 2013). For the abbreviations, see Appendix A.

5.3.2 Generalized Polytropic Relations

Space plasmas are, in their majority, weakly coupled particle systems. Then the equation of state is similar the ideal gas law $P = n\,k_\text{B}T$, where one of the three thermodynamic variables (P,n,T) is always dependent on the other two. Usually, we choose the two independent variables to be (n,T) or (n,P).

Given the polytropic relation in Eq. (5.1), the number of the independent variables reduces to one. The three variables (n,T,P) become dependent to each other. However, this dependence is not "global," namely, it is not characterizing the whole plasma, but only a certain streamline of the plasma flow. In particular, a polytropic relation is also dependent by a parameter that retains a fixed value along a streamline, that is, the parameter $\Pi = P \cdot (n/n_*)^{-a}$; this parameter remains constant along a streamline but differs among different streamlines. For this reason, any two parameters, e.g., (n,T) or (n,P), are not globally dependent parameters. This would be true only if Π was universal constant, i.e., independent of the streamline. In fact, there is not actual reduction of the degrees of freedom from two to one, since the one variable is always replaced by the new parameter Π, i.e., (n,T) or $(n, P) \rightarrow (n,\Pi)$. The only global relation is $P = n\,k_\text{B}T$, which really reduces the number of the degrees of freedom from three to two.

The invariant physical quantity that remains constant under a polytropic procedure is called the polytropic invariant pressure, and it is defined by

$$\Pi_a(P,n) \equiv P \cdot (n/n_*)^{-a}. \tag{5.3}$$

A generalization of polytropes comes from a synthetic thermodynamic process that can be expressed as a superposition of polytropic processes. This can be realized either as a superposition on both a or ν polytropic indices. Let us start with the first case, namely, with a density of polytropic indices $D(a)$. Then the generalized polytropic pressure $\Pi_D(P,n)$ describes the physical quantity that remains invariant under thermodynamic processes of the pressure–density relation that generalizes Eq. (5.3), that is

$$\Pi_D(P,n) \equiv P \cdot \int_{-\infty}^{\infty} D(a) \cdot (n/n_*)^{-a} da. \tag{5.4}$$

It has to be noted that the standard polytropic index is usually kept constant along a streamline, but it may be variable among streamlines (e.g., Nicolaou and Livadiotis, 2017). The concept here is that the polytropic index is variable along the streamlines.

As an example of $D(a)$, we consider a Gaussian distribution of polytropic indices about the most frequent polytropic index $\bar{a}$ with standard deviation σ_a, that is

$$D(a) = \frac{1}{\sqrt{2\pi}\sigma} \cdot \exp\left[-\frac{(a-\bar{a})^2}{2\sigma_a^2}\right]. \tag{5.5}$$

Then, according to Eq. (5.4), the generalized polytropic invariant pressure along the flow becomes

$$\Pi_D(P,n) = P \cdot \int_{-\infty}^{\infty} \frac{1}{\sqrt{2\pi}\sigma_a} \cdot \exp\left[-\frac{(a-\bar{a})^2}{2\sigma_a^2} - a \cdot \ln(n/n_*)\right] da$$

$$= P \cdot \int_{-\infty}^{\infty} \frac{1}{\sqrt{2\pi}\sigma_a} \cdot \exp\left\{-\frac{\{a - [\bar{a} - \sigma_a^2 \ln(n/n_*)]\}^2}{2\sigma_a^2}\right.$$

$$\left. - \bar{a} \cdot \ln(n/n_*) + \frac{1}{2}\sigma_a^2 \ln^2(n/n_*)\right\} da.$$

Therefore, the invariant polytropic pressure along the flow is given by

$$\Pi_D(P,n) \equiv P \cdot (n/n_*)^{-\bar{a}} \cdot \exp\left[\frac{1}{2}\sigma_a^2 \ln^2(n/n_*)\right], \tag{5.6}$$

and it remains constant along the plasma streamlines (note that when $\sigma_a \to 0$, Eq. (5.6) recovers the case of single polytrope for $a = \bar{a}$). Then, substituting $P = k_B n T$ and $\Pi_D = n_* k_B T_*$, we obtain

$$\ln(T/T_*) = (\bar{a} - 1) \cdot \ln(n/n_*) - \frac{1}{2}\sigma_a^2 \cdot \ln^2(n/n_*). \tag{5.7}$$

Livadiotis and McComas (2012, 2013b) found negative correlation between the temperature T and density n values of the proton plasma along the equatorial streamline from the heliospheric nose towards the heliotail. More precisely, it was shown that the variations of T and n follow an isobaric polytropic index, i.e., $a \approx 0$ (or $\nu \approx -1$). Livadiotis (2016c) showed that the parabolic model, Eq. (5.7), is better fit on the log—log scale of the observed values of T and n (Fig. 5.2). The average polytropic index is still isobaric, $\bar{a} \approx 0$, while we also derive the values of the polytropic standard deviation σ and the characteristic density scale n_*.

The superposition of polytropic processes can be also realized on ν polytropic indices, which correspond to the effective degrees of freedom $d_{\text{eff}} \equiv 2\nu = 2/(a - 1)$. Then, given a density of $D(\nu)$ (where the various densities are related as $D(\nu) = 2D(d_{\text{eff}}) = (a - 1)^2 D(a)$), we have the invariant density given by

$$N_D(n, T) \equiv n \cdot \int_{-\infty}^{\infty} D(\nu) \cdot (T/T_*)^{-\nu} d\nu. \tag{5.8}$$

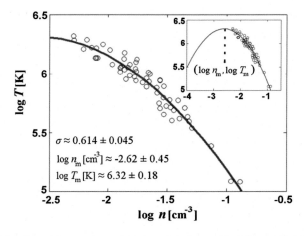

FIGURE 5.2

The parabolic fit between the density and temperature values (on a log—log scale) of the equatorial data of the inner heliosheath (Livadiotis et al., 2011). The fit shows that there is negative correlation between density and temperate values only for densities $\log n > \log n_m \cong -2.62 \pm 0.45$ (density units: cm^{-3}); for smaller densities, their correlation becomes positive. Adapted from Livadiotis (2016c).

As an example of $D(\nu)$, we consider again a Gaussian distribution of polytropic indices about the most frequent polytropic index $\bar{\nu}$ and standard deviation σ_ν, that is

$$D(\nu) = \frac{1}{\sqrt{2\pi}\sigma_\nu} \cdot \exp\left[-\frac{(\nu - \bar{\nu})^2}{2\sigma_\nu^2}\right]. \tag{5.9}$$

Then, according to Eq. (5.8), the generalized polytropic invariant density along the flow becomes

$$N_D(n, T) = n \cdot \int_{-\infty}^{\infty} \frac{1}{\sqrt{2\pi}\sigma_\nu} \cdot \exp\left[-\frac{(\nu - \bar{\nu})^2}{2\sigma_\nu^2} - \nu \cdot \ln(T/T_*)\right] d\nu$$

$$= n \cdot \int_{-\infty}^{\infty} \frac{1}{\sqrt{2\pi}\sigma_\nu} \cdot \exp\left\{-\frac{\{\nu - [\bar{\nu} - \sigma_\nu^2 \ln(T/T_*)]\}^2}{2\sigma_\nu^2}\right.$$

$$\left. - \bar{\nu} \cdot \ln(T/T_*) + \frac{1}{2}\sigma_\nu^2 \ln^2(T/T_*)\right\} d\nu.$$

Therefore, the invariant polytropic density along the flow is given by

$$N_D(n, T) \equiv n \cdot (T/T_*)^{-\bar{\nu}} \cdot \exp\left[\frac{1}{2}\sigma_\nu^2 \ln^2(T/T_*)\right]. \tag{5.10}$$

Then, substituting $N_D = n_*$, we obtain

$$\ln(n/n_*) = \bar{\nu} \cdot \ln(T/T_*) - \frac{1}{2}\sigma_\nu^2 \cdot \ln^2(T/T_*). \tag{5.11}$$

Next, we show that in the presence of a potential energy, the polytropic index depends on the kappa index and the form of the potential energy. While simple cases of kappa distributions may lead to the power law in Eq. (5.1), other, more complicated cases lead to the superposition polytropic relations as in Eq. (5.11).

5.3.3 Connection With the Kappa Index

The polytropic relation, Eq. (5.1), allows any arbitrary polytropic index a or ν, independently of the values of other thermodynamic parameters. However, the presence of a potential energy assigns specific dependence of the polytropic index on the kappa index or other parameters.

5.3.3.1 Thermal Equilibrium, $\kappa_0 \to \infty$

The positional dependence of thermal variables may be caused by the existence of a potential energy. At thermal equilibrium, this relation is derived from the Boltzmann–Gibbs distribution of the Hamiltonian function $H(\vec{r}, \vec{u}) = \varepsilon_K(\vec{u}) + \Phi(\vec{r})$, where $\varepsilon_K(\vec{u}) = \frac{1}{2}m(\vec{u} - \vec{u}_b)^2$ is the (velocity-dependent) kinetic energy and $\Phi(\vec{r})$ is the (position-dependent) potential energy,

$$P(\vec{r}, \vec{u}\,; T) \propto \exp\left[-\frac{H(\vec{r}, \vec{u})}{k_B T}\right] \propto \exp\left[-\frac{\varepsilon_K(\vec{u})}{k_B T}\right] \cdot \exp\left[-\frac{\Phi(\vec{r})}{k_B T}\right]. \tag{5.12}$$

The marginal distributions provide the positional and velocity distributions, which are, respectively,

$$P(\vec{r}; T) = \int_{-\infty}^{\infty} P(\vec{r}, \vec{u}; T)d\vec{u} \text{ and } P(\vec{u}; T) = \int_{-\infty}^{\infty} P(\vec{r}, \vec{u}; T)d\vec{r}, \tag{5.13}$$

where $d\vec{r}$ and $d\vec{u}$ denote, respectively, the infinitesimal 3-D volume of the regular space and the d_K-dimensional volume of the velocity space. Hence, because of the factorization identity of the exponential function (Chapter 1; Livadiotis and McComas, 2011b; Livadiotis, 2015c), we find

$$P(\vec{r}; T) \propto \exp\left[-\frac{\Phi(\vec{r})}{k_B T}\right] \text{ and } P(\vec{u}; T) \propto \exp\left[-\frac{\varepsilon_K(\vec{u})}{k_B T}\right]. \tag{5.14}$$

The local density is derived using the normalization

$$n(\vec{r})/n(\vec{r}_*) = P(\vec{r}; \kappa_0; T)/P(\vec{r}_*; \kappa_0; T), \tag{5.15}$$

where $\vec{r}_*$ indicates the position of zero potential, $\Phi(\vec{r}_*) = 0$:

$$n(\vec{r}) = n(\vec{r}_*) \cdot \exp\left[-\frac{\Phi(\vec{r})}{k_B T}\right]. \tag{5.16}$$

The local temperature is derived as the local mean kinetic energy $\langle \varepsilon_K \rangle (\vec{r}) \equiv \frac{1}{2} d_K k_B T(\vec{r})$, that is,

$$\frac{1}{2} d_K k_B T(\vec{r}) = \frac{\int_{-\infty}^{\infty} P(\vec{r}, \vec{u}; T)\varepsilon_K(\vec{u})d\vec{u}}{\int_{-\infty}^{\infty} P(\vec{r}, \vec{u}; T)d\vec{u}}, \tag{5.17}$$

and because of the factorization of the exponential in Eq. (5.12),

$$\frac{1}{2} d_K k_B T(\vec{r}) = \frac{\int_{-\infty}^{\infty} \exp\left[-\frac{\varepsilon_K(\vec{u})}{k_B T}\right]\varepsilon_K(\vec{u})d\vec{u}}{\int_{-\infty}^{\infty} \exp\left[-\frac{\varepsilon_K(\vec{u})}{k_B T}\right]d\vec{u}} = \frac{1}{2} d_K k_B T, \text{ or, } T(\vec{r}) = T.$$

$$\tag{5.18}$$

Therefore, the potential energy may cause some positional dependence on the density, while the temperature remains constant. The only way to have these two features is via an isothermal procedure, i.e., $[T(\vec{r})/T(\vec{r}_*)] \propto [n(\vec{r})/n(\vec{r}_*)]^{a-1}$, with $a \to 1$.

5.3.3.2 Out of Thermal Equilibrium, $\kappa_0 < \infty$

Out of thermal equilibrium, the kappa distribution of the Hamiltonian is given by

$$P(\vec{r}, \vec{u}; \kappa, T) \propto \left[1 + \frac{1}{\kappa} \cdot \frac{H(\vec{r}, \vec{u}) - \langle H \rangle}{k_B T} \right]^{-\kappa - 1}, \tag{5.19}$$

where $\langle H \rangle$ is the ensemble phase space average of the Hamiltonian function. When using the invariant kappa index κ_0, the phase space distribution, Eq. (5.19), becomes

$$P(\vec{r}, \vec{u}; \kappa_0, T) \propto \left[1 + \frac{1}{\kappa_0} \cdot \frac{\varepsilon_K(\vec{u}) + \Phi(\vec{r})}{k_B T} \right]^{-\kappa_0 - 1 - \frac{1}{2}d_K - \frac{1}{2}d_\Phi} \quad \text{in terms of velocity } \vec{u},$$

$$\tag{5.20a}$$

$$P(\vec{r}, \varepsilon_K; \kappa_0, T) \propto \left[1 + \frac{1}{\kappa_0} \cdot \frac{\varepsilon_K + \Phi(\vec{r})}{k_B T} \right]^{-\kappa_0 - 1 - \frac{1}{2}d_K - \frac{1}{2}d_\Phi} \varepsilon_K^{\frac{1}{2}d_K - 1} \quad \text{in terms of kinetic energy } \varepsilon_K,$$

$$\tag{5.20b}$$

where we set the kappa index, degenerated by both the kinetic and potential energy,

$$\kappa_0 \rightarrow \kappa_0 + \frac{1}{2}d_\Phi. \tag{5.21}$$

Therefore, the marginal distribution of the potential energy (called positional distribution) is given by

$$P(\vec{r}; \kappa_0; T) \propto \left[1 + \frac{1}{\kappa_0} \cdot \frac{\Phi(\vec{r})}{k_B T} \right]^{-\kappa_0 - 1 - \frac{1}{2}d_\Phi}, \tag{5.22}$$

where the positional degrees of freedom are given by

$$\frac{1}{2}d_\Phi \equiv \frac{\langle \Phi \rangle}{k_B T} = \frac{\int_{-\infty}^{\infty} \left[1 + \frac{1}{\kappa_0} \cdot \frac{\Phi(\vec{r})}{k_B T} \right]^{-\kappa_0 - 1 - \frac{1}{2}d_\Phi} \frac{\Phi(\vec{r})}{k_B T} d\vec{r}}{\int_{-\infty}^{\infty} \left[1 + \frac{1}{\kappa_0} \cdot \frac{\Phi(\vec{r})}{k_B T} \right]^{-\kappa_0 - 1 - \frac{1}{2}d_\Phi} d\vec{r}}. \tag{5.23}$$

The local density and temperature are calculated as for the equilibrium case, Eqs. (5.15 and 5.17). Indeed, first, from the marginal distribution of the potential energy, we derive the density $n(\vec{r})$. Then we derive the local mean kinetic energy $\langle \varepsilon_K \rangle (\vec{r}) \equiv \frac{1}{2}d_K k_B T(\vec{r})$,

$$\frac{1}{2}d_K k_B T(\vec{r}) = \frac{\int_0^\infty \left[1 + \frac{1}{\kappa_0} \cdot \frac{\varepsilon_K + \Phi(\vec{r})}{k_B T}\right]^{-\kappa_0 - 1 - \frac{1}{2}d_K - \frac{1}{2}d_\Phi} \varepsilon_K^{\frac{1}{2}d_K} d\varepsilon_K}{\int_0^\infty \left[1 + \frac{1}{\kappa_0} \cdot \frac{\varepsilon_K + \Phi(\vec{r})}{k_B T}\right]^{-\kappa_0 - 1 - \frac{1}{2}d_K - \frac{1}{2}d_\Phi} \varepsilon_K^{\frac{1}{2}d_K - 1} d\varepsilon_K}$$

$$= \frac{\int_0^\infty \left[1 + \frac{X}{\kappa_0 + \frac{\Phi(\vec{r})}{k_B T}}\right]^{-\kappa_0 - 1 - \frac{1}{2}d_K - \frac{1}{2}d_\Phi} X^{\frac{1}{2}d_K} dX}{\int_0^\infty \left[1 + \frac{X}{\kappa_0 + \frac{\Phi(\vec{r})}{k_B T}}\right]^{-\kappa_0 - 1 - \frac{1}{2}d_K - \frac{1}{2}d_\Phi} X^{\frac{1}{2}d_K - 1} dX} \cdot k_B T$$

$$= \frac{\kappa_0 + \frac{\Phi(\vec{r})}{k_B T}}{\kappa_0 + \frac{1}{2}d_K + \frac{1}{2}d_\Phi} \cdot \frac{\Gamma_{\kappa_0 + \frac{1}{2}d_K + \frac{1}{2}d_\Phi}\left(\frac{1}{2}d_K + 1\right)}{\Gamma_{\kappa_0 + \frac{1}{2}d_K + \frac{1}{2}d_\Phi}\left(\frac{1}{2}d_K\right)} \cdot k_B T = \frac{\kappa_0 + \frac{\Phi(\vec{r})}{k_B T}}{\kappa_0 + \frac{1}{2}d_\Phi} \cdot \frac{1}{2}d_K \cdot k_B T,$$

where we set $\chi \equiv \varepsilon_K / (k_B T)$, and because the q-Gamma functions (Chapter 1; Livadiotis and McComas, 2009; Appendix A) give

$$\frac{\Gamma_{\kappa_0 + \frac{1}{2}d_K + \frac{1}{2}d_\Phi}\left(\frac{1}{2}d_K + 1\right)}{\Gamma_{\kappa_0 + \frac{1}{2}d_K + \frac{1}{2}d_\Phi}\left(\frac{1}{2}d_K\right)} = \frac{1}{2}d_K \cdot \left(\kappa_0 + \frac{1}{2}d_K - \frac{1}{2}d_\Phi\right) \cdot \frac{\Gamma\left(\kappa_0 + \frac{1}{2}d_\Phi\right)}{\Gamma\left(\kappa_0 + 1 + \frac{1}{2}d_\Phi\right)}$$

$$= \frac{1}{2}d_K \cdot \frac{\kappa_0 + \frac{1}{2}d_K - \frac{1}{2}d_\Phi}{\kappa_0 + \frac{1}{2}d_\Phi}.$$

Hence, we have

$$n(\vec{r}) = n(\vec{r}_*) \cdot \left[1 + \frac{1}{\kappa_0} \cdot \frac{\Phi(\vec{r})}{k_B T}\right]^{-\kappa_0 - 1 - \frac{1}{2}d_\Phi}, \quad T(\vec{r}) = T(\vec{r}_*) \cdot \left[1 + \frac{1}{\kappa_0} \cdot \frac{\Phi(\vec{r})}{k_B T}\right].$$

$$(5.24)$$

Combining the above mentioned relations, we find

$$\left[\frac{n(\vec{r})}{n(\vec{r}_*)}\right] = \left[\frac{T(\vec{r})}{T(\vec{r}_*)}\right]^{-\kappa_0 - 1 - \frac{1}{2}d_\Phi} \equiv \left[\frac{T(\vec{r})}{T(\vec{r}_*)}\right]^\nu, \quad \text{with } \nu \equiv (a - 1)^{-1}$$

$$= -\kappa_0 - 1 - \frac{1}{2}d_\Phi. \qquad (5.25a)$$

For small values of $\frac{1}{2}d_\Phi$, we have $\nu \equiv (a-1)^{-1} \cong -\kappa_0 - 1 \cong -\left(\kappa - \frac{1}{2}\right)$ (see Chapter 11). For example, in plasmas characterized by kappa index near the limiting case of $\kappa_0 \cong 0$ (e.g., common spectrum, see Chapter 15), we find $\nu \cong -1$ or $a \cong 0$, which corresponds to the isobaric polytrope (constant thermal pressure), frequently observed in heliospheric and magnetospheric plasmas (Table 5.1).

Note that we have used the simplest formulation of phase space distribution, that of positive potential. In Chapter 3, we have seen how Eq. (5.25a) can be generalized for both positive and negative attractive potentials. Namely, setting $d_\Phi \equiv |\Phi|/(k_B T)$, we have

$$\left[\frac{n(\overrightarrow{r})}{n(\overrightarrow{r}_*)}\right] = \left[\frac{T(\overrightarrow{r})}{T(\overrightarrow{r}_*)}\right]^{-\kappa_0 - 1 \mp \frac{1}{2}d_\Phi} \equiv \left[\frac{T(\overrightarrow{r})}{T(\overrightarrow{r}_*)}\right]^\nu, \text{ with } \nu \equiv (a-1)^{-1}$$

$$= -\kappa_0 - 1 \mp \frac{1}{2}d_\Phi. \tag{5.25b}$$

where $\mp$ corresponds to the sign of $-\Phi$.

In addition, the prepotential factors $n(\overrightarrow{r}_*)$ and $T(\overrightarrow{r}_*)$ do not depend on the position $\overrightarrow{r}$, while there was no superposition of kappa distributions. In the following subsection, we examine how the polytropic relations can be modified when these complications are met.

5.3.4 Complicated Relations Between Polytropic and Kappa Indices

There are three possible reasons for which the relation between polytropic and kappa indices may be more complicated: (1) variety of kappa distribution formulae describing the potential energy; (2) positional dependence of the prepotential factor; and (3) superposition of kappa distributions:

1. Different formulae of the kappa distributions of the potential energy may lead to different relations between polytropic and kappa index, i.e., $\nu = -1 - \kappa_0 - \frac{1}{2}d_\Phi$ or $\nu = -1 + \kappa_0$. For example, for $\Phi(\overrightarrow{r}) \leq 0$, we may use the formulation (Table 3.1 in Chapter 3; Livadiotis, 2015):

$$P(\overrightarrow{r}, \overrightarrow{u}; \kappa_0, T) \propto \left[\frac{|\Phi(\overrightarrow{r})| - \varepsilon_K(\overrightarrow{u})}{\left(\frac{1}{2}d_\Phi - \kappa_0\right)k_B T} - 1\right]_+^{\kappa_0 - 1 - \frac{1}{2}d_K}. \tag{5.26}$$

For simplicity, we consider the relations of Table 3.1 with the kappa index degenerated only by the kinetic (and not the potential) degrees of freedom. We recall that the symbol "+" means that the distribution becomes zero when the base is nonpositive (see Eq. (1.13) in Chapter 1).

The marginal distribution of the potential energy is

$$
P(\overrightarrow{r}; \kappa_0; T) \propto \left[\frac{|\Phi(\overrightarrow{r})|}{\left(\frac{1}{2}d_\Phi - \kappa_0\right)k_B T} - 1 \right]^{\kappa_0 - 1}.
\tag{5.27}
$$

We derive the local density $n(\overrightarrow{r})$ using the normalization

$$
n(\overrightarrow{r}) \Big/ \left\{ n(\overrightarrow{r}_*) \left[\frac{|\Phi(\overrightarrow{r}_*)|}{\left(\frac{1}{2}d_\Phi - \kappa_0\right)k_B T} - 1 \right]^{\kappa_0 - 1} \right\} = P(\overrightarrow{r}; \kappa_0; T)/P(\overrightarrow{r}_*; \kappa_0; T),
\tag{5.28}
$$

where $\overrightarrow{r}_*$ now indicates an arbitrary position with potential $\Phi(\overrightarrow{r}_*) > (\frac{1}{2}d_\Phi - \kappa_0)k_B T$. The local temperature $n(\overrightarrow{r})$ is derived from the local mean kinetic energy

$$
\frac{1}{2}d_K k_B T(\overrightarrow{r}) = \frac{\displaystyle\int_{-\infty}^{\infty} P(\overrightarrow{r}, \varepsilon_K; T) \varepsilon_K^{\frac{1}{2}d_K} d\varepsilon_K}{\displaystyle\int_{-\infty}^{\infty} P(\overrightarrow{r}, \varepsilon_K; T) \varepsilon_K^{\frac{1}{2}d_K - 1} d\varepsilon_K}
$$

$$
= \frac{\displaystyle\int_{-\infty}^{\infty} \left[\frac{|\Phi(\overrightarrow{r})| - \varepsilon_K}{\left(\frac{1}{2}d_\Phi - \kappa_0\right)k_B T} - 1 \right]_+^{\kappa_0 - 1 - \frac{1}{2}d_K} \varepsilon_K^{\frac{1}{2}d_K} d\varepsilon_K}{\displaystyle\int_{-\infty}^{\infty} \left[\frac{|\Phi(\overrightarrow{r})| - \varepsilon_K}{\left(\frac{1}{2}d_\Phi - \kappa_0\right)k_B T} - 1 \right]_+^{\kappa_0 - 1 - \frac{1}{2}d_K} \varepsilon_K^{\frac{1}{2}d_K - 1} d\varepsilon_K}.
\tag{5.29}
$$

We find

$$
n(\overrightarrow{r}) = n(\overrightarrow{r}_*) \cdot \left[\frac{\dfrac{|\Phi(\overrightarrow{r})|}{\left(\frac{1}{2}d_\Phi - \kappa_0\right)k_B T} - 1}{\dfrac{|\Phi(\overrightarrow{r}_*)|}{\left(\frac{1}{2}d_\Phi - \kappa_0\right)k_B T} - 1} \right]^{\kappa_0 - 1},
$$

$$
T(\overrightarrow{r}) = T(\overrightarrow{r}_*) \cdot \left[\frac{\dfrac{|\Phi(\overrightarrow{r})|}{\left(\frac{1}{2}d_\Phi - \kappa_0\right)k_B T} - 1}{\dfrac{|\Phi(\overrightarrow{r}_*)|}{\left(\frac{1}{2}d_\Phi - \kappa_0\right)k_B T} - 1} \right], \text{ or}
\tag{5.30a}
$$

$$\left[\frac{n(\vec{r})}{n(\vec{r}_*)}\right] = \left[\frac{T(\vec{r})}{T(\vec{r}_*)}\right]^{\kappa_0-1} \equiv \left[\frac{T(\vec{r})}{T(\vec{r}_*)}\right]^{\nu}, \text{ with } \nu = \kappa_0 - 1. \tag{5.30b}$$

As we have seen in Chapter 3, the potential energy is causing degeneration to the kappa index so that $-\kappa_0 + \frac{1}{2}d_\Phi \to \kappa_0$, or $\kappa_0 \to -\kappa_0 + \frac{1}{2}d_\Phi$, Eq. (3.136a). Thus we end up with

$$n(\vec{r}) = n(\vec{r}_*) \cdot \left[\frac{\frac{|\Phi(\vec{r})|}{\kappa_0 k_B T} - 1}{\frac{|\Phi(\vec{r}_*)|}{\kappa_0 k_B T} - 1}\right]^{-\kappa_0 - 1 + \frac{1}{2}d_\Phi},$$

$$T(\vec{r}) = T(\vec{r}_*) \cdot \left[\frac{\frac{|\Phi(\vec{r})|}{\kappa_0 k_B T} - 1}{\frac{|\Phi(\vec{r}_*)|}{\kappa_0 k_B T} - 1}\right], \text{ or} \tag{5.31a}$$

$$\left[\frac{n(\vec{r})}{n(\vec{r}_*)}\right] = \left[\frac{T(\vec{r})}{T(\vec{r}_*)}\right]^{-\kappa_0 - 1 + \frac{1}{2}d_\Phi} \equiv \left[\frac{T(\vec{r})}{T(\vec{r}_*)}\right]^{\nu}, \text{ with } \nu = -\kappa_0 - 1 + \frac{1}{2}d_\Phi. \tag{5.31b}$$

2. The "prepotential" factors in Eq. (5.24) may also depend on the position:

$$n(\vec{r}) = n(\vec{r}_*) \cdot \frac{g_n(\vec{r})}{g_n(\vec{r}_*)} \cdot \left[1 + \frac{\Phi(\vec{r})}{\kappa_0 k_B T}\right]^{-\kappa_0 - 1 + \frac{1}{2}d_\Phi},$$

$$T(\vec{r}) = T(\vec{r}_*) \cdot \frac{g_T(\vec{r})}{g_T(\vec{r}_*)} \cdot \left[1 + \frac{\Phi(\vec{r})}{\kappa_0 k_B T}\right], \tag{5.32}$$

where $g_n(\vec{r})$ and $g_T(\vec{r})$ are some position functions. We may rewrite these as

$$n(X) = n(1) \cdot G_n(X) \cdot X^{-\kappa_0 - 1 + \frac{1}{2}d_\Phi}, \quad T(X) = T(1) \cdot G_T(X) \cdot X, \quad X \equiv 1 + \frac{\Phi(\vec{r})}{\kappa_0 k_B T}, \tag{5.33}$$

with $G_n(1) = G_T(1) = 1$, where the relation is parameterized with X, with $X = 1$ for $\vec{r} = \vec{r}_*$.

As an example of a function $g_n(\vec{r})$, let the case of plasma moving spherically outward, e.g., solar wind, where $g_n(\vec{r}) = 1/r^2$ and $\Phi(r) = -GMm/r$. We have (see Eq. (3.158a))

$$n(r) = n_0 \cdot \frac{r_0^2}{r^2} \cdot \left(\frac{\frac{3}{\kappa_0} \cdot \frac{r_0}{r} - 1}{\frac{3}{\kappa_0} - 1} \right)^{-\kappa_0+2} \quad , \quad T(r) = T_0 \cdot \left(\frac{\frac{3}{\kappa_0} \cdot \frac{r_0}{r} - 1}{\frac{3}{\kappa_0} - 1} \right), \text{ with}$$

$$r < \frac{3}{\kappa_0} \cdot r_0 = \frac{GMm}{\kappa_0 k_B T}, \quad r_0 \equiv \frac{GMm}{3 k_B T}, \tag{5.34}$$

where (n_0, T_0) are the density and temperature at radius $r = r_0$. Hence,

$$\frac{n(r)}{n_0} = \left(\frac{\kappa_0}{3} \right)^2 \left[1 + \frac{T(r)}{T_0} \cdot \left(\frac{3}{\kappa_0} - 1 \right) \right]^2 \cdot \left[\frac{T(r)}{T_0} \right]^{-\kappa_0+2}. \tag{5.35a}$$

For example,

$$\frac{n(r)}{n_0} \cong \left[\frac{T(r)}{T_0} \right]^4, \quad r \ll r_0, \ \kappa_0 \ll 3, \text{ or } \nu = 4, \ a = 1.25. \tag{5.35b}$$

By expanding Eq. (5.35),

$$\frac{n(r)}{n_0} = \left(\frac{\kappa_0}{3} \right)^2 \cdot \left\{ 1 + 2 \left(\frac{3}{\kappa_0} - 1 \right) \cdot \left[\frac{T(r)}{T_0} \right] + \left(\frac{3}{\kappa_0} - 1 \right)^2 \cdot \left[\frac{T(r)}{T_0} \right]^2 \right\} \cdot \left[\frac{T(r)}{T_0} \right]^{-\kappa_0+2},$$

$$\tag{5.35c}$$

we observe that

$$\frac{n(r)}{n_0} \propto \begin{cases} \left[\dfrac{T(r)}{T_0} \right]^{-\kappa_0+2} & \text{for } T(r) \ll T_0, \\[4mm] \left[\dfrac{T(r)}{T_0} \right]^{-\kappa_0+4} & \text{for } T(r) \gg T_0. \end{cases} \tag{5.36}$$

We define the effective polytropic index ν_{eff} as the slope of temperature–density diagram on a log–log scale:

$$\nu_{\text{eff}}(r) \equiv \frac{\partial \log n(r)}{\partial \log T(r)}, \tag{5.37}$$

so that Eq. (5.38) leads to

$$\nu_{\text{eff}} \cong \begin{cases} -\kappa_0 + 2 & \text{for } T(r) \ll T(r_*), \\ -\kappa_0 + 4 & \text{for } T(r) \gg T(r_*). \end{cases} \tag{5.38}$$

Fig. 5.3 plots $\log n(r)$ and $\nu_{\text{eff}}(r)$ as a function of $\log T(r)$. The effective polytropic index ν_{eff} is $\sim -\kappa_0 + 2$ at low and $\sim -\kappa_0 + 4$ at high temperature; thus $\nu_{\text{eff}} \sim -\kappa_0 + 3$ on average.

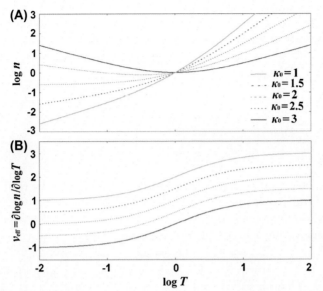

FIGURE 5.3

(A) The density–temperature relation in Eq. (5.35c) caused by the solar wind outward plasma and the gravitational potential. (B) The effective polytropic index ν_{eff} is $\sim -\kappa_0 + 2$ at low and $\sim -\kappa_0 + 4$ at high temperatures.

We may rewrite Eq. (5.35a) as

$$\ln\left[\frac{n(r)}{n_0}\right] = 2\ln\left(\frac{\kappa_0}{3}\right) + 2\ln\left[1 + \frac{T(r)}{T_0}\cdot\left(\frac{3}{\kappa_0}-1\right)\right] + (2-\kappa_0)\ln\left[\frac{T(r)}{T_0}\right]. \qquad (5.39)$$

Using the approximation

$$\ln[1 + (a-1)e^x] \cong \ln(a) + \frac{a-1}{a}\cdot x + \frac{a-1}{2a^2}x^2, \text{ i.e.,} \qquad (5.40a)$$

$$\begin{aligned}\ln\left[1 + \frac{T(r)}{T_0}\cdot\left(\frac{3}{\kappa_0}-1\right)\right] \cong{} & \ln\left(\frac{3}{\kappa_0}\right) + \left(1 - \frac{\kappa_0}{3}\right)\cdot\ln\left[\frac{T(r)}{T_0}\right] \\ & + \frac{1}{2}\frac{\kappa_0}{3}\left(1 - \frac{\kappa_0}{3}\right)\cdot\ln^2\left[\frac{T(r)}{T_0}\right],\end{aligned} \qquad (5.40b)$$

we find

$$\ln\left[\frac{n(r)}{n_0}\right] \cong \left(4 - \frac{5}{3}\kappa_0\right)\cdot\ln\left[\frac{T(r)}{T_0}\right] + \frac{\kappa_0}{3}\left(1 - \frac{\kappa_0}{3}\right)\cdot\ln^2\left[\frac{T(r)}{T_0}\right]. \qquad (5.41)$$

where the effective polytropic index is written as

$$\nu_{\text{eff}}(r) \equiv \frac{\partial\ln[n(r)/n_0]}{\partial\ln[T(r)/T_0]} = \left(4 - \frac{5}{3}\kappa_0\right) + \frac{\kappa_0}{3}\left(1 - \frac{\kappa_0}{3}\right)\cdot\ln\left[\frac{T(r)}{T_0}\right], \qquad (5.42)$$

We observe that Eq. (5.41) is parabolic, similar to the superposition of polytropic indices in Eq. (5.11), but with positive curvature.

3. Superposition of kappa distributions. Given the distribution density (discrete or continuous) of the kappa indices, $D(\kappa_0)$, the phase space distributions in Eqs. (5.20a) and (5.26) become

$$P(\vec{r}, \vec{u}) \sim \sum_{\kappa_0} D(\kappa_0) \left[1 + \frac{1}{\kappa_0} \cdot \frac{\varepsilon_K(\vec{u}) + \Phi(\vec{r})}{k_B T} \right]^{-\kappa_0 - 1 - \frac{1}{2}d_K - \frac{1}{2}d_\Phi},$$

$$P(\vec{r}, \vec{u}) \sim \sum_{\kappa_0} D(\kappa_0) \left[\frac{|\Phi(\vec{r})| - \varepsilon_K(\vec{u})}{\left(\frac{1}{2}d_\Phi - \kappa_0\right) k_B T} - 1 \right]_+^{\kappa_0 - 1 - \frac{1}{2}d_K}.$$

(5.43)

Then, following similar steps as previously, we can derive the relation between polytropic and kappa indices. After several simplifications, we may end up with a relation analogous to Eq. (5.34). Indeed, in the approximation of large potential energy $|\Phi(\vec{r})| \gg \frac{1}{2}d_\Phi k_B T$, we have

$$\frac{n(\vec{r})}{n(\vec{r}_*)} = \sum_{\kappa_0} D(\kappa_0) \left[\frac{\frac{|\Phi(\vec{r})|}{\kappa_0 k_B T} - 1}{\frac{|\Phi(\vec{r}_*)|}{\kappa_0 k_B T} - 1} \right]^{-\kappa_0 - 1 + \frac{1}{2}d_\Phi} \cong \sum_{\kappa_0} D(\kappa_0) \left[\frac{|\Phi(\vec{r})|}{|\Phi(\vec{r}_*)|} \right]^{-\kappa_0 - 1 + \frac{1}{2}d_\Phi},$$

(5.44a)

and

$$\frac{T(\vec{r})}{T(\vec{r}_*)} = \frac{\sum_{\kappa_0} D(\kappa_0) \left[\frac{\frac{|\Phi(\vec{r})|}{\kappa_0 k_B T} - 1}{\frac{|\Phi(\vec{r}_*)|}{\kappa_0 k_B T} - 1} \right]^{-\kappa_0 + \frac{1}{2}d_\Phi}}{\sum_{\kappa_0} D(\kappa_0) \left[\frac{\frac{|\Phi(\vec{r})|}{\kappa_0 k_B T} - 1}{\frac{|\Phi(\vec{r}_*)|}{\kappa_0 k_B T} - 1} \right]^{-\kappa_0 - 1 + \frac{1}{2}d_\Phi}} \cong \frac{\sum_{\kappa_0} D(\kappa_0) \left[\frac{|\Phi(\vec{r})|}{|\Phi(\vec{r}_*)|} \right]^{-\kappa_0 + \frac{1}{2}d_\Phi}}{\sum_{\kappa_0} D(\kappa_0) \left[\frac{|\Phi(\vec{r})|}{|\Phi(\vec{r}_*)|} \right]^{-\kappa_0 - 1 + \frac{1}{2}d_\Phi}}$$

$$= \left[\frac{|\Phi(\vec{r})|}{|\Phi(\vec{r}_*)|} \right],$$

(5.44b)

or

$$T(\vec{r}) \cong |\Phi(\vec{r})|.$$

(5.44c)

Then the polytropic relation is

$$\frac{n(\vec{r})}{n(\vec{r}_*)} \cong \sum_{\kappa_0} D(\kappa_0) \left[\frac{T(\vec{r})}{T(\vec{r}_*)} \right]^{-\kappa_0 - 1 + \frac{1}{2}d_\Phi},$$

(5.45)

or, given the case of $g_n(\vec{r}) = 1/r^2$,

$$\frac{n(\vec{r})}{n(\vec{r}_*)} \cong \frac{g_n\left\{ |\Phi|^{-1}[T(\vec{r})] \right\}}{g_n\left\{ |\Phi|^{-1}[T(\vec{r}_*)] \right\}} \cdot \sum_{\kappa_0} D(\kappa_0) \left[\frac{T(\vec{r})}{T(\vec{r}_*)} \right]^{-\kappa_0 - 1 + \frac{1}{2}d_\Phi}.$$

(5.46)

5.4 Correlation Between Particle Energies

The physical meaning of the kappa index is interwoven with the correlation of particles in systems out of thermal equilibrium (Chapter 1). One of the crucial limitations of the classical Boltzmann-Gibbs (BG) statistical approach is that correlations between particles are not included. By contrast, the nonextensivity of statistical mechanics naturally captures the correlations between the particles and the degrees of freedom. In particular, the correlation is mathematically modeled by the specific formulation of the kappa distribution and cannot be factored. If $P(\varepsilon_{(1)}, \varepsilon_{(2)})$ is the joint probability distribution, and $P(\varepsilon_{(1)})$, $P(\varepsilon_{(2)})$ are the individual marginal probability distributions of two particles with energies ε_1 and ε_2, then the factorization of the joint distribution is $P(\varepsilon_{(1)}, \varepsilon_{(2)}) = P(\varepsilon_{(1)}) \cdot P(\varepsilon_{(2)})$. For example, the exponential distribution can always be factored. This mathematical property is equivalent to the absence of correlation in systems described by the Maxwell distribution. On the other hand, the kappa distribution cannot be factored, $P(\varepsilon_{(1)}, \varepsilon_{(2)}) \neq P(\varepsilon_{(1)}) \cdot P(\varepsilon_{(2)})$, and there is a nonzero correlation coefficient R between the energies of any two particles of a system. Livadiotis and McComas (2011b) showed that $R = \frac{3}{2}/\kappa$, producing the "kappa spectrum" shown in Fig. 5.4.

Next we examine the correlation coefficient between the moments of any order of two particles, which is expressed as a function of the kappa index. Starting from the N-particle energy distribution function, that is, the joint probability distribution of N particles having energies $\left\{ \varepsilon_{(n)} \right\}_{n=1}^{N}$,

$$
P\left(\varepsilon_{(1)}, \ldots, \varepsilon_{(N)}; T, \kappa_0, d\right) = \kappa_0^{-\frac{d}{2}N} \cdot \frac{\Gamma\left(\kappa_0 + 1 + \frac{d}{2}N\right)}{\Gamma(\kappa_0 + 1) \cdot \Gamma\left(\frac{d}{2}\right)^N} \cdot (k_B T)^{-\frac{d}{2} \cdot N}
$$

$$
\times \left(1 + \frac{1}{\kappa_0} \cdot \frac{1}{k_B T} \sum_{n=1}^{N} \varepsilon_{(n)}\right)^{-\kappa_0 - 1 - \frac{d}{2}N} \cdot \prod_{n=1}^{N} \varepsilon_{(n)}^{\frac{d}{2} - 1},
$$

$$
(5.47)
$$

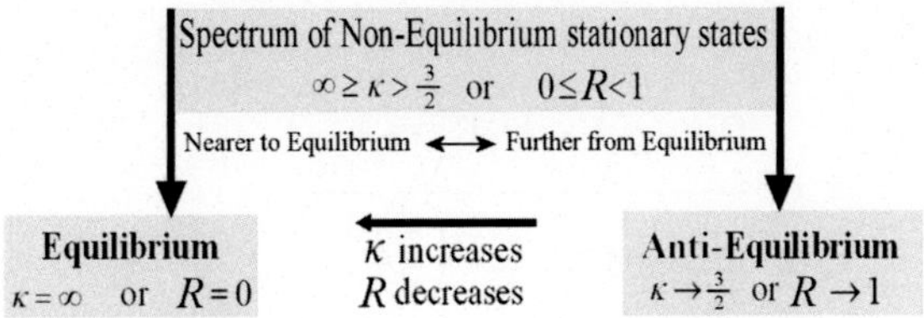

Spectrum of Non-Equilibrium stationary states

$\infty \geq \kappa > \frac{3}{2}$ or $0 \leq R < 1$

Nearer to Equilibrium ⟷ Further from Equilibrium

| Equilibrium | κ increases | Anti-Equilibrium |
| $\kappa = \infty$ or $R = 0$ | R decreases | $\kappa \to \frac{3}{2}$ or $R \to 1$ |

FIGURE 5.4

The spectrum of kappa indices. For $\kappa \to \infty$, the system resides at thermal equilibrium; for $\kappa \to \frac{3}{2}$, the furthest possible stationary state from thermal equilibrium is attained, the antiequilibrium. This is shown comparing with the correlation coefficient R. The classical case of systems residing in thermal equilibrium ($\kappa \to \infty$) corresponds to zero correlation ($R \to 0$), while the other extreme state of "antiequilibrium" $\left(\kappa \to \frac{3}{2}\right)$ indicates a maximum correlation ($R \to 1$).

we generate the two-particle distribution

$$P\big(\varepsilon_{(1)}, \varepsilon_{(2)}; T, \kappa_0, d\big) = \kappa_0^{-d} \cdot \frac{\Gamma(\kappa_0 + 1 + d)}{\Gamma(\kappa_0 + 1) \cdot \Gamma\left(\dfrac{d}{2}\right)^2} \cdot (k_B T)^{-d}$$

$$\times \left(1 + \frac{1}{\kappa_0} \cdot \frac{\varepsilon_{(1)} + \varepsilon_{(2)}}{k_B T}\right)^{-\kappa_0 - 1 - d} \cdot \varepsilon_{(1)}^{\frac{d}{2}-1} \varepsilon_{(2)}^{\frac{d}{2}-1}.$$

(5.48)

The (m_1, m_2)-th statistical moment of the two-individual distribution is given by

$$\left\langle x_{(1)}^{m_1} x_{(2)}^{m_2} \right\rangle = \frac{\Gamma(\kappa_0 + 1 + d)}{\Gamma(\kappa_0 + 1)\Gamma\left(\dfrac{d}{2}\right)^2} \kappa_0^{-d} \cdot \int_0^\infty \left[1 + \frac{1}{\kappa_0} \cdot \left(x_{(1)} + x_{(2)}\right)\right]^{-\kappa_0 - 1 - d}$$

$$\times x_{(1)}^{m_1 + \frac{d}{2} - 1} x_{(2)}^{m_2 + \frac{d}{2} - 1} dx_{(1)} dx_{(2)},$$

(5.49)

where we set $x_{(n)} \equiv \varepsilon_{(n)}/(k_B T)$. Hence,

$$\left\langle x_{(1)}^{m_1} x_{(2)}^{m_2} \right\rangle = \frac{\kappa_0^{m_1 + m_2} \Gamma(\kappa_0 + 1 + d)}{\Gamma(\kappa_0 + 1)\Gamma\left(\dfrac{d}{2}\right)^2 (\kappa_0 + d)^{m_1 + \frac{d}{2}} \left(\kappa_0 + \dfrac{d}{2} - m_1\right)^{m_2 + \frac{d}{2}}}$$

$$\times \Gamma_{\kappa_0 + d}\left(m_1 + \frac{d}{2}\right) \Gamma_{\kappa_0 + \frac{d}{2} - m_1}\left(m_2 + \frac{d}{2}\right),$$

(5.50)

where $\Gamma_{q=1+1/\kappa}(a)$ is the q-Gamma function (Chapter 1, Eqs. (1. Eqs.(1.16a–c)). Then,

$$\left\langle x_{(1)}^{m_1} x_{(2)}^{m_2} \right\rangle = \kappa_0^{m_1 + m_2} \frac{\Gamma(\kappa_0 + 1 - m_1 - m_2)\Gamma\left(m_1 + \dfrac{d}{2}\right)\Gamma\left(m_2 + \dfrac{d}{2}\right)}{\Gamma(\kappa_0 + 1)\Gamma\left(\dfrac{d}{2}\right)^2}.$$

(5.51)

Using Eq. (5.51) for $(m_1 = m, m_2 = 0)$, $(m_1 = 2m, m_2 = 0)$, and $(m_1 = m, m_2 = m)$, we derive the mean, variance, and covariance of the quantities $x_{(1)}^m$ and $x_{(2)}^m$,

$$\left\langle x_{(1)}^m \right\rangle = \kappa_0^m \frac{\Gamma(\kappa_0 + 1 - m)\Gamma\left(m + \dfrac{d}{2}\right)}{\Gamma(\kappa_0 + 1)\Gamma\left(\dfrac{d}{2}\right)},$$

(5.52a)

$$\sigma^2_{(1)(1)} = \left\langle x^{2m}_{(1)} \right\rangle - \left\langle x^m_{(1)} \right\rangle^2 = \frac{\kappa_0^{2m}}{\Gamma(\kappa_0 + 1)\Gamma\left(\frac{d}{2}\right)}$$

$$\times \left[\Gamma(\kappa_0 + 1 - 2m)\Gamma\left(2m + \frac{d}{2}\right) - \frac{\Gamma(\kappa_0 + 1 - m)^2\Gamma\left(m + \frac{d}{2}\right)^2}{\Gamma(\kappa_0 + 1)\Gamma\left(\frac{d}{2}\right)} \right],$$

$$(5.52b)$$

$$\sigma^2_{(1)(2)} = \left\langle x^m_{(1)} x^m_{(2)} \right\rangle - \left\langle x^m_{(1)} \right\rangle \left\langle x^m_{(2)} \right\rangle$$

$$= \frac{\kappa_0^{2m}\Gamma\left(m + \frac{d}{2}\right)^2}{\Gamma(\kappa_0 + 1)^2\Gamma\left(\frac{d}{2}\right)^2} \left[\Gamma(\kappa_0 + 1)\Gamma(\kappa_0 + 1 - 2m) - \Gamma(\kappa_0 + 1 - m)^2 \right],$$

$$(5.52c)$$

and thus, the (Pearson's) correlation coefficient R between the random variables $x^m_{(1)}$ and $x^m_{(2)}$ is

$$R = \frac{\sigma^2_{(1)(2)}}{\sigma_{(1)(1)}\sigma_{(2)(2)}} = \frac{\Gamma(\kappa_0 + 1 - 2m)\Gamma(\kappa_0 + 1) - \Gamma(\kappa_0 + 1 - m)^2}{\Gamma(\kappa_0 + 1 - 2m)\Gamma(\kappa_0 + 1)\dfrac{\Gamma\left(2m + \frac{d}{2}\right)\Gamma\left(\frac{d}{2}\right)}{\Gamma\left(m + \frac{d}{2}\right)^2} - \Gamma(\kappa_0 + 1 - m)^2}$$

$$= \frac{\dfrac{\Gamma(\kappa_0 + 1 - 2m)\Gamma(\kappa_0 + 1)}{\Gamma(\kappa_0 + 1 - m)^2} - 1}{\dfrac{\Gamma(\kappa_0 + 1 - 2m)\Gamma(\kappa_0 + 1)}{\Gamma(\kappa_0 + 1 - m)^2} \dfrac{\Gamma\left(2m + \frac{d}{2}\right)\Gamma\left(\frac{d}{2}\right)}{\Gamma\left(m + \frac{d}{2}\right)^2} - 1}$$

$$= \left[1 + \frac{\dfrac{\Gamma\left(2m + \frac{d}{2}\right)\Gamma\left(\frac{d}{2}\right)}{\Gamma\left(m + \frac{d}{2}\right)^2} - 1}{1 - \dfrac{\Gamma(\kappa_0 + 1 - m)^2}{\Gamma(\kappa_0 + 1 - 2m)\Gamma(\kappa_0 + 1)}} \right]^{-1}.$$

$$(5.53)$$

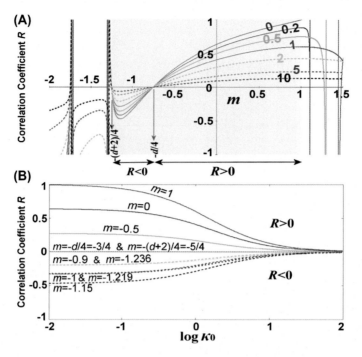

FIGURE 5.5

The correlation coefficient R between the value of the random variables ε^{-m} of two different particles is depicted as function of (A) the moment order m and for various kappa indices from $\kappa_0 = 0$ to $\kappa_0 = 10$, (B) the kappa index for orders $m = -(d+2)/4$ to $m = 1$. In order to avoid singularities, the allowed range of m values is $m \in \left(-\frac{d+2}{4}, 1\right)$ (*yellow shaded* region in panel (A)). The correlation coefficient is positive, $R > 0$, for values of the moment order $-d/4 < m < 1$, and negative $R < 0$ (anticorrelation), for $-(d+2)/4 < m < -d/4$. The correlation coefficient is zero at thermal equilibrium ($\kappa_0 \to \infty$), while it is maximized at antiequilibrium ($\kappa_0 \to 0$); however, only for $m = 1$, the maximum correlation is $R = 1$.

Fig. 5.5 plots Eq. (5.53) for various values of the order m and the kappa index κ_0.

For $m = 1$, we obtain the correlation coefficient R between particle energies as obtained in (Livadiotis and McComas, 2011b; Livadiotis, 2015c)

$$R_1(\kappa_0; d) = \left[1 + \frac{\dfrac{\Gamma\left(2 + \dfrac{d}{2}\right)\Gamma\left(\dfrac{d}{2}\right)}{\Gamma\left(1 + \dfrac{d}{2}\right)^2} - 1}{1 - \dfrac{\Gamma(\kappa_0)^2}{\Gamma(\kappa_0 - 1)\Gamma(\kappa_0 + 1)}}\right]^{-1} = \left(1 + \dfrac{\dfrac{1 + \dfrac{d}{2}}{\dfrac{d}{2}} - 1}{1 - \dfrac{\kappa_0 - 1}{\kappa_0}}\right)^{-1} = \dfrac{\dfrac{d}{2}}{\dfrac{d}{2} + \kappa_0}.$$

$$(5.54)$$

We also examine the correlation of the zero-th moment of the energy distribution $(m \to 0)$. We observe that

$$\frac{\Gamma\left(2m + \frac{d}{2}\right)\Gamma\left(\frac{d}{2}\right)}{\Gamma\left(m + \frac{d}{2}\right)^2} \cong 1 + \Psi_1\left(\frac{d}{2}\right) \cdot m^2 + O(m^3),$$

$$\frac{\Gamma(\kappa_0 + 1 - m)^2}{\Gamma(\kappa_0 + 1 - 2m)\Gamma(\kappa_0 + 1)} \cong 1 - \Psi_1(\kappa_0 + 1) \cdot m^2 + O(m^3),$$

hence,

$$R_0(\kappa_0; d) = \left[1 + \frac{\Psi_1\left(\frac{d}{2}\right)}{\Psi_1(\kappa_0 + 1)}\right]^{-1}, \text{ or,}$$

$$R_0(\kappa_0; 3) = \frac{1}{1 + \left(\frac{1}{2}\pi^2 - 4\right)\Big/\Psi_1(\kappa_0 + 1)} \text{ for } d = 3, \tag{5.55}$$

where $\Psi_1(x) \equiv \frac{d^2}{dx^2}\ln\Gamma(x)$ is the trigamma function. Note that $\Psi_1(x \to \infty) \to 0$ and $\Psi_1(x \to 1) = \frac{1}{6}\pi^2$. Hence, the correlation coefficient is zero for $\kappa_0 \to \infty$ (equilibrium), while its maximum value is at $\left[4(1 - 6/\pi^2)\right]^{-1} \approx 0.64$ for $\kappa_0 \to 0$.

As another example, we derive the correlation for $m = \frac{1}{2}$:

$$R_{\frac{1}{2}}(\kappa_0; d) = \left[1 + \frac{\dfrac{\Gamma\left(1 + \frac{d}{2}\right)\Gamma\left(\frac{d}{2}\right)}{\Gamma\left(\frac{d+1}{2}\right)^2} - 1}{1 - \dfrac{\Gamma\left(\kappa_0 + \frac{1}{2}\right)^2}{\Gamma(\kappa_0)\Gamma(\kappa_0 + 1)}}\right]^{-1} \leq R_{\frac{1}{2}}(0; d) = \frac{\Gamma\left(\frac{d+1}{2}\right)^2}{\Gamma\left(1 + \frac{d}{2}\right)\Gamma\left(\frac{d}{2}\right)},$$

$$\tag{5.56}$$

that approaches $\cong 1 - 1/(2d)$ as $d \to \infty$. Therefore, the moments with $m = \frac{1}{2}$ cannot reach correlation $R = 1$. In fact, the maximum correlation equals $R = 1$ only in the case of $m = 1$, Eq. (5.53). Then for $\kappa_0 \to 0$, the correlation becomes $R = 1$ only when $\Gamma(1 - m)^2\big/\Gamma(1 - 2m) \to \pm\infty$, which is true only for $m = 1$. The correlation coefficient R decreases monotonically from its maximum value $R = 1$ to $R = 0$, as the kappa index increases from $\kappa_0 \to 0$ (or $\kappa \to \frac{3}{2}$) to $\kappa_0 \to \infty$ (or $\kappa \to \infty$) (see Fig. 5.4).

Finally, we note that correlation, e.g., as given by Eq. (5.54), can be related to the Debye length λ_D and the number of particles in a Debye sphere N_D (see next section).

5.5 Debye Length in Equilibrium and Nonequilibrium Plasmas

5.5.1 General Aspects

Here we examine the electrostatic shielding in plasmas out of thermal equilibrium described by kappa distributions. We solve the Poisson equation of Gauss' law of electrodynamics and show the detailed derivation of the Debye length for three different dimensionalities of the potential: 1-D or linear symmetry for planar charge density, 2-D or cylindrical symmetry for linear charge density, and 3-D or spherical symmetry for a point charge. Then we provide three interpretations of the Debye length.

Once the formulation of the phase space distribution function is known (Livadiotis, 2015b), the Poisson equation for Gauss' law in electrodynamics can be solved to derive the exact potential configuration. At thermal equilibrium, the phase space distribution is given by the exponential Maxwell—Boltzmann distribution, while plasmas in stationary states out of thermal equilibrium are typically described by the kappa distribution (Chapter 1). In order to proceed from equilibrium to the nonequilibrium plasmas, it is critical to understand the concept of temperature for systems in stationary states that are out of thermal equilibrium. Fortunately, the temperature was shown to be well-defined for these nonequilibrium systems described by kappa distributions (for details, see Chapter 1; Livadiotis and McComas, 2009, 2010a, 2011b, 2013a, 2014a; see also the early work of Treumann, 1999a, 1999b; Treumann et al., 2004).

5.5.2 Poisson Equation for the Electrostatic Potential

Out of thermal equilibrium, the kappa distribution of the potential energy, Eq. (5.22), is derived as the marginal kappa distribution of the Hamiltonian, Eq. (5.19). Then we derive the density using the normalization $n(\vec{r})/n(\vec{r_*}) = P(\vec{r};\kappa_0;T)/P(\vec{r_*};\kappa_0;T)$, where $\vec{r_*}$ indicates now the position of zero potential $\Phi(\vec{r_*}) = 0$. For a charge perturbation, we expect this to be at infinity, $n_\infty \equiv n(\vec{r_*} \to \infty)$; this is called equilibrium density by means of the mechanical equilibrium where the potential becomes zero and should be not confused with thermal equilibrium. Hence, we have

$$n(\vec{r}) = n(\vec{r_*}) \cdot \left[1 + \frac{\Phi(\vec{r})}{\kappa_0 k_B T}\right]^{-\kappa_0 - 1 - \frac{1}{2}d_\Phi}, \text{ for } \Phi(\vec{r}) \geq 0,\ 0 \leq \kappa_0 \leq \infty, \quad (5.57a)$$

$$n(\vec{r}) = n(\vec{r_*}) \cdot \left[1 - \frac{|\Phi(\vec{r})|}{\kappa_0 k_B T}\right]^{-\kappa_0 - 1 + \frac{1}{2}d_\Phi}, \text{ for } \Phi(\vec{r}) \leq 0,\ 0 \leq \kappa_0 \leq \infty, \quad (5.57b)$$

Note that for large potential energies, different formalisms may be applied (as shown in Chapter 3). Both formulae, Eqs. (5.57a and b), must be used for small potential energies $|\Phi(\overrightarrow{r})|/(\kappa_0 k_B T) \ll 1$ and when the system's volume does not diverge to infinity ($r < \infty$). If we do consider that the volume expands from zero to (practically) infinity, then the density must converge; however, formulae, Eqs. (5.57a and b), diverge at $r \to \infty$. The potential is decreasing with the distance, becoming zero at $r \to \infty$. Then, in order for the density to converge, different formulation must be used. Table 3.1 (Chapter 3) included only the degeneration of the kappa index caused by the kinetic degrees of freedom. If we include also the degeneration caused by the potential degrees of freedom, $\kappa_0 \to -\kappa_0 + \frac{1}{2}d_\Phi$, as given in Eq. (3.136a), we obtain:

$$n(\overrightarrow{r}) = n(\overrightarrow{r_*}) \cdot \left[\frac{\Phi(\overrightarrow{r})}{\kappa_0 k_B T} - 1\right]^{\kappa_0 - 1 - \frac{1}{2}d_\Phi}, \text{ for } \Phi(\overrightarrow{r}) \geq 0, \; 0 \leq \kappa_0 < \frac{1}{2}d_\Phi, \quad (5.58a)$$

$$n(\overrightarrow{r}) = n(\overrightarrow{r_*}) \cdot \left[\frac{|\Phi(\overrightarrow{r})|}{\kappa_0 k_B T} - 1\right]^{-\kappa_0 - 1 + \frac{1}{2}d_\Phi}, \text{ for } \Phi(\overrightarrow{r}) \leq 0, \; 0 \leq \kappa_0 < \frac{1}{2}d_\Phi.$$
$$(5.58b)$$

For plasmas at thermal equilibrium, the ion/electron densities are given by the Boltzmann distribution of energy, where n_∞ denotes again the ion or electron equilibrium density.

$$n_i(\overrightarrow{r}) = n_\infty \cdot \exp\left[-\frac{\Phi_i(\overrightarrow{r})}{k_B T_i}\right], n_e(\overrightarrow{r}) = n_\infty \cdot \exp\left[-\frac{\Phi_e(\overrightarrow{r})}{k_B T_e}\right]. \quad (5.59)$$

For plasmas in stationary states out of thermal equilibrium, the ion and electron densities are described by the kappa distributions, Eqs. (5.57a and b), i.e.,

$$n_i(\overrightarrow{r}) = n_\infty \cdot \left[1 + \frac{\Phi_i(\overrightarrow{r})}{\kappa_0 k_B T_i}\right]^{-\kappa_0 - 1 - \frac{1}{2}d_{\Phi\,i}}, \; n_e(\overrightarrow{r}) = n_\infty \cdot \left[1 - \frac{|\Phi_e(\overrightarrow{r})|}{\left(\kappa_0 + \frac{1}{2}d_{\Phi\,e}\right)k_B T_e}\right]^{-\kappa_0 - 1}.$$
$$(5.60)$$

The ion and electron potential energies are related to the electric potential $V(\overrightarrow{r})$, i.e., $\Phi_i(\overrightarrow{r}) = eV(\overrightarrow{r})$ and $\Phi_e(\overrightarrow{r}) = -eV(\overrightarrow{r})$, respectively,

$$n_i(\overrightarrow{r}) = n_\infty \cdot \left[1 + \frac{eV(\overrightarrow{r})}{\kappa_0 k_B T_i}\right]^{-\kappa_0 - 1 - \frac{1}{2}d_{\Phi i}},$$

$$n_e(\overrightarrow{r}) = n_\infty \cdot \left[1 - \frac{eV(\overrightarrow{r})}{\kappa_0 k_B T_e}\right]^{-\kappa_0 - 1 + \frac{1}{2}d_{\Phi e}}, \text{ with } \frac{1}{2}d_{\Phi i,e} \equiv \frac{e\langle V\rangle_{i,e}}{k_B T_{i,e}},$$
$$(5.61)$$

where the common kappa index spans the usual interval of $\kappa_0 \in [0, \infty]$.

The total charge density is given by $\rho_e = e(n_i - n_e)$, i.e.,

$$
\rho_e(\vec{r}) = e[n_i(\vec{r}) - n_e(\vec{r})]
$$

$$
= e\, n_\infty \cdot \left\{ \left[1 + \frac{eV(\vec{r})}{\kappa_0 k_B T_i} \right]^{-\kappa_0 - 1 - \frac{1}{2} d_{\Phi i}} - \left[1 - \frac{eV(\vec{r})}{\kappa_0 k_B T_e} \right]^{-\kappa_0 - 1 + \frac{1}{2} d_{\Phi e}} \right\}, \qquad (5.62a)
$$

and in the approximation of small potential over kinetic energy $|\Phi(\vec{r})|/(\kappa_0 k_B T) \ll 1$, we have

$$
\frac{1}{\varepsilon_0} \rho_e(\vec{r}) \cong -\frac{1}{\lambda_D^2} \cdot V(\vec{r}), \qquad (5.62b)
$$

where T_0 denotes the mean inverse temperature as $T_0^{-1} = T_i^{-1} + T_e^{-1}$; λ_D is the kappa-dependent Debye length,

$$
\lambda_D = \lambda_{D\infty} \cdot \sqrt{\frac{\kappa_0}{\kappa_0 + 1}} \cdot g, \; \lambda_{D\infty} \equiv \sqrt{\frac{\varepsilon_0 k_B T_0}{e^2\, n_\infty}}, \qquad (5.63a)
$$

and the factor $g \approx 1$ is given by

$$
g = (1 - h)^{-\frac{1}{2}}, \; h \equiv \frac{T_0}{\kappa_0 + 1} \cdot \left(\frac{\frac{1}{2} d_{\Phi e}}{T_e} - \frac{\frac{1}{2} d_{\Phi i}}{T_i} \right) \ll 1. \qquad (5.63b)
$$

The number of particles in a Debye sphere $N_D \equiv n_\infty v_d \lambda_D^d$ is given by

$$
N_D = N_{D\infty} \cdot \left(\frac{\kappa_0}{\kappa_0 + 1} \right)^{\frac{d}{2}}, \; N_{D\infty} \equiv n_\infty v_d \lambda_{D\infty}^d. \qquad (5.64)
$$

where B_d and $v_d = \frac{1}{d} B_d = \pi^{\frac{d}{2}} / \Gamma\left(\frac{d}{2} + 1 \right)$ note the surface and volume of the unit d–D sphere so that $\int_{-\infty}^{\infty} f(r) d\vec{r} = B_d \int_0^\infty f(r) r^{d-1} dr$. Then by integrating both sides of Eq. (5.62b), we find

$$
\int_{-\infty}^{\infty} \rho_e(\vec{r}) d\vec{r} \cong -\frac{\varepsilon_0}{\lambda_D^2} \cdot \int_{-\infty}^{\infty} V(\vec{r}) d\vec{r}. \qquad (5.65)
$$

The charge excess is defined by

$$
e\Delta N = \int_{-\infty}^{\infty} \rho_e(\vec{r}) d\vec{r}, \qquad (5.66)
$$

and is equal to the number of charges of opposite sign, which are involved in the charge perturbation (Livadiotis and McComas, 2014a). Hence, the perturbation charge number ΔN is given by

$$
\Delta N = \frac{\varepsilon_0}{e\, \lambda_D^2} \cdot B_d \lambda_D^d \cdot V_0 \int_0^\infty e^{-r/\lambda_D} (r/\lambda_D)^{\frac{d-1}{2}} d(r/\lambda_D) = \frac{\varepsilon_0}{e\, \lambda_D^2} \cdot v_d \lambda_D^d d\, \Gamma\left(\frac{d+1}{2} \right) \cdot V_0,
$$

where we set the formula of the potential $d-$D $V(\vec{r})$ as

$$V(\vec{r}) = V_0 \cdot \frac{e^{-r/\lambda_D}}{(r/\lambda_D)^{\frac{d-1}{2}}}. \tag{5.67}$$

This relation will be shown further as follows, but we adopt here to find the constant V_0 that characterizes ΔN. Then the charge excess ΔN per Debye particles $N_D \equiv n_\infty v_d \lambda_D^d$ is given by

$$\frac{\Delta N}{N_D} = d\, \Gamma\left(\frac{d+1}{2}\right) \cdot \frac{e}{k_B T_0} \cdot V_0. \tag{5.68}$$

Thus the constant V_0 in Eq. (5.67) is given by

$$V_0 = \left[d\, \Gamma\left(\frac{d+1}{2}\right)\right]^{-1} \frac{k_B T_0}{e} \cdot \frac{\Delta N}{N_D}, \tag{5.69}$$

which is typically small because of the term $\Delta N/N_D$. The positional degrees of freedom are given by

$$\frac{1}{2}d_{\Phi\ i,e} \equiv \frac{e\langle V\rangle}{k_B T_{i,e}} = \frac{e}{k_B T_{i,e}} \cdot \frac{\int_{-\infty}^{\infty} \rho_e(\vec{r}) V(\vec{r}) d\vec{r}}{\int_{-\infty}^{\infty} \rho_e(\vec{r}) d\vec{r}} \cong \frac{e}{k_B T_{i,e}} \cdot \frac{\int_{-\infty}^{\infty} V^2(\vec{r}) d\vec{r}}{\int_{-\infty}^{\infty} V(\vec{r}) d\vec{r}}$$

$$\tag{5.70}$$

$$\cong \frac{eV_0}{k_B T_{i,e}} \cdot \frac{\int_0^{\infty} e^{-2r/\lambda_D} d(r/\lambda_D)}{\int_0^{\infty} e^{-r/\lambda_D} (r/\lambda_D)^{\frac{d-1}{2}} d(r/\lambda_D)} = \frac{eV_0}{k_B T_{i,e}} \cdot \left[2\Gamma\left(\frac{d+1}{2}\right)\right]^{-1}.$$

Therefore, the factor g in Eq. (5.63b) is given by

$$g = (1-h)^{-\frac{1}{2}}, \quad h \equiv \frac{1}{\kappa_0 + 1} \cdot \left[2d\, \Gamma^2\left(\frac{d+1}{2}\right)\right]^{-1} \cdot \frac{\Delta N}{N_D} \frac{1-t}{1+t}. \tag{5.71a}$$

where we set $t \equiv T_e/T_i$. We observe that $\frac{1}{2}d_{\Phi i,e}$ is also small because of $\frac{1}{2}d_{\Phi i,e} \propto V_0 \propto \Delta N/N_D$; thus

$$g \cong 1 + O(\Delta N/N_D). \tag{5.72}$$

Hence, the potential becomes

$$V(\vec{r}) = \left[d\, \Gamma\left(\frac{d+1}{2}\right)\right]^{-1} \frac{k_B T_0}{e} \cdot \frac{\Delta N}{N_D} \cdot \frac{e^{-r/\lambda_D}}{(r/\lambda_D)^{\frac{d-1}{2}}}, \tag{5.73}$$

with mean value

$$\frac{1}{2}d_{\Phi i,e} \cong \frac{eV_0}{k_B T_{i,e}} \cdot \left[2\Gamma\left(\frac{d+1}{2}\right)\right]^{-1} \cong \left[2d\, \Gamma\left(\frac{d+1}{2}\right)\right]^{-2} \frac{T_0}{T_{i,e}} \cdot \frac{\Delta N}{N_D} \ll 1. \tag{5.74}$$

Next, by substituting the relation between the electric potential and field $\vec{E} = -\vec{\nabla}V$ into the Gauss's law $\vec{\nabla} \cdot \vec{E} = \frac{1}{\varepsilon_0}\rho_e$ that connects the electric field with the charge density ρ_e, we derive the Poisson's equation

$$\nabla^2 V(\vec{r}) = -\frac{1}{\varepsilon_0}\rho_e(\vec{r}). \tag{5.75}$$

Given the total charge density $\rho_e = e(n_i - n_e)$, the Poisson's equation becomes

$$\nabla^2 V(\vec{r}) = -\frac{en_\infty}{\varepsilon_0} \cdot \left\{ \left[1 + \frac{eV(\vec{r})}{\kappa_0 k_B T_i}\right]^{-\kappa_0 - 1 - \frac{1}{2}d_{\Phi i}} - \left[1 - \frac{eV(\vec{r})}{\kappa_0 k_B T_e}\right]^{-\kappa_0 - 1 + \frac{1}{2}d_{\Phi e}} \right\}. \tag{5.76a}$$

This can be rewritten as

$$\nabla^2 \Psi(\vec{r}) - \frac{1}{\lambda_D^2} \cdot f[\Psi(\vec{r})] = 0, \text{ with } \Psi(\vec{r}) \equiv \frac{eV(\vec{r})}{\kappa_0 k_B T_0}, \quad \lambda_D \equiv \lambda_{D\infty} \cdot \sqrt{\frac{\kappa_0}{\kappa_0 + 1}}, \tag{5.76b}$$

where the function f is defined by

$$f(\Psi) \equiv -\frac{1}{\kappa_0 + 1} \cdot \left[\left(1 + \frac{T_0}{T_i}\Psi\right)^{-\kappa_0 - 1 - \frac{1}{2}d_{\Phi i}} - \left(1 - \frac{T_0}{T_e}\Psi\right)^{-\kappa_0 - 1 + \frac{1}{2}d_{\Phi e}}\right]. \tag{5.77}$$

This is approximated for $\Psi(\vec{r}) \ll 1$ as follows:

$$f(\Psi) \cong (1 - h) \cdot \Psi + O(\Psi^2) \cong \Psi + O(\Delta N/N_D), \tag{5.78}$$

(because of Eq. 5.71a) so that the Poisson's Eq. (5.76b) becomes

$$\nabla^2 \Psi(\vec{r}) - \frac{1}{\lambda_D^2} \cdot \Psi(\vec{r}) \cong 0 \text{ or } \nabla^2 V(\vec{r}) - \frac{1}{\lambda_D^2} \cdot V(\vec{r}) \cong 0. \tag{5.79}$$

(note that using $r \to r/\lambda_D$, we may rescale Eq. (5.79) and become dimensionless).

5.5.3 Symmetric Poisson Equation and Solutions

Next, we solve the Poisson differential Eq. (5.79) for different symmetries and dimensionalities, as shown in Fig. 5.6. Namely, 1-D (planar charge density and linear field symmetry), 2-D (linear charge density and cylindrical field symmetry), and 3-D (point charge and spherical field symmetry).

5.5.3.1 1-D Potential: Planar Charge Perturbation

For a planar charge perturbation at the y-z plane, the 1-D Poisson equation along x-axis is written as

$$V''(x) - \frac{1}{\lambda_D^2} \cdot V(x) \cong 0, \tag{5.80}$$

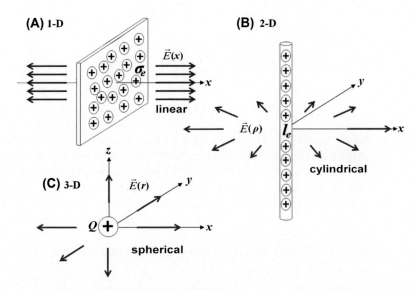

FIGURE 5.6
Three simple geometries of charge perturbation, with (A) linear, (B) cylindrical, (C) spherical symmetry, and respective
dimensionalities $d = 1,2,3$ (the electric field notations $\vec{E}(x)$, $\vec{E}(\rho)$, and $\vec{E}(r)$ correspond to the surface density σ_e, linear density
l_e, and point charge Q). Adapted by Livadiotis and McComas (2014a).

having the solution $V(x) \cong B \cdot \exp(-x/\lambda_{D\infty})$, where B is a constant expressed in
terms of the surface charge density σ_e; this is related to the electric field at $x = 0$,
i.e., $\frac{1}{2}(\sigma_e/\varepsilon_0) = E(x = 0) = -\Phi'(x = 0) \cong B/\lambda_D$, or $B \cong \frac{1}{2}(\sigma_e/\varepsilon_0)\lambda_D$; hence, the
solution becomes

$$V(x) \cong \frac{\sigma_e}{2\varepsilon_0}\lambda_D \cdot \exp(-x/\lambda_D), \tag{5.81}$$

5.5.3.2 2-D Potential: Linear Charge Perturbation

For a linear charge perturbation of large length L (practically $L \gg \lambda_D$) along the
z-axis, the symmetry of the electric potential and field is cylindrical. The 2-D
Poisson equation along x-y plane is

$$\frac{1}{\rho}\frac{d}{d\rho}\left[\rho\frac{dV(\rho)}{d\rho}\right] - \frac{1}{\lambda_D^2}\cdot V(\rho) \cong 0, \text{ with } \rho \equiv \sqrt{x^2 + y^2}, \tag{5.82}$$

that leads to the Bessel differential equation

$$V''(\rho) + \frac{1}{\rho}V'(\rho) - \frac{1}{\lambda_D^2}\cdot V(\rho) \cong 0, \tag{5.83}$$

with a solution given by the zero-th bounded modified Bessel function of the
second kind (or Macdonald function), $V(\rho) = B \cdot K_0(\rho/\lambda_D)$. The function $K_0(x)$
has the asymptotic behavior,

$$K_0(x) \cong \begin{cases} \dfrac{\sqrt{\pi}}{2} \cdot \dfrac{\exp(-x)}{\sqrt{x}}, & x \gg 1, \\[3mm] \dfrac{\exp(-x)}{\sqrt{x}}, & x \sim O(1), \\[3mm] -\ln\left(\dfrac{x}{2}\right) - \gamma_E, & x \ll 1, \end{cases} \qquad (5.84)$$

where $\gamma_E = 0.5772156649\ldots$ is the Euler constant; the asymptotic behavior for large x is sufficiently accurate ($\sim 90\%$) even for $x \sim 1$. In order to determine the constant B, we find the electric field near the perturbation, i.e., for $\rho/\lambda_D \ll 1$, where, according to Eq. (5.84), the potential is written as $V(\rho) \cong -B \cdot \ln(\rho)$, where we ignore the constant quantity $B \cdot \ln(2\lambda_D) - \gamma_E$. Hence, the corresponding electric field is $E(\rho) \cong B/\rho$; comparing this with the electric field of a linear charge perturbation of large length $E(\rho) \cong l_e/(2\pi\varepsilon_0\rho)$ with l_e, noting the linear charge density (e.g., Dash and Khuntia, 2010), we obtain that $B = l_e/(2\pi\varepsilon_0)$. Hence,

$$V(\rho) \cong \frac{l_e}{2\pi\varepsilon_0} \cdot K_0(\rho/\lambda_D) \cong \frac{l_e}{2\pi\varepsilon_0} \cdot \begin{cases} \sqrt{\dfrac{\pi}{2}} \cdot \dfrac{\exp(-\rho/\lambda_D)}{\sqrt{\rho/\lambda_D}}, & \rho \gg \lambda_D, \\[3mm] \dfrac{2}{\sqrt{\pi}} \cdot \dfrac{\exp(-\rho/\lambda_D)}{\sqrt{\rho/\lambda_D}}, & \rho \sim O(\lambda_D), \\[3mm] \ln(\lambda_D/\rho) + \text{const.}, & \rho \ll \lambda_D. \end{cases} \qquad (5.85)$$

5.5.3.3 3-D Potential: Point Charge Perturbation

For a point charge perturbation Q, the symmetry of the electric potential and field is spherical, and the 3-D Poisson equation is written in terms only of the spherical radius $r = \sqrt{x^2 + y^2 + z^2}$,

$$\frac{1}{r^2} \frac{d}{dr}\left[r^2 \frac{dV(r)}{dr} \right] - \frac{1}{\lambda_D^2} \cdot V(r) \cong 0 \qquad (5.86)$$

that leads to the differential equation

$$V''(r) + \frac{2}{r} V'(r) - \frac{1}{\lambda_D^2} \cdot V(r) \cong 0, \qquad (5.87)$$

with the solution given by the potential $V(r) = B \exp(-r/\lambda_D)/r$ (called Yukawa or Debye−Hückel). For $r \to 0$, this describes the potential of the charge Q without the plasma shielding, i.e., $V(r \to 0) = Q/(4\pi\varepsilon_0 r) = B/r$, or $B = Q/(4\pi\varepsilon_0)$. Hence,

$$V(r) \cong \frac{Q}{4\pi\varepsilon_0} \cdot \frac{1}{r} \exp(-r/\lambda_D). \qquad (5.88)$$

Table 5.2	**Charge/Field Dimensions and Symmetry**			
Charge Dimension	**Charge Distribution ρ_e**	**Field Dimension**	**Field Symmetry**	**Potential $V(\vec{r})$**
2-D	Surface density σ_e	1-D	Linear (x)	$\dfrac{\sigma_e}{2\varepsilon_0}\cdot\lambda_D\cdot\exp(-x/\lambda_D)$
1-D	Linear density l_e	2-D	Cylindrical (ρ)	$\sqrt{\dfrac{2}{\pi}}\,\dfrac{l_e}{2\pi\varepsilon_0}\cdot\dfrac{\exp(-\rho/\lambda_D)}{\sqrt{\rho/\lambda_D}}$
0-D	Point charge Q	3-D	Spherical (r)	$\dfrac{Q}{4\pi\varepsilon_0}\cdot\lambda_D^{-1}\cdot\dfrac{\exp(-r/\lambda_D)}{r/\lambda_D}$

5.5.3.4 d–D Potential

Eqs. (5.81) and (5.85) for $\rho \sim O(\lambda_D)$ and Eq. (5.88) show the shielding potential for dimensionalities $d = 1$, 2, and 3, respectively. The case of d–D potential can be derived inductively, as shown in Eqs. (5.67) and (5.72) (nevertheless, the whole space is still considered to be 3-D, i.e., $d \leq 3$). If we compare the constant V_0 in potential formula Eq. (5.72) with those of Eqs. (5.81), (5.85), and (5.88), one can easily show a trivial equivalency; for instance, for $d = 3$, we find that the number of the perturbed charge ions is $\Delta N = Q/e$ (!).

Table 5.2 gathers the derived formulations of the potential for all three dimensionalities, i.e., 1-D (planar charge density on y-z plane and linear field symmetry along x-axis), 2-D (linear charge density on z-axis and cylindrical field symmetry on x-y plane), and 3-D (point charge and spherical field symmetry).

5.5.4 The Case of Large Potential Energy

Next we investigate the Poisson's equation and its solutions when the volume expands from zero to practically infinity; then in order for the density to converge, the following formulation must be used:

$$n(\vec{r}) = n(\vec{r}_*)\cdot\left[\frac{\Phi(\vec{r})}{\kappa_0 k_B T} - 1\right]^{\kappa_0 - 1 - \frac{1}{2}d_\Phi}, \text{ for } \Phi(\vec{r}) \geq 0,\ 0 \leq \kappa_0 < \frac{1}{2}d_\Phi, \qquad (5.89a)$$

$$n(\vec{r}) = n(\vec{r}_*)\cdot\left[\frac{|\Phi(\vec{r})|}{\kappa_0 k_B T} - 1\right]^{-\kappa_0 - 1 + \frac{1}{2}d_\Phi}, \text{ for } \Phi(\vec{r}) \leq 0,\ 0 \leq \kappa_0 < \frac{1}{2}d_\Phi. \qquad (5.89b)$$

Thus for plasmas in stationary states out of thermal equilibrium, the ion and electron densities are

$$n_i(\vec{r}) = n_\infty\cdot\left[\frac{\Phi_i(\vec{r})}{\kappa_0 k_B T_i} - 1\right]^{\kappa_0 - 1 - \frac{1}{2}d_{\Phi i}},\ n_e(\vec{r}) = n_\infty\cdot\left[\frac{|\Phi_e(\vec{r})|}{\kappa_0 k_B T_e} - 1\right]^{-\kappa_0 - 1 + \frac{1}{2}d_{\Phi e}}.$$

$$(5.90)$$

Applying $\Phi_i(\vec{r}) = eV(\vec{r})$ and $\Phi_e(\vec{r}) = -eV(\vec{r})$, we derive

$$n_i(\vec{r}) = n_\infty \cdot \left[\frac{eV(\vec{r})}{\kappa_0 k_B T_i} - 1\right]^{\kappa_0 - 1 - \frac{1}{2}d_{\Phi i}},$$

$$n_e(\vec{r}) = n_\infty \cdot \left[\frac{eV(\vec{r})}{\kappa_0 k_B T_e} - 1\right]^{-\kappa_0 - 1 + \frac{1}{2}d_{\Phi e}}, \text{ with } \frac{1}{2}d_{\Phi i,e} \equiv \frac{e\langle V \rangle_{i,e}}{k_B T_{i,e}},$$

(5.91)

where the common kappa index spans the usual interval of $0 \leq \kappa_0 < \frac{1}{2}d_\Phi$. Given the total charge density $\rho_e = e(n_i - n_e)$, the Poisson's equation becomes

$$\nabla^2 V(\vec{r}) = -\frac{en_\infty}{\varepsilon_0} \cdot \left\{ \left[\frac{eV(\vec{r})}{\kappa_0 k_B T_i} - 1\right]^{\kappa_0 - 1 - \frac{1}{2}d_{\Phi i}} - \left[\frac{eV(\vec{r})}{\kappa_0 k_B T_e} - 1\right]^{-\kappa_0 - 1 + \frac{1}{2}d_{\Phi e}} \right\}. \quad (5.92)$$

For large potential energies, Eq. (5.92) is approximately expressed by

$$\nabla^2 V(\vec{r}) \cong -\frac{en_\infty}{\varepsilon_0} \cdot \left\{ \left[\frac{eV(\vec{r})}{\kappa_0 k_B T_i}\right]^{\kappa_0 - 1 - \frac{1}{2}d_{\Phi i}} - \left[\frac{eV(\vec{r})}{\kappa_0 k_B T_e}\right]^{-\kappa_0 - 1 + \frac{1}{2}d_{\Phi e}} \right\} \text{ or}$$

$$\nabla^2 \Psi(\vec{r}) - \frac{1}{\lambda_D^2} \cdot f[\Psi(\vec{r})] \cong 0,$$

(5.93)

where

$$f[\Psi(\vec{r})] \equiv -\frac{1}{\kappa_0 + 1} \cdot \left[(1 + 1/t)^{-\kappa_0 + 1 + \frac{1}{2}d_{\Phi i}} \Psi(\vec{r})^{\kappa_0 - 1 - \frac{1}{2}d_{\Phi i}} \right.$$
$$\left. - (1 + t)^{\kappa_0 + 1 - \frac{1}{2}d_{\Phi e}} \Psi(\vec{r})^{-\kappa_0 - 1 + \frac{1}{2}d_{\Phi e}} \right],$$

(5.94a)

e.g., for $t = 1$,

$$f\left[\frac{1}{2}\Psi(\vec{r})\right] \equiv -\frac{\frac{1}{2}}{\kappa_0 + 1} \cdot \left\{ \left[\frac{1}{2}\Psi(\vec{r})\right]^{\kappa_0 - 1 - \frac{1}{2}d_{\Phi i}} - \left[\frac{1}{2}\Psi(\vec{r})\right]^{-\kappa_0 - 1 + \frac{1}{2}d_{\Phi e}} \right\}. \quad (5.94b)$$

We may rescale Eq. (5.93) via $r \to r/\lambda_D$ to become dimensionless,

$$\nabla^2 \Psi(\vec{r}) - f[\Psi(\vec{r})] \cong 0,$$

(5.95)

which for the three cases of 1-D, 2-D, and 3-D potential, is respectively reduced to

$$\Psi''(x) - f[\Psi(x)] \cong 0, \quad \Psi''(\rho) + \frac{1}{\rho}\Psi'(\rho) - f[\Psi(\rho)] \cong 0,$$

$$\Psi''(r) + \frac{2}{r}\Psi'(r) - f[\Psi(r)] \cong 0.$$

(5.96a)

Eq. (5.96) is generalized types of the Lane—Emden differential equation (e.g., Chapter 3; Lane, 1870; Zwillinger, 1997). This is a dimensionless form of Poisson's equation derived for the gravitational potential of a Newtonian self-gravitating system.

The 1-D case of Eq. (5.96) can be multiplied by $\Psi'(x)$ to find

$$\Psi'(x)^2 - F[\Psi(x)] \cong \text{const.,} \quad \text{with} \tag{5.96b}$$

$$F[\Psi(\vec{r})] \equiv -\frac{1}{\kappa_0 + 1} \cdot \left[\frac{(1 + 1/t)^{-\kappa_0 + 1 + \frac{1}{2}d_{\Phi i}}}{\kappa_0 - \frac{1}{2}d_{\Phi i}} \Psi(x)^{\kappa_0 - \frac{1}{2}d_{\Phi i}} \right.$$
$$\left. + \frac{(1 + t)^{\kappa_0 + 1 - \frac{1}{2}d_{\Phi e}}}{\kappa_0 - \frac{1}{2}d_{\Phi e}} \Psi(\vec{r})^{-\kappa_0 + \frac{1}{2}d_{\Phi e}} \right]. \tag{5.96c}$$

Nevertheless, more complicated solutions may be derived in the 2-D/3-D cases.

5.5.5 Main Interpretations

The Debye length represents the physical scale of the transition from plasma collectivity to individual particle behavior. It interfaces between the physics of micro and macro scales and its technical definition and detailed attributes are crucial for theoretical and experimental plasma–physics research.

There is currently no strict definition of the term Debye length, while particular properties of electrostatic shielding have been used for the "interpretation" or "definition" of the "Debye length." There are three main interpretations or definitions of the Debye length in plasmas: (1) the distance at which the potential energy from a charge perturbation is equal to the thermal energy (e.g., Kallenrode, 2004; Baumjohann and Treumann, 1997); (2) the distance at which the potential energy from a charge perturbation has fallen to 1/e of its unshielded value (e.g., Montgomery and Tidman, 1964; Chen, 1974); and (3) the standard deviation of the positional charge distribution (Livadiotis and McComas, 2014a).

Interpretation 1 is a physical property of Debye shielding; it involves the distance where a sort of equilibrium is established between the "source" of the shielding that is the restoring electric potential and the "sink" of the shielding that is the disturbing thermal energy. In contrast, interpretation 2 is a mathematical property of the shielding, marking the distance where the shielded potential falls to a certain fraction of its unshielded value.

While the two interpretations are equivalent at thermal equilibrium, they are different for plasmas out of thermal equilibrium, and the validity of interpretation 1 is not certain for such plasmas. On the other hand, interpretation 2 is valid for systems both at and out of thermal equilibrium. This is because, similar to the Boltzmann distribution at thermal equilibrium, the kappa distribution can lead to a linearized Poisson equation with an exponential solution. Therefore, both the interpretations of the Debye length have been used for thermal equilibrium plasmas, but only interpretation 2 has been used for plasmas out of thermal equilibrium that are described by kappa distributions (Bryant, 1996; Rubab and Murtaza, 2006; Gougam and Tribeche, 2011). Indeed, interpretation 1 needs to be modified for nonequilibrium systems.

Even stable plasmas constantly undergo charge perturbations via the thermal motion of their particles. Whenever such a charge perturbation occurs in a plasma, positive and negative free charges respond by moving in opposite directions around the perturbation, producing a shielding effect that preserves the charge quasineutrality of the plasma (equidistribution of charge density) at large distances. Hence, the shielding electric potential energy recovers the local plasma's stability and restores its quasineutrality. On the other hand, thermal motions compete with the potential and make it more difficult for free charges to shield the charge perturbation. While the potential energy is larger than the thermal energy at distances near the charge perturbation, the thermal energy prevails at farther distances. The specific distance for which the potential and thermal energies are equal specifies the first interpretation of the Debye length λ_D (Kallenrode, 2004).

The potential energy is caused by the presence of ΔN ions that contribute to the charge perturbation. On the other hand, the thermal energy applies to all the N_D particles that could be available to shield the charge perturbation, i.e., included in a Debye length. Hence, the ratio R of the potential to thermal energy must be normalized by the quantity $\Delta N/N_D$, i.e.,

$$R \equiv \frac{N_D e \, V}{\Delta N \frac{1}{2} k_B T_0}. \tag{5.97}$$

The ratio of the total potential to thermal energy at the Debye length is a constant of the order of unity. This constitutes interpretation 1 of the Debye length. Indeed, in the 1-D case, we have $R(x = \lambda_D) \cong 1/e$. In addition, we show that for any dimensionality d of the potential, we have $R(r = \lambda_D) \cong c(d) \cdot e^{-1} \sim O(1)$, where the constant $c(d)$ differs for each dimensionality $d = 1, 2, 3$, but remains in the order of unity,

$$R(r = \lambda_D) \cong \left[\frac{d}{2} \, \Gamma\left(\frac{d+1}{2}\right) \right]^{-1} \cdot e^{-1} \sim O(1) \text{ or } \left[\frac{d}{2} \, \Gamma\left(\frac{d+1}{2}\right) e \right] \cdot R(r = \lambda_D) \cong 1. \tag{5.98}$$

Fig. 5.7A demonstrates this interpretation of Debye length through Eq. (5.98), that is, the approximate equality of the potential energy by the thermal energy at this distance.

Another frequently used interpretation of the Debye length (e.g., Montgomery and Tidman, 1964; Chen, 1974) involves the distance where the exponential factor, which is included in the spatial function of the electric potential, falls to $\sim 1/e$ of its unshielded value, i.e.,

$$\left. \frac{V(r; \lambda_D)}{V(r; \lambda_D \to \infty)} \right|_{r=\lambda_D} \sim e^{-1} \sim O(1), \quad R \cong \left[\frac{d}{2} \, \Gamma\left(\frac{d+1}{2}\right) \right]^{-1} \cdot e^{-1} \sim O(1) \text{ or}$$

$$\left. \frac{V(r; \lambda_D)}{e^{-1} V(r; \lambda_D \to \infty)} \right|_{r=\lambda_D} = 1. \tag{5.99}$$

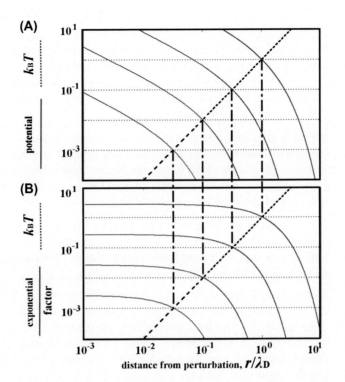

FIGURE 5.7

Demonstration of interpretations 1 and 2 of the Debye length in plasmas (for $\kappa_0 \to \infty$). (A) Interpretation 1. The 3-D potential energy (*red*) and the thermal energy (*blue*) are co-plotted for four values of thermal energy, $k_B T/[J] = 10^{-3}, 10^{-2}, 10^{-1}$, and 1, which correspond to four values of the potential energy because the potential's formulation involves also the temperature (through the Debye length). We plot the normalized potential energy $(3eN/\Delta N) \cdot eV$ and the thermal energy $k_B T$ at a distance r/λ_D from the perturbation. (B) Interpretation 2. The exponential factor $\exp(-r/\sqrt{k_B T})$ (*red*) and the thermal energy $k_B T$ (*blue*) are co-plotted for the same four values of thermal energy (note that any involved parameter is normalized to one except the temperature). The exponential factor is multiplied by $k_B T/e$ in order to be exactly equal to the thermal energy $k_B T$ at a distance equal to the Debye length. In both panels, the geometric locus of the intersections (connected with vertical dash–*dot lines*) is given by the parabola $k_B T = \lambda_D^2$ (on a log–log scale). Adapted by Livadiotis and McComas (2014).

This constitutes interpretation 2 of the Debye length, demonstrated in Fig. 5.7B.

One way to avoid including a threshold can be realized by using the differential equation rather than its solution, i.e., $\lim\limits_{V \to 0} (V/\nabla^2 V) = \lambda_D^2$. Here, the threshold $1/e$ that determines the Debye length is hided within the differential equation, Eq. (5.79), and the exponential form of its solution. However, we may substitute the Laplacian $\nabla^2 V$ with its equivalent form from the Poisson equation. Then a more general way to express interpretation 2 is via the ratio of the "outcome," the potential V, to its "cause," the charge distribution ρ_e; namely, from Eq. (5.72), we obtain

$$\lim_{V \to 0} \left(\frac{V}{\frac{1}{\varepsilon_0}\rho_e} \right) = \lambda_D^2. \tag{5.100}$$

The two Debye length interpretations, namely, (1) via the competing thermal and potential energies and their ratio, Eq. (5.98), and (2) via the exponential form of the potential, Eq. (5.100), are physically different, but they result to the same Debye length, and thus, they are mathematically equivalent (note that only interpretation 2 has been used for plasmas out of thermal equilibrium that are described by kappa distributions, e.g., Bryant, 1996; Rubab and Murtaza, 2006; Gougam and Tribeche, 2011).

We have seen that interpretation 2 is not restricted to any specific forms of the ion/electron distribution functions. The only requirement is that the specific forms of the distributions have to be known (either analytically or numerically). Here we present a third interpretation of Debye length for which only the second moment of the ion/electron distribution functions is necessary.

The concept of organization of plasmas by their Debye shielding within clusters of locally correlated particles (Debye spheres) has been shown to be interwoven with a new type of phase space quantization constant. According to this, the Debye length assigns a type of large-scale uncertainty in position $\sqrt{\langle \Delta r^2 \rangle} \sim \lambda_D$ that leads to a large-scale uncertainty in time $\delta t \sim \lambda_D / \theta$, where $\theta \equiv \sqrt{2 k_B T / m}$ is the characteristic thermal speed of the particle with mass m. Given this time uncertainty and the least energy of a particle in a Debye sphere, a large-scale quantization of phase space, some 12 orders of magnitude larger than the Planck constant was found (Section 5.9; Livadiotis and McComas, 2013b, 2014b; Livadiotis, 2015d).

Indeed, we have the positional statistical moments

$$\langle r^m \rangle = \frac{\int_{-\infty}^{\infty} \rho_e(\vec{r}) r^m d\vec{r}}{\int_{-\infty}^{\infty} \rho_e(\vec{r}) d\vec{r}} \cong \frac{\int_{-\infty}^{\infty} V(\vec{r}) r^m d\vec{r}}{\int_{-\infty}^{\infty} V(\vec{r}) d\vec{r}}$$

$$\cong \frac{\int_0^{\infty} e^{-r/\lambda_D} r^{m+(d-1)/2} dr}{\int_0^{\infty} e^{-r/\lambda_D} r^{(d-1)/2} dr} = \frac{\Gamma\left(m + \dfrac{d+1}{2}\right)}{\Gamma\left(\dfrac{d+1}{2}\right)} \cdot \lambda_D^m. \tag{5.101}$$

Hence, the first and second statistical moments are

$$\langle r \rangle \cong \frac{d+1}{2} \cdot \lambda_D, \quad \langle r^2 \rangle \cong \left(\frac{d+1}{2} + 1\right) \frac{d+1}{2} \cdot \lambda_D^2, \tag{5.102}$$

leading to the positional variance and standard deviation

$$\langle \Delta r^2 \rangle = \langle r^2 \rangle - \langle r \rangle^2 \cong \frac{d+1}{2} \cdot \lambda_D^2, \quad \sqrt{\langle \Delta r^2 \rangle} \cong \sqrt{\frac{d+1}{2}} \cdot \lambda_D. \tag{5.103}$$

#	Description	Restrictions and Requirements
1	$\lim\limits_{V \to 0}\left[d\,\Gamma\left(\dfrac{d+1}{2}\right)e\cdot\dfrac{N_D}{\Delta N}\cdot\dfrac{eV(\lambda_D)}{k_B T_0}\right]=1$	1. Poisson equation must be solved to find the potential. 2. Ions and electrons are described by kappa distributions. 3. These distributions must be known (analytically/numerically).
2	$\lim\limits_{V \to 0}\left(\dfrac{V}{\dfrac{1}{\varepsilon_0}\rho_e}\right)=\lambda_D^2$	1. The form of the potential is not necessary. 2. Any form of the ion/electron distribution functions. 3. These functions must be known (analytically/numerically).
3	$\lim\limits_{V \to 0}\langle\Delta r^2\rangle=\dfrac{d+1}{2}\cdot\lambda_D^2$	1. The form of the potential is not necessary. 2. Any form of the ion/electron distribution functions. 3. Only the second statistical moment is necessary.

Table 5.3 The Three Interpretations of the Debye Length

Specifically for $d=1$, we have $\sqrt{\langle\Delta x^2\rangle}\cong\lambda_D$. Note also that the previous equations are for small perturbations and thus can be written in terms of the limit of $V \to 0$, e.g.,

$$\lim_{V \to 0}\langle\Delta r^2\rangle=\frac{d+1}{2}\cdot\lambda_D^2,\ \lim_{V \to 0}\langle\Delta x^2\rangle=\lambda_D^2. \tag{5.104}$$

The three interpretations of the Debye length are compared in Table 5.3 and illustrated in Fig. 5.8.

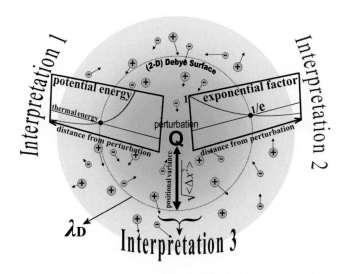

FIGURE 5.8
Schematic diagram of the three interpretations of the Debye length. Adapted by Livadiotis and McComas (2014a).

5.6 Electrical Conductivity

We derive the plasma electrical conductivity using the methodology of Gurnett and Bhattacharjee (2005, Chapter 11.4), where electrons are assumed to collide with fixed immobile ions (see also Goldston and Rutherford, 1995). As long as a plasma is undisturbed, it is stabilized into kappa distributions (that is, into a single kappa distribution or combination of multiple kappa distributions). For simplicity, we consider an ion/electron plasma, taking into account only the motion of electrons while we ignore the motion of the heavier and less mobile ions. Then the undisturbed electron distribution function is given by

$$f_0(\vec{u}) = n_0 \cdot P(\vec{u}; T_e; \kappa_0), \tag{5.105}$$

where n_0 is the average number density of ions and electrons, while $P(\vec{u}; T_e; \kappa_0)$ is the electron ($d_K = 3$)-dimensional kappa distribution of the Hamiltonian in the approximation of zero potential energy, that is,

$$P(\vec{u}; T_e; \kappa_0) = \left(\pi\theta_e^2\right)^{-\frac{3}{2}} \cdot \kappa_0^{-\frac{3}{2}} \cdot \frac{\Gamma\left(\kappa_0 + \frac{5}{2}\right)}{\Gamma(\kappa_0 + 1)} \cdot \left(1 + \frac{1}{\kappa_0} \cdot \frac{u^2}{\theta_e^2}\right)^{-\kappa_0 - \frac{5}{2}}, \tag{5.106}$$

in the flow reference frame $\vec{u}_b = 0$, recovering the Maxwell distribution for $\kappa_0 \to \infty$,

$$P(\vec{u}; T_e; \kappa_0 \to \infty) = \left(\pi\theta_e^2\right)^{-\frac{1}{2}d} \cdot \exp\left(-\frac{\vec{u}^2}{\theta_e^2}\right). \tag{5.107}$$

The previously mentioned kappa distribution is expressed in terms of the invariant kappa index κ_0, which is independent of the degrees of freedom; $\theta_e \equiv \sqrt{2k_B T_e/m_e}$ is the electron thermal speed.

Once the plasma is disturbed by some external force $\vec{F}$, then the distribution $f(\vec{u})$ is the solution of the Fokker–Planck equation

$$\left[\frac{\partial}{\partial t} + \vec{u} \cdot \vec{\nabla} + \frac{1}{m_e} \vec{F} \cdot \vec{\nabla}_u\right] f(\vec{u}) = \frac{1}{2} n_0 \Gamma_{ei} \cdot L_{F-P} f(\vec{u}), \tag{5.108}$$

where the constant $\Gamma_{ei} = \ln \Lambda \cdot e^4 / \left(4\pi \varepsilon_0^2 m_e^2\right)$ is characteristic of the electron–ion collisions, and $\Lambda \equiv 12\pi n_0 \lambda_{De}^3 = \ln(9N_{De})$; $\vec{\nabla}_u = (\partial/\partial u_x, \partial/\partial u_y, \partial/\partial u_z)$ is the gradient in the velocity space. The Fokker–Planck collision operator is

$$L_{F-P} = \vec{\nabla}_u \cdot \left(\frac{u^2 \overset{\leftrightarrow}{1} - \vec{u} \otimes \vec{u}}{u^3} \cdot \vec{\nabla}_u\right), \tag{5.109}$$

where $\overset{\leftrightarrow}{1}$ is the unit tensor; $\otimes$ is the tensor product (dyadic).

Consider now that a small electric field is applied, $\vec{E} = E \cdot \hat{z}$, $\vec{F} = -e\vec{E} = -eE \cdot \hat{z}$. After passing sufficient time so that all the transients have damped, the plasma evolves to a stationary $\partial f/\partial t = 0$ and spatially homogeneous $\vec{\nabla} f = 0$ state. Then Eq. (5.108) becomes

$$\frac{1}{m_e} \vec{F} \cdot \vec{\nabla}_u f(\vec{u}) = -\frac{eE}{m_e} \cdot \frac{\partial}{\partial u_z} f(\vec{u}) = \frac{1}{2} n_0 \Gamma_{ei} \cdot L_{F-P} f(\vec{u}). \tag{5.110}$$

We use spherical coordinates, while the direction of $\vec{E}$ implies azimuthally symmetric $\partial f/\partial \varphi = 0$, or $f(\vec{u}) = f(u, \vartheta)$. Hence, the operators in Eq. (5.110) become

$$\frac{1}{m_e} \vec{F} \cdot \vec{\nabla}_u = -\frac{eE}{m_e} \hat{z} \cdot \left(\frac{\partial}{\partial u} \hat{r} + \frac{1}{u} \frac{\partial}{\partial \vartheta} \hat{\vartheta} \right) = -\frac{eE}{m_e} \left(\cos \vartheta \frac{\partial}{\partial u} - \frac{\sin \vartheta}{u} \frac{\partial}{\partial \vartheta} \right), \text{ and}$$

$$\tag{5.111a}$$

$$L_{F-P} = \frac{1}{\sin \vartheta} \frac{1}{u^3} \left[\cos \vartheta \frac{\partial}{\partial \vartheta} + \sin \vartheta \frac{\partial^2}{\partial \vartheta^2} \right]. \tag{5.111b}$$

Then, we have

$$-\frac{eE}{m_e} \cdot \left[\cos \vartheta \frac{\partial f(u, \vartheta)}{\partial u} - \frac{\sin \vartheta}{u} \frac{\partial f(u, \vartheta)}{\partial \vartheta} \right]$$

$$= n_0 \Gamma_{ei} \cdot \frac{1}{2 \sin \vartheta} \frac{1}{u^3} \left[\cos \vartheta \frac{\partial f(u, \vartheta)}{\partial \vartheta} + \sin \vartheta \frac{\partial^2 f(u, \vartheta)}{\partial \vartheta^2} \right]. \tag{5.112}$$

The distribution function may be written as an expansion in terms of E,

$$f(u, \vartheta) = f_0(u) + \sum_{k=1} f_k(u) \cdot \Theta_k(\vartheta), \text{ with } f_k(u) \propto E^k. \tag{5.113}$$

For a small electric field E, we can keep up to the linear term,

$$f(u, \vartheta) = f_0(u) + f_1(u) \cdot \Theta(\vartheta), \tag{5.114}$$

where we set that the undisturbed distribution $f_0(u)$ is a kappa velocity distribution, as shown in Eq. (5.106), while the small perturbation of the distribution is proportional to its cause, the electric field $f_1(u) \propto E$,

$$-\frac{eE}{m_e} \cdot \left[\cos \vartheta \cdot f'_0(u) + \cos \vartheta \, \Theta(\vartheta) \cdot f'_1(u) - \sin \vartheta \frac{f_1(u)}{u} \cdot \Theta'(\vartheta) \right]$$

$$= n_0 \Gamma_{ei} \cdot \frac{f_1(u)}{u^3} \cdot \frac{\cos \vartheta \, \Theta'(\vartheta) + \sin \vartheta \, \Theta''(\vartheta)}{2 \sin \vartheta}, \tag{5.115}$$

Since $f_1 \propto E$, we keep only the first term in the left-hand side brackets; thus

$$-\frac{eE}{m_e} \cdot \cos \vartheta \cdot f'_0(u) \cong n_0 \Gamma_{ei} \cdot \frac{f_1(u)}{u^3} \cdot \frac{\cos \vartheta \, \Theta'(\vartheta) + \sin \vartheta \, \Theta''(\vartheta)}{2 \sin \vartheta}. \tag{5.116}$$

Then we require $\cos \vartheta \, \Theta'(\vartheta) + \sin \vartheta \, \Theta''(\vartheta)]/(2 \sin \vartheta \cos \vartheta) = \text{constant} \equiv a$ or $[\sin \vartheta \, \Theta'(\vartheta)]' = 2a \sin \vartheta \cos \vartheta$; thus $\Theta'(\vartheta) = a \sin \vartheta + \tilde{a}/\sin \vartheta$, or

$$\Theta(\vartheta) = -a \cos \vartheta + \tilde{a} \ln \tan \left(\frac{1}{2}\vartheta\right). \tag{5.117}$$

The solution $\ln \tan\left(\frac{1}{2}\vartheta\right)$ is not physically meaningful for $\vartheta = 0$ and π; thus, we set $\tilde{a} = 0$; hence, $\Theta(\vartheta) = \cos \vartheta$, and the right-hand side of Eq. (5.116) gives $[\cos \vartheta \, \Theta'(\vartheta) + \sin \vartheta \, \Theta''(\vartheta)]/(2 \sin \vartheta) = -\cos \vartheta$, or

$$f_1(u) \cong \frac{eE}{m_e n_0 \Gamma_{ei}} \cdot u^3 f'_0(u). \tag{5.118}$$

The derivative of the kappa distribution, Eq. (5.106), is $f'_0(u) = n_0 \cdot P'(u; T_e; \kappa_0)$, with

$$P'(u; T_e; \kappa_0) = -\frac{\kappa_0 + \frac{5}{2}}{\kappa_0} \cdot \frac{2u}{\theta_e^2}\left(1 + \frac{1}{\kappa_0}\cdot\frac{u^2}{\theta_e^2}\right)^{-1} \cdot P(u; T_e; \kappa_0), \text{ or} \tag{5.119}$$

$$f'_0(u) = -\frac{\kappa_0 + \frac{5}{2}}{\kappa_0} \cdot \frac{2u}{\theta_e^2}\left(1 + \frac{1}{\kappa_0}\cdot\frac{u^2}{\theta_e^2}\right)^{-1} \cdot f_0(u). \tag{5.120}$$

Therefore, substituting Eq. (5.120) into Eq. (5.118), we obtain

$$f_1(u) \cong -\frac{eE}{n_0 k_B T_e \, \Gamma_{ei}} \cdot \frac{\kappa_0 + \frac{5}{2}}{\kappa_0} \cdot \left(1 + \frac{1}{\kappa_0}\cdot\frac{u^2}{\theta_e^2}\right)^{-1} \cdot u^4 f_0(u). \tag{5.121}$$

The disturbed, linearized distribution function $f_1(u) \cos \vartheta$ causes a linear current density along E (z-axis),

$$J = -e \, n_0 \langle u_z \rangle = -e \cdot \int f_1(u) \cos \vartheta \cdot u \cos \vartheta \, d^3 \vec{u}$$

$$= -2\pi \, e \cdot \int_0^\infty f_1(u) \, u^3 \, du \cdot \int_{-1}^1 \cos^2 \vartheta \, d \cos \vartheta$$

$$= -\frac{4}{3}\pi \, e \cdot \int_0^\infty f_1(u) \, u^3 \, du$$

$$= \frac{\frac{4}{3}\pi \, e^2 E}{n_0 k_B T_e \, \Gamma_{ei}} \cdot \frac{\kappa_0 + \frac{5}{2}}{\kappa_0} \cdot \int_0^\infty f_0(u) \cdot \left(1 + \frac{1}{\kappa_0}\cdot\frac{u^2}{\theta_e^2}\right)^{-1} \cdot u^7 \, du. \tag{5.122}$$

Substituting the kappa distribution, Eq. (5.106), we derive

$$
J = \frac{\frac{4}{3}\pi e^2 E}{k_B T_e \, \Gamma_{ei}} \theta_e^5 \cdot \frac{\kappa_0 + \frac{5}{2}}{\kappa_0} \cdot \int_0^\infty P(u = w\theta_e; \kappa_0)\theta_e^3 \cdot \left(1 + \frac{1}{\kappa_0}\cdot w^2\right)^{-1} \cdot w^7 \, dw, \text{ or,}
$$

(5.123)

$$
J = \frac{\frac{4}{3}\pi e^2 E}{k_B T_e \, \Gamma_{ei}} \theta_e^5 \cdot \left[\int_0^\infty P(u = w\,\theta_e; \kappa_0 \to \infty)\theta_e^3 w^7 \, dw\right] \cdot I(\kappa_0),
$$

(5.124)

where we set $w \equiv u/\theta_e$, and

$$
\int_0^\infty P(u = w\theta_e; \kappa_0 \to \infty)\theta_e^3 w^7 \, dw = \pi^{-\frac{3}{2}}\int_0^\infty e^{-w^2} w^7 \, dw = 3\pi^{-\frac{3}{2}}.
$$

(5.125)

Thus

$$
J/E = 16(2/\pi)^{\frac{1}{2}}e^2(k_B T_e)^{\frac{3}{2}}m_e^{-\frac{5}{2}}\Gamma_{ei}^{-1}\cdot I(\kappa_0), \text{ or,}
$$

(5.126)

$$
J/E = \frac{64(2\pi)^{\frac{1}{2}}(k_B T_e)^{\frac{3}{2}}\varepsilon_0^2}{e^2 m_e^{\frac{1}{2}} \ln \Lambda}\cdot I(\kappa_0).
$$

(5.127)

The dependence on the kappa index is enclosed in $I(\kappa_0)$, where $I(\kappa_0 \to \infty) = 1$. We derive

$$
I(\kappa_0) \equiv \frac{\kappa_0 + \frac{5}{2}}{\kappa_0} \cdot \frac{\displaystyle\int_0^\infty P(u = w\theta_e; \kappa_0)\theta_e^3 \cdot \left(1 + \frac{1}{\kappa_0}w^2\right)^{-1} w^7 \, dw}{\displaystyle\int_0^\infty P(u = w\theta_e; \kappa_0 \to \infty)\theta_e^3 w^7 \, dw}
$$

$$
= \frac{\Gamma\left(\kappa_0 + \frac{7}{2}\right)}{3\kappa_0^{\frac{5}{2}}\Gamma(\kappa_0 + 1)}\cdot \int_0^\infty \left(1 + \frac{1}{\kappa_0}w^2\right)^{-\kappa_0 - \frac{7}{2}} w^7 \, dw
$$

$$
= \frac{1}{6}\left(\kappa_0 + \frac{5}{2}\right)^{-3}\kappa_0^{\frac{3}{2}}\frac{\Gamma\left(\kappa_0 + \frac{5}{2}\right)}{\Gamma(\kappa_0 + 1)}\cdot \int_0^\infty \left(1 + \frac{1}{\kappa_0 + \frac{5}{2}}W\right)^{-\left(\kappa_0 + \frac{5}{2}\right) - 1} W^{4-1} \, dW
$$

$$
= \left(\kappa_0 + \frac{5}{2}\right)^{-3}\kappa_0^{\frac{3}{2}}\frac{\Gamma\left(\kappa_0 + \frac{5}{2}\right)}{\Gamma(\kappa_0 + 1)}\cdot \frac{\Gamma_{\kappa_0 + \frac{5}{2}}(4)}{\Gamma(4)},
$$

(5.128)

(where we set $W \equiv \dfrac{\kappa_0 + \frac{5}{2}}{\kappa_0} w^2$), leading to

$$I(\kappa_0) = \kappa_0^{\frac{3}{2}} \frac{\Gamma\left(\kappa_0 - \frac{1}{2}\right)}{\Gamma(\kappa_0 + 1)}, \tag{5.129}$$

where we used the notation of the q-Gamma function (Chapter 1),

$$\Gamma_{\kappa_0 + \frac{5}{2}}(4) = \Gamma(4)\left(\kappa_0 + \frac{5}{2}\right)^3 \frac{\Gamma\left(\kappa_0 - \frac{1}{2}\right)}{\Gamma\left(\kappa_0 + \frac{5}{2}\right)}. \tag{5.130}$$

In an isotropic plasma, the conductivity is defined by the proportionality between the current J and a small electric field E, $\sigma = \lim_{E \to 0} J/E$ (Ohm's law). Hence,

$$\sigma = \frac{64(2\pi)^{\frac{1}{2}}(k_B T_e)^{\frac{3}{2}}\varepsilon_0^2}{e^2 m_e^{\frac{1}{2}} \ln \Lambda} \cdot I(\kappa_0). \tag{5.131}$$

At this point, we recall that the kappa index depends not only on the dimensionality of the velocity space, but also on the dimensionality of the positional space (Chapter 3; Livadiotis, 2015b),

$$\kappa = \kappa_0 + \frac{\langle H \rangle}{k_B T_e} = \kappa_0 + \frac{\langle \varepsilon_K \rangle}{k_B T_e} + \frac{\langle \Phi \rangle}{k_B T_e} = \kappa_0 + \frac{\langle \varepsilon_K \rangle}{k_B T_e} + \frac{eE\langle z \rangle}{k_B T_e} = \kappa_0 + \frac{3}{2} + 1. \tag{5.132}$$

The invariant kappa index κ_0 is independent of the positional and kinetic degrees of freedom, with a range of values $\kappa_0 \in (0, \infty)$. This is similar to the definition of $\kappa_0 = \kappa - \frac{1}{2}d_K$ (Chapter 1), but now we have included also the effect of the potential energy $\Phi(z) = eEz$ (because $\overrightarrow{F} = -e\overrightarrow{E} = -eE \cdot \hat{z} = -(\partial\Phi/\partial z) \cdot \hat{z}$, and setting $\Phi(0) = 0$). Hence, instead of $\kappa = \kappa_0 + \frac{1}{2}d_K$, which is the case for zero potential energy, we have $\kappa = \kappa_0 + \frac{1}{2}d_K + \frac{1}{2}d_\Phi$, where $\frac{1}{2}d_\Phi = \frac{d_r}{b} = 1$ (position dimensionality $d_r = 1$, power exponent $b = 1$, according to Eq. (4.119)). The undisturbed isotropic distribution must be considered as of potential dependence, i.e.,

$$\underbrace{f(u, \vartheta)}_{\substack{\text{small } E \\ \kappa = \kappa_0 + \frac{1}{2}d_K + \frac{1}{2}d_\Phi}} = \underbrace{f_0(u)}_{\substack{E \to 0 \\ \kappa = \kappa_0 + \frac{1}{2}d_K + \frac{1}{2}d_\Phi}} + \underbrace{f_1(u)}_{\substack{\text{small } E \\ \kappa = \kappa_0 + \frac{1}{2}d_K + \frac{1}{2}d_\Phi}} \cdot \Theta(\vartheta), \tag{5.133a}$$

so that the distribution of velocities in Eq. (5.106) can be rewritten as

$$P(\overrightarrow{u}; T_e; \kappa_0) \propto \left[1 + \frac{1}{\kappa_0} \cdot \frac{\varepsilon_K(\overrightarrow{u})}{k_B T_e}\right]^{-\kappa_0 - 1 - \frac{1}{2}d_K}$$

$$\rightarrow P(\overrightarrow{u}; T_e; \kappa_0) \propto \left[1 + \frac{1}{\kappa_0} \cdot \frac{\varepsilon_K(\overrightarrow{u}) + \Phi(E \rightarrow 0)}{k_B T_e}\right]^{-\kappa_0 - 1 - \frac{1}{2}d_K - \frac{1}{2}d_\Phi} . \tag{5.133b}$$

Therefore, the kappa index must increase by 1,

$$\kappa = \kappa_0 + \frac{3}{2} \rightarrow \kappa = \kappa_0 + \frac{5}{2}, \tag{5.134}$$

while the difference $\kappa_0 = \kappa - \frac{\langle H \rangle}{k_B T_e}$ remains constant:

$$\kappa_0 = \kappa - \frac{3}{2} \rightarrow \kappa_0 = \kappa - \frac{5}{2}. \tag{5.135}$$

Thus we apply

$$\kappa_0 \rightarrow \kappa_0 + 1 \text{ and } T \rightarrow T \cdot \kappa_0/(\kappa_0 + 1), \tag{5.136}$$

so that

$$T_e^{\frac{3}{2}} I(\kappa_0) = T_e^{\frac{3}{2}} \kappa_0^{\frac{3}{2}} \frac{\Gamma\left(\kappa_0 - \frac{1}{2}\right)}{\Gamma(\kappa_0 + 1)} \rightarrow T_e^{\frac{3}{2}} I(\kappa_0) = T_e^{\frac{3}{2}} \kappa_0^{\frac{3}{2}} \frac{\Gamma\left(\kappa_0 + \frac{1}{2}\right)}{\Gamma(\kappa_0 + 2)}, \tag{5.137}$$

Finally, we find that

$$I(\kappa_0) = \kappa_0^{\frac{3}{2}} \frac{\Gamma\left(\kappa_0 + \frac{1}{2}\right)}{\Gamma(\kappa_0 + 2)}. \tag{5.138}$$

Note that there is secondary dependence on the kappa index of the plasma conductivity, which is also included in the term $\ln \Lambda = \ln\left(12\pi n_0 \lambda_{De}^3\right)$. This dependence becomes weaker for larger values of Λ, i.e., of κ_0 or $N_{De\infty}$. Given the dependence of the Debye length on the kappa index, we have

$$\Lambda \equiv 9N_{De}, \ln \Lambda = \ln \Lambda_\infty + \frac{3}{2} \ln\left(\frac{\kappa_0}{\kappa_0 + 1}\right), \tag{5.139}$$

$$\ln \Lambda_\infty \equiv \ln\left(12\pi n_0 \lambda_{De\,\infty}^3\right) = \ln(9 N_{De\infty}),$$

thus

$$\sigma = \sigma_\infty \cdot \frac{I(\kappa_0)}{1 + \frac{3}{2}(\ln \Lambda_\infty)^{-1} \cdot \ln\left(\frac{\kappa_0}{\kappa_0 + 1}\right)}, \quad \sigma_\infty \equiv \frac{64(2\pi)^{\frac{1}{2}}\varepsilon_0^2}{e^2 m_e^{\frac{1}{2}}} \cdot \frac{(k_B T_e)^{\frac{3}{2}}}{\ln \Lambda_\infty}, \tag{5.140}$$

where the quantities with subscript ∞ correspond to thermal equilibrium, $\kappa_0 \rightarrow \infty$. Fig. 5.9 shows the dependence of the conductivity on the kappa index and for various value of $N_{De\infty}$.

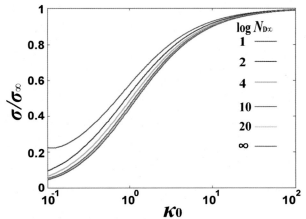

FIGURE 5.9

Plasma conductivity as a function of the kappa index κ_0. The dependence on $N_{De\,\infty}$ is also shown, which has weaker effect for larger values of κ_0.

5.7 Collision Frequency and Mean Free Path

The plasma electrical conductivity σ is related to the collision frequency for electrons colliding with ions ν_{ei}. The larger the collision frequency, the more scattering of the electron bulk flow, and thus, the weaker the resultant current density. Thus, for a given electric field, weaker resultant current density means smaller electrical conductivity. Hence, we expect $\sigma \propto \nu_{ei}^{-1}$. The exact relation can be obtained from a dimension analysis.

The current density and electric field dimensions are $[J] = [Q] \cdot [\ell]^{-2} \cdot [t]^{-1}$ and $[E] = [Q] \cdot [\ell]^{-2} \cdot \varepsilon_0^{-1}$. Then for the electrical conductivity, we have $[\sigma] = [J]/[E] = \varepsilon_0 [t]^{-1}$. Hence, the conductivity normalized to the permittivity ε_0 may describe a frequency $[\sigma/\varepsilon_0] = [t]^{-1}$. A characteristic time scale in plasmas is the inverse of the plasma frequency $\omega_{pl} \cong e\sqrt{n_0/(\varepsilon_0 m_e)}$, i.e., $[\omega_{pl}] = [t]^{-1}$; thus the conductivity over permittivity, normalized to the plasma frequency $\sigma/(\varepsilon_0 \omega_{pl})$, is dimensionless. In addition, the collision frequency is normalized to the plasma frequency becoming dimensionless, ν_{ei}/ω_{pl}. Then the inverse proportionality $\sigma \propto \nu_{ei}^{-1}$ can be written as equality between the normalized respective quantities, i.e., $\sigma/(\varepsilon_0 \omega_{pl}) = (\nu_{ei}/\omega_{pl})^{-1}$, that is, $\nu_{ei} = \varepsilon_0 \omega_{pl}^2 \big/ \sigma$, or

$$\nu_{ei} = \omega_{pl}^2 \cdot (\varepsilon_0/\sigma), \quad \omega_{pl} \cong e\sqrt{n_0/(\varepsilon_0 m_e)}, \quad \text{or,} \quad \nu_{ei} \cong n_0 e^2 /(m_e \sigma). \tag{5.141}$$

The Drude model for the electrical conduction (Drude, 1900) leads to the same relation. According to this, the electrons are accelerated by the electric field, but their acceleration is dumped by collisions,

$$m_e \frac{d\vec{u}_b}{dt} \cong eE - \nu_{ei} m_e \vec{u}_b, \tag{5.142}$$

where $\vec{u}_b \equiv \langle \vec{u} \rangle$ is the electron bulk velocity. The stationary solution is given by $d\vec{u}_b/dt = 0$, leading to the terminal speed

$$u_{b\,fin} \cong \frac{eE}{\nu_{ei}m_e},$$

(5.143)

that corresponds to a stationary state current density

$$J = en_0 u_{b\,fin} \cong \frac{e^2 n_0}{\nu_{ei}m_e} \cdot E, \text{ or } \sigma = J/E \cong \frac{e^2 n_0}{\nu_{ei}m_e},$$

(5.144)

that is, Eq. (5.141). Then, using Eq. (5.140), we find the collision frequency

$$\nu_{ei} \cong \nu_{ei\,\infty} \cdot \frac{1 + \frac{3}{2}(\ln\Lambda_\infty)^{-1} \cdot \ln\left(\dfrac{\kappa_0}{\kappa_0 + 1}\right)}{I(\kappa_0)}, \text{ with}$$

$$\nu_{ei\,\infty} \equiv \frac{e^2 n_0}{m_e \sigma_\infty} = \frac{e^4 n_0}{64(2\pi)^{\frac{1}{2}} m_e^{\frac{1}{2}} \varepsilon_0^2} \cdot \frac{\ln\Lambda_\infty}{(k_B T_e)^{\frac{3}{2}}}, \text{ or}$$

(5.145)

$$\nu_{ei}/\omega_{pl} \cong \frac{3}{16}\left(\frac{\pi}{2}\right)^{\frac{1}{2}} \cdot \frac{\ln(9N_{De\infty})}{9N_{De\infty}} \cdot \frac{1 + \frac{3}{2}\left[\ln(9N_{De\infty})\right]^{-1} \cdot \ln\left(\dfrac{\kappa_0}{\kappa_0 + 1}\right)}{I(\kappa_0)}.$$

(5.146)

We may describe collisions not only via the average time between successive collisions or its inverse, the collision frequency, but also via the average distance a particle covers between successive collisions, that is, the mean free path L_m. The average time between collisions is $\tau \sim 1/\nu_{ei}$. In the reference system of the bulk motion ($\vec{u}_b = 0$), the characteristic speed is given by the velocity standard deviation, i.e.,

$$L_m = \sqrt{\langle(\vec{u} - \vec{u}_b)^2\rangle} \cdot \tau = \sqrt{3k_B T_e/m_e}\Big/\nu_{ei} \cong L_{m\infty} \cdot \frac{I(\kappa_0)}{1 + \frac{3}{2}\left[\ln(9N_{De\infty})\right]^{-1} \cdot \ln\left(\dfrac{\kappa_0}{\kappa_0 + 1}\right)},$$

$$L_{m\,\infty} \equiv 16\sqrt{\frac{2}{3\pi}} \cdot \frac{9N_{De\infty}\,\lambda_{De\infty}}{\ln(9N_{De\infty})},$$

(5.147)

where we have used the identity

$$\sqrt{\frac{1}{3}\langle(\vec{u} - \vec{u}_b)^2\rangle}\Big/\omega_{pl\,e} = \sqrt{k_B T_e/m_e}\Big/\sqrt{e^2 n_0/(m_e\varepsilon_0)} = \sqrt{\varepsilon_0 k_B T_e/(e^2 n_0)}$$

$$= \lambda_{De\infty}, \text{ or } L_m\nu_{ei} \cong \sqrt{3}\,\lambda_{De\infty}\,\omega_{pl}.$$

(5.148)

We may also write the ratio

$$L_m/\lambda_{De} \cong 16\sqrt{\frac{2}{3\pi}} \cdot \frac{9N_{De\infty}}{\ln(9N_{De\infty})} \cdot \frac{\left(\frac{\kappa_0}{\kappa_0+1}\right)^{-\frac{1}{2}} I(\kappa_0)}{1 + \frac{3}{2}[\ln(9N_{De\infty})]^{-1} \cdot \ln\left(\frac{\kappa_0}{\kappa_0+1}\right)} > 1, \text{ or,}$$

(5.149a)

$$L_m/\lambda_{De} \cong 16\sqrt{\frac{2}{3\pi}} \cdot \frac{9N_{De}}{1 + \ln(9N_{De})} \cdot \left(\frac{\kappa_0}{\kappa_0+1}\right)^{-2} I(\kappa_0) > 1. \qquad (5.149b)$$

Fig. 5.10A plots L_m/λ_D (on a log–log scale) as a function of $N_{D\infty}$ and for various values of the kappa index κ_0, from $\kappa_0 = 0.01$ to $\kappa_0 = 100$ (practically corresponds to thermal equilibrium) (note that L_m/λ_D is deduced from Eq. (5.149a) by setting $T_i = T_e$, leading to $\lambda_{De} = 2^{\frac{1}{2}} \lambda_D$ and $N_{De} = 2^{\frac{3}{2}} N_D$). We observe that the

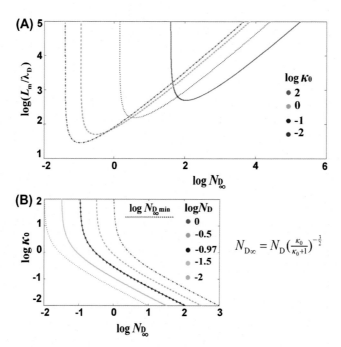

FIGURE 5.10
(A) Plot of the ratio L_m/λ_D (on a log–log scale) as a function of $N_{D\infty}$ for log $\kappa_0 = -2$ (*red*), -1 (*blue*), 0 (*green*), and 2 (*purple*).
(B) Plot of the value of $N_{D\infty}$ that corresponds to the minimum of L_m/λ_D as a function of κ_0 (on a log–log scale) (*red dotted line*);

the function $N_{D\infty}(\kappa_0) = N_D \cdot \left(\frac{\kappa_0}{\kappa_0+1}\right)^{-\frac{3}{2}}$ is co-plotted for log $N_D = -2$ (*brown*), -1.5 (*light blue*), -0.97 (*blue*), -0.5 (*green*), and

0 (*purple*).

ratio is always larger than 1, or $L_m > \lambda_D$. However, there is a minimum of this ratio at $N_{D\infty\,min}(\kappa_0)$. The latter is plotted in Fig. 5.10B as a function of κ_0 (red dotted line). We also co-plot $N_{D\infty}(\kappa_0) = N_D \cdot \left(\frac{\kappa_0}{\kappa_0+1}\right)^{-\frac{3}{2}}$ (trivially derived from Eq. (5.64) for $d=3$) for various values of N_D. We observe that the geometric locus of the minimum $N_{D\infty\,min}(\kappa_0)$ coincides with $N_{D\infty}(\kappa_0) = N_D \cdot \left(\frac{\kappa_0}{\kappa_0+1}\right)^{-\frac{3}{2}}$ for the specific value of $\log N_D \cong -0.97$ or $N_D \cong 0.107$ (or $\Lambda \approx 1$, given in Eq. (5.139)). The lower is the value of the ratio L_m/λ_D, the more collisional (less collisionless) should be the plasma (the rate of collisions increases). Consequently, the most collisional plasma state corresponds to the minimum of this rate, which is indicated by $\log N_D \approx -1$. For even lower values, the plasma collective behavior disappears and Eq. (5.149a) is rather meaningless; hence, collisionless plasmas are characterized by $\log N_D \geq -1$.

5.8 Magnetization: The Curie Constant

Let a system of paramagnetic particles with freely rotating magnetic moments $\overrightarrow{\mu}$ in the presence of an external magnetic field $\overrightarrow{B}$. The particle potential energy is given by

$$\Phi(\vartheta) = -\overrightarrow{\mu} \cdot \overrightarrow{B} = -\mu B \cos \vartheta, \tag{5.150}$$

where we set the z-axis on the direction of $\overrightarrow{B}$ with unit vector $\hat{z} \equiv \overrightarrow{B}/B$; ϑ is the polar angle between the vectors of the magnetic field $\overrightarrow{B}$ and moment $\overrightarrow{\mu} = \mu \cos \vartheta \cdot \hat{z}$ so that each particle is characterized by a certain value of the polar angle ϑ (with $\vartheta = 0$ set to be parallel to the field) (Fig. 5.11).

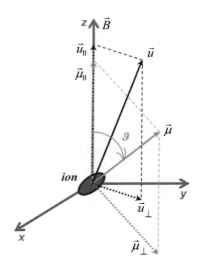

FIGURE 5.11
Schematic representation of the magnetic moment (*green*) and velocity (*red*) vectors and their projection components, parallel and perpendicular to z axis, that is, the direction of the magnetic field.

The magnetization M of a paramagnetic material is proportional to the strength B of an external magnetic field and inverse proportional to the temperature T in the approximation of high temperature, that is, $\beta \equiv \mu B/(k_B T) \ll 1$ (small magnetic energy over large thermal energy). This behavior constitutes the Curie's law, which is expressed using the Curie constant C as the proportionality coefficient,

$$M \cong C \cdot B/T, \tag{5.151}$$

while the per particle magnetization is given by the average magnetic moment $\vec{M} = \langle \vec{\mu} \rangle$. The magnetic moment is a positional function $\vec{\mu} = \vec{\mu}(\vec{r})$, or more precisely, is a function of the angle ϑ. Hence, $\langle \vec{\mu} \rangle$ is determined using the positional probability distribution $P(\cos \vartheta)$,

$$\langle \vec{\mu} \rangle = \mu \cdot \langle \cos \vartheta \rangle \cdot \hat{z}, \quad \langle \cos \vartheta \rangle = \int_{-1}^{1} P(\cos \vartheta) \cos \vartheta \, d\cos \vartheta. \tag{5.152}$$

The classical phase space distribution function is constructed by the BG energy distribution with the energy substituted by the Hamiltonian function, Eq. (5.12). Within the theory of kappa distributions (Livadiotis, 2015a) and its statistical framework of nonextensive statistical mechanics (Livadiotis and McComas, 2009), Eq. (5.19) becomes the kappa distribution of the Hamiltonian (Chapter 3; Livadiotis, 2015b). This distribution, in the case of negative potential energy (e.g., see Eqs. (4.125, 4.126) in Chapter 4), is expressed in terms of the velocity

$$P(\vartheta, \vec{u}) \propto \left\{ 1 + \frac{1}{\kappa_0} \cdot \left[\frac{(\vec{u} - \vec{u}_b)^2}{\theta^2} + \frac{\Phi(\vartheta)}{k_B T} \right] \right\}^{-\kappa_0 - 1 - \frac{1}{2} d_K \mp \frac{1}{2} d_\Phi}, \tag{5.153a}$$

or the kinetic energy, $\varepsilon_K = \frac{1}{2} m (\vec{u} - \vec{u}_b)^2$,

$$P(\vartheta, \varepsilon_K) \propto \left[1 + \frac{1}{\kappa_0} \cdot \frac{\varepsilon_K + \Phi(\vartheta)}{k_B T} \right]^{-\kappa_0 - 1 - \frac{1}{2} d_K \mp \frac{1}{2} d_\Phi} \varepsilon_K^{\frac{1}{2} d_K - 1}, \tag{5.153b}$$

where the upper/lower sign corresponds to positive/negative potential. The kinetic d_K and potential d_Φ degrees of freedom are given by $\frac{1}{2} d_K = \langle \varepsilon_K(\vec{u}) \rangle / (k_B T)$ and $\frac{1}{2} d_\Phi = |\Phi(\vartheta)| / (k_B T)$, respectively, while the kappa index degeneration is $\kappa = \kappa_0 + \frac{1}{2} d_K \pm \frac{1}{2} d_\Phi$. The normalization of the distributions is given by

$$\int_{-\infty}^{\infty} \int_{-1}^{1} P(\vartheta, \vec{u}; \kappa_0, T) d\cos \vartheta d\vec{u} = 1, \quad \int_{0}^{\infty} \int_{-1}^{1} P(\vartheta, \varepsilon_K; \kappa_0, T) d\cos \vartheta d\varepsilon_K = 1. \tag{5.154}$$

while the positional probability distribution is derived from

$$P(\vartheta; \kappa_0, T) = \int_{0}^{\infty} P(\vartheta, \varepsilon_K; \kappa_0, T) d\varepsilon_K. \tag{5.155}$$

The analysis of the kappa distribution of the Hamiltonian $H(\vartheta, \vec{u}) = \varepsilon_K(\vec{u}) + \Phi(\vartheta)$, with potential given by Eq. (5.150), has been studied in Livadiotis (2015b) and (2016a) in detail. There, for the sake of simplicity, it was assumed that the kappa index is degenerated only by the kinetic degrees of freedom, i.e., $\kappa = \kappa_0 + \frac{1}{2}d_K$. Here, we will also adopt this assumption and use κ_0, but one has to remember that the potential degrees have not considered in the kappa index degeneration so that the total kappa index is not given by $\kappa = \kappa_0 + \frac{1}{2}d_K - \beta\langle\cos\vartheta\rangle$ (as explained in Eq. (4.135a) in paragraph 4.5.11.5), but by $\kappa = \kappa_0 + \frac{1}{2}d_K$.

Hence, the analytical derivation of the phase space kappa distribution (in terms of $\cos\vartheta$ and ε_K) is given by

$$P(\cos\vartheta, \varepsilon_K) = \frac{\Gamma\left(\kappa_0 + \frac{1}{2}d_K + 1\right)(\kappa_0 + \beta\langle\cos\vartheta\rangle + \beta)^{\kappa_0}(1 + \beta\langle\cos\vartheta\rangle - \beta)}{2\Gamma(\kappa_0)\Gamma\left(\frac{1}{2}d_K\right)(\kappa_0 + \beta\langle\cos\vartheta\rangle)^{\kappa_0 + 1 + \frac{1}{2}d_K}}$$

$$\times (k_B T)^{-\frac{1}{2}d_K} \cdot \left[1 + \frac{1}{\kappa_0 + \beta\langle\cos\vartheta\rangle} \cdot \left(\frac{\varepsilon_K}{k_B T} - \beta\cos\vartheta\right)\right]^{-\kappa_0 - 1 - \frac{1}{2}d_K} \varepsilon_K^{\frac{1}{2}d_K - 1}.$$

$$(5.156)$$

The marginal distribution of the kinetic energy (velocity space) is

$$P(\varepsilon_K) = \frac{\Gamma\left(\kappa_0 + \frac{1}{2}d_K\right)(\kappa_0 + \beta\langle\cos\vartheta\rangle + \beta)^{\kappa_0}(1 + \beta\langle\cos\vartheta\rangle - \beta)}{2\beta\,\Gamma(\kappa_0)\Gamma\left(\frac{1}{2}d_K\right)(\kappa_0 + \beta\langle\cos\vartheta\rangle)^{\kappa_0 + \frac{1}{2}d_K}}$$

$$(5.157)$$

$$\times \left\{\left[1 + \frac{\varepsilon_K - \beta\,k_B T}{k_B T(\kappa_0 + \beta\langle\cos\vartheta\rangle)}\right]^{-\kappa_0 - \frac{1}{2}d_K}\right.$$

$$\left. - \left[1 + \frac{\varepsilon_K + \beta\,k_B T}{k_B T(\kappa_0 + \beta\langle\cos\vartheta\rangle)}\right]^{-\kappa_0 - \frac{1}{2}d_K}\right\}(k_B T)^{-\frac{1}{2}d_K} \cdot \varepsilon_K^{\frac{1}{2}d_K - 1},$$

while the marginal distribution of the potential energy is

$$P(\cos\vartheta) = \frac{\kappa_0(\kappa_0 + \beta\langle\cos\vartheta\rangle + \beta)^{\kappa_0}(1 + \beta\langle\cos\vartheta\rangle - \beta)}{2(\kappa_0 + \beta\langle\cos\vartheta\rangle - \beta\cos\vartheta)^{\kappa_0 + 1}}. \qquad (5.158)$$

The mean value of the potential energy is $\langle\Phi\rangle/(k_B T) = -\beta\langle\cos\vartheta\rangle$. The mean value of the $\cos\vartheta$ is a function of the kappa index and the inverse temperature, i.e., $\langle\cos\vartheta\rangle = f(\beta, \kappa_0)$, where f is implicitly given by

$$\frac{1 + \beta\langle\cos\vartheta\rangle + \beta}{1 + \beta\langle\cos\vartheta\rangle - \beta} = \left(\frac{\kappa_0 + \beta\langle\cos\vartheta\rangle + \beta}{\kappa_0 + \beta\langle\cos\vartheta\rangle - \beta}\right)^{\kappa_0}. \qquad (5.159a)$$

In addition, Eq. (5.159a) can be also expressed via the generalized Langevin function (Chakrabarti and Chandrashekar, 2010; Livadiotis, 2016a), i.e.,

$$\langle \cos \vartheta \rangle = L_{q=1+1/\kappa_0} \left(\frac{\beta}{1 + \frac{1}{\kappa_0}\beta\langle \cos \vartheta \rangle} \right), \quad \text{where} \tag{5.159b}$$

$$L_q(x) \equiv (2 - q)^{-1} \cdot \left[\coth_q(x) - \frac{1}{x} \right] = \frac{\kappa_0}{\kappa_0 - 1} \cdot \left[\coth_{q=1+1/\kappa_0}(x) - \frac{1}{x} \right], \tag{5.159c}$$

recovering the standard Langevin function for $q \to 1$ or $\kappa_0 \to \infty$, $L(x) = \coth(x) - x^{-1}$, which uses the q-hyperbolic cotangent, shown in Chapter 1, Section 1.4.2.

Fig. 5.12A and B plot the positional distribution $P(\cos \vartheta)$ for $\beta = 1$ and various kappa indices where we observe the monotonic and abrupt increase of the probability distributions as $\cos \vartheta$ approaches 1. This increase at angles near parallel to the magnetic field is more intense for smaller kappa indices. In order to measure this increase, we define the percentage ratio of $0.9 \leq \cos \vartheta \leq 1$ over $0 \leq \cos \vartheta \leq 0.9$:

$$C_{90}(\kappa_0, \beta) \equiv \int_{0.9}^{1} P(\cos \vartheta)d\cos \vartheta \bigg/ \int_{0}^{0.9} P(\cos \vartheta)d\cos \vartheta. \tag{5.160}$$

This ratio is depicted in panel (C), where we observe the abrupt increase of the ratio values as κ_0 decreases approaching zero, while there is similar behavior for

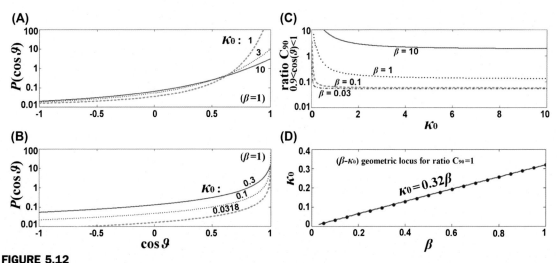

FIGURE 5.12
Positional distribution and magnetization. The positional distribution is plotted for $\beta = 1$ and various kappa indices, either ≥ 1 (A), or <1 (B). (C) The percentage ratio of $0.9 \leq \cos \vartheta \leq 1$ over $0 \leq \cos \vartheta \leq 0.9$ is depicted as a function of the kappa index and for various temperatures β. (D) The geometric locus of temperature and kappa index for $C_{90} = 1$, which fits to the line $\kappa_0 \cong 0.32\beta$.

any temperature β. Panel (D) shows the relation between temperature and kappa index for a constant ratio, e.g., equal to 1; the geometric locus of $C_{90}(\kappa_0,\beta) = 1$ is plotted, which fits to the line $\kappa_0 \cong 0.32\beta$.

Therefore, kappa indices smaller than 0.32β lead the majority of the particle moments to be near parallel to the magnetic field; thus high magnetization is expected. On the contrary, kappa indices larger than 0.32β lead the majority of the particle moments to be dispersed and not parallel to the magnetic field; then low magnetization is expected.

The per particle magnetization is $M = \mu\cdot\langle\cos\vartheta\rangle$; thus $\frac{1}{\mu}M = \langle\cos\vartheta\rangle$ is implicitly given by Eq. (5.159). This is plotted in Fig. 5.13, where we observe that the magnetization tends to zero either for high temperatures (small β) or for large kappa index κ_0. On the contrary, the magnetization tends to its largest value either for low temperatures (large β) or for small κ_0. Therefore, both temperature T and kappa index κ_0 have the same impact, that is, increasing the magnetization as T and/or κ_0 decrease. It is remarkable that strong magnetization at high temperatures can be obtained in systems far from thermal equilibrium ($\kappa_0 < 1$).

The Taylor expansion of the per particle magnetization M in terms of the dimensionless $\beta \equiv \mu B/(k_B T)$,

$$\frac{1}{\mu}M \cong \frac{1}{3}\left(1+\frac{1}{\kappa_0}\right)\cdot\left(\frac{\mu B}{k_B T}\right)\left[1+\frac{1}{15}\left(1+\frac{5}{\kappa_0}+\frac{1}{\kappa_0^2}\right)\cdot\left(\frac{\mu B}{k_B T}\right)^2\right]+O\left(\frac{\mu B}{k_B T}\right)^4,$$

$$(5.161)$$

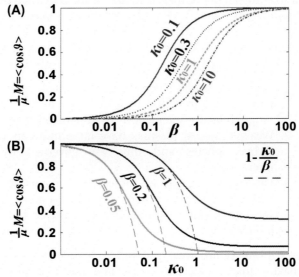

FIGURE 5.13

Magnetization depicted as a function of (A) the inverse temperature $\beta \equiv \mu B/(k_B T)$, and (B) the invariant kappa index κ_0; the limiting behavior $M/\mu \sim 1-\kappa_0/\beta$ for small kappa indices is also shown, *black dash*). Adapted by Livadiotis (2016a).

where the second order term is zero for any κ_0. The first order expansion is a good approximation,

$$M \cong \frac{1}{3} k_B^{-1} \mu^2 \cdot \left(1 + \frac{1}{\kappa_0}\right) \cdot \frac{B}{T}. \tag{5.162}$$

The Curie constant C_κ depends on the kappa index,

$$C_\kappa \equiv \lim_{(B/T) \to 0} \frac{\partial M}{\partial (B/T)} \cong C_\infty \cdot \left(1 + \frac{1}{\kappa_0}\right), \quad C_\infty \equiv \frac{1}{3} k_B^{-1} \mu^2, \tag{5.163}$$

while the paramagnetic susceptibility χ_κ becomes

$$\chi_\kappa \equiv \lim_{B \to 0} \frac{\partial M}{\partial B} \cong \chi_\infty \cdot \left(1 + \frac{1}{\kappa_0}\right), \quad \chi_\infty \equiv \frac{1}{3} \frac{\mu^2}{k_B T}. \tag{5.164}$$

Note that the magnetization density $\vec{M}_V = n \cdot \vec{M} = n \langle \vec{\mu} \rangle$ includes the inverse square Debye length, i.e., $M_V \propto (n/T) \cdot \left(1 + \kappa_0^{-1}\right) \cdot B \propto \lambda_D^{-2} \cdot B$ (see Eq. 5.64). Hence, the magnetization density per magnetic field depends only on the Debye length $\left[M_V/\left(\mu^2 B\right)\right] \cdot \lambda_D^2 \cong \frac{1}{3} \varepsilon_0 e^{-2} \cong 1.15 \times 10^{26} \text{J} \cdot \text{m}$. Therefore we may determine the Debye length if we measure the magnetization density caused by some small magnetic field.

5.9 Large-Scale Quantization Constant
5.9.1 The Role of Correlations

Numerous independent analyses of a broad range of space data led to the conclusion that in contrast to many Earthy and man-made plasmas, most space plasmas are out of thermal equilibrium and described by kappa distributions (or combinations thereof). The origin of the non-Maxwellian statistical behavior of these plasmas is the existence of local correlations between the plasma particles (Livadiotis and McComas, 2011b). Indeed, space plasmas are (1) collisionless, (2) weakly coupled particle systems, and (3) exhibiting local correlations:

1. Collisionless plasmas are weakly coupled, i.e., the coupling between the particles due to collisions is generally negligible, and instead, the long-range collective interactions bind particles together.
2. Weakly coupled plasmas have large numbers of particles within their Debye spheres and thus are governed by collective long-range interactions.
3. Particles in plasmas are subject to local correlations with other particles within their Debye sphere; correlations among particles can persist for substantial times without being destroyed by collisions or short-range interactions.

Space plasmas are weakly coupled particle systems with strong collective behavior that characterize correlated particles within Debye spheres. The presence of local correlations between space plasma particles produces not only the statistical behavior of these systems, which is different from classical systems in thermal equilibrium, but also has critical implications for fundamental physics.

Recent analyses of space, geophysical, and other plasmas revealed the existence of a new quantization constant $\hbar^*$, similar to the Planck constant $\hbar$, but ~ 12 orders of magnitude larger. Planck's constant constitutes the smallest possible phase space parcel for individual and uncorrelated particles, while the new quantization constant describes the smallest possible phase space parcel for collisionless particle systems characterized by local correlations (where correlations may be caused by Debye shielding, wave–particle long-range interactions, etc.).

Local correlations are manifested by the presence of a correlation length between particles. This divides the system into an ensemble of clusters of correlated particles. The particles within each of these "correlation clusters" participate altogether to this new type of quantization. The Debye length in plasmas expresses a large-scale positional uncertainty among particles, assigning thus, the smallest correlation length in plasmas (Section 5.5).

The Planck quantization constant $\hbar$ may characterize interactions among individual particles, and the large-scale quantization constant $\hbar^*$ applies only to clusters of correlated particles. The scheme in Fig. 5.14 demonstrates how the

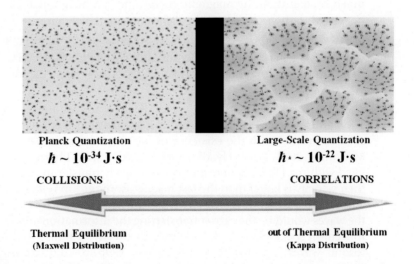

Planck Quantization
$$\hbar \sim 10^{-34}\,\text{J·s}$$
COLLISIONS

Large-Scale Quantization
$$\hbar^* \sim 10^{-22}\,\text{J·s}$$
CORRELATIONS

Thermal Equilibrium
(Maxwell Distribution)

out of Thermal Equilibrium
(Kappa Distribution)

FIGURE 5.14

(Upper panel) The Planck quantization constant characterizes particles without local correlations, while large-scale quantization constant applies in systems with local correlations between their particles. For example, the Debye length gives a measure of the smallest correlation length, and particles within the Debye length are always correlated, forming Debye spheres. The correlation length divides the system into an ensemble of clusters of correlated particles, e.g., Debye spheres. Particles within "correlation clusters" participate altogether to the large-scale quantization. (Lower panel) The nonequilibrium statistical behavior of space plasmas is related to the large-scale quantization constant. Both are caused by the presence of local correlations that survive in the collisionless plasma. Adapted by Livadiotis (2015d).

large-scale quantization is related to the nonequilibrium statistics of space plasmas and the competition between correlations and collisions. The existence of correlations between particles shifts plasmas away from thermal equilibrium (Livadiotis, 2015d) in stationary states described by kappa or kappa-like distributions (Livadiotis, 2015c). These cannot embody the Boltzmann–Gibbs statistical mechanics, but the generalized framework of nonextensive statistical mechanics (Tsallis, 2009b). The stronger the correlation, the furthest the plasma resides from thermal equilibrium. On the other hand, collisions destroy correlations recovering plasmas back at thermal equilibrium. Therefore, the kappa distributions and the large-scale quantization both appear in collisionless plasmas, namely, in plasmas where the correlation length is smaller than the mean free path (Livadiotis, 2014a). Since correlations provide evidence for collisionless plasma (collisions would have destroyed correlations), the presence of kappa distributions in plasma suggests that $\hbar^*$ characterizes this plasma (Livadiotis, 2015a); however, it is not known if a certain kappa value exists as a threshold between Planck and large-scale quantization.

5.9.2 The Smallest Phase Space Parcel in Plasmas

In a thermal (nonmagnetized) plasma, thermal motions introduce charge perturbations. Then the required work for removing a charge perturbation beyond the correlation length, that is, the Debye length λ_D, is around the thermal energy $\sim k_B T$. This may be expressed in terms of the ion acoustic (sound) speed V_S, i.e., of the order of $\sim \frac{1}{2} m V_S^2$. In a magnetized plasma where the thermal energy is much smaller than the magnetic energy and can be ignored, perturbations can be related to the magnetic energy of a single particle, that is, $\sim \frac{1}{2} B^2/(\mu_0 n)$, or in terms of the Alfvén speed $\left(V_A \equiv B/\sqrt{\mu_0(m_i + m_e)n}\right)$, $\sim \frac{1}{2} m V_A^2$. In the general case of plasmas, where both the thermal and magnetic energy may be important, then the information propagates via the fast magnetosonic speed

$$V_{ms} = \sqrt{V_A^2 + V_S^2}.$$

The particle–wave energy transfer in space plasmas has been phenomenologically connected with the existence of fast magnetosonic waves (Livadiotis and McComas, 2013, 2014; Livadiotis and Desai, 2016). Plasma particles are correlated within a Debye sphere and can effectively transfer information only when their velocity is larger than the fast magnetosonic group velocity, V_{ms}, the highest among ion plasma waves; otherwise, if the particle velocity is smaller than V_{ms}, the waves would be the carrier transferring the information and not the particles. Indeed, in space plasmas, information is effectively transferred via fast magnetosonic waves (e.g., Kivelson and Russell, 1995, p. 132; Gosling, 1998, p. 95; Treumann and Baumjohann, 2001, p. 168, 340; Hartquist et al., 2004, p. 50; Font, 2007). There are, of course, numerous other plasma waves and relevant information speeds, some of which are faster than the fast magnetosonic speed, even reaching the speed of light. However, these wave modes do not carry information about correlated particles, both ions and electrons, but information about individual particles. Therefore, the magnetosonic energy

$E_{ms} = \frac{1}{2}(m_i + m_e)V_{ms}^2$ is the smallest energy ε_C that can be transferred by the correlated particles of a Debye sphere to the plasma primary oscillation modes of frequency ω_{pl}.

As we have seen in Section 5.5, the Debye length expresses the positional standard deviation in plasmas, assigning a large-scale positional uncertainty. Therefore, the smallest length x_C, which is physically meaningful for interactions between the plasma-correlated ion/electron particles is the Debye length, i.e., $x_C \sim \frac{1}{2}\lambda_D$ (Livadiotis and McComas, 2013b, 2014a; Livadiotis, 2017a). The "lifetime" of a Debye sphere $t_C = x_C/\theta \sim \frac{1}{2}\lambda_D/\theta$ is the average time that a particle resides within a Debye length λ_D before leaving because of its thermal motion, i.e., when the particle travels on average $\frac{1}{2}\lambda_D$ owing to its thermal motion, with thermal speed $\theta = \sqrt{k_B T/m}$. The electrons, as the most mobile particles in the plasma, specify the smallest lifetime $t_C = x_C/\theta_e \sim \frac{1}{2}\lambda_D/\theta_e$, with $\theta_e = \sqrt{k_B T_e/m_e}$.

Combining the smallest energy exchange ε_C with the smallest lifetime t_C, we derive the smallest possible phase space portion that characterizes plasmas, noted by $\hbar_*$

$$\Delta\varepsilon \, \Delta t \geq \varepsilon_C \, t_C = \frac{1}{2}\hbar_*. \tag{5.165}$$

The lifetime and the plasma frequency are inversely proportional, $t_C \omega_{pl} \cong \frac{1}{2}$ (because of $\omega_{pl} = \theta_e/\lambda_D$ and $t_C = \frac{1}{2}\lambda_D/\theta_e$). This leads to the Planck's energy frequency equation. Indeed, starting from $t_C \omega_{pl} \cong \frac{1}{2}$, we multiply by the quantization constant to obtain $(\hbar\omega_{pl})t_C \cong \frac{1}{2}\hbar$, and comparing with $\varepsilon_C t_C = \frac{1}{2}\hbar$, we derive $\varepsilon_C \sim \hbar\omega_{pl}$. The ratio of the smallest energy ε_C over the plasma frequency ω_{pl} gives the quantization constant, but is this always given by the Planck constant?

The smallest phase space portion that characterizes typical plasmas, Eq. (5.165), is expected to be equal to the Planck's quantization constant. For example, laboratory helium plasma ($T \sim 4000K$, $n \sim 2.4\cdot 10^{25}$ m^{-3}; Jonkers and Van Der Mullen, 1999) gives the Planck constant. Indeed, in this case, we have $E_{ms} \approx k_B T$ (nonmagnetized plasma); hence the quantization constant is estimated by $k_B T/\omega_{pl} \approx 10^{-34}$J·s$\approx \hbar$. However, for typical space plasmas, the same ratio of energy over plasma frequency gives $\hbar_* \approx 10^{-22}$J·s. It is remarkable that the same value characterizes a large variety of geophysical/space plasmas (Livadiotis and McComas, 2013b, 2014b; Livadiotis, 2015d, 2016c, 2017a).

The smallest exchange of energy of a correlated particle ε_C is transformed into a plasmon of energy $\varepsilon_C \sim \hbar_*\omega_{pl}$. The ratio of the smallest energy ε_C over the characteristic frequency, the plasma frequency ω_{pl}, gives a new type of quantization constant $\hbar_*$. This is similar to the Planck energy–frequency relation $\varepsilon_{Pk} \sim \hbar\omega_{ph}$, where the least particle energy ε_{Pk} is transformed into a photon of frequency ω_{ph} (Fig. 5.15).

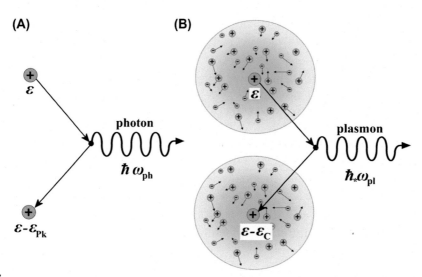

FIGURE 5.15

(A) The smallest energy that can be transferred from an individual particle is $\hbar\,\omega$, carried by a quasiparticle, i.e., a photon, of frequency ω_{ph}. (B) For a correlated particle in a plasma, the smallest energy that can be transferred is $\hbar_*\,\omega$, which can be carried by a plasmon having the plasma frequency ω_{pl}. Adapted by Livadiotis and McComas (2014b).

5.9.3 Estimation of $\hbar_*$ for Space Plasmas

The smallest energy ε_C that can be transferred from a plasma particle, the magnetosonic energy $E_{ms} = \frac{1}{2}(m_i + m_e)V_{ms}^2$, is related to the plasma frequency ω according to the old-known type of energy–frequency relation, $\varepsilon_C \sim \hbar_*\,\omega$. The plasma frequency ω is typically given by the primary plasma frequency ω_{pl}. As a consequence, the value of $\hbar_*$ can be approximated by $\hbar_* \sim \frac{1}{2}(m_i + m_e)V_{ms}^2/\omega_{pl}$. Analytically, the ratio of the magnetosonic energy to the plasma frequency is given by:

$$\hbar_* \cong E_{ms}/\omega_{pl} = \frac{1}{2}e^{-1}\mu_0^{-1}\sqrt{m_e\varepsilon_0}\,n^{-\frac{3}{2}}(B^2 + 2\mu_0 k_B\,a\,nT)$$

$$= 0.0705\,n^{-\frac{3}{2}}B^2 + 2.447\,a\,n^{-\frac{1}{2}}T, \qquad (5.166)$$

with units: E_{ms}/ω_{pl} $[10^{-22}\,\text{J s}]$, B $[\text{nT} = 10\,\mu\text{G}]$, n $[\text{cm}^{-3}]$, T $[\text{MK}]$; also, $T \equiv T_e \approx T_i$, $a \equiv a_e \approx a_i$ (polytropic index).

Using solar wind plasma parameters (density, temperature, and polytropic index) and the interplanetary magnetic field strength obtained during the $\sim$20-year-long Ulysses mission that covered a broad range of heliocentric distances $1.1 \le r_{sun} \le 5.5$ AU and heliolatitudes $-80° \le \vartheta_{sun} \le +80°$ (see Fig. 5.16), we have verified that the transfer of energy between plasma particles and waves is governed by a unique relationship: the ratio between the magnetosonic energy E_{ms} and the plasma frequency ω_{pl} is constant (e.g., see Livadiotis and McComas, 2013b, 2014b; Livadiotis, 2015d, 2016c, 2017a).

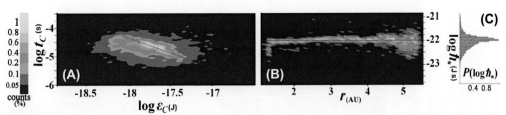

FIGURE 5.16

Phase space portion $\hbar_*$ calculated for the solar wind ion–electron plasma measurements using Eq. (5.166). (A) Diagram $(\varepsilon_C; t_C)$ (on a log–log scale), constructed from Ulysses daily measurements. (B) The product $\hbar_* = 2\varepsilon_C t_C$ is depicted as a function of heliocentric distance r. (A) and (B) are 2-D normalized histograms. (C) Normalized histogram of the values of the values of $\log \hbar_*$. The *fitted line* in (A) has slope -1 and intercept ~ -22.22 $\left(= \log\left(\frac{1}{2}\hbar_*\right)\right)$. The weighted mean of $\log \hbar_*$ values in (B) is found to be $\log(\hbar_*/[\mathrm{J \cdot s}]) \cong -21.92 \pm 0.15$. Adapted by Livadiotis and McComas (2013b).

The 1-D histogram in Fig. 5.16C shows that the value of this nearly constant ratio is $\log(\hbar^*/[\mathrm{J \cdot s}]) \approx -22$.

Nearly constant values for the ratio $E_{\mathrm{ms}}/\omega_{\mathrm{pl}}$ also appear to exist in other plasma environments, such as the Earth's magnetosphere, the inner heliosheath, and the local interstellar medium, perhaps pointing to physical mechanisms that govern the efficiency with which energy is transferred between particles and waves in a wide variety of space plasma environments (e.g., see Livadiotis and Desai, 2016). We also observe in Fig. 5.16B that significant deviations from the most probable value $\hbar^*$ for this ratio occurred when Ulysses was near ~ 1.1 and ~ 5 AU, i.e., during the fast latitude scans, and the slow descent toward and ascent away from the ecliptic plane near the orbit of Jupiter, respectively. Further evidence for significant variability and substantial departure from the value of $\log(\hbar^*/[\mathrm{J \cdot s}]) \approx -22$ is also found in other observations (cf, Figure 4 in Livadiotis and McComas, 2014b).

The ratio $E_{\mathrm{ms}}/\omega_{\mathrm{pl}}$ transitions remarkably smoothly and continuously from the slow solar wind with values $E_{\mathrm{ms}}/\omega_{\mathrm{pl}} \lesssim 10^{-23}$ to the fast solar wind, where $E_{\mathrm{ms}}/\omega_{\mathrm{pl}}$ tends toward the constant, $\hbar^* \sim 10^{-22}$ (Fig. 5.17). The same behavior appears from small to large values of the thermal and/or Alfvén speed (thus, of the magnetosonic speed or energy). Therefore the conditions where the ratio $E_{\mathrm{ms}}/\omega_{\mathrm{pl}}$ tends to its upper limit $\hbar^* \sim 10^{-22}$ are the following: (1) faster bulk plasma, in the case of solar wind, for example, the solar wind speed V_{SW} must be larger than ~ 550 km/s at 1 AU; and (2) large values of magnetosonic energy, while plasma beta is independent of the value of the ratio $E_{\mathrm{ms}}/\omega_{\mathrm{pl}}$. These conditions are useful in the case we intend to use the relation $E_{\mathrm{ms}}/\omega_{\mathrm{pl}} = \hbar^*$ in order to generate missing values of plasma parameters or magnetic field (e.g., see Livadiotis, 2015d).

In contrast to the 100% efficiently transferred E_{ms} in the fast solar wind, the nonefficiency of the slow solar wind can be caused by dispersion of fast

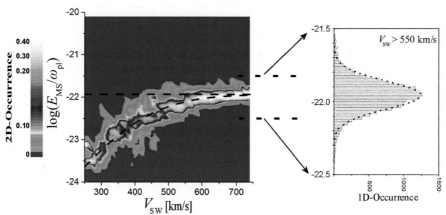

FIGURE 5.17

(Left): Normalized 2-D histograms of $\log(E_{ms}/\omega_{pl})$ versus solar wind speed V_{SW}. The *dashed curve* plots an empirical model and its limit at $\log(\hbar*)$. (Right): 1-D histogram of $\log(E_{ms}/\omega_{pl})$ for $V_{SW} > 550$ km/s. Adapted by Livadiotis and Desai (2016).

magnetosonic waves (Livadiotis and Desai, 2016). The constancy of the ratio E_{ms}/ω_{pl} is expected when the group speed is used instead of the phase speed. The fast solar wind is almost dispersion-less, characterized by quasiconstant values of the fast magnetosonic speed independently of the wavenumber k. The slow wind is subject to dispersion, which is less effective for larger wind speed and fast magnetosonic energy. Therefore, we focus on the fast solar wind values to estimate $\hbar* \approx E_{ms}/\omega_{pl}$. In Figs. 5.18A and B, we show the values of the ratio E_{ms}/ω_{pl} arranged with increasing speed V_{SW} in the case of the fast solar wind, while 5.18(c) plots their 1-D histogram, indicating a higher precision

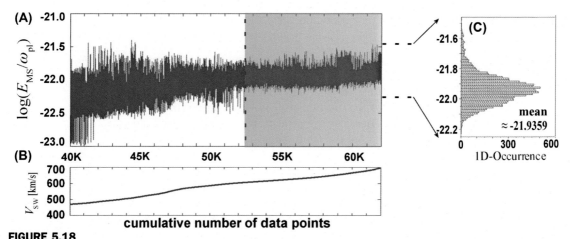

FIGURE 5.18

Fast solar wind: the values of the ratio E_{ms}/ω_{pl} (A) arranged with increasing solar wind speed V_{SW} (B), while panel (C) plots the one-dimensional histogram. Adapted by Livadiotis and Desai (2016).

value of $\hbar_*$. We find $\log(\hbar_*/[J \cdot s]) = -21.936 \pm 0.033$, or $\hbar_* = (1.160 \pm 0.083) \times 10^{-22} J \cdot s$, reaching the error limits induced by the data uncertainties.

The adaptation of the parameters involved in Eq. (5.166) in the case of kappa-distributed plasmas is as follows. The Alfvén speed remains as is, while the sound speed becomes

$$V_S \cong \sqrt{\left(a_e k_B T_e \frac{\kappa_{0_e}}{\kappa_{0_e} + 1} + a_i k_B T_i \frac{\kappa_{0_i}}{\kappa_{0_i} + 1}\right) \Big/ m_i} \cong \sqrt{\left(2a k_B T \frac{\kappa_0}{\kappa_0 + 1}\right) \Big/ m_i}$$

$$= V_{S \infty} \cdot \sqrt{\frac{\kappa_0}{\kappa_0 + 1}}, \quad V_{S \infty} \equiv \sqrt{(2a k_B T)/m_i},$$

(5.167)

where $T \equiv T_e \approx T_i$, $\kappa_0 \equiv \kappa_{0e} \approx \kappa_{0i}$, and $a \equiv a_e \approx a_i$ (see Chapter 8). The plasma frequency does not depend on the kappa index, while Eq. (5.63a) provides the kappa dependence on the Debye length, $\lambda_D = \lambda_{D \infty} \cdot \sqrt{\frac{\kappa_0}{\kappa_0 + 1}}$, $\omega_{pl} = \omega_{pl \infty}$. Then, considering that the electron contribution is more important and because of $a_e \approx 1$, we obtain

$$\lambda_{De} \omega_{pl} = \lambda_{D \infty e} \omega_{pl \infty} \cdot \sqrt{\frac{\kappa_0}{\kappa_0 + 1}}, V_{Se} \equiv \sqrt{\frac{\kappa_0}{\kappa_0 + 1}} k_B T_e / m_e = V_{S \infty e} \cdot \sqrt{\frac{\kappa_0}{\kappa_0 + 1}},$$

with $V_{S \infty e} \equiv \sqrt{k_B T_e / m_e}$, or

$$\lambda_{De} \omega_{pl} / V_{Se} = \lambda_{D \infty e} \omega_{pl \infty} / V_{S \infty e} = 1, \text{ or, } \omega_{pl} = V_{Se} / \lambda_{De} \text{ and } t_C = \frac{1}{2} \lambda_D / V_{Se},$$

(5.168)

where V_{Se} may be interpreted as the electron contribution to the total sound speed. Namely, the previous relations are still valid even for finite kappa index. The magnetosonic energy is also kappa dependent because of the involved sound speed, while the plasma frequency remains as is; their ratio becomes:

$$\hbar_* \cong E_{ms}/\omega_{pl} = \frac{1}{2} e^{-1} \mu_0^{-1} \sqrt{m_e \varepsilon_0} n^{-\frac{3}{2}} \left(B^2 + 2\mu_0 k_B a \frac{\kappa_0}{\kappa_0 + 1} nT\right)$$

$$= 0.0705 \, n^{-\frac{3}{2}} B^2 + 2.447 \, a \frac{\kappa_0}{\kappa_0 + 1} n^{-\frac{1}{2}} T.$$

(5.169)

In Chapter 2, we derive the entropy of a plasma described by kappa distributions, while in Chapter 4, we present the concept of waiting time (kappa) distributions. Both the entropy and the waiting time distribution have been used to estimate the value of $\hbar_*$ in Livadiotis and McComas (2013b). There are four main papers that estimate the value of $\hbar_*$ (Table 5.4), and their weighted mean is $\log \hbar_*[J \cdot s] = -21.924 \pm 0.020$ or $\hbar_* = (1.19 \pm 0.05) \times 10^{-22} [J \cdot s]$ (note that the logarithm values $\log \hbar_{*i}$ are normally distributed around their mean, and thus we use these values instead of $\hbar_{*i}$ for deriving average and error).

Table 5.4 Estimation of ℏ∗ and Its Weighted Mean

References	log $\hbar_*$[J·s]	δ log $\hbar_*$[J·s]
Livadiotis and McComas (2013)	−21.93	0.14
Livadiotis (2016)	−21.84	0.08
Livadiotis and Desai (2016)	−21.94	0.03
Livadiotis (2017)	−21.92	0.03
Weighted mean	−21.924	0.020

5.9.4 Application: Missing Plasma Parameters

We can exploit the constancy of $\hbar^*$ in space plasmas throughout the heliosphere and use the approximation $\hbar^* \sim \varepsilon_C/\omega_{pl}$ to derive unknown plasma parameters, e.g., magnetic field strength, density, and ion and electron temperature, which is expanded in Eq. (5.169). This relation can be reversed and expressed in terms of the unknown parameter, e.g., the magnetic field strength,

$$
\begin{aligned}
B &\cong \sqrt{2\hbar_* e \,\mu_0(m_e\varepsilon_0)^{-\frac{1}{2}}n^{\frac{3}{2}} - \mu_0 n\, k_B\left(a_i T_i \frac{\kappa_{0_i}}{\kappa_{0_i}+1} + a_e T_e \frac{\kappa_{0_e}}{\kappa_{0_e}+1}\right)} \\
&\cong \sqrt{(16.88 \pm 0.71)n^{\frac{3}{2}} - 34.699a\, nT \frac{\kappa_0}{\kappa_0+1}},
\end{aligned}
\tag{5.170}
$$

with the same units as in Eq. (5.167).

Fig. 5.19A plots the graph of the magnetic field and its error $B \pm \delta B$ as a function of density n, and for $T = 0.5$ MK, $a = 5/3$, and $\kappa_0 = 2.5$ (or $\kappa = 4$). This can be used to derive missing values of the magnetic field in space plasma datasets.

First, the method of deriving missing values of magnetic field was applied in the Ulysses measurements of the solar wind over a broad range of heliocentric distances 1.1 AU $< r <$ 5.5 AU (Livadiotis and McComas, 2013b). In Fig. 5.19B, the observational values of B are co-plotted with the modeled ones. Both sets of values follow similar distributions. The modeled B values appear with some larger variance around the statistical mode (most frequent value) at $B \sim 0.5-0.6$ nT.

Having verified the method, it was straightforward to apply it in space plasmas with unknown magnetic fields. As an example, it was used the inner heliosheath. The temperature, density, and kappa and polytropic indices have been estimated by Livadiotis et al. (2011, 2013; see also Livadiotis and McComas, 2012). The magnetic field strength was estimated by Livadiotis (2015d) using Eq. (5.170). Fig. 5.19C shows the results, that is, a histogram of all the derived values of the magnetic strength. The most frequent value is ~ 1.7 µG, that is, an average of the whole inner heliosheath. The derived value is consistent with the in situ measurements taken by Voyager 1 and 2 (note that all the plasma parameters have

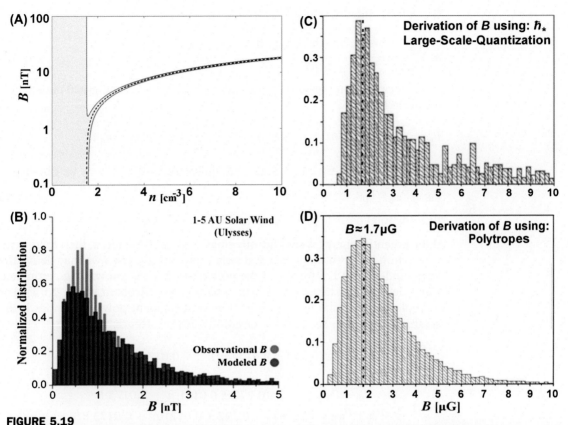

FIGURE 5.19
(A) Plot of the graph of the magnetic field and its error $B \pm \delta B$ as a function of density n. Then the magnetic field strength values are estimated for (B) the solar wind using Ulysses measurements and comparison with the respective observational values, and (C) the inner heliosheath using IBEX measurements. (D) The magnetic field values in the inner heliosheath are estimated using a different method. (B—D) panels are Adapted by Livadiotis (2015d, 2016c).

been estimated as radial averages for each sky direction, e.g., see Livadiotis et al., 2011; McComas et al., 2009). Finally, the magnetic field in the inner heliosheath was also estimated by Livadiotis (2016c) using a different method (superposition of polytropes), where the derived distribution of the magnetic field values is plotted in Fig. 5.19D. The two distributions in panels (C) and (D) have similar features (e.g., the same mode).

Eq. (5.169) can be reversed to derive any other plasma parameters. For example,

$$a\frac{\kappa_0}{\kappa_0+1} \cong (k_B\ nT)^{-1}\left[\hbar_* e(m_e\varepsilon_0)^{-\frac{1}{2}}n^{\frac{3}{2}} - \frac{1}{2}\mu_0^{-1}B^2\right]$$

$$\cong (0.486 \pm 0.021)\cdot n^{\frac{1}{2}}T^{-1} - 0.0288\ B^2\ (nT)^{-1}.$$

(5.171)

It is interesting that Eq. (5.171) may be written as

$$a\frac{\kappa_0}{\kappa_0+1} \cong \frac{\hbar*e(m_e\varepsilon_0)^{-\frac{1}{2}}n^{\frac{3}{2}} - \frac{1}{2}\mu_0^{-1}B^2}{k_B\,nT} = \frac{P_{pl}-P_{mg}}{P_{th}},$$ (5.172)

where the plasmon pressure P_{pl}, magnetic pressure P_{mg}, and thermal pressure P_{th} are given by

$$P_{pl} = \hbar*e(m_e\varepsilon_0)^{-\frac{1}{2}}n^{\frac{3}{2}} = n\cdot(\hbar*\omega_{pl}), \quad P_{mg} = \frac{1}{2}\mu_0^{-1}B^2, \quad P_{th} = k_B\,nT.$$ (5.173)

For systems with $\nu \cong \kappa_0$, Eq. (5.42), or equivalently, $\frac{\kappa_0}{\kappa_0+1} = a^{-1}$, we obtain

$$P_{th} + P_{mg} \cong P_{pl},$$ (5.174)

thus the total particle pressure gives the plasmon pressure.

If the potential energy is significantly small and can be ignored, then there is no connection between the kappa and polytropic indices. Then we may derive the kappa index from Eq. (5.171) if the polytropic index is known (e.g., estimated using density and temperature sequential values, Nicolaou et al., 2014). For example, for the solar wind plasma, the most frequent polytropic index is the adiabatic $a \cong \frac{5}{3}$ (e.g., Section 5.3; Livadiotis, 2016c). Hence, we have

$$\frac{\kappa_0}{\kappa_0+1} \cong \frac{P_{pl}-P_{mg}}{aP_{th}} \cong (0.292\pm0.012)\cdot n^{\frac{1}{2}}T^{-1} - 0.0173\,B^2(nT)^{-1} \Rightarrow$$ (5.175)

$$\kappa_0 \cong \frac{P_{pl}-P_{mg}}{aP_{th}+P_{mg}-P_{pl}} \cong \frac{(0.292\pm0.012)\cdot n^{\frac{3}{2}} - 0.0173\,B^2}{nT - (0.292\pm0.012)\cdot n^{\frac{3}{2}} + 0.0173\,B^2}.$$

5.10 Concluding Remarks

This chapter presented basic parameters and concepts involved in statistical plasma physics. We examined the polytropes and the polytropic index, the correlations between particle energies and moments, the Debye shielding and length scale, the electrical conductivity, the collision frequency and the mean free path, the plasma magnetization and the Curie constant, and the large-scale quantization constant (Fig. 5.20). We specified the effect of the kappa distributions on each of these parameters.

Table 5.5 summarizes the kappa dependence of the basic parameters presented in this chapter. It also provides the interparticle distance, Landau length, and entropic length, which were presented in Chapter 2. At thermal equilibrium, the distance of closest approach, also called the Landau length x_L, the electric potential becomes equal to the thermal energy, $e^2/(4\pi\varepsilon_0 x_L) \approx k_B T$. This is where the

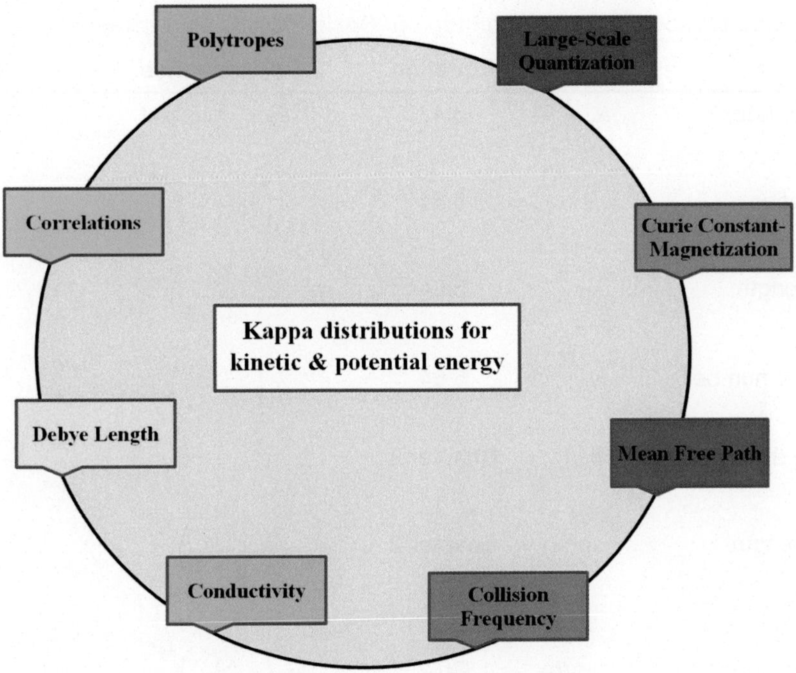

FIGURE 5.20
Infogram of the basic parameters adapted for plasmas described by kappa distributions.

electric charge density is maximized, as implied by Eq. (5.72). However, out of thermal equilibrium, the thermal energy is replaced by the term $[(\kappa_0 + 1)/\kappa_0] \cdot k_B T$, which is derived from the term λ_D^{-2} in Eq. (5.72).

It is now straightforward to apply these developments to study space, geophysical, laboratory, or other plasmas out of thermal equilibrium. The developed ideas and results will be useful and evaluated in general plasma physics research, as well as in the context of the following chapters of Parts B and C.

5.11 Science Questions for Future Research

Future analyses and observations need to address the following questions:

1. What is the solution of Lane–Emden equations for nonsmall potential?
2. How can the high-temperature plasma magnetization be applied?
3. Is there a threshold between Planck and large-scale quantization?

Acknowledgments

The work in Section 5.9 was supported in part by the project NNX17AB74G of NASA's HGI Program.

Table 5.5 Dependence on the Kappa Index of Basic Plasma Parameters

Quantity	Symbol	Equation	κ-dependence	Relations
Polytropic index	α, ν	5.42	[b] $\kappa_0 \pm$ constant	—
Correlation	R	5.54	$\propto \left(\kappa_0 + \dfrac{d}{2} \right)^{-1}$	—
Debye length	λ_D	5.64	$\propto \left(\dfrac{\kappa_0}{\kappa_0 + 1} \right)^{\frac{1}{2}}$	$\propto (T/n)^{\frac{1}{2}}$
Debye particle number	N_D	5.64	$\propto \left(\dfrac{\kappa_0}{\kappa_0 + 1} \right)^{\frac{d}{2}}$	$= V_d n\, \lambda_D^d$
Interparticle distance	b_p	Chapter 2	—	$= v_d^{-\frac{1}{d}} n^{-\frac{1}{d}} = \lambda_D \cdot N_D^{-\frac{1}{d}}$
Landau length	x_L	Chapter 2	$\propto \left(\dfrac{\kappa_0}{\kappa_0 + 1} \right)^{\frac{1-d}{2}}$	$= \frac{1}{3}\, \lambda_D \cdot N_D^{-1}$
Plasma coupling	$\Gamma \equiv x_L/b_p$	Ch.2	$\propto \left(\dfrac{\kappa_0}{\kappa_0 + 1} \right)^{\frac{1-d}{2}}$	$= \frac{1}{3} N_D^{\frac{1}{d}-1}$
Plasma frequency	ω_{pl}	5.141	—	$\propto n^{\frac{1}{2}}$
Conductivity	σ	5.140	[a] $\propto I(\kappa_0) \equiv \kappa_0^{\frac{3}{2}} \dfrac{\Gamma\left(\kappa_0 + \frac{1}{2}\right)}{\Gamma(\kappa_0 + 2)}$	$\propto T^{\frac{3}{2}}$
Collision frequency	ν_{ei}	5.146	[a] $\propto 1/I(\kappa_0)$	$= \omega_{pl}^2 \cdot (\varepsilon_0/\sigma)$
Mean free path	L_m	5.147	[a] $\propto I(\kappa_0)$	$\propto T^2/n$
Curie constant	C	5.163	$\propto \left(\dfrac{\kappa_0}{\kappa_0 + 1} \right)^{-1}$	—
Sound speed	V_S	5.167	$\propto \left(\dfrac{\kappa_0}{\kappa_0 + 1} \right)^{\frac{1}{2}}$	$\propto T^{\frac{1}{2}}$
Entropic length	σ_L	Chapter 2	$\propto \left(\dfrac{\kappa_0}{\kappa_0 + 1} \right)^{\frac{1}{2}}$	$\propto \lambda_D$

[a]approximately
[b]under conditions

CHAPTER 6

Superstatistics: Superposition of Maxwell—Boltzmann Distributions

C. Beck [1], E.G.D. Cohen [2,3]
[1]Queen Mary University of London, London, United Kingdom; [2]Rockefeller University, New York, NY, United States; [3]University of Iowa, Iowa City, IA, United States

Chapter Outline

6.1 Summary   313
6.2 Introduction: Dynamical Creation of Kappa Distributions   314
6.3 Timescale Separation in Nonequilibrium Situations   317
6.4 Typical Universality Classes for $f(\beta)$ 318
6.5 Asymptotic Behavior for Large Energies   319
6.6 Universality for Not Too Large Energies ε 321
6.7 From Measured Time Series to Superstatistics   322

6.8 Some Examples of Applications   324
  6.8.1 Classical Lagrangian Turbulence: Acceleration Statistics   324
  6.8.2 Quantum Turbulence   325
  6.8.3 Kappa Distributions in High-Energy Scattering Processes   326
6.9 Concluding Remarks   328
6.10 Science Questions for Future Research   328

6.1 Summary

In this chapter we think about dynamical reasons why kappa distributions occur so frequently in plasmas and other systems. In fact many complex driven nonequilibrium systems in statistical physics and in other areas of science (including plasma physics) are effectively described by a superposition of several statistics on different timescales, in short "superstatistics." A simple example is a tracer particle moving in a spatially inhomogeneous medium with temperature fluctuations on a large scale or some other parameter fluctuations on a large scale, but the concept is much more general. Superstatistical systems typically have marginal distributions that exhibit fat tails, for example, power law tails or stretched exponentials. Kappa distributions in space plasma physics are a particular example. In most cases observed for a variety of applications, one finds that there are three relevant universality classes: log-normal superstatistics, chi-square superstatistics, and inverse chi-square superstatistics. These can be effectively described by methods of nonextensive statistical mechanics. We outline the underlying theory; comment on the timescale separation assumption, on universality aspects, and the behavior of the tails of the distributions; and we briefly describe how to extract superstatistical parameters from a measured experimental time series. We finally describe some examples where kappa distributions (and their superstatistical descriptions) play an important role, namely, classical and quantum turbulence as well as scattering processes in high-energy physics.

Science Question: How can a superposition of Maxwellians lead to a kappa distribution?

Keywords: Lagrangian turbulence; Scattering processes; Superstatistics; Time-scale separation; Turbulence.

6.2 Introduction: Dynamical Creation of Kappa Distributions

Kappa distributions can be regarded as maximizing more general entropy measures subject to suitable constraints. These more general entropy measures can, for example, be the q-entropy introduced by Tsallis (1988) or other types of entropies such as the Renyi entropy (see Chapter 1; for a review, see Beck and Schlögl, 1993; Beck, 2009; Tsallis, 2009b).

However, while this entropy maximization based on more general entropy measures is certainly a valid mathematical approach, the important question that arises in this context is for the physical interpretation of kappa distributions. What could be a dynamical reason that a physical system is described by such a distribution? In particular, for kappa distributions observed in space plasmas this is a highly important question. Are these kappa distributions just fitting tools, or is there a deeper reason that they are observed?

At this point let us comment on an important development in the field of nonextensive statistical mechanics. It became clearer that the reason for the occurrence of generalized "power law" Boltzmann factors (so-called q-exponentials, equivalent to kappa distributions) can often be a driven nonequilibrium situation with local fluctuations of the environment. The starting point to understand this is the following formula:

$$\int_0^\infty d\beta\, f(\beta) e^{-\beta E} = [1 + (q-1)\beta_0 \varepsilon]^{-1/(q-1)}, \tag{6.1}$$

where

$$f(\beta) = \frac{(q-1)^{-\frac{1}{q-1}}}{\beta_0 \Gamma\left(\frac{1}{q-1}\right)} \cdot \left(\frac{\beta}{\beta_0}\right)^{\frac{1}{q-1}-1} e^{-\frac{\beta}{(q-1)\beta_0}}, \tag{6.2}$$

is a particular probability density, the so-called gamma distribution (or an χ^2 distribution, where $\frac{1}{2}\chi^2 \equiv \beta[(q-1)\beta_0]^{-1}$). The above formula is simply a mathematical fact, valid for the arbitrary energy level ε, but it can be given a deep physical interpretation. Ordinary Boltzmann factors $e^{-\beta\varepsilon}$ with fluctuating inverse temperatures, denoted by β, averaged in an inhomogeneous temperature environment, effectively give rise to the generalized Boltzmann factors of nonextensive statistical mechanics, which occur on the right-hand side of Eq. (6.1).

Traditionally, in nonextensive statistical mechanics one works with the parameter q, which in a sense describes how far away the complex system under consideration is from ordinary equilibrium statistical mechanics, which corresponds to $q = 1$. In plasma physics, the parameter κ is used instead:

$$\kappa = \pm\frac{1}{q-1}, \tag{6.3}$$

where the sign depends on the version of nonextensive statistical mechanics being used, that is, whether the so-called escort distributions (Beck and Schlögl, 1993; Tsallis et al., 1998) are used or not in implementing energy constraints (see Chapter 1; Beck, 2001; Livadiotis and McComas, 2009). The kappa index also depends on the dimensionality of the problem, that is, the kinetic degrees of freedom for particle systems (Livadiotis and McComas, 2011b; Livadiotis, 2015a). We will not go into detail here on the technical details; the important fact is that we now have a nice effective physical interpretation for why kappa distributions can occur in inhomogeneous nonequilibrium systems. Nonextensive statistical mechanics is relevant as an effective description for systems with temperature fluctuations on a large scale. But, more generally, it's not just temperature fluctuations; it could be other relevant system parameters as well that fluctuate, for example, the energy dissipation in a turbulent medium.

Clearly, the above example can be generalized in various ways, and that is the main scope of the so-called superstatistics approach, introduced by Beck and Cohen (2003) (see also: Cohen, 2004). Here the idea is that complex driven nonequilibrium systems often exhibit a dynamics that is a superposition of several dynamics on different timescales. For example, we may consider a tracer particle in a turbulent flow or in some other complicated environment, where the environmental conditions change, either spatially, temporally, or both. Assume that the environment exhibits temperature fluctuations or fluctuations of some other relevant parameter on a large spatiotemporal scale, large as compared to local relaxation times. Then there is a relatively fast dynamics given by the local velocity process of the tracer particle (or the local statistical mechanics system in a given cell) and also a slow one given by the parameter changes of the environment, which is spatio-temporally inhomogeneous. The two effects produce a superposition of two statistics, or in short, a "super-statistics" (Beck and Cohen, 2003; Tsallis and Souza, 2003a,b; Touchette, 2004; Beck and Cohen, 2004; Beck et al., 2005; Chavanis, 2006; Vignat et al., 2005; Rajagopal, 2006).

While this concept was introduced in a statistical mechanics setting in Beck and Cohen (2003), in the meantime many applications for a variety of complex systems have been pointed out (Ohtaki and Hasegawa, 2003; Reynolds, 2003; Daniels et al., 2004; Rizzo and Rapisarda, 2004; Beck, 2004; Baiesi et al., 2006; Abul-Magd, 2006; Porporato et al., 2006). The stationary probability distributions of a superstatistical system typically exhibit non-Gaussian behavior with fat tails, which can decay, e.g., with a power law, as a stretched exponential, or in an even more complicated way. As said before, what is important for this approach is the existence of an intensive parameter β, which fluctuates on a large spatiotemporal scale. For the simple example of a Brownian particle exhibiting a diffusive motion, β would be just the fluctuating inverse temperature of the environment, but in general β can also be an effective friction constant in a plasma, some other plasma parameter that fluctuates on a large scale, a changing mass parameter of a composed bigger object, a changing amplitude of Gaussian white noise in mathematical models, the fluctuating energy dissipation in turbulent flows, a fluctuating volatility in finance, an environmental parameter for biological systems, or simply a local variance parameter extracted from a given experimental signal. One can show that certain types of superstatistical models generate anomalous transport (Klages, 2008). Most superstatistical models are somewhat "less anomalous" than Levy-type models, in the sense that more higher moments exist than for Levy processes. The superstatistics concept is very general and has been successfully applied to a variety of complex systems, including hydrodynamic turbulence (Beck, 2003; Reynolds, 2003; Beck et al., 2005); pattern forming systems (Daniels et al., 2004); cosmic rays (Beck, 2004); solar flares (Baiesi et al., 2006); space plasmas (Schwadron et al., 2010; Ourabah et al., 2015); mathematical finance (Bouchard and Potters 2003; Ausloos and Ivanova 2003; Ohtaki and Hasegawa 2003); random matrix theory (Abul-Magd, 2006); networks (Abe and Thurner, 2005); quantum systems at low temperatures (Rajagopal, 2006); wind velocity fluctuations (Rizzo and Rapisarda, 2004) and hydroclimatic fluctuations (Porporato et & al., 2006); dynamics of cancer cells (Metzner et al., 2015); rainfall statistics (Yalcin et al., 2016); sea-level fluctuations (Rabassa and Beck, 2015); and population dynamics of species (Livadiotis et al., 2015, 2016). Kappa distributions observed in plasma physics can be produced by this mechanism as well. (See also the derivation of the Student's t distribution, Student, 1908).

In the present chapter, we emphasize the importance of the superposition and inhomogeneity aspect in understanding the origin of kappa distributions and their generalizations, thus illustrating the general concept of superstatistics. In Section 6.3, we discuss the necessary condition of timescale separation that is required for the superstatistics approach to work. Section 6.4 deals with typical superstatistical universality classes, one of them leading to kappa distributions. The asymptotic decay of the marginal probability distributions for large kinetic energies is analyzed in Section 6.5. Some universal behavior for sharply peaked temperature distributions is discussed in Section 6.6. A problem of interest for the experimentalist is discussed in Section 6.7: How does one extract super-statistical parameter distributions from a given experimentally measured time series? Section 6.8 then deals with some example data from classical and quantum turbulence and high-energy scattering processes. Finally, the concluding remarks are given in Section 6.9, while three general science questions for future analyses are posed in Section 6.10.

6.3 Timescale Separation in Nonequilibrium Situations

Consider a complex system in a stationary nonequilibrium state that is driven by some external forces. Usually we think here of a physical system (e.g., a turbulent flow), but we may easily apply similar techniques to economic, biological, and social systems, where just the meaning of the mathematical variables is different. Generally, such a complex system will be inhomogeneous in space and in time. Effectively, it is indeed useful to think of many spatial cells (or temporal window slices for time series) where there are different values of some relevant system parameter β (see Fig. 6.1).

The cell size is effectively determined by the experimental conditions, and it can be measured in terms of the correlation length of the continuously varying β-field. Superstatistical systems are characterized by a simplifying effect, namely the fact that the relaxation time is short so that each cell can be assumed to be in local equilibrium (in a certain approximation at least). Sometimes this property will be satisfied for a given complex system, sometimes not.

In the long term we need to average the spatial/temporal inhomogeneity, and the stationary distributions of the superstatistical inhomogeneous system arise as a superposition of Boltzmann factors $e^{-\beta\varepsilon}$ (or analogues of Boltzmann factors describing the local behavior of the system under consideration) weighted with the probability density $f(\beta)$ to observe some value β in a randomly chosen cell:

$$p(\varepsilon) = \int_0^\infty f(\beta) \cdot \frac{1}{Z(\beta)} \rho(\varepsilon) e^{-\beta\varepsilon} d\beta \tag{6.4}$$

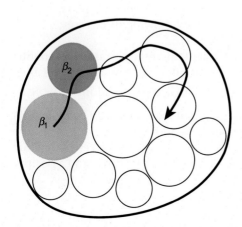

FIGURE 6.1
Concept of superstatistics. The complex systems under consideration consist of many different spatial cells, each having a different inverse temperature parameter β. The distribution of β is given by some probability density $f(\beta)$. Locally in each cell, ordinary statistical mechanics is valid, which is superimposed for the system as a whole if an average over the different cells is performed. Source: Beck (2011).

Here ε is an effective energy for each cell, $Z(\beta)$ is the normalization constant of $\rho(\varepsilon)e^{-\beta\varepsilon}$ for a given β, and $\rho(\varepsilon)$ is the density of states.

In the simplest case, we may just consider a Brownian particle of mass m moving through a changing environment in d dimensions. For a given particle velocity $\vec{u}$ one has the local Langevin equation

$$\frac{d\vec{u}}{dt} = -\gamma\vec{u} + \sigma\vec{L}(t), \tag{6.5}$$

where $\vec{L}(t)$ denotes a d-dimensional Gaussian white noise, which becomes superstatistical due to the fact that for a fluctuating environment the parameter $\beta = \frac{2}{m}\gamma\sigma^{-2}$ becomes a random variable as well. It varies from cell to cell on a rather large spatiotemporal scale $\tilde{T}$ (this symbol $\tilde{T}$ should not be confused with the temperature T, which is here denoted as $k_B T = \beta^{-1}$). Of course, for this example $E = \frac{1}{2}m\vec{u}^2$, and while the local stationary distribution in each cell is Gaussian

$$p(\vec{u}|\beta) = \left(\frac{1}{2\pi}\beta\right)^{\frac{1}{2}d} e^{-\frac{1}{2}\beta m \vec{u}^2}, \tag{6.6}$$

the marginal distribution describing the long-time behavior of the particle

$$p(\vec{u}) = \int_0^\infty f(\beta)p(\vec{u}|\beta)d\beta, \tag{6.7}$$

exhibits nontrivial behavior. The large-u tails of this distribution depend on the behavior of $f(\beta)$ for $\beta \to 0$ (Touchette and Beck, 2005).

6.4 Typical Universality Classes for $f(\beta)$

The distribution $f(\beta)$ is determined by the spatiotemporal dynamics of the driven nonequilibrium system under consideration. By construction, β is positive, so $f(\beta)$ cannot be a Gaussian. Let us here consider important examples of what to expect in typical experimental situations. In fact, analysis of the past 15 years shows that almost all complex systems that were investigated using this approach were well approximated by one of the following three cases:

1. There may be many (nearly) independent microscopic random variables $(\xi_j, j = 1,..., J)$ contributing to β in an additive way. For large J their rescaled sum $\frac{1}{\sqrt{J}}\sum_{j=1}^{J}\xi_j$ will approach a Gaussian random variable X_1 due to the Central Limit Theorem (CLT). In total, there can be many subsystems consisting of such microscopic random variables, leading to n Gaussian random variables $X_1,..., X_n$ due to various degrees of freedom in the system. As mentioned before, β needs to be positive, and a positive

β is obtained by squaring these Gaussian random variables. The resulting $\beta = \sum_{i=1}^{N} X_i^2$ is χ^2 distributed with degree n, i.e.,

$$f(\beta) = \frac{\left(\frac{n}{2}\right)^{\frac{n}{2}}}{\beta_0 \Gamma\left(\frac{n}{2}\right)} \cdot \left(\frac{\beta}{\beta_0}\right)^{\frac{n}{2}-1} e^{-\frac{n}{2} \cdot \frac{\beta}{\beta_0}}, \quad \beta_0 \geq 0, \ n > 1. \tag{6.8}$$

The resulting superstatistics generates kappa distributions, equivalent to Tsallis statistics. It exhibits power law tails for large particle speeds u. This statistics arises as a universal limit dynamics, i.e., the details of the microscopic random variables ξ_j (e.g., their probability densities) are irrelevant.

2. The same consideration as above may apply to the "temperature" β^{-1} rather than β itself. β^{-1} may be the sum of several squared Gaussian random variables arising out of many microscopic degrees of freedom ξ_j. The resulting $f(\beta)$ is the inverse χ^2 distribution, given by

$$f(\beta) = \frac{\left(\frac{n}{2}\right)^{\frac{n}{2}}}{\beta_0 \Gamma\left(\frac{n}{2}\right)} \cdot \left(\frac{\beta}{\beta_0}\right)^{-\frac{n}{2}-2} e^{-\frac{n}{2} \cdot \frac{\beta_0}{\beta}}. \tag{6.9}$$

It generates distributions that have exponential decays in u (Touchette and Beck, 2005). Again, this superstatistics is universal; details of the ξ_j are irrelevant.

3. Instead of β being a sum of many contributions, for other systems (in particular turbulent ones) the random variable β may be generated by multiplicative random processes. We may have a local cascade random variable $X_1 = \prod_{j=1}^{J} \xi_j$, where J is the number of cascade steps, and the ξ_i are positive microscopic random variables. Due to the CLT, for large J, $\log X_1 = \sum_{j=1}^{J} \log \xi_j$ becomes Gaussian if it is properly rescaled. Hence X_1 is log-normally distributed. In general there may be n such product contributions to the superstatistical variable $\beta = \prod_{i=1}^{n} X_i$. Then $\log \beta = \sum_{i=1}^{n} \log X_i$ is a sum of Gaussian random variables, hence it is Gaussian as well. Thus β is log-normally distributed, i.e.,

$$f(\beta) = \frac{a}{\beta} \exp\left[-c(\ln \beta - b)^2\right]. \tag{6.10}$$

The result is again independent of the details of the microscopic cascade random variables ξ_j, hence, there is universality again.

6.5 Asymptotic Behavior for Large Energies

Superstatistical probability densities, as given by Eqs. (6.4) or (6.7), typically exhibit "fat tails" for large ε, but what is the precise functional form of this large energy behavior? The answer depends on the distribution $f(\beta)$ and can be obtained from a variational principle. Details are described in Touchette and Beck (2005); here we just summarize some results. For a large ε we may use the saddle point approximation and write

$$B(\varepsilon) = \int_0^\infty f(\beta)e^{-\beta\varepsilon}d\beta = \int_0^\infty e^{-\beta\varepsilon+\ln f(\beta)}d\beta \sim e^{\sup_\beta\{-\beta\varepsilon+\ln f(\beta)\}} = e^{-\beta_E\cdot\varepsilon+\ln f(\beta_E)}$$

$$= f(\beta_E)e^{-\beta_E\varepsilon},$$

(6.11)

where the expression

$$\beta_E \equiv \sup_\beta\{-\beta\varepsilon + \ln f(\beta)\},$$

(6.12)

is a Legendre transform of $\ln f(\beta)$. The result of this transform is a function of ε, which can be thought of as representing a kind of entropy function if we consider $\ln f(\beta)$ to represent a free energy function. This entropy function, however, is different from other entropic formulations used in generalized versions of statistical mechanics. It describes properties related to the fluctuations of inverse temperature.

In the case where $f(\beta)$ is smooth and has only a single maximum, we can obtain the supremum by differentiating, i.e.,

$$\sup_\beta\{-\beta\,\varepsilon + \ln f(\beta)\} = -\beta_E\varepsilon + \ln f(\beta_E),$$

(6.13)

where β_E satisfies the differential equation

$$0 = -\varepsilon + (\ln f(\beta))' = -\varepsilon + \frac{f'(\beta)}{f(\beta)}.$$

(6.14)

By taking into account the next-order contributions around the maximum, Eq. (6.11) can be improved to

$$B(\varepsilon) \sim \frac{f(\beta_E)}{\sqrt{-(\ln f(\beta_E))''}} \cdot e^{-\beta_E\,\varepsilon}.$$

(6.15)

Let us illustrate all this by looking at a few examples. Suppose $f(\beta)$ is of the power law form $f(\beta) \sim \beta^\gamma$ and $\gamma > 0$ for small β. An example is an χ^2 distribution of n degrees of freedom, mentioned in Eq. (6.8). This behaves for $\beta \to 0$ as

$$f(\beta) \sim \beta^{\frac{1}{2}n-1}, \text{ or, } \gamma = \frac{1}{2}n - 1.$$

(6.16)

Other examples exhibiting this power law form are the so-called F distributions (Sattin and Salasnich, 2002; Beck and Cohen, 2003). With the above formalism, one obtains from Eq. (6.14)

$$\beta_E = \frac{\gamma}{\varepsilon}, \text{ and}$$

(6.17)

$$B(\varepsilon) \sim \varepsilon^{-\gamma-1}.$$

(6.18)

These types of $f(\beta)$ form the basis for power law generalized Boltzmann factors (Tsallis, 1988, 1999; Tsallis et al., 1998; Abe and Okamoto, 2001).

Another example would be a function $f(\beta)$, which for small β behaves as $f(\beta) \sim e^{-c/\beta}$, $c > 0$. Then

$$\beta_E = \sqrt{\frac{c}{\varepsilon}}, \quad \text{and} \tag{6.19}$$

$$B(\varepsilon) \sim \varepsilon^{-\frac{3}{4}} e^{-2\sqrt{c\,\varepsilon}}. \tag{6.20}$$

The above example can be generalized to stretched exponentials. For $f(\beta)$ of the form $f(\beta) \sim e^{-c\,\beta^{\delta}}$ one obtains after a short calculation

$$\beta_E = \left(\frac{\varepsilon}{c|\delta|}\right)^{1/(\delta-1)}, \quad \text{and} \tag{6.21}$$

$$B(\varepsilon) \sim \varepsilon^{\frac{1-\delta/2}{\delta-1}} e^{a\,\varepsilon^{\delta/(\delta-1)}}, \tag{6.22}$$

where a is some factor depending on δ and c.

6.6 Universality for Not Too Large Energies ε

One surprising effect is that the marginal distributions $p(\varepsilon) \sim B(\varepsilon)$ generated by integration over different distributions $f(\beta)$ of the parameter β often look very similar to kappa distributions, even if $f(\beta)$ is not exactly equal to the χ^2 distribution. This fact was emphasized as an important type of universality in Beck and Cohen (2003) and is of course relevant to understanding why kappa distributions, or approximations of them, are so often observed in space plasma physics and other areas of science.

In order to explain this fact, we essentially follow the argument given in Beck and Cohen (2003) and start from general distributions $f(\beta)$, but we assume that they are sharply peaked around some maximum value near the average value β_0. We write the generalized Boltzmann factor $B(\varepsilon)$, which is a superposition of ordinary Boltzmann factors $e^{-\beta\varepsilon}$ with a suitable weight function $f(\beta)$, as follows:

$$B(\varepsilon) = \int f(\beta) e^{-\beta\,\varepsilon} d\beta = \left\langle e^{-\beta\,\varepsilon} \right\rangle = e^{-\beta_0\varepsilon} \left\langle e^{-(\beta-\beta_0)\varepsilon} \right\rangle. \tag{6.23}$$

$\beta_0 = \langle\beta\rangle$ is the average of the fluctuating environmental parameter β, and $\langle\cdots\rangle$ denotes taking the average with the weight function $f(\beta)$. The above can now be written in a power series expansion as

$$B(\varepsilon) = e^{-\beta_0\,\varepsilon} \sum_{j=0}^{\infty} \frac{(-1)^j}{j!} \left\langle (\beta - \beta_0)^j \right\rangle \varepsilon^j. \tag{6.24}$$

The term $j = 0$ in this expansion equals 1, and the next term $j = 1$ vanishes, whereas the term $j = 2$ can be used to define quite generally (for any super-statistics) the parameter q by $(q - 1)\beta_0^2 = \sigma^2$, which is equivalent to

$$q \equiv \frac{\langle \beta^2 \rangle}{\langle \beta \rangle^2}. \tag{6.25}$$

Here we denoted by $\sigma^2 = \left\langle (\beta - \beta_0)^2 \right\rangle$ the variance of the fluctuating parameter β. The parameter q, or equivalently, the κ-index $\kappa = 1/(q - 1)$, describes deviations from the equilibrium state. Of course, if there are no temperature fluctuations, then $q = 1$ or $\kappa \to \infty$. This is the equilibrium situation. Generally, however, there are fluctuations of β around β_0, and all moments of this fluctuation enter in the above series expansion. The most important moment, however, is the variance σ^2, which is used to define the parameter q. In this way, we end up with the result that in this context of sharply peaked $f(\beta)$ generally *any* distribution generated by a superstatistics can be written as a generalized Boltzmann factor of the form

$$B(\varepsilon) = e^{-\beta_0 \varepsilon} \cdot \left[1 + \frac{1}{2}(q - 1)\beta_0^2 \varepsilon^2 + g(q)\beta_0^3 \varepsilon^3 + \dots \right], \tag{6.26}$$

where slight differences only arise via the nonuniversal function $g(q)$. These differences can be neglected if $\beta_0 \varepsilon$ is small enough. In this case all distributions given by $B(\varepsilon)$ look more or less the same for small energies ε. In practice, it turns out that if one looks at events that are less in size than a few standard deviations, then typical distributions generated by a superstatistics look very similar to kappa distributions. Only if one goes to quite extreme events and has sufficient statistics for them can one distinguish the power laws generated by the kappa distribution from other asymptotic behaviors. The other asymptotics can be determined using the techniques described in Section 6.5.

6.7 From Measured Time Series to Superstatistics

We now want to be more practically orientated and apply superstatistical techniques to some complex systems where we do not know the equations of motion but do have some information in the form of a measured time series. Suppose an experimentally measured time series $v(t)$ is given. This could, for example, be the velocity of a test body in an astrophysical plasma. Our goal is to test the hypothesis that the dynamics of this time series is well approximated by a superstatistical dynamics where a variance parameter fluctuates on a rather large timescale $\widetilde{T}$. Also, we want to numerically extract $f(\beta)$. First we have to determine the superstatistical timescale $\widetilde{T}$. For this we divide the time series into N equal time intervals of size τ. The total length of the signal is $t_{max} = N\tau$. We then define an averaged kurtosis function $K(\tau)$ by

$$K(\tau) = \frac{1}{t_{max} - \tau} \cdot \int_0^{t_{max} - \tau} dt_0 \frac{\left\langle (v - \bar{v})^4 \right\rangle_{t_0, \tau}}{\left\langle (v - \bar{v})^2 \right\rangle_{t_0, \tau}^2}. \tag{6.27}$$

Here $\langle\cdots\rangle_{t_0,\tau} = t_0 \int_{t_0}^{t_0+\tau} (\cdots)dt$ denotes an integration over an interval of length τ starting at t_0. The integration result fluctuates for each value of t_0 and is averaged by the integral over t_0. We are then looking for the special value $\tau = \widetilde{T}$, where

$$K(\widetilde{T}) = 3. \tag{6.28}$$

Clearly this condition defining the superstatistical timescale $\widetilde{T}$ simply reflects the fact that we are looking for locally Gaussian behavior in the time series, which implies a local kurtosis of $K = 3$. If τ is so small that only one constant value of the signal is observed in this interval, then of course $K(\tau) = 1$. On the other hand, if τ is so large that it includes the entire time series, then we obtain the flatness of the distribution of the entire signal, which is larger than 3, since superstatistical distributions are generically fat-tailed.

This analysis can be done for any given time series. As an example, Fig. 6.2 shows the time series given by share price return data (of the company Alcoa Inc.) on small (minute) timescales. (For more details, see Xu and Beck (2016).) Clearly, one sees from the figure that the superstatistical timescale is extracted from these data as $\widetilde{T} \approx 10$.

Next, we are interested in the analysis of the slowly varying stochastic process $\beta(t)$, in particular its probability distribution. Since the variance of local Gaussians $\sim \exp\left(-\frac{1}{2}\beta v^2\right)$ is given by β^{-1}, we can determine the process $\beta(t)$ from the time series as

$$\beta(t_0) = \frac{1}{\langle v^2 \rangle_{t_0,\tau} - \langle v \rangle_{t_0,\tau}^2}. \tag{6.29}$$

We can then easily make a histogram of $\beta(t_0)$ for all values of t_0, using time slices of length $\widetilde{T}$, thus obtaining the probability density $f(\beta)$. In this way one can see which type of superstatistics is relevant for a given complex system, using data

kurtosis

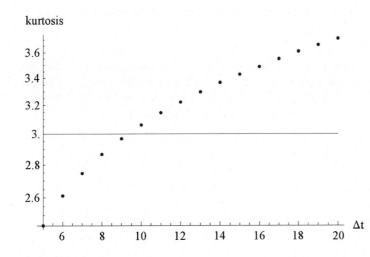

FIGURE 6.2
Example analysis for a time series given by share price returns on a timescale of minutes. The average kurtosis $K = 3$ is obtained for time slices of length $\tau = \Delta t \approx 10$. This defines the typical timescale $\widetilde{T}$ on which superstatistical parameter changes take place. Source: Xu and Beck (2016).

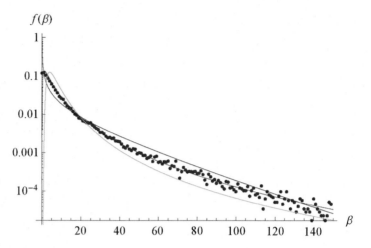

FIGURE 6.3
Extracted probability density $f(\beta)$ for the example of Alcoa share price returns on a small timescale. One sees that in this case a log-normal distribution (*red line*) yields a good fit of the data, better than the χ^2 distribution (*green line*) or inverse χ^2 distribution (*blue line*). Source: Xu and Beck (2016).

from a measured time series. For the above example of share price returns, the result is shown in Fig. 6.3. In this case one clearly sees that the log-normal distribution yields the best fit of the observed distribution $f(\beta)$. Thus this time series is an example of log-normal superstatistics. Of course, a similar analysis can be done for any other time series, for example, a measured time series for a given observable in a space plasma.

6.8 Some Examples of Applications

Our main example in this book will be kappa distributions in space plasma physics and more general distributions that are closely resembled by kappa distributions. This will be worked out in a lot of detail in the following chapters. But before concentrating on that topic, it is worthwhile to mention a few other examples of systems where superstatistical techniques have been successfully applied in the past.

6.8.1 Classical Lagrangian Turbulence: Acceleration Statistics

This field was particularly inspired by the measurements of Bodenschatz et al., published in the paper of La Porta et al. (2001). The main result was that the acceleration statistics of a tracer particle embedded in a fully developed turbulent flow is not at all Gaussian but rather exhibits a fat-tailed distribution. A typically observed probability distribution is shown in Fig. 6.4.

A theory for these distributions was developed in Beck (2007). The typical prediction of these types of models is log-normal superstatistics, i.e., a formula of the form

$$p(a_x) = \frac{\tau}{2\pi s} \int_0^\infty \beta^{-\frac{1}{2}} \exp\left[-\frac{1}{2s^2}\log^2\left(\frac{1}{m}\beta\right) \right] \cdot e^{-\frac{1}{2}\beta \, \tau^2 a_x^2} d\beta, \tag{6.30}$$

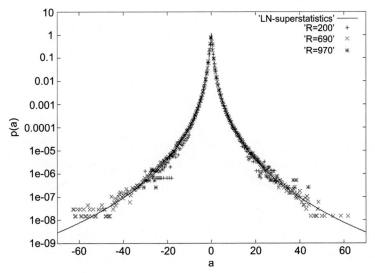

FIGURE 6.4

Measured probability density $p(a)$ of a single acceleration component (a) for various Reynolds numbers (R) and a fit by the log-normal (LN) superstatistics distribution given be Eq. (6.36), which provides an excellent fit of the data. Source: La Porta et al. (2001).

for the probability density of a single component a_x of the test particle. This formula agrees very well with the experimental data; see the above plot.

The measurements for accelerations presented in La Porta et al. (2001) are of very high quality and include very rare events, going up to accelerations of 40 times the standard deviation and more. Thus the data are precise enough to analyze the particular class of superstatistics that is relevant. In this case it turns out that log-normal superstatistics yields by far the best fits of the data, much better than a simple fit by χ^2 superstatistics (equivalent to kappa distributions). The log-normally distributed fluctuating parameter can be motivated by Kolmogorov's K62 theory (Castaing et al., 1990; Beck, 2007).

6.8.2 Quantum Turbulence

New and interesting data have been obtained for cold quantum liquids in a turbulent state. This topic comes under the general heading of "quantum turbulence." In these very cold quantum liquids, tracer particles can again be experimentally tracked. The result is surprising; whereas in classical turbulence the velocity distribution of the tracer particle is almost Gaussian, in quantum turbulence one observes power law distributions (Paoletti et al., 2008; La Mantia and Skrbek, 2014). In fact, as shown in Beck and Miah (2013) and Miah and Beck (2014), kappa distributions yield an excellent fit of the data.

The behavior of a turbulent quantum liquid is quite complicated; differently to the classical turbulence circulation, the turbulent quantum liquid is quantized, having a complex pattern of fluctuating vortex tangles. If the tracer particle comes close to such a vortex line it rapidly rotates and gets a high velocity. In the superstatistical model introduced in Beck and Miah (2013), the fluctuating

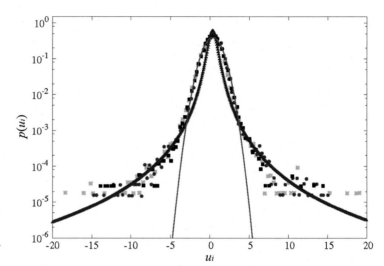

FIGURE 6.5
Measured velocity distribution of a test particle in a quantum turbulent flow and a fit with a superposition of a kappa distribution (*blue line*) and a Gaussian core.
Source: Miah and Beck (2014).

superstatistical parameter is identified as the distance of the tracer particle to the nearest vortex filament. This ultimately leads to the prediction that a kappa distribution of the form

$$p(\vec{u}) \propto \left[1 + \beta_0 \frac{1}{2} m (\vec{u} - \vec{u}_b)^2 \right]^{-\frac{3}{2}} \sqrt{\beta_0}, \tag{6.31}$$

should be a good fit of the velocity histogram data, and it is indeed, as shown in Beck and Miah (2013); $\vec{u}_b$ denotes the average or bulk velocity. The relevant nonequilibrium parameter predicted by the model is $q \to 5/3$, which is equivalent to $\kappa \to 3/2$ (that is, close to the furthest stationary state from thermal equilibrium, the antiequilibrium; see Chapter 1). Actually, what is being measured in the experiments and predicted in the theory is just the probability distribution of a single component of the velocity.

In quantum turbulence, one typically has a mixture of a normal liquid and a superfluid, and the normal liquid basically leads to a Gaussian core in the velocity distribution, whereas the superfluid leads to the above kappa distribution (Miah and Beck, 2014). A fit of experimental data is shown in Fig. 6.5.

6.8.3 Kappa Distributions in High-Energy Scattering Processes

The kappa distributions also yield an excellent fit for scattering data in high-energy physics obtained at particle colliders, for example, for differential cross-sections as a function of the transverse momenta. An example is shown in Fig. 6.6.

The connection between nonextensive statistical mechanics and scattering data in high-energy processes was first emphasized in the papers of Bediaga et al.

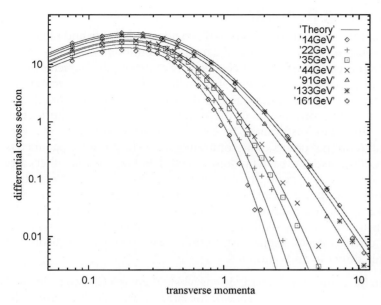

FIGURE 6.6

Measured differential cross-section as a function of observed transverse momenta for e^+e^- scattering for various center of mass energies and a fit with formulas based on kappa distributions and a phase space factor. Source: Beck (2000).

(2000) and Beck (2000), which used e^+e^- scattering data. Similar ideas were also applied to cosmic rays (Tsallis and Souza, 2003b; Beck, 2004) and many other scattering processes, such as p^+p^- at very high energies (Deppman, 2012; Wong et al., 2015).

In fact, what is underlying the fittings in Fig. 6.6 (and other experimental data) is kappa distributions of the form

$$p(\varepsilon) \propto [1 + (q-1)\beta_{eff}\varepsilon]^{-1/(q-1)}\varepsilon^2, \tag{6.32}$$

where ε is the energy of the particles

$$\varepsilon = \sqrt{c^2 p_x^2 + c^2 p_y^2 + c^2 p_z^2 + m^2 c^4}, \tag{6.33}$$

and $\beta_{eff} = (k_B T_{eff})^{-1}$ is an effective inverse temperature variable, given essentially by the Hagedorn (1965) temperature (a kind of boiling temperature of nuclear matter). For relativistic particles the rest mass m can be neglected, and one has $\varepsilon \approx c|\vec{p}|$. The distribution (Eq. 6.32) is a q-generalized relativistic Maxwell–Boltzmann distribution in the formalism of superstatistics. The kappa distribution is multiplied with ε^2, taking into account the available phase space volume. For high-energy scattering processes, the entropic index that fits the experimental data well is typically in the region of $q = 1.1...1.2$. The Hagedorn temperature parameter is of the order of 100 MeV. An upper bound $q \leq 11/9 = 1.2222$ was derived in Beck (2004), taking as an underlying model inverse temperature fluctuations given by a χ^2 superstatistics.

6.9 Concluding Remarks

In conclusion, the superstatistical approach provides a plausible and physically convincing reason for the occurrence of kappa distributions in a variety of circumstances. A primary reason is the heterogeneity of the physical systems under consideration, so that a superposition of many local Maxwell–Boltzmann distributions leads to kappa distributions globally when averaged over the fluctuating parameter. This corresponds to the simplest case, where the fluctuating parameter is chi-square distributed. In fact, chi-square distribution is a very natural distribution that occurs quite commonly if one has positive parameters that are sums of squared Gaussian random variables. But we saw that other cases are possible as well. For example, in cascade-like processes (like the Richardson cascade in turbulence), it is more plausible to have a log-normally distributed parameter leading to different statistics than kappa distributions when the parameter fluctuations are integrated out. Still, these different distributions are numerically often close to kappa distributions; we provided the theoretical reason for this similarity in Section 6.6.

One may also have mixtures of distributions where there is a Gaussian core and more heavy-tailed non-Gaussian behavior for large velocities (for example, the Lagrangian quantum turbulence considered in Fig. 6.5). Intrinsic long-range correlations and multifractal phase space structures may further modify the distributions observed. Finally, besides having spatial fluctuations in real space leading to superstatistical distribution functions, sometimes the parameter under consideration may simply fluctuate due to the sampling process involved in the way that the measurements are done. An example is the collection of high-energy scattering data described in Section 6.8. Here, all data were collected from a large number of scattering experiments where in each experiment the local collision temperature was different.

6.10 Science Questions for Future Research

Future analyses and observations need to address the following questions:

1. What classes of superstatistics are applicable in different plasmas?
2. Is the superposition of Maxwellians applied in space plasmas?
3. Can the superposition of Maxwellians describe the entropy?

CHAPTER 7

Linear Kinetic Waves in Plasmas Described by Kappa Distributions

A.F. Viñas [1]**, R. Gaelzer** [2]**, P.S. Moya** [3]**, R. Mace** [4]**, J.A. Araneda** [5]

[1]NASA Goddard Space Flight Center, Greenbelt, MD, United States; [2]Federal University of Rio Grande do Sul, Porto Alegre, Brazil; [3]University of Chile, Santiago, Chile; [4]University of KwaZulu-Natal, Durban, South Africa; [5]University of Concepción, Concepción, Chile

Chapter Outline

7.1 Summary   330
7.2 Introduction   330
7.3 Plasma Dielectric Tensor and the Dispersion Relation   336
7.4 Kappa Velocity Distribution Plasma Waves at Parallel Propagation ($\vartheta = 0$)   339
   7.4.1 General Aspects   339
   7.4.2 Low-Frequency Electromagnetic Alfvén Cyclotron Plasma Waves   342
   7.4.3 Low-Frequency Instabilities   343
      7.4.3.1 Alfvén Cyclotron Instability   343
      7.4.3.2 Proton Firehose Instability   345
   7.4.4 High-Frequency Electromagnetic Whistler Cyclotron Plasma Waves   349

   7.4.5 High-Frequency Instabilities   349
      7.4.5.1 Whistler Cyclotron Instability   349
      7.4.5.2 Electron Firehose Instability   351
   7.4.6 High-Frequency Electromagnetic Langmuir Plasma Waves   352
   7.4.7 Low-Frequency Electromagnetic Ion-Acoustic Plasma Waves   355
7.5 Kappa Velocity Distribution Plasma Waves at Oblique Propagation ($\vartheta \neq 0$)   359
7.6 Concluding Remarks   361
7.7 Science Questions for Future Research   361

Kappa Distributions. http://dx.doi.org/10.1016/B978-0-12-804638-8.00007-3

7.1 Summary

In this chapter we present an overview of the excitation, propagation, and absorption of linear plasma waves in collisionless, spatially uniform, multi-component, magnetized non-Maxwellian plasma characterized by a kappa velocity distribution based on a Vlasov kinetic description. Although traditional plasma physics texts are replete with examples of linear waves and instabilities in plasmas whose charged particles are modeled by the thermal Maxwellian velocity distribution, very few works exist that bring together in one place the analogous results for plasmas modeled by a kappa distribution—a more appropriate and versatile model for space and other collisionless plasmas. The treatment, which uses Vlasov kinetic theory coupled with Maxwell's equations, covers the range from low-frequency waves, where ion dynamics dominates the dispersion properties of the waves, up to frequencies in excess of the plasma frequency, where the electron physics plays the dominant role. For reasons of tractability, the primary focus is on waves and instabilities that propagate parallel to the ambient magnetic field, but both electrostatic as well as electromagnetic plasma waves are considered. The chapter begins with the fundamental concepts of nonequilibrium statistical mechanics and electrodynamics. The established kinetic theory of waves in a multispecies plasma with arbitrary velocity distribution is used to introduce the general dielectric tensor, which characterizes the plasma response to fluctuating electromagnetic fields. The general dielectric tensor is then derived using a drifting bi-kappa distribution for the special case of parallel propagation. Using this dielectric tensor in the Fourier-transformed wave equation, various fundamental parallel propagating longitudinal (electrostatic) and transverse (electromagnetic) plasma wave modes are discussed within the context of the bi-kappa velocity distribution. Both analytical and numerical results for the excitation (growth) and absorption (damping) of plasma waves in plasmas with a kappa velocity distribution will be discussed.

Science Question: What are the effects of kappa distributions on linear plasma waves and instabilities?

Keywords: Alfvén cyclotron instability; Electrostatic instabilities; Kinetic instabilities; Langmuir and ion-acoustic waves; Plasma waves; Proton and electron firehose instabilities; Whistler cyclotron instabilities.

7.2 Introduction

The long-range nature of the Coulomb force in conjunction with the propensity for plasmas to shield out short-range electric fields in an effort to remain charge neutral gives rise to a situation where interactions between charged particles are mediated primarily by the mean fields in the plasma, rather than by the microfields of their near neighbors. This leads to collective behavior (the focus of this chapter) with the possibility for oscillations, waves, and instabilities. Being collective in nature, oscillations, waves, and instabilities are insensitive to the precise positions and velocities of all of the particles. Instead, they depend on statistical properties of the ensemble, such as the distribution of particle velocities, which directly affect the likelihood of dynamical processes that can lead to an energy exchange between particles and waves.

In fact, a very important aspect of this many-particle description is that these particles can interact and exchange energy with waves, driving dynamical processes that produce both heating and particle acceleration as well as contribute to the excitation and damping of collective plasma waves. The rates of these processes, however, are governed by the statistical properties of the ensemble, in particular, the distribution of particle velocities. This collective description, which is described by the combination of the plasma kinetic Vlasov equation (or the collisionless Boltzmann equation), subject to the Lorentz electromagnetic forces and Maxwell's equations, predicts a definite set of plasma waves and oscillations, also called normal modes, which are inherent to the plasma system. They propagate through the plasma medium and interact self-consistently with the particles constituting the plasma.

In this chapter we shall mainly be concerned with the propagation of small-amplitude (linear) plane waves in a spatially homogeneous medium and their interaction with the plasma particles, whose statistical behavior is described by a long-tail kappa particle velocity distribution. Such linear waves are simple to analyze and demonstrate many of the essential features of the plasma behavior. The analysis presented here will be more relevant to space plasmas in which particles have a large mean free path (i.e., long-time correlation) between interparticle collisions, for which collective plasma behavior plays a more immediate role in their dynamics. The idea that the wave dispersion characteristics are not only dependent on the physical parameters (e.g., density, temperature, magnetic field, etc.) but that they also depend on the shape of the distribution probably extends back to the early work of Bernstein modes (Bernstein, 1966). However, to the best of our knowledge, it seems to have been first explicitly postulated by Abraham-Shrauner and Feldman (1977a, 1977b) in application to whistler and electromagnetic ion cyclotron waves in the solar wind.

The study of the properties of non-Maxwellian plasma particle distributions composed of suprathermal particle species has been a major focus of research in space and laboratory plasmas (Bornatici et al., 1983; Meyer-Vernet et al., 1995; Maksimovic et al., 1997; Leubner, 2005; Nieves-Chinchilla and Viñas, 2008; Zouganelis, 2008; Livadiotis and McComas, 2009, 2013a; Livadiotis, 2015a; Kaeppler et al., 2014; Viñas et al., 2015; among many others). Many of these studies have utilized the so-called kappa distribution, whose profile exhibits a power law tail. Variants of this distribution can also feature other free-energy sources such as a drifts with a speed $U_{\|s}$ relative to the average magnetic field (Summer and Thorne, 1991; Mace and Hellberg, 1995) and temperature anisotropy $\Gamma_s = T_{\perp s}/T_{\|s}$, where $T_{\|s}$ and $T_{\perp s}$ measure, respectively, the temperature of the distribution in the parallel and perpendicular directions relative to the mean ambient magnetic field, and the subscript s denotes the particle species. The kappa distribution has traditionally been expressed by

$$
f_s\left(u_\|, u_\perp\right) = n_s \frac{\pi^{-\frac{3}{2}} g(\kappa_s)}{w_{\|s} w_{\perp s}^2} \cdot \left\{ 1 + \frac{1}{\kappa_s} \cdot \left[\frac{\left(u_\| - U_{\|s}\right)^2}{w_{\|s}^2} + \frac{u_\perp^2}{w_{\perp s}^2} \right] \right\}^{-(\kappa_s+1)} , \quad \text{for } \kappa_s > \frac{3}{2},
$$

(7.1)

where we have defined $g(\kappa_s) = \kappa_s^{-\frac{3}{2}} \Gamma(\kappa_s + 1) \Big/ \Gamma\left(\kappa_s - \frac{1}{2}\right)$ solely in terms of the

parameter κ_s for each particle species s, and $\Gamma(x)$ is the Gamma function. Eq. (7.1) presents a type of anisotropic kappa distribution (Chapters 4 and 10; Lazar et al., 2012; Livadiotis, 2015a), also called a "drifting bi-kappa distribution"; it represents a non-Maxwellian distribution, and it is the focus of this chapter. Here, n_s is the particle number density, and $U_{\|s}$ is the particle species parallel (fluid) drift velocity relative to the average magnetic field direction B_{0z}.

The quantities $w_{\|s}$ and $w_{\perp s}$ are proportional to the parallel and perpendicular thermal speeds $(\theta_{\|s}, \theta_{\perp s})$, respectively, but they are also a function of the κ_s parameter and are given by

$$
w_{\|(\perp)s}^2 = \theta_{\|(\perp)s}^2 \cdot \frac{\kappa_s - \frac{3}{2}}{\kappa_s}, \text{ for } \theta_{\|(\perp)s}^2 \equiv \frac{2 k_B T_{\|(\perp)s}}{m_s},
$$

(7.2)

where $T_{\|(\perp)s}$ are the parallel and perpendicular temperatures (kinetically defined; see Chapter 1), m_s is the mass of species-s, and k_B is the Boltzmann constant. Then, the distribution function can be rewritten (e.g., see Summer and Thorne, 1991; Mace and Hellberg, 1995; Livadiotis and McComas, 2009; Livadiotis, 2015a), as

$$
f_s\left(u_\|, u_\perp\right) = n_s \frac{\pi^{-\frac{3}{2}} A(\kappa_s)}{\theta_{\|s} \theta_{\perp s}^2} \cdot \left\{ 1 + \frac{1}{\kappa_s - \frac{3}{2}} \cdot \left[\frac{\left(u_\| - U_{\|s}\right)^2}{\theta_{\|s}^2} + \frac{u_\perp^2}{\theta_{\perp s}^2} \right] \right\}^{-(\kappa_s+1)} ,
$$

(7.3)

where we have defined $A(\kappa_s) = \left(\kappa_s - \frac{3}{2}\right)^{-\frac{3}{2}} \Gamma(\kappa_s + 1) \Big/ \Gamma\left(\kappa_s - \frac{1}{2}\right)$ for each

particle species s solely in terms of the parameter κ_s. These definitions are important in order to define, in the case of an isotropic plasma, a single temperature moment distribution (see, for example, Mace and Hellberg, 1995; Livadiotis, 2015a; among others). Fig. 7.1 displays the profile of Eqs. (7.1) and (7.3) versus normalized velocity for an isotropic, nondrifting, and 1-D kappa distribution for different values of the kappa index.

The parameter kappa, κ, controls the shape of the distribution function, i.e., the level of suprathermal tail profile of the distribution. Smaller κ-indices represent

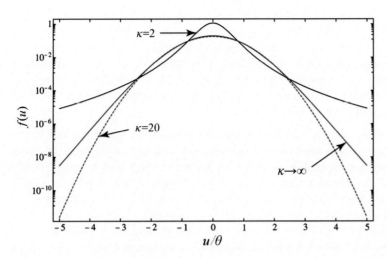

FIGURE 7.1
Normalized kappa distribution for different values of the κ-index.

higher degrees of suprathermal particles, whereas larger κ-indices indicate a smaller level of suprathermal components. The parameter κ, like the temperature T, is a statistical property of each charged particle species for a plasma out of thermal equilibrium. So, a plasma system can contain a mixture of plasma species that may have different κ-indices (i.e., $\kappa = \kappa_s$).

All of these types of kappa distributions, isotropic or anisotropic (for the explicit forms, see Chapter 4; see also Summers and Thorne, 1991; Livadiotis and McComas, 2013a; Livadiotis, 2015a), distinguish themselves by the presence of a suprathermal tail $\left(\text{at} \quad u^2 \gg w_{\parallel,\perp}^2 \right)$ that decays according to a power law instead of the exponential decay characteristic of Maxwellian distributions.

Olbert (1968), in collaboration with his former students Binsack (1966) and Vasyliūnas (1968) at MIT, first proposed this kind of distribution as a fitting model function to explain electron spectrometer measurements on the IMP-1 spacecraft and observations of magnetospheric electron velocity distributions on the OGO-1 satellite at the Earth's magnetotail, respectively. However, they did not provide for a first principle foundation as to the physical nature and origin of these velocity distributions. The origin of this kind of velocity distribution has been proposed from a fundamental set of postulates based on the principle of nonadditive entropy by Tsallis (1988). According to this representation, many particle physical systems that evolve subject to long-scale correlations and nonlinear effects can reach a quasistationary state in which the probability distribution functions of a physical quantity are not described by a Maxwell−Boltzmann distribution but rather by Eqs. (7.1)−(7.3). For such systems, the velocity distribution function arises from the maximization of a nonadditive q-entropy proposed by Tsallis (1988) subject to the constraints of the canonical ensemble. One review that accounts for the emergence of kappa distributions as a consequence of the Tsallis nonextensive statistical mechanics and the resulting implications for space plasmas has been given by Livadiotis and McComas (2009) and Livadiotis (2015a). A similar conceptual explanation about the origin of kappa distributions has been formulated by Treumann (1999a, 1999b). (For more details, see Chapter 1.) Nonetheless, here we will only be concerned with the electromagnetic dispersion effects that such velocity distributions will have on the plasma medium.

Non-Maxwellian particle velocity distributions with suprathermal tail extensions decrease as a power law of the velocity and are ubiquitous in space plasmas. They are especially inherent to the solar wind and magnetospheric electron and ion distributions (Binsack 1966; Olbert, 1968; Vasyliunas, 1968; Montgomery, 1968; Feldman et al., 1975; Pilipp et al., 1987; Maksimovic et al., 1997; Nieves-Chinchilla and Viñas, 2008; Zouganelis, 2008; among others). These distributions are remarkably well modeled by kappa-type distributions of the kinds discussed above as a fitting function. Such class of distributions are the result of a more general universal mechanism that generates these metastable distributions in many systems, such as astrophysical, quantum, biological, economic, and for certainty in space plasma systems (Gell-Mann and Tsallis, 2004; Tsallis and Brigatti, 2004). These distributions play an important role in the description of particle acceleration, in defining the thermodynamic or kinetic temperature, and in the wave–particle interactions of any plasma system. Among many fundamental challenging problems of laboratory, space, and astrophysical plasmas, three of the most important are the understanding of the following: (1) local sources of plasma instabilities by free energy in the particle velocity distributions, such as temperature anisotropy, suprathermal tails, and differential streaming, among other free-energy sources; (2) the relaxation of collisionless plasmas to nearly isotropic of non-Maxwellian velocity distribution functions; and (3) the resultant state of nearly energy-density equipartition with electromagnetic plasma turbulence.

Even in the absence of free energy for plasma instabilities, Maxwellian and non-Maxwellian plasma sustains a small but detectable spontaneous fluctuation in electromagnetic fields. Such fluctuations arise from the discreteness of the plasma particles, which produces fluctuations in the charge and current fluctuations (the sources for field fluctuations in the plasma). These electromagnetic field fluctuations, in the presence of a source of free energy, such as temperature anisotropy in the velocity distribution, provide the "seed" perturbations that are necessary for wave growth in processes called microinstabilities that amplify waves in time. The response of non-Maxwellian plasmas, due to these emissions, is intimately associated to the electric and magnetic field fluctuations, which can be described by the dielectric response function of the plasma. The response of plasmas to such electromagnetic disturbances is self-consistently determined by the equivalent dielectric tensor $\overleftrightarrow{\varepsilon}$ (Stix, 1992, p. 3), whose components are $\varepsilon_{ij}\left(\overrightarrow{k}, \omega\right)$ in an orthonormal frame of reference, where $\overrightarrow{k}$ is the wave vector, and

ω is the angular frequency of the wave. The spectral properties of the waves, which are strongly dependent on the dielectric response $\varepsilon_{ij}\left(\overrightarrow{k},\omega\right)$, provide substantial information about the state of a plasma, and in fact it is one of the most efficient methods of plasma diagnostics used today in laboratory fusion research devices and in space plasma measurement (e.g., Lund et al., 1994; Issautier et al., 2001; Viñas et al., 2005; Chapman and Gericke, 2011).

The analysis of observed emissions in space plasmas in terms of linear plasma wave solutions obtained using the equivalent dielectric tensor for Maxwellian-type charged particle populations is common practice in the space plasma community, despite the non-Maxwellian statistical behavior of space plasmas. As we have mentioned, this non-Maxwellian character is better modeled, within the framework of linear Vlasov–Maxwell theory, by a kappa-type distribution, with the "type" or "variant" characterizing the source of anisotropy or free energy in the system. This source of free energy, which can be any feature of the velocity distribution causing a deviation from an isotropic nonmonotonic decrease in particle energy with speed, can be converted into plasma waves via microinstabilities that amplify the seed fluctuations present because of particle discreteness effects. In general, plasma instability is a process whereby a small departure from equilibrium becomes the cause of a further deviation. A microinstability is an instability that depends on the microscopic details of the velocity distribution, rather than the bulk plasma conditions. These amplified waves react back on the particles, and in doing so they diminish or quench the nonequilibrium feature of the velocity distribution that is their source. This nonlinear effect (not included in the linear theory, which uses particle trajectories in the unperturbed electromagnetic fields) leads to eventual quenching of the microinstability and a new state of plasma equilibrium (or metaequilibrium). In the next section, we outline the calculation of the equivalent dielectric tensor for a general drifting bi-kappa distribution. This velocity distribution is particularly liable for space plasmas, for reasons already mentioned, and has the additional benefit that the results can be reduced to the well-known Maxwellian equivalents in the limit as $\kappa \rightarrow \infty$.

In this chapter we present an overview of linear plasma waves and instabilities in plasmas described by kappa velocity distributions. The main objective of this chapter is to provide a coherent and self-contained account of the fundamental theory of linear plasma waves, not as an isolated subject, but rather as an integral part of statistical plasma physics and its relevance to laboratory, geophysical, solar, space, and astrophysical plasmas from an academic viewpoint. This approach will assists young researchers as well as experienced investigators interested in the physics of linear plasma waves and instabilities in kappa-distributed plasmas.

In Section 7.3, we introduce the concept of plasma dielectric tensor and the general dispersion relation for the arbitrary initial equilibrium distribution function. Section 7.4 follows with a discussion of parallel propagating linear plasma waves for a kappa velocity distribution plasma. We also include specialized analytic and numerical solutions for various linear electromagnetic and electrostatic plasma waves at low and high frequencies. These represent typical general examples of plasma waves in both space and laboratory plasmas. In Section 7.5 we discuss the state of the field of linear plasma waves for a kappa velocity distribution at oblique propagation. Finally, the concluding remarks are given in Section 7.6, while three general science questions for future analyses are posed in Section 7.7.

7.3 Plasma Dielectric Tensor and the Dispersion Relation

The dielectric tensor describes the macroscopic response function to electromagnetic disturbances of a plasma particle system in the absence of external disturbances (e.g., Montgomery and Tidman, 1964; Krall and Trivelpiece, 1973; Ichimaru, 1973; Melrose, 1986; Stix, 1992; Swanson, 2003). The plasma dielectric tensor contains all of the essential information about the electromagnetic properties of the collective plasma oscillations and their linear interaction with the ensemble of particles. Since our interest is in electromagnetic and electrostatic phenomena, the main objective of this section is to establish the theoretical foundation for the electrodynamic description of plasmas as a continuous dielectric media. To do this self-consistently it is necessary to express the plasma current density $\vec{J}$ in terms of the induced electric field $\vec{E}$. One approach to achieve this is to introduce a conductivity tensor $\overset{\leftrightarrow}{\sigma}$, with components $\sigma_{ij}\left(\vec{k},\omega\right)$ in an orthonormal frame, which, in turn, can be related to the dielectric tensor. The plasma current density depends on the velocity distributions of the charged particle species, but we proceed generally at first, with an arbitrary velocity distribution $f_s\left(u_{\parallel},u_{\perp}\right)$ for each charged particle species s. The starting point of our analysis is the system of linearized Vlasov and Maxwell's equations. We assume that, prior to the imposition of any disturbances (perturbations), the equilibrium plasma parameters satisfy $\vec{E}_0 = 0$; $\vec{B}_0 = B_{0z}\hat{z}$; $\rho_0 = 0$ (no net external charge); $\vec{J}_0 = 0$ (no net external current); and that the particle velocity distributions are spatially uniform and time stationary $f_{0s} = f_s\left(u_{\parallel},u_{\perp}\right)$. The subscripts "0" denote equilibrium (unperturbed) values. Applying Fourier–Laplace transforms to our system of linearized Vlasov–Maxwell equations and eliminating $\vec{B}\left(\vec{k},\omega\right)$ in favor of the induced (perturbed) electric field $\vec{E}\left(\vec{k},\omega\right)$, we obtain the following system of algebraic equations:

$$\overset{\leftrightarrow}{D}\left(\vec{k},\omega\right)\cdot\vec{E}\left(\vec{k},\omega\right) = 0, \tag{7.4}$$

where $\overset{\leftrightarrow}{D}\left(\overrightarrow{k},\omega\right)$ is called the dispersion tensor in the Fourier domain, and the dispersion tensor is closely related to the dielectric tensor $\overset{\leftrightarrow}{\varepsilon}\left(\overrightarrow{k},\omega\right)$ and has the following explicit form:

$$\overset{\leftrightarrow}{D}\left(\overrightarrow{k},\omega\right) \equiv \overset{\leftrightarrow}{\varepsilon}\left(\overrightarrow{k},\omega\right) - \left(\frac{kc}{\omega}\right)^2\left(\overset{\leftrightarrow}{1} - \hat{k}\otimes\hat{k}\right), \tag{7.5}$$

where $\overset{\leftrightarrow}{1}$ is the unit tensor, $\otimes$ is the tensor (dyadic) product, and $\hat{k} = \overrightarrow{k}/k$ is the unit wave vector in cylindrical coordinates:

$$\overrightarrow{k} = k_\parallel\hat{z} + k_\perp\hat{x} = k\cos\vartheta\hat{z} + k\sin\vartheta\hat{x},$$

defined without a loss of generality, relative to the mean magnetic field $\overrightarrow{B}_0 = B_{0z}\hat{z}$ in terms of the propagation angle ϑ, ω is the wave frequency, and $\overset{\leftrightarrow}{\varepsilon}\left(\overrightarrow{k},\omega\right)$ is the dielectric tensor that contains the basic electromagnetic response of the plasma particle medium. It is defined as (Ichimaru, 1973, p. 51)

$$\overset{\leftrightarrow}{\varepsilon}\left(\overrightarrow{k},\omega\right) = \left(1 - \sum_s\frac{\omega_{ps}^2}{\omega^2}\right)\overset{\leftrightarrow}{1} + \sum_s\sum_{n=-\infty}^{n=+\infty}\frac{\omega_{ps}^2}{\omega^2}$$

$$\times\int\frac{\chi_{s,n}\left(\overrightarrow{k},\omega;u_\parallel,u_\perp\right)}{\omega - k_\parallel u_\parallel - n\Omega_s}\overset{\leftrightarrow}{T}_{s,n}\left(\overrightarrow{k},\omega;u_\parallel,u_\perp\right)d^3\overrightarrow{u}, \tag{7.6}$$

where we have defined the density normalized distribution $F_s\left(u_\parallel,u_\perp\right) = f_s\left(u_\parallel,u_\perp\right)/n_s$,

$$\chi_{s,n}\left(\overrightarrow{k},\omega;u_\parallel,u_\perp\right) \equiv k_\parallel\frac{\partial F_s}{\partial u_\parallel} + \frac{n\Omega_s}{u_\perp}\cdot\frac{\partial F_s}{\partial u_\perp}, \tag{7.7a}$$

and

$$\overset{\leftrightarrow}{T}_{s,n}\left(\overrightarrow{k},\omega;u_\parallel,u_\perp\right) \equiv \begin{bmatrix} \dfrac{n^2\Omega_s^2}{k_\perp^2}J_n^2 & iu_\perp\dfrac{n\Omega_s}{k_\perp}J_nJ_n' & u_\parallel\dfrac{n\Omega_s}{k_\perp}J_n^2 \\[2ex] -iu_\perp\dfrac{n\Omega_s}{k_\perp}J_nJ_n' & u_\perp^2J_n'^2 & -iu_\parallel u_\perp J_nJ_n' \\[2ex] u_\parallel\dfrac{n\Omega_s}{k_\perp}J_n^2 & iu_\parallel u_\perp J_nJ_n' & u_\parallel^2J_n^2 \end{bmatrix}. \tag{7.7b}$$

$J_n(x)$ and $J_n'(x)$ are the Bessel function of order n and argument $x = k_\perp u_\perp/\Omega_s$ and its derivative with respect to x. Here, $\omega_{ps} = \sqrt{n_sq_s^2/(\varepsilon_0 m_s)}$, ε_0 is the free-space permittivity, and $\Omega_s = q_s B_0 / m_s$ are the plasma and particle cyclotron frequencies, respectively. The volume integral in the velocity space is

$$\int d^3 \vec{u} = 2\pi \int_0^\infty u_\perp \, du_\perp \int_{-\infty}^{+\infty} du_{||}$$

in cylindrical coordinates when the velocity distribution function is assumed to be gyrotropic. Finally $F_s\left(u_{||}, u_\perp\right)$ represents the density normalized particle velocity distribution function for particle s-species that describes the statistical ensemble of particles. Eq. (7.4) has nontrivial solutions only if the determinant of the tensor vanishes, so that

$$\left\| \overset{\leftrightarrow}{D}\left(\vec{k}, \omega\right) \right\| = 0. \tag{7.8}$$

Eq. (7.8) determines an implicit relation between the frequency ω and wave vector $\vec{k}$, called the dispersion relation. It determines the wave frequency–wave vector relation for general linear electromagnetic and electrostatic waves propagating at an arbitrary angle ϑ relative to the average magnetic field in a homogeneous plasma medium. Eq. (7.8) involves the dielectric tensor $\overset{\leftrightarrow}{\varepsilon}\left(\vec{k}, \omega\right)$ in Eq. (7.5), which, in turn, depends on the species velocity distributions $F_s\left(u_{||}, u_\perp\right)$ through Eq. (7.6). Thus by defining the equilibrium particle velocity distributions $F_s\left(u_{||}, u_\perp\right)$ and solving Eq. (7.8) we obtain the general dispersion relation species s to plasmas having charged particles whose statistical behavior is governed by $F_s\left(u_{||}, u_\perp\right)$. This equation is the fundamental expression that will be exploited in this chapter to describe the basic properties of electromagnetic waves propagating in a plasma medium, and the dielectric tensor $\overset{\leftrightarrow}{\varepsilon}\left(\vec{k}, \omega\right)$ is the response of the plasma medium whose properties are characterized by the statistical particle distribution function $F_s\left(u_{||}, u_\perp\right)$. Thus by defining the equilibrium plasma model $F_s\left(u_{||}, u_\perp\right)$ and solving Eq. (7.8) we will obtain the microscopic response characteristics of the wave–particle interaction. The solution of Eq. (7.8) has been generally obtained for the traditional Maxwell–Boltzmann velocity distribution (Montgomery and Tidman, 1964; Stix, 1992; Krall and Trivelpiece, 1975; Ichimaru, 1973; Swanson, 2003; among others). These texts have provided solutions for obliquely propagating as well as particular parallel and perpendicular propagating of electromagnetic and electrostatic waves of the general dispersion relation for a Maxwellian plasma. Therefore we will not describe these thermal solutions here; we refer the reader to the extensive work in the literature and texts on the subject. Instead, in the next sections we will show particular solutions of Eq. (7.8) for the non-Maxwellian drifting bi-kappa distribution defined in Eqs. (7.1) or (7.3). These solutions will emphasize particular wave modes that have been observed in both laboratory as well as astrophysical space plasmas. The emphasis in these next sections is to provide guidance about the effect of suprathermal tails characterized by the κ-index on the dispersive properties and propagation characteristics of low- and high-frequency plasma waves. (The results we provide come from many research papers on this topic, e.g., Summers and Thorne (1991), Mace and Hellberg (1995), Mace (1996, 2003, 2004), Mace et al. (1998, 1999), Viñas et al. (2005, 2014, 2015), Lazar (2012), Lazar and Poedts (2009, 2014), Lazar et al. (2011a,b), Mace and Sydora (2010), Gaelzer and Ziebell (2014, 2016), dos Santos et al. (2014, 2015, 2016) and among many others.)

Unlike the thermal bi-Maxwellian distribution, the bi-kappa distribution cannot be factorized (unlike the Maxwellian exponential that can be factorized); this nonseparability is a nonextensive feature that can be traced back to the nonextensive statistical mechanics that underpins its existence (e.g., Livadiotis and McComas, 2011b). Owing to this nonseparability, the integrations over $u_{||}$ and $u_\perp$ in Eq. (7.6) do not decouple as they do for a bi-Maxwellian. This increases the mathematical complexity of the calculations, and at present there is no known closed form for the dielectric tensor elements in bi-kappa plasma for general oblique wave propagation. In what follows, therefore, we consider the cases of wave propagation parallel and perpendicular to the ambient magnetic field separately, as special cases, concluding with a discussion of the current state of research on the general oblique case.

7.4 Kappa Velocity Distribution Plasma Waves at Parallel Propagation ($\vartheta = 0$)

7.4.1 General Aspects

In general, the exact solutions of Eq. (7.8) can be obtained only numerically. Analytical solutions can be obtained under the assumption of small damping or growth rates γ for each mode, e.g., the Alfvén cyclotron (left-handed) or magnetosonic (right-handed) mode, considering $|\gamma| << |\omega_r| = \mathrm{Re}(\omega)$ to obtain the following relationship (Krall and Trivelpiece, 1973; Melrose, 1986):

$$D_r\left(k_{||}, \omega_r\right) = \mathrm{Re}\left\|\overleftrightarrow{D}\left(k_{||}, \omega_r\right)\right\| = 0,$$

$$\gamma = -\mathrm{Im}\left\|\overleftrightarrow{D}\left(k_{||}, \omega_r\right)\right\| \times \left\{\frac{\partial \mathrm{Re}\left\|\overleftrightarrow{D}\left(k_{||}, \omega\right)\right\|}{\partial \omega}\Bigg|_{\omega=\omega_r}\right\}^{-1}. \quad (7.9)$$

In this section, we focus only on the solution of Eq. (7.8) for a plasma modeled by the kappa velocity distribution function, Eqs. (7.1) or (7.3), and waves propagating parallel (i.e., $\vartheta = 0$) to the mean magnetic field. These solutions are the only current results that have been studied numerically and for which analytical results have been obtained. We consider a multicomponent plasma where each species s has a drifting bi-kappa velocity distribution, given by Eqs. (7.1) or (7.3). We consider only parallel wave propagation so that $k_\perp = 0$ and $k = |k_{||}|$. Under these conditions the arguments of the Bessel functions in T_{sn} in Eq. (7.7) are zero, Eq. (7.6) simplifies, and the dispersion tensor component D_{ij} is reduced to (Krall and Trivelpiece, 1973; Ichimaru, 1973), i.e.,

$$D_{xx}\left(k_{||}, \omega\right) = D_{yy}\left(k_{||}, \omega\right) = 1 + \frac{1}{2}\sum_s \frac{\omega_{ps}^2}{\omega^2}\left\{2(\Gamma_s - 1) + \Gamma_s\right.$$
$$\left. \times\left[\widehat{\xi}_s^- \, Z_\kappa(\xi_s^-) + \widehat{\xi}_s^+ \, Z_\kappa(\xi_s^+)\right]\right\} - \left(\frac{k_{||}c}{\omega}\right)^2$$

$$D_{xy}\left(k_{||}, \omega\right) = -D_{yx}\left(k_{||}, \omega\right) = -\frac{i}{2}\sum_s \frac{\omega_{ps}^2}{\omega^2}\Gamma_s\left[\widehat{\xi}_s^- Z_\kappa(\xi_s^-) - \widehat{\xi}_s^+ Z_\kappa(\xi_s^+)\right]$$

$$D_{zz}\left(k_{||}, \omega\right) = 1 + 2\sum_s \frac{\kappa - \frac{1}{2}}{\kappa}\frac{\omega_{ps}^2}{k_{||}^2 w_{||s}^2}\left[1 + \xi_s^\kappa Z_{\kappa+1}(\xi_s^\kappa)\right]$$

$$D_{xz}\left(k_{||}, \omega\right) = D_{zx}\left(k_{||}, \omega\right) = D_{yz}\left(k_{||}, \omega\right) = D_{zy}\left(k_{||}, \omega\right) = 0, \tag{7.10}$$

where we have defined the thermal anisotropy $\Gamma_s = T_{\perp s}/T_{||s}$ and also the parameters

$$\xi_s = \frac{\omega - k_{||}U_{||s}}{k_{||}w_{||s}}; \xi_s^\pm = \xi_s \mp \frac{\Omega_s}{k_{||}w_{||s}}; \widehat{\xi}_s^\pm = \xi_s \mp \frac{\Omega_s(1 - \Gamma_s^{-1})}{k_{||}w_{||s}}; \xi_{s\kappa} = \sqrt{\frac{\kappa_s + 1}{\kappa_s - \frac{3}{2}}}\xi_s;$$

$$\xi_{s\kappa}^\pm = \sqrt{\frac{\kappa_s}{\kappa_s - \frac{3}{2}}}\xi_s^\pm; \widehat{\xi}_{s\kappa}^\pm = \sqrt{\frac{\kappa_s}{\kappa_s - \frac{3}{2}}}\widehat{\xi}_s^\pm, \tag{7.11}$$

and where the kappa plasma dispersion function $Z_\kappa(\xi)$ has been defined as (Summers and Thorne, 1991)

$$Z_\kappa(\xi) = \frac{\pi^{-\frac{1}{2}}\kappa^{-\frac{1}{2}}\Gamma(\kappa)}{\Gamma\left(\kappa - \frac{1}{2}\right)} \cdot \int_{-\infty}^{+\infty}\left(1 + \frac{1}{\kappa}t^2\right)^{-\kappa}\frac{dt}{t - \xi}, \quad \text{for Im}(\xi) > 0, \tag{7.12}$$

For $\kappa > \frac{1}{2}$, the function $Z_\kappa(\xi)$ can be expressed in terms of the Gauss hypergeometric function $_2F_1(a, b; c; \xi)$ as (Mace and Hellberg, 1995)

$$Z_\kappa(\xi) = i \, \kappa^{-\frac{3}{2}} \left(\kappa - \tfrac{1}{2} \right) \cdot {}_2F_1 \left[1, 2\kappa; \kappa + 1; \frac{1}{2} \left(1 - \frac{\xi}{i\sqrt{\kappa}} \right) \right]. \tag{7.13}$$

These results will be used to illustrate the effects of non-Maxwellian tails represented by the κ-index on the dispersion and propagation properties of various types of plasma waves that represent intrinsic normal mode solutions of a homogeneous plasma.

The dispersion relation for electromagnetic plasma waves propagating parallel to the magnetic field can also be represented in a diagonal form using Eq. (7.4) with tensor elements in Eq. (7.10) to obtain (Krall and Trivelpiece, 1973; Ichimaru, 1973)

$$\left[D_{xx}\left(k_{\parallel}, \omega \right) + i D_{xy}\left(k_{\parallel}, \omega \right) \right] \cdot \left[D_{xx}\left(k_{\parallel}, \omega \right) - i D_{xy}\left(k_{\parallel}, \omega \right) \right]$$
$$= 0 \Leftrightarrow D_{xx}\left(k_{\parallel}, \omega \right) \pm i D_{xy}\left(k_{\parallel}, \omega \right) = 0, \tag{7.14}$$

where these expressions are written in a rotating coordinate system $D_\pm \equiv D_{xx}\left(k_{\parallel}, \omega \right) \pm i D_{xy}\left(k_{\parallel}, \omega \right)$, resulting in the following dispersion relation:

$$D_\pm\left(k_{\parallel}, \omega \right) \equiv 1 - \frac{c^2 k_{\parallel}^2}{\omega^2} + \sum_s \frac{\omega_{ps}^2}{\omega^2} \left[\Gamma_s - 1 + \Gamma_s \xi_s^\mp Z_\kappa(\xi_s^\mp) \right], \tag{7.15}$$

which is a simpler diagonal form. Here, the sign $+(-)$ represents right-R (left-L) circularly polarized waves, respectively. In other words, the left-handed is the Alfvén cyclotron wave mode, and the right-handed is the magnetosonic wave mode.

From the previous expressions it is clear that the properties of the kappa plasma dispersion function will affect the properties of the dispersion relation for the different plasma wave modes to be described below. In seeking analytical approximations to the results based on Eqs. (7.10) or (7.15) above, it is useful to use the power series (small argument) and asymptotic (large argument) expansions of Z_κ given in Eq. (7.13). These are, respectively (Hellberg and Mace, 2002; Mace and Hellberg, 2009; Gaelzer and Ziebell, 2014, 2016),

$$Z_\kappa(\xi) = i \, \pi^{\frac{1}{2}} \frac{\kappa^{-\frac{1}{2}} \Gamma(\kappa)}{\Gamma\left(\kappa - \frac{1}{2} \right)} \left(1 + \frac{1}{\kappa} \xi^2 \right)^{-\kappa} - 2 \left(\frac{\kappa - \frac{1}{2}}{\kappa} \right) \xi \left[1 - \frac{2}{3} \left(\kappa + \frac{1}{2} \right) \frac{\xi^2}{\kappa} \right.$$
$$\left. + \frac{2}{3} \left(\kappa + \frac{1}{2} \right) \frac{2}{5} \left(\kappa + \frac{3}{2} \right) \frac{\xi^4}{\kappa^2} - \cdots \right], \quad |\xi| < 1, \tag{7.16}$$

and

$$Z_\kappa(\xi) = \pi^{\frac{1}{2}} \frac{\kappa^{-\frac{1}{2}}\Gamma(\kappa)}{\Gamma\left(\kappa - \frac{1}{2}\right)} [i - \tan(\pi \kappa)] \cdot \left(1 + \frac{1}{\kappa}\xi^2\right)^{-\kappa}$$

$$- \frac{1}{\xi}\left[1 + \frac{1}{2}\left(\kappa - \frac{3}{2}\right)^{-1}\kappa\xi^{-2} + \frac{1}{2^2}\left(\kappa - \frac{3}{2}\right)^{-1}\left(\kappa - \frac{5}{2}\right)^{-1}\kappa^2\xi^{-4} + \cdots\right],$$

$$\kappa^{-\frac{1}{2}}|\xi| \gg 1. \tag{7.17}$$

It is important to mention that the imaginary terms in both expansions correspond only to the contribution of resonant particles, and the rest of the expansion includes both resonant and nonresonant particles. Eq. (7.15) is valid only for $\text{Im}(\xi) > 0$. The analytic continuation across the $\text{Im}(\xi) = 0$ has been considered, among others, by Mace and Hellberg (1995) and Gaelzer and Ziebell (2016). These results will be used to illustrate the effects of non-Maxwellian tails represented by the κ-parameter on the dispersion and propagation properties of various types of plasma waves that represent intrinsic normal mode solutions of a homogeneous plasma.

7.4.2 Low-Frequency Electromagnetic Alfvén Cyclotron Plasma Waves

The dispersion relation for parallel propagating waves follows from Eq. (7.8), with the D_{ij} given by Eq. (7.10). The general dispersion relation for parallel propagating waves in plasmas with a bi-kappa distribution has been obtained before (e.g., Summers and Thorne, 1991; Mace, 1998; Mace and Sydora, 2010; Mace et al., 2011; Navarro et al., 2015; among others). Here, we specialize to the case of Alfvén cyclotron waves that are driven unstable by thermal anisotropy in the proton bi-kappa distribution.

We consider the stationary plasma of single electron and proton components. The proton component has a stationary bi-kappa distribution given by Eq. (7.1) with $U_{\|p} = 0$ and a finite thermal anisotropy $\Gamma_p \neq 1$. The electron component is assumed to be thermalized, with a stationary ($U_{\|e} = 0$) isotropic (i.e., $\Gamma_e = 1$) Maxwellian distribution. For the electron terms in Eq. (7.10), we can take the limit $\kappa_e \to \infty$ yielding Z (Fried and Conte, 1961) in place of Z_κ. Alfvén cyclotron and magnetosonic waves are low-frequency waves with $\omega < \Omega_p \ll |\Omega_e|$, and their phase velocities are subluminal, i.e., $|\omega/k_\||| < c$. This enables us to obtain approximate results when appropriate. To ensure charge neutrality and zero parallel current conditions both species share equal densities $n_p = n_e$ and no parallel drift $U_{\|p} = U_{\|e} = 0$. Under these conditions, the dispersion relation obtained from the transverse dispersion tensor components (Mace and Sydora, 2010; Mace et al., 2011) in the rotating coordinate system (Eq. 7.15) becomes

$$D_\pm\left(k_\|, \omega\right) \cong 1 - \frac{c^2 k_\|^2}{\omega^2} + \frac{\omega_{pp}^2}{\omega^2}\left[\Gamma_p - 1 + \Gamma_p \xi_{p\kappa_p}^\pm Z_{\kappa_p}\left(\xi_{p\kappa_p}^\pm\right)\right] + \frac{\omega_{pe}^2}{\omega^2}\frac{\omega}{k_\| \theta_{\|e}} Z_e\left(\xi_e^\mp\right) = 0. \tag{7.18}$$

Depending on the local conditions, plasmas can become unstable and trigger microinstabilities (for $\text{Im}(\omega) = \gamma > 0$), associated with plasma wave modes. In the case of circularly polarized waves, the instabilities are particularly sensitive to the temperature anisotropy of the species. This will be examined in the next subsection.

7.4.3 Low-Frequency Instabilities

7.4.3.1 Alfvén Cyclotron Instability

The Alfvén cyclotron mode can be excited by resonant protons when the distribution contains free energy in the form of temperature anisotropy $\Gamma_p > 1$. Assuming cold plasmas, the argument of the dispersion function is large for protons and electrons (provided that the wave frequency is not near a cyclotron harmonic of either species), so that $|\xi_e| > 1$ and $|\xi_p| > 1$, and we can use the expansion for the large argument given by Eq. (7.17). In this limit we can consider protons to be resonant and electrons to be nonresonant; then, we expand $Z(\xi_e) \approx -1/\xi_e$ for electrons, and

$$Z_{\kappa_p}\left(\xi_{p\kappa_p}^{\pm}\right) = \pi^{\frac{1}{2}}\frac{\kappa_p^{-\frac{1}{2}}\Gamma(\kappa_p)}{\Gamma\left(\kappa_p - \frac{1}{2}\right)}\cdot\left[i - \tan\left(\pi\,\kappa_p\right)\right]\cdot\left(1 + \frac{\xi_{p\kappa_p}^{\pm 2}}{\kappa_p}\right)^{-\kappa_p} - \frac{1}{\xi_{p\kappa_p}^{\pm}}, \tag{7.19}$$

for protons (Gaelzer and Ziebell, 2014, 2016). Then, replacing the expressions in Eq. (7.19) into Eq. (7.18) and solving for the left-handed mode, we obtain the dispersion relation for the Alfvén-Cyclotron mode:

$$\omega_r\left(k_{\parallel}\right) = \omega_{AC}\left(k_{\parallel}\right) \cong \frac{1}{2}\Omega_p\left[\sqrt{1 + 4\frac{\omega_{pp}^2}{c^2k_{\parallel}^2}} - 1\right]\cdot\left(\frac{c^2k_{\parallel}^2}{\omega_{pp}^2}\right)$$

$$= \frac{1}{2}\Omega_p k_{\parallel}^2\lambda_p^2\left[\sqrt{1 + 4\frac{1}{k_{\parallel}^2\lambda_p^2}} - 1\right], \tag{7.20}$$

where $\lambda_p = c/\omega_{pp}$ is the proton inertial length (see also Baumjohann and Treumann, 1997a,b). Note that at this level of approximation the frequency of the Alfvén cyclotron mode does not depend on the proton anisotropy or the kappa parameter. In addition, it is easy to show that in the limit $|k_{\parallel}\lambda_p| \gg 1$ this mode is asymptotic to the proton gyro frequency $\omega_{AC}(k_{\parallel} \to \infty) = \Omega_p$.

Similarly, for the imaginary part of the frequency we can compute the damping/growth rate for Alfvén cyclotron waves in an electron–proton plasma. After some algebraic manipulations, we obtain

$$\frac{\gamma_{AC}}{\Omega_p} \cong \pi^{\frac{1}{2}}\frac{\left(\kappa_p - \frac{3}{2}\right)^{-\frac{1}{2}}\Gamma(\kappa_p)}{\Gamma\left(\kappa_p - \frac{1}{2}\right)}\left(1 + \frac{\xi_p^2}{\kappa_p - \frac{3}{2}}\right)^{-\kappa_p}\cdot\frac{(\omega_{AC} - \Omega_p)^2}{\omega_{AC}(\omega_{AC} - 2\Omega_p)}\cdot\frac{\Gamma_p(\omega_{AC} - \Omega_p) + \Omega_p}{k_{\parallel}\,\theta_{\parallel p}},$$

$$\tag{7.21}$$

where $\xi_p = (\omega_{AC} - \Omega_p)/(k_{\parallel}\theta_{\parallel p})$. From Eq. (7.21) we note that $\gamma > 0$ is possible solely for $\Gamma_p > 1$ and only in a limited range in wavenumber due to the κ_p-dependence. This expression reduces to the well-known solution (Krall and Trivelpiece, 1973) for the damping/growth rates of bi-Maxwellian distribution in the limit $\kappa_p \to \infty$, namely,

$$\lim_{\kappa_p \to \infty}\left[\frac{\gamma_{AC}\left(k_{\parallel}\right)}{\Omega_p}\right] = \pi^{\frac{1}{2}}\frac{\left(\omega_{AC} - \Omega_p\right)^2}{\omega_{AC}\left(\omega_{AC} - 2\Omega_p\right)} \cdot \frac{\Gamma_p\left(\omega_{AC} - \Omega_p\right) + \Omega_p}{k_{\parallel}\theta_{\parallel p}} \cdot e^{-\xi_p^2}. \qquad (7.22)$$

The results (see Fig. 7.2) for the frequency $\omega_{AC}(k_{\parallel})$ and growth/damping rate γ_{AC} of Alfvén cyclotron waves show clearly that the real frequency is not (strongly) dependent on the κ-index, whereas the growth and damping rates are more sensitive to kappa. Numerical results showing these solutions and the exact solutions are shown in Fig. 7.2 for different κ_p values. We have considered typical solar wind plasma parameters ($\omega_{pe}/\Omega_e = 100$), and $\Gamma_p = 3$ and $\beta_{\parallel} = 2\mu_0 n_0 k_B T_{\parallel}/B_0^2 = 0.1$ for both protons and electrons. In the figures, the solid and dashed lines correspond to the exact and approximate solutions, respectively.

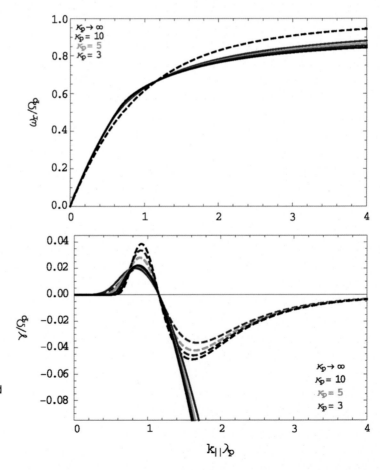

FIGURE 7.2

Frequency (top) and growth/damping rate (bottom) of Alfvén cyclotron waves, depicted as a function of wavelength $k_{\parallel}\lambda_p$ for a proton–electron plasma at different κ_p-indices with a proton temperature anisotropy $\Gamma_p = 3$.

7.4.3.2 Proton Firehose Instability

A second solution, the magnetosonic mode, can be excited by nonresonant particles in high-beta plasmas when the distribution contains free energy in the form of temperature anisotropy $\Gamma_p < 1$. In this case the instability is known as the proton firehose (PF) instability. To obtain an analytical expression for the frequency and growth/damping rate we will assume that the protons are slightly resonant $\left|\xi_p^-\right| \sim O(1)$ but use the large-argument expansion (to the second order) for their contribution. The electrons are considered to be nonresonant $|\xi_e| \gg 1$. They neglect the contribution of resonant electrons and consider only the large argument (nonresonant) asymptotic expansion (up to the second order, neglecting third-order terms) for electrons. Thus from Eq. (7.15) we obtained the dispersion relation for the PF instability, given by

$$D_+\left(k_{||}, \omega\right) \cong -\frac{c^2 k_{||}^2}{\omega^2} + \frac{\omega_{pp}^2}{\omega^2}\left[\Gamma_p - 1 + \Gamma_p \xi_{pK_p}^- Z_{K_p}\left(\xi_{pK_p}^-\right)\right] - \frac{\omega_{pe}^2}{\omega^2}\frac{\omega}{\omega - |\Omega_e|} = 0,$$

(7.23)

where we have assumed that the electron term is given by

$$Z_{K_e}\left(\xi_{eK_e}^-\right) \cong \lim_{K_e \to \infty}\left\{ -\frac{1}{\xi_{eK_e}^-}\cdot\left[1 + \frac{K_e}{K_e - \frac{3}{2}}\cdot\frac{1}{2}\left(\xi_{eK_e}^-\right)^{-2}\right]\right\} \cong -\frac{1 + \frac{1}{2}\left(\xi_e^-\right)^{-2}}{\xi_e^-}, \quad (7.24)$$

and assumed that $\omega \ll |\Omega_e|$ always. For the protons we assumed that

$$Z_{K_p}\left(\xi_{pK_p}^-\right) \cong -\frac{1}{\xi_{pK_p}^-}\cdot\left[1 + \frac{K_p}{K_p - \frac{3}{2}}\cdot\frac{1}{2}\left(\xi_{pK_p}^-\right)^{-2}\right] \cong -\frac{k_{||}w_{||p}}{\omega + \Omega_p}\cdot\left[1 + \frac{\frac{1}{2}k_{||}^2\theta_{||}^2}{(\omega + \Omega_p)^2}\right].$$

(7.25)

Then, replacing these expressions into the proton contribution of the dispersion relation, and after some tedious algebraic manipulations, we obtain an accurate approximate expression in terms of a fourth-order polynomial whose roots give the PF solution:

$$\left[\left(\frac{m_e}{m_p}g^2 + 1\right)z_r^2 - g^2(z_r + 1)\right](z_r + 1)^2 + \frac{1}{2}\beta_{||p}g^2\left[\Gamma_p(z_r + 1) - 1\right]\left(\frac{m_e}{m_p}z_r - 1\right) = 0,$$

(7.26)

where $z_r \equiv \omega_r/\Omega_p$ and $g \equiv k_{||}\lambda_p = k_{||}c/\omega_{pp}$. The solutions of this polynomial have been investigated, and there is only one real physical solution:

$$\omega_r\left(k_{||}\right) = \omega_{PF}(k)_{||} \cong \Omega_p \cdot \left\{\sqrt{1 - \frac{1}{2}(1 - \Gamma_p)\beta_{||p}}\cdot g + \frac{1}{2}\left[1 - \frac{1}{2}(1 - \Gamma_p)\beta_{||p}\right]\cdot g^2\right.$$

$$\left. + \frac{\beta_{||p}^2\left[16 - 5\Gamma_p(4 - \Gamma_p)\right] - 4\beta_{||p}(4 - \Gamma_p) + 4}{32\sqrt{1 - \frac{1}{2}(1 - \Gamma_p)\beta_{||p}}}\cdot g^3 + \frac{1}{4}\beta_{||p}\left\{1 - 5\beta_{||p}\left[1 - (5 - \Gamma_p)\Gamma_p\right]\right\}\cdot g^4\right\}.$$

(7.27)

This solution is the correct expression to order $O(g^4)$ in the quasifluid theory. It also demonstrates that the firehose mode is really a quasikinetic mode with fluid-collective behavior, rather than a fully kinetic resonant mode. Thus this property justifies the nonresonant asymptotic approximation used for protons. Note that in the limit, as g approaches infinity (i.e., $k_\| \to \infty$), the approximate solution of Eq. (7.27) is indistinguishable from the numerically exact right-handed solution:

$$\omega\left(k_\|\right) = \lim_{k_\| \to \infty} \left[\omega_r\left(k_\|\right)\right] \cong |\Omega_e| \left(1 - \frac{\omega_{pe}^2}{k_\|^2 c^2}\right). \tag{7.28}$$

There are two possible branches to the physical solution in Eq. (7.27). The first branch is when $\beta_{\|p} \geq 2\big/(1 - \Gamma_p)$. In this case the odd terms of Eq. (7.27) are imaginary, and the dispersion relation is quadratic for small values of g, yielding the result

$$\omega_{PF}\left(k_\|\right) \cong \Omega_p \cdot g^2 \cdot \frac{c_1 + g}{1 + g + \dfrac{m_e}{m_p} c_1 g^2 + \dfrac{m_e}{m_p} g^3},$$

$$c_1 = \frac{1}{2}\left[1 + \beta_{\|p}\left(1 - \frac{1}{2}\Gamma_p\right)\right] + \frac{1}{4}\frac{m_e}{m_p}\beta_{\|p}(1 - \Gamma_p). \tag{7.29}$$

The second branch is when $\beta_{\|p} < 2\big/(1 - \Gamma_p)$, and now the dispersion relation is linear for small g values, yielding the result

$$\omega_{PF}\left(k_\|\right) \cong \Omega_p \cdot g \cdot \frac{c_2 + g}{1 + \dfrac{m_e}{m_p} c_2 g + \dfrac{m_e}{m_p} g^2}, \quad c_2 = \sqrt{1 - \frac{1}{2}\beta_{\|p}(1 - \Gamma_p)}. \tag{7.30}$$

These results have been obtained by performing a Padé approximant expansion (Baker and Engelmann, 1975) on Eq. (7.27) for each branch case. With these approximate solutions we can now resolve for the imaginary part of the frequency, i.e.,

$$\frac{\gamma\left(k_\|\right)}{\Omega_p} = \frac{\gamma_{PF}\left(k_\|\right)}{\Omega_p} \cong \pi^{\frac{1}{2}} \frac{(\omega_{PF} + \Omega_p)^2}{\beta_{\|e}^{\frac{1}{2}} k_\| \lambda_p} \cdot \frac{1 - (1 + \omega_{PF}/\Omega_p)\Gamma_p}{\omega_{PF}(\omega_{PF} + 2\Omega_p)} \cdot \frac{\left(\kappa_p - \dfrac{3}{2}\right)^{-\frac{1}{2}} \Gamma(\kappa_p)}{\Gamma\left(\kappa_p - \dfrac{1}{2}\right)}$$

$$\left[1 + \frac{1}{\kappa_p - \dfrac{3}{2}} \cdot \frac{(1 + \omega_{PF}/\Omega_p)^2}{\beta_{\|p} k_\|^2 \lambda_p^2}\right]^{-\kappa_p}.$$

$$\tag{7.31}$$

This expression allows unstable (positive) mode solutions when

$$\left(\omega_{PF} + \Omega_p\right)\Gamma_p - \Omega_p < 0 \Rightarrow \frac{\omega_{PF}\left(k_{\|}\right)}{\Omega_p} < \Gamma_p^{-1} - 1. \tag{7.32}$$

If the conditions $\beta_{\|p} \geq 2 / \left(1 - \Gamma_p\right)$ and $\Gamma_p < 1$ are satisfied, then instabilities will be constrained to the range

$$k_{\|}\lambda_p \leq \sqrt{\frac{2\left(1 - \Gamma_p\right)}{\Gamma_p\left[1 + \beta_{\|p}\left(1 - \frac{1}{2}\Gamma_p\right)\right]}}. \tag{7.33}$$

On the other hand, for the low-β_p conditions $\beta_{\|p} < 2 / \left(1 - \Gamma_p\right)$ and $\Gamma_p < 1$, the domain of instabilities will occur for wavelengths in the range

$$k_{\|}\lambda_p \leq \sqrt{\frac{1 - \Gamma_p}{\Gamma_p\left[1 - \frac{1}{2}\beta_{\|p}\left(1 - \Gamma_p\right)\right]}}. \tag{7.34}$$

Thus our results predict that the PF instability exists in two possible regimes, and it can be excited only for $\Gamma_p < 1$. In the limit $\kappa_p \to \infty$ these expressions yield the threshold condition for PF instability in bi-Maxwellian plasmas:

$$\lim_{\kappa_p \to \infty}\left[\frac{\gamma_{PF}\left(k_{\|}\right)}{\Omega_p}\right] = \pi^{\frac{1}{2}}\frac{\left(1 + \omega_{PF}/\Omega_p\right)^2}{\beta_{\|p}^{\frac{1}{2}}k_{\|}\lambda_p} \cdot \frac{1 - \left(1 + \omega_{PF}/\Omega_p\right)\Gamma_p}{\left(\omega_{PF}/\Omega_p\right) \cdot \left(2 + \omega_{PF}/\Omega_p\right)}$$

$$\times \exp\left[-\frac{\left(1 + \omega_{PF}/\Omega_p\right)^2}{\beta_{\|p}k_{\|}^2\lambda_p^2}\right]. \tag{7.35}$$

The results for the frequency $\omega_{PF}(k_{\|})$ and growth/damping rate γ_{PF} for the PF instability exist for both the high- and low-β_p regimes and depends on the κ-indices. A comparison of both the analytical and exact numerical solutions for both regimes is shown in Fig. 7.3 (left) and (right) for different κ_p-indices. The left panel contains the dispersion relation solution for the case of $\beta_{\|p} \geq 2 / \left(1 - \Gamma_p\right)$ and $\Gamma_p < 1$, and the right panel contains the dispersion properties for the case of $\beta_{\|p} < 2 / \left(1 - \Gamma_p\right)$ and $\Gamma_p < 1$. The figure also shows a comparison of the exact numerical solution (solid lines) in Eq. (7.23) with the approximate polynomial solution via Padé approximants (dotted lines) in Eqs. (7.27) and (7.30). The results clearly show that at least within the range of $0 \leq k_{\|}\lambda_p \leq 3$, the Padé approximant solution provides a fairly good approximation when compared with the exact numerical solution. The PF instability in this regime shows both fluid collective and kinetic behavior

(Melrose, 1986). The imaginary part of the solution (dotted lines) clearly shows that at least for long wavelengths (small $k_\parallel \lambda_p < 1$), the fluid collective behavior is apparent, whereas for short wavelengths (large $k_\parallel \lambda_p \geq 1$), the kinetic behavior is present. The dispersion results also show that the frequency of the firehose instability is weakly dependent on the κ_p-index, but the imaginary part shows a slight enhancement as the κ_p-index decreases. The approximate solutions in both plots show similar spectral ranges for instability as the exact numerical solution.

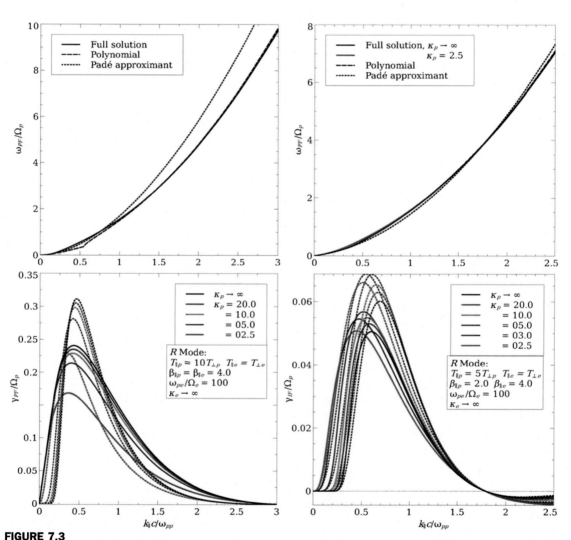

FIGURE 7.3

Proton firehose instability frequency (top) and growth/damping rate (bottom) as a function of wavelength $k_\parallel \lambda_p$ for a proton–electron plasma at different κ_p values. Left: high-β_p regime with a proton temperature anisotropy $\Gamma_p = 0.1$. Right: low-β_p regime with an anisotropy $\Gamma_p = 0.2$.

7.4.4 High-Frequency Electromagnetic Whistler Cyclotron Plasma Waves

In this case we will consider plasma composed of a single component of electrons and protons whose velocity distribution functions are modeled by equations by a drifting bi-kappa distribution, Eq. (7.1). The source of the instability is the electron thermal anisotropy in the form of $\Gamma_e > 1$, and the electrons' velocity distributions are represented by a small but finite kappa index, κ_e. Protons, on the other hand, will be assumed to be essentially Maxwellian-like particles, isotropic ($\Gamma_p = 1$), and with κ_p values approaching infinity, so that the modified dispersion function Z_κ reduces to the Maxwellian Z-function (Fried and Conte, 1961). To ensure charge neutrality and zero parallel current conditions, both species share a density $n_p = n_e$ and a null parallel drift $U_{\parallel p} = U_{\parallel e} = 0$. Here, we also consider high-frequency transverse electromagnetic waves, called whistler cyclotron (WC) waves and magnetosonic waves. As in Section 7.3 for the Alfvén cyclotron and PF waves, and under these aforementioned conditions, the dispersion relation for WC waves is obtained from the transverse dispersion tensor components (e.g., Mace, 1998; Mace and Sydora, 2010; Viñas et al., 2014, 2015) $D_\pm = D_{xx}\left(k_\parallel, \omega\right) \pm i D_{xy}\left(k_\parallel, \omega\right)$ in Eqs. (7.16) and (7.17):

$$D_\pm\left(k_\parallel, \omega\right) \cong -\frac{c^2 k_\parallel^2}{\omega^2} + \frac{\omega_{pe}^2}{\omega^2} \cdot \left[\Gamma_e - 1 + \Gamma_e \xi_{e\kappa_e}^\pm Z_{\kappa_p}\left(\xi_{e\kappa_e}^\pm\right)\right] + \frac{\omega_{pp}^2}{\omega^2} \frac{\omega}{k_\parallel \theta_{\parallel p}} Z_p\left(\xi_p^\pm\right) = 0. \tag{7.36}$$

Again, the sign $+(-)$ represents right (left) circularly polarized waves, respectively. This equation exhibits two different instabilities: the resonant right-handed WC instability and the nonresonant left-handed electron firehose (EF), whose exact solutions can be obtained numerically, as shown in Figs. 7.3 and 7.4. An analytical expression can be obtained under some assumptions, as in the previous section.

7.4.5 High-Frequency Instabilities

7.4.5.1 Whistler Cyclotron Instability

Here, we consider cold plasma and right-handed waves with frequencies in the vicinity of $\omega \approx |\Omega_e| >> \Omega_p$. Thus the proton contribution can be neglected (i.e., $Z(\xi_p) \approx 0$), and for electrons we consider the large argument expansion for $Z_{\kappa e}$, including the resonant particles. Using Eq. (7.36) and considering small growth/damping rates, we obtain the following approximate relationships for the frequency and growth/damping rate of the WC instability:

$$\omega_r\left(k_\parallel\right) = \omega_{WC}\left(k_\parallel\right) \cong |\Omega_e| \cdot \frac{k_\parallel^2 \lambda_e^2}{1 + k_\parallel^2 \lambda_e^2}, \tag{7.37}$$

and

$$\frac{\gamma_{WC}(k_\parallel)}{|\Omega_e|} \cong -\pi^{\frac{1}{2}} \frac{\left(\kappa_e - \frac{3}{2}\right)^{-\frac{1}{2}} \Gamma(\kappa_e)}{\Gamma\left(\kappa_e - \frac{1}{2}\right)} \left(1 + \frac{\xi_e^2}{\kappa_e - \frac{3}{2}}\right)^{-\kappa_e} \cdot \left(\frac{\omega_{WC}}{|\Omega_e|} - 1\right)^2 \frac{\Gamma_e(\omega_{WC} - |\Omega_e|) + |\Omega_e|}{k_\parallel \theta_{\parallel e}},$$

(7.38)

where $\xi_e = (\omega_{WC} - |\Omega_e|)/(k_\parallel \theta_{\parallel e})$. From Eq. (7.38) we note that $\gamma_{WC} > 0$ is possible solely when $\Gamma_e > 1$ and

$$\omega_{WC} < |\Omega_e|\left(1 - \Gamma_e^{-1}\right) \Rightarrow k_\parallel^2 \lambda_e^2 < \Gamma_e - 1$$

and only in a limited range in wavenumber due to the κ_e dependence, similar to the Alfvén cyclotron instability case. This expression reduces to the well-known

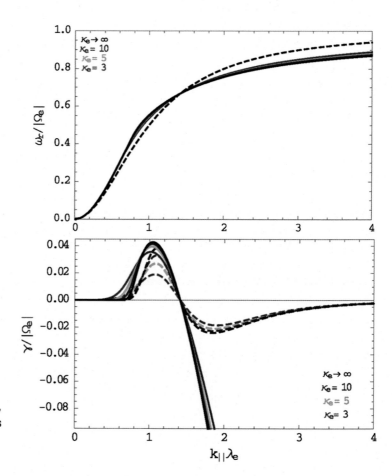

FIGURE 7.4

Electron whistler cyclotron instability frequency (top) and growth/damping rate (bottom) as a function of wavelength $k_\parallel \lambda_e$ for a proton–electron plasma with various κ_e-indices and temperature anisotropy $\Gamma_e = 3$.

solution for the damping/growth rates of bi-Maxwellian distribution in the limit $\kappa_e \to \infty$, namely,

$$\lim_{\kappa_e \to \infty} \left[\frac{\gamma_{WC}(k_{||})}{|\Omega_e|} \right] = -\pi^{\frac{1}{2}} \left(\frac{\omega_{WC}}{|\Omega_e|} - 1 \right)^2 \left[\frac{\Gamma_e(\omega_{WC} - |\Omega_e|) + |\Omega_e|}{k_{||}\,\theta_{||e}} \right] e^{-\xi_e^2}. \qquad (7.39)$$

The results (Fig. 7.4) for the frequency $\omega_{WC}(k_{||})$ and growth/damping rate γ_{WC} of the WC waves show clearly that the real frequency is not strongly dependent on the κ-index, whereas the growth and damping rates are more sensitive to kappa. Numerical results show these solutions, while the exact solutions are shown in Fig. 7.4 (for various κ_e values). We have considered typical solar wind plasma parameters ($\omega_{pe}/\Omega_e = 100$), and $\Gamma_e = 3$ and $\beta_{||} = 0.1$ for both protons and electrons. In the figures, the solid and dashed lines correspond to the exact and approximate solutions, respectively.

Mace and Sydora (2010) conducted a thorough survey of the whistler instability driven by electron thermal anisotropy. They found a rather complex dependence of the instability growth rate on κ_e-indices that either increases or decreases with this parameter, depending on the electron $\beta_{||e}$. While Fig. 7.4 depicts a case where the maximum growth rate of the Maxwellian plasma exceeds that of the kappa plasmas, Mace and Sydora (2010) delineate regions in the parameter space where the reverse is true. The precise behavior of the growth rate with κ_e is strongly dependent on the electron $\beta_{||e}$.

7.4.5.2 Electron Firehose Instability

Similar to the PF, EF instability is a nonresonant instability in high-beta plasmas, but in this case it corresponds to a left-handed mode that is unstable when $\Gamma_e < 1$. To obtain analytical expressions we begin with the dispersion relation:

$$D_-\left(k_{||}, \omega\right) \cong -\frac{c^2 k_{||}^2}{\omega^2} + \frac{\omega_{pe}^2}{\omega^2}\left[\Gamma_e - 1 + \Gamma_e \xi_{e\kappa_e}^+ Z_{\kappa_p}\left(\xi_{e\kappa_e}^+\right)\right] + \frac{\omega_{pp}^2}{\omega^2}\frac{\omega}{k_{||}\theta_{||p}}Z_p\left(\xi_p^-\right) = 0. \qquad (7.40)$$

Here, we neglect the contribution of nonresonant protons and resonant electrons, and we consider the large argument (nonresonant) asymptotic expansion for electrons. Then, inserting the asymptotic expressions for a large argument up to ξ^{-2} for electrons in Eq. (7.24) and considering the low frequencies of $\omega \ll \Omega_p$, that one obtains after some algebraic manipulations,

$$D_-\left(k_{||}, \omega\right) \simeq \frac{c^2 k_{||}^2}{\omega^2} + \frac{\omega_{pe}^2}{\omega^2}\left[\frac{\omega}{|\Omega_e|} - \frac{1}{2}(1 - \Gamma_e)\frac{k_{||}^2 \theta_{||e}^2}{\Omega_e^2}\right] - i\,\pi^{\frac{1}{2}}\frac{\omega_{pp}^2}{\omega^2}\frac{\omega}{k_{||}\theta_{||p}} = 0, \qquad (7.41)$$

resulting in the dispersion frequency for the EF instability (assuming ω is not close to a harmonic of $|\Omega_p|$):

$$\frac{\omega_r}{\Omega_p} = \frac{\omega_{EF}\left(k_{||}\right)}{\Omega_p} \cong -\left(1 + \frac{\pi}{k_{||}^2 \theta_{||p}^2}\frac{\Omega_p^2}{}\right)^{-1} \cdot \left[1 - \frac{1}{2}\beta_{||e}(1 - \Gamma_e)\right] \cdot \frac{k_{||}^2 c^2}{\omega_{pp}^2}, \tag{7.42}$$

and the growth/damping rate

$$\frac{\gamma\left(k_{||}\right)}{\Omega_p} = \frac{\gamma_{EF}\left(k_{||}\right)}{\Omega_p} \cong -\frac{\pi^{\frac{1}{2}}\Omega_p}{k_{||}\,\theta_{||p}}\left(1 + \frac{\pi}{k_{||}^2 \theta_{||p}^2}\frac{\Omega_p^2}{}\right)^{-1} \cdot \left[1 - \frac{1}{2}\beta_{||e}(1 - \Gamma_e)\right] \cdot \frac{k_{||}^2 c^2}{\omega_{pp}^2}. \tag{7.43}$$

Eq. (7.43) allows positive values (unstable modes) when

$$\beta_{||e} > \frac{2}{1 - \Gamma_e}. \tag{7.44}$$

Thus the EF can be excited only for $\Gamma_e < 1$ and large enough $\beta_{||e}$. The instability has a very weak dependence on κ_e and is equivalent to the PF. The results (Fig. 7.5) for the frequency $\omega_{EF}(k_{||})$ and growth/damping rate γ_{EF} for the EF instability show clearly that for this mode, both the real frequency and the damping/growth are somewhat weakly dependent on the κ-indices. This behavior remains even in the general case obtained by numerical integrations of Eq. (7.40), as shown in Fig. 7.5 for different κ_e-indices. We have considered typical solar wind plasma parameters ($\omega_{pe}/\Omega_e = 100$), $\Gamma_e = 0.25$, $\beta_{||e} = 4$ for electrons, and $\beta_{||p} = 0.1$ for protons. In the figures, the solid and dashed lines correspond to the exact and approximate solutions, respectively.

7.4.6 High-Frequency Electromagnetic Langmuir Plasma Waves

In this subsection we will consider a plasma composed of two component electrons whose velocity distribution function is given by Eq. (7.1) and a single Maxwellian proton component. One electron component, called a core, and is characterized by plasma properties such as n_{ec} and T_{ec}, which is isotropic (i.e., $\Gamma_{ec} = 1$, $T_{ec} = T_{||ec} = T_{\perp ec}$) and has no relative parallel drift $U_{||ec} = 0$. The other electron component is called the beam with plasma properties n_{eb} and T_{eb}, which is isotropic (i.e., $\Gamma_{eb} = 1$, $T_{eb} = T_{||eb} = T_{\perp eb}$) and has finite relative parallel drift $U_{||eb}$. The protons, on the other hand, will be assumed to be infinitely massive and relatively cold with Maxwellian-like particles and κ-indices approaching infinity. Here, we have only considered a class of high-frequency, longitudinal, electrostatic collective oscillations called Langmuir waves (Mace and Hellberg, 1995; Mace et al., 1998) whose frequency ω lies above ω_{pe}. Under these aforementioned conditions, the dispersion relation for electrostatic Langmuir waves is obtained from the longitudinal dispersion tensor component ($D_{zz} = 0$) in Eq. (7.10). Using Eqs. (7.11)–(7.15), we get

$$
D_{zz}\left(k_{\|},\omega\right) \cong 1 + 2\left(\frac{\kappa_{ec} - \frac{1}{2}}{\kappa_{ec} - \frac{3}{2}}\right)\frac{\omega_{pec}^2}{k_{\|}^2\theta_{\|ec}^2}\left[1 + \xi_{\kappa ec}Z_{\kappa+1}(\xi_{\kappa ec})\right]
$$

$$
+ 2\left(\frac{\kappa_{eb} - \frac{1}{2}}{\kappa_{eb} - \frac{3}{2}}\right)\frac{\omega_{peb}^2}{k_{\|}^2\theta_{\|eb}^2}\left[1 + \xi_{\kappa eb}Z_{\kappa+1}(\xi_{\kappa eb})\right] = 0,
$$

(7.45)

where we defined for the electron core (ec) and electron beam (eb) components:

$$
\xi_{ec} = \frac{\omega}{k_{\|}\theta_{\|ec}};\ \xi_{ec\kappa} = \left(\frac{\kappa_{ec} + 1}{\kappa_{ec} - \frac{3}{2}}\right)^{\frac{1}{2}}\xi_{ec};\ \xi_{eb} = \frac{\omega - k_{\|}U_{\|eb}}{k_{\|}\theta_{\|eb}};\ \xi_{eb\kappa} = \left(\frac{\kappa_{eb} + 1}{\kappa_{eb} - \frac{3}{2}}\right)^{\frac{1}{2}}\xi_{eb}.
$$

(7.46)

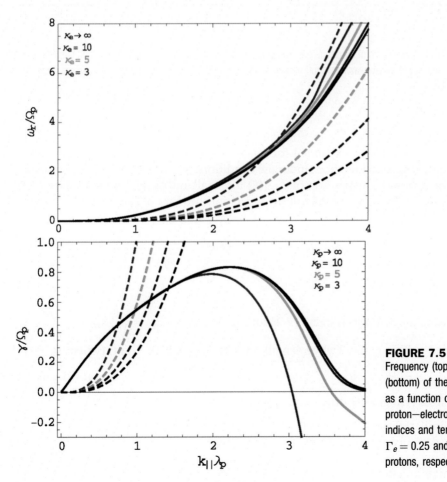

FIGURE 7.5
Frequency (top) and growth/damping rate (bottom) of the electron firehose instability as a function of wavelength $k_{\|}\lambda_p$ for a proton–electron plasma at different κ_e-indices and temperature anisotropies of $\Gamma_e = 0.25$ and $\Gamma_p = 0.1$ for electrons and protons, respectively.

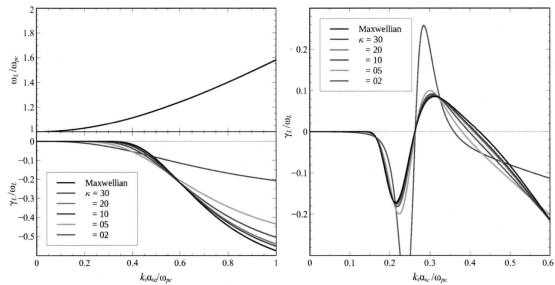

FIGURE 7.6

Langmuir waves versus normalized wavelength of a proton–electron plasma for the following cases: Left: dispersion curves (top) and growth/damping (bottom) of nonelectron beam plasma. Right: growth/damping of a drifting electron beam. All plots are depicted for various κ-indices (including Maxwellian).

Eq. (7.45) can be expanded for $|\gamma| \ll \omega_r$ to obtain a small damping approximate dispersion relation under the conditions $|\xi_{ec}| \gg 1$ and $|\xi_{eb}| \le 1$ and using the asymptotic expansion of $Z_{\kappa+1}(\xi)$ for small and large arguments, Eqs. (7.16) and (7.17), to obtain the following relationship for the Langmuir (L) frequency (Mace and Hellberg, 1995; Mace et al., 1998):

$$\omega_r\left(k_{\|}\right) \equiv \omega_L\left(k_{\|}\right) \cong \omega_{pec}\sqrt{1 + \frac{3}{2}\frac{\kappa_{ec}}{\kappa_{ec} - \frac{3}{2}}\frac{k_{\|}^2 w_{\|ec}^2}{\omega_{pec}^2}} = \omega_{pec}\sqrt{1 + 3k_{\|}^2\lambda_{De\infty}^2}, \quad (7.47)$$

using the definitions in Eq. (7.2) for the velocities $w_{\|ec} = \sqrt{\left(\kappa_{ec} - \frac{3}{2}\right)\Big/\kappa_{ec}}\cdot\theta_{\|ec}$ and defining the Debye length in the Maxwellian limit (when $\kappa_e \to \infty$), $\lambda_{De\infty}^2 = \frac{1}{2}\theta_{\|ec}^2\Big/\omega_{pec}^2 = \sqrt{\varepsilon_0 e^{-2}k_B T_{\|ec}\Big/n_{ec}}$. Here, $\omega_L\left(k_{\|}\right)$ denotes the Bohm-Gross (Langmuir) dispersion relation.

Similarly, we can compute the damping/growth rate for Langmuir waves in a core beam κ-plasma system, using the small damping/growth approximation in Section 7.3 (Eq. 7.9) to give, after some tedious algebraic manipulations,

$$\frac{\gamma}{\omega_L} \cong -\sqrt{\pi}\,\frac{\omega_L^2}{k_{\parallel}^2}\left\{\frac{\omega_L}{k_{\parallel}\theta_{\parallel ec}^3}A(\kappa_{ec})\left[1+\frac{\omega_L^2}{\left(\kappa_{ec}-\frac{3}{2}\right)k_{\parallel}^2\theta_{\parallel ec}^2}\right]^{-(\kappa_{ec}+1)}\right.$$

$$\left.+\frac{n_{eb}}{n_{ec}}\cdot\frac{\omega_L-k_{\parallel}U_{\parallel eb}}{k_{\parallel}\theta_{\parallel eb}^3}A(\kappa_{eb})\left[1+\frac{\left(\omega_L-k_{\parallel}U_{\parallel eb}\right)^2}{\left(\kappa_{eb}-\frac{3}{2}\right)k_{\parallel}^2\theta_{\parallel eb}^2}\right]^{-(\kappa_{eb}+1)}\right\}.$$

(7.48)

This expression reduces to the well-known solution for the damping/growth rates of the Maxwellian distribution in the limit as κ_{ec} and κ_{eb} approach infinity, yielding the solution (Krall and Trivelpiece, 1973)

$$\lim_{\kappa_{ec},\kappa_{eb}\to\infty}\left(\frac{\gamma}{\omega_L}\right)=-\sqrt{\pi}\,\frac{\omega_L^2}{k_{\parallel}^2}\left\{\frac{\omega_L}{k_{\parallel}\theta_{\parallel ec}^3}\exp\left(-\frac{\omega_L^2}{k_{\parallel}^2\theta_{\parallel ec}^2}\right)+\frac{n_{eb}}{n_{ec}}\cdot\frac{\omega_L-k_{\parallel}U_{\parallel eb}}{k_{\parallel}\theta_{\parallel eb}^3}\exp\right.$$

$$\left.\times\left[-\frac{\left(\omega_L-k_{\parallel}U_{\parallel eb}\right)^2}{k_{\parallel}^2\theta_{\parallel eb}^2}\right]\right\}.$$

(7.49)

The analytical results in Eqs. (7.47) and (7.48) for the frequency $\omega_L\left(k_{\parallel}\right)$ and a growth/damping rate γ/ω_L of the Langmuir oscillations show clearly, to the order of approximation used, that there is no κ_e dependence visible for the real frequency, whereas there is a κ_e dependence for the approximate growth/damping rate analytical expression. Numerical results that show Langmuir wave solutions are shown in Fig. 7.6.

Fig. 7.6 (left) shows plots of the Langmuir waves damping rate as a function of $k_{\parallel}$ for a kappa proton–electron plasma system with no beam (i.e., $n_{eb}=0$) or no free energy of any kind. These plots can be compared with Fig. 7.6 (right) that displays plots of the growth/damping rate of the Langmuir waves as a function of $k_{\parallel}$ for a kappa/Maxwellian proton–electron–electron beam plasma system with a parallel drift $U_{\parallel eb}/\theta_{\parallel ec}=4$ and for several values of κ. The real frequency is not shown since it is essentially the same as the plot to the left.

7.4.7 Low-Frequency Electromagnetic Ion-Acoustic Plasma Waves

In this case we will consider a plasma composed of single electrons and two proton components (three populations) whose velocity distribution functions are modeled by Eqs. (7.1) or (7.3). The ion (protons) inertia is also important; however, all components contribute to the dynamics. The instability is now due to a proton beam population, whereas the electrons are assumed to be described by a single velocity distribution function. We will assume the proton core (pc) population to be at rest, the proton beam (pb) to be drifting, and the electrons (e) also to be drifting so that the total parallel current vanishes. As in Section 7.4.5 for the Langmuir waves and under these aforementioned conditions, the dispersion relation for electrostatic ion-acoustic waves is obtained from the longitudinal dispersion tensor component D_{zz} in Eq. (7.9) using Eqs. (7.11)–(7.15) (Mace et al., 1999; Mace, 2003). The solution can be similarly expanded to obtain the following relationship from Eq. (7.10):

$$D_{zz}\left(k_{||}, \omega\right) \cong 1 + \frac{\kappa_{pc} - \frac{1}{2}}{\kappa_{pc} - \frac{3}{2}} \cdot \frac{2\omega_{ppc}^2}{k_{||}^2 \theta_{||pc}^2}\left[1 + \xi_{\kappa pc} Z_{\kappa+1}\left(\xi_{\kappa pc}\right)\right]$$

$$+ \frac{\kappa_{pb} - \frac{1}{2}}{\kappa_{pb} - \frac{3}{2}} \cdot \frac{2\omega_{ppb}^2}{k_{||}^2 \theta_{||pb}^2}\left[1 + \xi_{\kappa pb} Z_{\kappa+1}\left(\xi_{\kappa pb}\right)\right] + \frac{\kappa_e - \frac{1}{2}}{\kappa_e - \frac{3}{2}} \cdot \frac{2\omega_{pe}^2}{k_{||}^2 \theta_{||e}^2}\left[1 + \xi_{\kappa e} Z_{\kappa+1}(\xi_{\kappa e})\right] = 0,$$

$$(7.50)$$

where we defined the following arguments for the proton core, proton beam, and electron components:

$$\xi_{pc} = \frac{\omega}{k_{||}\theta_{||pc}}; \ \xi_{pc\kappa} = \sqrt{\frac{\kappa_{pc} + 1}{\kappa_{pc} - \frac{3}{2}}}\xi_{pc}; \ \xi_{pb} = \frac{\omega - k_{||}U_{||pb}}{k_{||}\theta_{||pb}}; \ \xi_{pb\kappa} = \sqrt{\frac{\kappa_{pb} + 1}{\kappa_{pb} - \frac{3}{2}}}\xi_{pb};$$

$$\xi_e = \frac{\omega - k_{||}U_{||e}}{k_{||}\theta_{||e}}; \ \xi_{e\kappa} = \sqrt{\frac{\kappa_e + 1}{\kappa_e - \frac{3}{2}}}\xi_e.$$

$$(7.51)$$

In the low-frequency regime ($|\omega| < \omega_{pe}$), there are two modes that result from Eq. (7.43) that can become unstable. The instabilities are (e.g., Gary, 1985, 1993) the electron/ion-acoustic instability, which is driven by the relative drift between the ion core and the electrons, and the ion/ion-acoustic instability, which is driven by the relative ion core beam. Another instability, which arises when the zero current condition is not met, is the Buneman instability (Melrose, 1986); this instability will not be further discussed here.

In order to find an approximate solution to Eq. (7.43), we will assume that the phase velocity of the unstable modes is fast with respect to the protons' thermal speed but is slow with respect to the electron thermal speed, so they satisfy the conditions $\theta_{||pc,pb} < |\omega/k_{||}| < \theta_{||e}$, $n_{pb} < n_{pc}$, $n_{pc} \cong n_e$, $U_{||e} = (n_{pb}/n_e)U_{||pb} << U_{||pb}$, and $U_{||e} << \theta_{||e}$. These approximations enable us to use the large argument expansion in Eq. (7.17) of the Z_κ function in the proton beam and core component terms and the small argument expansion in Eq. (7.16) of the Z_κ function in the electron component term in Eq. (7.50). Using the small damping/growth approximation $|\gamma| << \omega_r = \text{Re}(\omega)$ and $\omega_r \approx k_{||}c_{Sk}$, we obtained the frequency of the ion/electron acoustic modes:

$$\omega_r^2\left(k_{||}\right) = \omega_{IA}^2\left(k_{||}\right) \cong \frac{k_{||}^2 c_{Sk}^2}{1 + k_{||}^2 \lambda_D^2} \cdot \left(1 + 3\frac{\kappa_e - \frac{1}{2}}{\kappa_e - \frac{3}{2}}\frac{T_{||pc}}{T_{||e}}\right), \qquad (7.52)$$

assuming that the beam can be neglected since this component contributes negligibly to the real frequency and ignoring the thermal effects of the core ions, since $k_{\parallel}\theta_{\parallel pc} \ll \omega_r$. In Eq. (7.52) we have also defined (e.g., see also Bryant, 1996; Rubab and Murtaza, 2006; Gougam and Tribeche, 2011; Livadiotis and McComas, 2014a)

$$
c_{S\kappa}^2 = \frac{1}{2}\frac{m_e}{m_p}\left(\frac{\kappa_e - \frac{3}{2}}{\kappa_e - \frac{1}{2}}\right)\theta_{\parallel e}^2, \quad \lambda_{De}^2 = \left(\frac{\kappa_e - \frac{3}{2}}{\kappa_e - \frac{1}{2}}\right)\cdot\frac{1}{2}\frac{\theta_{\parallel e}^2}{\omega_{pe}^2} = \left(\frac{\kappa_e - \frac{3}{2}}{\kappa_e - \frac{1}{2}}\right)\cdot\lambda_{De\infty}^2,
$$

(7.53)

where $c_{S\kappa}$ and λ_D are the κ-dependent ion-acoustic speed and Debye length, respectively. These expressions reduce to the standard ion-acoustic speed $c_S^2 = k_B T_{\parallel e}/m_p$ and the Debye length $\lambda_{De\infty}^2 = \frac{1}{2}\theta_e^2/\omega_{pe}^2 = \sqrt{\varepsilon_0 e^{-2} k_B T_e/n_{ec}}$ in the Maxwellian limit (when $\kappa_e \to \infty$). Here, $\omega_L\left(k_{\parallel}\right)$ denotes the Bohm-Gross. Similarly, the damping/growth rate is obtained, after some algebraic manipulations, to be (Mace et al., 1998)

$$
\frac{\gamma_{IA}\left(k_{\parallel}\right)}{\omega_{IA}} \cong -\frac{\sqrt{\pi}}{8}\frac{\kappa_e - \frac{3}{2}}{\kappa_e - \frac{1}{2}}\left(1 + k_{\parallel}^2\lambda_D^2\right)^{-\frac{3}{2}}\left\{\left(\frac{T_{\parallel e}}{T_{\parallel pc}}\right)^{\frac{3}{2}}A(\kappa_{pc})\left(1 + \frac{1}{\kappa_{pc} - \frac{3}{2}}\cdot\frac{\kappa_e - \frac{3}{2}}{\kappa_e - \frac{1}{2}}\frac{\frac{1}{2}T_{\parallel e}/T_{\parallel pc}}{1 + k_{\parallel}^2\lambda_D^2}\right)^{-\left(\kappa_{pc}+1\right)}\right.
$$

$$
\left. + \frac{n_{pb}}{n_e}\left(\frac{T_{\parallel e}}{T_{\parallel pb}}\right)^{\frac{3}{2}}\left(1 - k_{\parallel}U_{\parallel pb}/\omega_{IA}\right)A(\kappa_{pb})\left[1 + \frac{1}{\kappa_{pb} - \frac{3}{2}}\frac{\left(\omega_{IA} - k_{\parallel}U_{\parallel pb}\right)^2}{k_{\parallel}^2\theta_{\parallel pb}^2}\right]^{-\left(\kappa_{pb}+1\right)}\right\},
$$

(7.54)

using $U_{\parallel e} \ll U_{\parallel pb}$ and $|\omega_r - k_{\parallel}U_{\parallel p}| \leq 1$, since there are two possible separate frequency domains where instabilities can occur. Here, we show mostly the ion/ion-acoustic mode domain. In the Maxwellian limit, this becomes

$$
\frac{\gamma_{IA}\left(k_{\parallel}\right)}{\omega_{IA}} \cong -\frac{\sqrt{\pi}}{8}\left(1 + k_{\parallel}^2\lambda_D^2\right)^{-\frac{3}{2}}\cdot\left[\left(\frac{T_{\parallel e}}{T_{\parallel pc}}\right)^{\frac{3}{2}}\exp\left(-\frac{\frac{1}{2}T_{\parallel e}/T_{\parallel pc}}{1 + k_{\parallel}^2\lambda_D^2}\right) + \sqrt{\frac{m_e}{m_p}}\right],
$$

(7.55)

which is the standard result. From Eq. (7.54) it follows that instability is only possible if

$$
\omega_{IA} - k_{\parallel}U_{\parallel pb} < 0 \Leftrightarrow U_{\parallel pb} > \frac{c_{S\kappa}}{\sqrt{1 + k_{\parallel}^2\lambda_D^2}},
$$

(7.56)

which happens whenever $U_{\parallel pb} > c_{S\kappa}$, or

$$
k_{\parallel}\lambda_D > \sqrt{\frac{c_{S\kappa}^2}{U_{\parallel pb}^2} - 1}, \quad \text{when } U_{\parallel pb} \leq c_{S\kappa}.
$$

(7.57)

Since $c_{S\kappa} < c_S$, the instability should be operative for lower values of $U_{||pb}$ in a kappa plasma. This necessary condition can also be written as

$$\frac{U_{||pb}^2}{\theta_{||pb}^2} > \frac{\kappa_e - \frac{3}{2}}{\kappa_e - \frac{1}{2}} \cdot \frac{T_{||e}}{2T_{||pb}} \cdot \frac{1}{1 + k_{||}^2 \lambda_D^2}. \tag{7.58}$$

Returning to Eq. (7.50), the second domain of the instability (namely the electron/ion-acoustic mode) occurs due to the electron drift relative to the proton core, when

$$\left| \frac{\kappa_e}{\kappa_e - \frac{3}{2}} \cdot \frac{\omega_{IA} - k_{||}U_{||e}}{k_{||}\theta_{||e}} \right| \leq 1. \tag{7.59}$$

Also, since $U_{||pb} > U_{||e}$, we can neglect the beam contribution to the dispersion relation Eq. (7.50) and write

$$\frac{\gamma_{IA}(k_{||})}{\omega_{IA}} \cong -\frac{\sqrt{\pi}}{8} \frac{\kappa_e - \frac{3}{2}}{\kappa_e - \frac{1}{2}} \left(1 + k_{||}^2 \lambda_D^2\right)^{-\frac{3}{2}} \cdot \left\{ \left(\frac{T_{||e}}{T_{||pc}}\right)^{\frac{3}{2}} A(\kappa_{pc}) \left(1 + \frac{1}{\kappa_{pc} - \frac{3}{2}} \frac{\kappa_e - \frac{3}{2}}{\kappa_e - \frac{1}{2}} \frac{\frac{1}{2}T_{||e}/T_{||pc}}{1 + k_{||}^2 \lambda_D^2}\right)^{-(\kappa_{pc}+1)} \right.$$

$$\left. + \sqrt{\frac{m_e}{m_p}} \left(1 - k_{||}U_{||e}/\omega_{IA}\right) A(\kappa_e) \left[1 + \frac{1}{\kappa_e - \frac{3}{2}} \frac{\left(\omega_{IA} - k_{||}U_{||e}\right)^2}{k_{||}^2 \theta_{||e}^2}\right]^{-(\kappa_e+1)} \right\}. \tag{7.60}$$

For this case, the necessary condition for instability is

$$\omega_{IA} - k_{||}U_{||e} < 0 \Leftrightarrow U_{||e} > \frac{c_{S\kappa}}{\sqrt{1 + k_{||}^2 \lambda_D^2}}. \tag{7.61}$$

When the zero-current condition $U_{||e} = (n_{pb}/n_e)U_{||pb}$ is imposed, this implies that

$$U_{||e} > \frac{n_e}{n_{pb}} \cdot \frac{c_{S\kappa}}{\sqrt{1 + k_{||}^2 \lambda_D^2}}, \tag{7.62}$$

which is a much more restrictive condition than that for the ion/ion-acoustic instability. Consequently, the electron/ion-acoustic instability is only efficient when there is a net current in the plasma. In this case, it is called the Buneman instability.

The plots/panels on Fig. 7.7 show solutions of the dispersion relation of the ion-acoustic mode and of the ion/ion-acoustic damping/growth rate obtained from the dispersion Eq. (7.52) for a beam-kappa plasma. Fig. 7.7 (left) shows plots of γ versus $k_{||}$ of the dispersion relations of the kappa-ion-acoustic mode for several values of $\kappa = \kappa_c = \kappa_b = \kappa_e$ and for fixed $T_{||e}/T_{||pc} = T_{||e}/T_{||pb} = 10$ and $n_{pb}/n_{pc} = 0.1$.

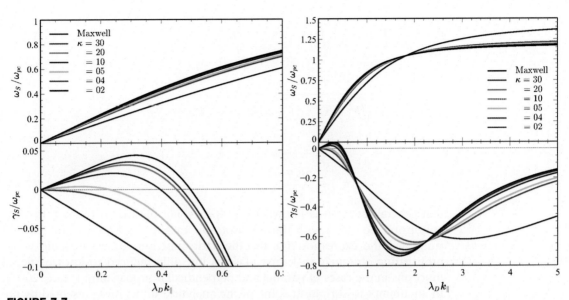

FIGURE 7.7

Ion-acoustic waves versus a normalized wavelength of a proton core, proton beam, and electron plasma for the following cases: Left: dispersion curves (top) and growth/damping rate (bottom) of a kappa plasma with a proton beam drift for various κ-indices. Right: similar plots and parameters as the left panel but for a larger $k_\parallel$ range.

In Fig. 7.7 (right) we showed similar plots of γ versus $k_\parallel$ (ion/ion-acoustic mode) as for the left panel but in a larger $k_\parallel$ domain. The parameters used are: $U_{\parallel pb}/\theta_{\parallel pc} = 3$, $T_{\parallel e}/T_{\parallel pc} = T_{\parallel e}/T_{\parallel pb} = 10$, and $n_{pb}/n_{pc} = 0.1$.

7.5 Kappa Velocity Distribution Plasma Waves at Oblique Propagation ($\vartheta \neq 0$)

The study of obliquely propagating plasma waves in a kappa velocity distribution plasma medium represents the current state-of-the-art frontier in plasma kinetic physics. At present, there is no close form analytical representation of the dielectric tensor for a kappa velocity distribution function at an arbitrary angle of propagation available for the investigation of plasma waves. Perhaps the first attempt at a general expression for the dielectric tensor for a multispecies plasma with isotropic kappa distributions was undertaken by Mace (1996). He showed that the dielectric tensor for such plasma could be written in a form similar to that obtained by Trubnikov for relativistic Maxwellian plasma. However, as in the relativistic Maxwellian case, the final expressions for the dielectric tensor elements were written in terms of integrals, which meant that they could not be expressed in closed form in terms of known functions. Later, Mace (2003) introduced a Gordeyev integral for the study of electrostatic waves propagating at an arbitrary angle relative to B_0. The Gordeyev integral introduced

$$G = \frac{\omega}{2^{\kappa-\frac{1}{2}}} \Gamma\left(\kappa + \frac{1}{2}\right) \int\limits_0^\infty \exp(i\omega\tau) \; z^{\kappa+\frac{1}{2}} K_{\kappa+\frac{1}{2}}(z) d\tau, \tag{7.63}$$

where

$$z(\tau) = \sqrt{2\kappa} \cdot \sqrt{k_\perp^2 w^2 \Omega^{-2} \cdot [1 - \cos(\Omega\tau)] + \frac{1}{2} k_\parallel^2 w^2 \tau^2}, \tag{7.64}$$

$$w^2 = \theta^2 \cdot \frac{\kappa - \frac{1}{2}}{\kappa}, \quad \theta^2 \equiv \frac{2k_B T}{m}, \tag{7.65}$$

that generalizes what is known for a Maxwellian plasma. (Compare the 3-D anisotropic Eq. (7.2) with the 1-D isotropic case in Eq. (7.65).) This integral simplified the expression for the dielectric function for electrostatic waves propagating obliquely in kappa plasmas but only reduced to known special functions in the cases of parallel and perpendicular propagation (Mace, 2003). In an attempt to surmount some of the analytical difficulties presented by the isotropic kappa distribution when evaluating the dielectric tensor elements, Cattaert et al. (2007) instead used a hybrid velocity distribution that was the product of a 1-D kappa distribution for the velocity component in the direction of B_0 and a 1-D Maxwellian distribution for the velocity component perpendicular to it. They showed that the dielectric tensor elements could be written in closed form in terms of the Z_κ function defined in Eq. (7.13) and the familiar sums of Bessel functions. However, this progress came at the expense of a true kappa representation.

Gaelzer and Ziebell (2014, 2016) have resumed work on this problem, having provided the most complete representation of the formalism of obliquely propagating waves. However, such a general solution still requires some considerable degree of numerical calculations in order to produce a description of the properties of the various plasma waves at oblique propagation angles. There are some other initial attempts toward this goal in the scientific literature such as Summers et al. (1994), Basu (2009), Liu et al. (2014), Astfalk et al., (2015), and Gaelzer and Ziebell (2014, 2016). Many of these works treat the dielectric tensor components via the numerical quadrature of Eqs. (7.6)–(7.8) or assumed power series expansions that do not provide for most of the physical and mathematical properties of the dielectric tensor, nor do they properly treat the analytical continuation of the results as the solution of the dispersion relation crosses the $\text{Im}(\omega) = 0$ boundary. Astfalk et al. (2015) developed an obliquely propagating dispersion solver, DSHARK, for kappa velocity distributions. This approach could provide many interesting results, but the code carried out the integrals numerically, and therefore the linear computational results require some considerable degree of numerical calculations in order to produce a description of the properties of the various plasma waves at oblique propagation angles. While a close form general solution of oblique propagation linear plasma waves in kappa distribution plasma still eludes treatment, there is considerable progress toward this objective.

7.6 Concluding Remarks

The results shown here are expected to provide sufficient motivation for researchers to investigate the behavior of collective plasma oscillations of low- and high-frequency electrostatic and electromagnetic waves in a kappa distribution plasma. They also aim to help researchers that are studying the stability of plasma systems with non-Maxwellian long-tail kappa velocity distributions, providing them with the physical effects of various types of plasma waves and giving them the mathematical tools needed to simplify and characterize their respective physical behavior. Thus this chapter provides academic guidance in the study of collective plasma oscillations in kappa velocity distribution plasma.

The instabilities we have chosen to discuss here, namely, Alfvén and WC waves, proton and EF waves, Langmuir oscillations, and ion and electron acoustic waves, are the most common and representative samples in space and laboratory plasma applications. Any new formalism that is able to integrate analytically the terms in the dielectric tensor for kappa distributions at arbitrary propagation angles should be able to recover, under small-angle approximation, the solutions presented here. This would also require the proper analytical continuation across the $Im(\omega) = 0$ contour in the complex domain.

The research on kappa velocity distribution driving linear plasma waves continues, but the trend is toward their effect at oblique propagation angles relative to the mean magnetic field. This is currently the forefront of linear plasma waves of kappa-distributed plasmas. The correct treatment and simplifications of the mathematical complexity of this frontier still elude us, but there are some very promising approximate solutions that may provide some light into the behavior of plasma collective oscillations with kappa particle velocity distributions.

7.7 Science Questions for Future Research

Future analyses and observations need to address the following questions:

1. What are the kappa dependent, obliquely propagated plasma waves?
2. What is the dispersion relation for other kappa distribution types?
3. How are the dispersion relations modified for a nonzero potential energy?

CHAPTER 8

Nonlinear Wave–Particle Interaction and Electron Kappa Distribution

P.H. Yoon [1,2], **G. Livadiotis** [3]

[1]University of Maryland, College Park, MD, United States; [2]Kyung Hee University, Yongin, Korea; [3]Southwest Research Institute, San Antonio, TX, United States

Chapter Outline

8.1 Summary 363
8.2 Introduction 364
8.3 Plasma Weak Turbulence Theory 365
  8.3.1 General Formulation 365
  8.3.2 Weak Turbulence Theory for Plasmas Described by Kappa Distributions 380
    8.3.2.1 Induced and Spontaneous Emissions 386
    8.3.2.2 Induced and Spontaneous Decay Processes 386
    8.3.2.3 Induced and Spontaneous Scattering Processes 388
8.4 Turbulent Quasiequilibrium and Kappa Electron Distribution 389

  8.4.1 Steady-State Particle Distribution 389
  8.4.2 Steady-State Langmuir Turbulence 392
    8.4.2.1 Balance of Induced and Spontaneous Emission 393
    8.4.2.2 Absence of Steady-State Decay Processes 394
    8.4.2.3 Balance of Induced and Spontaneous Scattering 395
8.5 Concluding Remarks 398
8.6 Science Questions for Future Research 398
Acknowledgments 398

8.1 Summary

In this chapter, we present the physics of nonlinear wave–particle interactions that lead to the electron kappa velocity distribution function as a stationary

Kappa Distributions. http://dx.doi.org/10.1016/B978-0-12-804638-8.00008-5

asymptotic state. The theoretical framework is based upon the plasma weak turbulence theory, which does not invoke nonextensive thermodynamics at all, but the final solution shows that the kappa distribution is the only allowable state in a weakly turbulent plasma. This finding implicates a profound interrelationship between the turbulent equilibrium state in plasmas and the nonextensive thermodynamic description.

Science Question: How the interaction of plasma particles with self-generated electromagnetic turbulence may lead to a kappa distribution?

Keywords: Electron velocity distribution; Langmuir waves; Plasma kinetic theory; Turbulence.

8.2 Introduction

Plasma is an ionized classical gas interacting through a well-defined long-range electromagnetic force. Consequently, it is a perfect nonequilibrium statistical mechanical system for studying the nonextensive behavior from first principles. The basic kinetic equation that governs the plasma processes is analogous to Boltzmann equation for dilute gas except that instead of binary collisions, it describes plasma particles collectively interacting through long-range electromagnetic fields. Consequently, the equation naturally lends itself for the study of nonextensive statistical property.

Plasmas do not naturally exist on Earth. Laboratory plasmas are prone to interference from the device wall, etc. In this regard, the space is an excellent natural laboratory for plasma physics. Since the 1960s, various spacecraft measured charged particle distributions and showed that they are distributed in velocity space according to none other than the kappa function (e.g., see Binsack, 1966; Olbert, 1968; Vasyliunas, 1968; Feldman et al., 1975; and the bibliography in Chapter 1). Recently, it was rigorously demonstrated that the electron kappa distribution is a solution of the plasma weak turbulence equation (Yoon, 2014). The theoretically derived value of $\kappa = 2.25$ means that for large u, the electron distribution is characterized by an inverse power law $f \propto u^{-2\kappa-2} \approx u^{-6.5}$. Observation shows that solar wind is characterized by $f \propto u^{-6.69}$ on average (Wang et al., 2012). Given observational uncertainties and inherent assumptions in the theory, the agreement is remarkable. This is a strong indication that space plasmas are indeed governed by nonextensive statistical mechanics.

In the present chapter, we show how the plasma turbulence theory for electrons may lead to the electron kappa velocity distribution function. A brief outline of this chapter is as follows: in Section 8.3, we present the basic derivation and formulation of plasma weak turbulence theory. Section 8.4 is devoted to the application of the formal weak turbulence theory to kappa plasmas. We also discuss the steady-state solution of the fundamental equations, which demonstrates that the kappa distribution is a unique solution for turbulent plasma in dynamic equilibrium. Finally, the concluding remarks are given in Section 8.5, while three general science questions for future analyses are posed in Section 8.6.

8.3 Plasma Weak Turbulence Theory

8.3.1 General Formulation

The plasma weak turbulence theory was formulated largely by scientists from the former Soviet Union in the early 1960s and continued on through the 1980s (Kadomtsev, 1965; Davidson, 1972; Akhiezer et al., 1975; Tsytovich, 1977; Melrose, 1980; Sitenko, 1982). The theory has been in use for many applications of plasma phenomena, which is ongoing to this date (Ziebell et al., 2015; Tigik et al., 2016). Let us describe the formal development. The starting point is the N-body probability distribution function in phase space $(\vec{r}, \vec{p})$, called the Klimontovich function, $N_a(\vec{r}, \vec{p}, t)$, defined by

$$N_a(\vec{r}, \vec{p}, t) = \sum_{j=1}^{N} \delta\left(\vec{r} - \vec{r}_j^a(t)\right) \delta\left(\vec{p} - \vec{p}_j^a(t)\right), \tag{8.1}$$

where $\vec{r}_j^a(t)$ and $\vec{p}_j^a(t)$ are exact particle orbits for the j-th particle of species a ($e_a = -e$ for electrons and $e_a = +Ze$ for ions, or $+e$ for protons),

$$\vec{u}_j^a(t) = \frac{d\vec{r}_j^a(t)}{dt}, \; \frac{d\vec{p}_j^a(t)}{dt} = e_a \vec{E}\left(\vec{r}_j^a(t), t\right) + e_a \vec{u}_j^a(t) \times \vec{B}\left(\vec{r}_j^a(t), t\right). \tag{8.2}$$

The electric $\vec{E}$ and magnetic $\vec{B}$ field vectors satisfy Maxwell's equation. The kinetic equation for the Klimontovich function is equivalent to the Liouville's equation for exact phase space mapping. Together with the Maxwell's equation, the fundamental equations for plasmas are given by

$$\left[\frac{\partial}{\partial t} + \vec{u} \cdot \vec{\nabla} + e_a\left(\vec{E}(\vec{r}, t) + \vec{u} \times \vec{B}(\vec{r}, t)\right) \cdot \frac{\partial}{\partial \vec{p}}\right] N_a(\vec{r}, \vec{p}, t) = 0,$$

$$\vec{\nabla} \times \vec{E}(\vec{r}, t) + \frac{\partial \vec{B}(\vec{r}, t)}{\partial t} = 0, \tag{8.3}$$

$$\vec{\nabla} \times \vec{B}(\vec{r}, t) - \frac{1}{c^2}\frac{\partial \vec{E}(\vec{r}, t)}{\partial t} - \mu_0 \sum_a e_a \int \vec{u} N_a(\vec{r}, \vec{p}, t) d\vec{p} = 0.$$

where c is the speed of light in vacuum; $\partial/\partial\vec{p}$ denotes the gradient in the momentum space (also noted with $\vec{\nabla}_p$). In Eq. (8.3), the other two equations of Maxwell's equations, $\nabla\vec{B} = 0$ and $\nabla\vec{E} = \varepsilon_0^{-1} \sum_a e_a \int N_a(\vec{r}, \vec{p}, t) d\vec{p}$, are implicit. The previously mentioned exact dynamical equation for $N_a(\vec{r}, \vec{p}, t)$ contains no microscopic dissipation. The dissipation (called also irreversibility), be it collective or collisional, arises as a result of the loss of information from the ensemble average procedures. The one-particle distribution function, $f_a(\vec{r}, \vec{p}, t)$, is the ensemble averaged Klimontovich function,

$$f_a(\vec{r}, \vec{p}, t) = \langle N_a(\vec{r}, \vec{p}, t)\rangle. \tag{8.4}$$

The Vlasov equation is formally identical to Eq. (8.3) except that the equation describes the dynamical path of the smoothed ensemble averaged function $f_a(\vec{r}, \vec{p}, t)$ instead of the exact phase space distribution, Eq. (8.1).

In what follows, let us assume that the plasma under consideration is immersed in a field-free environment, and the predominant mode of interaction is the electrostatic force. Under such an approximation, the basic Eq. (8.3) reduces to

$$\left(\frac{\partial}{\partial t} + \vec{u} \cdot \vec{\nabla} + e_a \vec{E}(\vec{r}, t) \cdot \frac{\partial}{\partial \vec{p}}\right) N_a(\vec{r}, \vec{p}, t) = 0,$$

$$\vec{\nabla} \cdot \vec{E}(\vec{r}, t) - \varepsilon_0^{-1} \sum_a e_a \int d\vec{p} N_a(\vec{r}, \vec{p}, t) = 0. \tag{8.5}$$

It is useful to consider the Klimontovich function, Eq. (8.1), describing the phase space evolution of free particles (ideal gas) that do not interact with each other,

$$N_a^0(\vec{r}, \vec{p}, t) = \sum_{j=1}^{N} \delta\left[\vec{r} - \vec{r}_j^{a0}(t)\right] \cdot \delta\left[\vec{p} - \vec{p}_j^{a0}(t)\right], \tag{8.6}$$

where $\vec{r}_j^{a0}(t)$ and $\vec{p}_j^{a0}(t)$ are exact orbits of free streaming particles satisfying, $d\vec{p}_j^{a0}(t)/dt = 0$, and $d\vec{r}_j^{a0}(t)/dt = \vec{u}_j^{a0}$. The corresponding Klimontovich equation for free particles is

$$\left(\frac{\partial}{\partial t} + \vec{u} \cdot \vec{\nabla}\right) N_a^0(\vec{r}, \vec{p}, t) = 0. \tag{8.7}$$

Let us denote the deviation of the Klimontovich function $N_a(\vec{r}, \vec{p}, t)$ from its average $f_a(\vec{r}, \vec{p}, t) = \langle N_a(\vec{r}, \vec{p}, t)\rangle$, that is, the fluctuation, $\delta N_a(\vec{r}, \vec{p}, t)$,

$$\delta N_a(\vec{r}, \vec{p}, t) = N_a(\vec{r}, \vec{p}, t) - \langle N_a(\vec{r}, \vec{p}, t)\rangle, \text{ or,}$$

$$N_a(\vec{r}, \vec{p}, t) = f_a(\vec{r}, \vec{p}, t) + \delta N_a(\vec{r}, \vec{p}, t). \tag{8.8}$$

Assuming random phase for the fluctuations, the ensemble average of $\delta N_a(\vec{r}, \vec{p}, t)$ is zero, $\langle \delta N_a(\vec{r}, \vec{p}, t)\rangle = 0$. Since the medium is free of average field, by definition, the electric field is only made of fluctuations, $\vec{E}(\vec{r}, t) = \delta \vec{E}(\vec{r}, t)$. Then, substituting Eq. (8.8) in the first equation in Eq. (8.5), we obtain

$$\left(\frac{\partial}{\partial t} + \vec{u} \cdot \vec{\nabla}\right) f_a(\vec{r}, \vec{p}, t) + e_a \frac{\partial}{\partial \vec{p}} \cdot \langle \delta \vec{E}(\vec{r}, t)\delta N_a(\vec{r}, \vec{p}, t)\rangle$$

$$+ \left(\frac{\partial}{\partial t} + \vec{u} \cdot \vec{\nabla}\right) \delta N_a(\vec{r}, \vec{p}, t) + e_a \delta \vec{E}(\vec{r}, t) \cdot \frac{\partial f_a(\vec{r}, \vec{p}, t)}{\partial \vec{p}} \tag{8.9a}$$

$$+ e_a \frac{\partial}{\partial \vec{p}} \cdot \left[\delta \vec{E}(\vec{r}, t)\delta N_a(\vec{r}, \vec{p}, t) - \langle \delta \vec{E}(\vec{r}, t)\delta N_a(\vec{r}, \vec{p}, t)\rangle\right] = 0,$$

while its fluctuation gives

$$\left(\frac{\partial}{\partial t} + \vec{u} \cdot \vec{\nabla}\right) f_a(\vec{r}, \vec{p}, t) + e_a \frac{\partial}{\partial \vec{p}} \cdot \langle \delta \vec{E}(\vec{r}, t) \delta N_a(\vec{r}, \vec{p}, t) \rangle = 0 \qquad (8.9b)$$

because $\langle \delta N_a(\vec{r}, \vec{p}, t) \rangle = \langle \delta \vec{E}(\vec{r}, t) \rangle = 0$. Subtracting Eq. (8.9b) from Eq. (8.9a), we derive

$$\left(\frac{\partial}{\partial t} + \vec{u} \cdot \vec{\nabla}\right) \delta N_a(\vec{r}, \vec{p}, t) + e_a \delta \vec{E}(\vec{r}, t) \cdot \frac{\partial f_a(\vec{r}, \vec{p}, t)}{\partial \vec{p}}$$
$$+ e_a \frac{\partial}{\partial \vec{p}} \cdot \left[\delta \vec{E}(\vec{r}, t) \delta N_a(\vec{r}, \vec{p}, t) - \langle \delta \vec{E}(\vec{r}, t) \delta N_a(\vec{r}, \vec{p}, t) \rangle \right] = 0. \qquad (8.9c)$$

Also, substituting Eq. (8.8) in the second equation in Eq. (8.5), we obtain

$$\vec{\nabla} \cdot \delta \vec{E}(\vec{r}, t) = \varepsilon_0^{-1} \sum_a e_a \int d\vec{p} \, \delta N_a(\vec{r}, \vec{p}, t). \qquad (8.9d)$$

We also define the fluctuation of the free-streaming Klimontovich function,

$$\delta N_a^0(\vec{r}, \vec{p}, t) = N_a^0(\vec{r}, \vec{p}, t) - \langle N_a^0(\vec{r}, \vec{p}, t) \rangle. \qquad (8.10)$$

Upon assuming that the ensemble average of the free streaming Klimontovich function is, in fact, the same as the one particle distribution function,

$$f_a(\vec{r}, \vec{p}, t) = \langle N_a^0(\vec{r}, \vec{p}, t) \rangle, \qquad (8.11)$$

the equation for $\delta N_a^0(\vec{r}, \vec{p}, t)$ is simply given by

$$\left(\frac{\partial}{\partial t} + \vec{u} \cdot \vec{\nabla}\right) \delta N_a^0(\vec{r}, \vec{p}, t) = 0. \qquad (8.12)$$

We may subtract Eq. (8.12) from the equation for δN_a in Eq. (8.9c) to obtain

$$\left(\frac{\partial}{\partial t} + \vec{u} \cdot \vec{\nabla}\right) \left[\delta N_a(\vec{r}, \vec{p}, t) - \delta N_a^0(\vec{r}, \vec{p}, t) \right] + e_a \delta \vec{E}(\vec{r}, t) \cdot \frac{\partial f_a(\vec{r}, \vec{p}, t)}{\partial \vec{p}}$$
$$+ e_a \frac{\partial}{\partial \vec{p}} \left[\delta \vec{E}(\vec{r}, t) \delta N_a(\vec{r}, \vec{p}, t) - \langle \delta \vec{E}(\vec{r}, t) \delta N_a(\vec{r}, \vec{p}, t) \rangle \right] = 0.$$

$$(8.13)$$

In Eq. (8.13), we are not interested in $\delta N_a^0(\vec{r}, \vec{p}, t)$ per se, but rather in the ensemble average of the product of two quantities $\delta N_a^0(\vec{r}, \vec{p}, t)$ and $\delta N_b^0(\vec{r}', \vec{p}', t')$, that is, the two-body correlation function for the fluctuations of the free-streaming Klimontovich function, $\langle \delta N_a^0(\vec{r}, \vec{p}, t) \delta N_b^0(\vec{r}', \vec{p}', t'') \rangle$.

We may compute this quantity directly from the definition Eq. (8.6), which is explicitly written as

$$N_a^0(\vec{r}, \vec{p}, t) = \sum_{i=1}^{N} \delta(\vec{r} - \vec{r}^{a0} - \vec{u}_i^a t)\delta(\vec{p} - \vec{p}_i^a), \tag{8.14}$$

to obtain

$$\left\langle \delta N_a^0(\vec{r}, \vec{p}, t)\delta N_b^0(\vec{r}', \vec{p}', t') \right\rangle$$
$$= \delta_{ab}\delta(\vec{r} - \vec{r}' - \vec{u}(t - t'))\delta(\vec{p} - \vec{p}')f_a(\vec{r}, \vec{p}, t), \tag{8.15}$$

(where δ_{ab} is the delta of Kronecker, $\delta_{ab} = 1$ if $a = b$, and $\delta_{ab} = 0$ if $a \neq b$).

To summarize, the set of self-consistent equations, which forms the basis for nonlinear plasma turbulence theory including the effects of single-particle fluctuations, are given by the formal particle kinetic equation, i.e., the equation for $f_a(\vec{r}, \vec{p}, t)$, Eq. (8.9), the equation for perturbed distribution function $\delta N_a^0(\vec{r}, \vec{p}, t)$, Eq. (8.13), the wave equation, that is, Eq. (8.9d), and the definition for the source fluctuation, Eq. (8.15). It is convenient to work in the spectral representation. We assume that the perturbed quantities are functions of two distinct timescales, the fast timescale of the fluctuations, t', and the slow timescale of the evolution of the distribution function and the wave amplitude, t. This means that we may express $\delta N_a(\vec{r}, \vec{p}, t) \to \delta N_a(\vec{r}, \vec{p}, t, t')$, $\delta N_a^0(\vec{r}, \vec{p}, t) \to \delta N_a^0(\vec{r}, \vec{p}, t, t')$, and $\delta\vec{E}(\vec{r}, t) \to \delta\vec{E}(\vec{r}, t, t')$. We are henceforth interested in spatially uniform system, $f_a(\vec{r}, \vec{p}, t) = f_a(\vec{p}, t)$. We assume that the perturbations can be decomposed in the sense of the customary Fourier–Laplace transformation over the fast timescale of the fluctuations, t', while the amplitudes of the spectra vary in slow timescale, t,

$$\delta N_a(\vec{r}, \vec{p}, t, t') = \int d\vec{k} \int_L d\omega \delta N_{\vec{k},\omega}^a(\vec{p}, t)e^{i\vec{k}\cdot\vec{r} - i\omega t'}, \delta\vec{E}(\vec{r}, t, t')$$

$$= \int d\vec{k} \int_L d\omega \delta\vec{E}_{\vec{k},\omega}(t)e^{i\vec{k}\cdot\vec{r} - i\omega t'} \tag{8.16a}$$

$$\delta N_{\vec{k},\omega}^a(\vec{p}, t) = \frac{1}{(2\pi)^4} \int d\vec{r} \int_0^\infty dt' \delta N_a(\vec{r}, \vec{p}, t, t')e^{-i\vec{k}\cdot\vec{r} + i\omega t'},$$

$$\delta\vec{E}_{\vec{k},\omega}(t) = \frac{1}{(2\pi)^4} \int d\vec{r} \int_0^\infty dt' \delta\vec{E}(\vec{r}, t, t')e^{-i\vec{k}\cdot\vec{r} + i\omega t'} \tag{8.16b}$$

where the integration $\int_L d\omega$ is taken along the path L stretching from $\omega = -\infty + i\sigma$ to $\omega = +\infty + i\sigma$, where $\sigma > 0$ and $\sigma \to 0$ in order to ensure causality. In the Fourier–Laplace transformation defined previously, the time dependence of the spectral amplitudes $\delta N_{\vec{k},\omega}^a(\vec{u}, t)$ and $\delta\vec{E}_{\vec{k},\omega}(t)$ is assumed to be slow and adiabatic. These quantities are calculated as if they are independent of time on the fast wave scale $(t' \sim \omega^{-1})$. We also write the electrostatic field in terms of the potential,

$$\delta \vec{E}_{\vec{k},\omega} = -i\delta\Phi_{\vec{k},\omega}\vec{k}. \qquad (8.17)$$

Then, in the two-time step approximation, after applying the Fourier–Laplace transformation defined previously to Eqs. (8.9b), (8.13), and (8.9d), the relevant equations become, respectively,

$$\frac{\partial f_a(\vec{p},t)}{\partial t} = ie_a \int d\vec{k} \int d\omega \int d\vec{k}' \int d\omega' \vec{k}'$$

$$\cdot \frac{\partial}{\partial \vec{p}} \left\langle \delta\Phi_{\vec{k}',\omega'}(t)\delta N^a_{\vec{k},\omega}(\vec{p},t) \right\rangle e^{i\left(\vec{k}+\vec{k}'\right)\vec{r}-i(\omega+\omega')t'}, \qquad (8.18a)$$

$$\left(\omega - \vec{k}\vec{u} + i\frac{\partial}{\partial t}\right)\left[\delta N^a_{\vec{k},\omega}(\vec{p},t) - \delta N^{a0}_{\vec{k},\omega}(\vec{p},t)\right] = -e_a\delta\Phi_{\vec{k},\omega}(t)\vec{k}\cdot\frac{\partial f_a(\vec{p},t)}{\partial \vec{p}}$$

$$- e_a \int d\vec{k}' \int d\omega' \vec{k}'\cdot\frac{\partial}{\partial \vec{p}}\left[\delta\Phi_{\vec{k}',\omega'}(t)\delta N^a_{\vec{k}-\vec{k}',\omega-\omega'}(\vec{p},t)\right.$$

$$\left.-\left\langle \delta\Phi_{\vec{k}',\omega'}(t)\delta N^a_{\vec{k}-\vec{k}',\omega-\omega'}(\vec{p},t)\right\rangle\right], \qquad (8.18b)$$

$$\delta\Phi_{\vec{k},\omega}(t) = \varepsilon_0^{-1}\sum_a \frac{e_a}{k^2}\int d\vec{p}\,\delta N^a_{\vec{k},\omega}(\vec{p},t). \qquad (8.18c)$$

The Fourier transformation of the free streaming particle fluctuations, $\left\langle \delta N_a^0(\vec{r},\vec{p},t,t')\delta N_b^0(\vec{r}',\vec{p}',t,t'')\right\rangle$, in fast timescale $\tau = t' - t''$ is given by

$$\left\langle \delta N_a^0(\vec{p},t)\delta N_b^0(\vec{p}',t)\right\rangle_{\vec{k},\omega} = \delta_{ab}\delta(\vec{p}-\vec{p}')(2\pi)^{-4}\int_0^\infty d\tau f_a(\vec{p},t)e^{-i\vec{k}\vec{u}\tau+i(\omega+\Delta)\tau}$$

$$+ \delta_{ab}\delta(\vec{p}-\vec{p}')(2\pi)^{-4}\int_{-\infty}^0 d\tau f_a(\vec{p},t)e^{-i\vec{k}\vec{u}\tau+i(\omega-\Delta)\tau}, \qquad (8.19)$$

where $\Delta \to 0^+$. The previously mentioned definition reduces to the customary Fourier transformation for $\Delta = 0$ exactly. From this, we have

$$\left\langle \delta N_a^0(\vec{p},t)\delta N_b^0(\vec{p}',t)\right\rangle_{\vec{k},\omega} = \delta_{ab}\delta(\vec{p}-\vec{p}')(2\pi)^{-3}\delta(\omega-\vec{k}\vec{u})f_a(\vec{p},t). \qquad (8.20)$$

For stationary and homogeneous turbulence, which means that the correlation of two fluctuating quantities depends only on the relative time and distance, not on absolute values—of course, such an assumption is violated when the underlying turbulence exhibits intermittent behavior—the cumulants of two fluctuating quantities $A(\overrightarrow{r}, t)$ and $B(\overrightarrow{r}, t)$ are a function of the relative distance and relative time only,

$$\left\langle A(\overrightarrow{r}, t) B(\overrightarrow{r}', t') \right\rangle = \left\langle AB \right\rangle_{\overrightarrow{r} - \overrightarrow{r}', t - t'} \equiv G(\overrightarrow{r} - \overrightarrow{r}', t - t'). \tag{8.21}$$

In spectral representation, this implies

$$\langle AB \rangle_{\overrightarrow{k}, \omega} = \left\langle A_{\overrightarrow{k}, \omega} B_{-\overrightarrow{k}, -\omega} \right\rangle = \left\langle A_{-\overrightarrow{k}, -\omega} B_{\overrightarrow{k}, \omega} \right\rangle, \tag{8.22}$$

or more generally,

$$\left\langle A_{\overrightarrow{k}, \omega} B_{\overrightarrow{k}', \omega'} \right\rangle = \langle AB \rangle_{\overrightarrow{k}, \omega} \, \delta\left(\overrightarrow{k} + \overrightarrow{k}' \right) \delta(\omega + \omega'). \tag{8.23}$$

This simplifies the formal particle kinetic equation in Eq. (8.18a,b,c).

At this point, we make the approximation by formally absorbing the slow time derivatives into the definition for the angular frequency,

$$\omega \to \omega + i \frac{\partial}{\partial t}, \tag{8.24}$$

and suppress the slow time t henceforth, until we reintroduce it at an appropriate later stage. Employing the above approximations, we now have a set of equations that forms the basis of subsequent weak turbulence analysis,

$$\frac{\partial f_a(\overrightarrow{p})}{\partial t} = i e_a \int d\overrightarrow{k} \int d\omega \, \overrightarrow{k} \cdot \frac{\partial}{\partial \overrightarrow{p}} \left\langle \delta\Phi_{\overrightarrow{k} \, \omega} \, \delta N^a_{-\overrightarrow{k}, -\omega}(\overrightarrow{p}) \right\rangle,$$

$$\delta N^a_{\overrightarrow{k} \, \omega}(\overrightarrow{p}) = \delta N^{a0}_{\overrightarrow{k} \, \omega}(\overrightarrow{p}) - \frac{e_a}{\omega - \overrightarrow{k} \overrightarrow{u} + i0} \, \overrightarrow{k} \cdot \frac{\partial f_a(\overrightarrow{p})}{\partial \overrightarrow{p}} \delta\Phi_{\overrightarrow{k} \, \omega} - \frac{e_a}{\omega - \overrightarrow{k} \overrightarrow{u} + i0}$$

$$\times \int d\overrightarrow{k}' \int d\omega' \, \overrightarrow{k} \cdot \frac{\partial}{\partial \overrightarrow{p}} \left[\delta\Phi_{\overrightarrow{k}' \, \omega'} \, \delta N^a_{\overrightarrow{k} - \overrightarrow{k}', \omega - \omega'}(\overrightarrow{p}) - \left\langle \delta\Phi_{\overrightarrow{k}' \, \omega'} \, \delta N^a_{\overrightarrow{k} - \overrightarrow{k}', \omega - \omega'}(\overrightarrow{p}) \right\rangle \right],$$

$$\delta\Phi_{\overrightarrow{k} \, \omega} = \varepsilon_0^{-1} \sum_a \frac{e_a}{k^2} \int d\overrightarrow{p} \, \delta N^a_{\overrightarrow{k} \, \omega}(\overrightarrow{p}),$$

$$\left\langle \delta N^0_a(\overrightarrow{p}) \delta N^0_b(\overrightarrow{p}') \right\rangle = (2\pi)^{-3} \delta_{ab} \delta(\overrightarrow{p} - \overrightarrow{p}') \delta(\omega - \overrightarrow{k} \overrightarrow{u}) f_a(\overrightarrow{p}). \tag{8.25}$$

(Note that it is useful to check our equations with a dimensional analysis. For example, given $[N_a] = \left[N^a_{\overrightarrow{k}, \omega} \right] \cdot [t] \cdot [r]^3$ (from Eq. 8.16a) and that $[f_a] = [N_a]$ (from Eq. 8.8), then, the first of Eq. (8.25) implies $[t]^{-1} = [e\Phi] \cdot ([r] \cdot [p])^{-1}$, which is obviously true.) Let us now introduce the following shorthand notations:

$$q \equiv (\overrightarrow{k}, \omega), \quad \overrightarrow{g}_q = \overrightarrow{g}^a_{k\,\omega} = -\frac{e_a}{\omega - \overrightarrow{k}\,\overrightarrow{u} + i0} \cdot \frac{\partial}{\partial \overrightarrow{p}}. \tag{8.26}$$

We also omit δ for the perturbed quantities. Then the nonlinear equation for the perturbed distribution can be expressed compactly as

$$N^a_q = N^{a0}_q + \overrightarrow{k} \cdot \overrightarrow{g}_q f_a \Phi_q + \int dq' \overrightarrow{k} \cdot \overrightarrow{g}_q \Big[\Phi_{q'} N^a_{q-q'} - \big\langle \Phi_{q'} N^a_{q-q'} \big\rangle \Big]. \tag{8.27}$$

To solve the nonlinear equation for N^a_q, we employ iterative means,

$$N^a_q = N^{a(1)}_q + N^{a(2)}_q + N^{a(3)}_q + \cdots, \tag{8.28}$$

under the assumption that each order in the perturbative expansion is of the similar magnitude with the electric field perturbation of the same order, $N^{a(n)}_q \propto \Phi^n_q$. Under this assumption, let us write down the iterative solution order by order. Then we have to the third order,

$$N^{a(1)}_q = N^{a0}_q + \overrightarrow{k}\,\overrightarrow{g}_q f_a \Phi_q,$$

$$N^{a(2)}_q = \int dq' \, \overrightarrow{k}\,\overrightarrow{g}_q \Big[\Phi_{q'} N^{a(1)}_{q-q'} - \big\langle \Phi_{q'} N^{a(1)}_{q-q'} \big\rangle \Big], \tag{8.29}$$

$$N^{a(3)}_q = \int dq' \, \overrightarrow{k}\,\overrightarrow{g}_q \Big[\Phi_{q'} N^{a(2)}_{q-q'} - \big\langle \Phi_{q'} N^{a(2)}_{q-q'} \big\rangle \Big].$$

The truncation of iterative solution up to third order (or for that matter, any finite order) implies that the perturbation expansion in electric field wave amplitude is valid. This assumption is one of the key ingredients of the weak turbulence theory, where the wave energy density is assumed to be sufficiently lower than the particle thermal energy. For highly turbulent system, the perturbation expansion fails. Theories that attempt to partially sum the infinite series such as Eq. (8.28) is known as the "renormalized kinetic theory" (for details, see, e.g., Kadomtsev, 1965; Krommes, 1984; or a recent monograph by Diamond et al., 2010).

Making successive use of iterative solution, we obtain

$$N^{a(1)}_q = N^{a0}_q + \overrightarrow{k}\,\overrightarrow{g}_q f_a \Phi_q,$$

$$N^{a(2)}_q = \int dq' \left(\overrightarrow{k}\,\overrightarrow{g}_q \right) \left(\overrightarrow{k} - \overrightarrow{k'} \right) \overrightarrow{g}_{q-q'} f_a [\Phi_{q'} \Phi_{q-q'} - \langle \Phi_{q'} \Phi_{q-q'} \rangle], \tag{8.30}$$

$$N^{a(3)}_q = \int dq' \int dq'' \left(\overrightarrow{k}\,\overrightarrow{g}_q \right) \left(\overrightarrow{k''}\,\overrightarrow{g}_{q-q'} \right) \left(\overrightarrow{k} - \overrightarrow{k'} - \overrightarrow{k''} \right) \overrightarrow{g}_{q-q'-q''} f_a$$

$$\times [\Phi_{q'} \Phi_{q''} \Phi_{q-q'-q''} - \Phi_{q'} \langle \Phi_{q''} \Phi_{q-q'-q''} \rangle - \langle \Phi_{q'} \Phi_{q''} \Phi_{q-q'-q''} \rangle],$$

(where we used $\langle \Phi_q \rangle = 0$). Inserting this "net solution" to the wave equation while symmetrizing various terms with respect to the dummy integral variable, we have

$$
0 = k^2 \varepsilon(q) \Phi_q - \varepsilon_0^{-1} \sum_a e_a \int d\vec{p} N_q^{a0}(\vec{p}) - \sum_{q_1} \sum_{q_2} ikk_1 k_2 \chi^{(2)}(q_1|q_2) [\Phi_{q_1} \Phi_{q_2}
$$
$$
\substack{(q_1+q_2=q)}
$$
$$
- \langle \Phi_{q_1} \Phi_{q_2} \rangle] - \sum_{q_1} \sum_{q_2} \sum_{q_3} kk_1 k_2 k_3 \chi^{(3)}(q_1|q_2|q_3) [\Phi_{q_1} \Phi_{q_2} \Phi_{q_3} - \Phi_{q_1} \langle \Phi_{q_2} \Phi_{q_3} \rangle
$$
$$
\substack{(q_1+q_2+q_3=q)}
$$
$$
- \langle \Phi_{q_1} \Phi_{q_2} \Phi_{q_3} \rangle],
$$

$$(8.31)$$

where we have defined various response functions,

$$
\varepsilon(q) = 1 + \chi(q), \chi(q) = \sum_a \chi_a(q) = \varepsilon_0^{-1} \sum_a \frac{e_a^2}{k^2} \int d\vec{p} \frac{\vec{k} \cdot \partial f_a / \partial \vec{p}}{\omega - \vec{k} \vec{u} + i0}, \quad (8.32a)
$$

$$
\chi^{(2)}(q_1|q_2) = \sum_a \chi_a^{(2)}(q_1|q_2)
$$
$$
= \sum_a \frac{1}{2}(-ie_a) \frac{\varepsilon_0^{-1} e_a^2}{k_1 k_2 |\vec{k}_1 + \vec{k}_2|} \int d\vec{p} \frac{1}{\omega_1 + \omega_2 - (\vec{k}_1 + \vec{k}_2) \cdot \vec{u} + i0}
$$
$$
\times \left[\vec{k}_1 \cdot \frac{\partial}{\partial \vec{p}} \left(\frac{\vec{k}_2 \cdot \partial f_a / \partial \vec{p}}{\omega_2 - \vec{k}_2 \vec{u} + i0} \right) + \vec{k}_2 \cdot \frac{\partial}{\partial \vec{p}} \left(\frac{\vec{k}_1 \cdot \partial f_a / \partial \vec{p}}{\omega_1 - \vec{k}_1 \vec{u} + i0} \right) \right],
$$

$$(8.32b)$$

$$
\chi^{(3)}(q_1|q_2|q_3) = \sum_a \chi_a^{(3)}(q_1|q_2|q_3) = \sum_a \frac{1}{2}(-ie_a)^2 \frac{\varepsilon_0^{-1} e_a^2}{k_1 k_2 k_3 |\vec{k}_1 + \vec{k}_2 + \vec{k}_3|}
$$
$$
\cdot \int d\vec{p} \times \frac{1}{\omega_1 + \omega_2 + \omega_3 - (\vec{k}_1 + \vec{k}_2 + \vec{k}_3) \cdot \vec{u} + i0} \cdot \vec{k}_1
$$
$$
\cdot \frac{\partial}{\partial \vec{p}} \left\{ \frac{1}{\omega_2 + \omega_3 - (\vec{k}_2 + \vec{k}_3) \cdot \vec{u} + i0} \right.
$$
$$
\times \left[\vec{k}_2 \cdot \frac{\partial}{\partial \vec{p}} \left(\frac{\vec{k}_3 \cdot \partial f_a / \partial \vec{p}}{\omega_3 - \vec{k}_3 \vec{u} + i0} \right) \right.
$$
$$
\left. \left. + \vec{k}_3 \cdot \frac{\partial}{\partial \vec{p}} \left(\frac{\vec{k}_2 \cdot \partial f_a / \partial \vec{p}}{\omega_2 - \vec{k}_2 \vec{u} + i0} \right) \right] \right\}.
$$

$$(8.32c)$$

The definitions and notations of the various dielectric susceptibilities are consistent with Sitenko (1982) or Yoon et al. (2012). Multiplying Eq. (8.31) with Φ_{-q} (that is, with Φ_q^*), we have

$$0 = k^2 \varepsilon(q) \Phi_q \Phi_{-q} - \varepsilon_0^{-1} \sum_a e_a \int d\vec{p}\, \Phi_{-q} N_q^{a0}(\vec{p}) - \sum_{q_1} \sum_{\substack{q_2 \\ (q_1+q_2=q)}} i k k_1 k_2 \chi^{(2)}(q_1|q_2)$$

$$\times \left[\Phi_{-q} \Phi_{q_1} \Phi_{q_2} - \Phi_{-q} \langle \Phi_{q_1} \Phi_{q_2} \rangle \right] - \sum_{q_1} \sum_{q_2} \sum_{\substack{q_3 \\ (q_1+q_2+q_3=q)}} k k_1 k_2 k_3 \chi^{(3)}(q_1|q_2|q_3)$$

$$\times \left[\Phi_{-q} \Phi_{q_1} \Phi_{q_2} \Phi_{q_3} - \Phi_{-q} \Phi_{q_1} \langle \Phi_{q_2} \Phi_{q_3} \rangle - \Phi_{-q} \langle \Phi_{q_1} \Phi_{q_2} \Phi_{q_3} \rangle \right],$$

$$(8.33a)$$

and taking the ensemble average,

$$0 = k^2 \varepsilon(q) \langle \Phi_q \Phi_{-q} \rangle - \varepsilon_0^{-1} \sum_a e_a \int d\vec{p} \left\langle \Phi_{-q} N_q^{a0}(\vec{p}) \right\rangle$$

$$- \sum_{q_1} \sum_{\substack{q_2 \\ (q_1+q_2=q)}} i k k_1 k_2 \chi^{(2)}(q_1|q_2) \left\langle \Phi_{-q} \Phi_{q_1} \Phi_{q_2} \right\rangle$$

$$- \sum_{q_1} \sum_{q_2} \sum_{\substack{q_3 \\ (q_1+q_2+q_3=q)}} k k_1 k_2 k_3 \chi^{(3)}(q_1|q_2|q_3) \left[\left\langle \Phi_{-q} \Phi_{q_1} \Phi_{q_2} \Phi_{q_3} \right\rangle - \left\langle \Phi_{-q} \Phi_{q_1} \right\rangle \left\langle \Phi_{q_2} \Phi_{q_3} \right\rangle \right],$$

$$(8.33b)$$

(again, because $\langle \Phi_{\pm q} \rangle = 0$). Then, for homogeneous and stationary turbulence, we have

$$\langle \Phi_{q_1} \Phi_{q_2} \Phi_{q_3} \Phi_{-q} \rangle = \delta(q_1 + q_2 + q_3 - q) \times \left[\delta(q_1 + q_2) \langle \Phi^2 \rangle_{q_1} \langle \Phi^2 \rangle_{q_3} \right.$$

$$\left. + \delta(q_1 + q_3) \langle \Phi^2 \rangle_{q_1} \langle \Phi^2 \rangle_{q_2} + \delta(q_2 + q_3) \langle \Phi^2 \rangle_{q_2} \langle \Phi^2 \rangle_{q_3} \right].$$

$$(8.34a)$$

Note that $\delta(q_1 + q_2 + q_3 - q)$ is automatically satisfied within the context of Eq. (8.33b). Hence,

$$\langle \Phi_{q_1} \Phi_{q_2} \Phi_{q_3} \Phi_{-q} \rangle = \left[\delta(q_1 + q_2) + \delta(q_1 + q_3) \right] \langle \Phi^2 \rangle_{q_1} \langle \Phi^2 \rangle_q$$

$$+ \delta(q_2 + q_3) \langle \Phi^2 \rangle_{q_2} \langle \Phi^2 \rangle_{q_3} \qquad (8.34b)$$

Then, by changing the dumb index $q_2 \to q_3$, we find $\delta(q_1 + q_2) + \delta(q_1 + q_3) \to 2\delta(q_1 + q_2)$. We also have

$$\delta(q_2 + q_3) \langle \Phi^2 \rangle_{q_2} \langle \Phi^2 \rangle_{q_3} \to \delta(q_2 + q_3) \langle \Phi^2 \rangle_{q_2} \langle \Phi^2 \rangle_{-q_2} = \delta(q_2 + q_3) \langle \Phi^4 \rangle_{q_2},$$

$$(8.34c)$$

where $\langle\Phi^4\rangle$ is the irreducible four-body correlation, which is small compared to the two-body correlations. Hence, we have $\langle\Phi_{q_1}\Phi_{q_2}\Phi_{q_3}\Phi_{-q}\rangle \to 2\delta(q_1 + q_2)$ $\langle\Phi^2\rangle_{q_1}\langle\Phi^2\rangle_q$. Also, $q_3 = q - (q_1 + q_2) = q$, $q_2 = q_1$; thus $k_3 = |k - (k_1 + k_2)| = k$, $k_2 = k_1$, which gives $\chi^{(3)}(q_1|q_2|q_3) \to \chi^{(3)}(q_1|-q_1|q)$. Therefore,

$$\sum_{q_1}\sum_{q_2}\sum_{q_3}_{(q_1+q_2+q_3=q)} kk_1k_2k_3\chi^{(3)}(q_1|q_2|q_3)\langle\Phi_{-q}\Phi_{q_1}\Phi_{q_2}\Phi_{q_3}\rangle$$

$$= 2\sum_{q_1} k^2k_1^2\chi^{(3)}(q_1|-q_1|q)\langle\Phi^2\rangle_{q_1}\langle\Phi^2\rangle_q. \tag{8.35}$$

Now, we set $q_1 \to q'$, thus

$$2\sum_{q_1} k^2k_1^2\chi^{(3)}(q_1|-q_1|q)\langle\Phi^2\rangle_{q_1}\langle\Phi^2\rangle_q = 2\sum_{q'} k^2k_1^2\chi^{(3)}(q'|-q'|q)\langle\Phi^2\rangle_{q'}\langle\Phi^2\rangle_q$$

$$= 2\sum_{q'} k^2k'^2\chi^{(3)}(q'|q|-q')\langle\Phi^2\rangle_{q'}\langle\Phi^2\rangle_q. \tag{8.36}$$

Therefore,

$$0 = k^2\varepsilon(q)\langle\Phi^2\rangle_q - i\sum_{q'} kk'\left|\vec{k} - \vec{k'}\right|\chi^{(2)}(q'|q-q')\langle\Phi_{q'}\Phi_{q-q'}\Phi_{-q}\rangle - 2$$

$$\times \sum_{q'} k^2k'^2\chi^{(3)}(q'|-q'|q)\langle\Phi^2\rangle_{q'}\langle\Phi^2\rangle_q - \varepsilon_0^{-1}\sum_a e_a\int d\vec{p}\left\langle\Phi_{-q}N_q^{a\,0}(\vec{p})\right\rangle. \tag{8.37}$$

Next, replacing q by $-q$ in Eq. (8.31), multiplying $N_{-q}^{a0}(\vec{p})$, and taking the average, we obtain

$$0 = k^2\varepsilon^*(q)\left\langle\Phi_{-q}N_q^{a0}(\vec{p})\right\rangle - \varepsilon_0^{-1}\sum_a e_a\int d\vec{p'}\left\langle N_q^{a0}(\vec{p})N_{-q}^{a0}(\vec{p'})\right\rangle$$

$$- \sum_{q'} ik^2k'\chi^{(2)}(-q|q')\left\langle\Phi_{-q}\Phi_{q'}N_q^{a0}(\vec{p})\right\rangle + \sum_{q'}\sum_{q''} k^2k'k''\chi^{(3)}(-q|q'|q'')$$

$$\times \left[\left\langle\Phi_{-q}\Phi_{q'}\Phi_{q''}N_q^{a0}(\vec{p})\right\rangle - \left\langle\Phi_{-q}N_q^{a0}(\vec{p})\right\rangle\langle\Phi_{q'}\Phi_{q''}\rangle\right], \tag{8.38}$$

where we have made use of the symmetry property $\varepsilon(-q) = \varepsilon^*(q)$. Now, because of Eq. (8.20), the second term is written as $-\varepsilon_0^{-1}(2\pi)^{-3}e_a\delta(\omega - \vec{k}\,\vec{u})f_a(\vec{p})$. Also, rearranging the third term, we write $\sum_{q'} ikk'\left|\vec{k} - \vec{k'}\right|\chi^{(2)}(q'|q-q')$ $\left\langle\Phi_{-q'}\Phi_{-q+q'}N_q^{a0}(\vec{p})\right\rangle$. The fourth term, following steps similar to those leading to Eq. (8.37), becomes $\sum_{q'} k^2 k'^2 \chi^{(3)}(q'|-q'|q)\langle\Phi^2\rangle_{q'}\langle\Phi_{-q}N_q^{a0}(\vec{p})\rangle$. Hence, we obtain

$$k^2 \varepsilon^*(q)\langle\Phi_{-q}N_q^{a0}(\vec{p})\rangle - \varepsilon_0^{-1}(2\pi)^{-3}e_a\delta(\omega - \vec{k}\,\vec{u})f_a(\vec{p})$$

$$+ \sum_{q'} ikk'\left|\vec{k} - \vec{k'}\right|\chi^{(2)}(q'|q-q')\langle\Phi_{-q'}\Phi_{-q+q'}N_q^{a0}(\vec{p})\rangle \qquad (8.39)$$

$$+ \sum_{q'} k^2 k'^2 \chi^{(3)}(q'|-q'|q)\langle\Phi^2\rangle_{q'}\langle\Phi_{-q}N_q^{a0}(\vec{p})\rangle = 0.$$

Note that the term $\langle\Phi_{-q}N_q^{a0}(\vec{p})\rangle$ appears twice; the part that involves $\chi^{(3)}$ is treated as a small correction to the linear solution, and thus we may ignore it. Hence, solving in terms of this term, we have

$$\langle\Phi_{-q}N_q^{a0}(\vec{p})\rangle \cong \frac{\varepsilon_0^{-1}e_a\delta(\omega - \vec{k}\,\vec{u})f_a(\vec{p})}{(2\pi)^3 k^2 \varepsilon^*(q)} - \frac{i}{ke^*(q)}\sum_{q'} k'\left|\vec{k} - \vec{k'}\right|\chi^{(2)*}(q'|q-q')$$

$$\times \langle\Phi_{-q'}\Phi_{-q+q'}N_q^{a0}(\vec{p})\rangle.$$

$$(8.40)$$

Consequently, inserting Eq. (8.40) to the right-hand side of Eq. (8.37), we arrive at

$$0 = \varepsilon(q)\langle E^2\rangle_q - i\int dq' \chi^{(2)}(q'|q-q')kk'\left|\vec{k} - \vec{k'}\right|\langle\Phi_{q'}\Phi_{q-q'}\Phi_{-q}\rangle$$

$$- 2\int dq' \chi^{(3)}(q'|-q'|q)\langle E^2\rangle_{q'}\langle E^2\rangle_q - \sum_a \int d\vec{p}\, \frac{(\varepsilon_0^{-1}e_a)^2\delta\left(\omega - \vec{k}\,\vec{u}\right)f_a(\vec{p})}{(2\pi)^3 k^2 \varepsilon^*(q)}$$

$$+ i\sum_a \int d\vec{p}\, \frac{(\varepsilon_0^{-1}e_a)\chi^{(2)*}(q'|q-q')}{ke^*(q)} k'\left|\vec{k} - \vec{k'}\right|\langle\Phi_{-q'}\Phi_{-q+q'}N_q^{a0}(\vec{p})\rangle,$$

$$(8.41)$$

where we have made use of $\langle E^2\rangle_q = \langle k^2\Phi^2\rangle_q$. Note that here we use summation and integral over $q \equiv (\vec{k}, \omega)$ interchangeably, that is, $\sum_q = \int dq = \int d\vec{k}\int d\omega$.

Eq. (8.41) contains third-order cumulants, $\langle\Phi_{q'}\Phi_{q-q'}\Phi_{-q}\rangle$, and $\langle\Phi_{-q'}\Phi_{-q+q'}N_q^{a0}(\vec{p})\rangle$. We may construct these quantities from Eq. (8.31) by ignoring the third-order nonlinearity at the outset. The three-body cumulants are zero if nonlinear terms are neglected since the linear solutions are essentially plane waves such that all odd moments of the wave amplitudes vanish upon taking the ensemble average. Thus if we write the perturbed field as a sum of the plane wave solution that satisfies the linear wave equation plus the nonlinear correction term, $\Phi_q = \Phi_q{}^{(0)} + \Phi_q{}^{(1)}$, where $\Phi_q{}^{(0)}$ satisfies $\varepsilon(q)\Phi_q{}^{(0)} = 0$, then upon inserting this expression to Eq. (8.31) and retaining the lowest-order terms within the nonlinear terms on the right-hand side, we obtain

$$\Phi_{q_1}^{(1)} = \frac{1}{k_1^2 \varepsilon(q_1)} \sum_{q''} i k_1 k'' \left| \vec{k}_1 - \vec{k''} \right| \chi^{(2)}(q''|q_1 - q'') \left[\Phi_{q''}^{(0)} \Phi_{q_1 - q''}^{(0)} - \left\langle \Phi_{q''}^{(0)} \Phi_{q_1 - q''}^{(0)} \right\rangle \right]$$

$$+ \frac{\varepsilon_0^{-1}}{k_1^2 \varepsilon(q_1)} \sum_a e_a \int d\vec{p}\, N_{q_1}^{a0}(\vec{p}).$$

(8.42)

The quantity $\langle \Phi_{q'} \Phi_{q-q'} \Phi_{-q} \rangle$ is constructed by making use of Eq. (8.48) for each of $\Phi_{q'}$, $\Phi_{q-q'}$, and Φ_{-q},

$$\langle \Phi_{q'} \Phi_{q-q'} \Phi_{-q} \rangle = \left\langle \Phi_{q'}^{(1)} \Phi_{q-q'}^{(0)} \Phi_{-q}^{(0)} \right\rangle + \left\langle \Phi_{q'}^{(0)} \Phi_{q-q'}^{(1)} \Phi_{-q}^{(0)} \right\rangle + \left\langle \Phi_{q'}^{(0)} \Phi_{q-q'}^{(0)} \Phi_{-q}^{(1)} \right\rangle + \cdots$$

(8.43)

Then we make use of Eq. (8.42) and omit the superscript (0) on the right-hand side at the end. Next, we make use of the symmetry property, $\chi^{(2)}(-q_1|-q_2) = -\chi^{(2)*}(q_1|q_2)$, to simplify various coupling coefficients, and decompose the four-body cumulants as products of two-body cumulants while ignoring irreducible components, thereby closing the hierarchy of correlations,

$$\langle \Phi_{q_1} \Phi_{q_2} \Phi_{q_3} \Phi_{q_4} \rangle = \delta(q_1 + q_2 + q_3 + q_4) \cdot \left[\delta(q_1 + q_2)\langle \Phi_{q_1} \Phi_{q_2} \rangle \langle \Phi_{q_3} \Phi_{q_4} \rangle \right.$$
$$+ \delta(q_1 + q_3)\langle \Phi_{q_1} \Phi_{q_3} \rangle \langle \Phi_{q_3} \Phi_{q_4} \rangle + \delta(q_1 + q_4)\langle \Phi_{q_1} \Phi_{q_4} \rangle \langle \Phi_{q_2} \Phi_{q_3} \rangle \big]$$
$$= \delta(q_1 + q_2 + q_3 + q_4) \cdot \left[\delta(q_1 + q_2)\langle \Phi^2 \rangle_{q_1} \langle \Phi^2 \rangle_{q_3} \right.$$
$$+ \delta(q_1 + q_3)\langle \Phi^2 \rangle_{q_1} \langle \Phi^2 \rangle_{q_2} + \delta(q_2 + q_3)\langle \Phi^2 \rangle_{q_2} \langle \Phi^2 \rangle_{q_3} \big].$$

(8.44)

In the previous equation, we have made use of $\langle \Phi_q \Phi_{-q} \rangle = \langle \Phi^2 \rangle_q$. After some tedious but otherwise straightforward algebraic manipulations, we obtain

$$\langle \Phi_{q'} \Phi_{q-q'} \Phi_{-q} \rangle = 2ikk' \left| \vec{k} - \vec{k'} \right| \left[\left[\frac{\chi^{(2)}(q|-q+q')}{k'^2 \varepsilon(q')} \langle \Phi^2 \rangle_{q-q'} \langle \Phi^2 \rangle_q + \frac{\chi^{(2)}(q|-q')}{\left| \vec{k} - \vec{k'} \right|^2 \varepsilon(q-q')} \langle \Phi^2 \rangle_{q'} \langle \Phi^2 \rangle_q \right. \right.$$

$$\left. - \frac{\chi^{(2)*}(q'|q-q')}{k^2 \varepsilon^*(q)} \langle \Phi^2 \rangle_{q'} \langle \Phi^2 \rangle_{q-q'} \right] + \varepsilon_0^{-1} \sum_a e_a \int d\vec{p} \left[\frac{\left\langle \Phi_{q-q'} \Phi_{-q} N_{q'}^{a0}(\vec{p}) \right\rangle}{k'^2 \varepsilon(q')} \right.$$

$$\left. \left. + \frac{\left\langle \Phi_{q'} \Phi_{-q} N_{q-q'}^{a0}(\vec{p}) \right\rangle}{\left| \vec{k} - \vec{k'} \right|^2 \varepsilon(q-q')} + \frac{\left\langle \Phi_{q'} \Phi_{q-q'} N_{-q}^{a0}(\vec{p}) \right\rangle}{k^2 \varepsilon^*(q)} \right] \right].$$

(8.45)

The closure scheme introduced in Eq. (8.44), namely, ignoring the irreducible four-body cumulant, is the simplest closure, which, in the theory of neutral fluid turbulence, is known as the "quasinormal closure."

Let us insert Eq. (8.45) to Eq. (8.41). The result is

$$
0 = \varepsilon(q)\langle E^2\rangle_q - \sum_a \frac{\left(\varepsilon_0^{-1} e_a\right)^2}{(2\pi)^3 k^2 \varepsilon^*(q)} \int d\vec{p}\,\delta(\omega - \vec{k}\,\vec{u}) f_a(\vec{p})
$$

$$
+ 2\int dq' \left[\chi^{(2)}(q'|q-q')\left(\frac{\chi^{(2)}(q|-q+q')}{\varepsilon(q')}\langle E^2\rangle_{q-q'} + \frac{\chi^{(2)}(q|-q')}{\varepsilon(q-q')}\langle E^2\rangle_{q'}\right)\right.
$$

$$
\left. - \chi^{(3)}(q'|-q'|q)\langle E^2\rangle_{q'}\right]\langle E^2\rangle_q - 2\int dq' \frac{\left|\chi^{(2)}(q'|q-q')\right|^2}{\varepsilon^*(q)}\langle E^2\rangle_{q'}\langle E^2\rangle_{q-q'}
$$

$$
- i\sum_a \varepsilon_0^{-1} e_a \int dq' \int d\vec{p}\; kk'\left|\vec{k}-\vec{k}'\right|\left\{\chi^{(2)}(q'|q-q')\left[\frac{\left\langle \Phi_{q-q'}\Phi_{-q}N_{q'}^{a0}(\vec{p})\right\rangle}{k'^2 \varepsilon(q')}\right.\right.
$$

$$
+ \frac{\left\langle \Phi_{q'}\Phi_{-q}N_{q-q'}^{a0}(\vec{p})\right\rangle}{\left|\vec{k}-\vec{k}'\right|^2 \varepsilon(q-q')} + \frac{\left\langle \Phi_{q'}\Phi_{q-q'}N_{-q}^{a0}(\vec{p})\right\rangle}{k^2 \varepsilon^*(q)}\right]
$$

$$
\left. - \chi^{(2)*}(q'|q-q')\frac{\left\langle \Phi_{-q'}\Phi_{-q+q'}N_q^{a0}(\vec{p})\right\rangle}{k^2 \varepsilon^*(q)}\right\}.
$$

$$(8.46)$$

It is evident that we need to further evaluate the remaining third-order cumulants $\left\langle \Phi_{q-q'}\Phi_{-q}N_{q'}^{a0}(\vec{p})\right\rangle$, $\left\langle \Phi_{q'}\Phi_{-q}N_{q-q'}^{a0}(\vec{p})\right\rangle$, $\left\langle \Phi_{q'}\Phi_{q-q'}N_{-q}^{a0}(\vec{p})\right\rangle$, and $\left\langle \Phi_{-q'}\Phi_{-q+q'}N_q^{a0}(\vec{p})\right\rangle$. These quantities are but special cases of a generic form $\left\langle \Phi_{q_1}\Phi_{-q_1+q_2}N_{-q_2}^{a0}(\vec{p})\right\rangle$. Let us proceed to evaluate this quantity. Making use of Eq. (8.42) to evaluate Φ_{q1} and Φ_{-q1+q2} successively, we have

$$
\left\langle \Phi_{q_1}\Phi_{-q_1+q_2}N_{-q_2}^{a0}(\vec{p})\right\rangle = \frac{2(2\pi)^{-3} i\varepsilon_0^{-1} e_a}{k_1 k_2 \left|\vec{k}_1 - \vec{k}_2\right|\varepsilon(q_2)}
$$

$$
\times \left[\frac{\chi^{(2)}(q_2|q_1-q_2)}{\varepsilon(q_1)}\langle E^2\rangle_{q_1-q_2} + \frac{\chi^{(2)}(-q_1|q_2)}{\varepsilon(-q_1+q_2)}\langle E^2\rangle_{q_1}\right]
$$

$$
\delta(\omega_2 - \vec{k}_2\vec{u}) f_a(\vec{p})
$$

$$(8.47)$$

Identifying $q_1 = q - q'$ and $q_2 = -q'$, we may obtain the expression for $\left\langle \Phi_{q-q'}\Phi_{-q}N_{q'}^{a0}(\vec{p})\right\rangle$. Making the identification for $q_1 = q'$ and $q_2 = -q + q'$, we may also have $\left\langle \Phi_{q'}\Phi_{-q}N_{q-q'}^{a0}(\vec{p})\right\rangle$. Likewise, setting $q_1 = q'$ and $q_2 = q$ leads to $\left\langle \Phi_{q'}\Phi_{q-q'}N_{-q}^{a0}(\vec{p})\right\rangle$. Finally, identifying $q_1 = -q'$ and $q_2 = -q$ yields the expression for $\left\langle \Phi_{-q'}\Phi_{-q+q'}N_q^{a0}(\vec{p})\right\rangle$. In this way, the contributions from all the necessary third-order cumulants to Eq. (8.46) can be obtained. The result is the so-called nonlinear spectral balance equation,

$$
0 = \varepsilon(q)\langle E^2\rangle_q - \sum_a \frac{(\varepsilon_0^{-1}e_a)^2}{(2\pi)^3 k^2 \varepsilon^*(q)} \int d\vec{p}\,\delta(\omega - \vec{k}\,\vec{u})f_a(\vec{p})
$$

$$
- 2\int dq' \frac{|\chi^{(2)}(q'|q-q')|^2}{\varepsilon^*(q)}\langle E^2\rangle_{q'}\langle E^2\rangle_{q-q'}
$$

$$
+ 2\int dq' \left\{ \left[\chi^{(2)}(q'|q-q')\right]^2 \left(\frac{\langle E^2\rangle_{q-q'}}{\varepsilon(q')} + \frac{\langle E^2\rangle_{q'}}{\varepsilon(q-q')}\right) - \chi^{(3)}(q'|-q'|q)\right\}\langle E^2\rangle_{q'}
$$

$$
+ \sum_a \int dq' \frac{2(\varepsilon_0^{-1}e_a)^2}{(2\pi)^3 k'^2 |\varepsilon(q')|^2} \left\{ \frac{[\chi^{(2)}(q'|q-q')]^2}{\varepsilon(q-q')}\langle E^2\rangle_q \right.
$$

$$
\left. - \frac{|\chi^{(2)}(q'|q-q')|^2}{\varepsilon^*(q)}\langle E^2\rangle_{q-q'}\right\} \times \int d\vec{p}\,\delta\left(\omega' - \vec{k}'\vec{u}\right)f_a(\vec{p})
$$

$$
+ \sum_a \int dq' \frac{2(\varepsilon_0^{-1}e_a)^2}{(2\pi)^3 \left|\vec{k}-\vec{k}'\right|^2 |\varepsilon(q-q')|^2} \times \left\{ \frac{[\chi^{(2)}(q'|q-q')]^2}{\varepsilon(q')}\langle E^2\rangle_q \right.
$$

$$
\left. - \frac{|\chi^{(2)}(q'|q-q')|^2}{\varepsilon^*(q)}\langle E^2\rangle_{q'}\right\} \cdot \int d\vec{p}\,\delta\left(\omega - \omega' - \left(\vec{k}-\vec{k}'\right)\vec{u}\right)f_a(\vec{p}).
$$

$$(8.48)$$

At this point, we reintroduce the slow time derivative, which was formally absorbed into the definition for new angular frequency in Eq. (8.23). Such a procedure was an approximate and heuristic treatment since the angular frequency ω appears in many places without the slow time derivative. In fact, the only place where ω appears with the slow time derivative factor $i\partial/\partial t$ is on the left-hand side of the equation for perturbed distribution. Consequently, when we extract the slow time derivative from the angular frequency in Eq. (8.48), a care must be exercised to implement it only in the leading term, namely, the linear response function,

$$
\varepsilon(q)\langle E^2\rangle_q \to \varepsilon\left(\vec{k},\omega + i\frac{\partial}{\partial t}\right)\langle E^2\rangle_q \to \left(\varepsilon(q) + \frac{i}{2}\frac{\partial\varepsilon(q)}{\partial\omega}\frac{\partial}{\partial t}\right)\langle E^2\rangle_q. \qquad (8.49)
$$

A more rigorous method is to employ mathematical multiple timescale perturbation analyses, as discussed by Davidson (1972). The present paper adopts the previously mentioned heuristic approach. The reason for factor $1/2$ in the previous equation is because the time derivative $\partial/\partial t$ is supposed to operate only on E_q within the ensemble average $\langle E_q^2\rangle = \langle E_q E_{-q}\rangle$, but since both E_q and E_{-q} are affected by $\partial/\partial t$, we simply divided the net result by the factor 2. This leads to

$$0 = \frac{i}{2}\frac{\partial \mathrm{Re}\varepsilon\left(\vec{k},\omega\right)}{\partial \omega}\frac{\partial \langle \delta E^2\rangle_{\vec{k},\omega}}{\partial t} + \mathrm{Re}\varepsilon\left(\vec{k},\omega\right)\langle \delta E^2\rangle_{\vec{k},\omega} + i\mathrm{Im}\varepsilon\left(\vec{k},\omega\right)\langle \delta E^2\rangle_{\vec{k},\omega}$$

$$+ 2\int d\vec{k}'\int d\omega'\left[\left\{\chi^{(2)}\left(\vec{k}',\omega'\left|\vec{k}-\vec{k}',\omega-\omega'\right.\right)\right\}^2\right.$$

$$\times \left(\frac{\langle \delta E^2\rangle_{\vec{k}-\vec{k}',\omega-\omega'}}{\varepsilon\left(\vec{k}',\omega'\right)} + \frac{\langle \delta E^2\rangle_{\vec{k}',\omega'}}{\varepsilon\left(\vec{k}-\vec{k}',\omega-\omega'\right)}\right)$$

$$\left. - \chi^{(3)}\left(\vec{k}',\omega'\left|-\vec{k}',-\omega'\right|\vec{k},\omega\right)\langle \delta E^2\rangle_{\vec{k}',\omega'}\right]\langle \delta E^2\rangle_{\vec{k},\omega}$$

$$- 2\int d\vec{k}'\int d\omega'\frac{\left|\chi^{(2)}\left(\vec{k}',\omega'\left|\vec{k}-\vec{k}',\omega-\omega'\right.\right)\right|^2}{\varepsilon^*\left(\vec{k},\omega\right)}\langle \delta E^2\rangle_{\vec{k}',\omega'}\langle \delta E^2\rangle_{\vec{k}-\vec{k}',\omega-\omega'}$$

$$- \frac{\varepsilon_0^{-2}}{8\pi^3}\frac{1}{k^2\varepsilon^*\left(\vec{k},\omega\right)}\sum_a e_a^2\int d\vec{p}\,\delta(\omega - \vec{k}\vec{u})f_a(\vec{p})$$

$$+ \frac{\varepsilon_0^{-2}}{4\pi^3}\int d\vec{k}'\int d\omega'\frac{1}{k'^2\left|\varepsilon\left(\vec{k}',\omega'\right)\right|^2}\left(\frac{\left\{\chi^{(2)}\left(\vec{k}',\omega'\left|\vec{k}-\vec{k}',\omega-\omega'\right.\right)\right\}^2}{\varepsilon\left(\vec{k}-\vec{k}',\omega-\omega'\right)}\langle \delta E^2\rangle_{\vec{k},\omega}\right.$$

$$\left. - \frac{\left|\chi^{(2)}\left(\vec{k}',\omega'\left|\vec{k}-\vec{k}',\omega-\omega'\right.\right)\right|^2}{\varepsilon^*\left(\vec{k},\omega\right)}\langle \delta E^2\rangle_{\vec{k}-\vec{k}',\omega-\omega'}\right)\sum_a e_a^2$$

$$\times \int d\vec{p}\,\delta\left(\omega' - \vec{k}'\vec{u}\right)f_a(\vec{p}) + \frac{\varepsilon_0^{-2}}{4\pi^3}\int d\vec{k}'\int d\omega'\frac{1}{\left|\vec{k}-\vec{k}'\right|^2\left|\varepsilon\left(\left|\vec{k}-\vec{k}',\omega-\omega'\right.\right)\right|^2}$$

$$\times \left(\frac{\left\{\chi^{(2)}\left(\vec{k}',\omega'\left|\vec{k}-\vec{k}',\omega-\omega'\right.\right)\right\}^2}{\varepsilon\left(\vec{k}',\omega'\right)}\langle \delta E^2\rangle_{\vec{k},\omega}\right.$$

$$\left. - \frac{\left|\chi^{(2)}\left(\vec{k}',\omega'\left|\vec{k}-\vec{k}',\omega-\omega'\right.\right)\right|^2}{\varepsilon^*\left(\vec{k},\omega\right)}\langle \delta E^2\rangle_{\vec{k}',\omega'}\right)$$

$$\times \sum_a e_a^2\int d\vec{p}\,\delta\left[\omega - \omega' - \left(\vec{k}-\vec{k}'\right)\vec{u}\right]f_a(\vec{p}), \tag{8.50}$$

where we have resorted back to the longhand notation.

We now discuss formal particle kinetic equation. From Eq. (8.24), the particle kinetic equation is given by

$$\frac{\partial f_a}{\partial t} = -ie_a \int d\vec{k} \int d\omega \, \vec{k} \cdot \frac{\partial}{\partial \vec{p}} \left\langle \Phi_{-q} N_q^a \right\rangle. \tag{8.51}$$

For the particle kinetic equation, it is sufficient to retain only the linear solution for N_q^a, that is, $N_q^a = N_q^{a0} + \vec{k} \cdot \vec{g}_q f_a \Phi_q$; see Eq. (8.30). It is possible to retain corrections to the particle kinetic equation. In fact, Tsytovich (1977) and Yoon et al. (2003) considered just such a problem, but they found that the nonlinear modification only leads to a small correction to the velocity–space diffusion coefficient. We consequently restrict ourselves to the lowest-order particle kinetic equation, which upon making use of

$$\left\langle \Phi_{-q} N_q^a \right\rangle = -\frac{e_a \langle E^2 \rangle_q}{\omega - \vec{k}\,\vec{u} + i0} \cdot \frac{\vec{k}}{k^2} \cdot \frac{\partial f_a}{\partial \vec{p}} + \frac{e_a \delta(\omega - \vec{k}\,\vec{u}) f_a}{(2\pi)^3 \varepsilon_0 k^2 \varepsilon^*(q)}, \tag{8.52}$$

leads to the desired formal particle kinetic equation,

$$\frac{\partial f_a}{\partial t} = e_a^2 \int d\vec{k} \int d\omega \, \frac{\vec{k}}{k} \cdot \frac{\partial}{\partial \vec{p}} \left[-\frac{i\delta(\omega - \vec{k}\,\vec{u}) f_a}{(2\pi)^3 \varepsilon_0 k \varepsilon^*(q)} + \frac{i\langle E^2 \rangle_q}{\omega - \vec{k}\,\vec{u} + i0} \cdot \frac{\vec{k}}{k} \cdot \frac{\partial f_a}{\partial \vec{p}} \right]. \tag{8.53}$$

8.3.2 Weak Turbulence Theory for Plasmas Described by Kappa Distributions

The customary weak turbulence theory found in the literature assumes that the bulk plasma distribution function is specified by quasi-Maxwellian form. However, in anticipation of the final result where we will find that the asymptotically steady-state electron distribution function is given by the kappa model, we reformulate the standard weak turbulence theory under the implicit assumption that the electron distribution is given by the kappa model of the form

$$f_e(u) = n_e (\pi \kappa' u_{Te}^2)^{-\frac{3}{2}} \frac{\Gamma(\kappa + 1)}{\Gamma\left(\kappa - \frac{1}{2}\right)} \left(1 + \frac{u^2}{\kappa' u_{Te}^2}\right)^{-\kappa - 1}, \tag{8.54}$$

where $\Gamma(x)$ is the gamma function. Note that Eq. (8.54) is slightly different and more general than the classic kappa distribution function in that the parameter κ' is different from κ. The κ' and u_{Te}^2 parameters must be determined eventually from the mean kinetic energy that defines the temperature (Chapter 1; Livadiotis, 2015). The kinetic temperature for the model distribution, Eq. (8.54), is rigorously defined by

$$\frac{3}{2} k_B T_e = \frac{1}{n_e} \int d\vec{u} \frac{1}{2} m_e u^2 f_e(u) = \frac{3}{2} \left(\frac{1}{2} m_e u_{Te}^2\right) \frac{\kappa'}{\kappa - \frac{3}{2}}. \tag{8.55}$$

Hence, we determine u_{Te}^2 from κ and T_e, and substituting in Eq. (8.54), the distribution becomes

$$f_e(u) = n_e \left[\pi \left(\kappa - \frac{3}{2} \right) \frac{2k_B}{m_e} T_e \right]^{-\frac{3}{2}} \frac{\Gamma(\kappa + 1)}{\Gamma\left(\kappa - \frac{1}{2} \right)} \left[1 + \frac{u^2}{\left(\kappa - \frac{3}{2} \right) \frac{2k_B}{m_e} T_e} \right]^{-\kappa - 1} , \qquad (8.56)$$

where we observe that κ' has been canceled out. By setting $\theta_e \equiv \sqrt{2k_B T_e / m_e}$ or $\theta_e^2 = u_{Te}^2 \kappa' / \left(\kappa - \frac{3}{2} \right)$, we obtain

$$f_e(u) = n_e \left[\pi \left(\kappa - \frac{3}{2} \right) \theta_e^2 \right]^{-\frac{3}{2}} \cdot \frac{\Gamma(\kappa + 1)}{\Gamma\left(\kappa - \frac{1}{2} \right)} \left(1 + \frac{1}{\kappa - \frac{3}{2}} \cdot \frac{u^2}{\theta_e^2} \right)^{-\kappa - 1} . \qquad (8.57)$$

We must also prove eventually that the kappa model of the form Eq. (8.54) is indeed the legitimate solution of the asymptotically steady-state equations of the weak turbulence theory.

Consider the real part of Eq. (8.50), which has linear and nonlinear terms, as well as terms associated with discrete particle effects. The discrete particle terms in Eq. (8.50) are those terms associated with $f_a(\vec{p})$, and these terms will eventually be responsible for various spontaneous emission and scattering processes. As far as the real part of Eq. (8.50) is concerned, nonlinear terms and spontaneous emission terms are ignored, thus leading to

$$\mathrm{Re}\varepsilon(\vec{k}, \omega) \langle \delta E^2 \rangle_{\vec{k}, \omega} = 0. \qquad (8.58)$$

If we denote the dispersion relation $\mathrm{Re}\varepsilon\left(\vec{k}, \omega^{\alpha}_{\vec{k}} \right) = 0$, then we may express the electric field fluctuation in terms of the eigenmode intensity by

$$\langle \delta E^2 \rangle_{\vec{k}, \omega} = \sum_{\sigma = \pm 1} \sum_{\alpha = L, S} I_{\vec{k}}^{\sigma \alpha} \delta\left(\omega - \sigma \omega^{\alpha}_{\vec{k}} \right). \qquad (8.59)$$

where $I_{\vec{k}}^{\sigma \alpha}$ is the intensity for each eigenmode, $\alpha = L, S$ denoting the Langmuir (L) and ion-sound S (or ion-acoustic) modes, respectively. The Langmuir wave dispersion relation for kappa distribution Eq. (8.54) can be obtained on the basis of the definition for linear response function Eq. (8.32a). We make use of nonrelativistic definition for momentum–velocity relation, $\vec{p} = m_a \vec{u}$, henceforth. Upon approximating for $\omega \gg \vec{k} \vec{u} = k u_{\|}$ (i.e., the fast mode condition), which is appropriate for Langmuir waves, and ignoring ion response, we have

$$\mathrm{Re}\varepsilon\left(\vec{k}, \omega^{\alpha}_{\vec{k}} \right) = 0 = 1 + \frac{e^2}{\varepsilon_0 k \omega} \int_{-\infty}^{\infty} du_{\|} \frac{\partial f_e}{\partial u_{\|}} \left(1 + \frac{k u_{\|}}{\omega} + \frac{k^2 u_{\|}^2}{\omega^2} + \frac{k^3 u_{\|}^3}{\omega^3} + \cdots \right)$$

$$= 1 - \frac{\omega_{pe}^2}{\omega^2} \left(1 + \frac{3}{2} \cdot \frac{k^2 \theta_e^2}{\omega^2} \right), \text{ or,}$$

$$\qquad (8.60)$$

$$\omega^2 = \omega_{pe}^2\left(1 + \frac{3}{2}\cdot\frac{k^2\theta_e^2}{\omega^2}\right) \cong \omega_{pe}^2\left(1 + \frac{3}{2}\cdot\frac{k^2\theta_e^2}{\omega_{pe}^2}\right) = \omega_{pe}^2 + 3k^2\frac{k_B T_e}{m_e}, \tag{8.61}$$

where $\omega_{pe}^2 = e^2 n_e/(\varepsilon_0 m_e)$ is the square of the plasma frequency (that is near to $\omega_{pl}^2 = e^2 n_e/(\varepsilon_0 m_0)$, with $m_0^{-1} = m_e^{-1} + m_p^{-1}$, as used in Chapter 5), and $\lambda_{De\infty}^2 = \frac{1}{2}\theta_e^2/\omega_{pe}^2$,

$$\omega\frac{L}{k} \cong \omega_{pe}\cdot\sqrt{1 + 3k^2\lambda_{De\infty}^2}, \tag{8.62}$$

with the superscript L shown the Langmuir waves.

To obtain the ion-sound mode dispersion relation, we impose the fast mode condition for the ion response, but approximate the slow mode condition, $\omega \ll \vec{k}\,\vec{u} = ku_\parallel$, for the electrons,

$$0 = 1 - \frac{\omega_{pi}^2}{\omega^2}\left(1 + \frac{3}{2}\frac{k^2 u_{Ti}^2}{\omega^2}\right) - \frac{e^2}{\varepsilon_0 k}\int du_\parallel \frac{\partial f_e}{\partial u_\parallel}\frac{1}{ku_\parallel}$$

$$= 1 - \frac{\omega_{pi}^2}{\omega^2}\left(1 + \frac{3}{2}\frac{k^2 u_{Ti}^2}{\omega^2}\right) + \frac{\omega_{pe}^2}{k^2\frac{1}{2}\theta_e^2}\cdot\frac{\kappa - \frac{1}{2}}{\kappa - \frac{3}{2}}, \tag{8.63}$$

where we have assumed Maxwellian distribution for the ions; $\omega_{pi}^2 = e^2 n_i/(\varepsilon_0 m_i)$ is the square of the ion (proton) plasma frequency. Solving in terms of ω^2, we have

$$\omega^2 = \omega_{pi}^2\left(1 + \frac{3}{2}\frac{k^2 u_{Ti}^2}{\omega^2}\right) - \frac{\omega^2}{k^2\lambda_{De}^2} \text{ or } \omega^2 = \omega_{pi}^2 k^2\lambda_{De}^2\frac{1 + \frac{3}{2}\frac{k^2\theta_i^2}{\omega^2}}{1 + k^2\lambda_{De}^2}. \tag{8.64}$$

Setting $\omega^2 \cong \omega_{pi}^2 k^2\lambda_{De}^2 + O(k^4)$ for the frequency in the right-hand side, we obtain

$$\omega^2 = k^2\cdot\frac{\omega_{pi}^2\lambda_{De}^2 + \frac{3}{2}\theta_i^2}{1 + k^2\lambda_{De}^2}, \text{ with } \theta_i^2 \equiv \frac{2k_B T_i}{m_i}. \tag{8.65}$$

Given that $n_i \approx n_e$ (by virtue of quasineutrality), we have

$$\omega_{pi}^2\lambda_{De}^2 = \frac{\omega_{pi}^2}{\omega_{pe}^2}\frac{1}{2}\theta_e^2\cdot\frac{\kappa - \frac{3}{2}}{\kappa - \frac{1}{2}} = \frac{k_B T_e}{m_i}\cdot\frac{\kappa - \frac{3}{2}}{\kappa - \frac{1}{2}}, \tag{8.66}$$

thus we end up with

$$\omega^2 = k^2\frac{k_B T_e}{m_i}\cdot\frac{\frac{\kappa - \frac{3}{2}}{\kappa - \frac{1}{2}} + 3\frac{T_i}{T_e}}{1 + k^2\lambda_{De}^2}, \tag{8.67}$$

where we used the kappa-dependent Debye length (Chapter 5, Section 5.5; Bryant, 1996; Rubab and Murtaza, 2006; Gougam and Tribeche, 2011; Livadiotis and McComas, 2014a), given by

$$
\lambda_{De}^2 = \frac{\frac{1}{2}\theta_e^2}{\omega_{pe}^2} \cdot \frac{\kappa - \frac{3}{2}}{\kappa - \frac{1}{2}} = \lambda_{De\infty}^2 \cdot \frac{\kappa - \frac{3}{2}}{\kappa - \frac{1}{2}}, \quad \lambda_{De\infty}^2 \equiv \frac{1}{2}\theta_e^2 \big/ \omega_{pe}^2. \tag{8.68}
$$

Hence, the ion acoustic mode dispersion relation follows:

$$
\omega_{\vec{k}}^S \cong k c_S \cdot \sqrt{\frac{1 + 3\frac{T_i}{T_e}\frac{\kappa - \frac{1}{2}}{\kappa - \frac{3}{2}}}{1 + k^2\lambda_{De}^2}}, \quad c_S \equiv \sqrt{\frac{k_B T_e}{m_i} \cdot \frac{\kappa - \frac{3}{2}}{\kappa - \frac{1}{2}}}, \tag{8.69}
$$

where the superscript S stands for the sound (or ion-acoustic) waves, with maximum propagation speed c_S.

The imaginary part of Eq. (8.50) leads to the wave kinetic equation, which can be manipulated to yield

$$
0 = \sum_{\sigma=\pm1}\sum_{\alpha}\left[\frac{\partial \mathrm{Re}\varepsilon\left(\vec{k},\sigma\omega_{\vec{k}}^{\alpha}\right)}{\partial\sigma\omega_{\vec{k}}^{\alpha}}\cdot\frac{\partial I_{\vec{k}}^{\sigma\alpha}}{\partial t} + 2\mathrm{Im}\varepsilon\left(\vec{k},\sigma\omega_{\vec{k}}^{\alpha}\right)I_{\vec{k}}^{\sigma\alpha}\right]\delta\left(\omega - \sigma\omega_{\vec{k}}^{\alpha}\right)
$$

$$
+ \sum_{\sigma=\pm1}\sum_{\alpha}4\mathrm{Im}\int d\vec{k}'\left\{\sum_{\sigma''=\pm1}\sum_{\gamma}\frac{\left[\chi^{(2)}\left(\vec{k}',\sigma\omega_{\vec{k}}^{\alpha} - \sigma''\omega_{\vec{k}-\vec{k}'}^{\gamma}\bigg|\vec{k} - \vec{k}',\sigma''\omega_{\vec{k}-\vec{k}'}^{\gamma}\right)\right]^2}{\varepsilon\left(\vec{k}',\sigma\omega_{\vec{k}}^{\alpha} - \sigma''\omega_{\vec{k}-\vec{k}'}^{\gamma}\right)}\right.
$$

$$
\left.I_{\vec{k}-\vec{k}'}^{\sigma''\gamma} + \sum_{\sigma'=\pm1}\sum_{\beta}\frac{\left[\chi^{(2)}\left(\vec{k}',\sigma'\omega_{\vec{k}'}^{\beta}\bigg|\vec{k} - \vec{k}',\sigma\omega_{\vec{k}}^{\alpha} - \sigma'\omega_{\vec{k}'}^{\beta}\right)\right]^2}{\varepsilon\left(\vec{k} - \vec{k}',\sigma\omega_{\vec{k}}^{\alpha} - \sigma'\omega_{\vec{k}'}^{\beta}\right)}I_{\vec{k}'}^{\sigma'\beta}\right.
$$

$$
\left. - \sum_{\sigma'=\pm1}\sum_{\beta}\chi^{(3)}\left(\vec{k}',\sigma'\omega_{\vec{k}'}^{\beta}\bigg| - \vec{k}',-\sigma'\omega_{\vec{k}'}^{\beta}\bigg|\vec{k},\sigma\omega_{\vec{k}}^{\alpha}\right)I_{\vec{k}'}^{\sigma'\beta}\right\}I_{\vec{k}}^{\sigma\alpha}\delta\left(\omega - \sigma\omega_{\vec{k}}^{\alpha}\right)
$$

$$
- \sum_{\sigma',\sigma''=\pm1}\sum_{\beta,\gamma}4\mathrm{Im}\int d\vec{k}'\frac{\left|\chi^{(2)}\left(\vec{k}',\sigma'\omega_{\vec{k}'}^{\beta}\bigg|\vec{k} - \vec{k}',\sigma''\omega_{\vec{k}-\vec{k}'}^{\gamma}\right)\right|^2}{\varepsilon^*\left(\vec{k},\sigma'\omega_{\vec{k}'}^{\beta} + \sigma''\omega_{\vec{k}-\vec{k}'}^{\gamma}\right)}I_{\vec{k}'}^{\sigma'\beta}I_{\vec{k}-\vec{k}'}^{\sigma''\gamma}
$$

$$
\delta\left(\omega - \sigma'\omega_{\vec{k}'}^{\beta} - \sigma''\omega_{\vec{k}-\vec{k}'}^{\gamma}\right)
$$

$$
- \mathrm{Im}\frac{2\varepsilon_0^{-2}}{(2\pi)^3 k^2\varepsilon^*\left(\vec{k},\omega\right)}\sum_{a}e_a^2\int d\vec{u}\,\delta(\omega - \vec{k}\,\vec{u})f_a(\vec{u})
$$

$$+ 2\mathrm{Im} \sum_a \int d\vec{k'} \int d\omega' \frac{2\left(\varepsilon_0^{-1} e_a\right)^2}{(2\pi)^3 k'^2 \left|\varepsilon\left(\vec{k'}, \omega'\right)\right|^2}$$

$$\left\{ \sum_{\sigma=\pm 1} \sum_\alpha \frac{\left[\chi^{(2)}\left(\vec{k'}, \omega' \middle| \vec{k} - \vec{k'}, \sigma\omega^\alpha_{\vec{k}} - \omega'\right)\right]^2}{\varepsilon\left(\vec{k} - \vec{k'}, \sigma\omega^\alpha_{\vec{k'}} - \omega'\right)} I^{\sigma\alpha}_{\vec{k}} \delta\left(\omega - \sigma\omega^\alpha_{\vec{k}}\right) \right.$$

$$\left. - \sum_{\sigma''=\pm 1} \sum_\gamma \frac{\left|\chi^{(2)}\left(\vec{k'}, \omega' \middle| \vec{k} - \vec{k'}, \sigma''\omega^\gamma_{\vec{k} - \vec{k'}}\right)\right|^2}{\varepsilon^*\left(\vec{k'}, \omega' + \sigma''\omega^\gamma_{\vec{k} - \vec{k'}}\right)} I^{\sigma''\gamma}_{\vec{k} - \vec{k'}} \delta\left(\omega - \omega' - \sigma''\omega^\gamma_{\vec{k} - \vec{k'}}\right) \right\}$$

$$\times \int d\vec{u} \, \delta\left(\omega' - \vec{k'}\vec{u}\right) f_a(\vec{u})$$

$$+ 2\mathrm{Im} \sum_a \int d\vec{k'} \int d\omega' \frac{2\left(\varepsilon_0^{-1} e_a\right)^2}{(2\pi)^3 \left|\vec{k} - \vec{k'}\right|^2 \left|\varepsilon\left(\vec{k} - \vec{k'}, \omega - \omega'\right)\right|^2}$$

$$\times \left\{ \sum_{\sigma=\pm 1} \sum_\alpha \frac{\left[\chi^{(2)}\left(\vec{k'}, \omega' \middle| \vec{k} - \vec{k'}, \sigma\omega^\alpha_{\vec{k}} - \omega'\right)\right]^2}{\varepsilon\left(\vec{k'}, \omega'\right)} I^{\sigma\alpha}_{\vec{k}} \delta\left(\omega - \sigma\omega^\alpha_{\vec{k}}\right) \right.$$

$$\left. - \sum_{\sigma'=\pm 1} \sum_\beta \frac{\left|\chi^{(2)}\left(\vec{k'}, \omega' \middle| \vec{k} - \vec{k'}, \omega - \omega'\right)\right|^2}{\varepsilon^*\left(\vec{k}, \omega\right)} I^{\sigma'\beta}_{\vec{k'}} \delta\left(\omega' - \sigma'\omega^\beta_{\vec{k'}}\right) \right\}$$

$$\int d\vec{u} \, \delta\left[\omega - \omega' - \left(\vec{k} - \vec{k'}\right)\vec{u}\right] f_a(\vec{u}). \tag{8.70}$$

Since the linear eigenmodes satisfy the dispersion relation, $\varepsilon\left(\vec{k}, \sigma\omega^\alpha_{\vec{k}}\right) \approx 0$, $\alpha = L, S$, we have

$$\frac{1}{\varepsilon\left(\vec{k}, \omega\right)} = P\frac{1}{\varepsilon\left(\vec{k}, \omega\right)} - \sum_\sigma \sum_{\alpha=L,S} \frac{i\pi\delta\left(\omega - \sigma\omega^\alpha_{\vec{k}}\right)}{\varepsilon'\left(\vec{k}, \sigma\omega^\alpha_{\vec{k}}\right)},$$

$$\frac{1}{\varepsilon^*\left(\vec{k}, \omega\right)} = P\frac{1}{\varepsilon^*\left(\vec{k}, \omega\right)} + \sum_\sigma \sum_{\alpha=L,S} \frac{i\pi\delta\left(\omega - \sigma\omega^\alpha_{\vec{k}}\right)}{\varepsilon'\left(\vec{k}, \sigma\omega^\alpha_{\vec{k}}\right)},$$

$$\tag{8.71}$$

where a shorthand notation $\varepsilon'\left(\vec{k}, \sigma\omega^\alpha_{\vec{k}}\right) = \partial \mathrm{Re}\varepsilon'\left(\vec{k}, \sigma\omega^\alpha_{\vec{k}}\right)\big/\partial\sigma\omega^\alpha_{\vec{k}}$, is used; P denotes the principal value. The following relations hold for linear eigenmodes $\omega = \sigma\,\omega^\alpha_{\vec{k}}$, where $\alpha = L,S$:

$$\frac{1}{\varepsilon'\left(\vec{k}, \sigma\omega^L_{\vec{k}}\right)} = \frac{\sigma\omega^L_{\vec{k}}}{2}, \quad \frac{1}{\varepsilon'\left(\vec{k}, \sigma\omega^S_{\vec{k}}\right)} = \frac{\sigma\mu^L_{\vec{k}}\omega^L_{\vec{k}}}{2},$$

(8.72)

$$\mu^L_{\vec{k}} = k^3\lambda^3_{De}\sqrt{\frac{m_e}{m_i}}\sqrt{\frac{\kappa - \dfrac{3}{2}}{\kappa - \dfrac{1}{2}} + \frac{3T_i}{T_e}},$$

Eqs. (8.71) and (8.72) lead to the reduction of Eq. (8.70)

$$\frac{\partial I^{\sigma\alpha}_{\vec{k}}}{\partial t} = -\frac{2\mathrm{Im}\varepsilon\left(\vec{k}, \sigma\omega^\alpha_{\vec{k}}\right)}{\varepsilon'\left(\vec{k}, \sigma\omega^\alpha_{\vec{k}}\right)}I^{\sigma\alpha}_{\vec{k}} + \sum_a \frac{\varepsilon_0^{-2}e_a^2}{4\pi^2 k^2\left|\varepsilon'\left(\vec{k}, \sigma\omega^\alpha_{\vec{k}}\right)\right|^2}\int d\vec{u}\,\delta\left(\sigma\omega^\alpha_{\vec{k}} - \vec{k}\,\vec{u}\right)f_a(\vec{u})$$

$$-\frac{4}{\varepsilon'\left(\vec{k}, \sigma\omega^\alpha_{\vec{k}}\right)}\sum_{\sigma'=\pm 1}\sum_{\beta}\int d\vec{k'}\,\mathrm{Im}\Bigg\{P\frac{2\left[\chi^{(2)}\left(\vec{k'}, \sigma'\omega^\beta_{\vec{k'}}\Big|\vec{k}-\vec{k'}, \sigma\omega^\alpha_{\vec{k}} - \sigma'\omega^\beta_{\vec{k'}}\right)\right]^2}{\varepsilon\left(\vec{k}-\vec{k'}, \sigma\omega^\alpha_{\vec{k}} - \sigma'\omega^\beta_{\vec{k'}}\right)}$$

$$-\chi^{(3)}\left(\vec{k'}, \sigma'\omega^\beta_{\vec{k'}}\Big|-\vec{k'}, -\sigma'\omega^\beta_{\vec{k'}}\Big|\vec{k}, \sigma\omega^\alpha_{\vec{k}}\right)\Bigg\}I^{\sigma\alpha}_{\vec{k}}I^{\sigma'\beta}_{\vec{k'}}$$

$$-\sum_a\frac{\varepsilon_0^{-2}e_a^2}{\pi^2\varepsilon'\left(\vec{k}, \sigma\omega^\alpha_{\vec{k}}\right)}\sum_{\sigma'=\pm 1}\sum_{\beta}\int d\vec{k'}\frac{\left|\chi^{(2)}\left(\vec{k'}, \sigma'\omega^\beta_{\vec{k'}}\Big|\vec{k}-\vec{k'}, \sigma\omega^\alpha_{\vec{k}} - \sigma'\omega^\beta_{\vec{k'}}\right)\right|^2}{\left|\vec{k}-\vec{k'}\right|^2\left|\varepsilon\left(\vec{k}-\vec{k'}, \sigma\omega^\alpha_{\vec{k}} - \sigma'\omega^\beta_{\vec{k'}}\right)\right|^2}$$

$$\times\left[\frac{I^{\sigma\alpha}_{\vec{k}}}{\varepsilon'\left(\vec{k'}, \sigma'\omega^\beta_{\vec{k'}}\right)} - \frac{I^{\sigma'\beta}_{\vec{k'}}}{\varepsilon'\left(\vec{k}, \sigma\omega^\alpha_{\vec{k}}\right)}\right]\int d\vec{u}\,\delta\left[\sigma\omega^\alpha_{\vec{k}} - \sigma'\omega^\beta_{\vec{k'}} - \left(\vec{k}-\vec{k'}\right)\vec{u}\right]f_a(\vec{u})$$

$$-\frac{4\pi}{\varepsilon'\left(\vec{k}, \sigma\omega^\alpha_{\vec{k}}\right)}\sum_{\sigma',\sigma''=\pm 1}\sum_{\beta,\gamma}\int d\vec{k'}\left|\chi^{(2)}\left(\vec{k'}, \sigma'\omega^\beta_{\vec{k'}}\Big|\vec{k}-\vec{k'}, \sigma''\omega^\gamma_{\vec{k}-\vec{k'}}\right)\right|^2$$

$$\times\left(\frac{I^{\sigma''\gamma}_{\vec{k}-\vec{k'}}I^{\sigma\alpha}_{\vec{k}}}{\varepsilon'\left(\vec{k'}, \sigma'\omega^\beta_{\vec{k'}}\right)} + \frac{I^{\sigma'\beta}_{\vec{k'}}I^{\sigma\alpha}_{\vec{k}}}{\varepsilon'\left(\vec{k}-\vec{k'}, \sigma''\omega^\gamma_{\vec{k}-\vec{k'}}\right)} - \frac{I^{\sigma'\beta}_{\vec{k'}}I^{\sigma''\gamma}_{\vec{k}-\vec{k'}}}{\varepsilon'\left(\vec{k}, \sigma\omega^\alpha_{\vec{k}}\right)}\right)$$

$$\delta\left(\sigma\omega^\alpha_{\vec{k}} - \sigma'\omega^\beta_{\vec{k'}} - \sigma''\omega^\gamma_{\vec{k}-\vec{k'}}\right).$$

(8.73)

This is formally identical to the wave kinetic equation for conventional weak turbulence theory except that, as we have already seen, the linear dispersive properties are modified by the kappa model. The nonlinear coefficients will also be modified for kappa plasmas, as we shall see next. Before we move on, we note that the first term on the right-hand side corresponds to the induced emission; the second term depicts the spontaneous emission. The third term represents the induced scattering, the fourth term describes the spontaneous scattering, and the final term depicts the decay processes. Of the terms depicting the decay processes, the first two terms within the large parenthesis are responsible for the induced decay process, while the final term corresponds to the spontaneous decay. We next calculate specific forms of each term in the previous wave kinetic equation.

8.3.2.1 Induced and Spontaneous Emissions

The induced emissions terms are given by

$$
\left.\frac{\partial I^{\sigma L}_{\vec{k}}}{\partial t}\right|_{ind.em.} = -\frac{2\mathrm{Im}\varepsilon\left(\vec{k},\sigma\omega^{L}_{\vec{k}}\right)}{\varepsilon'\left(\vec{k},\sigma\omega^{L}_{\vec{k}}\right)}I^{\sigma L}_{\vec{k}} = \pi\sigma\omega^{L}_{\vec{k}}\frac{4\pi e^2}{m_e k^2}\int d\vec{u}\,\delta\left(\sigma\omega^{L}_{\vec{k}} - \vec{k}\,\vec{u}\right)\vec{k}\,\frac{\partial f_e}{\partial \vec{u}}I^{\sigma L}_{\vec{k}},
$$

$$
\left.\frac{\partial I^{\sigma S}_{\vec{k}}}{\partial t}\right|_{ind.em.} = -\frac{2\mathrm{Im}\varepsilon\left(\vec{k},\sigma\omega^{S}_{\vec{k}}\right)}{\varepsilon'\left(\vec{k},\sigma\omega^{S}_{\vec{k}}\right)}I^{\sigma S}_{\vec{k}} = \pi\mu_{\vec{k}}\sigma\omega^{L}_{\vec{k}}\frac{4\pi e^2}{m_e k^2}\int d\vec{u}\,\delta\left(\sigma\omega^{S}_{\vec{k}} - \vec{k}\,\vec{u}\right)
$$

$$
\times\,\vec{k}\,\frac{\partial}{\partial \vec{u}}\left(f_e + \frac{m_e}{m_i}f_i\right)I^{\sigma S}_{\vec{k}}. \tag{8.74}
$$

Spontaneous emission terms are

$$
\left.\frac{\partial I^{\sigma L}_{\vec{k}}}{\partial t}\right|_{spont.em.} = \sum_a \frac{4e_a^2}{k^2\left|\varepsilon'\left(\vec{k},\sigma\omega^{L}_{\vec{k}}\right)\right|^2}\int d\vec{u}\,\delta\left(\sigma\omega^{L}_{\vec{k}} - \vec{k}\,\vec{u}\right)f_a(\vec{u})
$$

$$
= \frac{4\pi n_e^2 e^4}{m_e k^2}\int d\vec{u}\,\delta\left(\sigma\omega^{L}_{\vec{k}} - \vec{k}\,\vec{u}\right)f_e(\vec{u}),
$$

$$
\left.\frac{\partial I^{\sigma S}_{\vec{k}}}{\partial t}\right|_{spont.em.} = \sum_a \frac{4e_a^2}{k^2\left|\varepsilon'\left(\vec{k},\sigma\omega^{S}_{\vec{k}}\right)\right|^2}\int d\vec{u}\,\delta\left(\sigma\omega^{S}_{\vec{k}} - \vec{k}\,\vec{u}\right)f_a(\vec{u})
$$

$$
= \mu^2_{\vec{k}}\frac{4\pi n_e^2 e^4}{m_e k^2}\int d\vec{u}\,\delta\left(\sigma\omega^{S}_{\vec{k}} - \vec{k}\,\vec{u}\right)(f_e + f_i). \tag{8.75}
$$

8.3.2.2 Induced and Spontaneous Decay Processes

Sitenko (1982) discusses the various approximate and limiting forms of the nonlinear susceptibilities. In the present author's earlier publications (Yoon, 2000, 2005), one also finds similar simplifications of various nonlinear susceptibilities. However, these customary results implicitly assume Maxwellian distribution. Upon assuming kappa distribution, however, one may generalize the approximate second-order susceptibility in a straightforward manner, the derivation of which will be omitted for the sake of space economy. The final result is as follows:

$$
\left| \chi^{(2)}\left(\vec{k'}, \sigma'\omega_{\vec{k'}}^{L} \middle| \vec{k} - \vec{k'}, \sigma\omega_{\vec{k}}^{S} - \sigma'\omega_{\vec{k'}}^{L} \right) \right|^2 \cong \left(\frac{\kappa - \frac{1}{2}}{\kappa - \frac{3}{2}} \right)^2 \frac{e^2}{4(k_B T_e)^2} \cdot \frac{\left| \vec{k'}\left(\vec{k} - \vec{k'} \right) \right|^2}{k^2 k'^2 \left| \vec{k} - \vec{k'} \right|^2}.
$$

$$(8.76)$$

Making use of Eq. (8.76), the decay processes can be obtained as follows:

$$
\frac{\partial I_{\vec{k}}^{\sigma L}}{\partial t}\bigg|_{decay} = -\frac{8\pi}{\varepsilon'\left(\vec{k}, \sigma\omega_{\vec{k}}^{L} \right)} \sum_{\sigma',\sigma''=\pm1} \int d\vec{k'} \left| \chi^{(2)}\left(\vec{k'}, \sigma'\omega_{\vec{k'}}^{L} \middle| \vec{k} - \vec{k'}, \sigma''\omega_{\vec{k}-\vec{k'}}^{S} \right) \right|^2
$$

$$
\times \left(\frac{I_{\vec{k}-\vec{k'}}^{\sigma''S} I_{\vec{k}}^{\sigma L}}{\varepsilon'\left(\vec{k'}, \sigma'\omega_{\vec{k'}}^{L} \right)} + \frac{I_{\vec{k'}}^{\sigma' L} I_{\vec{k}}^{\sigma L}}{\varepsilon'\left(\vec{k} - \vec{k'}, \sigma''\omega_{\vec{k}-\vec{k'}}^{S} \right)} - \frac{I_{\vec{k'}}^{\sigma' L} I_{\vec{k}-\vec{k'}}^{\sigma'' S}}{\varepsilon'\left(\vec{k}, \sigma\omega_{\vec{k}}^{L} \right)} \right)
$$

$$
\times \delta\left(\sigma\omega_{\vec{k}}^{L} - \sigma'\omega_{\vec{k'}}^{L} - \sigma''\omega_{\vec{k}-\vec{k'}}^{S} \right) = \left(\frac{\kappa - \frac{1}{2}}{\kappa - \frac{3}{2}} \right)^2 \frac{\varepsilon_0^{-2} e^2}{32\pi (k_B T_e)^2}
$$

$$
\times \sum_{\sigma',\sigma''=\pm1} \sigma\omega_{\vec{k}}^{L} \int d\vec{k'} \frac{\left(\vec{k}\,\vec{k'} \right)^2}{k^2 k'^2 \left| \vec{k} - \vec{k'} \right|^2} \delta\left(\sigma\omega_{\vec{k}}^{L} - \sigma'\omega_{\vec{k'}}^{L} - \sigma''\omega_{\vec{k}-\vec{k'}}^{S} \right)
$$

$$
\times \left(\sigma\omega_{\vec{k}}^{L} I_{\vec{k'}}^{\sigma' L} I_{\vec{k}-\vec{k'}}^{\sigma'' S} - \sigma'\omega_{\vec{k'}}^{L} I_{\vec{k}-\vec{k'}}^{\sigma'' S} I_{\vec{k}}^{\sigma L} - \sigma'' \mu_{\vec{k}-\vec{k'}} \omega_{\vec{k}-\vec{k'}}^{L} I_{\vec{k'}}^{\sigma' L} I_{\vec{k}}^{\sigma L} \right),
$$

$$(8.77a)$$

$$
\frac{\partial I_{\vec{k}}^{\sigma S}}{\partial t}\bigg|_{decay} = -\frac{8\pi}{\varepsilon'\left(\vec{k}, \sigma\omega_{\vec{k}}^{S} \right)} \sum_{\sigma',\sigma''=\pm1} \int d\vec{k'} \left| \chi^{(2)}\left(\vec{k'}, \sigma'\omega_{\vec{k'}}^{L} \middle| \vec{k} - \vec{k'}, \sigma''\omega_{\vec{k}-\vec{k'}}^{L} \right) \right|^2
$$

$$
\times \left(\frac{I_{\vec{k}-\vec{k'}}^{\sigma'' L} I_{\vec{k}}^{\sigma S}}{\varepsilon'\left(\vec{k'}, \sigma'\omega_{\vec{k'}}^{L} \right)} + \frac{I_{\vec{k'}}^{\sigma' L} I_{\vec{k}}^{\sigma S}}{\varepsilon'\left(\vec{k} - \vec{k'}, \sigma''\omega_{\vec{k}-\vec{k'}}^{L} \right)} - \frac{I_{\vec{k'}}^{\sigma' L} I_{\vec{k}-\vec{k'}}^{\sigma'' L}}{\varepsilon'\left(\vec{k}, \sigma\omega_{\vec{k}}^{S} \right)} \right) \delta
$$

$$\left(\sigma \omega^S_{\vec{k}} - \sigma' \omega^L_{\vec{k'}} - \sigma'' \omega^L_{\vec{k}-\vec{k'}} \right) = \left(\frac{\kappa - \frac{1}{2}}{\kappa - \frac{3}{2}} \right)^2 \frac{\varepsilon_0^{-2} e^2}{32\pi (k_B T_e)^2} \sum_{\sigma', \sigma'' = \pm 1} \sigma \mu_{\vec{k}} \omega^L_{\vec{k}}$$

$$\times \int d\vec{k'} \frac{\left[\vec{k'} \left(\vec{k} - \vec{k'} \right) \right]^2}{k^2 k'^2 \left| \vec{k} - \vec{k'} \right|^2} \delta \left(\sigma \omega^S_{\vec{k}} - \sigma' \omega^L_{\vec{k'}} \right.$$

$$\left. - \sigma'' \omega^L_{\vec{k}-\vec{k'}} \right) \times \left(\sigma \mu_{\vec{k}} \omega^L_{\vec{k}} I^{\sigma'L}_{\vec{k'}} I^{\sigma''L}_{\vec{k}-\vec{k'}} \right.$$

$$\left. - \sigma' \omega^L_{\vec{k'}} I^{\sigma''L}_{\vec{k}-\vec{k'}} I^{\sigma S}_{\vec{k}} - \sigma'' \omega^L_{\vec{k}-\vec{k'}} I^{\sigma'L}_{\vec{k'}} I^{\sigma S}_{\vec{k}} \right) \qquad (8.77b)$$

The first terms on the right-hand sides represent the spontaneous decay processes, while the remaining two terms correspond to induced decays.

8.3.2.3 Induced and Spontaneous Scattering Processes

For this process, Yoon (2000, 2005) discussed the derivation in detail. For L mode, it was shown that the most important terms are induced scattering involving two L modes. For the second-order nonlinear susceptibility of relevance to such a process, one may treat $\sigma \omega^L_{\vec{k}} - \sigma' \omega^L_{\vec{k'}}$ as the slow mode. One may also approximate the real part of the dielectric constant $\varepsilon \left(\vec{k} - \vec{k'}, \sigma \omega^L_{\vec{k}} - \sigma' \omega^L_{\vec{k'}} \right)$ by $\left[\left| \vec{k} - \vec{k'} \right| \lambda^2_{De} \right]^{-1}$. The imaginary part of the previously mentioned dielectric constant has both electron and ion contributions. Upon treating the imaginary part as a small quantity, one may expand the term $1 / \varepsilon \left(\vec{k} - \vec{k'}, \sigma \omega^L_{\vec{k}} - \sigma' \omega^L_{\vec{k'}} \right)$ as $\left[\left| \vec{k} - \vec{k'} \right| \lambda^2_{De} \right]^{-1} - i \mathrm{Im} \varepsilon \left(\vec{k} - \vec{k'}, \sigma \omega^L_{\vec{k}} - \sigma' \omega^L_{\vec{k'}} \right)$. The third-order nonlinear susceptibility can also be approximated by treating $\sigma \omega^L_{\vec{k}} - \sigma' \omega^L_{\vec{k'}}$ as arbitrary (slow mode) frequency. However, again, we note that all these approximations implicitly assume Maxwellian bulk electron distribution. One may follow the same procedure to generalize the results to kappa plasma. The resulting expression can be shown to be

$$\frac{\partial I^{\sigma L}_{\vec{k}}}{\partial t} \bigg|_{ind.sc.} = -\frac{4}{\varepsilon' \left(\vec{k}, \sigma \omega^\alpha_{\vec{k}} \right)} \sum_{\sigma' = \pm 1} \int d\vec{k'}$$

$$\times \mathrm{Im} \left\{ P \frac{2 \left[\chi^{(2)} \left(\vec{k'}, \sigma' \omega^L_{\vec{k'}} \middle| \vec{k} - \vec{k'}, \sigma \omega^L_{\vec{k}} - \sigma' \omega^L_{\vec{k'}} \right) \right]^2}{\varepsilon \left(\vec{k} - \vec{k'}, \sigma \omega^L_{\vec{k}} - \sigma' \omega^L_{\vec{k'}} \right)} \right.$$

$$
\left. -\chi^{(3)}\left(\vec{k'},\sigma'\omega^L_{\vec{k'}}\middle|-\vec{k'},-\sigma'\omega^L_{\vec{k'}}\middle|\vec{k},\sigma\omega^L_{\vec{k}}\right)\right\}\cdot I^{\sigma L}_{\vec{k}}I^{\sigma'L}_{\vec{k'}}
$$

$$
= \left(\frac{\kappa-\frac{1}{2}}{\kappa-\frac{3}{2}}\right)^2\frac{\pi\varepsilon_0\left(\sigma\omega^L_{\vec{k}}\right)}{n_e^2 m_i}\sum_{\sigma'=\pm 1}\int d\vec{k'}\int d\vec{u}\,\frac{\left(\vec{k}\,\vec{k'}\right)^2}{k^2 k'^2}\delta\left[\sigma\omega^L_{\vec{k}}-\sigma'\omega^L_{\vec{k'}}-\left(\vec{k}-\vec{k'}\right)\vec{u}\right]
$$

$$
\times\left(\vec{k}-\vec{k'}\right)\frac{\partial f_i}{\partial\vec{u}}I^{\sigma L}_{\vec{k}}I^{\sigma'L}_{\vec{k'}}.
$$

(8.78)

Spontaneous scattering terms discussed by (Yoon, 2005) can also be generalized to kappa plasma by

$$
\left.\frac{\partial I^{\sigma S}_{\vec{k}}}{\partial t}\right|_{spont.sc.} = -\left(\sigma\omega^L_{\vec{k}}\right)\frac{e^4}{(k_B T_e)^2}\lambda^4_{De}\sum_{\sigma'=\pm 1}\int d\vec{k'}\int d\vec{u}\,\frac{\left(\vec{k}\,\vec{k'}\right)^2}{k^2 k'^2}
$$

$$
\times\left(\sigma'\omega^L_{\vec{k'}}I^{\sigma L}_{\vec{k}}-\sigma\omega^L_{\vec{k}}I^{\sigma'L}_{\vec{k'}}\right)\delta\left[\sigma\omega^L_{\vec{k}}-\sigma'\omega^L_{\vec{k'}}-\left(\vec{k}-\vec{k'}\right)\vec{u}\right](f_e+f_i).
$$

(8.79)

This completes the reformulation of wave kinetic equation for kappa plasmas in the context of the plasma weak turbulence theory. The particle equation is rather trivial, so we will omit the discussion. We next make use of the results we have discussed thus far and consider the asymptotical steady state of turbulent plasma (that is, for $t \to \infty$). The notion of the so-called turbulent equilibrium physically implies a plasma state in which particles exchange momentum and energy mediated by long-range collective interactions, i.e., waves, while dynamically the entire system is in equilibrium. Such a system is an evidently clear example of a statistical system that naturally lends itself out of thermal equilibrium and is described by the framework of nonextensive statistical mechanics.

8.4 Turbulent Quasiequilibrium and Kappa Electron Distribution

8.4.1 Steady-State Particle Distribution

Let us consider the electron kinetic Eq. (8.53), which can be rewritten as follows, taking into account only the Langmuir wave intensity in the velocity space diffusion coefficient:

$$
\frac{\partial f_e}{\partial t} = \frac{\partial}{\partial u_i}\left(A_i f_e + D_{ij}\frac{\partial f_e}{\partial u_j}\right),\quad\binom{A_i}{D_{ij}}
$$

$$
= \frac{\pi e^2}{m_e^2}\int d\vec{k}\,\frac{k_i}{k^2}\sum_{\sigma=\pm 1}\delta\left(\sigma\omega^L_{\vec{k}}-\vec{k}\,\vec{u}\right)\cdot\binom{\frac{1}{4\pi^2}\sigma\omega^L_{\vec{k}}m_e}{k_j I^{\sigma L}_{\vec{k}}},
$$

(8.80)

using the Centimetre–Gram–Second system. In what follows, we assume that the forward- and backward-propagating wave intensities are identical, $I^{\pm L}_{\vec{k}} = I^{L}_{\vec{k}}$. Upon decomposing the wave vector $\vec{k}$ into components perpendicular and parallel to the velocity vector $\vec{u}$, one can show that the perpendicular component of the velocity friction (or drag) coefficient $\vec{A}$ vanishes upon summation over ± 1. We may thus write

$$A_i = u_i G(u), \text{ with } G(u) = \frac{e^2}{4\pi m_e u^2} \sum_{\sigma=\pm 1} \int \frac{d\vec{k}}{k^2} \left(\omega^{L}_{\vec{k}}\right)^2 \delta\left(\sigma\omega^{L}_{\vec{k}} - \vec{k}\vec{u}\right).$$

(8.81)

The diffusion coefficient tensor D_{ij}, when expressed in as elements perpendicular and parallel with respect to the velocity vector, can also be shown to be given by

$$D_{ij} = D_{\perp}\left(\delta_{ij} - \frac{u_i u_j}{u^2}\right) + D_{\parallel}\frac{u_i u_j}{u^2},$$

$$\begin{pmatrix} D_{\perp} \\ D_{\parallel} \end{pmatrix} = \frac{\pi e^2}{m_e^2 u^2} \sum_{\sigma=\pm 1} \int \frac{d\vec{k}}{k^2}\left[\frac{k^2 u^2 - \left(\omega^{L}_{\vec{k}}\right)^2}{\left(\omega^{L}_{\vec{k}}\right)^2}\right]\delta\left(\sigma\omega^{L}_{\vec{k}} - \vec{k}\vec{u}\right)I_L(\vec{k}).$$

(8.82)

Inserting Eqs. (8.81) and (8.82) to the electron kinetic Eq. (8.80) and expressing the right-hand side in velocity spherical coordinate system, we obtain

$$\frac{\partial f_e}{\partial t} = \frac{1}{u^2}\frac{\partial}{\partial u}\left[u^2\left(uGf_e + D_{\parallel}\frac{\partial f_e}{\partial u}\right)\right] + \frac{1}{u^2}\frac{\partial}{\partial\mu}\left[(1-\mu^2)D_{\perp}\frac{\partial f_e}{\partial\mu}\right],$$

(8.83)

where μ is the cosine of the pitch angle. From this, it follows that the steady-state solution is given by

$$uGf_e + D_{\parallel}\frac{\partial f_e}{\partial u} = 0, \frac{\partial f_e}{\partial\mu} = 0.$$

(8.84)

That is, an isotropic distribution function ($\partial f_e/\partial\mu = 0$) is a legitimate solution for the steady state. The generic formal solution is

$$f_e = C\exp\left(-\int\frac{uG}{D_{\parallel}}du\right).$$

(8.85)

Making use of the isotropy associated with the wave intensity, $I_L\left(\vec{k}\right) = I_L(k)$, which we have already assumed, the quantity $uG/D_{\parallel}$ is simplified by

$$\frac{uG}{D_{\parallel}} = \frac{m_e u}{4\pi^2}\frac{\int_{\omega_{pe}/u}^{\infty}k^{-1}dk}{\int_{\omega_{pe}/u}^{\infty}k^{-1}I_L(k)dk}.$$

(8.86)

In deriving the previous equation, we have approximated the Langmuir wave dispersion relation by $\omega_{\frac{L}{k}} \cong \omega_{pe}$. Upon defining

$$\mathcal{H}(u) = \int_{\omega_{pe}/u}^{\infty} \frac{dk}{k}, \mathcal{H}(u)J(u) = \int_{\omega_{pe}/u}^{\infty} \frac{dk}{k} I_L(k), \tag{8.87}$$

we have

$$\frac{uG}{D_{\parallel}} = \frac{m_e u}{4\pi^2} \frac{1}{J(u)}. \tag{8.88}$$

If we consider $J = k_B T_e/(4\pi^2)$, a constant for temperature T_e, then we obtain a Maxwellian distribution, $f_e = C \cdot \exp(-u^2/\theta_e^2)$. However, for the present purpose, we are interested in the kappa steady-state distribution function, Eq. (8.54). We will have to show that such a model is a unique solution, but for the moment, let us assume that it indeed is the only allowable solution. Then the question is, what functional form for $J(u)$ is necessary for the formal solution, Eq. (8.85), to reduce to the desired form, Eq. (8.54)? After some algebraic manipulations and direct comparison between the formal solution, Eq. (8.85), and kappa model, Eq. (8.54), it is becomes straightforward to show that the necessary form of the self-consistent reduced Langmuir fluctuation spectrum $J(u)$ must be given by

$$J(k) = \frac{k_B T_e}{4\pi^2} \frac{\kappa - \frac{3}{2}}{\kappa + 1} \left[1 + \frac{\frac{1}{2} m_e \omega_{pe}^2}{\left(\kappa - \frac{3}{2}\right) k^2 k_B T_e} \right], \tag{8.89}$$

where we have made use of the resonance condition $k = \omega_{pe}/u$. The resonance condition can again be invoked to rewrite Eq. (8.87) as

$$\mathcal{H}(k) = \int_{k}^{\infty} \frac{dk'}{k'}, \mathcal{H}(k)J(k) = \int_{k}^{\infty} \frac{dk'}{k'} I_L(k'). \tag{8.90}$$

Making use of Eq. (8.90), it follows that

$$\frac{k_B T_e}{4\pi^2} \frac{\kappa - \frac{3}{2}}{\kappa + 1} \int_{k}^{\infty} \frac{dk'}{k'} \left[1 + \frac{m_e \omega_{pe}^2}{2\left(\kappa - \frac{3}{2}\right) k'^2 k_B T_e} \right] = \int_{k}^{\infty} \frac{dk'}{k'} I_L(k'). \tag{8.91}$$

Upon taking the derivative d/dk of both sides, we obtain the expressed for the actual Langmuir fluctuation spectrum, $I_L(k)$,

$$I_L(k) = \frac{k_B T_e}{4\pi^2} \frac{\kappa - \frac{3}{2}}{\kappa + 1} \left\{ 1 + \frac{m_e \omega_{pe}^2}{2\left(\kappa - \frac{3}{2}\right) k'^2 k_B T_e} [1 + 2\mathcal{H}(k)] \right\}. \tag{8.92}$$

In summary, the steady-state electron velocity distribution function and the necessary Langmuir turbulence intensity are given by Eqs. (8.54), (8.89), and (8.92). It is important to note at this stage that the kappa electron distribution, Eq. (8.54), is simply a forced solution at this stage. Had we chosen constant wave spectrum, then, as we already discussed, we would have obtained a thermal equilibrium Maxwellian distribution. For that matter, had we chosen a different spectral form for $J(k)$, or equivalently, $I_L(k)$, then we would have obtained a different kind of electron distribution function other than the Maxwellian or kappa model. As we move onto the discussion of the wave kinetic equations and the associated steady-state asymptotic properties of the wave equations, it will be shown that the only allowable form of particle distribution and Langmuir wave intensity spectrum are none other than those specified by Eqs. (8.54), (8.89), and (8.92).

8.4.2 Steady-State Langmuir Turbulence

Let assume that the ion-sound turbulence intensity can be ignored when compared with the Langmuir wave intensity in the steady state. This means that the decay processes that involve ion-sound mode can be ignored. We will show that indeed this is the case. Then the steady-state wave kinetic equation is given by collecting spontaneous and induced emission terms in Eqs. (8.68) and (8.69), and induced and spontaneous scattering terms in Eqs. (8.72) and (8.73),

$$
\begin{aligned}
0 = {} & \frac{\omega_{pe}^2}{k^2} \int d\vec{u}\,\delta\!\left(\sigma\omega^L_{\vec{k}} - \vec{k}\,\vec{u}\right)\left(\frac{ne^2}{\pi}f_e + \sigma\omega^L_{\vec{k}}I^{\sigma L}_{\vec{k}}\,\vec{k}\cdot\frac{\partial f_e}{\partial\vec{u}}\right) \\
& - \left(\frac{\kappa - \frac{1}{2}}{\kappa - \frac{3}{2}}\right)^2 \frac{e^2}{2(\varepsilon_0 k_B T_e)^2}\cdot\sum_{\sigma',\sigma''=\pm1}\sigma\omega^L_{\vec{k}}\int d\vec{k'}\,\frac{\mu_{\vec{k}-\vec{k'}}\left(\vec{k}\,\vec{k'}\right)^2}{k^2 k'^2\left|\vec{k}-\vec{k'}\right|^2} \\
& \times \sigma''\omega^L_{\vec{k}-\vec{k'}}I^{\sigma'L}_{\vec{k'}}I^{\sigma L}_{\vec{k}}\delta\!\left(\sigma\omega^L_{\vec{k}} - \sigma'\omega^L_{\vec{k'}} - \sigma''\omega^S_{\vec{k}-\vec{k'}}\right) \\
& - \left(\frac{\kappa - \frac{1}{2}}{\kappa - \frac{3}{2}}\right)^2 \frac{e^2}{\varepsilon_0^2 m_e^2 \omega_{pe}^2}\sum_{\sigma'=\pm1}\int d\vec{k'}\int d\vec{u}\,\frac{\left(\vec{k}\,\vec{k'}\right)^2}{k^2 k'^2}\delta\!\left[\sigma\omega^L_{\vec{k}} - \sigma'\omega^L_{\vec{k'}}\right. \\
& \left. - \vec{k} - \vec{k'}u^-\right] \times \left[\frac{ne^2}{\pi\omega_{pe}^2}\sigma\omega^L_{\vec{k}}\left(\sigma'\omega^L_{\vec{k'}}I^{\sigma L}_{\vec{k}} - \sigma\omega^L_{\vec{k}}I^{\sigma' L}_{\vec{k'}}\right)f_i\right. \\
& \left. - \frac{m_e}{m_i}\sigma\omega^L_{\vec{k}}I^{\sigma' L}_{\vec{k'}}I^{\sigma L}_{\vec{k}}\left(\vec{k}-\vec{k'}\right)\cdot\frac{\partial f_i}{\partial\vec{u}}\right].
\end{aligned}
\tag{8.93}
$$

We already assumed that in the steady state, the forward- and backward-propagating Langmuir waves have the same spectral form, $I^{+L}_{\vec{k}} = I^{-L}_{\vec{k}} = I^{L}_{\vec{k}}$.

Since the wave intensity is symmetric in $\vec{k}$, it is sufficient to consider only $\sigma = 1$ in the wave kinetic equation. The ion distribution is assumed to remain stationary and Maxwellian. These assumptions lead to

$$
0 = \frac{\pi \omega_{pe}^2}{k^2} \int d\vec{u}\, \delta\!\left(\omega^L_{\vec{k}} - \vec{k}\cdot\vec{u}\right)\left[\frac{ne^2}{\pi}f_e + \omega^L_{\vec{k}} I_L(\vec{k})\,\vec{k}\cdot\frac{\partial f_e}{\partial \vec{u}}\right]
$$

$$
- \left(\frac{\kappa - \frac{1}{2}}{\kappa - \frac{3}{2}}\right)^2 \frac{\pi}{2}\frac{e^2}{(k_B T_e)^2}\cdot \sum_{\sigma',\sigma''=\pm 1}\omega^L_{\vec{k}}\int d\vec{k}'\,\frac{\mu_{\vec{k}-\vec{k}'}\left(\vec{k}\vec{k}'\right)^2}{k^2 k'^2\left|\vec{k}-\vec{k}'\right|^2}
$$

$$
\times \sigma''\omega^L_{\vec{k}-\vec{k}'} I_L(\vec{k}') I_L(\vec{k}) \delta\!\left(\omega^L_{\vec{k}} - \sigma'\omega^L_{\vec{k}'} - \sigma''\omega^S_{\vec{k}-\vec{k}'}\right)
$$

$$
- \left(\frac{\kappa - \frac{1}{2}}{\kappa - \frac{3}{2}}\right)^2 \frac{\omega^L_{\vec{k}}}{4\pi n k_B T_i}\sum_{+,-}\int d\vec{k}'\int d\vec{u}\,\frac{\left(\vec{k}\vec{k}'\right)^2}{k^2 k'^2}\delta\!\left[\omega^L_{\vec{k}} \mp \omega^L_{\vec{k}'} - \left(\vec{k}-\vec{k}'\right)\vec{u}\right]
$$

$$
\times \left\{\frac{k_B T_i}{4\pi^2}\left[\pm\omega^L_{\vec{k}'} I_L(\vec{k}) - \omega^L_k I_L(\vec{k}')\right] + I_L(\vec{k}') I_L(\vec{k})\left(\omega^L_{\vec{k}} \mp \omega^L_{\vec{k}'}\right)\right\}f_i.
$$

(8.94)

8.4.2.1 Balance of Induced and Spontaneous Emission

In this subsection, we address the issue of the balance of spontaneous and induced emissions by first ignoring the nonlinear term, and see whether the resulting solution is consistent with the earlier solution obtained for the particle (electron) distribution, Eq. (8.54), and the (Langmuir) wave intensity, Eq. (8.89) or (8.92). Recall that the steady-state intensities, Eq. (8.89) or (8.92), were derived simply by requiring that it leads to the necessary electron distribution, Eq. (8.54). We must now prove that Eq. (8.89) or (8.92) also satisfies the steady-state linear wave kinetic equation. Consider the balance of spontaneous and induced emissions in Eq. (8.94),

$$
0 = \frac{\pi \omega_{pe}^2}{k^2}\int d\vec{u}\,\delta\!\left(\omega^L_{\vec{k}} - \vec{k}\cdot\vec{u}\right)\left(\frac{ne^2}{\pi}f_e + \omega^L_{\vec{k}} I_L(\vec{k})\,\vec{k}\cdot\frac{\partial f_e}{\partial \vec{u}}\right)
$$

$$
\approx \frac{\pi \omega_{pe}^2}{k^2}\frac{ne^2}{\pi}\int d\vec{u}\,\delta(\omega_{pe} - ku\mu)\left(1 - \frac{\kappa + 1}{\kappa - \frac{3}{2}}\cdot\frac{4\pi}{k_B T_e}\cdot\frac{I_L(k)}{1 + \frac{1}{2}m_e u^2 \Big/\left[\left(\kappa - \frac{3}{2}\right)k_B T_e\right]}\right)f_e.
$$

(8.95)

Let us integrate the previously mentioned quantity in k space,

$$
0 = 4\pi \int_0^\infty du\, u \int_{\omega_{pe}/v}^\infty \frac{dk}{k} \left(1 - \frac{\kappa+1}{\kappa-\frac{3}{2}} \cdot \frac{4\pi}{k_B T_e} \cdot \frac{I_L(k)}{1+\frac{1}{2}m_e u^2 \Big/ \left[\left(\kappa-\frac{3}{2}\right)k_B T_e\right]} \right) f_e
$$

$$
= 4\pi \int_0^\infty du\, u \left(\mathcal{H}(u) - \frac{\kappa+1}{\kappa-\frac{3}{2}} \cdot \frac{4\pi}{k_B T_e} \cdot \frac{\mathcal{H}(u)J(u)}{1+\frac{1}{2}m_e u^2 \Big/ \left[\left(\kappa-\frac{3}{2}\right)k_B T_e\right]} \right) f_e,
$$

(8.96)

which shows that $J(k)$ given by Eq. (8.89) indeed satisfies the equality. We thus see that the balance of induced and spontaneous emissions leads to the same solution as before. This resolves one of the outstanding issues, namely, whether the electron distribution, Eq. (8.54), and the imposed wave intensity, Eq. (8.89) or (8.92), are also the self-consistent solutions of the linear wave kinetic equation or not. We continue with the remaining piece in the steady-state wave kinetic equation, namely, the nonlinear term in the steady-state wave kinetic, Eq. (8.94).

8.4.2.2 Absence of Steady-State Decay Processes

Before we move onto the steady-state nonlinear wave–particle interaction processes, let us address the issue of whether the three-wave interaction processes are indeed negligible or not. We have made an a priori assumption that it is true, but now it is necessary to provide the proof. Taking $\sigma = 1$ in the Langmuir wave decay terms in Eq. (8.94), upon explicitly summing over the indices σ' and σ'', we find that the decay terms for L mode are given by

$$
\frac{dI_L(\vec{k})}{dt}\bigg|_{decay} = \left(\frac{\kappa-\frac{1}{2}}{\kappa-\frac{3}{2}}\right)^2 \frac{\pi}{2} \frac{e^2}{(k_B T_e)^2}
$$

$$
\times \sum_{\sigma',\sigma''=\pm 1} \omega_{\vec{k}}^L \int d\vec{k'}\, \frac{\mu_{\vec{k}-\vec{k'}} \left(\vec{k}\cdot\vec{k'}\right)^2}{k^2 k'^2 |\vec{k}-\vec{k'}|^2} \left[\delta\left(\omega_{\vec{k}}^L - \omega_{\vec{k'}}^L - \omega_{\vec{k}-\vec{k'}}^S\right) \right.
$$

$$
\left. - \delta\left(\omega_{\vec{k}}^L - \omega_{\vec{k'}}^L + \omega_{\vec{k}-\vec{k'}}^S\right)\right] \omega_{\vec{k}-\vec{k'}}^L I_L\left(\vec{k}\right) I_L\left(\vec{k'}\right).
$$

(8.97)

where we have ignored those terms involving $\omega_{\vec{k}}^L = \omega_{\vec{k'}}^L$ in the three-wave resonance condition, since obviously such terms cannot satisfy the resonance condition.

Consider the wave–wave resonance condition delta function:

$$\delta\left(\omega^L_{\overrightarrow{k}} - \omega^L_{\overrightarrow{k'}} \pm \omega^S_{\overrightarrow{k}-\overrightarrow{k'}}\right) = \frac{1}{\omega_{pe}}\delta\left[\frac{3}{2}\frac{(k^2 - k'^2)k_B T_e}{m_e \omega^2_{pe}} \pm \sqrt{\frac{k_B T_e}{m_i \omega^2_{pe}}}\sqrt{\frac{\kappa - \frac{3}{2}}{\kappa - \frac{1}{2}}}\left|\overrightarrow{k} - \overrightarrow{k'}\right|\right]$$

$$\approx \frac{1}{\omega_{pe}}\frac{2}{3}\delta\left[\frac{(k^2 - k'^2)k_B T_e}{m_e \omega^2_{pe}}\right],$$

(8.98)

which shows that the two decay terms in Eq. (8.98) cancel each other upon ignoring terms of order $\sqrt{m_e/m_i}$. This proves that the three-wave decay resonance term can be ignored in the discussion of the steady-state wave kinetic equation.

8.4.2.3 Balance of Induced and Spontaneous Scattering

We now return to the problem of balance between the spontaneous and induced scattering terms in Eq. (8.94),

$$0 = -\left(\frac{\kappa - \frac{1}{2}}{\kappa - \frac{3}{2}}\right)^2 \frac{\omega^L_{\overrightarrow{k}}}{4\pi n k_B T_i}\sum_{+,-}\int d\overrightarrow{k'}\int d\overrightarrow{u}\frac{\left(\overrightarrow{k}\,\overrightarrow{k'}\right)^2}{k^2 k'^2}\delta\left[\omega^L_{\overrightarrow{k}} \mp \omega^L_{\overrightarrow{k'}} - \left(\overrightarrow{k} - \overrightarrow{k'}\right)\overrightarrow{u}\right]$$

$$\times \left\{\frac{k_B T_i}{4\pi^2}\left[\pm\omega^L_{\overrightarrow{k'}}I_L\left(\overrightarrow{k}\right) - \omega^L_{\overrightarrow{k}}I_L\left(\overrightarrow{k'}\right)\right] + I_L\left(\overrightarrow{k'}\right)I_L\left(\overrightarrow{k}\right)\left(\omega^L_{\overrightarrow{k}} \mp \omega^L_{\overrightarrow{k'}}\right)\right\}f_i.$$

(8.99)

Explicitly writing down the summation over two signs and ignoring those terms associated with the resonance conditions with the delta function argument $\omega^L_{\overrightarrow{k}} + \omega^L_{\overrightarrow{k'}}$, we have

$$0 = \int d\overrightarrow{k'}\int d\overrightarrow{u}\frac{\left(\overrightarrow{k}\,\overrightarrow{k'}\right)^2}{k^2 k'^2}\delta\left[\omega^L_{\overrightarrow{k}} - \omega^L_{\overrightarrow{k'}} - \left(\overrightarrow{k} - \overrightarrow{k'}\right)\overrightarrow{u}\right]$$

$$\times \left\{\frac{T_i}{4\pi^2}\left[\omega^L_{\overrightarrow{k'}}I_L\left(\overrightarrow{k}\right) - \omega^L_{\overrightarrow{k}}I_L\left(\overrightarrow{k'}\right)\right] + I_L\left(\overrightarrow{k'}\right)I_L\left(\overrightarrow{k}\right)\left(\omega^L_{\overrightarrow{k}} - \omega^L_{\overrightarrow{k'}}\right)\right\}f_i.$$

(8.100)

Since the resonant velocity determined by the nonlinear wave–particle interaction condition between the two Doppler-shifted Langmuir waves and the ions,

$u_{res} = \left(\omega^L_{\overrightarrow{k}} - \omega^L_{\overrightarrow{k'}}\right)\Big/\left|\overrightarrow{k} - \overrightarrow{k'}\right|$, must be sufficiently low, that is, it must be of

the order of the ion thermal speed, $u_{res} \leq \theta_i \equiv \sqrt{2k_B T_i/m_i}$, it is reasonable to expect that the resonance is possible only if $\overrightarrow{k} \sim \overrightarrow{k'}$. We thus write $\overrightarrow{k'} = \overrightarrow{k} + \delta\overrightarrow{k}$

and expand various terms in Eq. (8.94) with respect to small $\delta \vec{k}$. Thus we have

$$
\omega^L_{\vec{k}'} I_L(\vec{k}) - \omega^L_{\vec{k}} I_L(\vec{k}') \approx \delta \vec{k} \cdot \frac{d\omega^L_{\vec{k}}}{d\vec{k}} I_L(\vec{k}) - \omega^L_{\vec{k}} \delta \vec{k} \cdot \frac{dI_L(\vec{k})}{d\vec{k}},
$$

$$
I_L(\vec{k}') I_L(\vec{k}) \left(\omega^L_{\vec{k}} - \omega^L_{\vec{k}'} \right) \approx - \left[I_L(\vec{k}) \right]^2 \delta \vec{k} \cdot \frac{d\omega^L_{\vec{k}}}{d\vec{k}},
$$

(8.101)

which yields

$$
0 = \int d(\delta \vec{k}) \int d\vec{u} \frac{\left(\vec{k} \vec{k}' \right)^2}{k^2 k'^2} \delta \left[\omega^L_{\vec{k}} - \omega^L_{\vec{k}'} - \left(\vec{k} - \vec{k}' \right) \vec{u} \right]
$$

$$
\times \delta k \left[\omega_L(k) \frac{dI_L(k)}{dk} + \frac{4\pi^2}{T_i} \frac{d\omega_L(k)}{dk} I_L(k)^2 - \frac{d\omega_L(k)}{dk} I_L(k) \right] f_i.
$$

(8.102)

In the previous equation, we have made use of the fact that the turbulence intensity and the wave dispersion relation depend only on the modulus of the wave vector k. Making use of the Langmuir wave dispersion relation and its derivative with respect to k, we obtain the necessary condition for the equality in Eq. (8.102) to be satisfied, namely,

$$
0 = \omega_L(k) \frac{dI_L(k)}{dk} + \frac{4\pi^2}{k_B T_i} \frac{d\omega_L(k)}{dk} I_L(k)^2 - \frac{d\omega_L(k)}{dk} I_L(k)
$$

$$
= \frac{3k}{m_e \omega_{pe}} \left[\left(1 + \frac{3k^2 k_B T_e}{2m_e \omega_{pe}^2} \right) \frac{dI_L(k)}{I_L(k) \left(3k^2 / 2m_e \omega_{pe}^2 \right)} + \frac{4\pi^2 T_e}{T_i} I_L(k)^2 - k_B T_e I_L(k) \right],
$$

(8.103)

which enjoys an exact solution,

$$
I_L(k) = \frac{k_B T_i}{4\pi^2} \left(1 + \frac{2}{3} \frac{m_e \omega_{pe}^2}{k^2 k_B T_e} \right)
$$

(8.104)

This solution represents the steady-state Langmuir turbulence intensity that results from balancing the spontaneous and induced scattering processes only. It is important to note that the derivation did not invoke that the electron distribution is of the kappa form, Eq. (8.54), at all. Consequently, this is an independent outcome that arises out of the consideration of wave kinetic equation. Yet the basic mathematical form of the wave intensity is, in an overall sense, identical to that of Eq. (8.92). Considering the fact that the wave intensity, Eq. (8.92), is the result of imposing the electron kappa distribution, Eq. (8.54), and that Eqs. (8.92) and (8.104) are mathematically identical except for constant multiplicative factors, it thus follows that the kappa distribution is the only allowable solution consistent with the balance of nonlinear wave kinetic processes. This constitutes the proof of uniqueness theorem for the kappa model.

Had one imposed any other electron distribution, then the resulting wave intensity, i.e., one that would have replaced Eq. (8.92), would simply be incompatible with the solution, Eq. (8.104), which was derived independently.

Before we close, we must now reconcile the two basically identical but alternative forms of the wave intensity, Eqs. (8.92) and (8.104). Upon comparison between the two forms of wave intensity expressions, we can immediately identify

$$\kappa = \frac{9}{4} + \frac{3\mathcal{H}}{2} = 2.25 + 1.5\mathcal{H}, \left(\kappa - \frac{3}{2}\right)T_e = (\kappa + 1)T_i, \tag{8.105}$$

provided we treat the correction factor $\mathcal{H}$ as constant. Note that the second relation determines the ratio of the electron over ion temperature in terms of the kappa index. This brings us to the issue of $\mathcal{H}$,

$$\mathcal{H}(k) = \int_k^\infty \frac{dk'}{k'} = \ln k'|_{k'=k}^\infty \tag{8.106}$$

Note that the quantity $\mathcal{H}$ diverges for both $k' \to 0$ and $k' \to \infty$. To remedy the divergent situation, we introduce the lower and upper cutoffs, k_{min} and k_{max},

$$\mathcal{H} = \ln \frac{k_{max}}{k_{min}}. \tag{8.107}$$

We may leave k_{min} and k_{max} as undetermined and treat $\mathcal{H}$ as a freely adjustable parameter. This quantity was necessary in the derivation of the spectrum $I_L(k)$ from J, but once the proper intensity is obtained, we may simply ignore $\mathcal{H}$. If we adopt such an approach, then we have a final form of electron kappa distribution and the associated Langmuir fluctuation intensity

$$I_L(k) = \frac{T_e}{4\pi^2} \frac{\kappa - \frac{3}{2}}{\kappa + 1} \left(1 + \frac{1}{\kappa - \frac{3}{2}} \cdot \frac{2\pi ne^2}{k^2 k_B T_e}\right), \kappa = \frac{9}{4} = 2.25, \frac{T_i}{T_e} = \frac{\kappa - \frac{3}{2}}{\kappa + 1} = \frac{3}{13}. \tag{8.108}$$

Returning to the issue of the quantity $\mathcal{H}$, obviously, the value of $\mathcal{H}$ is positive as seen from Eq. (8.84), although, as we mentioned, the precise treatment of this quantity is uncertain. Assuming that $\mathcal{H}$ can be treated as an adjustable positive definite parameter, the present theory can predict not only the kappa index 2.25, but also the entire near-equilibrium region [2.5, ∞] (Livadiotis and McComas, 2010a, 2013a,d), who demonstrated that $\kappa \sim 2.5$ in the nonextensive statistical mechanics has a special significance in that it represents a demarcation between the far-equilibrium (κ below ~ 2.5) and near-equilibrium regions (with κ above ~ 2.5). It is difficult to understand the reason for the existence of such a distinction of space plasmas. The present turbulent equilibrium theory might shed some light on this issue by introducing one way of creating stationary states with kappa indices in the near-equilibrium region, i.e., via nonzero $\mathcal{H}$.

8.5 Concluding Remarks

The purpose of the present chapter had been to put forth a self-contained treatise on the problem of electron kappa distribution in a quasiequilibrium plasma where enhanced electrostatic fluctuations exist. We presented a self-contained discussion of the weak turbulence theory, on the basis of which, we discussed the time-asymptotic $(t \rightarrow \infty)$ steady-state solution. By solving the set of coupled electron-Langmuir turbulence equations for a steady-state situation, we have demonstrated that the steady-state kappa electron distribution with the kappa index $\kappa = 2.25$ emerges as a unique asymptotically steady-state solution. Such a distribution behaves as quasiscale free velocity power law distribution, $f_e \sim u^{-6.5}$, for high velocities.

In order to verify this theoretical prediction against observation, Wang et al. (2012) compared velocity power law spectral indices predicted by both theory and observation made during quiet-time solar wind condition. They found a reasonable agreement between the theory $(u^{-6.5})$ and observation $(\sim u^{-5.0}$ to $u^{-8.7}$ with average $u^{-6.69})$. This seems to show that the quasisteady state, solar wind electrons may indeed be characterized by nonextensive thermo-statistical state.

One final thought before we close this chapter is that the preceding discussion forces the temperature ratio of protons to electrons to be a fixed number, $T_i/T_e = 3/13$. While the actual temperature ratio T_i/T_e in the solar wind is generally less than unity (e.g., Montgomery et al., 1968), the ratio is certainly not constant, and T_i/T_e, in some instances, can get higher than unity. This is an unresolved issue at the present state. The present theory assumes steady state, while the actual solar wind, even for quiet times, is dynamic. Also, observations show that the actual solar wind electrons and ions are made of several distinct components. In contrast, the present treatise assumes single-component electrons and ions (protons). More realistic modeling of the solar wind plasma may be necessary in order to advance further, but it is beyond the scope of the present chapter.

8.6 Science Questions for Future Research

Future analyses and observations need to address the following questions:

1. Can the result be extended for any ion-electron temperatures?
2. What is the role and influence of ambient magnetic field?
3. Can the weak turbulence analysis be generalized to other wave modes?

Acknowledgments

P. Yoon acknowledges NSF grant AGS1550566 to the University of Maryland, and the BK21 plus grant from the National Research Foundation (NRF) of the Korean Ministry of Education to Kyung Hee University, Korea. He also acknowledges the Science Award Grant from the GFT Foundation to the University of Maryland. Part of this work was carried out while P.H.Y. was visiting the Ruhr University Bochum (RUB), Germany, which was made possible by support from the Ruhr University Research School PLUS, funded by Germany's Excellence Initiative (DFG GSC 98/3), and the Visiting International Professor Fellowship administered by RUB.

CHAPTER 9

Solitary Waves in Plasmas Described by Kappa Distributions

G.S. Lakhina, S. Singh
Indian Institute of Geomagnetism, New Panvel (W), Navi Mumbai, India

Chapter Outline
9.1 Summary 399
9.2 Introduction: Observations and Origin of Suprathermal Electrons 400
9.3 Model of Ion-Acoustic Solitons and Double Layers in Plasmas With Suprathermal Electrons 402
  9.3.1 Two-Component Plasmas: Cold Protons and Suprathermal Electrons 405
  9.3.2 Three-Component Plasmas: Protons, Heavier Ions, and Suprathermal Electrons 407
  9.3.3 Two-Component Magnetized Plasmas: Protons and Suprathermal Electrons 413
9.4 Model for Electron-Acoustic Solitons in Plasmas With Suprathermal Electrons 415
9.5 Concluding Remarks 417
9.6 Science Questions for Future Research 418
Acknowledgments 418

9.1 Summary

Plasmas encountered in space and astrophysical environments are usually far from the thermal equilibrium containing suprathermal particles. Electron distributions in such plasmas are best described by kappa distributions rather than Maxwellian distributions. This chapter presents an overview of ion/electron Solitary Waves (SWs) and Double Layers (DLs) in multicomponent plasmas, where the suprathermal electron component is described by kappa

Kappa Distributions. http://dx.doi.org/10.1016/B978-0-12-804638-8.00009-7

distributions. Types of solitons and DLs are relevant to the electrostatic SWs observed in the solar wind and planetary magnetospheres.

Science Question: What is the effect of kappa distributions on the expressions of solitary waves?

Keywords: Double layers; Ion/electron-acoustic solitons; Solar wind; Solitary waves; Suprathermal particles.

9.2 Introduction: Observations and Origin of Suprathermal Electrons

SWs are the localized symmetric potential structures with no net potential drop. Respectively, DLs are the localized asymmetric potential structures with a net potential drop. The first observation of SWs and DLs in electric field components parallel to the magnetic field was reported by Temerin et al. (1982) from the S3-3 spacecraft (S/C) in the auroral acceleration region between 6000 and 8000 km altitude. Later on, SWs and DLs were observed on the auroral field lines by the Viking (Boström et al., 1988; Koskinen et al., 1990). The electrostatic solitary structures observed by the S3-3 and Viking were interpreted as ion holes, as they carried negative potentials and propagated along the magnetic field with speeds from ~5 to ≥50 km/s, which were of the order of ion-acoustic or ion beam speeds. The parallel electric field amplitudes were typically ~15–20 mV/m with pulse durations of ~2–20 ms. Later on, similar negative-potential solitary structures associated with the upward-going ion beams were reported from Polar S/C by Bounds et al. (1999).

Matsumoto et al. (1994) were the first to report the presence of SW structures having a positive potential in the plasma sheet boundary layer (PSBL) by analyzing waveform data from the Geotail Plasma wave Instrument. They named these isolated pulse waveforms electrostatic solitary waves (ESWs) (Kojima et al., 1997; Omura et al., 1999).

Since then, observations of similar ESWs have been reported in various regions of the Earth's magnetosphere by several researchers using waveform data from Polar, FAST, WIND, and Cluster S/C. For example, ESWs have been observed in the auroral acceleration region by Polar (Mozer et al., 1997; Bounds et al., 1999) and FAST (Ergun et al., 1998a, 1998b), the high-altitude polar magnetosphere, including the polar cap boundary layer by Polar (Cattell et al., 1999; Franz et al., 1998, 2005; Tsurutani et al., 1998), and the magnetosheath by Geotail S/C (Kojima et al., 1997) and the Cluster (Pickett et al., 2004, 2005, 2008). The electrostatic solitary structures are found in the electric field component parallel to the background magnetic field and are usually bipolar (sometimes monopolar or tripolar). The electrostatic solitary structures can have either positive (electron holes) or negative (ion holes) potential, and their electric field amplitudes decrease with distance from the Earth (Pickett et al., 2004; Franz et al., 2005; Ergun et al., 1998a, 1998b). The field-aligned velocities and scale sizes of ESWs are found to increase with distance from the Earth (Cattell et al., 1999; Omura et al., 1999).

ESWs observed by S/C are characterized by an amplitude–width relationship where the amplitude of the electrostatic potential tends to increase with its width. This property of ESWs is opposite to that of Korteweg-de Vries (KdV)-type solitons where the soliton amplitude increases as its width decreases. This would indicate that the ESWs observed by S/C are not the usual KdV type of small-amplitude ion-acoustic or electron-acoustic solitons. On the contrary, because of the misconception in the space plasma community that all weak solitons should behave like KdV solitons, the generation mechanisms for ESWs based on ion- or electron-acoustic solitons were considered to be unfeasible (Ergun et al., 1998b; Pickett et al., 2004; Franz et al., 2005). The properties of the arbitrary amplitude solitons predicted by the models based on the Sagdeev pseudopotential techniques are quite different from the KdV-type solitons as, depending upon the parametric range, their amplitudes can either increase or decrease with an increase in their width (Ghosh and Lakhina, 2004). This has brought the soliton/DL models based on the Sagdeev pseudopotential method to the forefront of viable models for the generations of ESWs. In particular, the models based on arbitrary amplitude electron-acoustic SWs (Pottelette et al., 1990, 1999; Dubouloz et al., 1991, 1993; Kakad et al., 2007; Ghosh et al., 2008; Berthomier et al., 1998, 2000; Lakhina et al., 2008a, 2008b, 2009, 2011a, 2011b) are being considered as an alternative to the phase-space electron hole models (Omura et al., 1996; Kojima et al., 1997; Goldman et al., 1999, 2007; Singh, 2003; Muschietti et al., 1999; Chen et al., 2005; Tao et al., 2011; Norgren et al., 2015) for the generation of ESWs.

During the past 50 years, there have been several theoretical studies on the ion-acoustic (Sagdeev, 1966; Washimi and Taniuti, 1966; Buti, 1980; Baboolal et al., 1990; Reddy and Lakhina, 1991; Reddy et al., 1992; Ghosh et al., 1996; Ghosh and Iyengar, 1997; Berthomier et al., 1998; Ghosh and Lakhina, 2004; Lakhina et al., 2008a, 2008b, 2011a, 2011b; Maharaj et al., 2012a; Singh et al., 2013; Rufai et al. 2012, 2014; Kakad et al., 2016) and electron-acoustic (Pottelette et al., 1990, 1999; Dubouloz et al., 1991, 1993; Berthomier et al., 2000; Singh et al., 2001, 2011; Singh and Lakhina, 2001, 2004; Tagare et al., 2004; Cattaert et al., 2005; Kakad et al., 2007, 2009; Ghosh et al., 2008; Lakhina et al., 2008a, 2008b; Maharaj et al., 2012b; Mbuli et al., 2015) SWs and DLs in both magnetized and nonmagnetized plasmas. These studies considered multicomponent plasmas where various species are treated either as fluids or having Maxwellian particle distributions. However, particle distribution functions in space plasmas are often found to deviate from Maxwellian due to the presence of suprathermal particles having high-energy tails (e.g., Vasyliunas, 1968; Leubner, 1982; Marsch et al., 1982; Armstrong et al., 1983). Such an excess suprathermal electron population may arise due to velocity space diffusion, leading to an inverse power-law distribution at a velocity much higher than the electron thermal speed. The kappa distribution has been widely adopted to model the observed suprathermal particle distributions (Chapters 10–17 in part C; also, see: Summers and Thorne, 1991; Thorne and Summers, 1991a, 1991b; Thorne and Horne, 1994; Mace and Hellberg, 1995; Hellberg and Mace, 2002; Hapgood et al., 2011; see also

Livadiotis and McComas, 2009; Pierrard and Lazar, 2010; Livadiotis, 2015a; and references therein).

This chapter presents an overview of the work done on ion- and electron-acoustic solitons and DLs where hot electrons are characterized by kappa distributions. In Section 9.3, we present ion-acoustic solitons and DLs, propagating either parallel or at an oblique angle to the ambient magnetic field, in multicomponent plasmas. In Section 9.4, we discuss electron-acoustic solitons propagating parallel to the ambient magnetic field in plasmas with suprathermal electrons. Finally, the concluding remarks are given in Section 9.5, while three general science questions for future analyses are posed in Section 9.6.

9.3 Model of Ion-Acoustic Solitons and Double Layers in Plasmas With Suprathermal Electrons

There are a few studies that have examined ion-acoustic SWs with highly energetic kappa-distributed electrons. Saini et al. (2009) have studied arbitrary amplitude ion-acoustic SWs in nonmagnetized plasma composed of cold ions and kappa-distributed electrons. Saini and Kourakis (2010) extended the analysis to include an electron beam in this system. Lakhina and Singh (2015) have studied ion-acoustic solitons and DL in a three-component plasma consisting of fluid protons and α particles and suprathermal electrons having a kappa distribution to explain the generation of weak double layers (WDLs) and low-frequency coherent electrostatic waves observed by Wind S/C in the solar wind at 1 AU (e.g., Mangeney et al., 1999; Lacombe et al., 2002; Salem et al., 2003).

We shall describe the model of Lakhina and Singh (2015), which extends the Saini et al. (2009) model to the case of hot protons as a special case when α particles are absent.

Solar wind plasma is modeled by a homogeneous, collisionless, and magnetized three-component plasma consisting of protons (n_p, T_p), heavier ions (n_i, T_i, U_i), and suprathermal electrons (n_e, T_e) having a kappa distribution, where n_j, T_j, and U_j represent values of density, temperature, and drift velocity (along the direction of the ambient magnetic field) of the species j: p, i, and e for protons, heavier ions, and suprathermal electrons, respectively. Their electrostatic equilibrium values (for $\Phi = 0$) are noted as N_p, N_i, and N_e, respectively, with $N_e = N_p + Z_i N_i$. As an example of heavier ions, here we consider the heavier ions to be He^{++} ions, i.e., α particles.

We model solar wind electron velocities with a kappa distribution (e.g., Mace and Hellberg, 1995; Livadiotis and McComas, 2009; Livadiotis, 2015a):

$$f(\vec{u}) = N_e \pi^{-\frac{3}{2}} \kappa^{-\frac{3}{2}} \frac{\Gamma(\kappa + 1)}{\Gamma(\kappa - 1/2)} \theta_\kappa^{-3} \left(1 + \frac{1}{\kappa} \cdot \frac{\vec{u}^2}{\theta_\kappa^2}\right)^{-\kappa - 1}, \quad \theta_\kappa \equiv \sqrt{\frac{\kappa - \frac{3}{2}}{\kappa}} \cdot \theta_e \,, \qquad (9.1)$$

where $\theta_e \equiv \sqrt{2k_B T_e/m_e}$ is the characteristic electron thermal speed and k_B is the Boltzmann constant. For $\kappa \to \infty$, the kappa distribution approaches a

Maxwellian distribution, i.e., thermal equilibrium, whereas for $\kappa \to \frac{3}{2}$ it represents the so-called antiequilibrium, the farthest state from thermal equilibrium (Livadiotis, 2015a). The electron number density in the presence of ion-acoustic waves having an electrostatic potential Φ can be obtained by replacing u^2/θ_e^2 in the distribution function given by Eq. (9.1) by $u^2/\theta_e^2 - e\,\Phi/(k_B T_e)$ and integrating it over the velocity space (as explained in the theory of kappa distributions in the presence of a potential energy, e.g., see Chapter 3; Livadiotis, 2015b; Devanandhan et al., 2015), namely,

$$n_e = N_e \cdot \left(1 - \frac{1}{\kappa - \frac{3}{2}} \cdot \frac{e\Phi}{k_B T_e}\right)^{-\kappa + \frac{1}{2}} , \tag{9.2a}$$

where N_e denotes the density at $\Phi \to 0$. Hereafter, unless otherwise specified, all densities are normalized to the equilibrium electron density, $N_e = N_p + Z_i\,N_i$, and the electrostatic potential Φ by $k_B T_e/e$, hence,

$$n_e = \left(1 - \frac{\Phi}{\kappa - \frac{3}{2}}\right)^{-\kappa + \frac{1}{2}} . \tag{9.2b}$$

We will consider the nonlinear electrostatic waves propagating parallel to the ambient magnetic field. For this case, the dynamics of both protons and α particles are described by the multifluid equations of continuity, momentum, an energy equation derived from the equation of state of each species, and the Poisson equation (Lakhina et al., 2014; Lakhina and Singh, 2015):

$$\frac{\partial n_j}{\partial t} + \frac{\partial (u_j n_j)}{\partial x} = 0 , \tag{9.3}$$

$$\frac{\partial u_j}{\partial t} + u_j \frac{\partial u_j}{\partial x} + \frac{1}{n_j \mu_j} \frac{\partial P_j}{\partial x} + \frac{Z_j}{\mu_j} \frac{\partial \Phi}{\partial x} = 0 , \tag{9.4}$$

$$\frac{\partial P_j}{\partial t} + u_j \frac{\partial P_j}{\partial x} + 3P_j \frac{\partial u_j}{\partial x} = 0. \tag{9.5}$$

Here, j: p, i, $\mu_j = m_j/m_p$, and Z_j is the positive electronic charge of the jth ion species. For protons, $Z_p = 1$, and for α particles, $Z_i = 2$. We normalized velocities with the ion-acoustic speed, defined by the electron temperature and the proton mass, $c_s = (k_B T_e/m_p)^{1/2}$, the time with the inverse of the proton plasma frequency, $\omega_p = [N_e e^2/(\varepsilon_0 m_p)]^{1/2}$, the lengths with the hot-electron Debye length, $\lambda_{D\infty} = [\varepsilon_0 k_B T_e/(N_e e^2)]^{1/2}$, temperatures with the electron temperature, T_e, and the thermal pressure P_j with the electron thermal pressure $N_e k_B T_e$. For convenience, we used the electron Debye length as defined at thermal equilibrium $\lambda_{D\infty}$ (that is, independent of the κ-index) rather than for the kappa distribution

(Livadiotis and McComas, 2014a). Further, we have taken the adiabatic index, $\gamma = 3$ (considering 1-D space for the description of electrons, i.e., $\gamma = 1 + 2/d$ for $d = 1$, where d is the degree of freedom), in the energy equation for the ions as given by Eq. (9.5).

In order to study the nonlinear evolution of ion-acoustic waves, Eqs. (9.3)–(9.5) are transformed into a stationary frame moving with velocity V, the phase velocity of the wave, i.e., $\xi = x - Mt$, where $M = V/c_s$ is the Mach number with respect to the ion-acoustic speed. In this reference frame, all variables, e.g., density and pressure, tend to their undisturbed values, and potential Φ tends to zero at $\xi \rightarrow \pm\infty$. The transformed Eqs. (9.3)–(9.5) yield the following expressions for the density of protons and heavy ions (Maharaj et al., 2012a; Lakhina and Singh, 2015), respectively:

$$n_p = \frac{N_p}{2\sqrt{3T_p}} \cdot \left\{ \left[\left(M + \sqrt{3T_p} \right)^2 - 2\Phi \right]^{\frac{1}{2}} - \left[\left(M - \sqrt{3T_p} \right)^2 - 2\Phi \right]^{\frac{1}{2}} \right\}, \qquad (9.6)$$

$$n_i = \frac{N_i}{2\sqrt{3T_i/\mu_i}} \cdot \left\{ \left[\left(M - U_0 + \sqrt{3T_i/\mu_i} \right)^2 - 2Z_i\Phi/\mu_i \right]^{\frac{1}{2}} \right.$$
$$\left. - \left[\left(M - U_0 - \sqrt{3T_i/\mu_i} \right)^2 - 2Z_i\Phi/\mu_i \right]^{\frac{1}{2}} \right\}. \qquad (9.7)$$

In Eqs. (9.6) and (9.7), (n_p, T_p) and (n_i, T_i) are the normalized density and temperature of protons and α particles, respectively, and $U_0 = U_i/c_a$ is the normalized relative drift between the protons and the α particles. Further, the above density expressions (and the following Sagdeev potential expressions) are written in symbolic forms where the operation of a square root on a squared expression returns the same expression, e.g., $\sqrt{(A \pm B)^2} = A \pm B$. The above basic set of equations is closed by the transformed Poisson's equation:

$$\frac{\partial^2 \Phi}{\partial \xi^2} = n_e - n_p - Z_i n_i. \qquad (9.8)$$

On substituting the above expressions for n_e, n_p, and n_i in the transformed Poisson Eq. (9.8), multiplying it by $d\Phi/d\xi$ and integrating it using the boundary conditions $\Phi = 0$ and $d\Phi/d\xi = 0$ at $\xi \rightarrow \pm\infty$, one gets the following energy integral:

$$\frac{1}{2}\left(\frac{d\Phi}{d\xi}\right)^2 + S(\Phi, M) = 0, \qquad (9.9)$$

where $S(\Phi, M)$ is the Sagdeev pseudopotential (Sagdeev, 1966), given

$$S(\Phi, M) = \left[1 - \left(1 - \frac{\Phi}{\kappa - \frac{3}{2}}\right)^{-\kappa + \frac{3}{2}}\right] + \frac{N_p}{6\sqrt{3T_p}}$$

$$\cdot \left[\begin{array}{l} \left(M + \sqrt{3T_p}\right)^3 - \left[\left(M + \sqrt{3T_p}\right)^2 - 2\Phi\right]^{\frac{3}{2}} \\ -\left(M - \sqrt{3T_p}\right)^3 + \left[\left(M - \sqrt{3T_p}\right)^2 - 2\Phi\right]^{\frac{3}{2}} \end{array} \right] + \frac{N_i}{6\sqrt{3T_i/\mu_i}}$$

$$\cdot \left[\begin{array}{l} \left(M - U_0 + \sqrt{3T_i/\mu_i}\right)^3 - \left[\left(M - U_0 + \sqrt{3T_i/\mu_i}\right)^2 - 2Z_i\Phi/\mu_i\right]^{\frac{3}{2}} \\ -\left(M - U_0 - \sqrt{3T_i/\mu_i}\right)^3 + \left[\left(M - U_0 - \sqrt{3T_i/\mu_i}\right)^2 - 2Z_i\Phi/\mu_i\right]^{\frac{3}{2}} \end{array} \right].$$

$$(9.10)$$

The first, second, and third terms, on the left side of Eq. (9.10), describe the contribution from electrons, protons, and α particles to the Sagdeev pseudopotential, respectively. It is interesting to note that Eq. (9.9) describes the motion of a pseudoparticle of unit mass in a pseudopotential $S(\Phi, M)$, where Φ and ξ play the role of displacement, x, from the equilibrium and time, t, respectively.

Eq. (9.9) gives soliton solutions, provided that the Sagdeev potential satisfies the following conditions:

(1) $S(\Phi, M) = 0$, $\dfrac{dS(\Phi, M)}{d\Phi} = 0$, and $\dfrac{d^2 S(\Phi, M)}{d\Phi^2} < 0$, at $\Phi = 0$;

(2) $S(\Phi, M) = 0$, at $\Phi = \Phi_0$; and

(3) $S(\Phi, M) < 0$, for $0 < |\Phi| < |\Phi_0|$.

$$(9.11)$$

If one more condition is satisfied in addition to those listed in Eq. (9.11), then one can get a DL solution, if

(4) $\dfrac{dS(\Phi, M)}{d\Phi} = 0$, at $\Phi = \Phi_0$. $\qquad\qquad(9.12)$

We will now discuss the solution of Eq. (9.9) for some special cases.

9.3.1 Two-Component Plasmas: Cold Protons and Suprathermal Electrons

First, we shall discuss a special case of two-component plasmas consisting of cold protons ($T_p = 0$) and hot electrons having κ distributed electrons, which has been studied by Saini et al. (2009). We must point out that Eq. (9.10) for the Sagdeev potential has been derived assuming a finite temperature for protons (and also α particles). Therefore one cannot take $T_p = 0$ in Eqs. (9.6) and (9.10), as the proton terms will blow up. To handle the case of cold protons, one needs to follow the approach outlined in Saini et al. (2009) and Lakhina et al. (2014). Then, the proton density term becomes

$$n_p = \left(1 - \frac{2\Phi}{M^2}\right)^{-\frac{1}{2}},$$

(9.13)

and the Sagdeev potential is written as

$$S(\Phi, M) = 1 - \left(1 - \frac{\Phi}{\kappa - \frac{3}{2}}\right)^{-\kappa + \frac{3}{2}} + M^2 \cdot \left[1 - \left(1 - \frac{2\Phi}{M^2}\right)^{\frac{1}{2}}\right].$$

(9.14)

From Eq. (9.14), it is seen that the Sagdeev potential, $S(\Phi, M)$, and its first derivative with respect to Φ, vanish at $\Phi = 0$. The condition $d^2S(\Phi, M)/d\Phi^2 < 0$ at $\Phi = 0$ is satisfied, provided that the Mach number, M, exceeds a critical value M_0, given by

$$M_0 = \left(\frac{\kappa - \frac{3}{2}}{\kappa - \frac{1}{2}}\right)^{\frac{1}{2}} \leq 1.$$

(9.15)

For $\kappa \to \infty$, Eq. (9.15) gives $M_0 = 1$ as expected for the case of Boltzmann electron distribution (Sagdeev, 1966). When the κ-index decreases from large values to small values, M_0 decreases monotonically, and in the limit of $\kappa \to \frac{3}{2}$, $M_0 \to 0$. For $M > M_0$, the soliton conditions listed in Eq. (9.11) above can be satisfied only for $\Phi > 0$. Therefore only positive-potential ion-acoustic solitons are possible for this model. The DLs do not occur in this model as the condition given by Eq. (9.12) is not satisfied. Also, notice that the argument M_0 is exactly the additional multiplying factor of the kappa-dependent Debye length, i.e., $\lambda_D = M_0\lambda_{D\infty}$ (Chapter 5, Section 5.5; also see Eq. (6) in Livadiotis and McComas, 2014a).

From Eq. (9.13), it is clear that there is an upper limit on the potential Φ beyond which the ion density becomes imaginary. The upper limit is given by $\Phi = \Phi_{max} \equiv \frac{1}{2}M^2$, which corresponds to an infinite compression of proton density. The upper limit on the Mach number $M = M_{max}$ can be found from the solution of $S(\Phi_{max}, M_{max}) = 0$ or the root M_{max} from the following equation:

$$M_{max}^2 + 1 - \left(1 - \frac{\frac{1}{2}M_{max}^2}{\kappa - \frac{3}{2}}\right)^{-\kappa + \frac{3}{2}} = 0.$$

(9.16)

The positive-potential ion-acoustic solitons can exist only in the Mach number range $M_0 < M < M_{max}$. In Fig. 9.1, we have shown the variation of critical (M_0) and maximum (M_{max}) Mach numbers for various values of the κ-index. The existence domain, the area between the M_{max} and M_0 curves, of ion-acoustic solitons decreases as the κ-index decreases. Fig. 9.2 shows the variation of Sagdeev potential versus the normalized electrostatic potential Φ for a fixed

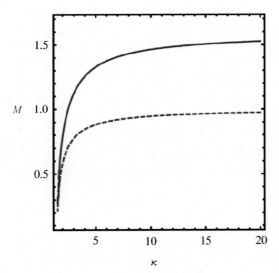

FIGURE 9.1
Ion-acoustic soliton existence domain in the parameter space of κ and the Mach number, M. Solitons occur in the region between the two curves. The *lower dashed curve* represents the critical Mach number, M_0, and the *upper solid curve* represents the maximum Mach number, M_{max}. Source: Saini et al. (2009).

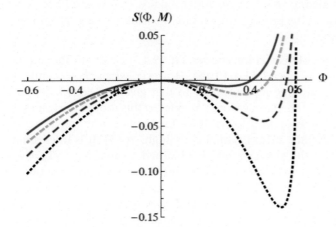

FIGURE 9.2
Variation of $S(\Phi, M)$ versus Φ for fixed $M = 1.1$ and different values of the κ-index: $\kappa = 3$ (*dotted curve*), $\kappa = 4$ (*dashed curve*), $\kappa = 6$ (*dot-dashed curve*), and $\kappa = 10$ (*solid curve*). Source: Saini et al. (2009).

value of $M = 1.1$ and for various values of the κ-index. It is clear that the positive-potential ion-acoustic soliton amplitude increases as the κ-index decreases. The effects of finite proton temperature on the ion-acoustic can be studied for the derived Sagdeev potential by setting $N_i = 0$ in Eq. (9.10).

9.3.2 Three-Component Plasmas: Protons, Heavier Ions, and Suprathermal Electrons

Three-component plasmas exist abundantly in nature, e.g., auroral field lines, PSBLs, solar wind, and in laboratory plasmas. We will consider the case of solar wind plasma, which can be modeled by hot fluid protons, hot fluid α particles drifting relative to protons along the magnetic field lines, and suprathermal electrons having a kappa distribution.

From Eq. (9.10), we observe that the Sagdeev potential and its first derivative with respect to Φ vanish at $\Phi = 0$. The soliton condition $d^2 S(\Phi, M)/d\Phi^2 < 0$ at $\Phi = 0$ demands that $M > M_0$, where M_0 satisfies the following equation:

$$\frac{N_p}{M_0^2 - 3T_p} + \frac{Z_i^2 N_i}{\mu_i \left[(M_0 - U_0)^2 - 3T_i/\mu_i \right]} - \frac{\kappa - \frac{1}{2}}{\kappa - \frac{3}{2}} = 0. \tag{9.17}$$

In the case of a proton–electron plasma ($N_i = 0$), Eq. (9.17) has a single positive root:

$$M_0 = \left(\frac{\kappa - \frac{3}{2}}{\kappa - \frac{1}{2}} + 3T_p \right)^{\frac{1}{2}}, \tag{9.18}$$

describing the ion-acoustic mode in a plasma with kappa-distributed electrons and hot protons, generalizing Eq. (9.15). Comparing Eqs. (9.15) and (9.18), we observe that an increase in the proton temperature, T_p, leads to a higher value of the critical Mach number.

Generally, in the presence of heavier ion species, Eq. (9.17) yields another mode with lower values of M_0, as given by Eq. (9.18). However, when $U_0 = 0$ and $T_i/T_p = \mu_i$, we still derive only one mode similar to that given by Eq. (9.18) but modified by the presence of the heavier ions. We refer to this root as the fast ion-acoustic mode and to the root with lower M_0 values as the slow ion-acoustic mode (Lakhina et al., 2008b, 2014; Lakhina and Singh, 2015). It is interesting to note that for $U_0 \neq 0$, the slower mode is obtained even when $T_i/T_p = \mu_i$. Therefore the slow ion-acoustic mode can always exist in solar wind except for the case of $U_0 = 0$ and $T_i/T_p = 4$. The slow ion-acoustic mode is an ion–ion hybrid mode that requires essentially two ion species having either different thermal velocities or a relative streaming. An interesting property of the new slow ion-acoustic mode is that it can support positive-/negative-potential solitons/DLs, depending on the plasma parameters.

Fig. 9.3 shows the variation of the Sagdeev potential versus Φ for various values of the Mach number in the case of the slow ion-acoustic mode. It shows four cases of double-layer (DL) occurrence under different parametric regimes that are representative of conditions occurring during the fast solar wind (panels A and B), the slow solar wind (panel C), and the intermediate solar wind (panel D) (Mangeney et al., 1999; Borovsky and Gary, 2014; Bourouaine et al. 2011; Maksimovic et al., 1997; Livadiotis, 2015a). The curves in Fig. 9.3A–D clearly show that the slow ion-acoustic soliton amplitude increases as M (above the critical value, M_0) increases (curves 1–3) until a DL (curve 4) is formed. Solitons do not exist for Mach numbers greater than the DL Mach number (curve 5). Here, the upper limit on the Mach number, $M_{\max}$, is provided by the

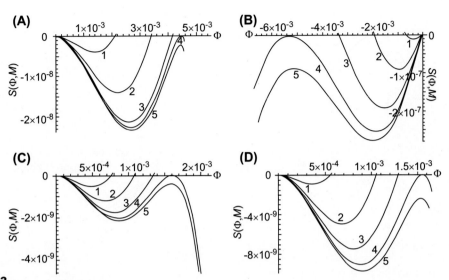

FIGURE 9.3

Sagdeev pseudopotential $S(\Phi, M)$ versus potential Φ for the slow ion-acoustic solitons and double layer (DL). The normalized parameters are as follows: (A) $N_i = 0.01$, $T_p = 1.0$, $T_i/T_p = 4$, $U_0 = 0.3$, $\kappa = 10$, and $M = 1.963$, 1.964, 1.9644, 1.96446 (DL), and 1.96449 for curves 1–5, respectively; (B) $N_i = 0.05$, $T_p = 2$, $T_i/T_p = 2$, $U_0 = 0.8$, $\kappa = 5$, and $M = 2.525$, 2.529, 2.531, 2.5317635 (DL), and 2.532 for curves 1–5, respectively; (C) $N_i = 0.05$, $T_p = 0.2$, $T_i/T_p = 4$, $U_0 = 0.2$, $\kappa = 2$, and $M = 0.9214$, 0.9215, 0.92155, 0.921569 (DL), and 0.921576 for curves 1–5, respectively; and (D) $N_i = 0.05$, $T_p = 0.5$, $T_i/T_p = 2$, $U_0 = 0.5$, $\kappa = 2$, and $M = 1.322$, 1.3225, 1.32265, 1.322724 (DL), and 1.32275 for curves 1–5, respectively. Source: Lakhina and Singh (2015).

DLs. Fig. 9.3A, C, and D shows examples of positively charged (i.e., $\Phi > 0$) solitons and a DL, whereas Fig. 9.3B shows negatively charged (with $\Phi < 0$) solitons and a DL. It is worth pointing out that the existence domain of the slow ion-acoustic soliton/DL is narrow, i.e., it exists just above the critical Mach numbers, M_0.

Fig. 9.4 shows the variation of the Sagdeev potential versus Φ for various values of the Mach number in the case of the fast ion-acoustic mode (for the same parameters as in Fig. 9.3). The fast ion-acoustic solitons occur for higher values of M than in Fig. 9.3. In this case, the soliton amplitude also increases with M until the upper limit of curve 4 is reached (curves 1–4). Beyond this, the soliton solution does not exist. The double layers do not occur, and the Mach upper limit M_{max} is provided by the restriction that the heavier ion number density must be real (Maharaj et al., 2012a). Hence, the solitons/double layers for both slow and fast ion-acoustic modes occur in a Mach number region of $M_0 < M \leq M_{max}$.

In Fig. 9.5, a variation of M_0, M_{max}, and the maximum value of the potential, Φ_{max}, with the κ-index are shown for both slow (Fig. 9.5A and B) and fast

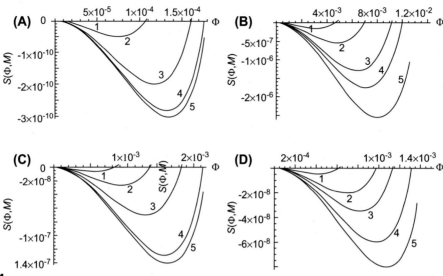

FIGURE 9.4

Sagdeev pseudopotential $S(\Phi, M)$ versus the potential Φ for the fast ion-acoustic solitons for the same plasma parameters as in Fig. 9.3, except that the Mach numbers are different in each panel. The parameters are as follows: (A) $M = 2.043, 2.044, 2.045, 2.04533$, and 2.0454 for curves 1–5, respectively; (B) $M = 2.62, 2.625, 2.63, 2.63205$, and 2.635 for curves 1–5, respectively; (C) $M = 1.014, 1.016, 1.018, 1.01962$, and 1.0199 for curves 1–5, respectively; and (D) $M = 1.398, 1.40, 1.401, 1.40223$, and 1.403 for curves 1–5, respectively. Source: Lakhina and Singh (2015).

ion-acoustic (Fig. 9.5C and D) modes for the parameters representing a slow solar wind stream case. From Fig. 9.5A, it is clear that as κ increases, both M_{max} and M_0 for the slow ion-acoustic case tend to increase. Fig. 9.5B shows that the maximum value of the positive potential of slow ion-acoustic solitons increases as κ increases. From Fig. 9.5C, it is clear that for the fast ion-acoustic mode, the range $(M_{max} - M_0)$ tends to increase as κ increases. Fig. 9.5D shows that Φ_{max} of the fast ion-acoustic mode also tends to increase together with the κ index.

Figs. 9.6 and 9.7 show the potential, Φ, and the electric field, E, profiles, respectively, for slow ion-acoustic solitons and the DL for the solar wind plasma parameters of Fig. 9.3. Curves 1–4 in each panel correspond to the Mach number in the corresponding panel in Fig. 9.3. Solitons (curves 1–3 in Fig. 9.6A–D) have symmetric profiles, and DLs (curve 4 in Fig. 9.6A–D) have asymmetric profiles. It is interesting to note that the electric fields for the solitons have bipolar structures, whereas those for the DLs have monopolar structures (Fig. 9.7A–D).

Observed WDLs in the solar wind at 1 AU are characterized by amplitudes $e\Phi/(k_B T_e) \sim 10^{-4} - 10^{-3}$ and widths W $\sim$ 5 to 60 $\lambda_{D\infty}$ with a peak around 25 $\lambda_{D\infty}$ (Mangeney et al., 1999; Lacombe et al., 2002; Salem et al., 2003). From Figs. 9.3 and 9.6, we note that slow ion-acoustic DLs have maximum amplitudes

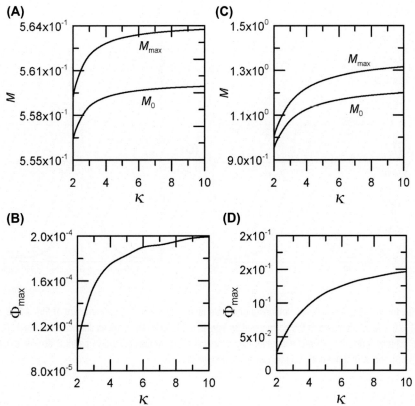

FIGURE 9.5

Variation of the critical Mach number M_0, maximum Mach number M_{max}, and maximum potential Φ_{max} with different values of the κ-index in a slow solar wind plasma with $N_i = 0.05$, $T_p = 0.2$, $T_i/T_p = 2$, and $U_0 = 0$. (A) and (B) correspond to the slow ion-acoustic case, while (C) and (D) correspond to the fast ion-acoustic case. Source: Lakhina and Singh (2015).

of $e\Phi_{max}/(k_BT_e) = -2.0 \times 10^{-4}$–0.0075 (negative DLs) and $e\Phi_{max}/(k_BT_e) =$ 0.001−0.007 (positive DLs) with widths $W \sim 12$−28 $\lambda_{D\infty}$. Hence, there is an excellent agreement with observed amplitudes and widths of WDLs. However, there is a disagreement between the observed and predicted shapes of the WDLs. Fusion of a positive DL and a negative-potential soliton seems to explain the WDL profiles!

We propose that both slow and fast ion-acoustic solitons can produce low-frequency electrostatic wave activity in the ion-acoustic frequency range, falling between the ion and electron plasma frequencies, as observed in the solar wind at 1 AU by the Wind S/C (Mangeney et al., 1999; Lacombe et al., 2002; Salem et al., 2003). First, this automatically explains the coherent nature of the waveforms. Second, the fast Fourier transform of the ion-acoustic solitons/DLs would produce a broadband spectrum with a main peak between 0.35 and 1.6 kHz and $E = (0.01$−0.7$)$ mV/m, which is in excellent agreement with the observed electric fields $\sim(0.0054$−0.54$)$ mV/m associated with the low-frequency waves observed in the solar wind at 1 AU (Mangeney et al., 1999).

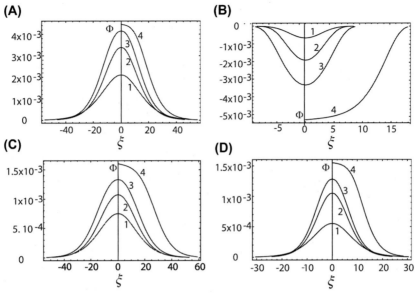

FIGURE 9.6

Potential Φ profiles for the slow ion-acoustic solitons and double layers for the plasma parameters of Fig. 9.3. Curves 1—4 in each panel correspond to the Mach numbers in the corresponding panels in Fig. 9.3. The profiles shown in panels A, C, and D are for the positive-potential solitons (curves 1—3) and a double layer (curve 4). The profiles shown in panel B are for the negative-potential solitons (curves 1—3) and a DL (curve 4). Source: Lakhina and Singh (2015).

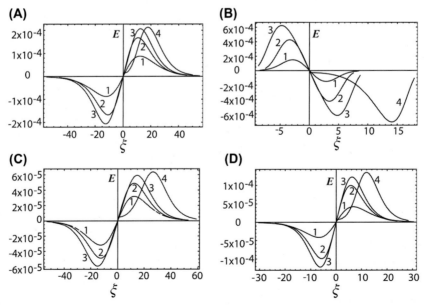

FIGURE 9.7

Electric field, E, profiles for the slow ion-acoustic solitons and double layers for the plasma parameters of Fig. 9.3. Curves 1—4 in each panel correspond to the Mach numbers in the corresponding panels in Fig. 9.3. The electric field profiles of the solitons have bipolar structures (curves 1, 2, and 3 in panels A, B, C, and D), and those of the double layers have monopolar structures (curve 4 in panels A, B, C, and D). Source: Lakhina and Singh (2015).

9.3.3 Two-Component Magnetized Plasmas: Protons and Suprathermal Electrons

In this section, we consider pure electron–proton, magnetized plasma, which consists of adiabatic warm ions and hot electrons. The dynamics of the protons are governed by fluid equations, and hot electrons are assumed to have a kappa distribution given by Eq. (9.1). The ion-acoustic solitons are considered to be propagating in the x–z plane, and the ambient magnetic field $\vec{B}_0$ is along the z direction (Singh et al., 2013). The electron density is given by Eq. (9.2). The set of normalized fluid equations for the protons in magnetized plasma is given below:

$$\frac{\partial n_p}{\partial t} + \frac{\partial(u_x n_p)}{\partial x} + \frac{\partial(u_z n_p)}{\partial z} = 0, \tag{9.19}$$

$$\frac{\partial u_x}{\partial t} + u_x\frac{\partial u_x}{\partial x} + u_z\frac{\partial u_x}{\partial z} + 3n_p T_p\frac{\partial n_p}{\partial x} + \frac{\partial \varphi}{\partial x} - Ru_y = 0, \tag{9.20}$$

$$\frac{\partial u_y}{\partial t} + u_x\frac{\partial u_y}{\partial x} + u_z\frac{\partial u_y}{\partial z} + Ru_x = 0, \tag{9.21}$$

$$\frac{\partial u_z}{\partial t} + u_x\frac{\partial u_z}{\partial x} + u_z\frac{\partial u_z}{\partial z} + 3n_p T_p\frac{\partial n_p}{\partial z} + \frac{\partial \varphi}{\partial z} = 0. \tag{9.22}$$

$R = \Omega_p/\omega_p$, where $\Omega_p = eB_0/m_p$ is the cyclotron proton frequency. In Eqs. (9.19)–(9.22) we used the same normalizations as in Eqs. (9.2)–(9.5); Eqs. (9.19)–(9.22) depend on the x, z, and t variables. In order to find a solution for Eqs. (9.19)–(9.22), they are first transformed to the single variable $\xi = ax + bz - Mt$, where $a = k_x/k = \sin\vartheta$ and $b = k_z/k = \cos\vartheta$ are the direction cosines along the x and z directions, respectively, satisfying the relation $a^2 + b^2 = 1$.

The inclusion of the ambient magnetic field introduces complexities in the system of equations and cannot be solved analytically. However, under the assumption that the quasineutrality condition, $n_e = n_p$, is satisfied, the transformed equations can be integrated. We make use of the boundary conditions, $n_{e,p} \to 1$, $u_{x,z} \to 0$, $\Phi \to 0$, and $d\Phi/d\xi \to 0$ at $\xi \to \pm\infty$, and arrive at the energy integral, given by Eq. (9.9), with the Sagdeev potential given by

$$S(\Phi, M) = R^2\frac{S_1(\Phi, M)}{S_2(\Phi, M)}, \quad \text{with} \tag{9.23}$$

$$S_1(\Phi, M) = \left(1 + \frac{\beta^2}{M^2}\right)\left[1 - \left(1 - \frac{\Phi}{\kappa - \frac{3}{2}}\right)^{-\kappa + \frac{3}{2}}\right] + (M^2 + \beta^2)\left[1 - \left(1 - \frac{\Phi}{\kappa - \frac{3}{2}}\right)^{\kappa - \frac{1}{2}}\right]$$

$$+ \left[1 - \beta^2\left(\frac{\kappa - \frac{1}{2}}{\kappa - \frac{3}{2}}\right)\right]\Phi - \frac{\beta^2}{2M^2}\left[1 - \left(1 - \frac{\Phi}{\kappa - \frac{3}{2}}\right)^{-2\kappa + 3}\right] - \frac{M^2}{2}\left[1 - \left(1 - \frac{\Phi}{\kappa - \frac{3}{2}}\right)^{2\kappa - 1}\right]$$

$$+ T_p\left\{ \begin{matrix} \frac{\beta^2}{M^2}\left[1 - \left(1 - \frac{\Phi}{\kappa - \frac{3}{2}}\right)^{-\kappa + \frac{3}{2}}\right] + \left(1 - \frac{\Phi}{\kappa - \frac{3}{2}}\right)^{-4\kappa + 3} - \left(1 - \frac{\Phi}{\kappa - \frac{3}{2}}\right)^{-3\kappa + \frac{3}{2}} \\[2mm] + \beta^2\left[1 - \left(1 - \frac{\Phi}{\kappa - \frac{3}{2}}\right)^{\kappa - \frac{1}{2}}\right] + \frac{(\beta^2 - 3)}{2}\left[1 - \left(1 - \frac{\Phi}{\kappa - \frac{3}{2}}\right)^{-2\kappa + 1}\right] \\[2mm] + \left(1 + \frac{T_p\beta^2}{M^2}\right)\left[1 - \left(1 - \frac{\Phi}{\kappa - \frac{3}{2}}\right)^{-3\kappa + \frac{3}{2}}\right] - \frac{T_p\beta^2}{2M^2}\left[1 - \left(1 - \frac{\Phi}{\kappa - \frac{3}{2}}\right)^{-6\kappa + 3}\right] \end{matrix} \right\}$$

$$S_2(\Phi, M) = \left[1 - M^2\left(\frac{\kappa - \frac{1}{2}}{\kappa - \frac{3}{2}}\right)\left(1 - \frac{\Phi}{\kappa - \frac{3}{2}}\right)^{2\kappa - 2} + 3T_p\left(\frac{\kappa - \frac{1}{2}}{\kappa - \frac{3}{2}}\right)\left(1 - \frac{\Phi}{\kappa - \frac{3}{2}}\right)^{-2\kappa}\right]^2.$$

Eq. (9.23) must satisfy the soliton conditions given by Eq. (9.11). The second derivative of the Sagdeev potential must satisfy at $\Phi = 0$:

$$\frac{d^2 S(\Phi, M)}{d\Phi^2} = \frac{R^2}{M^2}\left(\frac{M^2 - M_0^2}{M^2 - M_1^2}\right) < 0, \quad \text{with} \tag{9.24}$$

$$M_0 = \beta\left(\frac{\kappa - \frac{3}{2}}{\kappa - \frac{1}{2}} + 3T_p\right)^{\frac{1}{2}} \quad \text{and} \quad M_1 = \left(\frac{\kappa - \frac{3}{2}}{\kappa - \frac{1}{2}} + 3T_p\right)^{\frac{1}{2}}. \tag{9.25}$$

Before we proceed further, the importance of the second derivative along with Eq. (9.25) has to be understood properly. Here, $M_0 \leq M_1$, since $\beta = \cos \vartheta \leq 1$ and inequality in Eq (9.24) can be satisfied only when $M_0 \leq M \leq M_1$. Therefore M_0 and M_1 are the critical and upper limits of the Mach number, respectively, for soliton solutions to exist. For parallel propagation ($\vartheta = 0^0$), inequality in Eq (9.24) cannot be satisfied because both the critical (M_0) and the upper limit (M_1) of the Mach number will coincide with each other. Hence, a soliton solution will not be possible. Thus M_1 is the maximum possible Mach number beyond which soliton solutions do not exist. Further, the upper limit of the Mach number is independent of the angle of propagation. The amplitude of the ion-acoustic solitons increases with an increase in superthermality of electrons whereas the width decreases. Also, the critical Mach number as well as the upper Mach number decreases with an increase in superthermality (not shown here).

9.4 Model for Electron-Acoustic Solitons in Plasmas With Suprathermal Electrons

Over the last decade, electron-acoustic SWs have been widely studied, considering though, Maxwellian distributions for hot electrons. Relevant studies using kappa distributions for hot electrons have been very limited. For example, in three-component nonmagnetized plasma having stationary ions, cold inertial electrons, and hot suprathermal electrons, the existence of arbitrary amplitude electron-acoustic SWs has been studied by Younsi and Tribeche (2010). In a similar theoretical model, Sahu (2010) studied the existence of small-amplitude electron-acoustic DLs. Devanandhan et al. (2011a, 2011b) studied electron-acoustic solitons in nonmagnetized multicomponent plasma with two and three electron components and ions, respectively. In pure electron–ion magnetized plasma, Devanandhan et al. (2012) studied electron-acoustic solitons by considering ions to be much hotter than electrons. Sultana et al. (2012) studied arbitrary and small-amplitude electron-acoustic waves in a two-electron component magnetized, cold plasma by assuming a quasineutrality condition. Small-amplitude, weakly, nonlinear electron-acoustic waves have been studied by Javidan and Pakzad (2013) in a three-component magnetized plasma. Zakharov–Kuznetsov electrostatic solitons and modified Korteweg-de Vries (mKdV) solitons have been studied by Adnan et al. (2014) and Devanandhan et al. (2015) in a magnetized plasma having warm ions and suprathermal electrons.

We now consider the propagation of the electron-acoustic SWs parallel to the ambient magnetic field in a three-component homogeneous plasma consisting of cold (N_c, T_c) and hot (N_h, T_h) electrons, and protons (N_p, T_p), where N_j and T_j refer to the density and temperature of the jth species (at electrostatic equilibrium). The hot electrons are assumed to be following a kappa distribution function (Eq. 9.1), and a density expression given by Eq. (9.2) can be obtained upon integration of Eq. (9.1) over the velocity space (Chapter 3). The dynamics of cold electron and ion populations are described by the multifluid equations of continuity, momentum, an energy equation derived from the equation of state of each species, and the Poisson equation (Devanandhan et al., 2011a) and are given by Eqs. (9.3)–(9.5) and closed by Poisson Eq. (9.8), where n_e is replaced by $n_c + n_h$, and $n_p + Z_i n_i$ is replaced by n_p. Note that n_j, u_j, and P_j are the density, fluid velocity, and pressure of the jth species, with j: c and p represents cold electrons and protons, respectively. Also, in Eqs. (9.2)–(9.5) and (9.8), we normalized the densities with the total equilibrium electron $N_e = N_c + N_h = N_p$ or ion densities, the velocities with the thermal speed of hot electrons, $\theta_h \equiv \sqrt{2k_B T_h/m_e}$, the time with the inverse of the electron plasma frequency, $\omega_e = [N_e e^2/(\varepsilon_0 m_e)]^{1/2}$, the lengths with the hot-electron Debye length, $\lambda_{D\infty} = [\varepsilon_0 k_B T_h/(N_e e^2)]^{1/2}$, the potential with $k_B T_h/e$, and the thermal pressure P_j with $N_e k_B T_h$. As in Section 9.3.1, we have taken $\gamma = 3$ for both the electrons and the ions. The Sagdeev potential for electron-acoustic solitons is now written as

$$S(\Phi,M) = N_h \left[1 - \left(1 - \frac{\Phi}{\kappa - \frac{3}{2}} \right)^{-\kappa + \frac{3}{2}} \right]$$

$$+ \frac{N_c}{6\sqrt{3T_c}} \left\{ \begin{array}{l} (M + \sqrt{3T_c})^3 - \left[(M + \sqrt{3T_c})^2 + 2\Phi \right]^{\frac{3}{2}} \\ -(M - \sqrt{3T_c})^3 + \left[(M - \sqrt{3T_c})^2 + 2\Phi \right]^{\frac{3}{2}} \end{array} \right\}$$

$$+ \frac{\mu_p N_p}{6\sqrt{3T_p/\mu_p}} \left\{ \begin{array}{l} \left(M + \sqrt{3T_p/\mu_p} \right)^3 - \left[\left(M + \sqrt{3T_p/\mu_p} \right)^2 - 2\Phi/\mu_p \right]^{\frac{3}{2}} \\ -\left(M - \sqrt{3T_p/\mu_p} \right)^3 + \left[\left(M - \sqrt{3T_p/\mu_p} \right)^2 - 2\Phi/\mu_p \right]^{\frac{3}{2}} \end{array} \right\}.$$

$$(9.26)$$

Here $\mu_p = m_p/m_e$. The Sagdeev potential given by Eq. (9.26) satisfies the soliton condition $d^2 S(\Phi,M)/d\Phi^2 < 0$ at $\Phi = 0$ provided $M > M_0$, where M_0 is the critical Mach number given by the equation:

$$\frac{N_c}{(M_0^2 - 3T_c)} + \frac{1}{\mu_p \left(M_0^2 - 3T_p/\mu_p \right)} - N_h \left(\frac{\kappa - \frac{1}{2}}{\kappa - \frac{3}{2}} \right) = 0. \qquad (9.27)$$

If the contribution from the ions is neglected, the above equation can be approximated to

$$M_0 = \left[\frac{N_c}{N_h} \left(\frac{\kappa - \frac{3}{2}}{\kappa - \frac{1}{2}} \right) + 3T_c \right]^{\frac{1}{2}}, \qquad (9.28)$$

generalizing Eq. (9.18). In general, Eq. (9.27) will have four roots, but all the other roots are not physical. For numerical computations, we will consider only the real positive roots.

Fig. 9.8A shows the variation of Sagdeev potential versus Φ for different values of the Mach number M and the kappa index κ. The typical chosen values of the normalized parameters are cold electron density $N_c = 0.55$, hot electron density $N_h = 0.45$, and hot electron temperature $T_c = 0.01 = T_p$ (proton temperature). It can be seen from the figure that the amplitude of the electron-acoustic solitons increases with an increase in the Mach number. A soliton solution exists for the Mach numbers $0.6618 \le M \le 1.2878$ for $\kappa = 2$. Numerical computations show that the increase in the value of kappa shifts the range of Mach numbers to the higher values for which soliton solutions are obtained, e.g., for $\kappa = 4$, the value of the Mach number lies in the range $0.9508 \le M \le 1.3116$ in order to have the soliton solution. In Fig. 9.8B, the electric field amplitude of the electron-acoustic

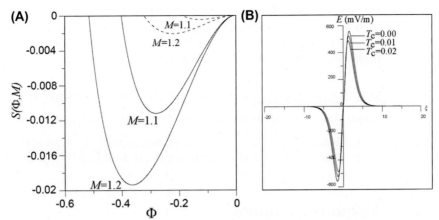

FIGURE 9.8
(A) Variation of the Sagdeev potential $S(\Phi,M)$ for $\kappa = 2$ (*solid line*), 4 (*dashed line*), and $M = 1.1$, 1.2, and the parameters $N_c = 0.55$, $N_h = 0.45$, and $T_c = 0.01 = T_p$. (B) Electric field amplitude for different values of T_c and the parameters $\kappa = 2$, $M = 1.1$, $N_c = 0.55$, $N_h = 0.45$, and $T_p = 0.01$.

solitons is depicted for different values of T_c for the parameters $\kappa = 2$, $M = 1.1$, $N_c = 0.55$, $N_h = 0.45$, and $T_i = 0.01$. It is observed that the bipolar electric field structures' amplitudes decrease with an increase in the T_c values. The maximum amplitude occurs for $T_c = 0$. Also, the Mach number regime, for which a soliton solution can be obtained, narrows down with an increasing value of T_c.

It is worth mentioning that having a finite temperature for the cold electrons modifies the regime for the existence of the electron-acoustic solitons. These nonlinear soliton structures will have a higher electric field amplitude for the case when the cold electron temperature is zero as compared to the finite temperature case.

9.5 Concluding Remarks

We have studied ion- and electron-acoustic solitons and DLs in multicomponent magnetized plasma containing suprathermal electrons governed by kappa distributions. In electron–proton plasmas, the ion-acoustic solitons have positive potentials. The DLs cannot exist in this case. As the κ-index increases, the existence domain of the ion-acoustic solitons increases, but their amplitudes decrease. In plasmas containing two ions and suprathermal electrons described by kappa distributions (e.g., solar wind approximated by populations of proton, α-particles, and electrons), the existence domain and the maximum amplitude of both the slow and fast ion-acoustic parallel propagating modes tend to increase as the kappa index increases. Three-component plasmas containing protons, cold electrons, and hot electrons governed by a kappa distribution are found to support electron-acoustic solitons having negative potentials. Larger κ-indices lead to solitons of higher Mach numbers (and smaller amplitudes of the electron-acoustic solitons).

The results of slow ion-acoustic DLs may explain the observations of WDLs in solar wind at 1 AU. The results of slow and fast ion-acoustic solitons can provide a plausible mechanism for the generation of coherent electrostatic turbulence in solar wind observed by Wind S/C at 1 AU.

9.6 Science Questions for Future Research

Future analyses and observations need to address the following questions:

1. How do kappa-distributed electrons control slow ion-acoustic DLs?
2. How do they control electrostatic waves excited by ion-acoustic modes?
3. Can acoustic modes exist in solar wind?

Acknowledgments

GSL thanks the National Academy of Sciences, India, for their support under the NASI-Senior Scientist Platinum Jubilee Fellowship Scheme. SS thanks Mr. B.I. Panchal and Dr. S. Devanandhan for their help in preparing the figures and typesetting the equations.

PART 3
Applications in Space Plasmas

10. Ion Distributions in Space Plasmas   421
    G. Livadiotis, D.J. McComas
11. Electron Distributions in Space Plasmas   465
    V. Pierrard, N. Meyer-Vernet
12. The Kappa-Shaped Particle Spectra in Planetary
    Magnetospheres   481
    K. Dialynas, C.P. Paranicas, J.F. Carbary, M. Kane,
    S.M. Krimigis, B.H. Mauk
13. Kappa Distributions and the Solar Spectra:
    Theory and Observations   523
    E. Dzifčáková, J. Dudík
14. Importance of Kappa Distributions to Solar
    Radio Bursts   549
    I.H. Cairns, B. Li, J.M. Schmidt
15. Common Spectrum of Particles Accelerated in
    the Heliosphere: Observations and a Mechanism   569
    L.A. Fisk, G. Gloeckler

16. Formation of Kappa Distributions at Quasiperpendicular
    Shock Waves   609
    G.P. Zank
17. Electron Kappa Distributions in Astrophysical Nebulae   633
    D.C. Nicholls, M.A. Dopita, R.S. Sutherland,
    L.J. Kewley

CHAPTER 10

Ion Distributions in Space Plasmas

G. Livadiotis[1], D.J. McComas[2]
[1]Southwest Research Institute, San Antonio, TX, United States;
[2]Princeton University, Princeton, NJ, United States

Chapter Outline

10.1 Summary 422
10.2 Introduction 422
10.3 Formulations of Ion Kappa Distributions 425
    10.3.1 Standard Kappa Distribution of Velocities 425
    10.3.2 "Negative" Kappa Distribution of Velocities 427
    10.3.3 Superposition of Kappa Distribution of Velocities 429
    10.3.4 Anisotropic Kappa Distribution of Velocities 430
    10.3.5 Phase Space Marginal Kappa Distribution of Velocities 432
10.4 Toward Antiequilibrium, the Farthest State From Thermal Equilibrium 432
10.5 Arrangement of the Stationary States 439
    10.5.1 A Measure of the Thermodynamic Distance From Thermal Equilibrium 439

    10.5.2 A Generalized Measure 441
    10.5.3 The Kappa Spectrum 443
10.6 Interpreting the Observations 444
    10.6.1 Observations and Measurements of Kappa Indices in the Heliosphere 444
    10.6.2 Observations in the Inner Heliosheath 451
        10.6.2.1 Far-Equilibrium Inner Heliosheath 451
        10.6.2.2 Anticorrelation Between Density and Temperature 452
        10.6.2.3 Isobaric Process 454
    10.6.3 Near Versus Far Equilibrium 456
    10.6.4 The Role of PickUp Ions in the Transitions of Kappa Distributions 458
10.7 Concluding Remarks 463
10.8 Science Questions for Future Research 463
Acknowledgment 463

10.1 Summary

In this chapter, we present the types and properties of kappa distributions of ion velocities that describe these populations in space plasmas. We also present and interpret the measurements of kappa indices of various ion populations in space plasmas throughout the heliosphere and provide a detailed discussion of the observations in the inner heliosheath. The kappa index is the primary parameter of kappa distributions, significant for identifying the nonequilibrium stationary states and characterizing their "thermodynamic distance" from thermal equilibrium and the furthest state from thermal equilibrium, or "antiequilibrium." We examine the detailed transitions of stationary states toward the extreme state of antiequilibrium. Observations show an empirical separation of the spectrum of kappa values in ion space plasmas. The kappa spectrum is divided in two regions: the near-equilibrium region with values of kappa, $2.5 < \kappa < \infty$ (e.g., Saturnian magnetosphere), and the far-equilibrium region with indices $1.5 < \kappa < 2.5$ (e.g., inner heliosheath). We discuss the role of PickUp Ions (PUIs) in the transitions of kappa distributions from near-equilibrium to far-equilibrium states. With the presented statistical formulations, properties, and observations of kappa distribution of ion populations in space plasmas, the chapter provides valuable and necessary tools for space physics education and research.

Science Question: What are the features of the ion kappa distributions in space plasmas?

Keywords: Antiequilibrium; Correlation; Heliosphere; Inner heliosheath; Ion distributions; Isobaric process; Near/far equilibrium; PickUp Ions; Solar wind; Thermodynamic distance.

10.2 Introduction

Kappa distributions of particle velocities were employed in various space plasma analyses: (1) the *inner heliosphere*, including solar wind (e.g., Collier et al., 1996; Maksimovic et al., 1997a,b, 2005; Pierrard et al., 1999; Mann et al., 2002; Marsch, 2006; Zouganelis, 2008; Štverák et al., 2009; Livadiotis and McComas, 2010a, 2011a, 2013a, 2013c; Yoon, 2014; Pierrard and Pieters, 2015; Pavlos et al., 2016), solar spectra (e.g., Chapter 13; Dzifčáková and Dudík, 2013; Dzifčáková et al., 2015), solar corona (e.g., Owocki and Scudder, 1983; Vocks et al., 2008; Lee et al., 2013; Cranmer, 2014), solar energetic particles (e.g., Xiao et al., 2008; Laming et al., 2013), corotating interaction regions (CIRs) (e.g., Chotoo et al., 2000), and solar-related (e.g., Mann et al., 2009; Livadiotis and McComas, 2013b; Bian et al., 2014; Jeffrey et al., 2016); (2) the *planetary magnetospheres*, including magnetosheath (e.g., Binsack, 1966; Olbert, 1968; Vasyliūnas, 1968; Formisano et al., 1973; Ogasawara et al., 2013), near magnetopause (e.g., Ogasawara et al., 2015), magnetotail (e.g., Grabbe, 2000), ring current (e.g., Pisarenko et al., 2002), plasma sheet (e.g., Christon, 1987; Wang et al., 2003; Kletzing et al., 2003), magnetospheric substorms (e.g., Hapgood et al., 2011), magnetospheres of giant planets (Chapter 12), such as Jovian (e.g., Collier and Hamilton, 1995; Mauk et al., 2004), Saturnian (e.g., Schippers et al., 2008; Dialynas et al., 2009; Livi et al., 2014; Carbary et al., 2014), Uranian (e.g., Mauk et al., 1987), Neptunian (Krimigis et al., 1989), magnetospheres of planetary moons, such as Io (e.g., Moncuquet et al., 2002), and Enceladus (e.g., Jurac et al., 2002), or cometary magnetospheres (e.g., Broiles et al., 2016a,b); (3) the *outer heliosphere and the inner heliosheath* (e.g., Decker and Krimigis, 2003; Decker et al., 2005; Heerikhuisen et al., 2008, 2010, 2014, 2015; Zank et al., 2010; Livadiotis et al., 2011, 2012, 2013; Livadiotis and McComas, 2011a, 2011b, 2012, 2013a, 2013c, 2013d; Livadiotis, 2014a; Fuselier et al., 2014; Zirnstein and McComas, 2015); and (4) *beyond the heliosphere*, including HII regions (i.e., regions of interstellar atomic hydrogen that is ionized; e.g., Nicholls et al., 2012), planetary nebula (e.g., Nicholls et al., 2012; Zhang et al., 2014), and supernova magnetospheres (e.g., Raymond et al., 2010). Maxwell distributions have also been used in space science (e.g., Hammond et al., 1996), especially due to their simplicity; for example, they are often used to fit the "core" of the observed distributions, that is, the part of the distribution around its maximum.

There are various physical mechanisms that can generate kappa distribution. Such mechanisms may be applied in different plasma environments and successfully explain the observed kappa distributions. Some of them are detailed described in this book, e.g., polytropic bevavior of plasmas (Chapter 5), "superstatistics," that is, the superposition of various Maxwell distributions of different temperature (Chapter 6), asymptotically stationary electron distribution functions caused by Langmuir turbulence (Chapter 8), diffusive shock acceleration or stochastic acceleration processes (Chapter 15), formation of kappa distributions across shock waves (Chapter 16). Here we present another mechanism that can decrease the kappa index of ion kappa distributions when bulk plasma incorporates with PUIs (Livadiotis and McComas, 2011a). Aside from their empirical successful usage and their extraction from different mechanisms, kappa distributions are naturally exported (Chapter 1; Treumann, 1997; Milovanov and Zelenyi, 2000; Leubner, 2002; Livadiotis and McComas, 2009; Livadiotis, 2015a) from the foundations of nonextensive statistical mechanics (Chapter 1; Tsallis, 1988, 2009b; Tsallis et al., 1999).

Prior to the connection of kappa distributions with nonextensive statistical mechanics, there was confusion in space physics community about the physical meaning of the temperature and kappa index. For instance, several modified versions of kappa distribution have been suggested (e.g., Hawkins et al., 1998, Mauk et al., 2004), with all having different definitions of the kappa index. The associated temperature was involved in even more confused scenarios and there were numerous publications defining different temperature-like parameters (c.f. Table 2 in Livadiotis and McComas, 2009). However, the exact definition of the nonequilibrium temperature is not something that can be simply chosen; rather, it must emerge from statistical mechanics. The nonextensive statistical mechanics not only offers a solid foundation of kappa distributions, but also provides a set of proven tools for understanding these distributions, including a consistent definition of temperature for nonequilibrium systems (e.g., see Abe, 2001; Abe et al., 2001; Rama, 2000; Livadiotis and McComas, 2009, 2010a) and the physical meaning of the kappa index as a measure of how far a stationary state that the system resides in is from equilibrium (Livadiotis and McComas, 2010a, 2010b).

In the pioneering work of Binsack (1966), Olbert (1968), and Vasyliūnas (1968), the temperature was given by its kinetic definition, that is, the variance of the distribution of velocities or the mean kinetic energy, $T \equiv \frac{2}{3} k_B^{-1} \cdot \langle \varepsilon_K \rangle$. Livadiotis and McComas (2009) showed the equality between this kinetic definition and the primary thermodynamic definition within the framework of nonextensive statistical mechanics. This equivalence is sufficient to provide a consistent and unique definition of nonequilibrium temperature (see references in Livadiotis and McComas, 2009). In addition, Eq. (10.2) ensures that the internal energy is invariant under variations of the kappa index. The internal energy per particle, $U/N = \langle \varepsilon_K \rangle$, is a characteristic of the system and cannot be different for different stationary states. All the stationary states must equivalently describe a system that has a particular internal energy U. Since the kappa index is a characteristic of stationary states (it differs for different states), the internal energy U must be independent of this index. Therefore, the temperature T, defined out of equilibrium, and the kappa index κ are the *two independent and controlling parameters* of nonequilibrium systems, such as space plasmas.

The significant role of the kappa index is that it identifies nonequilibrium stationary states and gives a measure of their "thermodynamic distance" from thermal equilibrium. The kappa distribution identifies a single stationary state, corresponding to a certain value of the kappa index. Different stationary states are indicated by different values of the kappa index. The temperature, on the other hand, is independent of the kappa index, and thus the characterization of the stationary state of a system is irrelevant to its temperature in contrast to the kappa index. Therefore, stationary states can be arranged according to their kappa indices so that the smaller the kappa index is, the further from equilibrium the system is. At equilibrium, the kappa index is $\kappa \to \infty$. In practice, for kappa indices larger than $\kappa \sim 10$, the kappa distribution is considered close to a Maxwellian (quasi-equilibrium). On the other hand, for $\kappa \to \frac{3}{2}$, the furthest possible stationary state from equilibrium is attained, called antiequilibrium. Therefore, the whole set of stationary states can be realized in a spectrum-like arrangement of the kappa index in the interval $\kappa \in \left(\frac{3}{2}, \infty \right]$.

This chapter presents the basic formulae of velocity kappa distributions that describe ions in space plasmas. The formulations are given in terms of the standard 3-D kappa index κ and the more fundamental modified kappa index κ_0, which is invariant of the particle degrees of freedom. Aside of the standard formula of multidimensional kappa distribution, we present and describe in the "negative" kappa distribution, the superposition of kappa distribution, the anisotropic kappa distribution, and the phase space marginal kappa distribution of velocities. These distributions can also describe electrons; however, this is the topic of Chapter 11. We will also discuss the consistent definitions of the "thermodynamic distance" between stationary states and the extreme stationary state of thermal equilibrium. This is based on the observation that the furthest state from thermal equilibrium, "antiequilibrium," has universal properties and is independent of the particle dimensionality or degrees of freedom; thus the two extreme states, equilibrium and antiequilibrium, have an invariant distance that can be set to 1, while all the other states in between can be expressed as a fraction of this distance. Finally, the chapter reviews the observations of the kappa indices of ion populations in space plasmas or ion-related systems throughout the heliosphere. With the statistical formulations, properties, and observations of kappa distribution of ion populations in space plasmas presented here, this chapter provides valuable and necessary tools for the space physics community.

The chapter is structured as follows: in Section 10.3, we present the basic formulae of velocity space kappa distributions, useful for describing ions in space plasmas. Section 10.4 examines the transitions of stationary states toward the extreme state of antiequilibrium. Section 10.5 presents the consistent definition of the measure of the thermodynamic distance from thermal equilibrium and then describes the arrangement of the stationary states into the kappa spectrum. Section 10.6 presents and interprets the observations of kappa indices of ion populations in space plasmas throughout the heliosphere and provides a detailed discussion of observations in the inner heliosheath. In particular, we present measurements of the kappa indices in the heliosphere, in the inner heliosheath, the distinction of the thermodynamic behavior between near- and far-equilibrium regions, and the role of PUIs in the transitions of kappa distributions. Finally, the concluding remarks are given in Section 10.7, while three general science questions for future analyses are posed in Section 10.8.

10.3 Formulations of Ion Kappa Distributions

Here we present the most frequently used formulae of kappa distributions for describing the ion populations in space plasmas.

10.3.1 Standard Kappa Distribution of Velocities

For a system of particles with d_K kinetic degrees of freedom per particle, the kappa distribution of velocities is given by

$$P(\vec{u};\kappa,T) = \left[\pi\,\theta^2\left(\kappa - \frac{d_K}{2}\right)\right]^{-\frac{1}{2}d_K} \cdot \frac{\Gamma(\kappa+1)}{\Gamma\left(\kappa - \frac{d_K}{2}+1\right)} \cdot \left[1 + \frac{1}{\kappa - \frac{d_K}{2}} \cdot \frac{(\vec{u} - \vec{u}_b)^2}{\theta^2}\right]^{-\kappa-1},$$

(10.1a)

where $\theta \equiv \sqrt{2k_B T/m}$ is the thermal speed of particle of mass m, that is, the particle temperature T expressed in speed units; both the particle $\vec{u}$ and bulk $\vec{u}_b$ velocities are in an inertial reference frame. The distribution is isotropic in the commoving system (where the fluid is in its rest frame, $\vec{u}_b = 0$) and in the absence of a potential energy (Chapter 3).

The kappa distribution, Eq. (10.1a), in terms of the kinetic energy, $\varepsilon_K = \frac{1}{2}m(\vec{u} - \vec{u}_b)^2$, is written as

$$P(\varepsilon_K; \kappa, T) = \frac{\left[\left(\kappa - \dfrac{d_K}{2}\right)k_B T\right]^{-\frac{1}{2}d_K}}{B\left(\dfrac{d_K}{2}, \kappa - \dfrac{d_K}{2} + 1\right)} \cdot \left(1 + \frac{1}{\kappa - \dfrac{d_K}{2}} \cdot \frac{\varepsilon_K}{k_B T}\right)^{-\kappa - 1} \varepsilon_K^{\frac{1}{2}d_K - 1}, \quad (10.1b)$$

where $B(x, y) \equiv \Gamma(x)\Gamma(y)/\Gamma(x + y)$ and $\Gamma(x)$ are the beta and gamma functions.

The kappa index κ depends on the kinetic degrees of freedom d_K, in contrast to the invariant kappa index κ_0 that is independent of d_K (Chapters 1, 3, 5; Livadiotis and McComas, 2011b; Livadiotis, 2015c). The physical meaning of the kappa index is interwoven with the reciprocal correlation coefficient of the energies of any two correlated kinetic degrees of freedom. In particular, the correlation coefficient is given by $\rho = \frac{1}{2}f/\kappa(f)$. The kappa index κ is dependent on the correlated degrees of freedom f, and can be related to an invariant kappa index κ_0 by

$$\kappa(f) = \kappa_0 + \frac{1}{2}f, \quad (10.2)$$

If N_C is the number of correlated particles and d is the number of degrees of freedom per particle, then $f = d_K N_C$ and the dependent kappa index is $\kappa = \kappa_0 + \frac{1}{2}d_K N_C$. Note that κ_0 is the actual kappa index that characterizes a stationary state, and it is invariant from the number of particles and degrees of freedom of the system. The "thermalization" of the system, the process of the system reaching thermal equilibrium, is realized for quite large values of the kappa index, i.e., $1 << N_C << \kappa_0 \to \infty$, that is, to approach infinity and the distribution to become Maxwellian. Using the invariant kappa index κ_0, the expression of the kappa distribution of velocities becomes

$$P(\vec{u}; \kappa_0, T) = (\pi \kappa_0 \theta^2)^{-\frac{1}{2}d_K} \cdot \frac{\Gamma\left(\kappa_0 + 1 + \dfrac{d_K}{2}\right)}{\Gamma(\kappa_0 + 1)} \cdot \left[1 + \frac{1}{\kappa_0} \cdot \frac{(\vec{u} - \vec{u}_b)^2}{\theta^2}\right]^{-\kappa_0 - 1 - \frac{1}{2}d_K}, \quad (10.3a)$$

or, in terms of kinetic energy,

$$P(\varepsilon_K; \kappa_0, T) = \frac{(\kappa_0 k_B T)^{-\frac{1}{2}d_K}}{B\left(\dfrac{d_K}{2}, \kappa_0 + 1\right)} \cdot \left(1 + \frac{1}{\kappa_0} \cdot \frac{\varepsilon_K}{k_B T}\right)^{-\kappa_0 - 1 - \frac{1}{2}d_K} \varepsilon_K^{\frac{1}{2}d_K - 1}. \quad (10.3b)$$

Fig. 10.1 shows the kappa distribution of energy $\varepsilon_K/(k_B T)$ for $f = 3$ degrees of freedom and various values of the kappa index. As the kappa index increases, the maximum of the distribution lowers, while the two "edges," for small and large x, raise, so that the mean value of x (energy) is preserved. Indeed, the

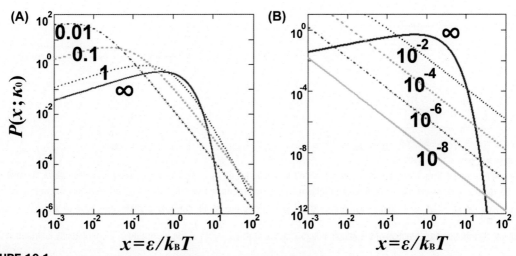

FIGURE 10.1

(A) The distribution is plotted in terms of the energy $x = \varepsilon/(k_B T)$ and for values of the invariant kappa index $\kappa_0 = 0.01, 0.1, 1$, and ∞. (B) The distribution is plotted for very small values of the invariant kappa index, $\kappa_0 = 10^{-2}, 10^{-4}, 10^{-6}, 10^{-8}$, indicating the power-low behavior of the distribution for these small indices; the Maxwellian ($\kappa_0 \to \infty$) is depicted in both panels for comparison (*red solid line*). Source: Livadiotis (2014a).

mean kinetic energy is independent of the kappa index; thus it remains constant under variations of the kappa index (for kinetic definition of temperature, see Chapter 1).

10.3.2 "Negative" Kappa Distribution of Velocities

The negative kappa distribution is obtained by setting $\kappa_0 \to -\kappa_0$ in the formalism of the standard kappa distribution, given in Eqs. (10.1a,b) (see Chapters 1, 3; Livadiotis, 2015b),

$$P(\vec{u}; \kappa_0, T) = \frac{\left(\pi \kappa_0 \theta^2\right)^{-\frac{1}{2}d_K} \Gamma(\kappa_0)}{\Gamma\left(\kappa_0 - \frac{d_K}{2}\right)} \cdot \left[1 - \frac{1}{\kappa_0} \cdot \frac{(\vec{u} - \vec{u}_b)^2}{\theta^2}\right]_+^{\kappa_0 - 1 - \frac{1}{2}d_K}, \qquad (10.4a)$$

or in terms of kinetic energy,

$$P(\varepsilon_K; \kappa_0, T) = \frac{\left(\kappa_0 k_B T\right)^{-\frac{1}{2}d_K}}{B\left(\kappa_0 - \frac{d_K}{2}, \frac{d_K}{2}\right)} \cdot \left(1 - \frac{1}{\kappa_0} \cdot \frac{\varepsilon_K}{k_B T}\right)_+^{\kappa_0 - 1 - \frac{1}{2}d_K} \varepsilon_K^{\frac{1}{2}d_K - 1}, \qquad (10.4b)$$

(where we used the cutoff condition noted by the subscript "+", namely, $[x]_+ = x$, if $\gamma \geq 0$ and $[x]_+ = 0$, if $x \leq 0$; see Chapter 1, Section 1.4.2). The distribution converges at $\varepsilon_K \to 0$ when $\kappa_0 > \frac{1}{2}d_K$.

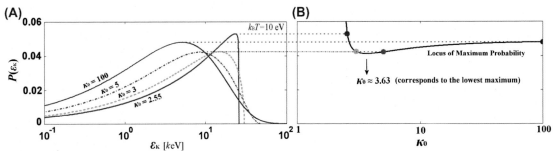

FIGURE 10.2

The negative kappa distribution of kinetic energy, plotted for kappa indices $\kappa_0 = 2.55$ (*red thick solid*), 3 (*green dash*), 5 (*dash–dot*), and 100 (*blue solid*); practically a Maxwell distribution). The temperature is fixed at 10 eV. (A) The distribution has a maximum $\varepsilon_{K_{max}} < \kappa_0 k_B T$ if $\kappa_0 > \frac{1}{2}d_K + 1$, that is, $\kappa_0 > \frac{5}{2}$ for three kinetic degrees of freedom per particle. (B) Locus of the maxima of the probability distribution as a function of the kappa index κ_0; at the limit $\kappa_0 \rightarrow \left(\frac{5}{2}\right)^+$, the maximum of the probability distribution peaks to infinity.

Fig. 10.2A depicts the negative kappa distribution, Eq. (10.4b), for various kappa indices (the temperature is kept fixed); the kappa index is chosen so that $\kappa_0 > \frac{1}{2}d_K + 1 = \frac{5}{2}$ (for $d_K = 3$); we observe that as κ_0 decreases, approaching this limit $\kappa_0 \rightarrow \left(\frac{5}{2}\right)^+$, the distribution becomes sharper at its maximum. Fig. 10.2B plots the locus of the maxima of the probability distribution as a function of the kappa index κ_0.

The maximum of the energy distribution, Eq. (10.4b), is located at

$$\varepsilon_{K_{max}} = \frac{\kappa_0}{\kappa_0 - 2} \cdot \left(\frac{1}{2}d_K - 1\right) \cdot k_B T. \tag{10.5}$$

Any energy with nonzero probability is $\varepsilon_K < \kappa_0 k_B T$. This is true also for the energy at the maximum of the distribution, $\varepsilon_{K_{max}}$. Hence, we have $\varepsilon_{K_{max}}/(\kappa_0 k_B T) = \left(\frac{1}{2}d_K - 1\right)\Big/(\kappa_0 - 2) \leq 1$ or $1 + \frac{1}{2}d_K \leq \kappa_0$. We obtain

$$0 < \frac{\varepsilon_{K_{max}}}{\kappa_0 k_B T} < 1 \text{ for } 1 + \frac{1}{2}d_K \leq \kappa_0, \frac{\varepsilon_{K_{max}}}{\kappa_0 k_B T} = 1 \text{ for } \frac{1}{2}d_K < \kappa_0 \leq 1 + \frac{1}{2}d_K. \tag{10.6}$$

The two cases are plotted in Fig. 10.3.

The standard kappa distribution corresponds to the q-exponential distribution for $q > 1$ under the transformation $\kappa = 1/(q-1)$. The negative kappa distribution corresponds to the same distribution for $q < 1$ under the transformation $\kappa = 1/(1-q)$. Both of these distributions were studied under the framework of nonextensive statistical mechanics; later, the properties of the negative kappa distribution were also studied in space physics (e.g., Leubner, 2004b), while the notion of temperature has been also corrected (Chapter 1; Livadiotis, 2015b).

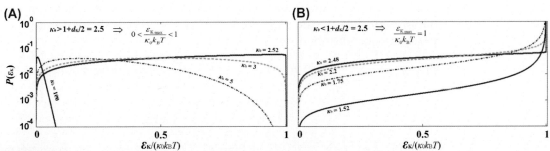

FIGURE 10.3

The negative kappa distribution of kinetic energy is plotted for kappa indices (A) $1 + \frac{1}{2}d_K < \kappa_0$, or $\frac{5}{2} < \kappa_0$ for $d_K = 3$, i.e., $\kappa_0 = 2.52$ (*red thick solid*), 3 (*green dash*), 5 (*dash–dot*), and 100 (*blue solid*) (practically a Maxwell distribution), and (B) $\frac{1}{2}d_K < \kappa_0 < 1 + \frac{1}{2}d_K$, or $\frac{3}{2} < \kappa_0 < \frac{5}{2}$ for $d_K = 3$, i.e., $\kappa_0 = 2.48$ (*red thick solid*), 2.2 (*green dash*), 1.75 (*dash–dot*), and 1.52 (*blue solid*). The temperature is kept constant at 10 eV.

The kappa indices may be different in the "positive" and "negative" kappa distributions, given in Eqs. (10.3b) and (10.4b), respectively, e.g., noted with κ_0^+ and κ_0^-. A linear superposition of these distributions is given by

$$P(\varepsilon_K; \kappa_0^+, \kappa_0^-, T, c) = cP(\varepsilon_K; \kappa_0^+, T) + (1 - c)P(\varepsilon_K; \kappa_0^-, T), \ 0 \le c \le 1, \qquad (10.7)$$

which can be a flexible description for space plasma particles. An equally weighted superposition is given by the distribution called bi-kappa introduced and studied by Leubner and Voros (2005),

$$P_{\frac{1}{2}}(\varepsilon_K; \kappa_0, T) = P\left(\varepsilon_K; \kappa_0, \kappa_0, T, c = \frac{1}{2}\right) = \frac{1}{2}[P(\varepsilon_K; \kappa_0, T) + P(\varepsilon_K; \kappa_0, T)].$$

$$(10.8)$$

A general expression of the distribution, Eq. (10.8), can be constructed by an arbitrary superposition of kappa distributions with different kappa indices (Chapter 4).

10.3.3 Superposition of Kappa Distribution of Velocities

A generalized distribution formula can be deduced as a superposition of kappa distributions where the system's density of kappa indices, $D(\kappa_0)$, is known. A linear superposition is given by

$$f(\varepsilon_K) = \sum_{\kappa_0} D(\kappa_0) P(\varepsilon_K; \kappa_0, T). \qquad (10.9a)$$

A more general, nonlinear, and nonadditive superposition is given in terms of

$$f(\varepsilon_K) = \phi^{-1}\left\{\sum_{\kappa_0} D(\kappa_0) \cdot \phi[P(\varepsilon_K; \kappa_0, T)]\right\}, \qquad (10.9b)$$

where ϕ is some monotonic function (see the concept of ϕ-averages in Livadiotis, 2007, 2012). The superposition may apply to both the positive and negative kappa distributions:

$$P(\varepsilon_K; T) = c\sum_{\kappa_0^+} D(\kappa_0^+)P(\varepsilon_K; \kappa_0^+, T) + (1-c)\sum_{\kappa_0^-} D(\kappa_0^-)P(\varepsilon_K; \kappa_0^-, T), \text{ or },$$

(10.10a)

$$P(\varepsilon_K; T) = \phi^{-1}\left\{ c\sum_{\kappa_0^+} D(\kappa_0^+)\cdot\phi[P(\varepsilon_K; \kappa_0^+, T)] \right.$$
$$\left. + (1-c)\sum_{\kappa_0^-} D(\kappa_0^-)\cdot\phi[P(\varepsilon_K; \kappa_0^-, T)] \right\}.$$

(10.10b)

The advantage of the linear superposition, Eq. (10.10a), is that the constructed superposed distribution has temperature identical to that of all the isothermal components $P(\varepsilon_K; \kappa_0^+, T)$ and $P(\varepsilon_K; \kappa_0^-, T)$. In the case where each component is characterized by different temperature,

$$P(\varepsilon_K; T_s) = c\sum_{\kappa_0^+} D(\kappa_0^+)P[\varepsilon_K; \kappa_0^+, T(\kappa_0^+)]$$
$$+ (1-c)\sum_{\kappa_0^-} D(\kappa_0^-)P[\varepsilon_K; \kappa_0^-, T(\kappa_0^-)],$$

(10.10c)

then the temperature of the superposed distribution T_s is equal to the average component temperature:

$$\frac{d_K}{2}k_B T_s = \int_0^\infty P(\varepsilon_K; T_s)\varepsilon_K d\varepsilon_K = c\sum_{\kappa_0^+} D(\kappa_0^+)\int_0^\infty P[\varepsilon_K; \kappa_0^+, T(\kappa_0^+)]\varepsilon_K d\varepsilon_K$$

$$+ (1-c)\sum_{\kappa_0^-} D(\kappa_0^-)\int_0^\infty P[\varepsilon_K; \kappa_0^-, T(\kappa_0^-)]\varepsilon_K d\varepsilon_K = c\sum_{\kappa_0^+} D(\kappa_0^+)\cdot\frac{d_K}{2}k_B T(\kappa_0^+)$$

$$+ (1-c)\sum_{\kappa_0^-} D(\kappa_0^-)\cdot\frac{d_K}{2}k_B T(\kappa_0^-) = c\frac{d_K}{2}k_B\langle T\rangle + (1-c)\frac{d_K}{2}k_B\langle T\rangle = \frac{d_K}{2}k_B\langle T\rangle$$

$$\text{or } T_s = \langle T\rangle.$$

(10.11)

The manipulation of the temperature is not trivial in the case of the nonlinear superposition, Eq. (10.10b); the result of Eq. (10.11) cannot be true without the consideration of special conditions for the kappa index densities $D(\kappa_0^+)$ and $D(\kappa_0^-)$.

10.3.4 Anisotropic Kappa Distribution of Velocities

Anisotropy between the particle kinetic degrees of freedom occurs when the kinetic energy is not equally distributed to the d_K kinetic degrees of freedom. The equipartition theorem is not valid for each of the kinetic degrees, but to all of them together. Consequently, the actual temperature T is still given by the mean of all the component temperatures,

$$\langle \varepsilon_K \rangle = \sum_{i=1}^{d_K} \langle \varepsilon_{K_i} \rangle = \frac{1}{2} k_B \sum_{i=1}^{d_K} T_i \equiv \frac{d_K}{2} k_B T, \text{ or, } T \equiv \frac{1}{d_K} \sum_{i=1}^{d_K} T_i, \quad (10.11a)$$

e.g., for $d_K = 3$,

$$\langle \varepsilon_K \rangle = \frac{1}{2} k_B T_x + \frac{1}{2} k_B T_y + \frac{1}{2} k_B T_z \equiv \frac{3}{2} k_B T, \text{ or } T = (T_x + T_y + T_z)/3. \quad (10.11b)$$

When there is anisotropy between the velocity projection at a certain direction $u_\parallel$ and the projection at the plane perpendicular to that direction, $\vec{u}_\perp$ (e.g., due to the presence of a local magnetic field), with respective temperature components $\theta_\parallel$ and $\theta_\perp$, then the temperature T is given by the average $T = \left(T_\parallel + 2T_\perp\right)/3$.

There are two main types of anisotropic kappa distributions. In the first type, each kinetic degree of freedom, e.g., the one corresponding to the velocity component u_i, is involved in a certain 1-D kappa distribution of different temperature, T_i or $\theta_i \equiv \sqrt{2k_B T_i/m}$, i.e., $P(\vec{u}, \kappa_0; \theta_i; d_K = 1)$. No correlation is considered between the degrees of freedom; thus the joint distribution for the velocity components is given by the product of the one-degree marginal distributions, that is,

$$P(\vec{u}) \propto \prod_{i=1}^{d_K} P(\vec{u}, \kappa_0; \theta_i; d_K = 1) = \prod_{i=1}^{d_K} \left[1 + \frac{1}{\kappa_0} \cdot \frac{(u_i - u_{bi})^2}{\theta_i^2}\right]^{-\kappa_0 - \frac{3}{2}},$$

$$(10.12)$$

(where the average particle velocity represents the bulk speed of the flow of particles, $\vec{u}_b \equiv \langle \vec{u} \rangle$). In the second type, all the degrees of freedom are considered to be correlated to each other, and the joint distribution is given by

$$P(\vec{u}) \propto \left[1 + \frac{1}{\kappa_0} \cdot \sum_{i=1}^{d_K} \frac{(u_i - u_{bi})^2}{\theta_i^2}\right]^{-\kappa_0 - 1 - \frac{1}{2} d_K}. \quad (10.13)$$

The second type is more frequently used is plasmas, where particles are always correlated (for more details on these formulations, see Chapter 4; Livadiotis, 2015a).

In the case of the anisotropy caused by a magnetic field, when the parallel and perpendicular components are noncorrelated, the distribution is

$$P(\vec{u}) \propto \left[1 + \frac{1}{\kappa_0} \cdot \frac{\left(u_\parallel - u_{b\parallel}\right)^2}{\theta_\parallel^2}\right]^{-\kappa_0 - \frac{3}{2}} \cdot \left[1 + \frac{1}{\kappa_0} \cdot \frac{\left(\vec{u}_\perp - \vec{u}_{b\perp}\right)^2}{\theta_\perp^2}\right]^{-\kappa_0 - 2}. \quad (10.14)$$

If all the degrees of freedom are correlated, then the distribution is

$$P(\vec{u}) \propto \left\{ 1 + \frac{1}{\kappa_0} \cdot \left[\frac{\left(u_\parallel - u_{b\parallel}\right)^2}{\theta_\parallel^2} + \frac{\left(\vec{u}_\perp - \vec{u}_{b\perp}\right)^2}{\theta_\perp^2} \right] \right\}^{-\kappa_0 - \frac{5}{2}} . \tag{10.15}$$

(For further reading, see also: Pierrard and Lazar, 2010; Lazar et al. 2012; Livadiotis 2015a.)

10.3.5 Phase Space Marginal Kappa Distribution of Velocities

In the presence of a potential energy $\Phi(\vec{r}, \vec{u})$ that depends on the position and velocity, the marginal kappa distribution of the velocity may cause anisotropy to the distribution of the velocity (kinetic energy).

As an example, consider the potential $\Phi(\vec{r}, \vec{u}) = \sum_{i=1}^{d_\kappa} \Phi_i(\vec{r}) \cdot u_i^2$. The Hamiltonian will be $H(\vec{r}, \vec{u}) = \sum_{i=1}^{d_\kappa} \left[\frac{m}{2} + \Phi_i(\vec{r}) \right] \cdot u_i^2$; the Hamiltonian factor $H(\vec{r}, \vec{u})/(k_B T)$ will be replaced by $\sum_i \left[\frac{m}{2} u_i^2 / (k_B T_i) \right]$, where $T_i \equiv T / \left[1 + \frac{2}{m} \Phi_i(\vec{r}) \right]$. Another example is the case of a potential that is independent of the velocity magnitude u but depends on the polar ϑ_u and azimuthal φ_u angles of the velocity space, compactly symbolized by $\Omega_u \equiv (\vartheta_u, \varphi_u)$, i.e., $\Phi(\vec{r}, \Omega_u)$ (see the example in Chapter 3; see also Livadiotis, 2015b).

10.4 Toward Antiequilibrium, the Farthest State From Thermal Equilibrium

The values of the kappa indices are determined from requiring the convergence of the integrals of normalization, mean value, and second statistical moment of velocities. The latter implies stronger conditions; hence the following integral must have finite value,

$$\int_{-\infty}^{\infty} P(\vec{u}; \kappa, \theta) d\vec{u} < +\infty, \quad \langle \vec{u} \rangle = \int_{-\infty}^{\infty} P(\vec{u}; \kappa, \theta) \vec{u} \, d\vec{u} < +\infty,$$

$$\langle u^2 \rangle = \int_{-\infty}^{\infty} P(\vec{u}; \kappa, \theta) u^2 d\vec{u} < +\infty, \tag{10.16}$$

from which we imply that $\left\langle (\vec{u} - <\vec{u}>)^2 \right\rangle < \infty$. The isotropic kappa distribution

$$P(u; \kappa, \theta) \propto \left[1 + \frac{1}{\kappa - \frac{f}{2}} \cdot \frac{(\vec{u} - \vec{u}_b)^2}{\theta^2} \right]^{-\kappa - 1}, \tag{10.17}$$

and the density of velocity states $g_V(u)$,

$$g_V(u) \equiv \int \frac{d\vec{u}}{du} = \int d\Omega_u \cdot u^{f-1} = B_f \cdot u^{f-1}, \qquad (10.18)$$

stand for the f-dimensional velocity space; $d\vec{u}$ is the infinitesimal volume, and $d\Omega_u$ is the infinitesimal spherical angle of an f-dimensional sphere (a product of $f-1$ infinitesimal angles, where its integration gives the f-dimensional spherical surface B_f).

In the classical case of the stationary state at equilibrium, $\kappa \to \infty$, the probability distribution is given by a Maxwellian, which decays exponentially, and thus the relevant integrals converge for any power-like expression of the density of velocity states, $g_V(u)$. However, in the case of the power-law−like decay of the kappa distribution (for $\kappa < +\infty$), the convergence is not trivial, as it depends on the value of the kappa index and the dimensionality f. In particular, the integrals converge as soon as the relevant integrands in the high-energy (H-E) limit of $u \to \infty$ attain at least a power-law decay of $1/u^r$, with $r > 1$. At this limit, the kappa distribution, Eq. (10.17), has the asymptotic behavior of $P(u;\kappa,\theta) \sim u^{-2(\kappa+1)}$, so that

$$u^2 \cdot P(u;\kappa,\theta) g_V(u) \sim u^{-2\kappa+f-1} \Rightarrow 1 < r = 2\kappa - f + 1. \qquad (10.19)$$

Hence, in order for the integral of the second statistical moment to converge, we obtain the condition

$$\kappa > \frac{1}{2}f, \text{ or }, \kappa_0 \equiv \kappa - \frac{1}{2}f > 0. \qquad (10.20)$$

The requirement of Eq. (10.20) is general for any f-dimensional system. All stationary states must have values of the kappa index within the interval $\kappa \in \left(\frac{f}{2}, \infty\right)$. The first boundary value, $\kappa \to \infty$, defines equilibrium, while the second boundary value, $\kappa \to \frac{f}{2}$, gives the furthest stationary state from equilibrium that can be approached (the antiequilibrium). All the attainable stationary states lie between these extreme states. The kappa index is dependent on the degrees of freedom, Eq. (10.2), and thus the invariant kappa index κ_0 is more convenient for characterizing the stationary states. Therefore, the two boundary states correspond to $\kappa_0 \to \infty$ and $\kappa_0 \equiv \kappa - \frac{f}{2} \to 0$.

In contrast to thermal equilibrium, which is trivially achievable, the other extreme stationary state of $\kappa_0 \equiv \kappa - \frac{f}{2} \to 0$ is not attainable, even though it can be approached arbitrarily closely. In Fig. 10.4A, we depict the kappa distribution $P(w;\kappa,\theta)$ in terms of the velocity $w \equiv |\vec{u} - \vec{u}_b|$ for various values of the kappa index (and $\theta = 1$), showing the statistical behavior of the particles as the extreme stationary state of $\kappa \to \frac{f}{2}$ is approached. The distribution is normalized by the

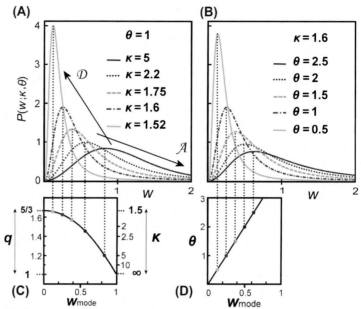

FIGURE 10.4

The kappa distribution is depicted for (A) constant temperature $\theta = 1$ and various kappa indices; and (B) constant kappa index $\kappa = 1.6$ or $\kappa_0 = 0.1$ and various temperatures θ. (A) By decreasing κ_0, the distribution is shifted to smaller velocities (or energies) toward the antiequilibrium state (along the indicated D array). By increasing κ_0, the distribution is shifted to larger velocities toward equilibrium (along the indicated A array). (B) The distribution is shifted to smaller (or larger) velocities due to the more familiar procedure of decreasing (or increasing) the temperature and staying at the same stationary state (the one corresponding to $\kappa = 1.6$). The shifting of distributions is also demonstrated by depicting the most possible velocity $w_{max}(\kappa_0, \theta)$: (C) for constant temperature, $\theta = 1$, and various kappa indices (the respective q-index is also given); and (D) for constant kappa index $\kappa = 1.6$ and various temperatures θ. Source: Livadiotis and McComas (2010a).

density of velocity states $g_V(w) = B_f w^{f-1}$, namely, $\int_0^\infty P_w(w; \theta, \kappa) dw = 1$, with the whole speed distribution $P_w(w; \kappa, \theta) \equiv P(w; \kappa, \theta) g_V(w)$ to be given by

$$P_w(w; \kappa_0, \theta) = \frac{2\kappa_0^{-\frac{1}{2}d_K} \theta^{-d_K}}{B\left(\dfrac{d_K}{2}, \kappa_0 + 1\right)} \cdot \left(1 + \frac{1}{\kappa_0} \cdot \frac{w^2}{\theta^2}\right)^{-\kappa_0 - 1 - \frac{1}{2}d_K} w^{d_K - 1}, \qquad (10.21)$$

where we have considered one-particle distribution ($N_C = 1$), i.e., $f = d_K$ kinetic degrees of freedom.

The maximum of the speed distribution, Eq. (10.21), is given by

$$w_{max}(\kappa_0, \theta) = \sqrt{\frac{\kappa_0}{\kappa_0 + \dfrac{3}{2}}} \cdot \sqrt{\frac{d_K - 1}{2}} \cdot \theta, \qquad (10.22)$$

attaining the maximum of the probability density, i.e.,

$$P_{w_{max}}(\kappa_0, \theta) = \frac{1}{\sqrt{\kappa_0 \cdot \theta^2}} \cdot \frac{2}{\Gamma\left(\frac{d_K}{2}\right)} \cdot \frac{\Gamma\left(\kappa_0 + 1 + \frac{d_K}{2}\right)}{\Gamma(\kappa_0 + 1)} \cdot \frac{\left(\kappa_0 + \frac{3}{2}\right)^{\kappa_0 + \frac{3}{2}} \left(\frac{d_K - 1}{2}\right)^{\frac{d_K - 1}{2}}}{\left(\kappa_0 + 1 + \frac{d_K}{2}\right)^{\kappa_0 + 1 + \frac{d_K}{2}}}.$$

$$(10.23)$$

Note that w_{max} is proportional to θ, but depends also on the kappa index. For any value of κ_0, it is always restricted to the interval $0 < w_{max} \leq \theta$ (for $d_K = 3$), with the equality holding at thermal equilibrium ($\kappa_0 \to \infty$). For smaller values of κ_0, the maximum w_{max} is also smaller and shifted closer to the zero. The suprathermal tails appear sufficiently far from this maximum, $w > w_{max}$, and for lower values of w_{max}, the tail is expected to become more dominant. This is consistent with Eq. (10.22), where with decreasing kappa values, w_{max} also decreases, generating suprathermal tails in the distribution.

Fig. 10.4A shows that as the kappa index increases and the stationary state becomes closer to equilibrium, the kappa distribution, Eq. (10.3b), is shifted to larger values of velocity. This can also be shown through the most likely velocity $w_{max}(\kappa_0, \theta)$, which is depicted in Fig. 10.4C, respectively, for the same values of kappa index (and $\theta = 1$).

The probability distributions for the two extreme stationary states are written as follows:

1. Equilibrium, $\kappa_0 \to \infty$: The speed distribution is given by a Maxwellian, $P(\vec{w}; \kappa_0 \to \infty, \theta) = \pi^{-\frac{1}{2}d_K} \cdot \theta^{-d_K} \cdot e^{-(w/\theta)^2}$; multiplying it by the density of states given in Eq. (10.18), we obtain

$$P_w(w; \kappa_0 \to \infty, \theta) \cong \Gamma\left(\frac{d_K}{2}\right)^{-1} \cdot \theta^{-d_K} \cdot e^{-(w/\theta)^2} \cdot w^{d_K - 1}. \qquad (10.24)$$

2. Antiequilibrium, $\kappa_0 \to 0$ (the furthest state from equilibrium): The distribution is given by $P(\vec{w}; \kappa_0 \to 0, \theta) = \pi^{-\frac{1}{2}d_K} \Gamma\left(\frac{d_K}{2} + 1\right) \cdot \kappa_0 \theta^2 \cdot w^{-d_K - 2}$, which is an approximation for $w \gg \sqrt{\kappa_0 \theta^2}$. The speed distribution is $P_w(w; \kappa_0 \to 0, \theta) = P(\vec{w}; \kappa_0 \to 0, \theta) \cdot g_V(w)$, i.e.,

$$P_w(w; \kappa_0 \to 0, \theta) \cong d_K \kappa_0 \theta^2 \cdot w^{-3}, \text{ for } w \gg \sqrt{\kappa_0 \theta^2}. \qquad (10.25a)$$

On the other hand, for $w \ll \sqrt{\kappa_0 \theta^2}$, a different approximation is produced,

$$P_w(w; \kappa_0 \to 0, \theta) \cong d_K \kappa_0^{-\frac{1}{2}d_K} \theta^{-d_K} \cdot w^{d_K - 1}, \text{ for } w \ll \sqrt{\kappa_0 \theta^2}. \qquad (10.25b)$$

The maximum of the distribution corresponds to energy and probability values

$$w_{\max}(\kappa_0, \theta) \propto \sqrt{\kappa_0 \theta^2}, P_{w_{\max}}(\kappa_0, \theta) \propto \frac{1}{\sqrt{\kappa_0 \theta^2}}, \tag{10.26}$$

thus their product is independent of the kappa index and temperature,

$$w_{\max}(\kappa_0, \theta) \cdot P_{w_{\max}}(\kappa_0, \theta) \cong C(d_K) \equiv \left(\frac{d_K - 1}{d_K + 2}\right)^{\frac{1}{2}d_K} \cdot \frac{3d_K}{d_K + 2}, C(d_K = 3) \cong \frac{36}{25\sqrt{10}}$$

$$\cong 0.455.$$

$$\tag{10.27}$$

For values of the kappa index taken arbitrary close to $\kappa_0 \to 0$, the product of the maximum probability density with the relevant velocity is independent of the kappa and the temperature, while it depends only on the dimensionality d_K; e.g., its constant value is $\cong 0.455$ for $d_K = 3$. Fig. 10.5 shows that the geometrical locus of all the modes (black solid line) is given by the function $P_{w_{\max}}(w_{\max}) \propto w_{\max}^{-1}$, verifying Eq. (10.27).

The previously mentioned considerations support the fact that for any small value of κ_0, the probability distribution of velocities $P_w(w; \kappa_0 \to 0, \theta)$ will never be equal to zero for nonzero velocities since its unimodal structure is always preserved. In other words, the most probable velocity will be given by the nonzero velocity at the maximum, i.e., $w_{\max}(\kappa_0 \to 0, \theta) \neq 0$, and thus the particles never slow down to exactly zero velocity in this statistical mechanics treatment.

It is clear that the marginal probability distribution of $\kappa_0 \to 0$ cannot be reached in a very similar way to how absolute zero temperature $\theta \to 0$ can only be approached, but is ultimately unattainable for any stationary state, even for the classical Maxwellian in equilibrium. In fact, it is remarkable that the behavior of approaching both the limits, $\kappa_0 \to 0$ or $\theta \to 0$, is so qualitatively similar. We may define the kappa-dependent thermal speed,

$$\theta_\kappa \equiv \sqrt{\kappa_0 \bigg/ \left(\kappa_0 + \frac{d_K}{2}\right)} \cdot \theta = \sqrt{\left(\kappa - \frac{d_K}{2}\right) \bigg/ \kappa \cdot \theta}, \text{ so that both the limits of } \kappa_0 \to 0$$

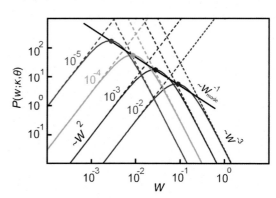

FIGURE 10.5
The kappa distribution, Eq. (10.3b), is depicted on a log—log scale for values of the kappa index near antiequilibrium $(\kappa_0 \to 0)$, i.e., $\kappa_0 = 10^{-2}, 10^{-3}, 10^{-4}, 10^{-5}$. We also verify the approximations in Eqs. (10.26) and (10.27). Source: Livadiotis and McComas (2010a).

or $\theta \to 0$ can be described by the same single limit; that is $\theta_\kappa \to 0$. Nonetheless, as we will see as follows, there is one significant difference when approaching the limit $\kappa_0 \to 0$ rather than the limit $\theta \to 0$: the internal energy and temperature remain constant.

In Fig. 10.4B, we depict the kappa distribution, Eq. (10.3b), for $\kappa = 1.6$ and various values of θ, while in Fig. 10.4D, we show the most frequent velocity $w_{\max}(\kappa, \theta)$, respectively, for the same values of θ (and $\kappa = 1.6$). As θ (or T) increases, the distribution, Eq. (10.3b), is shifted to larger values of velocity producing a phenomenological acceleration of the particles, while for θ (or T) decreasing, the distribution is shifted to smaller values of velocity, and a phenomenological deceleration of the particles is observed.

The phenomenological acceleration and deceleration of particles can be observed also by varying the kappa index instead of the temperature. The important difference is that by varying the temperature, the deduced acceleration and deceleration refer to particles of any energy; however, varying the kappa index may have the opposite effect on the higher energy particles than the lower energy particles.

First, we observe in Fig. 10.4A and C that as the kappa index increases, the stationary state becomes closer to thermal equilibrium and the probability distribution includes larger values of velocity. If the system attains a state of larger kappa index, its particles are characterized by larger values of velocity, producing a phenomenological acceleration. Similarly, for the reverse procedure of decreasing the kappa index and approaching the extreme state of anti-equilibrium at $\kappa_0 \to 0$, we have a phenomenological deceleration, with the particles accumulated at near zero velocity. On the other hand, the kappa index and the temperature are two independent parameters, and thus any change of the kappa index must not affect the temperature. Then we may ask, *how can we have acceleration or deceleration by varying the kappa index if the mean energy, that is, the temperature, remains constant?* Let's examine the case with an increasing kappa index. What really happens is that particles with near to zero velocities indeed increase their velocities (acceleration), but at the same time, particles with very high velocities decrease their velocities (deceleration). Acceleration and deceleration take place *at the same time*, preserving the balance in energy.

Figs. 10.6 and 10.7 illustrate both acceleration and deceleration. In Fig. 10.6, we depict the distribution $P(u; \kappa)$ for $\kappa_0 = 0.5$ and $\kappa_0 = 8.5$ (the corresponding 3-dim kappa indices are $\kappa = 2$ and $\kappa = 10$); also, $\theta = 1$ and $f = 3$. Assuming that the kappa index increases from $\kappa_0 = 0.5$ to $\kappa_0 = 8.5$, we observe that the probability distribution decreases for velocities $u < u_A$ and $u_D < u$, while it increases for the velocities between $u_A < u < u_D$. The boundaries u_A and u_D are the velocity values at the first and second intersection points of the two curves.

The same results can be shown with infinitesimal variations of the kappa index. In order to examine how the distribution changes upon variation of the kappa index, we study the derivation of the kappa distribution. In Fig. 10.7, we depict

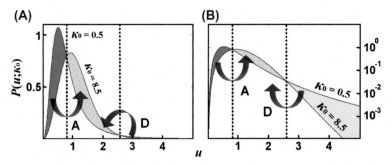

FIGURE 10.6
Illustration of the simultaneous acceleration and deceleration preserving the balance in mean energy (temperature). When the kappa index increases, (1) particles with low velocities from the region $u < u_A$ (*blue*) increase their velocities and depart to the region $u_A < u < u_D$ (*green*), leading to a phenomenological acceleration (A); and (2) particles with high velocities from the region $u_D < u$ (*yellow*) decrease their velocities and depart to the region $u_A < u < u_D$, leading to a phenomenological deceleration (D). The distribution is plotted in a linear scale (A) and semilog scale (B), showing its variation when the kappa index increases from $\kappa_0 = 0.5$ to $\kappa_0 = 8.5$ (region $u_A < u < u_D$ more visible in panel (B)). For these kappa indices, the region boundaries are $u_A/\theta \cong 0.76$, $u_D/\theta \cong 3.07$ (the distributions are depicted for $\theta = 1$ and $f = 3$). Source: Livadiotis and McComas (2011b).

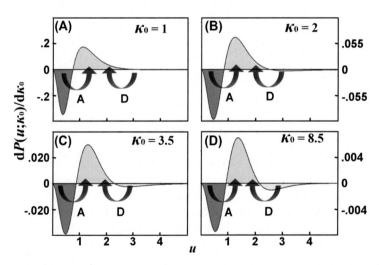

FIGURE 10.7
Phenomenological acceleration and deceleration illustrated by the distribution derivative in terms of the kappa index. The derivative $dP(u; \kappa_0)/d\kappa_0$ is depicted for $\kappa_0 = 1$, 2, 3.5, and 8.5. We observe that this is negative for velocities $u < u_A$ and $u_D < u$ and positive for the velocities between $u_A < u < u_D$ (see text). Therefore, particles with near to zero velocities in the region $u < u_A$ and extremely high velocities in the region $u_D < u$ depart to the mid-velocities region $u_A < u < u_D$ so that the balance in energy is preserved. Source: Livadiotis and McComas (2011b).

the derivative $dP(u; \kappa_0)/d\kappa_0$ for various kappa indices. We observe that this is negative for velocities $u < u_A$ and $u_D < u$, while it is positive for the velocities between $u_A < u < u_D$. At the boundaries u_A and u_D, the derivative $dP(u; \kappa_0)/d\kappa_0$ equals zero, with u_A corresponding to $d^2P(u; \kappa_0)/d\kappa_0^2 > 0$ and u_D to $d^2P(u; \kappa_0)/d\kappa_0^2 < 0$.

We conclude that the variation of the kappa index has opposite effect on the lower and higher energies of the kappa distribution. By increasing the kappa index, lower energy particles accelerate and higher energy particles decelerate to moderate energies. It is important to note that while the mean energy is always balanced (constant temperature), a larger fraction of particles is accelerated than decelerated, and thus the acceleration is dominant.

We may show the dominance of the acceleration of the low-velocity particles on the deceleration of the high-velocity particles with the following numerical example. We examine the case of increasing kappa index from $\kappa_0 = 1.00$ to $\kappa_0 = 1.01$ ($\Delta\kappa_0 = 0.01$) in a system of $N = 10^6$ particles. We have $\Delta N_A = \int_0^{u_A} \Delta P(u)du$ accelerating and $\Delta N_D = \int_{u_D}^{\infty} \Delta P(u)du$ decelerating particles, while $\Delta N_{acc} = \int_{u_A}^{u_D} \Delta P(u)du$ particles are accumulated in the mid velocities; we find $u_A/\theta \cong 0.76$, $u_D/\theta \cong 3.07$, and the probability distribution difference is given by $\Delta P(u) \cong (dP(u; \kappa_0)/d\kappa_0) \cdot \Delta\kappa_0$. Then we estimate that $\Delta N_A \cong 1427$ particles will accelerate, while only $\Delta N_D \cong 65$ particles will decelerate to the mid velocities, that is, $\Delta N_D/\Delta N_A \cong 0.046$. The ratio $\Delta N_D/\Delta N_A$ increases with the kappa index, attaining its maximum value ~ 0.39 for $\kappa_0 \to \infty$.

Finally, if by increasing the kappa index and shifting the system closer to thermal equilibrium, the majority of the particles are accelerated; then by decreasing the kappa index and shifting the system toward antiequilibrium, the majority of the particles are now decelerated. Then a new kind of freezing is defined by the process of decreasing the kappa index where the majority of particles are decelerated, while some particles are highly accelerated so that the mean energy is kept constant (for this reason, antiequilibrium was also called q-frozen state; e.g., see Livadiotis and McComas, 2013a). In the next session, we will describe the concept of thermodynamic distance and the approach to antiequilibrium in terms of correlation, which gives a better physical understanding of the process from my perspective.

10.5 Arrangement of the Stationary States

10.5.1 A Measure of the Thermodynamic Distance From Thermal Equilibrium

The kappa index can provide a measure of the departure of the stationary states from equilibrium. Indeed, the "thermodynamic distance" of a stationary state from equilibrium can be defined and expressed as a function of the kappa index (e.g., Burlaga and Viñas, 2005; Livadiotis and McComas, 2010a,b,c, 2011b; Livadiotis, 2015a). The consistency of this definition lies on the universal properties of the extreme states, equilibrium and antiequilibrium, e.g., their exponential and power-law functional behavior, respectively, which is independent of the bulk speed, temperature, magnetic field, or other properties of the system.

The universal behavior of the state at equilibrium is the exponential. Indeed, independently of the dimensionality f of the system, this will be always attained for $\kappa_0 \rightarrow \infty$, while the characteristic probability velocity distribution is a Maxwellian $P^{\text{EQ}}(w) \sim e^{-(w/\theta)^2}$. On the other hand, near antiequilibrium, we have $P^{\text{aEQ}}(w) \sim w^{-f-2}$. Then, because of $g_V(w) \sim w^{f-1}$, Eq. (10.18), we find that the whole distribution is $P_w^{\text{aEQ}}(w) \sim w^{-f-2} \cdot w^{f-1} = w^{-3}$, independently of the dimensionality f. The particle flux is also independent of the dimensionality, $j^{\text{aEQ}}(\varepsilon) \propto P^{\text{aEQ}}(\varepsilon)g(\varepsilon) \propto w^{-f-2} \cdot w^{f-1} = w^{-3} \propto \varepsilon^{-\frac{3}{2}}$. This universal behavior indicates an invariant measure and leads us to ascribe a fixed "thermodynamic distance" between the equilibrium and antiequilibrium. All the stationary states must lie between those states (see Section 10.5.3). Hence, the maximum "thermodynamic distance" is fixed, while each stationary state represents only a specific fraction of this total distance. In this way, antiequilibrium can be referred to as 100% away from equilibrium, while all the stationary states lie between the extreme stationary states of 0% (equilibrium) and 100% (antiequilibrium).

A mathematical description of the thermodynamic measure was previously developed by Livadiotis and McComas (2010a, 2010b). The formulated measure of metastability, noted by M_q, involves also the dimensionality f and is given by

$$M_q(\kappa_0; f) = \frac{\dfrac{f+1}{2}}{\kappa_0 + \dfrac{f+1}{2}}. \tag{10.28}$$

This measure M_q is bounded in the unity interval, $0 \leq M_q < 1$, and it may be referred as a percentage, i.e., $0\% \leq M_q < 100\%$.

As an application, we examine the transition of metastable stationary states toward equilibrium by the following example. Fig. 10.8A shows a hypothetical route of metastable stationary states leading toward the specific state at equilibrium. This simple demonstration shows a monotonic switching of a 3-D system over stationary states. These metastable stationary states are characterized by the kappa index gradually increasing from $\kappa_0 \rightarrow 0$ (antiequilibrium) to $\kappa_0 \rightarrow \infty$ (equilibrium). In particular, this figure maps a hypothetical sequential transition of the system through the stationary states with measures $M_q^A = 99\%$, $M_q^B = 90\%$, $M_q^C = 60\%$, $M_q^D = 30\%$, and $M_q^E = 0\%$ (equilibrium).

The particle flux $j(x)$ is given in the reference frame of the plasma's bulk flow, $\overrightarrow{u}_b = 0$, and expressed in terms of $x \equiv \varepsilon/(k_B T)$, while the convenient dimensionless flux $J(x)$ is defined by

$$J(x) \equiv 2\sqrt{\pi} \cdot \frac{k_B T}{n\theta} \cdot j(x), j(x) = \frac{n}{4m} \cdot P(w = \theta \cdot \sqrt{x}; \kappa_0; \theta). \tag{10.29}$$

In Fig. 10.8B the respective dimensionless flux $J(x)$ is depicted as a function of the dimensionless energy x for each of the five stationary states A–E. The flux is maximized for

$$x_{\text{Max}}(M_q) = 4 \cdot \frac{1 - M_q}{4 - M_q}. \tag{10.30}$$

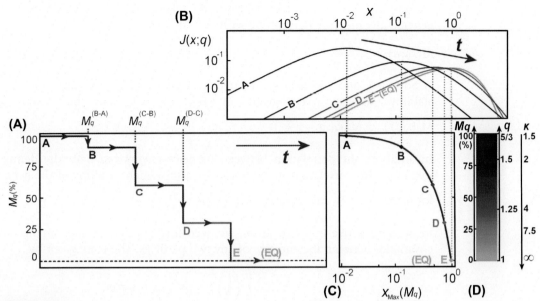

FIGURE 10.8
(A) A route of metastable stationary states from $\kappa_0 \to 0$ (antiequilibrium) to $\kappa_0 \to \infty$ (equilibrium). The states shown are $M_q^A = 99\%$, $M_q^B = 90\%$, $M_q^C = 60\%$, $M_q^D = 30\%$, $M_q^E = 0\%$. (B) The dimensionless flux $J(x)$ is depicted as a function of the dimensionless energy $x \equiv \varepsilon/(k_B T)$ for each of the five stationary states A–E. (C) The maximum energy $x_{Max}(M_q)$ is plotted as a function of M_q. (D) The stationary states are represented by a gradient color, from green for states near equilibrium to blue for those near antiequilibrium (q-indices are also given). Source: Livadiotis and McComas (2010c).

As shown in Fig. 10.8B and C, as the measure M_q decreases (i.e., as κ_0 increases), the flux $J(x)$ is shifted to larger values of energy. The respective maximum energy x_{Max} (that maximizes flux) for the A–E stationary states is $x_{Max}^A \cong 0.0133$, $x_{Max}^B \cong 0.129$, $x_{Max}^C \cong 0.471$, $x_{Max}^D \cong 0.757$, $x_{Max}^E = 1$. The maximum energy, Eq. (10.30), is depicted in Fig. 10.8C.

10.5.2 A Generalized Measure

The kappa index κ_0 itself can be a measure of how far a stationary state is from thermal equilibrium. In fact, the inverse, $1/\kappa_0$, may be naturally used as such a measure of the thermodynamic distance of a stationary state from equilibrium; that is denoted by $M(\kappa_0)$,

$$M(\kappa_0) = \frac{1}{\kappa_0}, 0 \le M(\kappa_0) < \infty. \tag{10.31}$$

Even though this does not constitute a bounded measure, the *escape* state equally separates the near- and far-equilibrium regions. Namely, $1/\kappa_0$ becomes 0 at equilibrium and ∞ at antiequilibrium, the farthest state from equilibrium. In general, any measure that can be written in the general form

$$M(\kappa_0; a) = \frac{a}{\kappa_0 + a},$$
(10.32)

is bounded, and can also be referred as a percentage. Obviously, the meta-stability measure M_q, given in Eq. (10.28), can be given for

$$M_q(\kappa_0; f) = M\left(\kappa_0; a = \frac{f+1}{2}\right).$$

In addition, the correlation between particles energy can naturally define a measure. This correlation measure can be also written in terms of the generic form, Eq. (10.32), e.g., $\rho(\kappa_0; f) = M\left(\kappa_0; a = \frac{f}{2}\right)$.

In order to derive the correlation, we need first to calculate the variance and covariance between the energies of any two particles. These are given by

$$\left\langle \vec{u}_{(n)}^2 \cdot \vec{u}_{(m)}^2 \right\rangle = \theta^4 \cdot \frac{\kappa_0}{\kappa_0 - 1} \times \begin{cases} \left(\frac{f}{2}\right)^2 & \text{if } n \neq m, \\ \left(\frac{f}{2}\right) \cdot \left(\frac{f}{2} + 1\right) & \text{if } n = m, \end{cases}$$
(10.33a)

or

$$\left\langle \varepsilon_{K(n)} \cdot \varepsilon_{K(m)} \right\rangle = (k_B T)^2 \cdot \frac{\kappa_0}{\kappa_0 - 1} \times \begin{cases} \left(\frac{f}{2}\right)^2 & \text{if } n \neq m, \\ \left(\frac{f}{2}\right) \cdot \left(\frac{f}{2} + 1\right) & \text{if } n = m. \end{cases}$$
(10.33b)

Then, the correlation of particle energies was found to be (Livadiotis and McComas, 2011b; Livadiotis, 2015c)

$$\rho(\kappa_0; f) = \frac{\left\langle \varepsilon_{(n)} \cdot \varepsilon_{(m)} \right\rangle - \left\langle \varepsilon_{(n)} \right\rangle \cdot \left\langle \varepsilon_{(m)} \right\rangle}{\sqrt{\left\langle \varepsilon_{(n)}^2 \right\rangle - \left\langle \varepsilon_{(n)} \right\rangle^2} \cdot \sqrt{\left\langle \varepsilon_{(m)}^2 \right\rangle - \left\langle \varepsilon_{(m)} \right\rangle^2}} = \frac{\frac{1}{2}f}{\kappa_0 + \frac{1}{2}f}.$$
(10.34)

For particles with different degrees of freedom, the respective relations are

$$\left\langle \vec{u}_{(n)}^2 \cdot \vec{u}_{(m)}^2 \right\rangle = \theta^4 \cdot \frac{\kappa_0}{\kappa_0 - 1} \times \begin{cases} \left(\frac{f_{(n)}}{2}\right) \cdot \left(\frac{f_{(m)}}{2}\right) & \text{if } n \neq m, \\ \left(\frac{f_{(n)}}{2}\right) \cdot \left(\frac{f_{(n)}}{2} + 1\right) & \text{if } n = m, \end{cases}$$
(10.35a)

or in terms of energy,

$$\langle \varepsilon_{(n)} \cdot \varepsilon_{(m)} \rangle = (k_B T)^2 \cdot \frac{\kappa_0}{\kappa_0 - 1} \times \begin{cases} \left(\dfrac{f_{(n)}}{2}\right) \cdot \left(\dfrac{f_{(m)}}{2}\right) & \text{if } n \neq m, \\[2ex] \left(\dfrac{f_{(n)}}{2}\right) \cdot \left(\dfrac{f_{(n)}}{2} + 1\right) & \text{if } n = m. \end{cases} \tag{10.35b}$$

Then the correlation between these different particles is given by

$$\rho\left(\kappa_0; f_{(n)}, f_{(m)}\right) = \sqrt{\frac{\frac{1}{2} f_{(n)}}{\kappa_0 + \frac{1}{2} f_{(n)}}} \cdot \sqrt{\frac{\frac{1}{2} f_{(m)}}{\kappa_0 + \frac{1}{2} f_{(m)}}} = \sqrt{\rho\left(\kappa_0; f_{(n)}\right) \cdot \rho\left(\kappa_0; f_{(m)}\right)}$$

$$(10.36)$$

For example, the correlation between the energies specified in one degree of freedom is

$$\rho(\kappa_0; f = 1) = \frac{1}{2\kappa_0 + 1}, \tag{10.37a}$$

while between the energies of a pair of degrees of freedom, the correlation is

$$\rho(\kappa_0; f = 2) = \frac{1}{\kappa_0 + 1}. \tag{10.37b}$$

This equation sets a physically meaningful measure: it is more fundamental to ask about the correlation of two Hamiltonian pairs of degrees of freedom instead of any other number of kinetic degrees or particles. Here we assume one kinetic degree and one squared potential degree, e.g., a 2-D phase space (x, p_x), (the result can be further generalized using the positive attractive power-law potential $\Phi(r) = \frac{1}{b} k \cdot r^b$). The notion of correlation between the energy of particles is described and studied in Chapter 5 (see also Livadiotis and McComas, 2011b; Livadiotis, 2015c). The statistics of this measure, $M = 1/(\kappa_0 + 1)$, and its relation with plasma density and temperature were studied in Livadiotis (2015a).

10.5.3 The Kappa Spectrum

The significance of the kappa index is primarily given by its role in identifying the nonequilibrium stationary states. The stationary states can be arranged according to their kappa index so that the smaller the kappa index is, the further from equilibrium and the nearer to antiequilibrium the system resides. The system resides at equilibrium for $\kappa_0 \to \infty$, while the antiequilibrium attained for $\kappa_0 \to 0$; this is the smaller possible kappa index, and thus this is the furthest possible stationary state from equilibrium. Therefore, the whole set of different stationary states can be realized in a spectrum-like arrangement of different kappa indices in the interval $\kappa_0 \in (0, \infty]$ (note that thermal equilibrium is attainable, while antiequilibrium can be reached as close as desired, but is unattainable).

	Equilibrium	Escape State	Anti-Equilibrium
κ_0	∞	1	0
q_0	1	2	∞
κ_3 or κ	∞	2.5	1.5
q_3	1	1.4	5/3
κ_f	∞	$1+f/2$	$f/2$
q_f	1	$(4+f)/(2+f)$	$1+1/f$
$1/\kappa_0$	0	1	∞
$M_q(\kappa_0,f)$	0	$(D+1)/(D+3)$	1
$\rho(\kappa_0,f)$	0	$D/(D+2)$	1
$\rho(\kappa_0,f=2)$	0	1/2	1

Near-Equilibrium Region   Far-Equilibrium Region

FIGURE 10.9

Characterization of the whole spectrum of stationary states via various measures. The basic states are shown, i.e., the boundary states of equilibrium and antiequilibrium, as well as the escape state that separates the spectrum in the near- and far-equilibrium regions. First used is the invariant zero-dimensional κ_0, which is the most fundamental index to characterize the stationary states and, through its inverse, $1/\kappa_0$, to give a measure of their thermodynamic distance from thermal equilibrium. We also give the f-dimensional kappa index, κ_f, that is not invariant but dependent on the degrees of freedom f, and the usual 3-D κ_3, which has been used in all the past publications (without the subscript "3"). The corresponding q-indices are also given. The last three rows show different quantities that may be used as bounded measures: the measure $M_q(\kappa_0;f)$, Eq. (10.28), the correlation $\rho(\kappa_0;f)$, Eq. (10.34), and the correlation for a pair of degrees of freedom $M(\kappa_0;a=1)=\rho(\kappa_0;2)$, Eqs. (10.32) and (10.37b). The first, $M_q(\kappa_0;f)$, obeys certain mathematical rules developed in Livadiotis and McComas (2010b). The second, $\rho(\kappa_0;f)$, has the unique physical meaning of correlation between particles, while the third measure is specifically the correlation between two Hamiltonian degrees of freedom, equally separating the near/far—equilibrium regions. Source: Livadiotis and McComas (2011b).

Using the invariant kappa index, its interpretation as inverse correlation, or any other measure of the thermodynamic distance between stationary states and thermal equilibrium, we can classify the types of systems in an arrangement called the kappa spectrum (Fig. 10.10). Therefore, the classical, single stationary state at equilibrium is generalized into a whole set of different nonequilibrium stationary states, each labeled by a separate kappa index, and collectively producing the kappa spectrum.

10.6 Interpreting the Observations

10.6.1 Observations and Measurements of Kappa Indices in the Heliosphere

Here we describe the kappa indices that have been detected by various observations of plasmas in the heliosphere. For convenience, we may use together with the invariant kappa index κ_0, the standard 3-D kappa index given by $\kappa = \kappa_0 + \frac{3}{2}$.

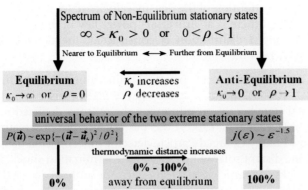

FIGURE 10.10
The kappa spectrum of stationary states (compare with Fig. 5.4). The stationary states of systems described by kappa distributions can be arranged according to the value of the invariant kappa index *that is related to the correlation*. The quantity κ_0 is invariant of the degrees of freedom of the system. In contrast to κ_0, the standard kappa index depends on the kinetic or potential degrees of freedom f, according to $\kappa = \kappa_0 + \frac{1}{2}f$. The whole spectrum of stationary states is described for $\kappa_0 \in (0, \infty)$, where the boundary states of equilibrium ($\kappa_0 \to \infty$) and antiequilibrium ($\kappa_0 \to 0$) correspond to distributions with universal behavior, e.g., the exponential distribution at equilibrium and the power-law flux of exponent -1.5 at the antiequilibrium. The invariant kappa index κ_0, the correlation coefficient $\rho = \frac{1}{2}f / \left(\kappa_0 + \frac{1}{2}f \right)$, Eq (10.34), and other quantities can be used as a measure of the "thermodynamic distance" of stationary states from thermal equilibrium nonequilibrium, and for the characterization of the κ-spectrum. (Compare with Fig. 5.4 in Chapter 5.) Source: Livadiotis and McComas (2013a).

First, we focus on values near the extreme states of equilibrium and anti-equilibrium. Particles at thermal equilibrium have velocities described by Maxwell distributions, that is, kappa distributions approaching the limit of $\kappa_0 \to \infty$. Maxwell distributions are observed in space plasmas (e.g., Hammond et al., 1996), but in general, they are less frequent than kappa distributions with smaller kappa indices, e.g., $\kappa_0 \approx 0.1-5$. On the other hand, particles at anti-equilibrium, the furthest state from thermal equilibrium, have velocities described by power-law distributions with exponent -5 in the velocity space or respective flux with exponent -1.5.

The antiequilibrium state was first detected in space plasma by Steffl et al. (2004) in the Io plasma torus. In particular, by analyzing the data from the Ultraviolet Imaging Spectrograph onboard the Cassini spacecraft, they found that the derived value of the kappa index that describes ions and electrons has a value around $\kappa \cong 5$ ($\kappa_0 \cong 3.5$) for the inner layers of $\sim 6 \, R_{\text{Io}}$ (Io's radius), and decreased with radial distance reaching asymptotically the antiequilibrium value $\kappa \cong 1.5$ ($\kappa_0 \cong 0$) for $\sim 9 \, R_{\text{Io}}$. Antiequilibrium was also detected in the solar wind protons by Fisk and Gloeckler (2006). This analysis was based on the observations of Ulysses and ACE Solar Wind Ion Composition Spectrometers that collected data on the suprathermal tails at distances ranging from 1 AU to 5.4 AU. In particular, by plotting the probability distribution of velocities

$P(\vec{u})$ where the contribution of the density velocity states $g_V(u) \sim u^2$ is excluded, they concluded that there is a "universal" power law in the H-E region, $P(u) \sim u^{-5}$. When the speed distribution is constructed, the derived power law is $P_w(u) \equiv P(u)g_V(u) \sim u^{-3}$.

Close to the kappa index $\kappa \cong 1.5$ ($\kappa_0 \cong 0$) is the value estimated by the analysis of Decker et al. (2005), that is, $\kappa \cong 1.63$ ($\kappa_0 \cong 0.13$) (this is sometimes named as fundamental state because it minimizes the entropy depicted in Fig. 10.17). The analysis of Decker et al. (2005) was based on recent observations from the Voyager 1, Low Energy Charged Particle detector that collected data in the inner heliosheath, the distant solar wind beyond the termination shock. In particular, by plotting the flux of particles versus energies, the authors detected a power-law suprathermal ($u \gg u_b$) tail for protons with a spectral index of about $\gamma \cong 1.6$, while the precise value of $\kappa \cong 1.63$ was estimated by the fitting of a kappa distribution (note that at high energies, the spectral index equals exactly the kappa index, Livadiotis and McComas, 2009.) This value of kappa index was subsequently used by other authors in several theoretical analyses and models (e.g., Prested et al., 2008; Heerikhuisen et al., 2008, 2010, 2014, 2015; Schwadron et al., 2009).

Another important stationary state is the one with kappa index $\kappa \cong 2.5$ ($\kappa_0 \cong 1$), which separates the near-equilibrium region with kappa indices $2.5 < \kappa$ (or $1 < \kappa_0$) from the far-equilibrium region with kappa indices $\kappa < 2.5$ (or $\kappa_0 < 1$). This particular kappa index is frequently observed in space plasmas. It was first noted in particles populations in the Uranian magnetosphere (Mauk et al., 1987) where the spectral indices at high energies (thus corresponding to kappa indices) were found to be distributed in the near-equilibrium region, $2.5 < \kappa$. Similar results were found by Dialynas et al. (2009) by examining particle distributions in the Saturnian magnetosphere from the Magnetospheric Imaging Instrument suite onboard Cassini (Chapter 12). The kappa indices were estimated for a large number of kappa distributions (both protons and O^+) that were organized by the L-shell of Saturn over 5–20 Saturnian radii. The estimated kappa indices were found to be large and distributed in a remarkable arrangement (c.f. Fig. 2 in Dialynas et al., 2009), ranging mostly between 2–2.5 and 11–13, corresponding to metastability measures M_q ranging between 15%–20% and 65%–80%. The histograms of kappa indices are shown in the insets of Fig. 10.11B,D, respectively, for the hydrogen and oxygen cations, where we observe that the most probable values are $\kappa_{H^+} \cong 3.5$ and $\kappa_{O^+} \cong 5$ (corresponding to $\kappa_0 \sim 2$ and ~ 3.5, and measures $M_{q,H^+} \cong 50\%$ and $M_{q,O^+} \cong 35\%$.

It is remarkable that the solar wind hydrogen ions are characterized by large deviations from equilibrium $M_q > 93.9\%$, while ions in the Saturnian magnetosphere were found to be characterized by the significantly smaller deviation of $M_{q,H^+} \cong 50\%$. This fact clearly supports the idea that the Saturnian magnetospheric plasma is approaching a quasiequilibrium status, namely, stationary states near equilibrium. In contrast to the sparse heliosphere, the dense

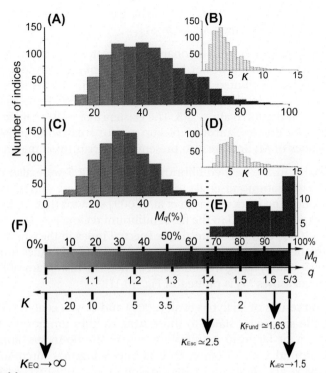

FIGURE 10.11

The histograms of the kappa indices of the near-equilibrium stationary states as detected by Dialynas et al. (2009) for the protons (B) and oxygen cations (D) in the Saturnian magnetosphere. The protons have spectral indices similar to that of O^+, but still, a significant difference exists between the most probable values of $\kappa_{H^+} \cong 3.5$ and $\kappa_{O^+} \cong 5$ that clearly implies $M_{q,H^+} \neq M_{q,O^+}$. This difference is recovered when we plot directly the respective histograms (A) and (C) of the metastability measure M_q, where the most probable values are $M_{q,H^+} \cong 40\%$ and $M_{q,O^+} \cong 35\%$. Panel (E) depicts the histogram of the metastability measures of the far-equilibrium (or near-antiequilibrium) states detected by Dayeh et al. (2009) in the inner heliosphere. The critical values of kappa index characterizing equilibrium $\kappa_{EQ} = \infty$, the escape state $\kappa_{Esc} \cong 2.5$, the fundamental state $\kappa_{Fund} \cong 1.63$, and the antiequilibrium state $\kappa_{aEQ} \cong 1.5$ are indicated in the whole interval $\kappa \in (3/2, \infty]$ (that is, $\kappa_0 \in (0, \infty]$ and $M_q \in [0,1)$ in panel (F). Source: Livadiotis and McComas (2010a).

hydrogen and oxygen in the Saturnian magnetosphere produces copious charge exchange between hydrogen ions and oxygen neutral atoms (and vice versa, $H^+ + O \rightleftarrows H + O^+$). This interaction leads the hydrogen ions to distribute in a specific way, inheriting the respective characteristics of oxygen ions, such as the same kappa index. Interestingly, the protons were found to have spectral indices similar to that of O^+, but still, a significant difference exists between the values of $\kappa_{H^+} \cong 3.5$ and $\kappa_{O^+} \cong 5$ that clearly implies $M_{q,H^+} \neq M_{q,O^+}$. Surprisingly, this difference is recovered when we plot directly the histograms of the q-metastability measure. In this case, we observe in Fig. 10.11A and C that the most probable values are $M_{q,H^+} \cong 40\%$ and $M_{q,O^+} \cong 35\%$. Given that the bin length in the histograms is $\delta M_q \cong 5\%$, then $M_{q,H^+} \cong M_{q,O^+}$.

The observed arrangement of stationary states near equilibrium is similar for both protons and O^+. In particular, about $\sim 99.5\%$ of the derived kappa indices were found to be distributed in the interval $\kappa > 2.5$. The same arrangement seems to be followed by the protons, where about $\sim 90\%$ of the kappa indices are distributed in the interval $\kappa > 2.5$. The existence of $\sim 10\%$ of kappa indices $\kappa < 2.5$ might be due to Saturnian pickup and/or solar wind protons, which have large values of M_q and are also present in the Saturnian magnetosphere.

Stationary states in the near-equilibrium region $\kappa > 2.5$ were also detected in solar wind (e.g., Hammond et al., 1996). However, these appear to be rather rare. For instance, the work of Dayeh et al. (2009) showed that all the estimated kappa indices were arranged in the far-equilibrium region, $\kappa < 2.5$. In this work, Dayeh et al. (2009) estimated directly the spectral index in the suprathermal tail spectra of the heavy ions CNO, Ne-S, and Fe during 1 AU quiet times of solar cycle 23 (1995–2007) using wind measurements. The histogram of the relevant metastability measures M_q is plotted in Fig. 10.11E.

The important results of Dialynas et al. (2009) and Dayeh et al. (2009) drive us to separate the stationary states to those near equilibrium and to those near antiequilibrium or far from equilibrium, where the boundary appears to be somewhere around $\kappa \cong 2.5$ ($\kappa_0 \cong 1$). It is interesting that interplanetary (IP) shocks (Desai et al., 2004) and CIRs (Mason et al., 2008) were found to be characterized by spectral indices in the far-equilibrium region, similar to the results of Dayeh et al. (2009). As we shall see as follows, the specific stationary state with kappa index $\kappa \cong 2.5$, called the escape state, has a key role in the transitions of stationary states. Stationary states can escape from the far-equilibrium region toward the near-equilibrium region, passing through the escape state.

This peculiar state has been observed in space plasmas. For example, the energy spectrum of cosmic ray Fe nucleus has been studied by Sato et al. (1985), where they found the spectral index of Fe being equal to $\gamma \cong -2.45 \pm 0.04$. Also, the most probable value of kappa index in the Uranian magnetosphere was found to be around the escape value $\kappa \cong 2.5$ that appears from $\sim 7\ R_{Io}$ to $\sim 8\ R_{Io}$ (Mauk et al., 1987). Finally, we point out that the escape value is consistent with the H-E power law of the energy flux of energetic neutral atoms (ENAs) observed by IBEX toward the heliospheric tail (McComas et al., 2009).

The previously described values of the kappa index were also found in other plasmas, or plasma-related, slowly driven systems that emit energy bursts. The emitted photon fluxes obey a power-law energy distribution, which can be explicitly derived from the asymptotic behavior of kappa distributions in the H-E regions. (Note that if $j(\varepsilon)$ is the emitted photon flux, then the emitted energy flux is $\int j(\varepsilon)\varepsilon d\varepsilon$, e.g., see Shimizu, 1995; Aschwanden et al., 2000.)

As an example, solar flares have a power-law energy distribution with reported spectral indices ranging from 1.5 to 2.6 (e.g., see Aschwanden et al., 2000; Charbonneau et al., 2001; Norman et al., 2001). However, if a consistent set of geometrical assumptions are made, the upper value may be reduced from 2.6 to approximately 2.1 (Parnell and Jupp, 2000; McIntosh and Charbonneau, 2001; McIntosh et al., 2002). On the other hand, the energy distributions of nanoflares are believed to have a different power-law index, namely, larger than 2, though with a large uncertainty (e.g., Benz, 2004; Pauluhn and Solanki, 2007; Bazarghan et al., 2008), and the same holds for flares that are not associated with coronal mass ejections (Yashiro et al., 2006). It can be conjectured that proper flares dominate the far-equilibrium region $\kappa < 2.5$, while nanoflares are mostly refer to the near-equilibrium region $\kappa > 2.5$ (we use nomenclature for proper, in contrast to micro-/nanoflares, e.g., see Reale, 2010). In addition, the statistical analysis of Crosby et al. (1993) of hard X-ray flares detected the spectral index $\gamma \cong 1.62$ for the flares recorded during 1980−1982, similar to the kappa index $\kappa \cong 1.63$ of Decker et al. (2005) near antiequilibrium. Kashyap et al. (2002) found that stellar flares are distributed with a power law of spectral index $\gamma \cong 2.5 \pm 0.4$, that is, around the escape value $\kappa \cong 2.5$. Shimizu (1995) studied the active region transient brightenings at the soft X-rays and found a power law with spectral index $\sim 1.5-1.6$. These values are near the antiequilibrium.

Near these kappa indices is also the spectral index $\gamma \cong 1.66$ characterizing the soft gamma repeater bursts (Göğüş et al., 1999). However, spectral indices close to the fundamental state ($\kappa \sim 1.63$) have been observed also in the very H-E gamma ray sources (Kifune et al., 1995), e.g., in blazar's spectrum, ranging from very low-frequency radio to extremely energetic gamma rays. On the other hand, the specific blazar studied in Schroedter et al. (2005) was found to be characterized by the spectral index $\gamma \cong 2.54 \pm 0.18$, that is, very close to the escape value $\kappa \cong 2.5$. In general, blazar γ-ray spectral indices were found to be distributed between 1.5 and 2.6 (Montigny et al., 1995; Venters and Pavlidou, 2007), which coincides with the far-equilibrium region, $\kappa < 2.5$. However, the majority of them are distributed near the escape value (Aharonian et al., 2006), e.g., the spectral indices that characterize Markarian 501, that is a bright optical blazar in the core of a giant elliptical galaxy (e.g., see Krennrich et al., 1999).

The antiequilibrium state was also detected in the X-ray spectra of the accretion disks around galactic and extragalactic black holes (e.g., Syunyaev et al., 1994). In addition, Ilovaisky et al. (1986) and Miyamoto et al. (1991) found the indices $\gamma \cong 2.27 \pm 0.27$ and $\gamma \cong 2.45 \pm 0.15$, respectively, which are around the escape state of $\kappa \cong 2.5$. It is interesting that when the accretion rate of the disk is very high, the spectral index asymptotically becomes 1.5 (Chakrabarti and Titarchuk, 1995) (we call up that antiequilibrium cannot be reached but asymptotically approached). Much earlier, Schreier et al. (1971) detected anti-equilibrium in the spectrum of the pulsating X-ray source Cygnus X-1. In particular, the spectral index $\gamma \cong 1.5$ found to characterize the spectra for energies above about 8−9 keV, while for lower energies, significant larger values of spectral index were observed, that is, $\gamma \geq 2.5$, lying thus on the near-equilibrium region. This is not strange, since in the case where the particles are separated into two parts with a different index each (Wang et al., 2003), e.g., $\gamma_1 = 1.5$ and $\gamma_2 = 2.5$, then at low energy values, the steeper spectrum of index $\gamma_2 = 2.5$ controls, while at sufficiently high energy values, the flatter spectrum of $\gamma_1 = 1.5$ index dominates. Finally, we add that the fundamental state was also detected in soft X-ray spectra of Seyfert galaxies, e.g., by Arnaud et al. (1985), where they analyzed observations of the Seyfert 1 galaxy MKN 841.

Table 10.1 shows several examples of critical kappa indices of stationary states out of equilibrium that characterize ions, electrons, and photons in space plasmas.

Table 10.1   Critical Indices of Ions, Electrons, and Photons in Space Plasmas

Stationary State(s)	Space Plasma	References	γ, κ
Antiequilibrium $\kappa \cong 1.5$	Pulsating X-ray source Cygnus X-1	Schreier et al. (1971)	γ^*
	X-rays spectra of accretion disks	Syunyaev et al. (1994) and Chakrabarti and Titarchuk (1995)	γ^*
	Io plasma torus	Steffl et al. (2004)	κ
$\kappa \cong 1.63$	Solar wind, inner heliosphere	Fisk and Gloeckler (2006)	γ
	Soft X-ray spectra of Seyfert galaxies	Arnaud et al. (1985)	γ
	Hard X-ray flares	Crosby et al. (1993)	γ
	Active region transient brightenings	Shimizu (1995)	γ
	Blazar gamma rays	Kifune et al. (1995)	γ
	Soft γ-rays repeater bursts	Göğüş et al. (1999)	γ
	Solar wind, inner heliosheath	Decker et al. (2005)	κ
Escape state $\kappa \cong 2.5$	Cosmic ray iron nucleus	Sato et al. (1985)	γ
	X-rays from accretion disks	Ilovaisky et al. (1986) and Miyamoto et al. (1991)	γ
	Extreme ultraviolet flares	Jeffrey et al. (2016)	κ
	Blazar γ-rays	Krennrich et al. (1999) and Schroedter et al. (2005) and Aharonian et al. (2006)	γ
	Stellar flares	Kashyap et al. (2002)	γ
	Io plasma torus	Steffl et al. (2004)	κ
	Heliospheric tail	McComas et al. (2009, 2012a, 2014) and Livadiotis et al. (2011)	γ
Far-equilibrium region, $\kappa < 2.5$	Blazar γ-rays	Montigny et al. (1995) and Venters and Pavlidou (2007)	γ
	Solar flares	Aschwanden et al. (2000) and Charbonneau et al. (2001) and Norman et al. (2001)	γ
	Interplanetary shocks	Desai et al. (2004)	γ
	Corotating interaction regions (CIRs)	Mason et al. (2008)	γ
	Quiet times of solar wind, inner heliosphere	Dayeh et al. (2009)	γ

Continued

Table 10.1 **Critical Indices of Ions, Electrons, and Photons in Space Plasmas—cont'd**

Stationary State(s)	Space Plasma	References	γ, κ
Near-equilibrium region, $\kappa > 2.5$	X-rays from accretion disks	Schreier et al. (1971)	γ
	Uranian magnetosphere	Mauk et al. (1987)	γ^*
	Solar wind, inner heliosphere	Collier et al. (1996)	κ
	Possible nanoflares	Benz (2004) and Pauluhn and Solanki (2007) and Bazarghan et al. (2008)	γ
	Terrestrial magnetosheath	Ogasawara et al., 2013	κ
	Saturnian magnetosphere	Dialynas et al. (2009)	κ

γ, A power law was fitted at high energies.
κ, A kappa distribution was used to model the data.
*, Possible coexistence of more than one stationary states. The steeper law reveals in law energies, while the flatter law dominates in high energies (see text).

10.6.2 Observations in the Inner Heliosheath

10.6.2.1 Far-Equilibrium Inner Heliosheath

The Interstellar Boundary Explorer (IBEX) mission was launched in October 2008 to observe ENA emissions from the outer boundary of the heliosphere, where the subsonic solar wind interacts with the interstellar neutrals (McComas et al., 2009, 2010). IBEX images the sky via two ultra-sensitive ENA detectors (IBEX-Hi, IBEX-Lo) and constructs $6^0 \times 6^0$ all-sky maps every ~ 6 months, covering a broad energy range from ~ 0.01 to ~ 6 keV. Using ENA observations from the IBEX-Hi sensor with energies above ~ 0.7 keV, Livadiotis et al. (2011) derived the sky maps of the (radially) average values of the kappa indices, temperature, and other thermodynamic parameters of the ENA source proton population in the inner heliosheath. This was developed by connecting the observed ENA energy flux to the velocity kappa distribution of the source protons (as shown in Livadiotis et al., 2012, 2013; Livadiotis and McComas, 2012).

We found that the kappa indices are mostly distributed in the interval 1.5–2.5, which coincides with the far-equilibrium region. The identification of the kappa index in a power-law energy flux spectrum is given by the spectral index. As it was shown, these indices are approximately equal (Livadiotis et al., 2011). Then the fact that all the (~ 1800) spectral indices were found to have a lower limit $\gamma_{MIN} \sim 1.5$, which is equal to the lower limit of the kappa index, supports the kappa distributions as the most likely model for describing the proton plasma (ENA source) in the inner heliosheath (McComas et al., 2012a, 2014). Fig. 10.13 shows the sky map of the spectral indices in the first 5 years of IBEX. We also point out that the escape value of $\kappa \sim 2.5$ is also consistent with the IBEX observations, as shown in Figs. 10.12 and 10.13; the upper limit of $\gamma \sim 2.5$ supports that the kappa indices are distributed in the far-equilibrium region, as noted by the theory of the kappa distributions.

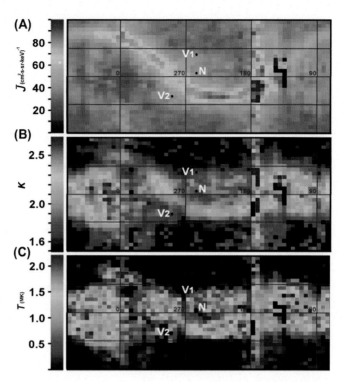

FIGURE 10.12

Flux (averaged over the observed energy range), the kappa index, and temperature sky maps of the source protons. The ribbon has five (numbered) relatively intense "knots" (the directions toward the helionose (N) and Voyager spacecraft (V_1 and V_2) are indicated). Source: Livadiotis et al. (2013).

The measured temperatures in most directions are about a million degrees, in agreement with most heliospheric models for the inner heliosheath. Thus the termination shock is stronger in most places than revealed by the Voyagers' observations at their two unique crossing points. The global sky maps indicate low-temperature regions in various directions, including toward Voyagers, where the temperature is an order of magnitude lower than the most frequent temperature of $\sim 10^6$ K, consistent with in situ Voyager observations.

Finally, we note that IBEX exposes the plasma properties through the detection of ENAs generated by charge exchange in the inner heliosheath. However, charge exchange modifies the plasma as it flows, making it difficult to ascertain the parent proton distribution. The evolution of proton distributions, initially represented by a kappa function, that experience losses due to charge exchange were investigated by Zirnstein and McComas (2015).

10.6.2.2 Anticorrelation Between Density and Temperature

Fig. 10.14 plots the values of the quadruplet (κ, n, T, P) of the proton plasma in the inner heliosheath from Livadiotis et al. (2011) and illuminates the thermodynamic processes that are taking place. In panel (b_1), we show the thermodynamic diagram $\kappa - \log T$ and find a relationship between κ and T fit by $T \cong Ax^2 \exp(-ax)$ (where T is given in MK $= 10^6$ K) with $x \equiv \kappa - 1.6$,

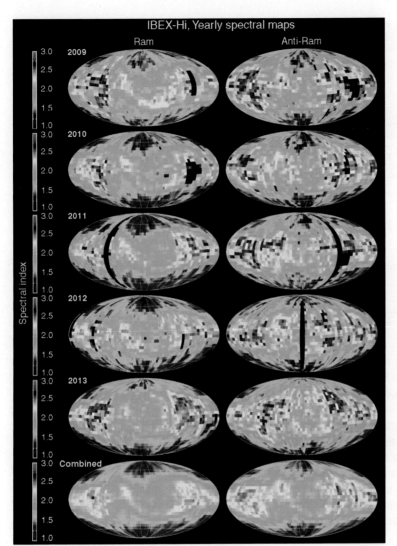

FIGURE 10.13
Sky maps of energy flux spectral index γ data for years 2009–2013 and all years combined (bottom) in annual ram (left) and antiram (right) directions. In general, spectral indices were found to have a lower limit $\gamma_{MIN} \sim 1.5$, which is equal to the lower limit of the kappa index that labels kappa distributions (note: these maps were selectively plotted by collecting data from the ram or antiram directions). The flux is enhanced or reduced in specified parts of the maps when the spacecraft is ramming into ENAs. Also, the energies were corrected to the solar frame). Source: McComas et al. (2014).

$A \cong 18 \pm 7$, and $a \cong 0.4 \pm 0.2$. This relation asymptotically becomes isothermal with $T \approx 10^6 K$ for large kappa ($\kappa > 2$) and iso-kappa (a process of fixed kappa) with $\kappa \approx 1.7$ for temperature lower than $T \sim 250{,}000K$. In panel (b$_3$), we show the thermodynamic diagram $\log n - P$ where the transition between the isothermal and iso-kappa processes is characterized by an isobaric process. This is $P \approx 3.2$ pdyn cm^{-2} for the ribbon (red) and $P \approx 2.1$ pdyn cm^{-2} for the globally distributed ENA flux from near the equator (green). In panel (b$_2$), we show the thermodynamic diagram $\log n - \log T$, which shows a possible power law between n and T, and the associated polytropic indices. Remarkably, the ribbon, having higher pressure, is naturally separated from the

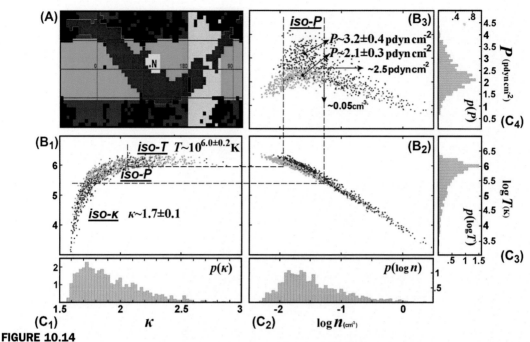

FIGURE 10.14

Thermodynamic processes of the space plasma in the inner heliosheath. (A) Separation of the sky map into the regions of the ribbon (*red*) and the globally distributed ENA flux that further separates into near-equator (*green*) and near-polar (*blue*) regions; also indicated are regions with high uncertainty (*light gray*) and the heliospheric nose (*yellow*) (N) (each point of the sky map corresponds to one of 1800 $6^0 \times 6^0$ pixels). (B) Thermodynamic diagrams: (b_1) κ–log T, (b_2) log n–log T (slope is $1/\nu$), (b_3) log $n - P$. Note that the values of the quadruplet (κ, n, T, P) are given for each pixel (from Livadiotis et al., 2011). Three thermodynamic processes have been detected: isothermal (iso-T) $T \approx 10^6 K$, isobaric (iso-P) $P \approx 3.2$ pdyn cm^{-2} (ribbon) and $P \approx 2.1$ pdyn cm^{-2} (globally distributed flux, mostly near the equator; green), and iso-kappa (fixed kappa index) with $\kappa \approx 1.7$ (globally distributed flux, mostly near the poles; blue). Panels (B_2) and (B_3) clearly show the thermodynamic separation of ribbon from the globally distributed ENA flux. In particular, panel (B_3) clarifies that ribbon has higher pressures with the boundary roughly given by $P \sim 2.5$ pdyn cm^{-2}. (C) The normalized histograms P of each variable of the quadruplet (κ, n, T, P) are given along the sides. Source: Livadiotis and McComas (2012).

globally distributed ENA fluxes; the boundary is roughly given by $P \sim 2.5$ pdyn cm^{-2}, and this separation is apparent in both panels (b_2) and (b_3). Further, the globally distributed ENA flux is separated into the near-equator region that is characterized by lower densities and the near-polar region, with the boundary roughly given by $n \sim 0.05$ cm^{-3}.

10.6.2.3 Isobaric Process

The solar wind flows radially out from the sun until it reaches the termination shock, where it is slowed and compressed. Beyond the termination shock, in the inner heliosheath, the solar wind flow bends back and ultimately flows around the sides of the termination shock and down the heliotail. This bending of the flow in the inner heliosheath may make it possible for the plasma (and consequent ENA emissions) from various directions in the sky to be correlated. Fig. 10.15A plots the temperature and density values along the equator, clearly showing their correlation.

The method of the maximized correlation was applied to the datasets of temperature and density to derive precisely the polytropic index (Livadiotis and McComas, 2013c). Namely, we examine the correlation r between the datasets n_i and T_i^ν, for $i = 1, ..., 37$. In Fig. 10.15C, we plot the modified correlation coefficient $R = 1-r^2$ as a function of the fitting parameter, which represents the polytropic index ν. Remarkably, we find $\nu^* \cong -0.96 \pm 0.06$, which is an excellent approximation of an isobaric thermodynamic process, that is, of polytropic index $\nu = -1$, i.e., constant thermal pressure $P \propto n \cdot T \sim$ constant (Chapter 5; Livadiotis and McComas, 2012; Livadiotis, 2016c).

We note that anticorrelations between density and temperature, consistent with constant or quasiconstant thermal pressure P, were also found in several other space plasmas; among others, we mention the boundary layer at the terrestrial magnetosheath (Sckopke et al., 1981; Hapgood and Bryant, 1992), the terrestrial central plasma sheet (Pang et al., 2015a, 2015b), the Jovian magnetosheath (Nicolaou et al., 2015), and the solar wind around Pluto magnetosphere (Elliot et al., 2016).

A superposition of polytropes was applied in the inner heliosheath proton plasma (Livadiotis, 2016c). Then the linear polytropic density–temperature relation on log–log scale was generalized to a concave-downwards parabola; the estimated mean polytropic index was $\nu \approx -1$ or $a \equiv 1+\nu^{-1} \approx 0$, indicating

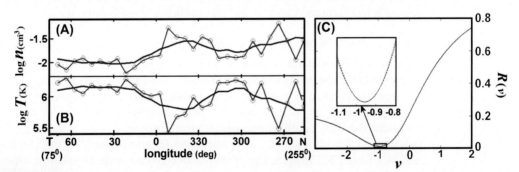

FIGURE 10.15
(A) Temperature T, and (B) number density n of the globally distributed flux in the inner heliosheath and along the equator from the heliosphere's upwind direction at the helionose (N) toward the heliotail (T) (*red*) (each point corresponds to a $6^0 \times 6^0$ pixel of the sky. The pristine globally distributed flux accounts for 37 pixels out of 60 across the equator). The *smoothed curve* (*blue*) is a seven-point moving average centered on each pixel. The temperature is clearly negatively correlated with density. The correlation coefficient is $r \cong -99.3\%$ (for the whole set of 60 points across the equator; $r \cong -89.0\%$). (C) The modified correlation coefficient $R = 1-r^2$ is depicted as a function of the fitting parameter ν that is the polytropic index, which is found to be $\cong -0.96 \pm 0.06$. Source: Livadiotis and McComas (2013c).

dominance of isobaric thermodynamic processes in the inner heliosheath, similar to other previously published analyses.

In Chapter 3 (Section 3.9.4), we have seen that in the presence of a central potential energy $\Phi(r)$, particles are subject to a polytropic relation between the local density $n(r)$ and temperature $T(r)$,

$$\left[\frac{n(\vec{r})}{n(\vec{r}_*)}\right] = \left[\frac{T(\vec{r})}{T(\vec{r}_*)}\right]^{-\kappa_0 - 1 \mp \frac{1}{2}d_\Phi} \equiv \left[\frac{T(\vec{r})}{T(\vec{r}_*)}\right]^{\nu}, \text{ with } \nu \equiv (a-1)^{-1}$$

$$= -\kappa_0 - 1 \mp \frac{1}{2}d_\Phi, \qquad (10.38)$$

where $n(\vec{r}_*)$ and $T(\vec{r}_*)$ are the local density and temperature at the position of zero potential $\vec{r} = \vec{r}_*$ (see Eq. 3.78b), where the sign $\mp$ is: $-$ if $\Phi > 0$ and $+$ if $\Phi < 0$. The potential degrees of freedom are defined by $\frac{1}{2}d_\Phi \equiv \langle|\Phi|\rangle \big/ (k_B T)$ (Eq. 3.10). For example, if the potential is small compared to the temperature (e.g., see the case of the Debye length in Chapter 5, Section 5.5), then we have $\nu \approx -\kappa_0 - 1$. If this is the case applied in the inner heliosheath, then when the plasma is close to the antiequilibrium state, $\kappa_0 \approx 0$, we obtain $\nu \approx -1$. If the potential energy, however, is given by a power-law relation $\Phi(r) = \pm\frac{1}{b}kr^{\pm b}$ (where the position vector dimensionality is d_r, e.g., $d_r = 3$ for $\vec{r} = (x, y, z)$), then we obtain $\frac{d_\Phi}{2} = \frac{1}{b}d_r$ (Eq. 3.141b). In this case, the isobaric process may be achieved when $\Phi(r) < 0$ and $\nu = \frac{1}{b}d_r - 1 - \kappa_0 \approx -1$ or $\kappa_0 \approx \frac{1}{b}d_r$. In the case of gravitational field, we have $b = 1$; if $d_r = 3$, then the isobaric process can be realized when the plasma is close to $\kappa_0 \approx 3$. See also the derivation of polytropic relations in Chapters 3 and 5, and the application is Chapter 11.

10.6.3 Near Versus Far Equilibrium

Space plasma observations have shown an empirical separation of the kappa spectrum. This spectrum is divided into near- and far-equilibrium regions with the boundary value at $\kappa \sim 2.5$, a state called "escape state": spectral or kappa indices $2.5 < \kappa \le \infty$ (e.g., Saturnian magnetosphere, Dialynas et al., 2009) represent the near-equilibrium region, while indices $1.5 < \kappa \le 2.5$ (e.g., inner heliosheath, Livadiotis et al., 2011) note the far-equilibrium region (Livadiotis and McComas, 2013d) (see Fig. 10.16). Indeed, various space plasmas appear to have indices distributed essentially over only one or the other region. For example, solar wind is characterized by kappa indices in the near-equilibrium region (e.g., Livadiotis and McComas, 2013c), while heavy ions during quiet times (Dayeh et al., 2009), IP shocks (Desai et al., 2004), and CIRs (Mason et al., 2008) are characterized by kappa indices in the far-equilibrium region. Even the concept of correlation can naturally separate these two regions; according to Eq. (10.37b), the near/far−equilibrium regions can be characterized by weak and strong correlation, respectively (Fig. 10.9).

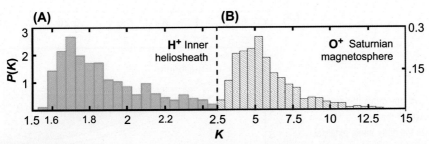

FIGURE 10.16
The distribution of kappa indices, $P(\kappa)$, constructed for (A) inner heliosheath protons (*green*) (Livadiotis et al., 2011), and (B) Saturnian magnetosphere oxygen O^+ (*red*) (kappa indices derived by Dialynas et al., 2009). While the kappa indices of the inner heliosheath are distributed only in the far-equilibrium region, the kappa indices of the Saturnian magnetosphere are distributed only in the near-equilibrium region. Source: Livadiotis and McComas (2013d).

One intriguing case of far- and near-equilibrium plasmas comes from comparing the inner heliosphere and the inner heliosheath proton plasmas. Indeed, the inner heliosphere is typically characterized by kappa indices in the near-equilibrium region (e.g., Dialynas et al., 2009; Livadiotis and McComas, 2013c; Livadiotis, 2015a), while the kappa indices in the inner heliosheath were found to be restricted to the far-equilibrium region (e.g., Livadiotis et al., 2011) (see Fig. 10.14c1). This may be surprising because the inner heliosheath is largely populated by solar wind plasma that has traveled out beyond the termination shock and because both the additional time for it to evolve and the action of shock heating might be expected to push the plasma closer to equilibrium. Livadiotis and McComas (2010a) anticipated this dissimilar behavior of kappa indices between the inner heliosphere and inner heliosheath by proposing that pickup protons, which are ions with highly organized phase space distributions, can reduce the entropy of the combined system of solar wind and pickup protons, and thus push the values of the kappa index toward the far-equilibrium region. This result was first shown based on analytical work (Livadiotis and McComas, 2011a) and then verified by analyzing IBEX ENA observations (Livadiotis et al., 2011).

The thermodynamics of the far-equilibrium region is more complicated than that of the near-equilibrium region. This may be shown using the entropy when expressed only in terms of the kappa index. Livadiotis and McComas (2010a, 2010c, 2013a) developed an entropic measure, which is independent of temperature, expressed solely in terms of the kappa index, $S(\kappa)$; it was shown how variations of entropy can gradually lead the system toward or away from equilibrium. The entropy $S(\kappa)$ is generally a nonmonotonic function with respect to the kappa index. In particular, it is monotonic in the near-equilibrium region, i.e., for $\infty > \kappa > 2.5$, while the nonmonotonicity (concave upwards) appears in the far-equilibrium region, i.e., for $2.5 \geq \kappa > 1.5$, exhibiting thus a more complicated behavior.

FIGURE 10.17
(A) Spontaneous increases of entropy move the system gradually toward equilibrium within the near-equilibrium region, $\kappa > 2.5$. On the other hand, external factors that may decrease entropy move the system away from equilibrium into the far-equilibrium region, $\kappa \leq 2.5$. Newly formed PickUp Ions can play this critical role in the case of solar wind and other space plasmas because of their highly ordered motion. They reduce the entropy of the system and move it toward the far-equilibrium region. (B) Distribution of kappa indices $P(\kappa)$ for the inner heliosheath protons H^+. (Compare with Fig. 2.6 in Chapter 2.) Source: Livadiotis and McComas (2013a).

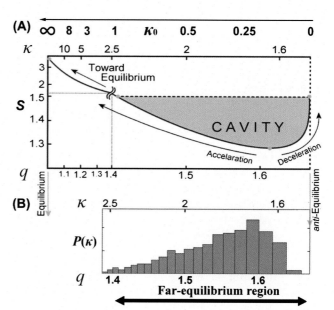

Spontaneous procedures that increase entropy move the system gradually toward equilibrium ($\kappa_0 \rightarrow \infty$), that is, the state with the maximum entropy (acceleration arrow in Fig. 10.17). On the other hand, external factors that may decrease entropy move the system back to states further from equilibrium and closer to antiequilibrium ($\kappa_0 \rightarrow 0$) (deceleration arrow in Fig. 10.17). Newly formed PUIs can play this critical role in the case of solar wind and other space plasmas because of their highly ordered motion.

10.6.4 The Role of PickUp Ions in the Transitions of Kappa Distributions

Livadiotis and McComas (2010a,c, 2011a) proposed that the pickup protons, which are ions with highly organized phase space distributions, reduce the entropy of the combined system of solar wind and pickup protons (incorporated solar wind) and thus push the values of the kappa index to be far from equilibrium and within the far-equilibrium region $\kappa_0 \in (0,1]$.

A PUI is generated when a neutral atom, for example, from the interstellar neutral gas that flows into the heliosphere, becomes ionized. The dominant ionization processes are charge exchange with solar wind ions and photoionization from solar photons although processes such as impact ionization from solar wind electrons can also contribute. The charge exchange ionization of a neutral atom is also a mechanism to generate ENAs. Because energy and momentum transfer is negligible in this process, the newly neutralized solar wind ion (usually hydrogen) is born from the local proton distribution, and thus this

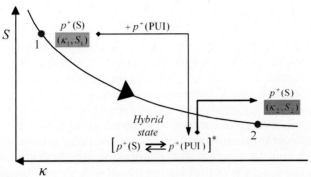

FIGURE 10.18
The transition of space plasmas in near-equilibrium stationary states is characterized by monotonic entropic behavior. As the entropy decreases, the kappa index also decreases and the system is diverted further from equilibrium. Solar wind protons, $p^+(S)$, residing in a stationary state with kappa index κ_1 and entropy S_1, are mixed with pickup protons, $p^+(PUI)$, forming the hybrid state of undistinguished protons $[p^+(S) \rightleftarrows p^+(PUI)]^*$, which finally degenerates into a new stationary state with kappa index κ_2 and entropy S_2. Source: Livadiotis and McComas (2011a).

ENA emission accurately reflects the properties (such as energy distribution and density) of the local source population of solar wind and PUIs. The initially circular motion of PUIs is superposed on the linear motion of the magnetic field, leading to a cycloidal motion in the inertia reference frame or a ring-shaped velocity distribution in the solar wind frame. This is quickly transformed into a distribution on a spherical shell by pitch angle scattering (Holzer and Axford, 1970). As the population of PUIs travels outward the heliosphere, it cools and the spherical shell velocity distribution shrinks. Then, newly formed PUIs replace the exterior shells, and gradually new shells fill the entire sphere. Finally, the local distribution of PUIs is composed of a filled spherical shell (Vasyliunas and Siscoe, 1976).

Fig. 10.17 in Section 10.6.3 shows that spontaneous increases of entropy gradually lead the system toward equilibrium. On the contrary, any exterior factor that decreases the entropy can redirect the plasma back to far from equilibrium stationary states. One specific mechanism that can cause decrease of entropy is related to the PUIs as soon as they can add "order" in the solar wind protons. Indeed, at least in the inner heliosheath, their highly ordered arrangement, which arises in the small thickness of their spherical shell distribution, decreases the average entropy and redirects the system away from equilibrium.

The addition and incorporation of highly ordered distributions of PUIs can increase the ordering of space plasmas, decreasing their entropy and driving them away from equilibrium (Fig. 10.18). Such a model was first presented by Livadiotis and McComas (2011a) in order to describe the entropy decrease in space plasma protons caused by the pickup protons and their highly ordered arrangement. The presented model was used for quasiequilibrium stationary states in order to describe the solar wind prior to incorporation of PUIs; for kappa indices larger than $\kappa \sim 10$, the kappa distribution is extremely close to a Maxwellian distribution.

Directly after the entrance of pickup protons in solar wind proton flow, a hybrid state is generated. The variation of the average entropy (entropy per particle) between the hybrid state and the original stationary state of the bulk solar wind protons, ΔS, depends on a dimensionless "fluctuation number" noted by Θ_d, that is, a measure of the organizing role of the pickup protons in the solar wind flow. This number characterizes and compares two competing fluctuations, the thermal motion of the solar wind protons, and the radial diffusion of the pickup protons that forms the spherical shell distribution; it is proportional to the ratio of the phase space density of the pickup protons with the relevant density of the bulk solar wind protons. The entropy change ΔS is proportional to the negative logarithm of Θ_d, $\Delta S \propto -\ln \Theta_d$: when $\Theta_d > 1$, we have strong thermal fluctuations of the bulk solar wind protons, and the entrance of the less disorganized pickup protons decreases the entropy, i.e., $\Delta S < 0$. When $\Theta_d < 1$, then the bulk solar wind protons are already well-organized and the entrance of the more disorganized pickup protons increases now the entropy, i.e., $\Delta S > 0$. When $\Theta_d = 1$, the two competing fluctuations balance each other, causing no change to the entropy, i.e., $\Delta S = 0$.

The solar wind protons, together with the pickup protons, are characterized by a probability distribution that is the normalized sum of the probability distributions of the bulk protons (kappa distribution) and the pickup protons (spherical shell distribution). Potentially, this hybrid describes a more organized state, having lower entropy than the stationary state assigned by the original probability distribution of the solar wind before the inclusion of pickup protons.

Calculating the possible values of Θ_d throughout the heliosphere and in the inner heliosheath, the authors found that upstream the termination shock the entropy generally increases; beyond the termination shock in the inner heliosphere, where the PUIs can be found in large fractions, the entropy decreases significantly. The decrease of entropy redirects the protons population back to far equilibrium.

Near equilibrium, the entropy $S(\kappa)$ increases with the kappa index (as shown in Fig. 10.17),

$$S(\kappa) \cong S_0 - 9.25\kappa^{-1} - 25.46\kappa^{-2} + O(\kappa^{-3}), \tag{10.39}$$

with $S_0 \equiv S(\kappa \to \infty) = \frac{3}{2}(1 + \ln\pi)$. Therefore, starting at equilibrium, which corresponds to the maximum κ-index ($\kappa \to \infty$) and a Maxwellian distribution, any decreases of entropy will lead to kappa distributions with smaller κ-indices, and thus the system will be directed further from equilibrium. Hence, $\Delta S \equiv S(\kappa) - S_0 = -9.25/\kappa - 25.46/\kappa^2$, and when $\Delta S < 0$, this gives a real solution for a finite kappa index. The initial Maxwellian distribution is replaced by a kappa distribution of (finite) kappa index, given by

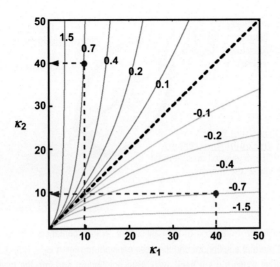

FIGURE 10.19
Nonequilibrium transitions caused by entropy variations. If the entropy of a system varies by ΔS, then its initial kappa distribution of index κ_1 will go to a kappa distribution of index κ_2, as given by Eq. (10.17), valid for sufficiently large kappa indices ($\kappa \gg 2.5$). Then the system transmits in a stationary state with κ_2 nearer (for $\Delta S > 0$) or further from equilibrium (for $\Delta S < 0$). In the case of the solar wind protons described initially by a kappa distribution of κ_1, PickUp Ions (PUIs) in the inner heliosheath can effectively reduce the system's entropy, leading to incorporated solar wind protons described by a kappa distribution of $\kappa_2 < \kappa_1$ (further from equilibrium). On the contrary, PUIs increase the system's entropy in the heliosphere sunward the termination shock, resulting to a kappa distribution of $\kappa_2 > \kappa_1$ (nearer to equilibrium). An example of decreasing entropy is shown by the *red arrow* (gray in print versions) where the initial index $\kappa_1 \cong 40$ leads to the index $\kappa_2 \cong 10$ for $\Delta S = -1$ (for even larger entropy decreases, we have even smaller kappa indices, i.e., $\kappa_2 \cong 5$ for $\Delta S = -2.5$, and $\kappa_2 \cong 3$ for $\Delta S = -5$). The *blue array* (light gray in print versions) shows the case of increasing entropy, where the index $\kappa_1 \cong 10$ leads to $\kappa_2 \cong 40$ for $\Delta S = 1$. Source: Livadiotis and McComas (2011a).

$$\kappa \cong 4.63 \cdot \left(1 + \sqrt{1 + 1.19|\Delta S|}\right) \Big/ |\Delta S|. \tag{10.40}$$

Starting from a quasiequilibrium stationary state that is corresponding to a large κ-index in the near-equilibrium region ($\kappa \gg 2.5$), the entropy variation between two stationary states with kappa indices $\kappa_1 \rightarrow \kappa_2$ is given by

$$\Delta S \equiv S(\kappa_2) - S(\kappa_1) \cong -9.25/\kappa_2 - 25.46/\kappa_2{}^2 + 9.25/\kappa_1 + 25.46/\kappa_1{}^2. \tag{10.41}$$

If the entropy of the system varies by ΔS, then its initial kappa distribution of index κ_1 will be replaced by a kappa distribution of index κ_2, given by

$$\kappa_2 \cong 4.63 \cdot \frac{1 + \sqrt{1 + 1.19\left(9.25/\kappa_1 + 25.46/\kappa_1{}^2 - \Delta S\right)}}{9.25/\kappa_1 + 25.46/\kappa_1{}^2 - \Delta S}, \tag{10.42}$$

and thus the system will transmit in a stationary state with κ_2 nearer to equilibrium (for $\Delta S > 0$), or further from equilibrium (for $\Delta S < 0$). The relation

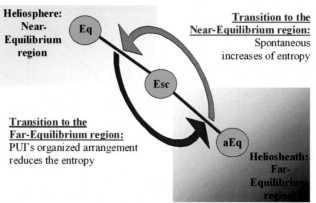

Heliosphere: Near-Equilibrium region

Transition to the Near-Equilibrium region: Spontaneous increases of entropy

Eq

Esc

Transition to the Far-Equilibrium region: PUI's organized arrangement reduces the entropy

aEq

Heliosheath: Far-Equilibrium region

FIGURE 10.20

Scheme of nonequilibrium transitions of the heliosphere. While spontaneous increases of entropy move the system toward thermal equilibrium and the near-equilibrium region $\kappa_0 \in (1, \infty]$, the impact of highly-organized PickUp Ions (PUIs) turns it back away from equilibrium and into the far-equilibrium region $\kappa_0 \in (0,1]$. As a result, the inner heliosphere, with comparatively fewer PUIs resides in near-equilibrium stationary states, while the much more distant inner heliosheath resides in far from equilibrium states (Eq, Esc, and aEq abbreviate equilibrium, escape, and antiequilibrium states, respectively.) Source: Livadiotis and McComas (2011b).

between the initial κ_1 and the resulting κ_2 kappa indices is shown in Fig. 10.19 for various values of the entropy variation, either positive $\Delta S > 0$ (green) or negative $\Delta S < 0$ (yellow). The red arrow indicates an example for $\Delta S \cong -1$, where the initial kappa distribution of $\kappa_1 \cong 40$ after the entropy reduction leads to the resulting kappa distribution of $\kappa_2 \cong 10$. The blue arrow shows the opposite case of increasing entropy $\Delta S \cong 1$, where the initial kappa distribution of $\kappa_1 \cong 10$ leads to the resulting kappa distribution of $\kappa_2 \cong 40$.

The model developed by (Livadiotis and McComas, 2011a) concluded that the inner heliosphere resides mostly in near-equilibrium stationary states, while the inner heliosheath resides mostly in far-equilibrium stationary states (Fig. 10.20), a result later verified by the analysis on IBEX-Hi ENA spectra, shown in Section 10.6.2 (Livadiotis et al., 2011; Livadiotis and McComas, 2012).

Other, more distant astrophysical systems have vastly different plasma parameter regimes, where the effects of both strongly enhancing entropy (driving the system toward equilibrium), and strongly decreasing entropy (driving the system toward more order and away from equilibrium) may have critical implications for evolution of the plasma environments and remotely observable emissions. For instance, Soker et al. (2010) showed that the presence of PUIs in some planetary nebulae and symbiotic astrophysical systems might reduce the postshock temperature of the fast wind and jets. These PUIs might also reduce the entropy of these systems, leading them into far-equilibrium transitions.

10.7 Concluding Remarks

This chapter presented the types of kappa distributions that describe the ion populations in space plasmas throughout the heliosphere. Specifically, we presented:

- formulations of ion kappa distributions: standard kappa distributions, "negative" kappa distributions, superposition of kappa distributions, anisotropic kappa distribution, and phase space marginal kappa distribution of velocities;
- description of the antiequilibrium, the furthest stationary state from thermal equilibrium, investigation of its statistical formulation and properties, and the transition of stationary states toward this extreme state;
- arrangement of the stationary states into the "kappa spectrum";
- development of the consistent measures describing the thermodynamic distance from thermal equilibrium;
- the measurements of kappa indices of various ion space plasmas throughout the heliosphere, while detailed discussion is given for the observations in the inner heliosheath;
- the different thermodynamic behavior between near- and far-equilibrium regions; and
- the role of PUIs in the transitions of kappa distributions.

The presented statistical formulations, properties, and observations of kappa distribution of ion populations in space plasmas provide valuable and necessary tools for space physics education and research.

10.8 Science Questions for Future Research

Future analyses and observations need to address the following questions:

1. What causes the distinction in near/far—equilibrium regions?
2. How do PUIs affect the kappa index, temperature, and entropy?
3. What mechanisms generate the observed ion kappa distributions?

Acknowledgment

This research work was partially funded by the IBEX mission as a part of NASA's Explorer Program.

CHAPTER 11

Electron Distributions in Space Plasmas

V. Pierrard[1,2], N. Meyer-Vernet[3]

[1]Royal Belgian Institute for Space Aeronomy, Brussels, Belgium; [2]Université Catholique de Louvain, Louvain-La-Neuve, Belgium; [3]LESIA, Observatoire de Paris, CNRS, PSL Research University, UPMC, Sorbonne University, Paris Diderot, Paris, France

Chapter Outline

11.1 Summary 465
11.2 Introduction: Observations and Origins of Suprathermal Electrons 466
11.3 Coronal Heating by Velocity Filtration Due to Suprathermal Electrons 470
11.4 Heat Flux 472

11.5 Influence of Suprathermal Electrons on the Acceleration of Escaping Particles 474
11.6 Concluding Remarks 478
11.7 Science Questions for Future Research 479
Acknowledgment 479

11.1 Summary

Space plasmas are essentially collisionless systems out of thermal equilibrium, where enhanced populations of suprathermal particles are observed. The typical distributions are generally better described by kappa distributions than by Maxwellians, especially for electrons, which has serious consequences since their small electron mass makes them major agents for plasma energy transport. This chapter presents suprathermal electrons and their critical role in the heating and acceleration of plasmas in several important space and astrophysical contexts. They affect the generation of the ambipolar electric field and contribute to the collisionless electron heat flux. In the solar corona, such electrons make a dominant contribution to the electron heat flux and play an important role in the coronal heating energy budget. They also support large ambipolar electric fields along open magnetic flux tubes in solar/stellar coronae and in planetary ionospheres and thus contribute significantly to solar and stellar wind acceleration, outflow from planetary ionospheres, and possibly even exoplanetary

Kappa Distributions. http://dx.doi.org/10.1016/B978-0-12-804638-8.00011-5

atmospheric loss. For the Earth's environment, it has been demonstrated that suprathermal electrons play a controlling role in the plasmasphere thermal structure, have a major effect on ionospheric outflows, and control the electron temperature and consequently the topside ionospheric scale height through the generation of heat flux.

Science Question: What are the features of electron kappa distributions in space plasmas?

Keywords: Halo; Heat flux; Strahl; Superhalo; Velocity filtration.

11.2 Introduction: Observations and Origins of Suprathermal Electrons

Nonthermal particle distributions are ubiquitously observed in space plasmas. Measurements confirmed by many interplanetary missions show enhanced populations of suprathermal electrons. The electron velocity distribution functions (VDFs) in space plasmas have non-Maxwellian suprathermal tails that decrease as a power law of the velocity. Such distributions can be approximated by the so-called kappa or generalized Lorentzian distributions:

$$f(\overrightarrow{u}) = n\pi^{-\frac{3}{2}}\kappa^{-\frac{3}{2}}\frac{\Gamma(\kappa+1)}{\Gamma\left(\kappa-\frac{1}{2}\right)}\theta_{\kappa}^{-3}\left(1+\frac{1}{\kappa}\cdot\frac{\overrightarrow{u}^2}{\theta_{\kappa}^2}\right)^{-\kappa-1}, \quad \theta_{\kappa} \equiv \sqrt{\frac{\kappa-\frac{3}{2}}{\kappa}}\cdot\theta, \qquad (11.1)$$

where $\theta \equiv \sqrt{2k_{B}T/m}$ is the characteristic thermal speed; n and T are the number density and temperature of the particle with mass m, respectively; and k_{B} is the Boltzmann constant (Chapter 1; see the reviews in Pierrard and Lazar, 2010; Livadiotis and McComas, 2013a; Livadiotis, 2015a; references therein). This distribution was first introduced by Binsack (1966), Olbert (1968), and Vasyliunas (1968) to fit electron observations in the magnetosphere. The value of the κ-index determines the slope of the energy spectrum of the suprathermal electrons forming the tail of the VDF. In the limit $\kappa \to \infty$, the kappa distribution degenerates into a Maxwellian, decreasing exponentially with the square velocity. A sum of two Maxwellians can also fit distributions with suprathermal tails, but that model yields weaker fits than kappa functions (Maksimovic et al., 1997b).

The electron VDFs are characterized by a thermal core (around 10 eV) that contains 90% of the density, an isotropic halo (50—80 eV) associated to the excess of suprathermal electrons, and an antisunward strahl parallel to the interplanetary magnetic field. An example of a typical solar wind electron VDF is presented in Fig. 11.1. Although the core population seems rather well described by a Maxwellian (in red on Fig. 11.1), a point that should be taken with caution because of the measurement uncertainties at low energies due to the

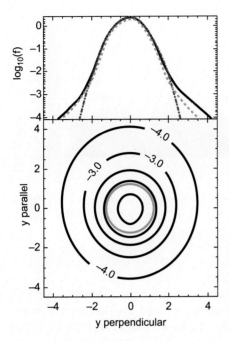

FIGURE 11.1

Typical velocity distribution function observed by WIND spacecraft at 1 AU in the slow speed solar wind. Top panel: Cross-section as a function of the parallel velocity (*black*) and perpendicular velocity (*blue*) normalized by the thermal velocity. (Maxwellian is in *red* for comparison.) Bottom panel: Contours of the observed distribution in the plane, defined by the parallel and perpendicular velocity normalized by the thermal velocity. The halo population (suprathermal electrons) is present in the parallel and in the perpendicular direction, in the slow speed solar wind, and with higher tails in the high speed solar wind. The anisotropy is low, with a velocity distribution function lightly focused in the direction of the interplanetary magnetic field (strahl component). The core is Maxwellian. Source: Pierrard (2012a).

photoelectrons and the positive spacecraft potential, there is a large excess of suprathermal electrons in the tail of the distribution: the halo population (e.g., Pierrard et al., 2001b). This halo population exists in all directions, thus also in the direction perpendicular to the interplanetary magnetic field (illustrated in blue on Fig. 11.1). In addition, strahl electrons increase the population in the direction of the magnetic field (the more so as the wind speed increases) (Marsch, 2006). The strahl population is clearly illustrated by the black line at positive velocities parallel to the magnetic field on the top panel of Fig. 11.1.

In addition to the core, halo, and strahl populations, solar wind electrons are observed to have a power law superhalo at energies above 2 keV (Lin, 1998). Wang et al. (2012) found that 2−20 keV superhalo electrons generally show a power law energy spectrum of differential flux, $j \sim \varepsilon^{-\gamma}$, with an average spectral index of $\gamma \sim 2.3$. The angular distribution of superhalo electrons is mainly isotropic (Yang et al., 2015).

Considering the core and the halo populations, the fits of the observed electron VDFs from the Ulysses spacecraft in the solar wind with kappa distributions give values of the kappa index κ ranging between 2 and 7 in general, with smaller averaged values in the fast wind than in the slow wind (Maksimovic et al., 1997b). The higher values of kappa obtained in the slow solar wind than in the fast wind suggest a link between the wind velocity and the suprathermal electrons.

Fig. 11.2 shows that the kappa index remains nearly constant with distance around $\kappa \approx 2$ between 1.5 and 2.3 AU in the fast wind (Le Chat et al., 2011); the

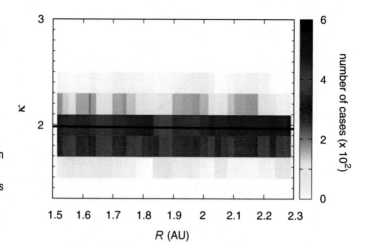

FIGURE 11.2
The kappa index of the whole electron distribution remains roughly constant with the heliocentric radial distance R above 1.5 AU in the fast wind, as measured by the method of quasithermal noise spectroscopy with Ulysses/URAP high-latitude data. Source: Le Chat et al. (2011).

results are derived from quasithermal noise spectroscopy (Meyer-Vernet and Perche, 1989) of Ulysses data at high latitudes, implemented with a kappa electron distribution (Le Chat et al., 2009). The kappa index found by fitting the whole distribution with a kappa distribution reveals its nonthermal character. The constancy of this kappa index is consistent with that of the relative density of the sum of the halo and strahl components found by a number of authors. Indeed, some authors have fitted the electron distributions not with just one kappa distribution but rather with a sum of a bi-Maxwellian distribution for the core population and a bi-kappa distribution for the halo (Maksimovic et al., 2005; Stverak et al., 2009; Pierrard et al., 2016). The sum of these two distributions is then removed from the observed VDF to deduce the characteristics of the strahl population (Stverak et al., 2009). The fit is then determined by many fitting parameters, e.g., $(n_c, T_{c\mathrm{II}}, T_{c\perp})$, $(n_h, T_{h\mathrm{II}}, T_{h\perp})$, and κ (where c is for the core and h is for the halo) as well as additional ones to characterize the strahl, while the fit with a global kappa VDF gives three parameters (n, T, κ), characterizing the global electron distribution. With an analysis of more than 240,000 samples from Helios (>1 AU), Cluster (1 AU), and Ulysses (>1.2 AU), it is found that below 1 AU, the halo increases with distance in relative density at the expense of the strahl, suggesting a possible diffusion of the strahl to the halo (Maksimovic et al., 2005; Stverak et al., 2009). The anticorrelation found for the core and halo temperatures is consistent with the radial evolution of the kappa model (Pierrard et al., 2016). Analyses of electron temperature anisotropies show that the bulk of solar wind electrons are constrained by Coulomb collisions, while the large departures from anisotropy are constrained by instabilities (Stverak et al., 2008).

Such suprathermal tails are observed in situ in solar wind and in the magnetosphere of the Earth and of the other planets (Pierrard and Lazar, 2010), including in the radiation belts (Pierrard and Lemaire, 1996b; Pierrard and

Borremans, 2012b). The presence of such distributions in many different space plasmas suggests a universal mechanism for the creation of suprathermal tails. Pierrard and Lazar (2010) reviewed several works analyzing the origin, effects, and consequences of the presence of nonthermal kappa distributions in space plasmas, not only for electrons but also for different ion species.

As first shown by Scudder and et al. (1979a, 1979b, 1983), the fast increase of the particle Coulomb free-paths with energy constitutes a basic mechanism for producing suprathermal tails in the presence of gradients (Shoub, 1983). This effect can be easily understood by noting that when the Knudsen number (the ratio of the mean free-path to the pressure scale-height) equals unity, the electrically charged particles moving, say, three times faster than the thermal ones have a free-path larger by a factor of 3^4, i.e., two orders of magnitude. More precisely, Scudder and Karimabadi (2013) showed that the nonthermal regime arises as soon as the Knudsen number exceeds 0.01, so that the whole solar corona and wind should exhibit nonthermal velocity distributions.

The formation of high-energy tails in the electron VDF was also studied with a Fokker-Planck model (Lie-Svendsen et al., 1997; Pierrard et al., 1999, 2001c). Due to the properties of the Coulomb collisions, the energetic particles are noncollisional even when thermal particles are submitted to collisions (e.g., Pierrard, 2012a). Turbulence can also play a role in the generation of such suprathermal tails as well as the long-range correlations supplied by the fields and plasma instabilities (Pierrard et al., 2011). The generation of suprathermal electrons by resonant interaction with whistler waves was suggested in the solar corona and wind (Vocks and Mann, 2003; Vocks et al., 2008). When large-amplitude waves are present, the nonlinear Landau damping can be responsible for the energization of plasma particles (Shizgal, 2007). In addition, coronal nanoflares generating kinetic Alfvén and whistler wave turbulence have been suggested to produce solar wind suprathermal electrons (Che and Goldstein, 2014).

Advances in space physics show that empirical kappa distributions naturally emerge from statistical mechanics and thermodynamics (Livadiotis and McComas, 2009, 2013a; Livadiotis, 2015a; references therein). Empirical velocity kappa distributions provide a straightforward replacement of the Maxwell distribution for systems out of thermal equilibrium such as space plasmas, while numerous applications of these distributions have been involved in the framework of particle acceleration.

In the present chapter, we show the importance of suprathermal electrons in the heating processes and in the escape of ions and electrons from planetary and solar/stellar exospheres. The enhanced population of energetic particles plays a crucial role in the heating and acceleration of plasma in several important space and astrophysical contexts. Such consequences are well evidenced by using kappa distributions in kinetic models where no closure requires the distributions to be nearly Maxwellian (e.g., Pierrard, 2012b). In Section 11.3, we show how the high temperature of the solar corona (and other ionized atmospheres) might be explained by the presence of an enhanced population of suprathermal particles. The important consequences of suprathermal electrons on the heat flux are described in Section 11.4. Section 11.5 describes the influence of the suprathermal electrons on the electric potential, leading to a higher escaping flux from planetary ion-exospheres and an increased acceleration of the solar wind. Finally, the concluding remarks are given in Section 11.6, while three general science questions for future analyses are posed in Section 11.7.

11.3 Coronal Heating by Velocity Filtration Due to Suprathermal Electrons

Suprathermal particles can be at the origin of the observed temperature increase with altitude in the stellar, solar, and planetary atmospheres. Scudder (1992a,b) pioneered the importance of these suprathermal particles in the heating of solar/ stellar atmospheres and demonstrated that the heating necessary to produce the steep temperature inversion in the solar transition region and corona can be achieved without the deposition of wave or magnetic energy. The suprathermal tails cause an increase of temperature in the transition region and low corona by velocity filtration, i.e., without any energy addition (Scudder, 1992b,c). Indeed, starting from isotropic kappa distributions, given by Eq. (11.1), at a given radial distance r_0, the Jeans theorem tells us that the distribution at radial distance r, where the (attractive) potential energy has increased by $\Delta W(r) = W(r) - W(r_0)$, is a kappa distribution of the same kappa index, with temperature increasing as

$$T(r) = T(r_0) \cdot \left[1 + \frac{\Delta W(r)}{\left(\kappa - \frac{3}{2} \right) k_B T(r_0)} \right], \tag{11.2}$$

and density decreasing in proportion of $T^{1/2-\kappa}$. In the solar corona, we have $\Delta W(r) = m\Phi_g(r) + Ze\Phi(r) - m\Phi_g(r_c) - Ze\Phi(r_c)$, where $\Phi_g(r) = -MG/r$ is the gravitational potential, $\Phi(r)$ is the electrostatic potential, and $r_0 = r_c$ corresponds to the top of the chromosphere. When kappa tends to infinity, the Maxwellian isothermal property is recovered: $T = T(r_c)$.

Electrons and protons are the major species. So they determine the electrostatic potential, which is simply given by the Pannekoek-Rosseland expression if the distributions are the same for electrons and protons. The temperature increase will concern the particle species for which an enhanced population of suprathermal particles is considered, and it is then given by Eq. (11.2). Let us take as an example the case of $T_0 = 10^4 K$ and kappa distributions with $\kappa = 5$ for all of the different particles (electrons and ions) in the solar corona. Fig. 11.3 illustrates the temperature increase obtained in such a case. The temperature increase is similar for electrons and protons, since they are the major species that determine the electrostatic potential. On the contrary, the minor ions have a higher temperature increase because the total potential W depends on the mass

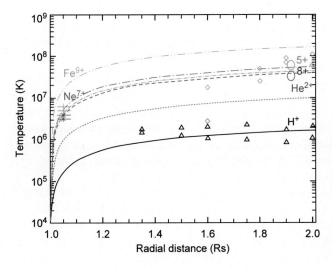

FIGURE 11.3
Temperature increase obtained for Fe^{9+} (*gray*), Ne^{7+} (*magenta*), O^{5+} (*green*), O^{8+} (*blue*), He^{2+} (*red*), and H^+ (*black*) in the solar corona when kappa distributions with $T \sim 10^4 K$ and $\kappa = 5$ are assumed at the top of the chromosphere. The temperature profiles are compared to Solar and Heliospheric Observatory (SOHO) temperature observations. Source: Pierrard and Lamy (2003).

m and the charge Z of the particles. Fig. 11.3 illustrates the temperature increase obtained for Fe^{9+}, Ne^{7+}, O^{5+}, O^{8+}, He^{2+}, and H^+, showing comparisons with temperature measurements deduced from Solar and Heliospheric Observatory (SOHO) (SUMER and UVCS) (Esser and Edgar, 2000; Tu et al., 1998). The same colors are used for the respective ion measurements.

Such temperature inversions can also apply to other stars (Scudder, 1992a, 1992b). A simple way of understanding the temperature increase via velocity filtration is to start with a velocity distribution made up of the superposition of a cold (c) and a hot (h) Maxwellian population, with $n_c >> n_h$ and $k_B T_c < \Delta W < k_B T_h$, so that the mean temperature at the base $T = (n_c T_c + n_h T_h)/(n_c + n_h) \approx T_c$ (Pierrard, 2012c). Because of Liouville's theorem, the temperature of each Maxwellian population remains constant with altitude. However, since the cold population does not have enough kinetic energy to overcome the attractive potential ΔW, it remains at low altitudes, whereas the hot population, which is barely affected by the potential, can reach high altitudes. As a result, the average temperature at high altitudes is roughly T_h (Meyer-Vernet, 2007).

This temperature increase with altitude via velocity filtration in an attractive potential was demonstrated graphically by Scudder (1992a) for an energy distribution whose logarithm has a slope that becomes flatter at a higher energy (contrary to the Maxwellian constant slope). This behavior can also be derived analytically for a distribution made of a sum of any number of Maxwellians (Meyer-Vernet et al., 1995). Furthermore, it can be shown that this property holds not only for the conventional kinetic temperature defined from the second moment of the velocity distribution but also for any normalized scalar moment of the distribution (Meyer-Vernet, 2001).

Similar properties also hold in planetary environments, where kappa velocity distributions are ubiquitous, for example, in the Earth's plasmasphere (e.g., Pierrard and Stegen, 2008) and the exospheres of Jupiter and Saturn (Pierrard, 2009).

In such fast rotating magnetized planets, centrifugal force plays an important role far from the planet, where the contribution of the centrifugal force projected along the magnetic field far exceeds the contribution of the gravitational force. In that case, the particles are pulled toward the centrifugal equator and form a plasma torus. The problem of calculating the torus latitudinal structure is similar to the calculation of the altitude profile of a gravitationally bound atmosphere, but in that case the gravitational term in the energy ΔW is replaced by $3m\Omega^2 z^2/2$, where Ω is the (angular) rotational speed, and z is the distance to the equator. With a kappa velocity distribution, velocity filtration thus produces in that case an increase in temperature with latitude (Chapters 3, 4, and 5; Meyer-Vernet, 2001; Livadiotis, 2015b):

$$
T(z) = T_0 \cdot \left[1 + \frac{\Delta W(z)}{\left(\kappa - \frac{3}{2} \right) k_B T_0} \right],
\tag{11.3}
$$

with an anticorrelation between density and temperature

$$
n \propto T^\nu, \text{ or } T \propto n^{a-1}, \quad \text{with} \quad \nu \equiv \frac{1}{a-1} \approx - \left(\kappa - \frac{1}{2} \right),
\tag{11.4}
$$

where a denotes the polytropic index. Such a temperature increase as density decreases with increasing latitude has been observed in the plasma torus associated to the satellite Io of Jupiter (Meyer-Vernet et al., 1995). The observed polytropic relation has been used to derive the value of $\kappa \approx 2.4$ for the electron distribution (Fig. 11.4), which was later confirmed by independent observations with Cassini (Steffl et al., 2004) and Hubble (Rutherford et al., 2003).

11.4 Heat Flux

As first noted by Shoub (1983), the nonthermal distributions associated to the weakness of collisions have major consequences on the plasma heat flux, which cannot be estimated from the classical collisional Spitzer-Härm value, even for small values of the Knudsen number typical of the transition region and the corona, $\geq 10^{-3}$ (e.g., Landi et al. 2014; references therein). The heat flux is mainly transported by electrons because of their small mass (and therefore large speeds), and even a weak power law tail in the electron VDF can allow heat to flow up a radially directed temperature gradient (Dorelli and Scudder, 1999). The electron temperature reaches a peak around $2R_s$, in good agreement with coronal brightness measurements obtained during solar eclipses (Pierrard et al., 2014).

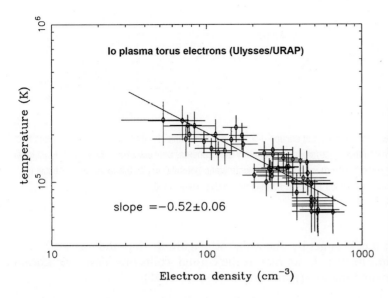

FIGURE 11.4
Electron density and temperature measured in situ in the Io plasma torus aboard Ulysses with the method of quasithermal noise spectroscopy (Meyer-Vernet and Perche, 1989). From the slope of the best-fit line, one deduces that
$\kappa = \frac{1}{2} - 1/\text{slope} \cong 0.5 + 1/0.52 \cong 2.4$.
(Here we ignore the small contribution of the potential degrees of freedom, as shown in Chapter 3.) Adapted from Meyer-Vernet et al. (1995).

Fig. 11.5 shows the total heat flux in the solar corona as a function of the kappa index, derived from a numerical simulation taking collisions into account (Landi and Pantellini, 2001). The heat flux is normalized to $10^{-3} m_e n_0 \theta_0^3$, where n_0 and $\theta_0 = (2k_B T_0/m_e)^{1/2}$ are the electron density and thermal speed at the base of the simulation. For $\kappa \to \infty$ (Maxwellian), the heat flux approaches the classical Spitzer-Härm value. As the power law tail increases (κ decreases), the heat flux changes, and for $\kappa < 4$ its sign becomes opposite, with heat flowing upward, from cold to hot regions. For $\kappa = 3$, heat flows not only from cold to hot, but

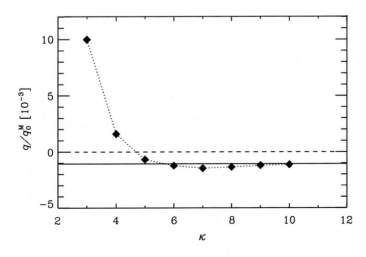

FIGURE 11.5
Heat flux in the solar corona from a numerical simulation with different values of κ (*dotted*), compared to the classical Spitzer-Härm heat flux (*solid line*) corresponding to Maxwellian distributions. Adapted from Landi and Pantellini (2001).

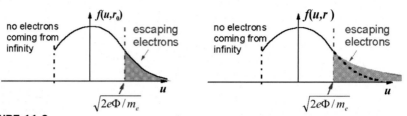

FIGURE 11.6

Suprathermal electrons increase the escape flux and thus the ambipolar electric field and the velocity of the particles escaping from solar/stellar or planetary atmospheres. Only electrons with a velocity u higher than the escape velocity determined from the electrostatic potential Φ_0 at the exobase can escape (escaping electrons are shown in *gray*). Adapted from Meyer-Vernet (2007).

also the heat flux is larger by one order of magnitude than the classical value. The simulation further shows that in the corona, collisions have a significant effect on velocity filtration (Landi and Pantellini, 2001).

Consider now the solar wind proper, where the Knudsen number is much larger (of the order of magnitude of unity). In this case, the heat flux is expected to be much closer to the so-called noncollisional value (Hollweg, 1974) than to the Spitzer-Härm one. The noncollisional heat flux Q_{NC} is produced by the tail of the electron velocity distribution that can escape outward, whereas no electron is coming from infinity (Fig. 11.6, left). With a Maxwellian velocity distribution, Q_{NC} is related to the electron density n, the mean speed u, and the temperature T by $Q_{NC} = \alpha\, n\, u \cdot \left(\frac{3}{2} k_B T\right)$, where α is of an order of magnitude unity in the solar wind at 1 AU (Hollweg, 1974).

This noncollisional heat flux is of a similar order of magnitude as the Spitzer-Härm value in the solar wind at 1 AU (Meyer-Vernet, 2007), but both vary very differently with the plasma properties since the noncollisional heat flux is proportional to n and u. It has been shown from data and simulations that the solar wind heat flux behaves as the noncollisional heat flux when the Knudsen number exceeds 0.01 (Landi et al., 2014). Fig. 11.6 shows how an excess of suprathermal electrons increases the heat flux and the wind speed. (This is explained in detail in Section 11.5.)

11.5 Influence of Suprathermal Electrons on the Acceleration of Escaping Particles

The simplest models applicable to winds that are not dominated by collisions are the exospheric models (Lemaire and Pierrard, 2001; Lemaire, 2010; references therein). The particle velocity distributions are assumed at the base of the wind, and their evolution with distance is calculated from the Jeans theorem, i.e., neglecting collisions. Contrary to the fluid models, the heat flux is not assumed a priori but rather is calculated from the third moment of the velocity distributions. The method is fully self-consistent since the electrostatic potential Φ is calculated everywhere from charge quasineutrality and zero charge flux (Lemaire and Scherer, 1971).

Exospheric models based on anisotropic kappa distributions have been developed to study the outflow of particles (electrons, protons, and other ions) from planetary and stellar exospheres (Pierrard and Lemaire, 1996a). Such models show that suprathermal electrons generate large ambipolar electric fields and heat flux along open magnetic flux tubes in solar/stellar coronae and in planetary ionospheres and thus contribute significantly to solar and stellar wind acceleration, outflow from planetary ionospheres, and possibly even exoplanetary atmospheric loss.

Electrons are always attracted to the planet or star by electrostatic potential, so that only energetic electrons can escape. An enhanced population of energetic electrons thus increases the escaping flux, without modifying much the density that is mainly contributed by the core electrons, as illustrated on Fig. 11.6. In order to keep the flux of protons everywhere equal to the electron flux, the electrostatic potential increases; this contributes to accelerating the particles to high-bulk velocities. On the contrary, suprathermal protons do not have much of an effect on the solar wind escaping flux and the electrostatic potential, because the positively charged particles are pushed outside by the electric force that already dominates at low radial distances. Thus most protons of the distribution (not only the suprathermal ones) contribute to the flux escaping in the solar wind. That is why the presence of suprathermal electrons has much more important consequences than suprathermal ions, concerning the escape of particles (not only of the electrons themselves but also of the protons and other ions) from planetary and solar atmospheres.

These models enable us to estimate analytically the value of the potential and wind speed as a function of the coronal temperature and suprathermal index κ as well as to explain the observed variation of electron temperature with distance (Pierrard and Lemaire, 1996a; Meyer-Vernet and Issautier, 1998; Meyer-Vernet, 1999; Issautier et al., 2001).

An example of solar wind profiles obtained with a Maxwellian (blue) and a kappa ($\kappa = 2$) electron distribution is illustrated in Fig. 11.7 for a temperature of 10^6K at the exobase. The electric potential is repulsive for the protons and other positive ions, while it is attractive for the electrons, so that the presence of suprathermal electrons has a high influence on the final bulk velocity of the solar wind in contrast to what would be obtained by assuming suprathermal protons.

In such models, the velocity at 1 AU depends on the suprathermal tail of the escaping population and thus on the kappa value. A low altitude of the exobase also accelerates the wind. This dependence is illustrated in Fig. 11.8 for solar wind models.

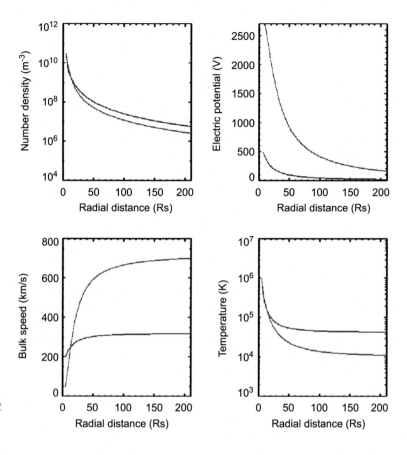

FIGURE 11.7

Density, electric potential, bulk speed, and temperature profiles of solar wind electrons for a Maxwellian velocity distribution function (*blue*) and a kappa velocity distribution function with $\kappa = 2$ (*red*). Source: Pierrard (2012a).

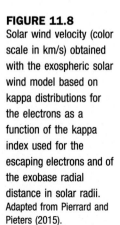

FIGURE 11.8

Solar wind velocity (color scale in km/s) obtained with the exospheric solar wind model based on kappa distributions for the electrons as a function of the kappa index used for the escaping electrons and of the exobase radial distance in solar radii. Adapted from Pierrard and Pieters (2015).

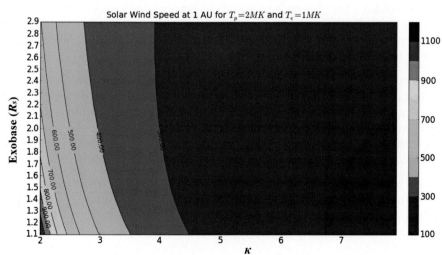

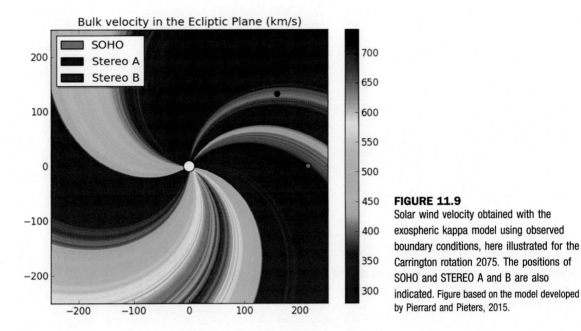

FIGURE 11.9
Solar wind velocity obtained with the exospheric kappa model using observed boundary conditions, here illustrated for the Carrington rotation 2075. The positions of SOHO and STEREO A and B are also indicated. Figure based on the model developed by Pierrard and Pieters, 2015.

The spiral shape of the magnetic field is also taken into account (Pierrard et al., 2001a), as illustrated in Fig. 11.9, and using realistic boundary conditions, the model can be used for the prediction of space weather (Pierrard and Pieters, 2015).

The kappa ion–exosphere model (Pierrard and Lemaire, 1996a) has been adapted and improved to study the solar wind (Maksimovic et al., 1997a; Pierrard et al., 2001a; Zouganelis et al., 2004, 2005), the solar corona (Pierrard and Lamy, 2003; Pierrard et al., 2004), the terrestrial ionosphere and plasmasheet (Khazanov et al., 1998), the polar wind of the Earth (Lemaire and Pierrard, 2001; Barghouthi et al., 2001; Tam et al., 2007) and of other planets like Jupiter and Saturn (Pierrard, 2009), the terrestrial auroral regions (Pierrard, 1996; Pierrard et al., 2007), the plasmasphere (Pierrard and Borremans, 2012a), and the radiation belts (Pierrard and Borremans, 2012b), among others, since suprathermal electrons are observed in all of these space plasmas. The exospheric model has been generalized to arbitrary potential energy distributions (Lamy et al., 2003). Anisotropic kappa distributions have also been used to model space plasmas (Chapters 4 and 10; Lazar et al., 2012c; Livadiotis, 2015a).

Fig. 11.10 illustrates the flux escaping from the Saturnian ionosphere for different kappa distributions and for a Maxwellian. While the flux is negligible at low temperatures for a Maxwellian, it becomes very high with kappa distributions due to the suprathermal particles. In this case, the polar wind

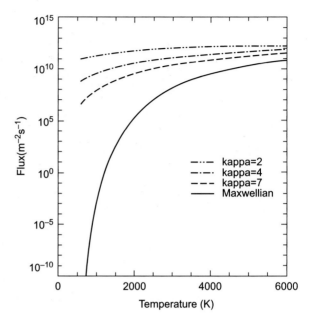

FIGURE 11.10
Example of particle fluxes (polar wind) escaping from the Saturnian ionosphere for different values of the kappa parameter. Source: Pierrard (2009).

escaping from the giant planets becomes an important source of their magnetosphere.

11.6 Concluding Remarks

In conclusion, the electron VDFs in space plasmas are observed to have ubiquitous suprathermal tails, well fitted by kappa distributions, which suggests a universal mechanism for their formation. The presence of such suprathermal electrons has major consequences for the heating and acceleration of space plasmas. These effects have to be taken into account in the models. This concerns coronal heating, acceleration of the solar wind, flux escape from planetary and solar exosphere, and characteristics of magnetospheric plasmas of the Earth and other planets. The anisotropy in the VDF (especially the strahl electrons in the solar wind) plays an important role in the escape flux, since the suprathermal population is even higher in the direction parallel to the magnetic field than in the perpendicular and other directions (halo electrons). This is directly related to the acceleration of particles escaping from stellar and planetary exospheres, especially the solar and polar winds. Superhalo electrons are also observed at energies >2 keV in the solar wind. Further studies of these different populations should help investigators to determine the origins and the physical processes generating these suprathermal electrons ubiquitously observed in space plasmas.

11.7 Science Questions for Future Research

Future analyses and observations need to address the following questions:

1. Do kappa-distributed electrons contribute to coronal heating?
2. Is the value of the kappa index affected by the Knudsen number?
3. Which mechanisms determine the kappa index of electrons?

Acknowledgment

This research was subsidized by the Scientific Federal Policy in the framework of the program Interuniversity Attraction Pole for the project P7/08 CHARM.

CHAPTER 12

The Kappa-Shaped Particle Spectra in Planetary Magnetospheres

K. Dialynas [1], C.P. Paranicas [2], J.F. Carbary [2], M. Kane [3], S.M. Krimigis [1,2], B.H. Mauk [2]

[1]Academy of Athens, Athens, Greece; [2]Johns Hopkins University Applied Physics Laboratory, Laurel, MD, United States; [3]Harford Research Institute, Bel Air, MD, United States

Chapter Outline

12.1 Summary 482
12.2 Introduction 483
12.3 Measuring and Interpreting the Kappa Distribution in Space Plasmas 487
12.4 Kappa Distribution in the Magnetospheres of the Gas Giant Planets 488
  12.4.1 Jupiter 488
    12.4.1.1 In General 488
    12.4.1.2 Energetic Ion Spectra 488
    12.4.1.3 Particle Energization Processes and Anisotropies 490
    12.4.1.4 Energetic Particle Moments 492
    12.4.1.5 Plasma Sources and Aurora 493
  12.4.2 Saturn 494
    12.4.2.1 In General 494
    12.4.2.2 Ion Spectra 494
    12.4.2.3 Neutrals 497
    12.4.2.4 Electron Spectra 498
  12.4.3 Uranus 500
    12.4.3.1 In General 500
    12.4.3.2 Plasma Energy Spectra 501
    12.4.3.3 Energetic Ion Spectra 501
    12.4.3.4 Whistler Waves 503
  12.4.4 Neptune 503
    12.4.4.1 In General 503
    12.4.4.2 Energetic Ions 504
12.5 Kappa Distribution in the Magnetospheres of the Terrestrial Planets 506
  12.5.1 Earth 506
    12.5.1.1 In General 506
    12.5.1.2 "Quiet" Plasma Sheet Spectra 507

Kappa Distributions. http://dx.doi.org/10.1016/B978-0-12-804638-8.00012-7

12.5.1.3 "Disturbed"
Plasma Sheet
Spectra 510
12.5.1.4 Middle
Magnetosphere
and
Substorms 511
12.5.1.5 High-Latitude
Spectra 512
12.5.1.6 Magnetosheath
Spectra 513
12.5.2 Mercury 514
12.5.2.1 In General 514

12.5.2.2 Energetic
Electron
Bursts 515
12.5.2.3 MESSENGER
Measurements
515
**12.6 Are Kappa Distributions
Useful for Magnetospheric
Research? 518**
12.7 Concluding Remarks 521
**12.8 Science Questions
for Future Research 521**
Acknowledgments 522

12.1 Summary

The kappa distribution function has become an essential form for analyzing trapped distributions of charged particles in space. It is loosely based on the Maxwellian distribution and can be thought of as a generalization of it, specifically with the capacity to describe higher fluxes at high energy, i.e., exhibiting a power law tail at the highest energies that roll over at lower energies to become Maxwellian at the lowest energies. Given that the charged particle populations themselves can have many different kinds of sources (solar wind, atmosphere, volcanic moons, etc.) and sinks (collisions with ambient neutrals or moon or ring surfaces, reconnection, etc.) and move under the influence of strong and dynamic planetary magnetic fields, it is advantageous to have a function flexible enough to capture the essential properties of entire particle distributions, e.g., plasma (<1 keV) and energetic ions (>5 keV). However, some distributions are well modeled by combinations (or modifications) of kappa distributions, since the most accurate quantitative characterizations often need even more flexibility than afforded by the standard kappa function. In this chapter we will review the use of kappa distributions to describe charged particles in the magnetospheres of the gas giant planets (Jupiter, Saturn, Uranus, and Neptune) and the inner planets (Earth and Mercury) that have allowed the determination of physical quantities of the plasma and energetic particle spectra such as temperatures, densities, pressures, spectral indices, and convection bulk velocities for one or more species, as available. These are critical quantities for understanding the structure and dynamics of magnetospheres. The dependence of these quantities on spatial and magnetic field parameters has rendered the kappa distribution a powerful diagnostic tool for understanding processes in space plasma physics. We also consider the degree to which the kappa distribution succeeds or fails to characterize planetary particle distributions under different conditions.

Science Question: What are the features of the ion/electron kappa distributions in magnetospheres?

Keywords: Energetic neutral atoms; Energetic particles; Magnetospheres; Particle detectors; Plasma processes.

12.2 Introduction

Some structures associated with the magnetospheres of the planets include the following: the bow shock (the first dynamical surface that the solar wind encounters as it flows toward the planet); the magnetosheath (the region primarily populated by shocked solar wind); the magnetopause (a surface that separates the solar wind particles from the magnetosphere's population); and the magnetotail (a reservoir of plasma and energy that is formed by the transfer of tangential stress between the solar wind and the planetary plasma environment (Southwood, 2015), extending the magnetosphere in the antisunward direction) (see Fig. 12.1). Kivelson and Russell (1995) is a comprehensive basic resource on magnetospheric systems. Sources of magnetospheric particles include the solar wind, atmospheres and ionospheres, cosmic ray and other particle interactions with surfaces such as rings and moons, and active satellites such as Io and Enceladus. Composition and spectral analyses in these diverse space environments are the key to establishing the sources and losses of magnetospheric particles as well as the physics that governs these particles motions. (Books that discuss both particle distributions and particle dynamics are: Baumjohann and Treumann (1997a,b), Kallenrode (1998), and Gurnett and Bhattacharjee (2005).)

Since it is impossible to solve the equations of motion for every single charged particle, a statistical approach is used to describe the plasma as a whole by taking advantage of the so-called "kinetic theory" (described in detail in Choudhuri, 1998) that leads to the development of velocity distributions. Among the observed/measured velocity distributions, Maxwellians have been widely used to describe the properties of space plasmas. However, Binsack (1966), Olbert (1968), and Vasyliunas (1968), in explaining the particle spectra from IMP-1, underscored the need for a different function that would resemble a Maxwellian but include an additional κ-parameter, providing a measure of the departure of the distribution from its Maxwellian character, forming a function that later became known as the kappa distribution function (Chapter 1; Vasyliunas, 1968, 1971).

If we assume that phase space density distribution f of a collection of particles depends only on the particle speed, u, so that $f = f(u)$, then the kappa distribution function in differential intensities, i.e., $j = f(u)u^2/m$, can be written following the notion of Vasyliunas (1968):

$$j(\varepsilon) = n \cdot w_0 \frac{2}{\sqrt{\pi}} \cdot \frac{\Gamma(\kappa + 1)}{\kappa^{3/2} \Gamma\left(\kappa - \frac{1}{2}\right)} \cdot \left[1 + \frac{1}{\kappa} \cdot \frac{\varepsilon}{E_0}\right]^{-\kappa - 1} \varepsilon, \qquad (12.1)$$

where m is the particle mass, $\varepsilon = \frac{1}{2} mu^2$ is the kinetic energy, n is the number density, w_0 is the most probable speed $\left(E_0 = \frac{1}{2} mw_0^2\right)$, κ is the exponent of the distribution at high velocities, and $\Gamma(x)$ is the gamma function; both w_0 and E_0 depend on the particle temperature, i.e., $E_0 = k_B T \cdot (\kappa - \frac{3}{2})/\kappa$.

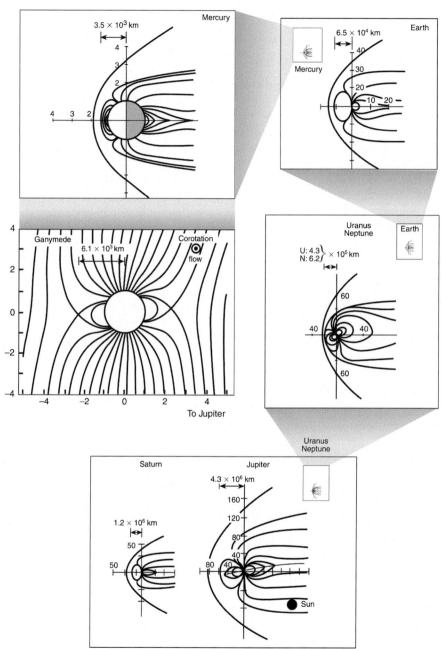

FIGURE 12.1
Comparative sketch of planetary magnetospheres. Adapted from Paranicas et al. (2005); reproduced from Williams et al. (1998).

The magnetospheric particle distributions are ideally described in a reference frame in which the plasma flow is zero. However, it is often necessary to describe the distribution in another frame, such as a spacecraft frame, using a convected kappa distribution function that, after some basic algebra, can be expressed in terms of the particle (u) and bulk plasma (u_b) speeds as well as the polar angle (ϑ) between the measured u and u_b (e.g., Krimigis et al., 1983; Kane et al., 1995; Livadiotis and McComas, 2013a). (Note that in this chapter, we will refer to the bulk flow of the plasma as seen in an inertial frame as the "convective motion," in keeping with the standard magnetospheric usage, thus "convected distribution" means its description in the inertial reference frame.)

$$j(u, \vartheta) = \frac{n\theta^{-3}}{m}\left[\pi\left(\kappa - \frac{3}{2}\right)\right]^{-\frac{3}{2}}\frac{\Gamma(\kappa + 1)}{\Gamma\left(\kappa - \frac{1}{2}\right)} \cdot \left[1 + \frac{1}{\kappa - \frac{3}{2}} \cdot \frac{u^2 - 2uu_b\cos\vartheta + u_b^2}{\theta^2}\right]^{-\kappa - 1}u^2,$$

$$(12.2)$$

where $\theta \equiv \sqrt{2k_BT/m}$ (see also Chapter 4).

When the particle speeds (e.g., measured in the reference frame where the plasma is at rest) are significantly higher than the convection bulk speed ($u \gg u_b$), a situation that is very common in magnetospheric plasmas, the spectrum approaches a power law in energy, $j \sim \varepsilon^{-\kappa}$ (Livadiotis et al., 2011). Typically, the kappa (κ)-index must take sufficiently large values, greater than the critical value $\kappa_c = 3/2$, where the distribution function collapses (Livadiotis and McComas, 2009, 2010a) and cannot be defined. All different forms of kappa distributions reduce to a Maxwellian form as $\kappa \to \infty$. The transition between the Maxwellian term that dominates at low energies and the power law that dominates at high energies occurs smoothly (both the distribution and first derivative are continuous) at $\sim (\kappa + 2)k_BT$ (Roelof et al., 1976; Krimigis et al., 1981a; Krimigis and Roelof, 1983). As shown in Fig. 12.2 and discussed later in this chapter, the kappa distribution can also be used to explore the particle anisotropies.

Paranicas et al. (1999) introduced a modified version of the kappa distribution function to characterize the energetic ion spectra near Ganymede, while Mauk et al. (2004) used the same function for the ion measurements within the Jovian magnetosphere, sampled by the Energetic Particle Detector (EPD) (Williams et al., 1992) experiment onboard Galileo:

$$j(\varepsilon) = C \cdot \frac{\left[(\gamma_1 + 1)k_BT_\gamma + \varepsilon_1\right]^{-\gamma_1 - 1}}{1 + (\varepsilon/e_t)^{\gamma_2}} \cdot \varepsilon,$$

$$(12.3)$$

where j is the differential intensity, ε is the measured energy, and $\varepsilon_1 \equiv \left(\sqrt{\varepsilon} + \sqrt{\varepsilon_b} - 2\sqrt{\varepsilon}\sqrt{\varepsilon_b}\cos\vartheta\right)^2$ is the energy measured with respect to the comoving reference frame of the plasma flow, and $\varepsilon_b = \frac{1}{2}mu_b^2$ ($\varepsilon_1 = \varepsilon$, if all measurements are made in the reference frame of the plasma flow). The numerator of this equation is a form similar to the standard kappa

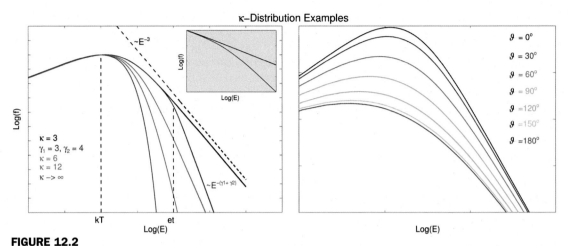

FIGURE 12.2

Left: The kappa distribution functions (numerator of Eq. (12.3) with $\varepsilon_1 = E$) for increasing κ. The *dashed line* represents a power law with $\gamma = 3$, which is shown for comparison. The *purple line* represents Eq. (12.3) where a softening break at e_t occurs. The inlay highlights that the transition of a kappa function with a slope γ_1 occurs smoothly to a tail with a slope $(\gamma_1 + \gamma_2)$. Right: Flux of kappa distributions for different polar angles ϑ using Eq. (12.2).

distribution function, in which T_γ connects to the actual temperature T of the particular system under consideration, $T_\gamma = T(\kappa - \frac{3}{2})/(\kappa + 1)$ (Livadiotis and McComas, 2009) and $\kappa = \gamma_1$ (e.g., Dialynas et al., 2009). The denominator is an additional term to characterize any softening break in slope, resulting in a $\gamma_1 + \gamma_2$ power law contribution at very high energies $(\varepsilon >> \varepsilon_1)$, with the transition energy given roughly by the parameter e_t. A comparative examination of radiation belt and energetic ion ring current particles has shown that the additional complexity afforded by equations like Eq. (12.3) is needed, particularly for the most intense observed spectra during active times (e.g., Mauk and Fox, 2010), when it might be expected that distributions have not achieved equilibrium states. Other functional forms of the kappa distribution function, as well as the results obtained by its use in planetary magnetospheres, will be explained in the following sections.

This chapter is structured as follows Section 12.3 includes a brief presentation on kappa distributions used to interpret the measured particle distributions. Following the work of Carbary et al. (2014), Section 12.4 describes the use of kappa distribution functions in the magnetospheres of the gas giant planets, while Section 12.5 deals with the use of kappa distributions on Mercury and Earth. Section 12.6 discusses the advantages and limitations of the kappa distribution and gives an overview of distributions beyond the classical kappa form. Finally, the concluding remarks are given in Section 12.7, while three general science questions for future analyses are posed in Section 12.8.

12.3 Measuring and Interpreting the Kappa Distribution in Space Plasmas

In the present chapter we discuss the kappa distribution functions that emerge in either the low-energy charged particle content of the magneto-spheres (plasma energies of a few eV up to ∼ 1 keV) and the energetic particles (typically >1 keV), or Energetic Neutral Atoms (ENAs), from the eV to the keV range. The kappa distribution function, represented by Eq. (12.2), is a function that has been used to represent a population of particles convecting past a spacecraft, assuming that particles are gyrotropic and isotropic in their rest frame. Those conditions are not always met in space plasmas. For example, if there is a density gradient present, neither condition will be met. In those cases, one must use caution interpreting the values of parameters calculated from the data.

Although most of the instruments onboard spacecraft that measure the energy spectrum of particles provide good spatial coverage of ∼ 2π sr or more, with high-energy resolution, usually consistent with Poissonian fluctuations ($\Delta\varepsilon/\varepsilon \sim \varepsilon^{-0.5}$), and high-time resolution at the order of ∼ 1 s, they have a limited range; this will place constraints on the calculation of the parameters of the distribution and must be taken into account when evaluating those parameters. Particle detectors such as the Ion and Neutral Camera (INCA, Cassini) (Krimigis et al., 2004; Mitchell et al., 1998) and the Low Energy Charged Particle (LECP, Voyager) (Krimigis et al., 1977) begin to become less efficient at the very low energy ends of their range. In principle, this can affect the quantity $k_B T$, so that in cases where low-energy channel efficiencies are not known with accuracy, such channels are excluded when calculating kappa distributions that fit the data, unlike the cases of LECP and INCA. Further, characteristic temperatures may fall outside the range of the detectors used, in which case they are extrapolations of trends seen in the detector data.

To properly determine the function given in Eq. (12.2), it is necessary to have measurements in a variety of look directions relative to the convection frame. However, this condition is frequently not met, so approximations need to be made. If there is a good sampling of the data in the direction that is nearly perpendicular to the flow direction, for example, it is possible to reconstruct the spectrum in the plasma rest frame with errors only at a lower order.

Detectors may be able to differentiate between species, but many particle detectors used in space missions cannot do so at all energies. Different species have their own unique distributions, with heavy ions tending to be hotter than lighter ions. Therefore detectors that cannot differentiate between ion species will, if there are multiple species present, be fit best by summing model spectra of the species known to exist. An extended discussion of particle detectors and their capabilities can be found in Carbary et al. (2014) and references therein.

12.4 Kappa Distribution in the Magnetospheres of the Gas Giant Planets

Researchers have used many different kinds of methods to garner scientific information about the outer planets. Jupiter's magnetosphere has been visited most often, and Saturn's has had more than a decade of Cassini data. Only the Voyager 2 spacecraft has visited Uranus and Neptune. Remote observations have added new knowledge, including the discovery of planetary moons. Here we describe the deviations from the Maxwellian plasma picture, taking note of processes that make the outer planet magnetospheres compelling to us. These include moons as significant sources of plasma, moons and rings as sinks of plasma, a range of obliquities and relationships between the spin and dipole axes, and great variations in radiation belts and their sources.

12.4.1 Jupiter

12.4.1.1 In General

Jupiter is the largest planet in the Solar System. It has a mean equatorial radius of 71,492 km and the strongest global magnetic field, with a dipole moment of $4.28\,G\,R_J{}^3$ (G: gravitation constant; R_J: Jovian mean equatorial radius). The magnetic (dipole) axis is tilted 9.6 degrees with respect to the spin axis. The planet spins with a period of 9.925 h (9 h, 55 min, 30 s), so the magnetic field and charged particles entrained in it exhibit this period when viewed by a stationary (or slowly moving) observer. The rapid spin of the planet imparts centrifugal forces on the trapped cold plasma that cause the magnetosphere to distort from a simple dipole to a magnetodisk that, driven by the dipolar tilt, oscillates up and down with the planetary period. Among Jupiter's four major moons, the innermost, Io, is tidally heated by gravitational interactions between Jupiter and another moon, Europa, which internally heat Io and generate massive volcanic activity. The Io volcanoes release volcanic products, mostly sulfur and oxygen, into the magnetosphere at a rate of ~ 1 Mg/s. Ionized by solar radiation and energetic particles, the sulfur and oxygen become the primary source of ions and electrons in the magnetosphere. Unlike Saturn, where neutral species dominate, the plasma dominates in the Jovian magnetosphere. By processes yet to be understood, the charged particles are accelerated to energies of tens of keV or higher. The energy distributions of these particles can be characterized as Maxwellian, power law, or kappa distribution.

12.4.1.2 Energetic Ion Spectra

Using Pioneer data (Van Allen et al., 1974), Baker and Van Allen (1976) were the first to use a kappa or kappa-like distribution to characterize the energetic particles, although Fillius and McIlwain (1974) and McIlwain and Fillius (1975) advocated a similar distribution function. Jupiter's very energetic electrons (50 keV $< \varepsilon <$ 50 MeV) were fitted to a distribution of the form:

$$dj/d\varepsilon = K \cdot [1 + \varepsilon/H]^{-r} \cdot \varepsilon^{-1.5}, \tag{12.4}$$

where K and H are constants, ε is energy, and r is a κ-like exponent. This type of spectra is characterized by a "bend" in the middle of the energy range. The energy scale H (which can be related to the temperature) varied from ~ 5 MeV in the inner magnetosphere to $\sim 10-35$ MeV in the outer magnetosphere, and the r index varied from ~ 2 (hard) at low magnetic latitudes to ~ 6 (soft) at higher latitudes. The oscillation of the magnetodisk causes the r index to vary periodically. The Pioneer plasma data were interpreted in terms of the classical Maxwellian (Frank et al., 1976), although some spectra in Jupiter's magnetosheath appeared to have non-Maxwellian characteristics (Mihalov et al., 1976).

Energetic particle observations made by the Voyager LECP instrument provided the next opportunity to use kappa-like distributions at Jupiter. The initial analyses of ion energy spectra revealed that they often exhibited a "bend" near ~ 100 keV, showing a convected Maxwellian form in energy below 100 keV and a power law at higher energies (Krimigis et al., 1979, 1981b; Hamilton et al., 1981). Densities, temperatures, and convection bulk speeds could be obtained by fitting the Maxwellian part of the spectrum. Such measurements revealed that a significant portion of the pressure lay in the high-energy tail of the spectrum, making Jupiter's plasma a "high-β" region in which the particle pressure dominated the magnetic pressure. The plasma beta β, that is, the ratio of plasma pressure divided by magnetic pressure, is important in establishing magnetospheric dynamics, especially in balancing internal pressure with external solar wind pressure and for accelerating plasma in the magnetotail. Notably, the thermal electrons at lower energies, well below those of LECP, also exhibited a high-energy, non-Maxwellian tail (Scudder et al., 1981; Sittler and Strobel, 1987). The kappa-like function of Eq. (12.4) was also employed to fit combined plasma and energetic particle data near Ganymede (Paranicas et al., 1999).

One of the first specific uses of a kappa function to describe the Jovian ion spectra was by Collier and Hamilton (1995), who also used Voyager LECP energetic ion (28 keV–60 MeV) observations. Fig. 12.3 illustrates the interpretation of these spectra as a Maxwellian-plus power law (left) and then as a kappa distribution (right). By employing the κ-fit results for both the nightside and dayside, the κ parameter was found to vary linearly with the proton temperature ($\kappa \approx b \cdot T_H + \kappa_0$, with $3 < \kappa < 6$, $5 < T_H < 35$ keV, and $b \sim 0.08$, $\kappa_0 \sim 2.86$). The observed increase of proton temperatures with decreasing distance from Jupiter is indicative of an inward radial diffusion of protons, conserving their magnetic moment (the first adiabatic invariant). (Note that there are many different relations between temperature and the κ-index found in space plasmas; as explained in Section 1.8.3.2 in Chapter 1, these do not constitute fundamental state equations, but they are generalized polytropic relations that depend on the local plasma streamlines.)

Kane et al. (1992, 1995, 1999) expanded the use of the kappa function to a form that assumes that all of the plasma particles are isotropic in a common frame of reference, Eq. (12.2), where the velocity of this frame of reference is

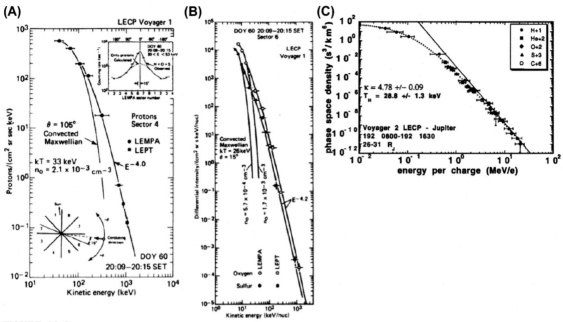

FIGURE 12.3

Sample ion spectra from the Voyager Low Energy Charged Particle (LECP) instrument at Jupiter. Typical (A) proton and (B) oxygen and sulfur spectra sampled in the outer Jovian magnetosphere, together with a convected Maxwellian fit (with temperatures $\sim 3\text{–}4 \times 10^8$ K) at low energies ($\varepsilon < 100$ keV) plus a power law at high energies. The detector's angular response was used to separate H^+ from heavy-ion components (abundance ratio for O/S ~ 3 at $\varepsilon > 100$ keV). *(From Krimigis et al. (1979).)* (C) The right spectrum incorporates both ideas into the first use of a kappa distribution to describe the particle spectra on Jupiter (Collier and Hamilton, 1995).

called the bulk plasma velocity. In this form, the particle velocity is measured in the observation (spacecraft) frame, whereas the temperature-like parameter of the distribution may be defined by the most probable particle speed (in the comoving frame), $E_0 = \frac{1}{2}mw_0^2$. (This is related to the actual temperature T with $E_0 = k_B T \cdot (\kappa - \frac{3}{2})/\kappa$.) The analysis included the dominant species in Jupiter's magnetosphere, namely, protons, oxygen ions, and sulfur ions, and was performed using both LECP data from Voyagers 1 and 2 and EPD data from the Galileo orbiter (Fig. 12.4). In this form, it was possible to determine that the bulk flow speeds were increasing with radial distance from the planet but were generally ~ 0.5 of the corotation speed (where $v_{cor} = \omega_J R_J$) (Kane et al., 1999).

12.4.1.3 Particle Energization Processes and Anisotropies

Using the convected distribution, we may also examine other parameters such as the temperature, density, and spectral index. In particular, the temperature has been shown to be ordered by radial distance in the magnetospheres of Jupiter and Saturn. In Fig. 12.5A, the radial dependence of the quantity $k_B T$ is shown for

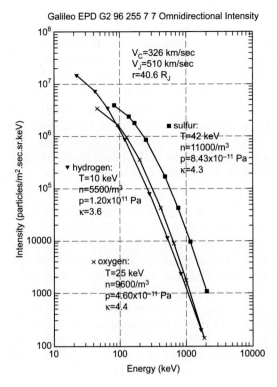

FIGURE 12.4
Sample ion spectra from the Galileo Energetic Particle Detector (EPD) instrument, showing kappa distribution fits to the three dominant species in Jupiter's magnetosphere. From Kane et al. (1999).

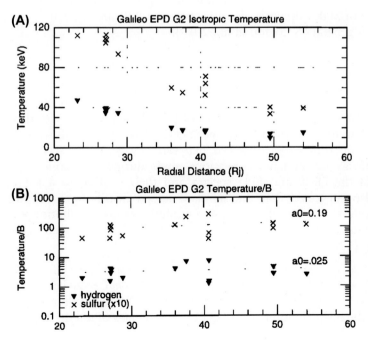

FIGURE 12.5
(A) Sulfur and hydrogen ion temperatures during the outbound pass of orbit G2 in the magnetosphere of Jupiter. Sulfur ion temperatures are significantly higher than those of hydrogen, and both species cool as they are transported outward. (B) The ratio T/B (in keV/nT), where B is the local magnetic field strength, is shown as a function of the radial distance from Jupiter in the planetary radii. The constancy of this quantity is indicative of adiabatic (no energization processes affecting the ions) transport. Adapted from Kane et al. (1999).

sulfur and hydrogen ions. Note that the sulfur temperatures are higher throughout this (nightside) region sampled by the Galileo EPD instrument during one of its orbits. Both species temperatures decline with increasing distance. In the absence of an energization source for the hot plasma ions, they should exhibit temperatures that are consistent with the conservation of the first adiabatic invariant $\mu = (1/2)mw_{0\perp}^2/B = E_0/B$ (where B is the local magnetic field strength, and, given our assumption of an isotropic distribution, $w_0 = w_{0\perp}$). For the same data, the quantity E_0/B is plotted for radial distances from Jupiter measured in planetary radii in Fig. 12.5B. For both species, this ratio is constant and was interpreted as evidence that the bulk plasma is adiabatically cooled upon outward transport (bulk flow analysis shows that the transport is predominantly outward).

Hawkins et al. (1998) conducted a similar multicomponent analysis using data from the Heliosphere Instrument for Spectrum, Composition, and Anisotropy at Low Energies on Ulysses (Lanzerotti et al., 1992) during its one pass by Jupiter. They found evidence that gradient anisotropies could be mimicking flow anisotropies. They suggested that observed flows were corotational near the plasma sheet but differed significantly at higher latitudes, although there were difficulties with using the Ulysses detectors (e.g., Krupp et al., 2001). The ultraviolet emissions of the Io plasma torus have also been interpreted in terms of a kappa distribution with $\kappa \sim 2.4$ to derive the composition of that region from ultraviolet observations from Cassini (Steffl et al., 2004). Finally, the convected kappa distribution has been applied to the ion data from the Soar Wind Around Pluto (McComas et al., 2008) detector on New Horizons when it traversed the distant Jovian magnetosheath, deriving speeds of 300–500 km/s, densities between 0.01 and 0.2 cm^{-3}, and $1.5 < \kappa < 1.8$ in general (Nicolaou et al., 2013).

12.4.1.4 Energetic Particle Moments

Using the kappa distribution, Mauk et al. (1996, 1998) have carried out simulations of the hot plasma at Jupiter and compared these with observations. Specifically, significantly lower fluxes of hot plasma were observed by Galileo's EPD detector compared to Voyager's LECP detector. Such differences could also explain differences observed between Voyager-era observations of Jupiter's aurora and similar observations made during the Galileo era. Examination of the LECP and EPD spectra and their relation to a kappa distribution also suggested that the standard kappa distribution was not adequate to model the observed spectra, so a modified kappa distribution was suggested by Mauk et al. (2004) to account for additional "bends" seen in the spectra; see Eq. (12.3).

In this new format, typical parameters (for the dominant ion S^+) were found to range from $k_B T \sim 700$ keV in the inner magnetosphere to $k_B T \sim 20$ keV in the outer, $e_t \sim 10^4$ keV, and $1.6 < \gamma_1 < 4.4$ and $0.5 < \gamma_2 < 3.1$, depending on the location. From the integral moments of Eq. (12.3), estimates of ion density and pressure were computed for the dominant ion species. Sulfur (from Io) completely dominated the ion densities within the magnetosphere, with n_S varying from ~ 0.5 cm^{-3} at 10 R_J to ~ 0.01 cm^{-3} at 40 R_J, and $n_S > n_O \geq n_H > n_{He}$ at all radial distances. From the pressure moment and the magnetic field observations, plasma β was computed to range from ~ 0.3 at 10 R_J to ~ 100 at 40 R_J, confirming the earlier measurements of this parameter and suggesting that the ion pressure actually "blows open" the outer Jovian magnetosphere, thus generating an outflowing "planetary wind."

12.4.1.5 Plasma Sources and Aurora

An anisotropic kappa distribution function has been used to model the electron distributions in the Io plasma torus and explain the latitudinal structure of the torus (Meyer-Vernet et al., 1995; Meyer-Vernet, 2001; Moncuquet et al., 2002). This distribution is sometimes called a bi-kappa distribution; however, the same name has also been used for other forms. A more accurate name may be a bidirectional kappa distribution (see Chapter 5). This distribution was originally developed to analyze microinstabilities in space plasmas (Summers and Thorne, 1991, 1992). The normalized bi-kappa has the form (e.g., Moncuquet et al., 2002; Livadiotis and McComas, 2014a; Livadiotis, 2015a):

$$
f(u_x, u_p) = \pi^{-\frac{3}{2}} \left(\kappa - \frac{3}{2} \right)^{-\frac{3}{2}} \frac{\Gamma(\kappa + 1)}{\Gamma\left(\kappa - \frac{1}{2} \right)} \theta_p^{-2} \theta_x^{-1} \left[1 + \frac{1}{\kappa - \frac{3}{2}} \left(\frac{u_x^2}{\theta_x^2} + \frac{u_p^2}{\theta_p^2} \right) \right]^{-\kappa - 1},
$$

(12.5)

where u denotes velocity, the "p" and "x" subscripts refer to components parallel and perpendicular to the magnetic field, respectively, and θ_x and θ_p are temperature-related parameters. This bidirectional implementation describes the nonequilibrium nature of the Io plasma torus wherein electrons moving parallel to the magnetic field can populate high latitudes, acting as if they were effectively "filtered" in direction (Meyer-Vernet, 2001; Moncuquet et al., 2002). These results reconcile the off-equatorial measurements of the Ulysses encounter with the equatorial observations of the Voyager.

The bidirectional kappa distribution has also found use in describing a number of other phenomena in the Jovian magnetosphere. The bi-kappa function explains how electrons can migrate from the equatorial regions to the polar regions and give rise to the aurora associated with Io (Ajello et al. 2001; Bondfond et al., 2009). This distribution was used to investigate various waves and instabilities on Jupiter as well as in other magnetospheres (Tripathi and Singhal, 2007; Pandey and Kaur, 2015a). The local stability of the Io torus against stratification-driven instabilities has been tested using the bidirectional kappa distribution, which can reproduce much of the observed data near the torus (André and Ferriére, 2008).

The exosphere of Jupiter (and Saturn) has been modeled using a kappa distribution, which enhances by several orders of magnitude the polar wind that escapes from the planet and populates the magnetosphere (Pierrard, 2009; see also Pierrard and Lazar, 2010).

12.4.2 Saturn

12.4.2.1 In General

Saturn's mean distance from the Sun is 9.58 AU (1 AU $\sim$ 150 $\times$ 10^6 km), nearly twice the distance as Jupiter. The mean equatorial radius is $R_S \sim$ 60,268 km, and the obliquity (angle between the planet's equatorial plane and its orbital plane) is 26.7 degrees, causing the planet to experience seasons. The planet has a magnetic dipole moment of 0.210 G $R_S{}^3$ or about 5% of Jupiter's dipole moment. The magnetic (dipole) axis is tilted <1 degree with respect to the spin axis and is likely offset with respect to the planet's gravitational center by $\sim$0.04 R_S (Connerney et al., 1983; Dougherty et al., 2005). The planet is a rapid rotator like Jupiter, with a fairly well-constrained rate of rotation. Periodicities measured by various instruments on the Cassini spacecraft suggest a rotation period of $\sim$10.7 h (Carbary and Mitchell, 2013). Both Jupiter and Saturn contain a magnetospheric plasma that partially rotates with the planet to great distances, thus allowing the rotation to strongly influence the dynamics of the plasma. As is the case at Jupiter, rapid rotation forces the rotating plasma to be confined near the centrifugal equator.

At Jupiter, the moon Io is the source of most of the magnetospheric sulfur and oxygen plasma, while at Saturn the ultimate source of water group plasma ions is the inner moon Enceladus (Dougherty et al., 2006; Hansen et al., 2006, 2011; Porco et al., 2006), although this generally involves a two-step process whereby plume neutrals become ionized. If the ionization processes occur in the magnetosphere, the new ions are accelerated by the planetary electric field that maintains corotation. They are described as being "picked up" by the corotating plasma. As is the case for Jupiter, the energy spectra of these particles may be Maxwellian at lower energies and power law at higher energies. In fact, the kappa distribution that includes these features is found in a variety of environments at Saturn, including the magnetospheric plasma (Krimigis et al., 1983; Dialynas et al., 2009); the neutral cloud (Jurac et al., 2002); and even the ions and electrons in Saturn's exosphere (Pierrard, 2009).

12.4.2.2 Ion Spectra

Spectral analysis of the hot plasma ions detected by LECP on Voyager 1 and 2 revealed that the spectra could be accurately represented by Maxwellians at low energies (tens of keV) and power laws at higher energies (hundreds of keV to several Mev) (Krimigis et al., 1981a, 1982). This led to a detailed analysis of energetic plasma spectra presented in Krimigis et al. (1983). The spectral form of the hot ion distributions was fit to the data by assuming that the kappa distribution in its convected form, Eq. (12.2), represented the actual observed spectra. Angular averaged data was used for the fits, and the bulk flow speed of the plasma being measured was assumed to be the rigid corotation speed $v_{cor} = \omega_S R_S$.

Angular averaged ion spectra (assumed to be protons) from LECP scans are shown in Fig. 12.6. While the fits are good in a wide variety of locations and conditions and yield a temperature of the distribution, densities of these hot ions are several orders of magnitude lower than those of the cold plasma, e.g., Frank et al. (1980). The temperature thus represents the hot ion distribution ($\sim$30 keV–2 MeV) and not the temperature of the cold ion plasma. The instrument used to obtain these spectra, LECP, measures hydrogen ions above $\sim$30 keV. Therefore interpretation of the resulting temperatures should include consideration of the low-end energy range limitations of the data.

Carbary et al. (1983) have also analyzed Voyager LECP data using the convected kappa distribution. Their approach was to use angular averaged data, and, assuming rigid corotation, derive through fitting to the data values for the distribution parameters $k_B T$ and κ. This determines the spectral shape (but not the overall scaling factor) from which the first order anisotropy may be derived. The first order anisotropy is a measure of how much the distribution in the rest frame of the plasma is altered when the distribution is measured in the spacecraft frame; it is characterized best when a large span of look directions relative to the flow are simultaneously sampled. They were able to conclude from this use of the convected kappa distribution that the observed first order anisotropies (and thus the convection speed of the plasma) were below predicted rigid corotation values inbound but at or near them outbound within Titan's orbit.

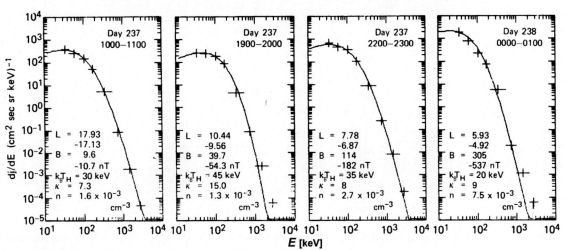

FIGURE 12.6
Sample angular averaged ion spectra, assumed to be hydrogen, from the Voyager 2 Low Energy Charged Particle (LECP) instrument in the magnetosphere of Saturn. The data points are fit to a kappa distribution (*solid curves*) for various distances from Saturn. Adapted from Krimigis et al. (1983).

In the Cassini era, particle detectors in the Magnetospheric Imaging Instrument (MIMI) instrument cluster (CHEMS, Low Energy Magnetospheric Measurement System, and INCA) have all measured energy spectra that can be well represented by the convected kappa distribution, Eq. (12.2). Measurements by the INCA detector have been used by Kane et al. (2008, 2014) to derive temperatures and the velocity of plasma in the magnetosphere of Saturn. Their approach utilized the full 3-D convected form of the distribution and fitted thousands of data points having different energies and viewing directions simultaneously to derive plasma flow velocities from anisotropies in the data. Spectra measured by INCA and other MIMI cluster instruments appear in Fig. 12.7. In the outer regions of the magnetosphere, the proton distributions are cold, so that in the energy range of the INCA detector, the distributions are nearly convected power laws, i.e., the tail of a cold kappa distribution is being sampled. Oxygen ions, the other dominant ion species measured by INCA, have a nonpower law spectrum well described by the kappa distribution.

Once particle data is fit using a kappa distribution, temperatures of the distributions are useful indicators of the nature of plasma transport. Dialynas et al. (2009) performed a statistical study of the spectra of H^+ and O^+ ions in the magnetosphere of Saturn using MIMI cluster instruments. Their results showed

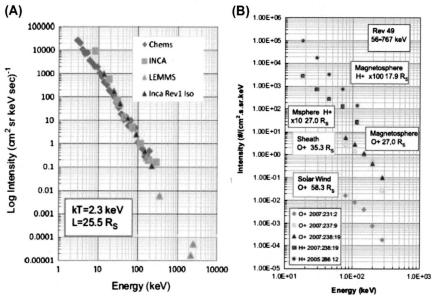

FIGURE 12.7

H^+ and O^+ spectra measured by instruments from the Magnetospheric Imaging Instrument (MIMI) cluster on the Cassini spacecraft, from Kane et al. (2014). (A) Hydrogen ions are nearly power law in the outer regions of Saturn's magnetosphere ($\sim 25\ R_S$). (B) H^+ and O^+ ion spectra, measured by the Ion and Neutral Camera (INCA) detector on Cassini. O^+ heavier ions have a higher temperature than that of H^+ ions, the uppermost spectrum in the figure, a trend seen in Galileo Energetic Particle Detector (EPD) measurements on Jupiter. Adapted from Kane et al. (2014).

(Fig. 12.8) that the H^+ temperatures increase with decreasing distance from Saturn, where the magnetic field increases in strength, consistent with energization by conservation of the first adiabatic invariant. The O^+ ions did not show any trend, so that they may be heated locally at each L-shell, a characteristic that can result from ion acceleration in rapid magnetic field reconfigurations, similar to what happens on Earth (see Section 12.5.1). Furthermore, owing to their greater lifetimes due to charge-exchange, H^+ can survive much more efficiently than O^+ throughout Saturn's magnetosphere, showing that Saturn's neutral cloud plays a key role in determining the shapes of the ion spectra, presenting a significant loss term of the $\sim 30-100$ keV ions.

The H^+ profile could also be interpreted as adiabatic energy loss with outward transport, since anisotropy studies (e.g., Kane et al., 2014) showed that radial transport is primarily outward. The kappa distribution has also been used to fit O^+ ions in the upstream region (Krimigis et al., 2009), identified by their spectra as originating in Saturn's magnetosphere and flowing upstream of the bow shock when the upstream region is magnetically connected to the magnetosphere. Brandt et al. (2010) used kappa distributions derived by Dialynas et al. (2009) to show that energetic ion injections (Mauk et al., 2005; Mitchell et al., 2009) that occur preferably in the postmidnight sector (Mitchell et al., 2005; Carbary et al., 2008) create a region of enhanced pressure that drifts around the planet and is sufficient to cause the observed magnetic field perturbations shown by Khurana et al. (2009) and explained by Andrews et al. (2008), Provan et al. (2009), and Southwood and Kivelson (2007). Furthermore, injections together with continuous heating/acceleration processes associated with the transition in the current sheet from a broad region on the dayside to a thin region on the nightside, as well as charge-exchange decay, are likely to be the cause for the observed day–night and dusk–dawn energetic ion asymmetries reported by Dialynas et al. (2013).

12.4.2.3 Neutrals

Saturn contains a large neutral (OH and other water group neutrals) cloud in the vicinity of the inner moon Enceladus and beyond (Shemansky et al., 1993;

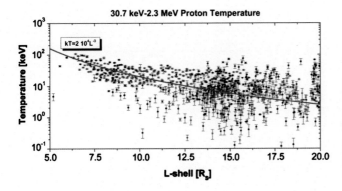

FIGURE 12.8
Radial profile of the temperature of hot H^+ ions in the magnetosphere of Saturn. The temperature increase with decreasing distance (and increasing magnetic field strength) from Saturn suggests that adiabatic transport of the hot plasma is occurring. Source: Dialynas et al. (2009).

Richardson et al., 1998). Before it was known that Enceladus is the source of this cloud, Jurac et al. (2002) used a complex model to simulate the mechanisms that sustain the observed cloud particles. Their model includes the effects of neutral ion collisions between the high-speed corotating plasma and the Keplerian motion of the neutral atoms and molecules. Neutral ion collisions can essentially transfer the corotation energy to the neutrals and cause a spatial inflation of the neutral gas torus. In Fig. 12.9, Jurac et al. (2002) found that not only is the z component (parallel to Saturn's spin axis) of the OH cloud particle velocities well fit to a kappa distribution, but also that the vertical (z-axis) density profile is similarly well represented.

A self-consistent model of the plasma and neutral cloud at Saturn was developed by Jurac and Richardson (2005). Their model required that a large amount of water originate from the orbit of Enceladus, although they noted that generating the rate needed from known processes (the Enceladus contribution had not been discovered at the time) such as sputtering from the E ring grains was unlikely. After the discovery of Enceladus as the needed source of water by the Cassini spacecraft, Dialynas et al. (2013) revisited the problem and developed a neutral gas distribution model using the kappa distribution-modeled ions from Dialynas et al. (2009). They concluded that the neutral gas vertical distribution at Saturn must be ~3 to 4 more extended than previously thought in order to accurately simulate the observed ENA images obtained by INCA.

12.4.2.4 Electron Spectra

The magnetosphere of Saturn contains copious fluxes of both thermal ($\varepsilon < 100$ eV) and suprathermal electrons ($\varepsilon > 1$ keV), and these electron fluxes

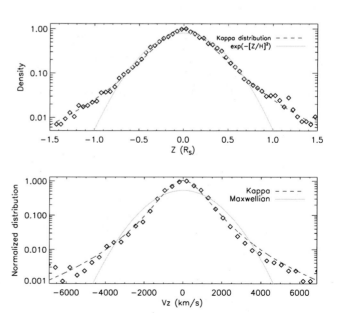

FIGURE 12.9

Density and velocity distribution of the OH cloud at Saturn. The vertical density profile is well fit to a kappa distribution (Livadiotis, 2015a). The velocity distribution of individual particles is non-Maxwellian and well fit to a kappa distribution. Source: Jurac et al. (2002).

are usually characterized by a bimodal distribution with a "cold" and a "hot" component (Rymer et al., 2007). This dual-spectral distribution is best fitted not with Maxwellians but rather with two kappa distributions (Schippers et al., 2008). The dual-kappa fits are integrated to obtain moments of electron density, temperature, and spectral (kappa) index as functions of location in the magnetosphere. Three distinct radial regimes were distinguished. In the inner regime inside 9 R_S, the suprathermal population densities dropped off, while outside the thermal population dropped off. In the outermost region outside 15 R_S, the thermal electrons became virtually absent. In the middle region between 9 and 15 R_S, suprathermal electrons dominated. Fig. 12.10 shows two examples of the dual-k distributions found in Saturn's magnetosphere. The appearance of dual-kappa distributions indicates that neither the thermal nor the suprathermal electrons are in equilibrium.

A similar approach using a single kappa distribution can be applied to electrons with energies exceeding ~ 20 keV (Carbary et al., 2011). For such electrons, the κ-index ranged from -3 to -4 in the outer magnetosphere ($>20\,R_S$), achieved minima near -4.5 between 10 and 15 R_S, and then hardened to values

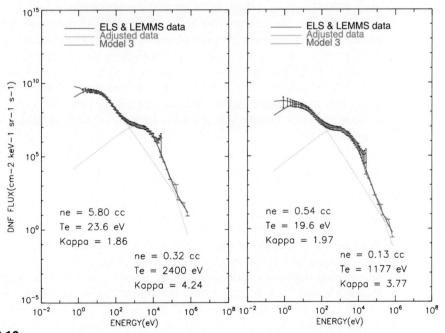

FIGURE 12.10

Left: A specimen from the inner ($\sim 9\,R_S$) and (right) outer Saturnian magnetosphere ($\sim 12\,R_S$), fitted with dual-kappa functions (the term "dual" refers to modeling with a linear superposition of two different kappa distributions). The ELS (Young et al., 2004) and Low Energy Magnetospheric Measurement System (LEMMS) refer to the two electron detectors that made the measurements, and "adjusted data" refers to an interpolation/smoothing that was required to connect the spectra from the two detectors. A solid blue line indicates the combined fit. Source: Schippers et al. (2008).

approaching -1.0 in the inner magnetosphere inside $5\,R_S$. Paranicas et al. (2014) also fit energetic electron spectra in the inner magnetosphere and found that simple functions, such as power laws in energy, would capture various portions of the energy range well. A dual-kappa distribution can also be applied to the electron plasma sheet in the magnetotail, where statistical analysis used a polytropic formalism to relate electron pressure to density, $P_e \propto n_e^a$ (Arridge et al., 2009). Because $a \approx 1$ for most of the 10-month survey, the electrons were shown to behave isothermally, which indicates that they must be in constant contact with an external heat reservoir.

Several theoretical investigations have arisen from the kappa spectra measured at Saturn (e.g., Henning et al., 2011). The first of these used Voyager measurements and a single κ function to model the mean electron flux precipitating on Titan's nightside, above the polar regions, finding that magnetospheric electrons produced rates between 1 and 5 electrons/cm^3/s at ~ 550 and 650 km (Galand et al., 1999). An electron kappa distribution was inserted into a standard Saturn Thermosphere Ionosphere Model to deduce possible ionospheric contributions to Saturn's plasmasphere (Moore and Mendillo, 2005). Electron densities up to ~ 200 cm^{-3} were predicted by this process. Later theoretical studies (e.g., Baluku et al., 2011) employed the electron kappa distributions derived by Schippers et al. (2008) and used them to assert that the electron acoustic waves might be observed in the outer magnetosphere ($>10\,R_S$), where they were expected to be weakly damped by dual-kappa distributions presented by the Cassini data. Koen et al. (2012) modeled the "cool" and "hot" electron velocities with kappa distributions (with low indices, $\kappa_c \sim 2$ and $\kappa_h \sim 4$) in the plasma sheet region to conclude that such electron acoustic waves are coupled with the electron plasma, while two-temperature kappa electron models (Annou, 2015) resulted in the deduction of ion acoustic waves for the inner, intermediate, and outer Saturn's magnetosphere.

12.4.3 Uranus

12.4.3.1 In General

The Voyager 2 encounter at Uranus revealed a very unusual magnetospheric configuration, with the spin axis pointing within ~ 8 degrees of the Uranus–Sun line and the magnetic dipole axis tilted by ~ 58 degrees with respect to the planet's spin axis (Ness et al., 1986). This configuration creates an unusual relationship between rotational and solar wind influence of the magnetosphere (Selesnick and Richardson, 1986; Vasyliunas, 1986; Selesnick and McNutt, 1987), allowing the solar wind to drive dynamical processes inside the system (Mauk et al., 1987, 1994; Sittler et al., 1987; Belcher et al., 1991; Kane et al., 1994). Furthermore, the planetary satellites are a combination of prograde bodies and likely captured bodies. The combinations of satellite inclinations and the tilt of the magnetic dipole make the interactions of moon and plasma complex (Krimigis et al., 1986; Stone et al., 1986; Cooper and Stone, 1991; Mauk et al., 1994). Composition analyses of the Uranian magnetosphere revealed the absence of any heavy-ion plasma torus (Bridge et al., 1986), while the particle composition at energies lower than ~ 6 keV (McNutt et al., 1987a, 1987b) and at radiation belt energies >0.6 MeV/nuc (Krimigis et al., 1986) were dominated by H^+ and electrons, with a minor ion H_2^+ component (shown as a subdominant peak in LECP pulse height measurements) at an abundance of 10^{-3} relative to H^+.

12.4.3.2 Plasma Energy Spectra

The Plasma science experiment observations of McNutt et al. (1987a, 1987b) showed that <6 keV proton distributions are generally characterized by Maxwellian cores (a "warm" component) with typical temperatures of $\sim 5-10$ eV and typical densities of $\sim$ few H^+/cm^3 but extend to variable power law nonthermal tails (a "hot" component) at higher energies that carry most of the plasma energy density (which is, however, small compared to the energy density of the planetary magnetic field). No α-particles are detected inside the Uranian magnetosphere (Krimigis et al., 1986), and the source of these "warm" protons is not the solar wind but most probably the Uranian ionosphere and the in situ ionization of the planet's extended hydrogen atmosphere (Fig. 12.11). By contrast, the nonthermal protons are possibly connected to solar wind–driven convection (Ye and Hill, 1994) from the magnetotail (or the auroral region of the ionosphere), pointing toward substorm-like events at Uranus.

12.4.3.3 Energetic Ion Spectra

The reported $\sim 0.6-1.2$ MeV electron and ion energy spectra (Krimigis et al., 1986) showed prominent nonthermal features that supplement the nonthermal characteristics of the low-energy plasma. Between the orbits of the moons Ariel and Umbriel, electron spectra are observed with power law tails at energies >590 keV, with a typical spectral index of $\gamma \sim 2.7$, but with subtle rolloff at lower energies, suggesting Maxwellian cores with temperatures $k_BT \sim 15$ keV. The spectra sampled inside the orbit of Miranda exhibited an overall shape consistent with a power law form in energy with a typical index $\gamma \sim 1.1$, possibly indicative of a cosmic ray albedo neutron decay source rather than inward diffusion. The maximum ion intensities were found to be comparable to those measured at Saturn (but lower than the intensities at Jupiter).

Although most of the H^+ spectra were consistent with a kappa distribution form in energy, Maxwellian fits with typical temperatures of $k_BT \sim 10-50$ keV were employed to describe the low-energy portion of their distribution (<200 keV) and power laws with $\gamma \sim 3-10$ for their high-energy nonthermal tails. The overall energetic particle pressure divided by the total magnetic field pressure (the "energetic" partial β) was found to be at most ~ 0.1 at the magnetospheric regions $<15\ R_U$ (Uranian mean equatorial radius: 1 $R_U = 25{,}600$ km), therefore lower than the maximum values found in the magnetospheres of Jupiter and Saturn. In situations where the magnetic field "pressure" is much greater than the particle pressure, the field lines can retain their dipole shape. Therefore unlike Saturn, where nightside-injected particles (local bundles of plasma)

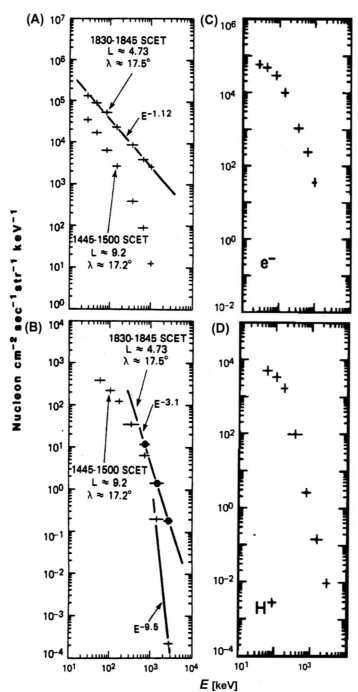

FIGURE 12.11

Left column: (A) electron, (B) proton spectra, measured within the Uranian magnetosphere. Right column: (C) electron, (D) proton spectra. Source: Krimigis et al. (1986); Mauk et al. (1987).

subsequently rotate around the planet and produce local diamagnetic depressions to the magnetic field (dipole lines become inflated) (Achilleos et al., 2010a, 2010b), at Uranus, even the low-energy plasma partial β is very small, leading researchers to conclude that the particle stress is insufficient to distort the Uranian magnetic field very profoundly.

Following the work of Krimigis et al. (1986), a reanalysis of the LECP energetic particle data by Mauk et al. (1987) provided evidence of the Uranian plasma sheet expansion during substorm-like injection of particles. They added that only in a few magnetospheric regions were the ion spectra adequately represented by kappa distributions. In some extreme cases, ion distributions showed Maxwellian shapes with $k_B T$ reaching as high as 125−250 keV, indicating the most intense energetic particle spectra measured in any magnetosphere to date. However, the electron spectral shapes inside the orbit of the moon Miranda were neither purely Maxwellian nor power law, a fact that raised questions about possible mechanisms driving such extreme local "thermalization." Despite the fact that the total plasma and particle pressures are low, the integral electron flux substantially exceeded the whistler mode stably trapped limit (Kennel and Petschek, 1966). Since such waves were observed by the plasma wave experiment onboard Voyager 2 (Gurnett et al., 1986), researchers connected these whistler mode waves and the local thermalization of particles inside the orbit of Miranda.

12.4.3.4 Whistler Waves

To model the nonthermal high-energy tails of the electron spectra that drive the whistler mode wave instability inside the Uranian magnetosphere, Pandey and Kaur (2015b) and Kurth (1992) used a kappa distribution function. Their work showed that the calculated temperature anisotropy, number density, and thermal energy of particles support the growth of electromagnetic electron cyclotron waves inside the magnetosphere, similar to the case of Earth (e.g., Pandey et al., 2008), despite the huge differences between magnetosphere sizes, plasma sources, etc. Furthermore, the relatively high wave number range in Uranus ($0 < k_U < 1$) pointed toward a high-energy range of plasma waves that provide the theoretical explanation for these high-energetic particles inside the Uranian magnetosphere.

12.4.4 Neptune

12.4.4.1 In General

The Voyager 2 encounter with Neptune revealed a large magnetosphere, possessing several characteristics of special interest. The large offset of its internal magnetospheric configuration ($\sim 0.55\, R_N$; Neptunian mean equatorial radius: $1\, R_N = 24{,}765$ km) (Ness et al., 1989) results in a unique radiation belt−planetary atmosphere interaction (Mauk et al., 1990; Cheng, 1990a; Selesnick and Stone, 1991), while the large dipole axis tilt by ~ 47 degrees (Ness et al., 1989) with respect to the planet's spin axis imposes significant periodicities driven by the solar wind and magnetosphere interaction (as the magnetic "cusp" points toward the solar wind every 16.1 h). Of particular interest is Neptune's satellite Triton, which forms a neutral atom torus in the vicinity of its orbit (Richardson et al., 1990). Despite the large H source rate at Triton, only a small fraction of these H atoms ($\sim 15\%$) have orbits sufficiently close to Neptune to get ionized within the magnetosphere, meaning that Triton is likely not a major source of magnetospheric plasma (Cheng, 1990b). However, composition analyses of the low-energy plasma (<6 keV) revealed the presence of H^+ and a heavy component presumed to be N^+ (Belcher et al., 1989), consistent with the species predicted to escape from Triton (Summers and Strobel, 1991). The energetic particles, on the other hand ($0.6 < \varepsilon < 1$ MeV/nuc) (Krimigis et al., 1989), are mainly dominated by protons, with a minor $H_2{}^+$ component and a substantially small fraction of Helium, with relative abundances of $H:H_2:He^4 \sim 1.3 \times 10^3{:}1{:}0.1$, while no statistically significant counts consistent with heavy ions were identified.

12.4.4.2 Energetic Ions

The energetic particle spectra (Krimigis et al., 1989) were found to be highly variable within the Neptunian magnetosphere. Albeit inside the orbit of the moon Triton the H^+ spectra showed a kappa distribution form in energy, Maxwellian fits with typical $k_B T \sim 55$ keV (number density of ~ 3–30×10^{-4} cm^{-3} and energy density of ~ 3–30 eV/cm^3) were employed. The high-energy tail of the distribution, e.g., Fig. 12.12B, exhibits a substantial flux excess (a power law with typical $\gamma \sim 4.3$) that deviates from a Maxwellian. Outside the orbit of Triton, protons form power law distributions with typical $\gamma \sim 4$ and even lower number densities (~ 2–8×10^{-6} cm^{-3}) and energy densities (~ 0.1–0.9 eV/cm^3). The energetic electrons were found to be consistent with power laws but showed prominent differences inside and outside Triton's orbit, with a slight bend at low energies accompanied by a softening break at >100 keV with $\gamma \sim 5.6$ (inside) and a single power law with typical $\gamma \sim 4.4$–4.7 (outside).

A detailed reanalysis of the LECP measurements (Mauk et al., 1991) presented particle pressures and densities comparable to those at Uranus (although by 10–100 times smaller for the inner magnetosphere) and a larger partial β value ($\beta \sim 0.2$) but still too small to produce magnetic field distortions. However, assuming a high N^+ contribution from Triton, β in the *trans*-Triton regions (toward the tail) can reach unity, therefore allowing local diamagnetic depressions in the magnetic field, such as the observed "tail events" presented by Ness et al. (1989).

Using a combination of Maxwellian fits and power laws, H^+ temperatures were found to range from ~ 10 keV to as high as 150 keV and electron temperatures of a few tenths of keV. A kappa distribution (although not used) would suffice to characterize the entire spectrum in the middle magnetosphere regions, while near Triton (and beyond) the spectra obtain a purely Maxwellian form. These high Maxwellian temperatures are correlated with Triton's position at its "inner

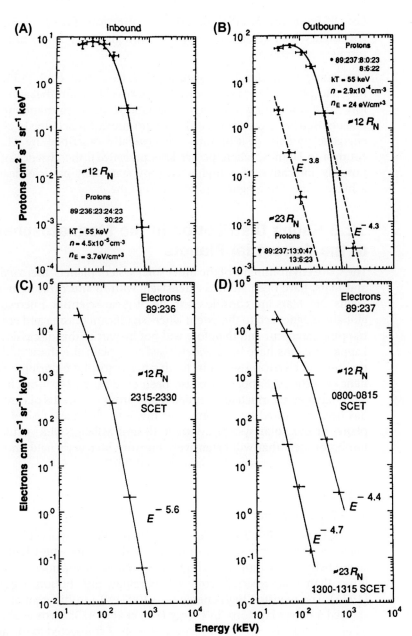

FIGURE 12.12
Energy spectra of (A, B) H$^+$ and (C, D) electrons at selected regions (inside and outside Triton's orbit) during the inbound and outbound Voyager 2 passes. Source: Krimigis et al. (1989).

L-shell" boundary, but their ordering as a function of the L-shell does not imply consistency with adiabatic transport (Krimigis et al., 1990) inside the magnetosphere of Neptune.

In contrast to the Uranian magnetosphere, e.g., Mauk et al. (1987), the middle magnetosphere of Neptune showed no noticeable azimuthal asymmetries in the ion flux, confirming previous expectations of a corotationally dominated magnetosphere, e.g., Hill (1984), with no substorm activity. In addition, as the energetic particle distributions are generally described by Maxwellians (in contrast to Uranus, where power laws prevailed), the absence of substorms is possibly indicative of a higher entropy state that particle distributions can achieve inside the magnetosphere of Neptune.

12.5 Kappa Distribution in the Magnetospheres of the Terrestrial Planets

The space environments of the inner solar system planets, Mercury, Venus, Earth, and Mars, have been studied extensively with in situ measurements. Although Venus and Mars are considered to be of great scientific interest, they are not globally magnetized to the extent where significant plasma and energetic particle trapping can occur and therefore will not be part of this discussion. By contrast, kappa functions have been widely used to explain the shape and properties of the ion and electron spectra in Earth's magnetosphere during various geomagnetic conditions. Kappas have also been used in the Hermean magnetosphere, although they are much less studied than Earth's to date. Its plasma environment has been measured by the only two missions that have visited the innermost planet of our solar system, Mariner 10 and MESSENGER, while the BebiColombo orbiters that will extensively measure Mercury's field and particle environment are not expected to launch until 2017.

12.5.1 Earth

12.5.1.1 In General

Dating back to the observations made by the IMP-7 spacecraft (1970s), it was suggested that energy spectra in the terrestrial plasma sheet had a nonthermal shape/behavior, comprising a high-energy tail with considerably higher particle fluxes than that expected from a Maxwellian distribution, e.g., Roelof et al. (1976). The reported high-energy ($\varepsilon > 100$ keV) ion differential energy spectra showed a power law form in energy (Fan et al., 1975; Sarris et al., 1976, 1981), where the flux $j(\varepsilon) \sim \varepsilon^{-\gamma}$ with typical $\gamma \sim 4-7$ connected to the acceleration of ions up to >290 keV and electrons up to >220 keV. Furthermore, Roelof et al. (1976) pointed out that the high-energy H^+ distributions would require an effective $\gamma > 30$ at ~300 keV in order to be consistent with a thermal plasma of $T < 10$ keV. The combination of thermal plasma and energetic ion measurements (~50 eV/e to 4 MeV; eV/e: energy in eV per charge) from the IMP-7 and IMP-8 spacecraft (Sarris et al., 1981), suggested that the high-energy portion of the distribution correlates with the low-energy one, i.e., the transition energy that separates thermals from energetics can be approximated by $\sim k_B T(\gamma + 1)$. In other words, these high-energy tails of the plasma sheet population are an integral part of the same distribution from eV to MeV ions that essentially behave as a single population.

This connection among particle populations was made clearer with data from the ISEE 1, 2, 3, and AMPTE-IRM/CCE spacecraft. The work of Christon et al. (1988, 1989) suggested that both ion and electron energy spectra are well represented by kappa distributions rather than a Maxwellian distribution. The statistical study of Artemyef et al. (2014) using Interball-Tail measurements in the $20 < \varepsilon < 600$ keV range confirmed that the magnetotail spectra are separated into two parts: a thermal core and a high-energy tail $\sim \varepsilon^{-\gamma}$ with $4 < \gamma < 5$. Solar wind particles are likewise described by a combination of distribution functions, each of which loosely dominates over one range of energies.

12.5.1.2 "Quiet" Plasma Sheet Spectra

According to Christon et al. (1988, 1989), during quiet magnetospheric conditions (Auroral Electroject index $AE < 100$ nT), the central plasma sheet-charged particle populations included ions with characteristic energies of $E_0 \sim 1.3$ keV and electrons with $E_0 \sim 0.2$ keV, while both included a persistent power law tail with a κ-index of $5 < \kappa < 6$. Note that the temperature $k_B T$ may be calculated from E_0 using $k_B T = E_0 \cdot \kappa \left/ \left(\kappa - \frac{3}{2} \right) \right.$. Although the shape of the ion spectra remains unaffected by the changes of the AE index for these quiet magnetospheric conditions, the characteristic energy of ions can increase with increasing geomagnetic activity. The undisturbed plasma sheet is supplied by the terrestrial ionosphere and the solar wind, typically dominated by relatively low-energy H^+ (<10 keV), which coexists with a low-density O^+ population of ionospheric origin ($0.01 < O^+/H^+ < 0.1$) and an even lower density He^{++} population from the solar wind ($0.02 < He^{++}/H^+ < 0.04$) that persists during all geomagnetic activity levels (Peterson et al., 1981; Lennartson and Shelley, 1986). According to the kappa fits, these quiet plasma sheet He^{++} distributions ($E_0 \sim 2.85$ keV, $\kappa \sim 5.1$, and $n \sim 0.013$ cm^{-3}) and H^+ distributions ($E_0 \sim 0.66$ keV, $\kappa \sim 4.1$ and, $n \sim 0.37$ cm^{-3}) yield a flux ratio of $H^+/He^{++} \sim 1$ for the >24 keV energy range, in addition to the overall consistency between the He^{++} and H^+ spectral shapes (Fig. 12.13, top panel).

Analyses using measurements from the THEMIS mission, initially five spacecraft in the Earth's orbit (Stepanova and Antonova, 2015), showed that when the ion and electron spectra are organized by radial distance around the central plasma sheet ($\sim 7–30\ R_E$; $R_E \sim 6378.1$ km: Earth's mean equatorial radius), the κ-values increase with distance down the tail, i.e., $5 < \kappa < 10$ for ions and $3 < \kappa < 5$ for electrons (Fig. 12.14). In principle, the tendency of the spectra to become "softer" (lower intensity with increasing energy) suggests the possibility of "aging," i.e., relaxation of the particle distributions to a Maxwellian

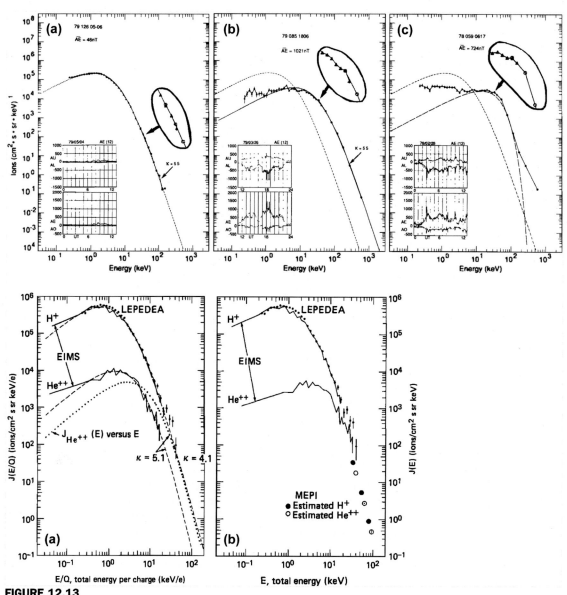

FIGURE 12.13

Top: Plasma sheet ion spectra (Christon et al., 1991), plotted with Auroral Electroject (*AE*) values from Kamei and Maieda (1981, 1982). *Circled insets* show the instruments overlapping the energy range ($\sim$24–45 keV). The quiet time data (A) is repeated as *short-dashed curves* in (B) and (C) while *long-dashed curves* in (C) represent a shifted Maxwellian. Bottom: Plasma sheet ion spectra sampled over a quiet interval from ISEE 1 (Christon et al., 1989). (A) H$^+$ and He^{++} (*solid curves*) from the EIMS *(From Peterson et al. (1981).)* and total ion flux from the LEPEDEA (*small points*). Kappa distribution fits to the EIMS data (*dashed curves*) are shown extrapolated up into the MEPI energy range. (B) EIMS H$^+$ and He^{++} (*solid curves*) and the LEPEDEA (*small points*) and MEPI (*large points*) total ion flux data are presented as flux versus total particle energy *E*.

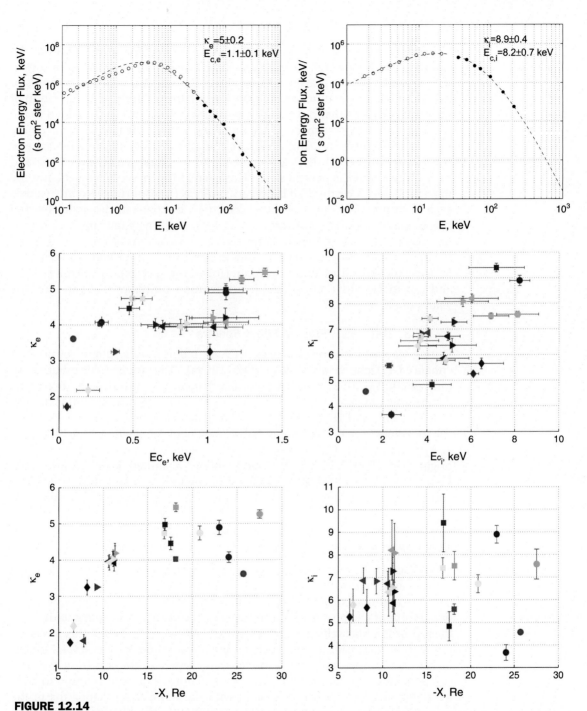

FIGURE 12.14

Left column: Average electron energy flux spectra (top) measured by ESA (*white circles*) and SST (*black circles*) instruments onboard the THEMIS-B satellite on February 22, 2008, and fitted by kappa distributions. The kappa index is plotted with the core energy (middle) and radial (bottom) distance. Right column: The same for ions. Colors indicate different events on February 14–26, 2008 (for details, see Stepanova and Antonova, 2015). (*Symbols* indicate different satellite used). Adapted from Stepanova and Antonova (2015).

("thermalization") due to diffusion in the velocity space, as described by Collier (1999). However, the tailward increase cannot be explained by any of the sources of plasma transport, unless an additional acceleration mechanism, like turbulent acceleration, or energization of particles due to dipolarization of the magnetic field (that can occur even at times of low activity) is added to the framework of velocity space diffusion. These authors where in favor of the second mechanism, i.e., the transport of particles from the inner magnetosphere toward the tail that can be related to the presence of turbulence and eddy diffusion in the plasma sheet. However, as this process is maybe not the most probable, if we think of the reverse process (transport toward the planet), even if, in principle, most of the energization mechanisms leave the spectral shape unaffected, e.g., Christon et al. (1991), rapid magnetic field reconfigurations that are well studied at Earth can cause violations of the first adiabatic invariant (mainly for heavier ions due to their long gyroperiods compared to H^+) and the second adiabatic invariant (for all ion species), e.g., Mauk (1986), and can in fact drive nonadiabatic acceleration processes that can alter the values of κ.

12.5.1.3 "Disturbed" Plasma Sheet Spectra

During geomagnetic disturbances ($AE \gg 100$ nT), ionospheric outflow is enhanced (Daglis et al., 1990, 1997, 1999a). The average contribution of ionospheric O^+ (newly injected particles) increases considerably, becoming comparable to the H^+ densities ($0.1 < O^+/H^+ < 0.6$). Gloeckler and Hamilton (1987), using AMPTE 1–315 keV/e measurements, showed that ions of ionospheric origin (e.g., O^+, H^+) follow a two-parameter kappa function with a high-energy index of ~ 5.5. However, the higher charge state ions, i.e., of solar wind origin (He^{++}, O^{++}, C^{6+}, O^{6+}, etc.), follow a shifted kappa distribution, $j \sim j_0 [1 + (\varepsilon^{1/2} - \varepsilon_s^{1/2})^2/3]^{-5.5}$, with a shift energy $\varepsilon_s \sim 4$ keV/amu (~ 8 keV/e) and an ion temperature ~ 0.5 keV/amu (~ 1 keV/e). Moreover, the substorm plasma sheet spectra, especially O^+ and He^+, showed a hardening in their high-energy tails (high-energy flux enhancements). By contrast, the substorm ions of solar wind origin (He^{++} and $CNO^{>3+}$) are depleted by a factor of ~ 2.

Consequently, despite the existence of several kappa-like portions in both the electron and ion distributions (Christon et al., 1991), the active time spectra are complex and may well include contributions from $Z > 2$ ions. A single functional form of a kappa distribution function is not always adequate to describe them (nor is a Maxwellian or a velocity exponential). The overall flux levels are found to be increasing at high energies, but a discernible knee at several to tens of keV (more often for ions than electrons) separates the hot from the cold components of the spectra. The partial temperature of the ions is found to be in the range of 2.8–12.5 keV, consistent with the more recent kappa fits to the energetic H^+ measurements from Cluster/RAPID (in the 1 keV to 4 MeV energy range), where ion temperatures are ~ 10.5 keV and $\kappa \sim 9.7$ during disturbed conditions (temperature ~ 3 keV and $\kappa \sim 4.4$ during quiet conditions) (Kronberg et al., 2010). The evolution of the particle spectra can be attributed to acceleration mechanisms, i.e., both ions and electrons participate in two different acceleration processes (betatron and cross-tail current sheet) with increasing geomagnetic activity (from low to high AE). During such conditions, the plasma sheet acts as a container of energy in the form of stretched magnetic field lines (by contrast to the dipole lines of the ring current), which is released tailward through plasmoids followed by a dipolarization of the magnetic field (Nose et al., 2010), responsible for dramatic ion heating toward the Earth (Delcourt, 2002; Mitchell et al., 2003).

The results of Christon et al. (1991) have been confirmed with the statistical analysis of Haaland et al. (2010) using Cluster particle (0.7 keV–2 MeV) measurements. According to their fits, the overall proton spectral shape is fairly consistent with a kappa distribution function with $\kappa \sim 3.5$ to 4 on an average for various magnetospheric conditions, over a broad range of solar wind pressures, while no direct response of the κ-index with the interplanetary magnetic field (IMF) changes is observed. Although deviations of the thermal part of the distributions can be attributed to some degree to the influence of ionospheric ions (increased ionospheric outflow), as explained earlier in this section, the shapes of the spectra (especially the high-energy power law tails) are due to internal magnetospheric processes, such as acceleration of particles by several possible mechanisms: nonadiabatic (adiabatic) acceleration of particles with large (small) gyro radii compared to the current sheet thickness (Speicer, 1965); acceleration by the induced electric field (the changing magnetic field) in reconnection (Grigorenko et al., 2009; Angelopoulos et al., 1994); nonadiabatic acceleration of large gyroradius particles (Delcourt et al., 1990) that occurs during dipolarization following substorms (Nose et al., 2000); or a combination of all these mechanisms.

12.5.1.4 Middle Magnetosphere and Substorms

No matter which mechanism dominates, a significant part of the stored magnetic energy in the tail that is released over the substorm expansion phase gets converted to particle energy ($\varepsilon > 100$ keV). Ring current and radiation belt particle spectra derived from a plasma sheet source retain their power law tails with Maxwellian-like roleoffs at lower energies (e.g., Pisarenko et al., 2002b). The most intense spectra run into limits, however, that distort the lower energy particles away from their purely Maxwellian ideals. The more energetic particles play a key role in both the ring current and radiation belts. The pressure-driven currents along with other magnetospheric electrical currents are linked to the overall closure of the terrestrial current system, thus providing the coupling of the inner magnetosphere with the ionosphere. Therefore although the quiet ring current region is generally dominated by >50 keV H^+ ($>75\%$ of total ion density), forming a current with an average ion current density of $1–4$ nA/m^2, the O^+ particle population ($>2\%$ that can reach up to 19%) can be significantly enhanced from the ionosphere during magnetic field reconfigurations (the storm time ion current density may exceed ~ 7 nA/m^2) (Daglis et al., 1999b). Other species, e.g., He^+, are much smaller components of the ring current density, while the solar wind ions, i.e., He^{++} and $CNO^{>3+}$, are only <1% of the total density (Gloeckler and Hamilton, 1987).

The region where dipole magnetic field lines evolve to stretched, i.e., near the "boundary of injection" of ring current particles (Mauk and Meng, 1983; Lopez et al., 1990), is where the expansion phase of substorms begins, and nonadiabatic acceleration processes are essential in determining the particle and ring current dynamics. Liu and Rostoker (1995) have simulated the effects on the evolution of an initial Maxwellian distribution of ions that over several substorm expansive phase cycles becomes a kappa distribution function. The Interball-Tail probe crossing of the outer ring current region (Pisarenko et al., 2002a, 2002b) showed that the 0.5 keV to 3 MeV ion spectra are fairly consistent with kappa distributions with typical temperatures of $6.8 < k_B T < 9.6$ keV and spectral slopes of $4.2 < \kappa < 7.1$. However, as shown earlier (Christon et al., 1991), although the high-energy tails are accurately described by power laws, an intensity mismatch was found for the thermal part of the distribution. Nevertheless, the smooth ring current spectral shapes support the basic conclusion of Christon et al. (1991) that both spectrum-preserving and spectrum-altering heating processes (betatron and cross-tail current sheet acceleration) participate in the overall particle energization during geomagnetic disturbances. This also indicates the persistent character of the transition from plasma sheet ions to the ring current (Pisarenko et al., 2002b).

Some of the pioneers in attempting the theoretical foundation of the kappa distribution function, Hasegawa et al. (1985), proved that a plasma, which is surrounded by a superthermal radiation field, undergoes velocity space diffusion that leads to the formation of a power law distribution at high energies. Consequently, the kappa shapes of particles inside the magnetosphere (e.g., in the magnetotail) can be due to the presence of low-frequency fluctuating radiation fields (Gurnett et al., 1976) that are spread through the central plasma sheet. The work of Grabbe (2000) showed that the use of kappa distribution functions with $6 < \kappa < 7$ leads to the formation of an unstable spectrum that exhibits minimal trapping in the tail and is driven by electron/ion beam instabilities. On the other hand, closer to Earth, measurements at the energy range of 75 keV to 1.5 MeV made by the Synchronous Orbit Particle Analyzer instrument onboard 1989-046 and LANL-01A satellites at geosynchronous orbit ($\sim 6.6 R_E$) have shown (Xiao et al., 2008) that the kappa distribution functions fit well the electron spectra at low energies but fail to capture the measured tails at >300 keV, because the kappa distribution falls as $\sim u^{-2\kappa-2}$, whereas the measurements imply a $\sim p^{-\kappa-1}$ dependence at the relativistic energy (where p is the relativistic particle momentum), which is inconsistent with the power law.

12.5.1.5 High-Latitude Spectra

The electron and ion densities are comparable in the plasma sheet, but owing to their greater ion mass compared to electrons, most of the pressure between the tail lobes and the central plasma sheet is attributed to ions ($\sim$85%). The Hydra instrument onboard the Polar mission showed that high-latitude electron distributions are very consistent with kappa distributions ($\kappa < 10$), with typical densities of $\sim$0.1 to 0.3 cm^{-3} (pole to equator) and characteristic energies of 0.4–0.9 keV, verifying a trend of increasing energy from the plasma sheet boundary to the central plasma sheet (Kletzing et al., 2003). While the near-Earth, solar wind electrons are also best approximated by kappa distributions (Maksimovic et al., 1997b), these lobe plasma electrons may retain their spectral form even before they are transported to the plasma sheet. Furthermore, due to the reported connection between high-latitude plasma sheet electrons to the auroral ionosphere (Kletzing and Scudder, 1999), all models of the auroral acceleration region should include the effects on acceleration profiles produced by kappa distributions, rather than Maxwellians (e.g., Dors and Kletzing, 1999).

12.5.1.6 Magnetosheath Spectra

In the outermost parts of the terrestrial magnetosphere, the magnetosheath was studied by Formisano et al. (1973) using Helios1 0.2–16 keV measurements, and the kappa distribution was employed for $\kappa = 2$, 3, 5, and $\kappa \to \infty$ to show that when the solar wind Mach number (M) and β are low ($M < 3$ and $\beta < 0.01$), the proton velocities in the magnetosheath are better described by Maxwellian. However, high-energy tails are formed with increased M and β in the absence of upstream waves. This is not inconsistent with the later composition analyses of Gloeckler and Hamilton (1987), who employed a superposition of two Maxwellians to fit the heated solar wind ions with a temperature $\sim$0.2 keV/amu in the energy range 0.3–1.3 keV and a temperature $\sim$0.8 keV/amu in the higher energy range 1.3–5.5 keV.

Magnetosheath spectra are complicated, as they are formed in the framework of shock acceleration and wave particle interactions. Such complications can be resolved partly by using remote sensing techniques, such as ENA imaging, which can link the upstream particle parameters to shock and solar wind conditions without reliance on a single spacecraft location (Fig. 12.15). The 0.1 eV to $\sim$6 keV ENA-derived H$^+$ magnetosheath spectra (Ogasawara et al., 2013) from the Interstellar Boundary Explorer High Energy imager (Funsten et al., 2009) showed enhanced high-energy tails and were successfully fitted with kappa distributions for both the dynamic and static solar wind phases. According to these fits, the H$^+$ partial temperatures were found to range from $\sim$0.1 to <0.7 keV, while the κ values of the magnetosheath H$^+$ were distributed within the near equilibrium, i.e., >2.5 (Livadiotis and McComas, 2010a; 2013d). Interestingly, some linear correlation between κ and T was found ($T \sim a \cdot \kappa + \delta$) in the magnetosheath data, while a correlation of the kappa tails with the solar wind angle points toward a perpendicular shock configuration that is able to produce these nonthermal tails.

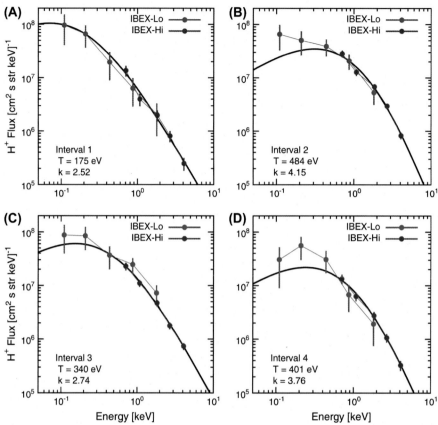

FIGURE 12.15

Combined magnetosheath proton spectra using Interstellar Boundary Explorer High Energy (IBEX)-Lo and IBEX-Hi data averaged for four intervals according to differences in upstream region properties and energetic neutral atoms (ENA) rates: (A) turbulent with low ENA rates, (B) turbulent with high ENA rates, (C) quiet with high solar wind pressure and quasiradial interplanetary magnetic field (IMF), and (D) quiet with moderate solar wind pressure and quasiparallel IMF. Solid lines represent the kappa distributions that are fitted only to the IBEX-Hi data. Adapted from Ogasawara et al. (2013).

12.5.2 Mercury

12.5.2.1 In General

The Hermean system mainly consists of planetary ions, namely Na^+, O^+, and He^+, due to photoionization of Mercury's exosphere (Raines et al., 2013) with an equatorial pressure that can be a substantial fraction of the H^+ pressure (Zurbuchen et al., 2011). Cheng et al. (1987) showed that heavy ions, mainly Na, from the Hermean exosphere can be accelerated to keV energies, making it an important contributor of mass and energy to the magnetosphere, e.g., Na^+ densities are on the order of 10% of the solar wind protons (Gershman et al., 2014). The Hermean magnetic moment is $\sim 2-7 \times 10^{-4}$ that of Earth's, but nevertheless, its intrinsic magnetic field dominates the magnetospheric cavity, deflecting the solar wind plasma flow and the IMF (solar wind pressure and IMF are stronger near Mercury), forming a neutral current sheet close to the planetary heliocentric orbital plane. Consequently, the general dynamics of the Hermean system (e.g., Caan et al., 1977; Kokubun and McPherron, 1981; Eraker and Simpson, 1986; Christon et al., 1987) were expected to be somewhat similar to Earth's (Siscoe et al., 1975).

12.5.2.2 Energetic Electron Bursts

The first measurements from the Mariner 10 mission (Simpson et al., 1974) showed foreground fluxes of H^+ and electrons with energies of ~ 550 and ~ 300 keV, respectively. The differential intensity spectra were found to be consistent with a power law form in energy ($\sim j_0 \cdot \varepsilon^{-\gamma}$) with a typical spectral index $\gamma \sim 5.5$ for >500 keV H^+ and $\gamma > 9$ for >170 keV electrons. The researchers found large-scale electron oscillations (period ~ 6 s) that were in phase with H^+ bursts and attributed these events to acceleration of particles in the magnetotail or the magnetosheath. Armstrong et al. (1975) confirmed the existence of electron bursts but questioned the >500 keV H^+, a fact that was confirmed by recent analyses of the MESSENGER Energetic Particle Spectrometer (EPS) measurements. For example, Ho et al. (2011, 2012) and Lawrence et al. (2015) showed that bursts of energetic particles were in fact due to energetic electrons (and not ions) that are continuously supplied by substorm-like processes (Fig. 12.16).

Analyses of the Mariner 10 data (Christon, 1987) revealed the nonthermal properties of the premidnight (cool plasma region) electron energy spectra, including a high-energy tail with considerably higher particle fluxes than that expected from a Maxwellian distribution. These electron spectra are well fitted with kappa distribution functions with typical characteristic energies of $E_0 \sim 28$ eV and $\kappa \sim 7$ and are very similar in shape with electron spectra sampled from the central plasmasheet of Earth's magnetotail but by a factor of ~ 7 lower characteristic energies than those found at Earth. Flux enhancements in the >35 keV electron spectra histories, combined with the measured magnetic field in the tail, suggest that these events are indicative of dipolarization events that occur after substorms, similar to what happens on Earth.

12.5.2.3 MESSENGER Measurements

These observations were corroborated by ~ 35 keV to 1 MeV electron measurements from the EPS onboard MESSENGER (Ho et al., 2012) that detected bursts of tens to hundreds of keV electrons within the Hermean system. Although in their statistical analysis the authors found several cases of electron spectra that are best represented by kappa distributions, they performed power law fits in the $36-100$ keV range with a mean spectral index $\gamma \sim 2.51 \pm 0.85$. The spectra sampled upstream of Mercury's bow shock had a slightly softer spectral index ($\gamma \sim 2.91 \pm 0.87$) than the remaining events ($\gamma \sim 2.44 \pm 0.87$), while the spectral index at energies >100 keV was found to be generally larger ($\gamma > 3$). Despite the fact that the foreground measurements in the equatorial region were

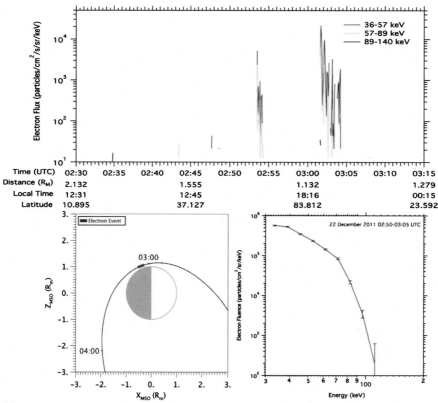

FIGURE 12.16

Top: Energetic electron intensity versus time and (bottom left) the location for a pair of two closely spaced high-latitude energetic electron bursts on December 22, 2011. Bottom right: Event-averaged electron energy spectrum, showing a kappa distribution form. Adapted from Ho et al. (2012).

generally weak (may be the high-energy tail of the 1−10 keV quasitrapped population of electrons in Mercury's equatorial regions, consistent with the hybrid simulations of Schriver et al. (2011)), both the high-latitude and night-side events showed similar spectral and pitch angle characteristics. This implies that an acceleration mechanism may be at work, similar to reconnection (possibly connected to substorms), with a high reconnection rate (due to the plethora of burst events per each orbit), consistent with the reported magnetic field measurements from Slavin et al. (2009, 2010). Owing to the planet's small size and its very efficient reconnection, substorms on Mercury occur frequently and on timescales as short as a few minutes.

Starr et al. (2012) reported the detection of fluorescent X-ray emissions due to the impact of 1−10 keV electrons on Mercury's nightside using X-ray Spectrometer measurements onboard MESSENGER. To assess their model sensitivity to the changes in the assumed excitation spectra, the researchers

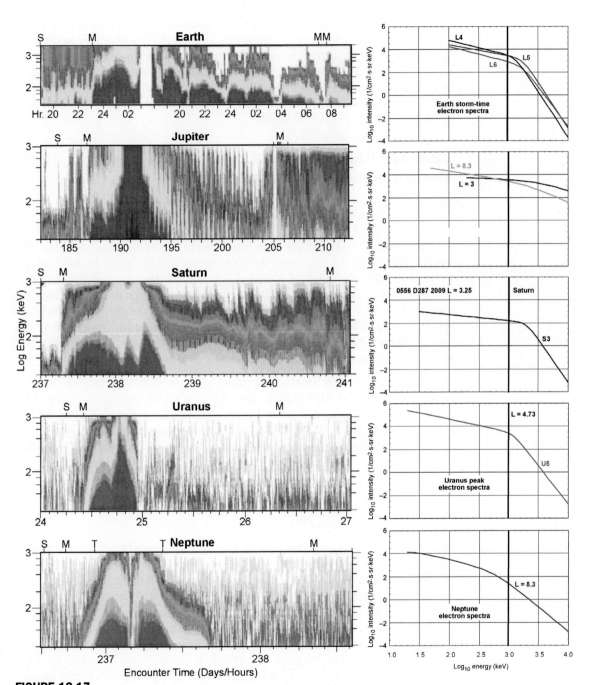

FIGURE 12.17

(Left column) Color-coded intensities (spectrograms) of energetic electrons as a function of energy (20 keV to ~1 MeV) and time using International Sun Earth Explorer (ISEE) mission data for Earth and Voyager measurements for all other planets (from Mauk et al., 1995, reproduced at Mauk and Fox, 2010). Magnetopause crossings are indicated by the "M" character above each plot. (Right column) Fits of the intense spectra within each different magnetosphere using Eq. 5.1. Taken from Mauk and Fox (2010).

employed the kappa distribution for four distinct energies (0.8, 1.0, 1.2, and 1.4 keV). A higher energy population of energetic electron events was reported by Lawrence et al. (2015) using MESSENGER's Gamma Ray and Neutron Spectrometer, but instead of a kappa distribution (or a power law), their modeling involved a Maxwellian with ~90 keV average energy. Their results showed a prominent dusk–dawn asymmetry, with the majority of the electron burst events located close to dawn.

12.6 Are Kappa Distributions Useful for Magnetospheric Research?

As explained earlier in this chapter and in Carbary et al. (2014), there are several cases where kappa distributions do a relatively poor job in capturing the low-energy rollover of energetic particle spectra in planetary magnetospheres (e.g., Christon et al., 1991). In other cases (e.g., Xiao et al., 2008; Mauk and Fox, 2010; Mauk, 2014) (Fig. 12.17), even the high-energy tails of particles have shown deviations from the classical kappa distribution form. At least for the more intense particle spectra, other functional forms have proven to be more useful in the description of plasma and energetic particles. For example, the multispecies (H, He, O, S) EPD spectra (Mauk et al., 2004) sampled in Jupiter's magnetosphere showed distinct "softening" breaks at high energies, Eq. (12.3), which could not be described by the convected kappa distribution of Eq. (12.2).

In describing the most intense spectral shapes of radiation belt electrons at energies ~0.1 to >1 MeV (Mauk and Fox, 2010) and ring current ions at energies ~100 keV to 1 MeV in the magnetospheres of Earth, Jupiter, Saturn, Uranus, and Neptune, Mauk (2014) used an empirical functional form that is flexible to capture the observed discontinuities (softening breaks at ~1 MeV) that cannot be explained by Maxwellian, power law, or single kappa distributions:

$$j(\varepsilon, \alpha) = C \cdot \frac{\left[(\gamma_1 + 1)k_B T_\gamma + \varepsilon_1\right]^{-\gamma_1 - 1}}{1 + (\varepsilon/e_t)^{\gamma_2}} \cdot \varepsilon \cdot \sin^{2S}(\alpha), \tag{12.6}$$

which is a generalization of Eq. (12.3). A similar functional form was used in the earliest observations of the Jovian radiation belts (Baker and Van Allen, 1976). The numerator of the fraction is the known kappa distribution; the denominator describes the spectral softening breaks (e.g., spectra at Uranus, Neptune, and Jupiter); and the α-dependent term adds a pitch angle dependence of the particle flux that relates to the anisotropy (S exponent) relative to the magnetic field. This function has been used to investigate the role of the differential so-called Kennel–Petschek limit (Kennel and Petschek, 1966), as moderated by whistler mode scattering for electrons and electromagnetic cyclotron waves for ions in the strongly magnetized planetary magnetospheres. Note that the temperature-like parameter T_γ can be related to the actual temperature T of the system, e.g., when $S = \gamma_2 = 0$, then $\kappa = \gamma_1$ and $T_\gamma = T\left(\kappa - \frac{3}{2}\right)\Big/(\kappa + 1)$.

All of the previous observations oblige us to ask the following question: "Why are kappa distribution functions used in so many cases to describe plasma in magnetospheric systems?" There have been several theories concerning the origins and theoretical developments of kappa distributions (e.g., Pierrard and Lazar, 2010). However, one can easily understand that all of these theories have one common puzzling factor: collisions. The Maxwellian velocity distribution function is inherently collisional, and if we think of the kappa as a generalized form of these distributions, then one can misinterpret the kappa distributions as also being collisional. But we know that space plasmas can be accelerated to produce nonequilibrium, suprathermal tails at high energies, generally described by power laws, because they are in fact collisionless. On the other hand, in magnetospheric contexts, plasma electrons can be heated due to their coexistence with ions. For instance, Young et al. (2005) showed that the corotation speeds of the ions organize both ion and electron data, while Rymer et al. (2007) suggested that given sufficiently long residence times (slow plasma transport), the electrons can significantly equilibrate with the ions through Coulomb collisions and get heated.

Therefore the question becomes, "How do these distributions even exist in magnetospheric plasmas?" Among several mechanisms, it has been proposed (e.g., Hasegawa et al., 1985) that electromagnetic field fluctuations can "replace" collisions and provide the necessary binding for plasmas, and velocity space diffusion enhanced by Coulomb-field fluctuations produces high-energy power law distributions. Consequently, the kappa distributions are maybe not intrinsically "universal laws," but they are important in the sense that they generalize the notion of equilibrium for collisionless plasmas that are not in thermal equilibrium. While the coupling between the particles due to collisions is generally negligible, they do show strong collective behavior owing mainly to wave–particle interactions. In collisionless plasmas, the independence of particles breaks down due to long-range correlations mediated by the plasma waves, and classical statistics cannot provide a basic derivation for kappa distributions (Lazar et al., 2012b). Whatever the mechanism, the significance of the kappa distribution function has been explained under its new theoretical development for nonextensive systems (Livadiotis and McComas, 2009, 2013a; Livadiotis, 2015a) by extending the work of Tsallis (1988, 1999), proving that the fundamental origins of the kappa distribution function lie in a new statistical approach that points toward a new thermodynamic description of plasma.

Therefore, we ask the following: "How has the kappa distribution function helped us to interpret magnetospheric physics results to date?" First of all, kappa distributions can be considered to be the "Maxwellians" for nonextensive systems and have been proven to be convenient tools to describe plasma systems not in thermodynamic equilibrium, such as magnetosphere plasmas. As shown earlier in this chapter, in a number of applications the kappa function has been successfully used to fit both the low-energy "thermal" parts and the flux excess of suprathermal particles at higher energies, giving the opportunity of calculating the power law index that has proven to be a universal characteristic of planetary space environments and corresponds to energization and/or equilibrating processes in planetary magnetospheres. Furthermore, the extraction of other vital parameters for given plasma systems, such as flow speeds, plasma moments, and partial temperatures and densities, was made possible by the kappa distribution function, since it encapsulates the fundamental physical properties of space plasmas.

The κ-index is a very important parameter, since it is a prime indicator for systems that are not in local thermodynamic equilibrium and can be characterized as a source of "free" energy that can drive different plasma processes (Hapgood et al., 2011). For example, a metric that describes how close a given system is to equilibrium can be achieved through the κ-index (Livadiotis and McComas, 2010a) by using the following equation: $M_q = 4(q-1)/(q+1)$, where $q = 1 + 1/\kappa$. As explained earlier in this chapter, in some cases where the ion distributions are fit to kappas, a useful relationship between the core temperature and the κ-index may emerge: $\kappa \approx bT_{core} + \kappa_0$ (b and κ_0 are some positive constants) (Collier, 1999), implying that as the distributions become hotter, they also become more Maxwellian ($\kappa \to \infty$) and less suprathermal. So, the plasma with such characteristics will get close to equilibrium ($q \to 1, M_q \to 0$). According to this analysis, more Maxwellian distributions inside magnetospheres will be identified as "older" in the sense that they have undergone more velocity space diffusion. However, we note that the above formula does not imply a physical law, i.e., the T and κ are not necessarily correlated as such. By contrast, this formula is one of the many generalized polytropic relations that depend on local plasma streamlines.

Characterizing the plasma's thermodynamic properties is critical for constructing sophisticated MHD models. For example, the ideal gas law connects the plasma temperature (T) with the pressure (P) and density (n) of a given system, and if we consider its definition we will find an important property explained by Livadiotis (2015a), i.e., that the thermodynamic definition of temperature and the kinetic temperature coincide, even when the system relaxes to stationary states that are away from equilibrium. Furthermore, we should not forget that both temperatures and densities are key in determining one of the most central properties of plasmas: the Debye length (e.g., Bryant, 1996; Rubab and Murtaza, 2006; Gougam and Tribeche, 2011; Livadiotis and McComas, 2014a). Although the suprathermal particles (the high-energy tails shown in the kappa distributions) do not contribute significantly in plasma shielding, they do contribute significantly in the calculation of the plasma temperature.

Another example comes directly from spacecraft measurements. Although the authors that followed the methods described below did not necessarily use the kappa distribution specifically, their work highlights the need for a distribution that is able to provide the aforementioned fundamental parameters (such as the κ-index). Baumjohann and Paschmann (1989) were able to fit the magnetotail plasma data using the polytropic expression $P \propto n^a$ to calculate the mean polytropic index (a) of the Earth's plasma sheet. As they note, the determination of the polytropic index is not only important for the determination of whether steady state convection of plasma is possible or not, but it also has implications for the stability of the tail. Their calculated index was $a \sim 1.66$ (i.e., $\sim 5/3$) on average, implying that the plasma sheet behaves nearly adiabatically. However, the $a \sim 1.4$ index for the quiet plasma sheet showed that the magnetotail is not in thermodynamic equilibrium but rather continuously cools until new energy enters from outside. Spence and Kivelson (1990), using a 2-D model of adiabatic convection, explored the differences on previous calculations of the polytropic index by Baumjohann and Paschmann (1989) and Huang et al. (1989). Similar methods were used (Pudovkin et al., 1997; Borovsky et al., 1998) for determining the polytropic index for the plasma sheet and the specific entropy (Borovsky and Cayton, 2011) for Earth and other space environments, such as the solar wind (Newbury et al., 1997) and the heliosheath (Livadiotis and McComas, 2012). Nicolaou et al. (2014) used the method of correlation maximization (Livadiotis and McComas, 2013a) between density and temperature datasets to estimate the polytropic index variability at 1 AU; they found that the distribution of the polytropic indices is best described by a kappa function with a mean ≈ 1.8 and standard deviation ≈ 2.4.

12.7 Concluding Remarks

We have argued that the kappa distribution function is extremely useful and flexible in describing the energy spectra of the charged particles in planetary magnetospheres. The high-energy tails of the observed particle distributions inside magnetospheres were shown to deviate from the Maxwellian form. The kappa distribution functions have provided the observational and theoretical framework for capturing the thermal portion of the particle distributions, together with the power law tails (remarkably well most of the time), giving us insights about plasma processes and magnetospheric dynamics of planetary systems. However, a plethora of distinct cases of particle distributions over intense magnetospheric conditions highlighted the need for further development of new functional forms based on the classical kappa distributions. The search for distribution functions that best fit the particle spectra and their theoretical development is becoming more topical and has proven to be of paramount importance for understanding the physics of the magnetospheres. To date, all of the different functional forms seem to share the same basic ingredients: the kappa distribution function and the physical parameters that largely depend on it. All we need to do is keep asking the right questions.

12.8 Science Questions for Future Research

Future analyses and observations need to address the following questions:

1. What magnetospheric plasma processes dictate kappa distributions?
2. What are the interpretations of the mapping of κ values in magnetospheres?
3. What is the general function for quiet/active processes in magnetospheres?

Acknowledgments

The authors would like to thank Timothy A. Cassidy (LASP, Colorado, USA), Elias Roussos (MPS, Gottingen, Germany), and Thomas S. Sotirelis (JHU/APL, Maryland, USA) for their invaluable help in reviewing parts of this manuscript.

CHAPTER 13

Kappa Distributions and the Solar Spectra: Theory and Observations

E. Dzifčáková, J. Dudík

Astronomical Institute of the Czech Academy of Sciences, Ondřejov, Czech Republic

Chapter Outline

13.1 Summary 523
13.2 Introduction 524
13.3 Synthetic Line and Continuum Intensities 526
   13.3.1 Ionization and Recombination Rates: Behavior of the Ionization Equilibrium 526
   13.3.2 Excitation Rates 529
   13.3.3 Continuum 531
   13.3.4 KAPPA Package 533
   13.3.5 Synthetic Spectra and the Atmospheric Imaging Assembly Response to Emissions 534
13.4 Plasma Diagnostics From Emission Line Spectra 535
   13.4.1 Density Diagnostics 536
   13.4.2 Single-Ion Diagnostics of the κ-index 537
   13.4.3 Diagnostics Involving Ionization Equilibrium 540
   13.4.4 Diagnostics From Transition Region Lines: Si III 542
13.5 Differential Emission Measures for Kappa Distributions 544
13.6 Concluding Remarks 546
13.7 Science Questions for Future Research 547

13.1 Summary

Emitted spectra reflect the physical conditions in the emitting medium, including the distribution of particle energies. The kappa distributions influence line intensities through the rates of collisional processes involving ionization, recombination, and excitation. Generally, these distributions lower and widen the ionization peaks, which can also be shifted to lower or higher temperatures.

Kappa Distributions. http://dx.doi.org/10.1016/B978-0-12-804638-8.00013-9

This can be used in diagnostics together with ratios involving lines with different excitation energies or different behavior of the excitation cross-section with energy. Kappa distributions also influence the continua. Especially at flare temperatures and X-ray energies, the bremsstrahlung emissivity is greatly increased and exhibits a strong high-energy tail. The height of the ionization edges is also increased, although details depend on the temperature of the ionization equilibrium. The behavior of the synthetic spectra with the κ-index is reflected in the derived quantities, such as the responses of various extreme-ultraviolet (EUV) and ultraviolet (UV) filters used to observe the Sun, and also the differential emission measure (EM). Theoretical methods for the diagnostics of plasma parameters are presented together with their applications to observed spectra of the solar corona, transition region, and flares.

Science Question: What is the effect of kappa distributions in the solar spectra?

Keywords: Radiation mechanisms: nonthermal; Stars: coronae; Sun: corona; Sun: transition region; Sun: ultraviolet radiation; Sun: X-rays.

13.2 Introduction

The detection of kappa distributions in solar wind (e.g., see Chapters 10 and 11) suggests a possibility that they could originate in the solar corona (see, e.g., Vocks and Mann, 2003). Indeed, it has been argued that the solar and stellar coronae are non-Maxwellian above 1.05 stellar radii (Scudder and Karimabadi, 2013). The transition region between the corona and the underlying chromosphere could also be non-Maxwellian due to the presence of large temperature gradients at relatively low densities (Roussel-Dupré, 1980; Shoub, 1983). A population of suprathermal particles in the solar corona could also be generated through coronal heating (Testa et al., 2014), currently thought to proceed via nanoflares, small-scale impulsive events of unknown nature releasing magnetic energy (e.g., Klimchuk et al., 2010; Viall and Klimchuk, 2011; Bradshaw et al., 2012; Cargill, 2014) either by reconnection or by wave–particle interactions (see Vocks et al., 2008).

The high temperatures characteristic of the solar corona make it impossible to perform any physical measurements in situ. Instead, the emitted radiation is the only source of information about the physical conditions therein. The coronal radiation is typically in the form of optically thin emission lines originating in the multiple ionized metal ions such as Fe, Ca, and Si. These lines are typically located in the X-ray, EUV, and UV parts of the spectrum, with a small number of predominantly forbidden lines being present at the optical and infrared wavelengths (e.g., Phillips et al., 2008).

That the formation of emission lines is affected by non-Maxwellian distributions has been recognized by Gabriel and Phillips (1979). These authors considered the influence of a power law high-energy tail at energies greater than about 10 keV on the excitation of the allowed Fe XXV resonance line in the solar flares.

Since the excitation threshold of this line is about 6 keV, the excess of high-energy electrons causes an increase in the excitation of the resonance line. Contrary to that, the Fe XXIVd dielectronic satellites are unaffected by the high-energy tail, so their intensities do not change. Thus the Fe XXIVd/Fe XXV ratio of line intensities would decrease.

The situation described by Gabriel and Phillips (1979) pertained to solar flares; however, the principle is the same for all optically thin lines. Namely, changes in the distribution function would lead to changes in the line intensities, where they can produce measurable effects. The solar flares are, however, unique, since the high-energy tail produced there can be directly observed via a bremsstrahlung continuum emission (Fig. 13.1; Lin and Hudson, 1971; Brown, 1971; also, see the review of Fletcher et al., 2011), and the coronal X-ray sources can be easily fitted with a kappa distribution (Kašparová and Karlický, 2009; Oka et al., 2013, 2015). The formation of a kappa distribution in solar flares has also been derived analytically (Bian et al., 2014) if turbulence with a diffusion coefficient inversely proportional to velocity is present.

Direct detection of the high-energy particles in the corona of a quiet Sun has only yielded upper limits on their EMs (Hannah et al., 2010). Therefore one has to rely on the use of emission lines to detect the kappa distributions in the solar

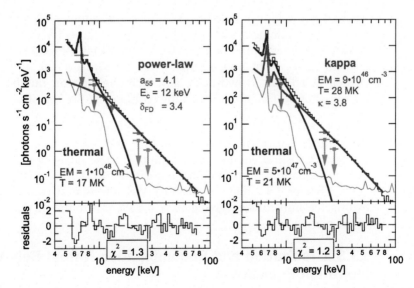

FIGURE 13.1

Reuven-Ramaty High-Energy Solar Spectroscopic Imager observations of a solar flare that occurred on December 31, 2007. Spatially integrated photon spectra are fitted with a thermal (Maxwellian) component and a power law tail (left). The slope of the power law tail is 3.4, and a low-energy cutoff E_c is imposed at 12 keV. The same spectrum can be fitted with a thermal component and a kappa distribution with $\kappa = 3.8$. The temperatures and emission measures of each fit component are indicated. The *gray curve* indicates the background (nonflaring) emission. The residuals in the lower panels are shown in the units of standard deviation as derived from the photon noise. Source: Oka et al. (2013).

corona. The methods of such diagnostics are described in this chapter. Although these methods have been developed for applications to the solar spectra, they are also applicable to the spectra of other active stars, nebulae, or galaxies, provided that the emitting medium is optically thin and excited by inelastic electron–ion collisions. A theoretical background on the formation of emission lines for kappa distributions is given in Section 13.2, while an overview of diagnostic methods and results is described in Section 13.3. Section 13.5 discusses multi-thermal analysis for kappa distributions. Finally, the concluding remarks are given in Section 13.6, while three general science questions for future analyses are posed in Section 13.7.

13.3 Synthetic Line and Continuum Intensities

Solar corona and transition regions are optically thin emitting environments. The observed radiation is therefore given by the sum of all contributions arising along the line of sight l of the observer. The intensity I_{ij} of a line λ_{ij} is then an integral product of the radiative transition probability A_{ij} from the upper level i to the lower level j, the energy of the emitted photon, and the density of emitters (e.g., Mason and Monsignori-Fossi, 1994; Phillips et al., 2008):

$$I_{ij} = \frac{A_{ij}}{4\pi} \frac{hc}{\lambda_{ij}} \int n\left(X_i^{+k}\right) dl = \frac{A_{ij}}{4\pi} \frac{hc}{\lambda_{ij}} \int \frac{1}{n_e} \frac{n\left(X_i^{+k}\right)}{n(X^{+k})} \frac{n(X^{+k})}{n(X)} A_X n_H n_e dl, \tag{13.1}$$

where h is the Planck constant, $n\left(X_i^{+k}\right)$ is the number density of the excited ions $+k$ of the element X with an electron on the upper level i, $n(X^{+k})$ is the total density of k-times ionized ions of element X, $n(X)$ is the total number density of element X, n_H is hydrogen density, $A_X = n(X)/n_H$ is the abundance of the element X relative to hydrogen, and n_e is the electron number density. The quantity $EM = \int n_H n_e dl$ is called the (column) EM of the plasma.

In Eq. (13.1), the fraction $n\left(X_i^{+k}\right)/n(X^{+k})$ describes the excitation state, while $n(X^{+k})/n(X)$ corresponds to the relative ion abundance. For the plasma-satisfying statistical equilibrium conditions, these quantities can be calculated from the equations of ionization and excitation equilibrium (e.g., Phillips et al., 2008, Chapter 4.3.2 and 4.5 therein). We note that at coronal temperatures, hydrogen is fully ionized, giving $n_H \approx 0.83\, n_e$ due to the chemical composition of the plasma.

13.3.1 Ionization and Recombination Rates: Behavior of the Ionization Equilibrium

In the solar corona, ionization and recombination proceed by inelastic electron–ion collisions. The ionization processes considered here are direct electron impact ionization and excitation–autoionization. The latter is a process where a previously doubly excited ion spontaneously ionizes without a radiative transition. The most important recombination processes are radiative recombination by electron capture and dielectronic recombination.

Dielectronic recombination is a resonance process inverse to autoionization. It involves a radiative transition during the deexcitation of the doubly excited state. Other ionization and recombination processes are usually not important in coronal conditions, although photoionization may be important for transition region ions. In general, the rate coefficient R of any collision process is given by

$$R = \int_{\varepsilon_0}^{\infty} v\sigma(v)f(v)dv = \int_{\varepsilon_0}^{\infty} \left(\frac{\varepsilon}{2m_e}\right)^{\frac{1}{2}} \sigma(\varepsilon)f(\varepsilon)d\varepsilon, \qquad (13.2)$$

where ε is the incident particle energy, ε_0 is the energy threshold, $\sigma(\varepsilon)$ is the cross-section of the process that becomes zero for $\varepsilon < \varepsilon_0$, and $f(\varepsilon)$ is the isotropic particle distribution function (Chapters 1 and 4; Livadiotis and McComas, 2009, 2013a; Livadiotis, 2015a).

The change of the shape of the distribution with κ influences both the ionization and recombination rates and therefore the ionization equilibrium for any element (Dzifčáková, 1992, 2002; Dzifčáková and Dudík, 2013). An example of the changes in the total ionization and recombination rates for Fe XI (that is, Fe^{+10}) is shown in Fig. 13.2. In general, the ionization rate is dominated by the high-energy particles. For this reason, the ionization rate increases with decreasing κ, especially at low temperatures. Compared to the Maxwellian distribution, the ionization rate can be increased by several orders of magnitude, depending on T. However, at the temperatures corresponding to the maximum of the relative ion abundance, the total ionization rates for the kappa distributions can be lower than the corresponding Maxwellian one. We note that the maximum of the relative ion abundance is attained at temperatures where the total ionization rate equals the total recombination

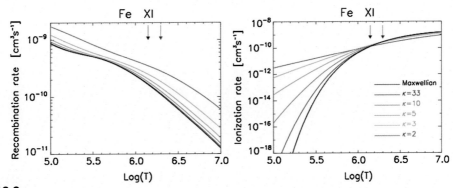

FIGURE 13.2
Dependence of the total recombination rates (left) and total ionization rates (right) on κ for the Fe XI ion. Individual line styles and colors stand for different values of κ. *The black and red arrows* denote the temperature of the maximum ion abundance for the Maxwellian and $\kappa = 2$ distributions, respectively.

rate, so that this temperature depends on the behavior of the recombination rates as well.

The radiative recombination rates are dominated by the low-energy electrons, since the corresponding cross-section decreases as a power law with electron energy. The dielectronic recombination rate reflects the number of particles in the distribution at the respective resonance energies. For Fe XI, the total recombination coefficient is dominated by the radiative recombination. At a constant temperature, it is enhanced for the kappa distributions compared to the Maxwellian one, this being a consequence of the increase of the number of low-energy electrons with κ.

These changes in the ionization and recombination rates are reflected in the resulting ionization equilibrium (see Fig. 13.3). The ionization peaks for the

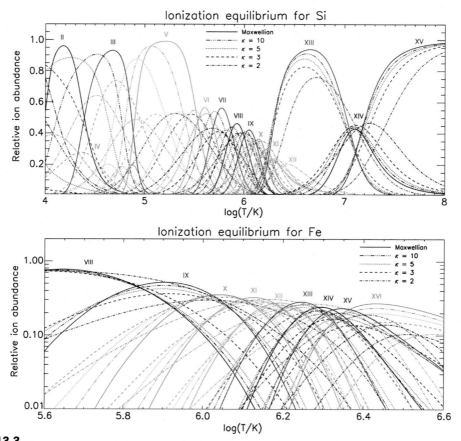

FIGURE 13.3

Collisional ionization equilibrium for Si (top) and Fe (bottom) for kappa distributions. Individual ion stages are denoted by different colors, while different values of κ are indicated by different line styles. Note that the ionization peaks for the kappa distributions are wider and flatter compared to the Maxwellian distribution, especially for Fe IX, and can be shifted to higher temperatures at $\log(T\,[\mathrm{K}]) > 6$, while at transition region temperatures, there are significant shifts toward lower temperatures. Note the logarithmic scale on the y axis in the bottom picture. Source: Dzifčáková and Dudík (2013).

kappa distributions are typically wider and flatter for lower κ. They can also be shifted in temperature. A significant shift toward lower temperatures occurs for transition region ions (formed below 1 MK). This shift can be almost an order of magnitude for $\kappa = 2$ compared to the Maxwellian distribution, e.g., for Si IV. For the coronal and flare temperatures, the shift can be in both directions, typically toward the low temperatures for large values of κ and in the opposite direction for small values of $\kappa < 3$ (Fig. 13.3). These changes in the ionization equilibrium offer possibilities to diagnose the distribution function directly, e.g., from measurements of ion composition in the solar wind (e.g., Owocki and Scudder, 1983) or from the ratios of emission line intensities originating in neighboring ionization stages (Section 13.3.3).

The first analysis of the effect of kappa distributions on the abundance ratios of O VII/O VIII and Fe XII/Fe XIII was performed by Owocki and Scudder (1983). These calculations allowed them to explain observations of the ionization state of the solar wind that differ from the classical ionization theory. Subsequently, Dzifčáková (1992) calculated the Fe ionization equilibrium for kappa distributions using the atomic data of Arnaud and Rothenflug (1985) and Shull and Van Steenberg (1982). In these calculations, the direct ionizaton and auto-ionization rates are calculated using a cross-section. For the radiative recombination rates, the approximate expressions for the cross-sections by Owocki and Scudder (1983) were used. Luhn and Hovestadt (1985) and Ko et al. (1996) presented mean charges for various elements, mainly for applications in the solar wind and solar energetic particles. Newer atomic data by Arnaud and Raymond (1992), Verner and Ferland (1996), and Mazzotta et al. (1998) were used by Dzifčáková (2002) and Wannawichian et al. (2003) to recalculate the ionization equilibria using kappa distributions. Finally, Dzifčáková and Dudík (2013) used the latest ionization cross-sections from Dere (2007) and recombination cross-sections from the work of Badnell et al. (2003). Again, the ionization rates were calculated directly from cross-sections, while the recombination rates were obtained using approximate methods. Hahn and Savin (2015) approximated the kappa distributions by a sum of Maxwellian distributions with different temperatures, which allowed them to calculate the corresponding rates for the kappa distributions as a sum of Maxwellian rates at corresponding temperatures. Their ionization equilibria are in good agreement with those of Dzifčáková and Dudík (2013). These authors also showed that electron impact multiple ionization may be important when using kappa distributions with a low κ.

13.3.2 Excitation Rates

The emission lines of the solar corona in the EUV and X-rays are dominantly excited by electron–ion collisions. It is customary to express the rate of the collisional excitation $j \rightarrow i$ in terms of the dimensionless, distribution-averaged, collision strength arguments Υ_{ji}, given by

$$\Upsilon_{ji}(T, \kappa) = \frac{\sqrt{\pi}}{2} \exp\left(\frac{\Delta\varepsilon_{ij}}{k_B T}\right) \int_{\Delta E_{ij}}^{\infty} \Omega_{ji}(\varepsilon) \left(\frac{\varepsilon}{k_B T}\right)^{\frac{1}{2}} f_\kappa(\varepsilon) d\varepsilon, \tag{13.3}$$

where $\Delta\varepsilon_{ij}$ is the energy difference between levels i and j, k_B is the Boltzmann constant, and $\Omega_{ji}(\varepsilon)$ is the dimensionless cross-section, called the collision strength. Note that the excitation can only occur by collisions with electrons having $\varepsilon \geq \Delta\varepsilon_{ij}$. The collisional rate corresponding to Υ_{ji} can be obtained by multiplication of Υ_{ji} by an appropriate constant and a $T^{-1/2}\exp(-\Delta\varepsilon_{ij}/(k_B T))$ factor. Fig. 13.4 (right-hand side) shows typical changes of Υ_{ji} with decreasing κ, a strong increase for the lower temperatures, and a small decrease for the higher ones in comparison with the Maxwellian distribution.

The deexcitation of level i can also occur via collisions as a reverse process; however, under most circumstances, deexcitation proceeds by spontaneous emission, described as the corresponding Einstein coefficient A_{ij}; see also Eq. (13.1). However, a small subset of excited levels has only forbidden or semiforbidden radiative transitions characterized by small values of A_{ij}. For such transitions, collisional deexcitation can be important. Furthermore, we note that for some transitions, especially in the forbidden ones in the UV and optical parts of the spectrum, the contribution to excitations from proton–ion collisions or from photoexcitation by the photospheric radiation of the Sun cannot be neglected.

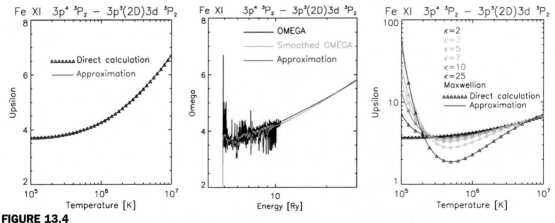

FIGURE 13.4

Left: Approximation of the Maxwellian-averaged collision strength Υ_{ji} for the Fe XI 188.216 Å line (transition 3s² 3p⁴ ³P₂–3s² 3p³ (2D) 3d ³P₂). The direct calculation that was obtained by integrating Ω_{ji} over the Maxwellian distribution is shown in the *triangles*, while the approximation is shown by the *red line*. Middle: The corresponding collision strength Ω_{ji} (*black*) together with the approximation obtained by fitting Υ_{ji} (left, *red color*) and a smoothed Ω_{ji} (*cyan*). The resonance region extends up to about 10 Ry (136 keV). Right: Distribution-averaged collision strengths Υ_{ji} for kappa distributions. Direct calculations from Dudík et al. (2014a) are shown by *triangles*, while those obtained by the approximate method of Dzifčáková et al. (2015) are shown by *full lines*.

The calculation of Υ_{ji} can, in principle, be done directly, i.e., using the collision strengths $\Omega_{ji}(\varepsilon)$ (Fig. 13.4). For kappa distributions, the integral in Eq. (13.3) can then be calculated numerically (Bryans, 2005; Dudík et al., 2014a). In practice, the coronal ions contain many thousands of energy levels, and each transition contains resonance contributions at lower incident energies (Fig. 13.4, middle), so Ω_{ji} must be calculated at a sufficiently dense grid of incident energies ε. The resulting files are prohibitively large, up to several tens of Gbytes, and thus are not publicly accessible. Υ_{ji} can then be calculated using an approximate method, first developed by Dzifčáková (2006). This approximate method assumes a functional form on $\Omega_{ji}(\varepsilon)$ (Mewe, 1972) in the form of a power law series in nondimensionalized energy plus an additional logarithmic component. The coefficients of such expansions are then determined by fitting the Maxwellian distribution-averaged collision strength Υ_{ji}. Subsequently, the Υ_{ji} for kappa distributions is calculated analytically using Eq. (13.3). An example of such approximation for the $3s^2\,3p^4\,{}^3P_2 - 3s^2\,3p^3\,(2D)\,3d\,{}^3P_2$ transition in Fe XI at 188.216 Å is shown in Fig. 13.4. The approximate method was validated by Dzifčáková et al. (2015) by comparison to direct calculations of Dudík et al. (2014a), who used the original $\Omega_{ji}(\varepsilon)$ calculated for Fe XI by Del Zanna et al. (2010) and for Fe XVII by Del Zanna (2011). A typical precision of 5% was found, with all cases being within 10%. Similar precisions were found by Hahn and Savin (2015) by using the Maxwellian decomposition technique, involving the decomposition of a kappa distribution into a series of Maxwellian distributions at different temperatures. (Note: This expression of kappa distribution is examined by "superstatistics", e.g., see Chapter 6; Beck and Cohen, 2003; Tsallis and Souza, 2003a,b; Touchette, 2004; Beck & Cohen, 2004; Beck et al., 2005; Chavanis, 2005; Vignat 2005; Rajagopal 2006, Tsallis 2009b; Schwadron et al., 2010; Livadiotis et al. 2016).

13.3.3 Continuum

In addition to line radiation, the plasma can radiate via processes resulting in continuous spectrum, such as the electron–ion bremsstrahlung and radiative recombination. These two processes are the dominant contributors to the continuum radiation of the solar corona, transition region, and flares (e.g., Phillips et al., 2008). Typically, the continuum intensity is important in X-rays at high temperatures in solar flares. The corresponding theoretical emissivities of the respective continua were obtained by Dudík et al. (2012), and an example is shown in Fig. 13.5. Due to the relatively low abundances of elements with $Z \geq 3$, the total continuum radiation is dominated by completely ionized hydrogen and helium.

The electron–ion bremsstrahlung spectrum for kappa distributions is characterized by a significant increase of emissivity by orders of magnitude at short wavelengths. Here, the contribution to bremsstrahlung radiation is dominated by the high-energy tail of the kappa distribution. The effect is already noticeable for $\kappa \sim 33$, where the increase at 1 Å is already nearly an order of magnitude for $T = 10$ MK, corresponding to a flare plasma. At shorter wavelengths and smaller κ, the effect is much more pronounced.

Near the peak of the emissivity at longer wavelengths, the bremsstrahlung spectrum is flatter for kappa distributions compared to Maxwellian distributions. It is increased again at long wavelengths, where the emissivity curves for different κ are nearly parallel to each other. This makes it impossible to detect kappa distributions, e.g., from radio radiation, where they lead only to modification of the brightness temperature of the radio source.

The strongly enhanced bremsstrahlung emission at short X-ray wavelengths is routinely observed during solar flares by the Reuven-Ramaty High-Energy Solar Spectroscopic Imager (Lin et al., 2002). It is usually represented as a non-Maxwellian power law tail in both the photon and electron spectra (e.g., Brown, 1971; Lin and Hudson, 1971; Holman et al., 2003; Veronig et al., 2010; Fletcher et al., 2011; Oka et al., 2013), which is also observed in microflares (Hannah et al., 2011). Kašparová and Karlický (2009) were the first to interpret this high-energy component with a kappa distribution. These authors showed that the spectra of partially occulted solar flares are well represented by a kappa distribution. The occultation by the solar disk means that only the emission of the coronal source region, believed to be close to the reconnection site, is observed and is not contaminated by the dominantly thermal emission of the bright flare loops. Following their work, Oka et al. (2013) fitted a spectrum of the December 31, 2007, flare with different models, including a power law tail, a kappa distribution (see Fig. 13.1), and a combination of the super-hot thermal and a power law component. It was shown that the power law fit produced a worse approximation than the kappa distribution and the combination of a thermal and a power law component. The power law component is also incompatible with the upper limit on the electron distribution at lower energies derived from observations (Oka et al., 2015) of the Atmospheric

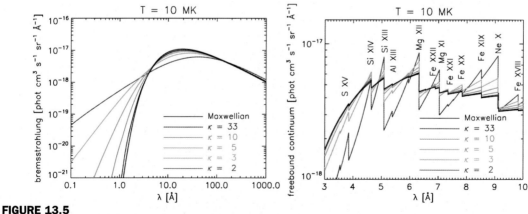

FIGURE 13.5
Free—free (bremsstrahlung) and free-bound continua for kappa distributions with temperature $T = 10$ MK. The bremsstrahlung is increased especially in the X-rays (left), while the free-bound continuum shows a significant increase in the height of the individual ionization edges (right). The ion species producing the individual ionization edges are indicated. The curves need to be multiplied by the emission measure to obtain the individual emissivities. Source: Dudík et al. (2012) reproduced with permission © ESO.

Imaging Assembly (AIA) onboard the Solar Dynamics Observatory (SDO, Pesnell et al. 2012). The interpretation of the kappa distribution is preferable, since it requires only three free parameters, compared to five terms needed in the case of the combination of thermal and power law components. Also, the power law requires an artificial low-energy cutoff (≈ 30 keV), which introduces some difficulties in interpreting the obtained temperatures and densities. Contrary to that, the kappa distribution extends seamlessly into a thermal core and allows for a natural estimation of the electron number densities and the relative energy carried by the accelerated particles (Oka et al., 2013). The interpretation of the flare X-ray emission using a kappa distribution was supported by the theoretical derivation of Bian et al. (2014), who assumed turbulence with a diffusion coefficient inversely proportional to velocity.

The free-bound continuum, arising due to the radiative recombination, is dominated by the low-energy electrons (Section 13.3.1). Near the ionization energy from a given level in an ion, the cross-section reaches its maximum values, leading to the appearance of ionization edges of various metal ions. The kappa distributions, characterized by an increase in the number of low-energy particles with decreasing κ at a given temperature, lead to a significant increase of the ionization edges. The ionization edges are strongly pronounced, especially for low κ, with some of the ionization edges being increased by a factor of several for $\kappa = 2$ compared to the Maxwellian distribution (Fig. 13.5). Details depend on the behavior of the relative ion abundance with κ and T, so that at a particular temperature, some ionization edges may actually be decreased.

In principle, the relative increase in the ionization edges with respect to the neighboring continuum can be utilized for the diagnostics of κ. In practice, however, the X-ray spectrum is crowded with a multitude of emission lines, so that a reliable measurement of the continuum with a sufficient signal-to-noise ratio would be difficult to perform. Some ionization edges, such as the Si XIV edge at 4.64 Å, are relatively unobscured by lines and could be used for diagnostics in the future if significantly sensitive instrumentation becomes available.

13.3.4 KAPPA Package

The KAPPA package (Dzifčáková et al., 2015) is a modification of the freely available CHIANTI 7.1 database and software (Dere et al., 1997; Landi et al., 2013), widely used for Maxwellian spectral synthesis. The KAPPA package contains the ionization equilibria as well as ionization and recombination rates for the values of $\kappa = 2, 3, 4, 5, 7, 10, 15, 25$, and 33. For these values of κ, Υs are also available for all transitions included in CHIANTI 7.1. Routines for the calculation of synthetic line and continuum spectra have been developed as well. These are based on the original CHIANTI routines, with the first parameter always being the value of κ (the KAPPA package is freely available at http://kappa.asu.cas.cz).

13.3.5 Synthetic Spectra and the Atmospheric Imaging Assembly Response to Emissions

An example of the synthetic spectra for a Maxwellian and a kappa distribution with $\kappa = 2$ is shown in Fig. 13.6 for $\log(T[K]) = 6.25$. There, a portion of the EUV spectrum near 193 Å is shown. These spectra were calculated using the CHIANTI and KAPPA packages. It can be seen that the spectra are dominated by emission lines. The continuum was included in the calculations, but its intensity is weak compared to the line intensities. The spectra look very different for the Maxwellian and $\kappa = 2$. For the Maxwellian distribution, the spectrum is dominated by the Fe XII 195.119 Å line, while for $\kappa = 2$, the intensity of this line is about a factor of 2 lower. This is a result of both the ionization and excitation equilibrium. The effect of the ionization equilibrium is the stronger one. At $\log(T [K]) = 6.25$, the relative ion abundance of Fe XI is significantly increased for $\kappa = 2$ compared to the Maxwellian distribution (Fig. 13.3, bottom), while the relative abundances of Fe XII and Fe XIII are decreased as their respective peaks are shifted to higher temperatures. This is reflected in the character of the spectra that shows an increase of the Fe XI lines, while the Fe XII and Fe XIII lines are decreased.

These wavelengths correspond to the bandpass of the 193 Å filter of the AIA (Lemen et al., 2012; Boerner et al., 2012), an instrument widely used to image the full Sun solar corona with a 0.6 arc sec spatial resolution. The filter transmissivity as a function of the wavelength λ is shown by the thick brown line. The integral product of the filter transmissivity and the emitted spectrum constitutes the contribution to the filter response to the plasma emission. Due to the behavior of the ionization equilibrium, the AIA 193 Å response, dominated by

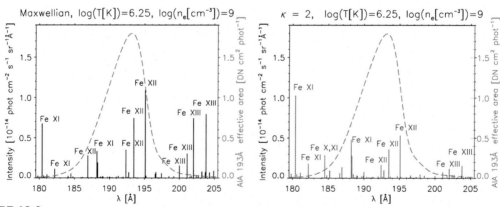

FIGURE 13.6

Synthetic spectra in the extreme-ultraviolet near 193 Å for the Maxwellian distribution (left) and the kappa distribution with $\kappa = 2$ (right). The temperature and density are the same in both images: $\log(T [K]) = 6.25$ and $\log(n_e[cm^{-3}]) = 9.0$. The pass-band of Atmospheric Imaging Assembly 193 Å is shown by the *dashed brown line*. AIA, Atmospheric Imaging Assembly. Source: Dzifčáková et al. (2015).

Fe XII, is shifted toward higher temperatures (Fig. 13.7). Secondary peaks due to O V, Ca XVII, and Fe XXIV (O'Dwyer et al., 2010) are present and behave similarly as the main one. An exception is the O V peak at $\log(T\,[K]) \approx 5.4$, which disappears for $\kappa \leq 3$ due to the widening of the O V relative ion abundance. The responses of other AIA bands show a similar behavior (Dzifčáková et al., 2015).

13.4 Plasma Diagnostics From Emission Line Spectra

The intensity of a spectral line, Eq. (13.1), arising in a given structure within solar corona is in principle a function of four plasma parameters: EM, temperature T, electron density n_e, and κ. The dependence on T and κ arises through the ionization and excitation equilibrium. The dependence on n_e arises through the relative level population, i.e., the excitation equilibrium, since the total number of electron−ion collisional excitations depends on the product of the rate coefficient and n_e, while deexcitation is radiative and does not depend on n_e. Details, however, are complex due to the statistical equilibrium involving many hundreds of levels within a single ion. EM is a multiplicative parameter that can be obtained by multiplying the synthetic spectrum (of a given T, n_e, and κ) with the observed intensity.

The diagnostics of T, n_e, and κ are usually evaluated by using ratios of the intensities. This is because the intensity ratios are independent of EM. Since the line intensity ratio is a function of three parameters, it is not possible to diagnose κ alone, but all three parameters must usually be diagnosed at the same time. A rare exception are ions such as Fe XVII (Section 13.4.2), for which the line ratios are independent of n_e. For other ions, the density can be diagnosed independently using the technique described in Section 13.4.1. The diagnostics of T and κ are always coupled, since they are both parameters of the electron distribution

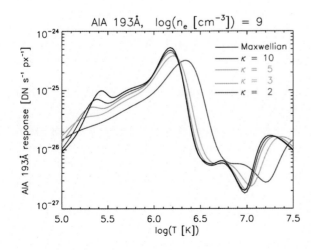

FIGURE 13.7

Atmospheric Imaging Assembly 193 Å response to emission for the kappa distributions for $\log(n_e[\mathrm{cm}^{-3}]) = 9.0$. Source: Dzifčáková et al. (2015).

function, requiring simultaneous diagnostics from the same lines. These lines should be formed from levels with significantly different excitation energies to maximize the effect of the high-energy tail on the electron excitation rates and therefore on the line intensity ratio. However, in practice this can be problematic because of the limited spectral bands of the existing EUV spectrometers.

The analysis of the solar spectra is furthermore complicated by the optically thin nature of the solar corona, resulting in the presence of a diffuse background emission or even multiple emitting structures along the line of sight. Then the situation is characterized by a differential emission measure (DEM) (Section 13.5). Furthermore, the analysis may be complicated by various instrumental problems, such as calibration uncertainties, degradation of the detectors (e.g., Del Zanna, 2013a), etc.

13.4.1 Density Diagnostics

In many cases, it is advantageous to diagnose the electron density prior to the diagnostics of T and κ. This is typically done by using an intensity ratio of two lines of the same ion, one being an allowed line and the other one originating in a metastable level. Iron ions emit many lines observed in the EUV by the Extreme-Ultraviolet Imaging Spectrograph (EIS) onboard the Hinode satellite (Culhane et al., 2007) that can be used for the diagnostics of n_e under the assumption of a Maxwellian distribution (e.g., Young et al., 2009; Watanabe et al., 2009; Del Zanna et al., 2012a). The dependency of such density-sensitive ratios on T is typically small. Dzifčáková and Kulinová (2010) were the first to note that many of the Fe density-sensitive line ratios do not depend strongly on κ as well as T and can thus be used as an independent density diagnostics. Suitable ratios of Fe lines were listed and tested. Mackovjak et al. (2013) explored density diagnostics from EUV line ratios of Si, S, Ar, and Ni ions. Dudík et al. (2014a) used state-of-art atomic data corresponding to CHIANTI (Del Zanna et al., 2015a) to find suitable ratios for independent density diagnostics across the entire wavelength range, from X-rays and EUV to the visible and infrared parts of the spectrum. Dudík et al. (2014b) investigated density diagnostics from the UV transition region O IV lines observed by the Interface Region Imaging Spacecraft (IRIS; DePontieu et al., 2014). They showed that some line ratios are independent of κ, but others are not. Indeed, in solar flares it is observed that some of the latter ratios are not consistent with a Maxwellian distribution (Polito et al., 2016).

An example of electron density diagnostics from the Fe XI 182.167 Å/188.216 Å and Fe XIII 203.8 Å sbl/202.044 Å is shown in Fig. 13.8. Here, the acronym "sbl" denotes that the Fe XIII 203.8 Å line is a self-blend of four separate transitions occurring close in wavelength (see Del Zanna and Storey, 2012; Dudík et al., 2014a, Table B.4). In Fig. 13.8, the black lines denote the line intensity ratio calculated for the Maxwellian distribution, while red lines correspond to $\kappa = 2$. Note that $\kappa = 2$ is the lowest value of κ available within the KAPPA package (Section 13.3.4). Full lines correspond to temperatures of the maximum of the

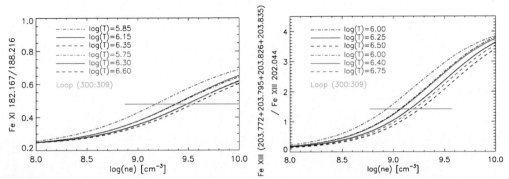

FIGURE 13.8

Density diagnostics from Fe XI and Fe XIII line ratios. *Black curves* stand for ratios calculated for the Maxwellian distribution, while *red curves* represent $\kappa = 2$. Individual line styles represent the temperature, while temperatures corresponding to the maximum of the peak of the ionization equilibrium are shown by *full lines*. The observed line ratio is indicated by a *horizontal azure line*. Source: Dudík et al. (2015).

relative ion abundance, while the dotted and dashed lines correspond to temperatures where the ion abundance has 1% of its maximum value. Thus these three lines denote the approximate temperature range where the ion is formed for each distribution. Their spread is directly proportional to the uncertainty in the resulting n_e due to the temperature dependence of the density-sensitive ratio. The dependence of the ratio on κ can be ascertained by the shift between the black and red curves. We see that for $\kappa = 2$, the diagnosed densities are typically lower by approximately 0.1 in the decadic logarithm.

The observed line ratios from Dudík et al. (2015) are denoted by azure lines in Fig. 13.8. These ratios were used for the diagnostics of electron density in a transient coronal loop that shows evidence for kappa distributions (see Section 13.4.3). We see that the plasma is consistent with $\log(n_e\ [\mathrm{cm}^{-3}])$ being in the range of 8.8–9.6 irrespective of T or κ. This result is a necessary prerequisite used in Section 13.4.3 to diagnose κ, albeit with a high uncertainty.

13.4.2 Single-Ion Diagnostics of the κ-index

Simultaneous diagnostics of T and κ using intensities of lines produced by the same ion is the most reliable theoretical diagnostic method. This is because such diagnostics avoid the use of the relative ion abundance factor $n(X^{+m})/n(X)$ that can be a source of uncertainty due to the uncertainties in the underlying ionization and recombination cross-sections as well as possible departures of the plasma from ionization equilibrium that could occur if the plasma has been rapidly heated or is cooling strongly.

Since two parameters are being diagnosed simultaneously, at least two line ratios are needed. The diagnostics is then typically done using the ratio–ratio diagram, involving a dependence of one ratio $r_y = I_3/I_4$ on the other ratio $r_x = I_1/I_2$ (see Fig. 13.9). Usually, it is sufficient to take I_4 equal to I_1 or I_2, so

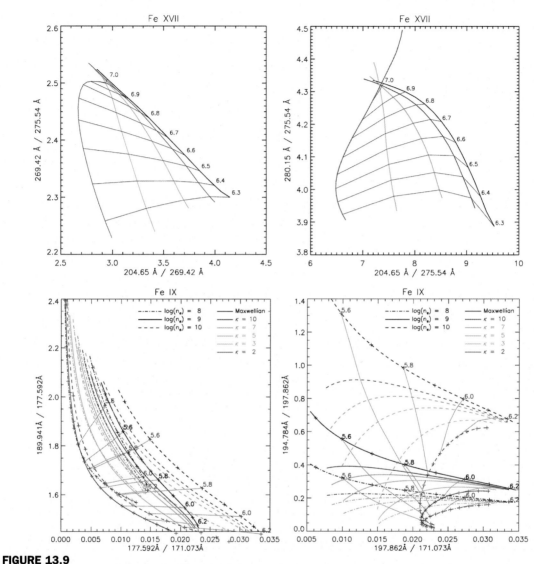

FIGURE 13.9

Theoretical single-ion diagnostics from Fe XVII (top) and Fe IX (bottom). Note that the Fe IX ratios involve only lines from the short-wavelength channel of the Hinode/ Extreme-Ultraviolet Imaging Spectrograph. Individual line styles stand for different densities: $\log(n_e[\text{cm}^{-3}]) = 8$ (*dash-dot*), 9 (full), and 10 (*dashed lines*). These densities cover the density range typical of the solar corona. Individual colors stand for different values of κ. The *thin gray lines* connect points of constant $\log(T[\text{K}])$ at the curves for different κ but constant n_e. Source: Dzifčáková and Kulinová (2010) and Dudík et al. (2014a).

that in principle only three lines are needed. The ratios will be sensitive to T and κ if the lines correspond to transitions with a strongly different dependence of the excitation cross-sections on energy or if the lines are widely separated in λ, so that the corresponding excitation thresholds occur at widely different energies, with the individual transitions being excited by different parts of the electron energy distribution.

Dzifčáková and Kulinová (2010) and Mackovjak et al. (2013) explored the possibility of diagnosing T and κ simultaneously from the Hinode/EIS spectra. They used the atomic datasets corresponding to the latest available versions of CHIANTI at the time and investigated both single-ion and multi-ion diagnostics. Several single-ion ratios of lines of Fe and O sensitive to κ were found. However, the single-ion O IV ratios sensitive to κ involved lines that are in practice too weak to be observed with EIS (Mackovjak et al., 2013). The sensitivities of the Fe VIII—XV single-ion ratios to κ were found to be weaker than their sensitivities to n_e, since the transitions are close in wavelength. For the Fe XVII, however, the ratios are independent of n_e (Fig. 13.9, top; see also Dzifčáková and Del Zanna, 2012), because this ion does not have any meta-stable levels at coronal and even flare densities for any κ-index. Examples of the Fe XVII ratio—ratio diagrams are shown in the top part of Fig. 13.9. Dzifčáková and Kulinová (2010), however, did not perform the diagnostics of κ using EIS observations and single-ion ratios.

The calculations of the collision strengths Ω_{ji} for the coronal Fe ions underwent a significant improvement (Del Zanna et al., 2010, 2012a, 2012b, 2014; Del Zanna and Storey, 2012, 2013). In addition to density diagnostics, Dudík et al. (2014a) used these calculations and direct integration of Υ_{ji} (Section 13.3.2) to investigate the single-ion diagnostics of T and κ from the Fe IX—XIII spectra. It was found that most EUV line ratios do not offer any sensitivity to κ. A small number of ratios were found to be sensitive to κ. However, in line with Dzifčáková and Kulinová (2010), these line ratios are dominantly sensitive to density. Additionally, the sensitivity of the EIS lines to κ for a known density was found to be typically of the order of $\sim$ few tens %. This is due to the relatively small wavelength range of EIS (171—211 Å and 250—290 Å) and the similar energy dependence of the cross-sections. Most of these transitions are to the lowest five energy levels. This level of sensitivity is similar to the EIS calibration uncertainty, which is 20% (Culhane et al., 2007), precluding such diagnostics.

Among those investigated, the Fe IX ion was found to be a notable exception. Several ratio—ratio diagrams offer a strong sensitivity to κ of the order of a factor of several. An example of the 177.592 Å/171.073 Å—189.941 Å/177.592 Å ratio—ratio diagram is shown in the bottom left panel of Fig. 13.9. In this figure, individual line styles stand for different densities, while colors stand for individual values of κ. We see that the ratio—ratio diagram is dominantly sensitive to κ and only weakly sensitive to n_e. In addition, Fe IX produces an observable transition between levels 13 and 148 at 197.862 Å ($3s^2\ 3p^5\ 3d\ ^1P_1$—$3s^2\ 3p^5\ 4p$ $^1S_0$). This transition involves different energy levels than all of the other Fe IX transitions observed by EIS. Ratio—ratio diagrams for Fe IX involving this line offer a strong sensitivity to κ. The best example is shown in Fig. 13.9 at the bottom right. The sensitivity to κ increases with increasing n_e and decreasing T.

The increase of sensitivity with T is not surprising, since at lower temperatures the high-energy tail of a kappa distribution can lead to a stronger increase in the excitation of the upper level.

Other notable exceptions are the forbidden lines in the UV, visible, or infrared parts of the spectrum. The Fe X 6378.26 Å line is one of the best examples. In combination with the EUV lines, these lines represent another potential diagnostic.

The scarcity of good diagnostics for κ from single-ion spectra has an important consequence. Unless a specifically designed observation and careful diagnostics are performed, the observed spectra can *always* be modeled with some values of T and n_e and an assumption of a Maxwellian distribution.

13.4.3 Diagnostics Involving Ionization Equilibrium

Because of the small sensitivity of the single-ion diagnostic techniques, it is necessary to resort to diagnostics involving multiple ionization stages. Such diagnostics can be performed using ratio—ratio technique as well (Dzifčáková and Kulinová, 2010). However, one of the ratios will now involve lines formed in the neighboring ionization stages. Since the relative ion abundance is always a peak-shaped function of T (Section 13.3.1, Fig. 13.3), such a ratio will be dominantly sensitive to T but will also be sensitive to κ. It is advantageous to combine it with a single-ion ratio sensitive to the κ-index, e.g., a ratio of lines separated in wavelength.

Mackovjak et al. (2013) found good examples of such ratio—ratio diagrams using O IV—V and S X—XI lines within the EIS wavelength range, e.g., O IV 207.2 Å sbl/O IV 279.93 Å—O V 271.04 Å/O IV 279.93 Å. Although indications of strong departures from the Maxwellian distribution were found, the lines used were very weak. Usually, these lines are unobservable with EIS. An instrument with a larger effective area would be needed to use these ratio—ratio diagrams for diagnostics.

The first successful diagnostics of κ from a coronal spectra was performed by Dudík et al. (2015) using the Hinode/EIS observations of the active region NOAA 11,704 on March 30, 2013, obtained as part of the Hinode Operational Program 226. During the last scan with 60-s exposures at each slit position, a transient, cooling coronal loop appeared within the field of view of EIS. The appearance of a loop followed a previous B-class solar flare 3 h earlier, and it was located in a neighborhood of several jets, indicating ongoing magnetic reconnection. A portion of the transient loop was located along a portion of the EIS slit at 13:19 UT (Fig. 13.10). Context imaging observations from the SDO/AIA are shown in Fig. 13.10 together with a portion of the EIS field of view used for analysis. The individual narrow-band AIA filters (Section 13.3.5) are dominated by Fe IX (171 Å), Fe XII (193 Å), Fe XIV (211 Å), and Fe XVI (335 Å), although significant contributions from other ions are present (e.g., O'Dwyer et al., 2010;

Del Zanna, 2013b). The AIA observations showed that the EIS captured the loop at its brightening phase as it cooled through the individual AIA bandpasses.

The EIS instrument obtained Fe XI–XIII spectra of the transient loop in multiple lines, allowing for density diagnostics (Section 13.4.1) and multi-ion diagnostics of T and κ from multiple combinations of Fe XI line ratios and ratios involving an Fe XII and an Fe XI line. Two examples of such ratio–ratio diagrams are shown in Fig. 13.11. The Fe XI r_x ratios are plotted on the x axis. They contain one line observed by a short-wavelength detection of EIS, either Fe XI 182.167 Å or 188.216 Å, combined with the Fe XI 257.554 Å or 257.772 Å self-blends observed by a long-wavelength detector of EIS. The Fe XII lines used in the r_y ratio are either the 186.887 Å sbl or 195.119 Å sbl and are combined with one of the Fe XI lines. Since these ratio–ratio diagrams are sensitive to n_e, the individual curves for different κ are plotted for the two extreme values of the density range diagnosed in Section 13.4.1. The observed ratios of background-subtracted line intensities are shown by the violet crosses representing observational uncertainties. The uncertainties are caused by photon statistics, calibration uncertainty (which is about 20%), and minor contributions from detector noise. In practice, the uncertainties are dominated by the calibration uncertainty. Nevertheless, we see that the ratios indicate $\kappa \leq 2$ independently of the combination of lines used.

We point out that the observational uncertainty is comparable to the spread of the diagnostic curves for the entire range of κ considered. This, in practice, precludes diagnostics of intermediate values of κ. Significantly lower calibration uncertainties, $\sim 5\%$, would be required to discern the situations with $\kappa = 3–10$, as even the 10% calibration uncertainty (shown by azure crosses in Fig. 13.11) is not feasible for this purpose. Even then, higher kappa indices would be practically unmeasurable from ratios of EIS lines.

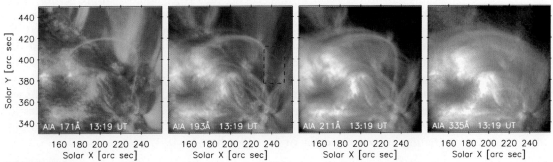

FIGURE 13.10

Solar Dynamics Observatory/Atmospheric Imaging Assembly observations of the transient coronal loop in the active region NOAA 11,704 on March 30, 2013. The time shown corresponds to the time of the Extreme-Ultraviolet Imaging Spectrograph observations of the loop. The *red* box in the Atmospheric Imaging Assembly 193 Å panel is the analyzed portion of the Extreme-Ultraviolet Imaging Spectrograph field of view. Source: Dudík et al. (2015).

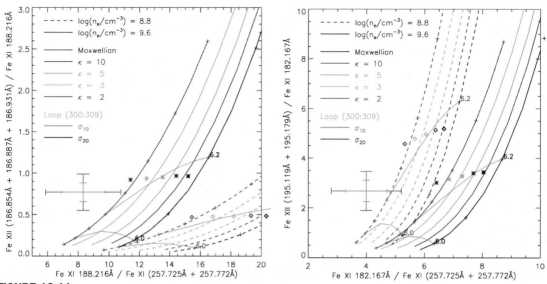

FIGURE 13.11

Diagnostics of the kappa distribution from observations using ratio–ratio diagrams. The x axis represents the single-ion Fe XI ratio, while the y axis contains a ratio of an Fe XII line to an Fe XI line. Individual colors stand for different κ, while the line style denotes the extrema of the density range diagnosed from Fig. 13.8. The observed values with their respective errors are shown in violet. The *azure color* represents the error bar if the calibration uncertainty was 10% instead of 20%. *Diamonds and asterisks* represent the differential emission measure-predicted ratios for individual κ and n_e. Source: Dudík et al. (2015).

13.4.4 Diagnostics From Transition Region Lines: Si III

The solar transition region can exhibit significant temperature gradients along the magnetic field lines. As a result, departures from the Maxwellian distribution can occur in the form of high-energy tails (e.g., Roussel-Dupré, 1980; Shoub, 1983; Ljepojevic and MacNiece, 1988; see also Testa et al., 2014). The presence of such particles was first indicated by Dufton et al. (1984) using the UV line intensities of Si III as observed by Skylab in the quiet Sun and in a solar flare. Pinfield et al. (1999) used the measurements of 10 Si III lines at 1100−1320 Å made by SOHO/SUMER (Wilhelm et al., 1995) and concluded that the observed line ratios are not compatible with a Maxwellian distribution. These results were based on the reported enhancement of the 1313 Å line, which originates in a level 3s4s $^1S_0$ having a higher excitation threshold than the other lines originating in levels 3s3d ^3D_J or 3p^2 ^3P_J, where J = 0, 1, or 2.

Dzifčáková and Kulinová (2011) used the line intensities observed by Pinfield et al. (1999) and calculated a grid of model line intensities for $\kappa = 2-34$ and a range of temperatures and densities. These authors found that a kappa distribution with $\kappa = 7$ provides the best match for all line intensities observed in a solar active region provided that the photoexcitation contribution by the photospheric blackbody radiation with a temperature of 6000 K is taken into account. The inconsistency of the observed line intensities with the Maxwellian distribution is demonstrated in Fig. 13.12 (left-hand side). The observed 1313 Å/1113 Å line intensity ratio in an active region spectrum is larger than the theoretically predicted ones by a factor of ~ 2 for all T and n_e considered. Contrary to that, the line intensities are consistent with $T = 10^4$ K if $\kappa = 7$ across the Si III spectrum (Fig. 13.10; Dzifčáková and Kulinová, 2011, Table 3). The low temperatures diagnosed for $\kappa = 7$ can be explained by the behavior of the ionization equilibrium, where the Si III peak is significantly shifted toward lower T (Fig. 13.3, top) in combination with the decreasing slope of the DEM with T. For this reason, Dzifčáková and Kulinová (2011) also considered the influence of the DEM (Section 13.5) on the diagnostics and showed that the observed ratios are not inconsistent with a range of typical slopes of the DEM (see Eq. 13.4; Section 13.5). This result makes the diagnostics more robust with respect to possible plasma inhomogeneities.

Similar modeling of the entire Si III spectrum was done for the quiet Sun and coronal hole observations, where consistency was found for $\kappa = 11$ and 13, respectively. However, these results were disputed by Del Zanna et al. (2015b), who calculated the new atomic data for Si III using large R-Matrix calculations. These authors showed that the observed line intensities are in general compatible with a Maxwellian distribution, if the new atomic data and the temperature sensitivity of individual lines are taken into account. They have also pointed out that in the SUMER observations of the active region, the 1313 Å line was not observed simultaneously with the other ones. This is potentially a problem due to the high temporal variability of the transition region, at least in some areas (e.g., Testa et al., 2013).

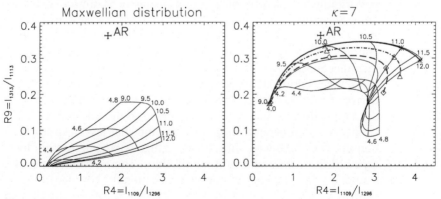

FIGURE 13.12

Si III temperature—density ratio—ratio diagrams for different values of κ: Maxwellian distribution (left) and $\kappa = 7$ (right). The theoretical positions of points corresponding to individual temperatures, $\log(T\,[\text{K}]) = 4.0$—4.8 as well as individual densities, $\log(n_e[\text{cm}^{-3}]) = 9.0$—12.0, are indicated. The observed ratios for the active region spectrum are indicated by the *red cross labeled* "AR." The *red curves* on the right image correspond to different slopes of differential emission measures (see Eq. 13.4). Source: Dzifčáková and Kulinová (2011).

13.5 Differential Emission Measures for Kappa Distributions

Diagnostics using optically thin emission lines is usually complicated with the possible presence of multiple emitting structures along the line of sight. Some coronal loops are known not to be isothermal, as is the unresolved loop emission of active region cores (e.g., Schmelz et al., 2009; Warren et al., 2012; Del Zanna, 2013b). In addition, the observations are fraught with the presence of unresolved, diffuse backgrounds that must be subtracted in order to separate the intensity of the emitting structure (e.g., Tripathi et al., 2009; Gupta et al., 2015).

In practice, the presence of many emitting structures along the line of sight results in contributions to the total intensity arising at different plasma temperatures. Eq. (13.1) for the total intensity can then be rewritten as (e.g., Phillips et al., 2008)

$$I_{ij} = \frac{A_{ij}}{4\pi} \frac{hc}{\lambda_{ij}} \int \frac{1}{n_e} \frac{n(X_i^{+k})}{n(X^{+k})} \frac{n(X^{+k})}{n(X)} A_X n_H n_e \frac{dl}{dT} dT, \qquad (13.4)$$

where the expression $n_H n_e \, dl/dT$ is the DEM. Inversions of Eq. (13.4) using many observed line intensities or even filter-imaging observations (Section 13.3.5) have been extensively applied to the analysis of the coronal, transition region, and flare observations (e.g., Schmelz et al., 2009, 2014; Hannah and Kontar, 2012, 2013; Warren et al., 2012; Del Zanna, 2013b; Dudík et al., 2015).

Mackovjak et al. (2014) were the first to study the behavior of the DEM for kappa distributions. This was done by supplying the line contribution functions (subintegral expressions in Eq. (13.1)) calculated for kappa distributions and performing the inversion of Eq. (13.4) using the Withbroe–Sylwester (Withbroe, 1975; Sylwester et al., 1980) and the regularized inversion methods (Hannah and Kontar, 2012). These authors studied the DEMs for several active region cores using Hinode/EIS intensities reported in a survey study by Warren et al. (2012) as well as the quiet Sun EIS observations of Landi and Young (2010). For the active region cores, the resulting DEMs were found to be multithermal for all κ with a similar degree of multithermality. Specifically, the low-temperature slopes of the DEMs for the kappa distributions were the same within the uncertainties, while the much steeper high-temperature slopes were somewhat less steep for the kappa distributions. The DEMs were shifted toward higher temperatures for low κ, a behavior expected from the ionization equilibrium. These results mean that theoretical studies on the frequency of nanoflares heating the coronae of solar active region cores (e.g., Cargill, 2014) are valid irrespective of the amount of accelerated particles.

For the quiet Sun, Mackovjak et al. (2014) found that the respective EM–loci curves, i.e., intensities divided by the line contribution functions, move toward a near-isothermal crossing point with decreasing κ (Fig. 13.13). The DEM(T) obtained were still found to be multithermal, with a decreasing degree of multithermality with κ. Although this result could be interpreted as the quiet Sun intensities originating in $\kappa = 2$ plasmas, it is not known which mechanism could accelerate such a high relative number of energetic particles.

Finally, Dudík et al. (2015) applied the technique developed by Mackovjak et al. (2014) to investigate the degree of multithermality of the observed transient non-Maxwellian coronal loop (Section 13.4.3). To do this, the DEM inversions for kappa distributions were performed using the regularized inversion method applied to AIA data (Hannah and Kontar, 2013), since the EIS instrument did not observe enough lines spanning a sufficiently wide temperature interval. The loop was found to be significantly multithermal even after background subtraction. However, the DEM-predicted line intensities calculated using Eq. (13.4)

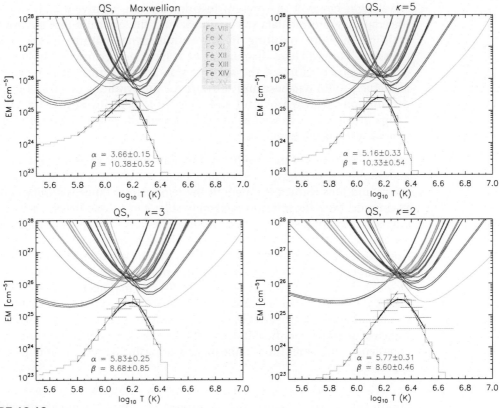

FIGURE 13.13

Emission measure distributions EM(T) for the Quiet Sun (QS) spectra observed by Landi and Young (2010) calculated for kappa distributions. The *azure* and *black colors* show the EM(T) obtained using the Withbroe—Sylwester method and the regularized inversion method, respectively. Emission measure—loci curves are shown in different colors, and each color represents a different Fe VIII—Fe XV ion. Low-temperature and high-temperature slopes α and β are indicated in each image. Source: Mackovjak et al. (2014).

and the resulting ratios (asterisks and diamonds in Fig. 13.11) show a convergence toward the observed values, thereby confirming the diagnostics of $\kappa \leq 2$.

13.6 Concluding Remarks

In this chapter, we discussed the method and applications of the diagnostics of non-Maxwellian kappa distributions from the optically thin observations of the solar corona, transition region, and flares. Kappa distributions represent a good approximation to the power law nature of the high-energy, hard X-ray tails observed in the continuum bremsstrahlung emission in solar flares. In the case of the solar corona and transition region, the diagnostics of kappa distributions is carried out using the line ratio technique. The determination of κ requires a simultaneous determination of T and n_e. While the electron density can be diagnosed independently of κ and T, the determination of κ from EUV and UV emission lines is in many instances hampered by the small wavelength range of the present instrumentation, relatively high calibration uncertainties, atomic data uncertainties, and the optically thin nature of the emitting medium resulting in the multithermal nature of the observed emission. Nevertheless, diagnostics of κ have been carried out successfully from the observations of the solar corona and transition region. The freely available KAPPA database and software for the synthesis of spectra for kappa distributions can be a helpful and effective tool to investigate the effects of kappa distributions as well as their diagnostics in optically thin, low-density plasma at high temperatures.

Future diagnostics of the κ-index will require the exploration of the single-ion and multi-ion diagnostic techniques using emission lines of different elements and ions formed throughout the temperature range spanning orders of magnitude, from the transition region to solar corona and flares. The single-ion diagnostic techniques are preferable since they avoid potential problems with uncertainties in the ionization and recombination rates together with departures from the ionization equilibrium. The latter may be important for lower densities or fast-intensity changes typical of the transition region (e.g., Testa et al., 2013; Doyle et al., 2013; Olluri et al., 2013a, 2013b) and may also be important for the solar corona (e.g., Bradshaw and Klimchuk, 2011). Of special importance is the combination of lines that can be observed with a single instrument, which offers sufficient sensitivity to the κ-index. X-ray flare lines may be a good choice, since an independent verification of the diagnostics using a bremsstrahlung emission is possible.

The effect of the accelerated particles, whether described by a kappa distribution or not, will also need to be studied in conjunction with possible departures from the ionization equilibrium. This is because kappa distributions strongly modify the ionization and recombination rates, which in turn will be reflected in the behavior of the transient ionization in response to sudden heating or cooling events. However, since transient ionization is susceptible to advective flows (e.g., Bradshaw and Mason, 2003a, 2003b), such investigation of nonequilibrium, non-Maxwellian ionization will require the utilization of the full hydrodynamic codes, treating radiation and plasma evolution in a self-consistent manner.

13.7 Science Questions for Future Research

Future analyses and observations need to address the following questions:

1. Can kappa distributions describe coronal nonequilibrium ionizations?
2. What are the effects of kappa distributions on solar magnetic topology?
3. Are kappa distributions in the solar corona caused by local heating?

CHAPTER 14

Importance of Kappa Distributions to Solar Radio Bursts

I.H. Cairns, B. Li, J.M. Schmidt
University of Sydney, Sydney, NSW, Australia

Chapter Outline

14.1 Summary   549
14.2 Introduction   550
14.3 Qualitative Aspects for the Generation and Damping of Plasma Waves and Radio Emissions   554
14.4 Type III Bursts, Electron Beams, and Langmuir Waves   557
14.5 Type II Bursts, Shocks, and Electron Reflections   561
14.6 Concluding Remarks   565
14.7 Science Questions for Future Research   567
Acknowledgment   567

14.1 Summary

Non-Maxwellian electron distributions are observed widely throughout the solar system and are often well described as kappa distributions. This chapter addresses the importance of kappa distributions to the growth and damping of plasma waves (specifically Langmuir waves), the reflection of electrons from shocks, the evolution of electron beams, and the generation of type II and III solar radio bursts. Kappa distributions extend to much higher speeds and have many more fast particles than Maxwellians. Accordingly, kappa distributions lead to a much larger spontaneous emission of plasma waves and radio emissions produced by incoherent processes. Similarly, shocks that accelerate the background electrons have much higher numbers of fast electrons upstream for kappa-distributed electrons than for Maxwellian electrons, leading to stronger electron beams and stronger Langmuir waves and radio emissions. This is demonstrated theoretically for type II solar radio bursts. The extension of kappa distributions to larger speeds than for Maxwellian distributions leads to larger damping rates for electron beam-generated Langmuir waves. It also limits the relaxation of the beam to higher

Kappa Distributions. http://dx.doi.org/10.1016/B978-0-12-804638-8.00014-0

speeds, and thus reduces energy transfers to the Langmuir waves and any radio emissions generated therefrom. These effects are demonstrated theoretically for type III solar radio bursts, allowing us to understand why kappa distributions for the background electrons lead naturally to faster beams with speeds of $\geq 0.5c$ and to the type III bursts being weaker, having faster frequency drift rates, and starting at lower frequencies for smaller values of the kappa index, κ.

Science Question: How do solar type II and III radio bursts depend on electron kappa distributions?

Keywords: Plasma waves; Radio emissions; Shocks; Solar physics; Type II bursts; Type III bursts.

14.2 Introduction

Kappa distributions of electrons appear to be ubiquitous in space plasma (e.g., Meyer-Vernet, 2001; Livadiotis and McComas, 2009, 2010a, 2011b, 2013a; Pierrard and Lazar, 2010; Livadiotis, 2015a). Specifically, the combination of the core and halo electrons observed in solar wind is well described as a kappa distribution (e.g., Maksimovic et al., 1997b). Strong theoretical arguments also exist that coronal electrons are likely better described using kappa distribution functions than Maxwellian distribution functions. For definiteness, the $d = 3$ dimensional kappa distribution $f(\vec{u})$ of electron velocities is

$$f(\vec{u};T;\kappa) = n_e \cdot \left(\pi \kappa \theta_\kappa^2\right)^{-\frac{3}{2}} \cdot \frac{\Gamma(\kappa+1)}{\Gamma\left(\kappa-\frac{1}{2}\right)} \cdot \left[1 + \frac{1}{\kappa} \cdot \frac{(\vec{u}-\vec{u}_b)^2}{\theta_\kappa^2}\right]^{-\kappa-1}, \quad (14.1)$$

normalized by the electron number density $n_e = \int f(\vec{u})d\vec{u}$, $\theta_\kappa \equiv \sqrt{(2\kappa-3)/(2\kappa)} \cdot \theta$, with $\theta \equiv \sqrt{2k_BT/m_e}$ noting the electron thermal speed. Both the particle velocity $\vec{u}$ and the mean velocity $\vec{u}_b$ are in an inertial reference frame; here, κ denotes the 3-D kappa index, which is related to the 1-D kappa index, κ_1, by $\kappa = \kappa_1 + 1$ and to the invariant kappa index κ_0 by $\kappa = \kappa_0 + \frac{3}{2}$. In comparison, the velocity distribution function for d kinetic degrees of freedom (Livadiotis, 2015a; Chapter 1), is

$$f(\vec{u};T;\kappa_0) = n_e \cdot \left(\pi \kappa_0 \theta^2\right)^{-\frac{1}{2}d} \cdot \frac{\Gamma\left(\kappa_0+1+\frac{1}{2}d\right)}{\Gamma(\kappa_0+1)} \cdot \left[1 + \frac{1}{\kappa_0} \cdot \frac{(\vec{u}-\vec{u}_b)^2}{\theta^2}\right]^{-\kappa_0-1-\frac{1}{2}d}. \quad (14.2)$$

For $\kappa \to \infty$, Eqs. (14.1) and (14.2) tend to the Maxwellian distribution:

$$f(\vec{u};T;\kappa_0 \to \infty) = n_e\left(\pi \theta^2\right)^{-\frac{3}{2}} \cdot \exp\left[-\frac{(\vec{u}-\vec{u}_b)^2}{\theta^2}\right]. \quad (14.3)$$

Usually, Eqs. (14.1)–(14.3) are written in the reference frame of the background plasma. Accordingly, when the difference between $\vec{u}_b$ and the velocity of the background plasma is much larger than the thermal speed θ, then $\vec{u}_b$ is often called a "beam velocity," and the particles modeled by Eqs. (14.1)–(14.3) represent a beam of fast particles moving through the background plasma with an average velocity $\vec{u}_b$. Typically, $\vec{u}_b$ is either parallel or antiparallel to the background magnetic field $\vec{B}$.

Plasma waves and radio emissions are important for multiple reasons. First, they contain information about the source plasma. For instance, a given wave mode may correspond to a resonance in the plasma, including those expected at the electron and ion gyro frequencies $f_{c\,a} = \Omega_{c\,a}/(2\pi) = ZeB/(2\pi\, m_a)$ or the electron and ion plasma frequencies $f_{p\,a} = \omega_{pa}/(2\pi) = Ze\sqrt{n_a/(\varepsilon_0 m_a)}/(2\pi)$, where a: e or i, corresponding to electrons or ions (of charge Z, which hereafter will be taken equal to 1), respectively. The Langmuir mode occurs at frequencies just above f_{pa}, with a dispersion relation $\omega_L^2 = \omega_{pe}^2 + \frac{3}{2}k^2\theta^2$. Similarly, free-space radio waves (in the so-called "o and x modes") also occur above f_{pe}, with a dispersion relation $\omega_T^2 = \omega_{pe}^2 + k^2 c^2$, where c is the speed of light. Often the term "plasma wave" is used to refer to a mode that cannot escape the source plasma, unlike a radio emission (in principle). Second, the following aspects all relate to the physics of the source plasma and associated dynamical physics: non-Maxwellian plasma waves and radio waves releases of signal energy in the source, the existence of free energy available for redistribution as the source plasma tends toward thermal equilibrium, and a source plasma that is not in thermodynamic equilibrium. Third, radio emissions that escape the source plasma are scattered, refracted, and even absorbed along the path, thereby containing information on the propagation physics and path.

The Sun is not in thermodynamic equilibrium but is instead subject to many important dynamical events such as solar flares, coronal mass ejections (CMEs), and shocks. These processes produce intense radio emissions, X-rays, and both ultraviolet (UV) and white-light emissions. Of primary interest here are the multiple classes of solar radio bursts identified in the 1950s (Wild and McCready, 1950; Wild and Smerd, 1972) and shown schematically in Fig. 14.1: types II and III are the most intense and frequent, with frequencies typically drifting from ~ 200 MHz to ~ 10 MHz in about 5 s (type III) to 5–10 min (type II), as sources move from the deep corona to ≈ 5 solar radii. Type II lasts for days in the solar wind, as they drift down to ~ 20 kHz near the Earth's orbit (1 AU). In contrast, type III lasts for 1–2 h in the solar wind, as their exciting electrons move to 1 AU. (Relevant reviews may include those of Nelson and Melrose, 1985; Suzuki and Dulk, 1985; Bastian et al., 1998; Cairns, 2011; Reid and Ratcliffe, 2014.)

Type II solar radio bursts are produced by shocks, probably all associated with CMEs (the converse is not true: not all CMEs have observable type II bursts), while type III bursts are associated with flares and reconnection events

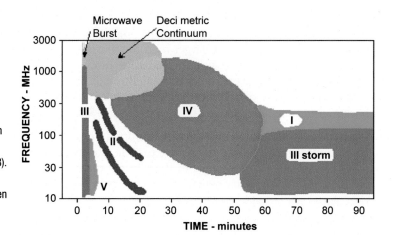

FIGURE 14.1

The zoo of solar radio emissions, based on earlier work (e.g., Wild and Smerd, 1972; Suzuki and Dulk, 1985; Bastian et al., 1998). Type II and III bursts are the focus here, associated with coronal mass ejection-driven shocks and flare-accelerated electrons, respectively. Source: Kennewell (2016).

(Fig. 14.2). Flares and CMEs both result from a reorganization of the Sun's magnetic fields via the process called "magnetic reconnection"; this process changes the connection of magnetic field lines and converts magnetic energy into bulk motion, heating of the plasma, and accelerated particles, especially electrons.

Flares typically involve two magnetic loops interacting, with bulk motions of plasma both toward the chromosphere (and photosphere) on closed field lines and also outward into the corona as well as accelerated particles moving both Sunward and anti-Sunward. The bulk plasma flows downward, leading to bursts of UV, white-light, and H-α emissions called flares, while energetic electrons accelerated Sunward produce bursts of X-rays via bremsstrahlung. In contrast, the outgoing bulk flows primarily enhance the solar wind, perhaps also leading to heating via stream–stream interactions. The outgoing fast electrons, in contrast, form electron beams that produce non-Maxwellian formulations of Langmuir waves and radio emissions, the latter called type III solar radio bursts.

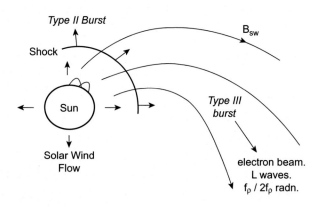

FIGURE 14.2

Schematic of the basic physics of type II and III bursts, associated with a coronal mass ejection-driven shock and flare-accelerated electrons, respectively. Both involve electron beams (with different origins and characteristics), Langmuir waves, and f_{pe} and $2f_{pe}$ radiation.

For CMEs, magnetic reconnection proceeds differently: typically a loop of solar magnetic field lines is magnetically disconnected at low altitudes, within less than 0.1 solar radii from the Sun's photosphere, and then reconnected into a self-contained ball of plasma and magnetic fields that moves outward at speeds between about 200 and 2000 km s^{-1}. The high densities and temperatures of CMEs make them glow brightly in white light and UV observations. Moreover, typically the CME speed is large enough to produce a shock upstream of the CME, analogous to the bow shock in front of a supersonic fighter jet.

There are multiple reasons why shocks are crucial in space physics, plasma physics, and astrophysics. Shocks result from nonlinear steepening of waves due to the group speed of a wave with a given wavenumber, depending on the amplitude, leading to larger amplitude portions of a disturbance catching up with slower portions to build a steep wave front; this steepens until the nonlinear steepening is balanced by dissipation. Most shocks are expected to be fast magnetosonic (or fast-mode) shocks. Shocks slow the upstream super-fast mode flow, compress and heat the plasma downstream, and amplify and rotate the magnetic field. A strong shock with a factor of 4 increases in density, magnetic field, and inverse flow speed going from upstream to downstream, with much larger increases in temperature. These are all observable signatures, both in situ and remotely in spectroscopic and white-light data (e.g., Vourlidas et al., 2003; Gopalswamy and Yashiro, 2011; Schmidt et al., 2016). Here, though, the importance for type II bursts is that a shock's magnetic mirror can reflect and accelerate electrons back upstream, leading to electron beams, Langmuir waves, and radio emissions.

Fig. 14.2 summarizes the basic physics of coronal and interplanetary type II and III bursts as well as their associations with CMEs, shocks, flares, electron beams, and Langmuir waves. These bursts involve fundamental plasma physics with widespread applications to astrophysical, laboratory, and space phenomena. The basic physics is as follows: (1) energetic electrons develop "beam" distributions as they travel away from the acceleration region (either the reconnection region or the shock), for instance because faster electrons outrun slower ones to form a localized peak in velocity at a given location; (2) the beams drive Langmuir waves, with frequencies just above the electron plasma frequency f_{pe} (that measures the electron number density, $n_e \propto f_{pe}^2$); (3) linear and nonlinear mechanisms convert Langmuir energy into radiation near f_{pe} and $2f_{pe}$; and (4) the radio frequency decreases with time due to n_e decreasing with increasing heliocentric distance as the type III electron beam or CME-driven type II-producing shock leaves the Sun.

The further structure of this chapter is as follows. Section 14.3 addresses the qualitative reasons why kappa distributions are important for the generation of plasma waves and radio emissions, addressing both incoherent (single-particle) emission processes and also collective, coherent emission processes like plasma wave instabilities and damping plus nonlinear wave–wave emission processes. Section 14.4 considers type III bursts in some detail, showing that kappa distributions of the background plasma particles have crucial effects. Type II bursts are addressed in Section 14.5, focusing on the reflection of kappa-distributed electrons from the shock and the associated differences by orders of magnitude in the predicted radio flux. Importantly, the observed and predicted radio spectra agree very well for $\kappa = 4$ (or $\kappa_0 = 2.5$ for $d = 3$), a value consistent with in situ solar wind data, demonstrating the importance of kappa distributions in explaining the observations. Finally, the concluding remarks are given in Section 14.6, while three general science questions for future analyses are posed in Section 14.7.

14.3 Qualitative Aspects for the Generation and Damping of Plasma Waves and Radio Emissions

Two vital qualitative points are made before directly addressing the growth of plasma waves and radio emissions. The first is that for speeds of u greater than several thermal speed θ (at the comoving system), (isotropic) kappa distributions yield higher values of $f(\vec{u})$ and more fast particles at a given u than a Maxwellian. This can be demonstrated directly using Eqs. (14.1) and (14.2). The second point is that the slope on a kappa distribution is much flatter than a Maxwellian at suprathermal speeds (much higher than the speed at the mode of the distribution), where the Maxwellian approaches zero at a very large rate (exponentially).

The relevance of these points to the incoherent growth of plasma waves and radio emissions, otherwise described as spontaneous emission via the Cerenkov or cyclotron resonances, is that such incoherent growth processes directly depend on $f(\vec{u})$. Specifically, the rate at which energy $W_M(\vec{k})$ is emitted incoherently into waves in the mode M with angular frequency $\omega = \omega_M(\vec{k})$ and phase velocity $\vec{u}_{ph} = (\omega/k)\hat{k}$ is

$$\frac{d}{dt}W_M(\omega, \vec{k}) = \int d\vec{u} \, A(\vec{u}, \omega, \vec{k}) f(\vec{u})\delta(\omega - \vec{k}\vec{u}) \propto A(\vec{u}_{ph})f(\vec{u}_{ph}).$$

(14.4)

Here the delta function enforces the Cerenkov resonance but can be replaced by the cyclotron resonance when magnetization effects are important; A is a function that relates to the probability of emission. The important point is that the power input is proportional to the value of $f(\vec{u}_{ph})$. Thus since a kappa distribution has more fast particles than a Maxwellian at the same suprathermal speed, i.e., a high speed beyond $\vec{u}_{ph}$, the power input will be larger. Put another way, a kappa distribution will lead to higher power inputs into waves and radio emissions via incoherent emission processes such as Cerenkov emission, bremsstrahlung, and also gyro resonant processes such as synchrotron radiation. The spectra can also be changed due to the different function form of $f(\vec{u}_{ph})$, but this aspect is not pursued here.

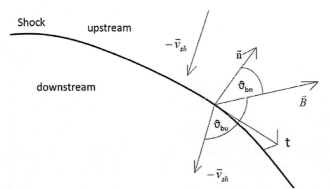

FIGURE 14.3
Geometric variables relevant to particle acceleration at a shock.

Another important single particle (or incoherent process) is particle reflection and associated energization at shocks (e.g., Filbert and Kellogg, 1979; Wu, 1984; Cairns, 1987, 2011). Consider the shock quantities in Fig. 14.3 (e.g., Schmidt and Cairns, 2012a). In this reference frame (the shock's rest frame) the plasma moves through the shock from upstream with velocity $-\vec{v}_{sh}$, the local upstream magnetic vector is $\vec{B}$, a convection electric field exists with $\vec{E} = \vec{v}_{sh} \times \vec{B}$, the local shock normal is $\vec{n}$, and the angles ϑ_{bn} and ϑ_{bu} are between $(\vec{B}, \vec{n})$ and $(\vec{B}, -\vec{v}_{sh})$, respectively. In this frame, ignoring the gyro motion, electrons move in straight lines defined by the $\vec{E} \times \vec{B}$ drift velocity $\vec{v}_d = \vec{E} \times \vec{B}/B^2$, their velocity is parallel to $\vec{B}$, they encounter the change in magnetic field strength at the shock, they attempt to move through the shock with constant magnetic moment, and they are reflected or transmitted based on their initial pitch angles, the shock's jump in magnetic field, and any electric fields encountered. The situation is very similar to that of particle motion in a magnetic bottle. In the plasma frame, electrons change their kinetic energy when reflected by the shock's magnetic mirror.

Analyzing this situation is easiest in the de Hoffman-Teller frame, which moves along the local shock surface with a speed of $v_d \cdot \tan(\vartheta_{bn})$ and has zero convection electric field on both sides of the shock. The speed with which the shock moves along $\vec{B}$ in the shock frame is $v_{sh} \cdot \sec(\vartheta_{bn})$. Electrons, then, obey conservation of the magnetic moment and energy (subject also to the cross-shock electrostatic (ES) potential), leading to a loss cone distribution of reflected electrons. For us, though, the important point is that the only electrons that are reflected are those that can escape upstream of the shock, corresponding to $|u_\parallel| > v_c = v_{sh} \cdot \sec(\vartheta_{bn})$ (Cairns, 1987). This cutoff effect leads to a beam feature in the distribution function of electrons reflected back upstream at speeds near v_c. In more detail, the reflected and initial parallel speeds (subscripts r and i, respectively) are related by

$$u_{\parallel,r} = 2v_{sh} \cdot \sec(\vartheta_{bn}) - u_{\parallel,i} + v_{sh} \cdot \cos(\vartheta_{bu}), v_c = v_{sh} \cdot \sec(\vartheta_{bn}), \tag{14.5}$$

where the term $2v_{sh} \cdot \sec(\vartheta_{bn})$ corresponds to the change in reference frame, and via Liouville's theorem

$$f\left(u_{\|,r}\right) = f\left(u_{\|,i}\right),\tag{14.6}$$

with $u_{\|,i} < v_c$ to encounter the shock initially. Eq. (14.5) also leads to reflected particles having $u_{\|,r} > v_c = v_{sh} \cdot \sin(\vartheta_{bu})$. Larger beam speeds v_c correspond to larger energies available for wave growth. These occur for larger u, for ϑ_{bu} close to 90 degrees, and, most importantly, for ϑ_{bn} closer to 90 degrees. Thus fast-moving shocks with magnetic fields close to perpendicular to the shock surface are ideal. Furthermore, via Eq. (14.6), it is important to have more initial fast particles in the upstream distribution for the shock to accelerate. Thus a shock encountering a plasma with kappa-distributed electrons will reflect more particles than a Maxwellian, since it has more fast particles at a given $u_{\|,i}$ above a few thermal speeds than the Maxwellian.

Turning now to collective, coherent wave processes, these typically depend on the gradients of the particle distribution function in the velocity space, e.g., on terms like $\partial f / \partial \overrightarrow{u}$. Specifically, the growth and damping of waves depend on expressions like

$$\frac{d}{dt} W_M\left(\omega, \overrightarrow{k}\right) = \int d\overrightarrow{u} \; \overrightarrow{A}\left(\overrightarrow{u}, \omega, \overrightarrow{k}\right) \cdot \frac{\partial f}{\partial \overrightarrow{u}} \cdot \delta\left(\omega - \overrightarrow{k}\,\overrightarrow{u}\right)$$
$$\times W_M\left(\omega, \overrightarrow{k}\right) \propto \overrightarrow{A}\left(\overrightarrow{u}_{ph}\right) \cdot \frac{\partial f\left(\overrightarrow{u}_{ph}\right)}{\partial \overrightarrow{u}} \cdot W_M\left(\omega, \overrightarrow{k}\right).\tag{14.7}$$

Clearly, Eq. (14.7) leads to exponential growth or damping of the waves. Typically, growth occurs if $\partial f / \partial u_{\|} > 0$, corresponding to an increase in $f(u_{\|})$ with increasing $u_{\|}$ (in some range of $u_{\|}$), the opposite of a thermal distribution, while damping occurs where $\partial f / \partial u_{\|} < 0$. An example of growth is shown in Fig. 14.4, looking at the (infinite) positive slope at speed v_c, with the beam of

FIGURE 14.4
Superposition of the distribution $F_r(u_{\|})$ of reflected electrons onto the background distribution $F_i(u_{\|})$ leads to a beam-like distribution (that rises abruptly from $F_i(u_{\|})$ at $u_{\|} = v_c$ to the peak of triangle A and then follows $F_i + F_r$ from $u_{\|} = v_c$ to c), with a cutoff at the escape speed v_c. The back reaction to the wave growth (quasilinear relaxation) leads to the plateau limited by shaded areas A and B (which conserves the particle number), with the lower limit $u_{\|} = a$, determined by both F_i and F_r. Source: Schmidt and Cairns (2012a).

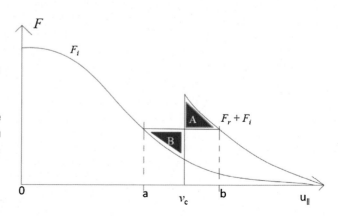

reflected electrons (superposed on the background distribution) defined by the heavy solid line. In contrast, for a kappa distribution we have $\partial f / \partial u_{\parallel} < 0$ for the entire range of $u_{\parallel}$, corresponding to damping for this entire range of parallel phase speeds. A crucial point here is that the Maxwellian does not extend over the whole range of $u_{\parallel}$, and so does not lead to significant damping for $u_{\parallel} \gtrsim 0.3c$. In this case, then, the kappa distribution leads to a much higher damping of waves at high phase speeds than for a Maxwellian.

When the background plasma has a larger damping rate for the waves, a stronger source of free energy is required to overcome the damping and leads to a net growth for the waves. For Langmuir waves driven by an electron beam, the electron beam needs to have a larger growth rate and be denser, faster, and/or narrower (in $u_{\parallel}$) if the background plasma is described by a kappa distribution. This is shown in the following paragraphs to be relevant for type III bursts.

It is also important to take into account the so-called quasilinear relaxation of the unstable particle distribution function. This is the back reaction to the wave growth and corresponds to electrons giving up energy to the wave via instability, moving to lower energy and different pitch angles. (Specifically, instabilities that are driven by gradients of $\partial f / \partial u_{\parallel}$ will have electrons decreasing their $u_{\parallel}$ and the gradient $\partial f / \partial u_{\parallel}$, increasing the pitch angle, whereas those driven by gradients $\partial f / \partial u_{\perp}$ will have electrons decreasing their $u_{\perp}$ and $\partial f / \partial u_{\perp}$, decreasing the pitch angle). The end state corresponds to a zero net growth rate for the waves. For electron beams driving Langmuir waves, the initial and end states are illustrated in Fig. 14.4: the end state is a plateau that from Eq. (14.7) has a growth rate of $\Gamma = dW_L/dt \cdot W_L^{-1} \propto \partial f / \partial u_{\parallel} = 0$, corresponding to a flat distribution that meets the background distribution at some speed $u_{\parallel} = a$, with the height corresponding to the conservation of the particle number. Clearly, if the background distribution is a kappa distribution rather than a Maxwellian, then for the same plasma number density the plateau will extend to a much higher value of $u_{\parallel}$ than for a Maxwellian. This effect is shown in the following paragraphs to be vital in constraining the average speed of the electrons driving the Langmuir waves (often called the beam speed) in type II and III sources (Schmidt and Cairns, 2012a, 2012b; Li and Cairns, 2013a, 2013b).

14.4 Type III Bursts, Electron Beams, and Langmuir Waves

The qualitative physics of type III bursts involves several steps (e.g., Suzuki and Dulk, 1985; Bastian et al., 1998; Ziebell et al., 2001; Kontar and Pecseli, 2002; Li and Cairns, 2013a; Li et al., 2003, 2008; Reid and Ratcliffe, 2014):

1. Electrons accelerated in reconnection regions (or elsewhere) develop a beam distribution function by time-of-flight effects, corresponding to fast electrons outrunning slower electrons along $\vec{B}$.

2. The electron beam drives high levels of Langmuir waves by the standard bump-on-tail or beam instability, leading to quasilinear relaxation of the beam.

3. Spatial inhomogeneities lead to burstiness in the growth of the Langmuir waves and a slower relaxation of the electron beam, often called "stochastic growth effects."

4. Langmuir waves are subject to several nonlinear wave–wave processes, including the following:

    a. the electrostatic (ES) decay $L \to L' + S$ of beam-generated Langmuir waves L into backscattered Langmuir waves L' and forward-going ion acoustic waves S;

    b. the production of radio waves T just above f_{pe} by the electromagnetic (EM) decay $L \to T(f_{pe}) + S'$, stimulated by the S waves from the ES decay; and

    c. the production of radio waves near $2f_{pe}$ by the coalescence process $L + L' \to T(2f_{pe})$.

5. Straight-line propagation to the observer from a plane-stratified three-dimensional source, within an allowed emission cone given by observations, is then used to predict the radio emission observable at Earth.

Quantitative descriptions of each of these steps are available, albeit with limitations. For instance, the electron–Langmuir wave physics is modeled quantitatively with the so-called quasilinear equations, almost invariably in one dimension (along $\pm \vec{B}$) (e.g., Kontar and Pecseli, 2002; Li et al., 2003, 2008):

$$\frac{\partial f(u, x, t)}{\partial t} + u \frac{\partial f}{\partial x} = \frac{\partial}{\partial u} [A(u, x, t)f] + \frac{\partial}{\partial u}\left[D(u, x, t)\frac{\partial f}{\partial u}\right] + S(u, x, t), \tag{14.8}$$

$$\frac{\partial N_L(k, x, t)}{\partial t} + u_g \frac{\partial N_L}{\partial x} - \frac{\partial \omega_L}{\partial x}\frac{\partial N_L}{\partial x} = \alpha_L(k, x, t) + \Gamma(k, x, t)N_L + R_L(k, x, t). \tag{14.9}$$

The first equation couples the electron distribution function to the waves, including changes in the distribution function and its gradient $\partial f/\partial u_\parallel$ due to resonant interactions with the wave electric fields. The second couples the growth and damping of the wave distribution function (in wavevector) to the electron distribution and its gradients. In more detail, A and D relate to spontaneous emission and quasilinear diffusion effects, respectively, and depend on the wave distribution as well as other wave and electron properties; S is a source/loss term; u_g is the group speed; the derivative in ω_L relates to refraction; α_L is the spontaneous emission term for waves; $\Gamma(k)$ is the linear growth/damping rate of waves due to the gradient $\partial f/\partial u_\parallel$; and R_L is the rate for nonlinear wave–wave processes. These equations allow for the prediction of the electron distribution function and the 1-D Langmuir wavevector distribution as functions of time, position, and either velocity or wavevector, respectively.

A 3-D wavevector distribution for the Langmuir waves is then obtained from the 1-D output from the quasilinear code by assuming an angular form (Li et al., 2003, 2008). Standard nonlinear plasma theory is used to obtain the nonlinear rates for the processes in step 4 (a–c) (above), leading to predictions for the 3-D wavevector distributions of all the waves as functions of position and time that are similar to Eq. (14.8). Spatial inhomogeneities can also be introduced into the plasma density and the ion and electron temperatures on spatial scales corresponding to macroscopic coronal/solar wind gradients and intermediate scales (Li et al., 2011a,b, 2012) and on small scales. The latter correspond to density turbulence and lead to stochastic growth effects (Robinson et al., 1993), which can be modeled quantitatively (Li et al., 2006a,b). Implementations allow the background and accelerated electrons to be modeled with Maxwellian, kappa or even a combination of these distributions (Li and Cairns, 2013a,b, 2014). An important point is that the nonlinear rates for the wave–wave equations are very robust against changes in the functional form of the background distributions (Layden et al., 2011; 2013) until the relevant thermal speeds approach c, since the basic approximation is that the Langmuir wave phase speeds are much less than c.

Fig. 14.5 shows the intensity as a function of frequency and time ("dynamic spectrum") of the f_{pe} radio emission predicted for a type III burst using Eqs. (14.8) and (14.9) and similar equations for the same accelerated electrons (e.g., using the same source term $S(u,x,t)$ in Eq. (14.8)), but different models for the background plasma distribution function. Specifically, in each case the background plasma has the same profiles for the plasma density and both the electron and ion temperatures but is either a kappa distribution (with one of two values of κ) or a Maxwellian distribution. The differences in the predicted radio emission are thus associated with differences in the electron–Langmuir wave interactions related to the background plasma's distribution function. With a Maxwellian corresponding to the limit $\kappa \to \infty$, it is clear that as κ decreases, the type III emission is predicted to be weaker, to drift faster in frequency (larger df/dt), to start at lower frequencies, and to start at larger distances from but end closer to the source of the accelerated electrons (Li and Cairns, 2013). Fig. 14.5 thus provides direct evidence that kappa distributions should be important in the production and interpretation of type III solar radio bursts.

The variations in Fig. 14.5 of the radiation flux and frequency drift rate with frequency are quantified in the left and right panels of Fig. 14.6, respectively. Considering the motion of an electron beam that moves radially outward with a constant speed u_b as it continually produces f_{pe} radiation, it is easy to show that the frequency drift rate df/dt is related to the plasma density profile dn_e/dr and u_b by (Wild and McCready, 1950)

$$\frac{df}{dt} = \frac{u_b}{2n_e} \frac{dn_e}{dr} f. \tag{14.10}$$

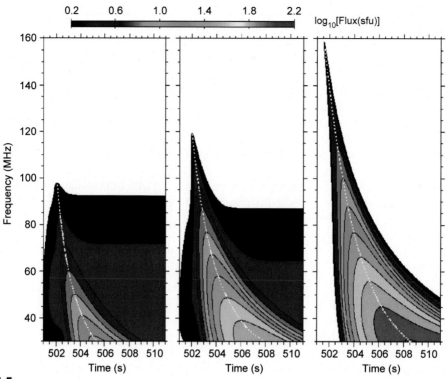

FIGURE 14.5
Dynamic spectra predicted for type III bursts using the same plasma profiles and input accelerated electrons but different distribution functions for the background plasma: kappa distributions with $\kappa = 5$, 7, or ∞ (Maxwellian) from left to right. The dynamic spectra provide clear evidence that the description of the background plasma with kappa distributions should have major observable consequences. *SFU*, solar flux unit. Source: Li and Cairns (2013a).

Accordingly, if larger values of df/dt are obtained for the same f and density profile, then u_b must be larger. Comparing the drift rates predicted by Eq. (14.10) for the simulated density profiles with the simulation results implies that u_b increases strongly as κ decreases, with $u_b \approx 0.25c$ for $\kappa \rightarrow \infty$ but $u_b \approx 0.58c$ for $\kappa = 5$. Direct analysis of the predicted electron distributions (not shown) demonstrates that these inferred beam speeds agree excellently with the average speed for the quasilinearly relaxed electron beams observed in the simulations. Thus the kappa index effects for the background plasma lead to major changes in the speed of the type III electron beam. This counterintuitive effect is explained using the concepts in Section 14.2.

Consider quasilinear relaxation of an electron beam superposed on a kappa distribution of background electrons or the corresponding Maxwellian (with the same number density and temperature). As demonstrated in Fig. 14.7, the quasilinearly relaxing beam distribution encounters the background kappa distribution at a significantly higher speed than for the Maxwellian, since the kappa

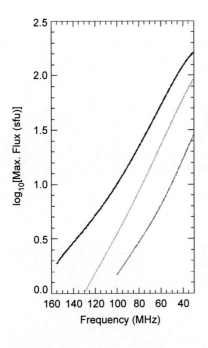
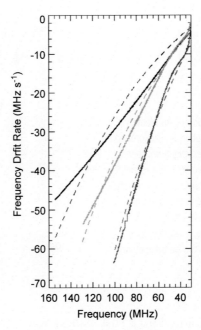

FIGURE 14.6
Quantification of the type III dynamic spectra in Fig. 14.5 (Li and Cairns, 2013a) for simulations with the same plasma profiles and input accelerated electrons but background plasma that are Maxwellian (*black solid curves*) or kappa distributions with $\kappa = 7$ (*green solid*) or $\kappa = 5$ (*red solid*). (Left) Radiation flux versus frequency and (right) frequency drift rate *df/dt* versus frequency. Predictions from Eq. (14.9) for *df/dt* with the imposed density profile for specific beam speeds u_b are shown with dashed curves for each simulation, providing direct evidence for u_b increasing greatly as κ decreases. (*SFU*, solar flux unit.) Source: Li and Cairns (2013).

distribution has many more fast electrons than the Maxwellian. Thus the plateau does not move as far toward low $u_{\parallel}$, so the beam loses less energy (and the Langmuir waves are significantly weaker) and has a higher average speed than for a Mawellian. Put another way, in the kappa case the nominal average beam speed is higher ($\sim 0.43c$) than for the Maxwellian case ($\sim 0.32c$), and the frequency drift rate should be higher according to Eq. (14.9). The Langmuir waves and the radio intensity should be weaker. These attributes are qualitatively consistent with Figs. 14.5 and 14.6 (Li and Cairns, 2013a). In addition, it is clear that kappa distributions are vital for both the accelerated and background plasmas to explain type III beams with speeds well above $0.3c$ (Li and Cairns, 2013a, 2013b), as sometimes observed (Poquerusse, 1994) as so-called type IIId bursts.

The foregoing analyses demonstrate that detailed kappa versus Maxwellian characteristics of the background plasma are vital in understanding the high speeds of type III electron beams. Not surprisingly, having a kappa (or power law) versus a Maxwellian distribution function for the accelerated electrons is also important. As shown by Li and Cairns (2014), type III beams with drift speeds above $0.5c$ require kappa distributions with $\kappa < 5$ for both the background and accelerated electrons.

14.5 Type II Bursts, Shocks, and Electron Reflections

Fig. 14.8 illustrates qualitatively the generation of type II bursts, extending the basic summary in Section 14.1 (e.g., Nelson and Melrose, 1985; Bastian et al.,

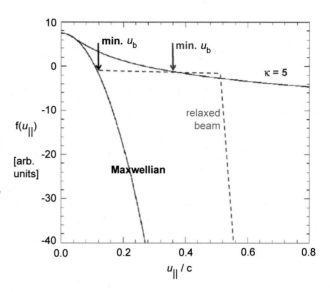

FIGURE 14.7
Illustration of how quasilinear relaxation of an electron beam superposed on a kappa distribution of background electrons (*red curve*) leads to a larger minimum speed, and so a higher average "beam speed," for the relaxed beam electrons (where the blue dashed curve meets the red curve) than for Maxwellian background electrons (*black curve*). The distributions $f(u_{\parallel})$ shown are 1-D distributions in the plasma rest frame, resulting from integrating the 3-D distributions over the perpendicular velocity phase space.

1998; Cairns, 2011). First, a shock wave, whether a "blast wave shock" due to a remote event or a shock driven directly by (and standing in front of) a CME, reflects and accelerates electrons. Second, the reflected electrons develop a beam distribution function due to two effects: the minimum speed required to escape upstream of the shock, and/or the competition between electron motion along $\pm \vec{B}$ from the shock and the $\vec{E} \times \vec{B}$ drift, which leads to a minimum speed $v_{\parallel}$ for any specified point upstream of the shock. Third, the electron beam drives Langmuir waves via the bump-on-tail instability. Fourth, the radio emission is

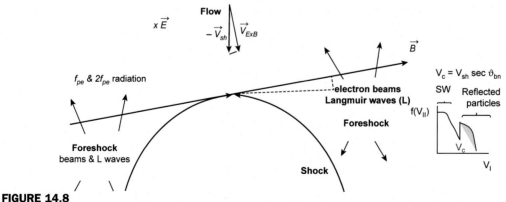

FIGURE 14.8
Schematic for type II bursts: the shock reflects and accelerates electrons back upstream of the shock, where they naturally form electron beams (the inset on the right shows the electron distribution function at the spatial location where the dashed lines meet) that drive Langmuir waves, some of whose energy is converted into f_{pe} and $2f_{pe}$ radio emissions. Reflected electrons are found downstream of the magnetic field line tangent to the shock. (*SW*, shock wave.)

produced near f_{pe} and $2f_{pe}$ by nonlinear processes. Fifth, the radio emission propagates to the observer, subject to refraction, scattering, and absorption. The third to fifth steps are the same as those for type III bursts, described further in Section 14.4, differing only in that the characteristic beam speeds for type II bursts are in the range of 3−10 (background) electron thermal speeds while those for type III bursts are in excess of 0.1c and are typically greater than 20 electron thermal speeds.

Each of the steps in the above type II theory can be described quantitatively using analytic theory (Fig. 14.9), a different approach than for the type III theory in Section 14.4. For instance, the electron distributions at each upstream point are calculated using Liouville's theorem combined with conservation of the magnetic moment and energy in the de Hoffman-Teller frame. Similarly, the energy flow into the Langmuir waves is predicted analytically by assuming a limitation of quasilinear relaxation of the bump-on-tail instability due to the Langmuir waves evolving in a locally inhomogenous plasma and entering a state described by stochastic growth theory (Robinson et al., 1993; Cairns and Robinson, 1999), in which the beam persists in a marginally stable state. Then, the predicted energy flows into the Langmuir waves are multiplied by the conversion efficiencies for the ES decay, EM decay, and Langmuir coalescence processes to predict the energy flows into f_{pe} and $2f_{pe}$ radiation everywhere upstream of the shock. Evidence exists for the ES decay process in particular (Graham and Cairns, 2015). Finally, these energy flows into radiation are integrated over the upstream volume and propagated to the observer (using straight-line propagation with a fall-off with inverse distance squared when the radiation frequency is above the local plasma frequency along the path but zero net emission when the radiation is unable to reach the observer due to the radiation frequency not always being above the local plasma frequency along the path) to predict the radiation intensity as a function of frequency and time. A crucial question is how to predict the properties of the shock and the upstream plasma as functions of position and time. One way to do this is to specify them for a model calculation while another is to use magnetohydrodynamic (MHD) simulation.

Fig. 14.10 takes the first approach (Cairns et al., 2003; Knock et al., 2003). The speed of a 3-D paraboloid shock is specified in a locally homogeneous background plasma, assumed to be represented by a kappa distribution of electrons, the Rankine-Hugoniot conditions are used to calculate the properties of the shock in a 3-D manner, and the flux of radio emission is predicted at a remote observer. Note that none of the Rankine-Hugoniot conditions involves the kappa index, because the conservation of mass, momentum, and energy equations are independent of the kappa index; thus the known Rankine-Hugoniot conditions remain the same even for kappa distributions (see Livadiotis, 2015b.) Fig. 14.10 shows clearly that the radio flux increases monotonically as the shock speed (the plasma flow speed in the shock rest frame) increases. Physically, this increase results from the cutoff speed and the local beam speed being proportional to the shock speed via Eq. (14.4), having more energy available to flow into the Langmuir waves and radio emissions. In addition, the

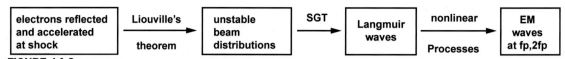

FIGURE 14.9
Steps and associated theoretical formalisms involved in the generation of type II bursts. (*EM*, electromagnetic; *SGT*, stochastic growth theory.)

multiple curves in Fig. 14.10 correspond to different κ, with smaller κ having a larger flux. This is due to the shock reflecting more electrons as κ decreases due to there being more fast electrons with suitable v to reflect, as explained in Section 14.3 and the papers of Cairns et al. (2003) and Knock et al. (2003). (Additional examples of this first approach may be found in the papers of Knock et al., 2001; Knock and Cairns, 2005; Hillan et al., 2012a, 2012b.)

The second approach is to predict the shock and plasma parameters using an MHD simulation (e.g., Schmidt and Gopalswamy, 2008; Schmidt & Cairns 2012a;b; 2014; 2016; Schmidt et al., 2013, 2014; Cairns and Schmidt, 2015). The most recent works use the 3-D MHD code called BATS-R-US (Powell et al., 1999; Toth et al., 2012) with realistic data-driven models for the corona, solar wind, and the CME (e.g., Roussev et al., 2003, 2004; Cohen et al., 2007, 2008; Downs et al., 2010; Schmidt et al., 2013, 2014; Schmidt and Cairns, 2014, 2016; Cairns and Schmidt, 2015). Moreover, the type II predictions use a kappa distribution for the background electrons, with $\kappa = 2.5$ and the plasma density and electron temperature (equal to the ion temperature) given by the MHD simulations. The dynamic spectra predicted for interplanetary type II bursts so far agree very well with the observations (Schmidt et al., 2013, 2014; Schmidt and Cairns, 2014, 2016; Cairns and Schmidt, 2015). Readers are referred to the foregoing papers for details. Here we focus on the specific importance of kappa

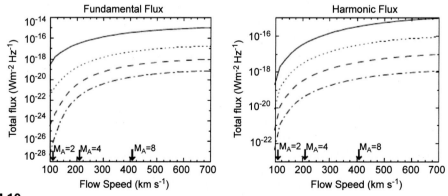

FIGURE 14.10
The predicted fluxes of fundamental and harmonic type II emissions increase as the shock speed and Alfven Mach number M_A increase and κ decreases (Cairns et al., 2003; Knock et al., 2003). From bottom to top, the curves are for $\kappa = 2, 3, 4$, and 5. Note that 10^{-22} Wm^{-2}Hz^{-1} equals one solar flux unit. The observer is 10^9 m upstream of the Earth's bow shock.

distributions to the theory for a specific observed coronal type II burst and do not address other aspects of the theory.

Fig. 14.11 shows the dynamic spectrum of a coronal type II burst observed by the Learmonth radio spectrograph in Western Australia near 0210 UT on September 7, 2014, with calibration by J. Harding personal communication), as well as several preceding type III bursts. Note that this type II burst consists of four bands of emission: the two higher frequency bands differ by about 20% in frequency, the two lower frequency bands differ by about 10% in frequency, and the two lower and two higher bands differ in frequency by a factor close to 2.0. These are so-called "split-bands" of both fundamental and harmonic radiation, corresponding to f_{pe} and $2f_{pe}$ radiation, respectively. Fig. 14.12 shows that the observed radio fluxes are in the approximate range of 10^{-23} to 10^{-21} Wm^{-2}Hz^{-1} or 0.1 to 10 solar flux units. It will be relevant for future work that this is the first type II observed by the Murchison Widefield Array, also in Western Australia, which has imaging capability, unlike the Learmonth radio spectrograph.

Carefully constructing the 3-D MHD configurations of the corona, solar wind, and CME using the foregoing techniques, Fig. 14.12 presents the predicted dynamic spectra for two values of κ for the background plasma electrons: $\kappa = 10$, which corresponds effectively to a Maxwellian distribution, and $\kappa = 2.5$, which is characteristic of the solar wind (e.g., Maksimovic et al., 1997b; see also the distinction between near-equilibrium and far-equilibrium regions at $\kappa \sim 2.5$, called the escape state, in Chapter 10; Livadiotis and McComas, 2010a, 2011b, 2013d). Several important results are apparent:

1. The predicted spectra have split-bands of both fundamental and harmonic radiation that look very similar to the observed dynamic spectrum. Moreover, the detailed morphologies are similar with the lower (upper) frequency fundamental split-band and higher (lower) frequency harmonic split-band, weaker in both the observations and predictions than the other two bands, with the lower (upper) frequency fundamental split-band disappearing at intermediate times.
2. The intensities of the four bands agree very well between the theory for $\kappa = 2.5$ and the observations, within better than a factor of 10.
3. The theoretical predictions for $\kappa = 2.5$ and 10 differ in flux by a factor of 100 and not in morphological structure, with the predictions for $\kappa = 2.5$ agreeing with the observations within less than a factor of 10.

These results provide strong evidence that the type II theory accounts very well for the observed coronal type II burst, including its split-band structure, and that the coronal plasma (along the shock's path) is well described as having a kappa electron distribution with $\kappa \approx 2.5$.

14.6 Concluding Remarks

In conclusion, kappa distributions are very important for the growth and damping of plasma waves (specifically Langmuir waves), the reflection of

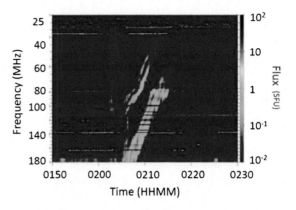

FIGURE 14.11
Learmonth dynamic spectrum of the type II burst observed on September 7, 2014, for the period 0205–0215 UT. Several type III bursts precede type II, which has both fundamental and harmonic split-bands.

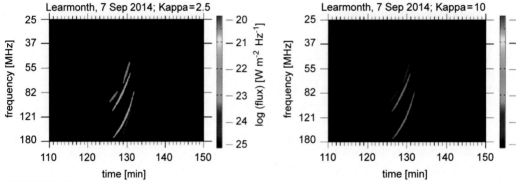

FIGURE 14.12
Dynamic spectra predicted for the type II burst of September 7, 2014, for $\kappa = 2.5$ and 10, the latter essentially corresponding to a Maxwellian.

electrons at shocks, the evolution of electron beams, and the generation of type II and III solar radio bursts. These relate to the extension of kappa distributions to much higher speeds, thus having many more fast particles, than a Maxwellian. Kappa distributions thus lead to a much larger spontaneous emission of plasma waves and radio emissions produced by incoherent processes. They also lead to shocks accelerating and reflecting many more of the plasma electrons back upstream of the shock, leading to stronger electron beams, Langmuir waves, and radio emissions. This was demonstrated theoretically for type II solar radio bursts, as is excellent agreement between observations and predictions for the coronal type II burst of September 7, 2014. This is the first such demonstration for coronal type II bursts and provides strong support for the associated theory/simulation capability. The extension of kappa distributions to larger speeds than for Maxwellian distributions leads to larger damping rates at high phase speeds for electron beam-generated Langmuir waves. It also limits the relaxation of the electron beam to higher speeds and reduces energy transfers to the Langmuir waves and any radio emissions generated therefrom. These effects are demonstrated theoretically for type III solar radio bursts, allowing us to show that the coronal electrons must have a kappa distribution with $\kappa \lesssim 5$ when fast type III beams with speeds $\gtrsim 0.5c$ are observed. These effects also lead to the type III bursts being weaker, having faster frequency drift rates, and starting at lower frequencies for smaller values of κ.

Regarding future work, we make several points, condensed into specific questions in Section 14.7. First, the kappa index is likely to vary from one type II or III burst to another, producing significantly different fluxes and burst properties. Indeed, the temporal and spatial variability of the solar active regions and corona suggests that kappa might well vary with position and time, leading to important changes in the location, number, and properties of source regions for even a single type II or III burst. Another possible cause of variability is preconditioning of the plasma by earlier shocks and reconnection events. Second, it is as yet unclear whether the foregoing theories can explain the detailed characteristics of a large sample of representative type II and III bursts (e.g., dynamic spectra, source locations, in situ electrons and Langmuir waves, fine structures in the radio emissions like split-bands), both in the corona and the solar wind. Only then will it be possible to claim that type II and III bursts are well understood.

Finally, preconditioning of the plasma by preceding type II or III events may well affect the predicted and observed properties, For instance, theoretical calculations of type II bursts for so-called "CME cannibalism" events (Gopalswamy et al., 2001), in which a second CME/shock pair follows in the path of an earlier CME/shock pair, catches up, and produces an enhanced radio emission in the interaction region, should be performed. Matters to address include whether the initial and enhanced radio emissions are both predicted well and whether agreement between the observed and predicted emissions requires smaller values of κ (and so more suprathermal electrons) and shock processing of the plasma by the first CME/shock pair. The analogue for type III bursts is when two successive electron beams follow the same field lines, with the second encountering remnant wave kappa distributions (generated by the first electron beam) that affect the evolution of the second beam. Li et al. (2002) have demonstrated this effect for Maxwellian plasmas, but the generalizations to kappa plasmas and to the post-2008 type III theory remain to be seen.

14.7 Science Questions for Future Research

Future analyses and observations need to address the following questions:

1. Does kappa vary significantly between different type II and III bursts?
2. Is preconditioning with κ distributions important for CME cannibalism?
3. How do shocks/flares/CMEs affect the kappa description of radio bursts?

Acknowledgment

This research work was supported in part by the Australian Research Council via grant DP140103933.

CHAPTER 15

Common Spectrum of Particles Accelerated in the Heliosphere: Observations and a Mechanism

L.A. Fisk, G. Gloeckler
University of Michigan, Ann Arbor, MI, United States

Chapter Outline

15.1 Summary   570
15.2 Introduction   571
15.3 Observations   574
    15.3.1 Common Spectrum in the Inner Heliosphere   575
        15.3.1.1 August 12, 2001, Local Acceleration Event   575
        15.3.1.2 October 25, 2001, Local Acceleration Event   577
    15.3.2 Common Spectrum of Anomalous Cosmic Rays Accelerated in the Nose of the Heliosheath   580
    15.3.3 Common Spectrum in the Fast, Polar Coronal Holes Solar Wind   582
15.4 Acceleration Mechanism That Yields the Common Spectrum   582

15.4.1 Conditions in Which the Acceleration Must Operate   583
15.4.2 Illustration of How the Acceleration Mechanism Works   583
15.4.3 Parker Equation: Describing the Behavior of Accelerated Particles   585
15.4.4 Why Is the Spectral Index in the Velocity Space Equal to −5?   586
15.4.5 Deriving an Equation for the Time Evolution of the Common Spectrum   587
15.4.6 Comparing Solutions for the Time Evolution of the Common Spectrum With Observations   590
15.4.7 Why Is Pump Acceleration the Dominant Acceleration Mechanism?   591

Kappa Distributions. http://dx.doi.org/10.1016/B978-0-12-804638-8.00015-2

15.4.8 Subtleties Associated With Solutions for Pump Acceleration 592

15.4.9 Source Particle Spectrum 594

15.5 Applications of the Pump Acceleration Mechanism 597

15.5.1 Acceleration of Energetic Particles at Shocks 598

15.5.1.1 Model for the Shock 598

15.5.1.2 Spectrum at the Shock 600

15.5.1.3 Role of Diffusive Shock Acceleration 603

15.5.2 Acceleration in the Solar Corona 605

15.5.2.1 Choice of the Diffusion Coefficient 605

15.5.2.2 Impulsive Solar Energetic Particle Events 606

15.6 Concluding Remarks 608

15.7 Science Questions for Future Research 608

Acknowledgments 608

15.1 Summary

Measurements of accelerated suprathermal particles in various locations of the heliosphere reveal a remarkably simple common spectrum. At low energies the spectrum is always the same, i.e., the velocity distribution is a power law with a spectral index of -5, which corresponds to the limiting kappa distribution with a κ-index $= 1.5$. At higher energies a rollover occurs, which can often be described as exponential. This common spectral shape cannot be accounted for by any conventional acceleration mechanism, such as diffusive shock acceleration or traditional stochastic acceleration. Thus a new simple "pump acceleration mechanism" has been constructed to account for these observations. In this mechanism, particles are pumped up in energy through a series of adiabatic compressions and expansions. The conditions under which the pump acceleration is the dominant acceleration mechanism are quite general and are likely to occur in other astrophysical plasmas. This chapter reviews relevant observations of the energy spectra of accelerated suprathermal tails throughout the solar system, revealing the first detection of the limiting kappa distribution. Also, the governing equation of the pump acceleration mechanism, responsible for generating this distribution, is derived in detail and applied in a couple of representative examples: acceleration downstream from locally observed shocks and acceleration in impulsive solar energetic particle events.

Science Question: How can a pump acceleration mechanism shift plasmas to the limiting state of $\kappa \sim 1.5$?

Keywords: Common spectrum, Mechanisms of kappa distributions, Particle acceleration, Suprathermal particles.

15.2 Introduction

Detailed observations have been made of the spectra of lower energy particles accelerated both in the inner heliosphere, from the Ulysses and the Advanced Composition Explorer (ACE) spacecraft, and in the heliosheath beyond the termination shock of the solar wind, from the Voyager spacecraft. In both cases, in specific events in the inner heliosphere and throughout the heliosheath, the spectra are the same: a power law in particle speed with a spectral index of -5, when the spectrum is expressed as a distribution function (also referred to as phase space density). Equivalently, in the differential intensity j (sometimes mentioned simply by the term "flux") energy spectrum, this is a power law, $j \propto \varepsilon^{-\gamma}$, where $-\gamma$ denotes the spectral index, with $\gamma \sim 1.5$. This common spectrum generally has an exponential rollover at higher energies, indicating the maximum energy particles acquired in the acceleration process.

In the inner heliosphere at 1 astronomical unit (AU), the common spectrum occurs at quite low energies, which makes it most evident in the observations of the Solar Wind Ion Composition Spectrometer (SWICS) instruments on ACE and Ulysses (Gloeckler et al., 1992); the rollover or maximum energy of accelerated particles is in all cases less than ~ 1 MeV/nucleon and in many cases below ~ 100 keV/nucleon. These observations are summarized in Fisk and Gloeckler (2012a). In the heliosheath, the common spectrum is observed over the full energy range of the Voyager Low Energy Charged Particle and Cosmic Ray Subsystem instruments (Stone et al., 1977; Krimigis et al., 1997). Shortly after both Voyager 1 and Voyager 2 crossed the termination shock, the common spectrum (an intensity spectrum with a spectral index of $-\gamma$ with $\gamma \sim 1.5$) was observed in particles accelerated to about a few MeV/nucleon (Decker et al., 2006; Gloeckler et al., 2008). Not only were the spectral indices the same at both Voyager 1 and Voyager 2, but the absolute intensities were also the same at both spacecraft, even though the two Voyagers were more than 150 AU apart. Furthermore, these common spectra remained essentially unchanged for years after the termination shock crossings. Moreover, when Voyager 1 reached the prime acceleration region of the anomalous cosmic rays (ACRs) at ~ 117 AU, the full ACR oxygen spectrum had the common spectrum (Fisk and Gloeckler, 2013; references therein).

It is remarkable that the spectra of accelerated particles in the inner heliosphere and in the heliosheath are the same. It is also remarkable that no traditional acceleration mechanism, such as diffusive shock acceleration or stochastic acceleration, can account for the common spectrum. Diffusive shock acceleration does yield power law spectra, but in general the spectral index depends on the compression ratio of the shock, with no preference for a spectral index of -5. Traditional stochastic acceleration often yields exponential spectra, and if power laws result, there is again no reason for a spectral index of -5.

The energy-flux spectral index γ is trivially related to the κ-index $= \gamma$ (Table 15.1 in Livadiotis and McComas, 2009; Livadiotis et al. 2011). The common spectrum has the features of a certain kappa distribution, corresponding to the limiting

Table 15.1 Shock and Particle Acceleration Results for Local Acceleration Events in August 2001

Date of LAE	Averaging Interval (h)	Location	Com/Sion Ratio, R	θ_{Bn} (Deg)	Tail γ	Peak Tail n (cm^{-3})	u_{roll} (cm/s)	U_{mod} (cm/s)
Aug. 3	06:00—07:00	Shock a	2.9 ± 0.7	70 ± 3	5.0 ± 0.4[c]	$5.85 \cdot 10^{-3}$	$2 \cdot 10^8$	$\sim 5 \cdot 10^7$
Aug. 3	08:00—09:00	Peak b	—	—	5.03 ± 0.09	$6.30 \cdot 10^{-3}$	$2 \cdot 10^8$	—
Aug. 12	10:00—11:00	Shock c	3.85 ± 0.9	76 ± 3	4.7 ± 0.35[c]	$2.69 \cdot 10^{-3}$	$5 \cdot 10^8$	$2.45 \cdot 10^8$
Aug. 12	11:00 —12:00	Peak d	—	—	5.03 ± 0.09	$1.24 \cdot 10^{-2}$	$\sim 3 \cdot 10^8$	—
Aug. 17	10:00—11:00	Shock e	4.3 ± 0.3[c]	62 ± 6	4.2 ± 0.3[c]	$7.7 \cdot 10^{-3}$	$> 3.5 \cdot 10^8$	$\sim 5 \cdot 10^7$
Aug. 17	12:00—14:00	Peak f	—	—	4.98 ± 0.06	$3.15 \cdot 10^{-2}$	$\sim 3 \cdot 10^8$	—
Aug. 24	7:00—22:00[a]	Diffuse g	—	—	4.99 ± 0.09	$3.5 \cdot 10^{-4}$	$> 3.5 \cdot 10^8$	—
Aug. 26	5:00—18:00	Prompt h	—	—	4.94 ± 0.07	$8.5 \cdot 10^{-4}$	$2 \cdot 10^8$	—
Aug. 27	20:00—21:00	Shock i	2.8 ± 0.6	89 ± 6	4.9 ± 0.4[c]	$8.4 \cdot 10^{-3}$	$> 3.5 \cdot 10^8$	$\sim 3 \cdot 10^7$
Aug. 27	21:00—2:00[b]	Peak j	—	—	5.03 ± 0.09	$3.15 \cdot 10^{-2}$	$\sim 2 \cdot 10^8$	—

LAE, local acceleration event.
[a] August 25, 2001.
[b] August 28, 2001.
[c] Complex spectra fit by power law.

value of $\kappa = \gamma \sim 1.5$. At this limit, the energy density is logarithmically divergent, hence the requirement of a rollover at higher energies. The value of the κ-index coincides with the spectral index that determines the slope of the energy-flux spectrum (on a log–log scale) of the suprathermal electrons forming the tail of the velocity distribution. In the limit $\kappa \to \infty$, the kappa distribution degenerates into a Maxwellian, decreasing exponentially with the square velocity. On the other limit, however, $\kappa \to 1.5$, the kappa distribution degenerates into a power law with exponent $\kappa \to 1.5$. In a Maxwellian limit, the plasma is at thermal equilibrium, while in the other limit, the plasma resides in the furthest stationary state from thermal equilibrium (also called antiequilibrium). (For more details, see Chapter 1; Livadiotis and McComas, 2009, 2013a; Livadiotis, 2015a.)

These features provide a guide to the origin of the common spectrum, as we discuss in detail in this chapter. The common spectrum is most pronounced in regions of the heliosphere where there is extensive compressive turbulence (e.g., downstream from shocks or compression waves) and throughout the heliosheath. These regions also have ample supplies of particles to be accelerated: particles heated at the shock front and PUIs in the heliosheath. We can justify these regions as being thermally isolated, in which particles are accelerated locally, not diffusing in or out from adjacent regions. In a thermally isolated volume, energy is conserved. In a thermally isolated volume with an irreversible acceleration due to compressions and expansions coupled with spatial diffusion, the particles will tend toward a state of maximum entropy, which, as we shall demonstrate, requires the common spectrum: a power law with a spectral index of -5 when expressed as a distribution function.

The concept that the common spectrum results from acceleration of a low-energy source of particles in compressive turbulence, in a thermally isolated volume, is supported by earlier theoretical work by Bykov (2001), who found the common spectrum in superbubbles of the Galaxy before the common spectrum was known in the heliosphere. Bykov (2001, 2014) performed a nonlinear numerical simulation in which there is a reaction of the accelerated particles on the turbulence, such that energy is conserved between the accelerated particles and the turbulence. The requirement that energy is conserved is the equivalent of requiring that the acceleration occurs in a thermally isolated system, with no external source of energy. The resulting nonrelativistic distribution function of the accelerated particles found by Bykov (2001, 2014) is a power law with a spectral index of -5.

The thermodynamic argument for the common spectrum is interesting, but what is required is a detailed mechanism for accomplishing the acceleration, from which we can determine the rate of acceleration and apply the mechanism in many different astrophysical settings. In a series of papers, Fisk and Gloeckler (2007, 2008, 2012a, 2014) and Fisk et al. (2010) developed a new acceleration mechanism that yields the common spectrum: a pump mechanism that accelerates particles in compressive turbulence in regions of space that can be considered to be thermally isolated, i.e., no external sources of energy. We will discuss the pump acceleration mechanism in detail in this chapter and show that it can account for all of the observations.

We begin by summarizing the observations of the common spectrum in the heliosphere, emphasizing those observations that best reveal the conditions in which the acceleration must operate. We then build upon the earlier work of Fisk and Gloeckler (2012a, 2014) and Fisk et al. (2010) and provide a complete derivation of a pump acceleration mechanism that yields the common spectrum, discussing the various subtleties associated with this derivation. We also present two representative applications of the pump acceleration mechanism: acceleration downstream from shocks and acceleration in impulsive solar energetic particle (SEP) events. References to other applications are provided, such as to the acceleration of ACRs in the heliosheath and to the acceleration of galactic cosmic rays in the Galaxy.

In Section 15.3, we present observations of particles that best reveal the conditions in which the acceleration of the common spectrum occurs. In Section 15.4 we present the acceleration mechanism that yields the observed common spectrum. Several applications of the pump acceleration mechanism are discussed in Section 15.5. Finally, the concluding remarks are given in Section 15.6, while three general science questions for future analyses are posed in Section 15.7.

15.3 Observations

The best and arguably the only way to obtain definitive information on where particles are accelerated and on the acceleration mechanism is through observations of differential intensity spectra or velocity distributions of several species with different charge-to-mass (Q/A) ratios over a broad energy range (from <1 keV/nucleon to >100 MeV/nucleon) in the location where the particles are accelerated.

In the inner heliosphere, there are instances of substantial enhancement of particle intensities over a wide suprathermal energy range, typically below a few MeV, lasting from hours to days. We refer to these events as local acceleration events (LAEs) and the location in which they occur as local acceleration regions (LARs). This nomenclature does not prejudice the acceleration process responsible for the LAEs, since, as we shall discuss, sometimes the LAEs are accompanied by locally recorded shocks, (e.g., SEP and CIR events); however, the main acceleration is generally not coincident with a shock. Beyond the termination shock, the entire nose region of the heliosheath can be considered to be a large LAR.

Within the LARs, there are often embedded prime acceleration regions (PARs), in which accelerated tail particle densities are at local maximum values. In these PARs, contributions from remote acceleration regions are small. There, the operating acceleration mechanism produces energy spectra that are not contaminated by particles accelerated remotely.

In this section, we present observations of particles that best reveal the conditions in which the acceleration of the common spectrum must occur: (1) spectra of protons accelerated downstream from shocks in the slow wind at 1 AU, (2) the acceleration of ACRs in the heliosheath, and (3) the acceleration of protons in the fast solar wind from the polar coronal holes. All of the spectra that we will show have been transformed to the solar wind frame using the duty cycle approximation. It is essential to make this transformation to the solar wind frame to observe the true spectral shapes, especially in the SWICS energy range, below ~ 100 keV/nucleon.

With knowledge of the required conditions in which the acceleration of the common spectrum must occur, we can develop an acceleration mechanism that can yield the common spectrum in Section 15.4.

15.3.1 Common Spectrum in the Inner Heliosphere

In Fig. 15.1 we show variations with the time of the hourly values of the density of suprathermal tails observed with SWICS on ACE during a 42-day period in 2001, computed by integrating the solar wind frame distribution function of protons from $u_{proton} = V_{sw}$ (the solar wind speed) to $u_{proton} = 9V_{sw}$. The tail density is highly variable, from low values of $\sim 5 \cdot 10^{-5}$ cm^{-3} to values exceeding 10^{-1} cm^{-3}. Local peak tail densities are often (but not always) associated with shocks. However, half of the shocks produce no local peaks, and several local peaks are not accompanied by shocks. This result is consistent with the results of a statistical study of numerous shocks and SEP events by Lario et al. (2005).

15.3.1.1 August 12, 2001, Local Acceleration Event

We select the second of the most intense of the four LAEs shown in Fig. 15.1. It was accompanied by a strong shock ($R = 3.85 \pm 0.05$, $V_{sh} = 409 \pm 29$ km/s,

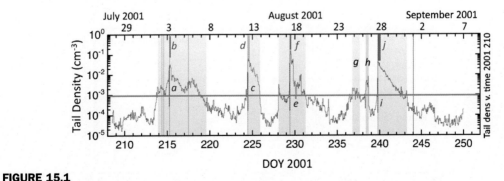

FIGURE 15.1

Variations with the time of hourly values of the density of tail particles during a CME-dominated 42-day period in 2001. Tail densities varied by about three orders of magnitude. Six local acceleration events (LAEs) are indicated by the shaded areas. LAEs are defined to be time periods when the tail density exceeds 10^{-3} cm^{-3} (*green horizontal line*). Eight shocks (*indicated by the vertical red lines*) were recorded during this period at times. Solar wind frame velocity distributions (from 10^6 to $\sim 3 \cdot 10^9$ cm/s) were computed during time intervals labeled *a* through *j* and are listed in Table 15.1. *DOY*, Day of the year.

$\theta_{Bn} = 76 \pm 3°$, $M_A = 2.8$), observed by the MAG and the Solar Wind Electron, Proton, and Alpha Monitor on ACE on August 12, 2001, at 10:50. The time profiles of the solar wind bulk speed and its thermal speed, along with the densities of the solar wind and tail particles, are displayed in Fig. 15.2. The tail density (density of protons with speeds between V_{sw} and $9V_{sw}$) is ramped up by a factor of ~ 7, from a low value of $\sim 2 \cdot 10^{-3}$ cm^{-3} to $\sim 1.3 \cdot 10^{-2}$ cm^{-3} at the shock, reaching its peak value of $\sim 5 \cdot 10^{-2}$ cm^{-3} 1 h downstream of the shock.

In Fig. 15.3 (left panel) we show the 12-min averaged proton spectrum during which the shock passed ACE. The solar wind frame velocity distribution has three components: (1) the slightly heated bulk solar wind, (2) a low-energy tail component at a speed between $\sim 10^7$ and $\sim 10^8$ cm/s with a complex shape, and (3) a high-energy tail component above $\sim 10^8$ cm/s. The high-energy tail spectrum is a modulated -5 common spectrum of particles accelerated not at the shock itself but most likely downstream of the shock. This is evident from the right panel of Fig. 15.3 that shows the solar wind frame proton spectrum averaged over 8.4 h starting right after the shock passage. The common tail spectrum starts at 10^7 cm/s, which is about four times the average thermal speed of the solar wind during the 8.4-h time period and is well fit with Eq. (15.1):

$$f(u) = f_0 \left(\frac{u}{10^7}\right)^{-5} \exp\left[-\left(\frac{u}{u_{roll}}\right)^{1.05}\right], \quad u > 10^7 \text{cm/s}. \tag{15.1}$$

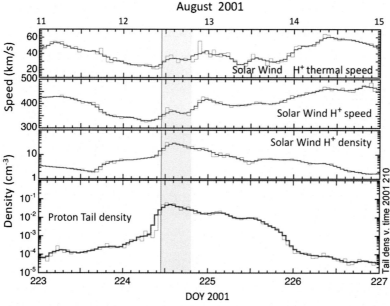

FIGURE 15.2

Variations with time during DOY 223 to 227 of 2001 of hourly values of (from top to bottom) the solar wind thermal speed, solar wind bulk speed, solar wind proton density, and the density of tail particles. The solar wind and tail density as well as the solar wind bulk and thermal speeds were higher several hours downstream of the shock then they were during the 1 h of shock passage.

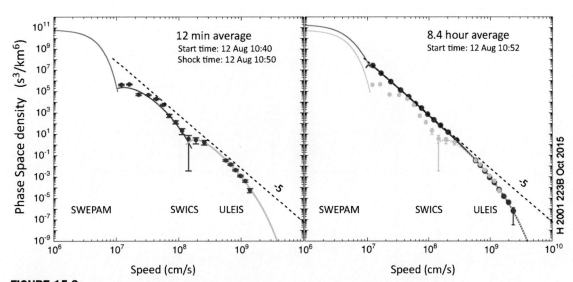

FIGURE 15.3

Left panel: 12-min averaged solar wind frame velocity distribution functions of protons during the passage of a strong shock (compression ratio of $R = 3.85 \pm 0.15$) on August 12, 2001, at 10:50 UT (universal time). The *blue curve* is the Maxwellian velocity distribution computed from the Solar Wind Electron, Proton, and Alpha Monitor bulk solar wind parameters. Right panel: 8.4-h averaged solar wind frame velocity distribution functions of protons just downstream of the shock (*shaded time period* in Fig. 15.2). The spectrum at the shock passage from the left panel is shown as *gray symbols* and a *curve*. The spectra are identical in the Ultra Low Energy Isotope Spectrometer energy range. The *dashed lines* are power laws with a spectral index of -5.

Here, u is the proton speed in the solar wind frame, $f_0 = 7 \cdot 10^7$ s^3/km^6, and $u_{roll} = 5 \cdot 10^8$ cm/s is the e-folding rollover speed.

The results for the August 12, 2001, LAE as well as five other LAEs in August 2001 are summarized in Table 15.1. The most striking result is that during the peak tail density of each LAE a pure -5 power law is observed, in most cases from the end of the bulk solar wind segment ($u_{proton} \approx 3u_{th}$, where $u_{th} \equiv \sqrt{k_B T / m_p}$, with k_B noting the Boltzmann's constant and m_p the proton mass) to the end of the SWICS energy range. The tail spectra above $\sim 10^8 \cdot$cm/s at the four shocks with compression ratios ranging from ~ 2.5 to 4 were modulated -5 power law spectra of particles accelerated downstream of the shocks. At speeds between $\sim 10^7$ and $\sim 10^8$ cm/s the spectra were complex. Fitting these segments with a power law yielded power law indices listed in column six of Table 15.1.

It should be noted that diffusive shock acceleration applied to realistic shocks, which have, for example, nonplanar shock geometries and time variations, can yield complex spectra at the shock front and downstream.

15.3.1.2 October 25, 2001, Local Acceleration Event

Associated with the so-called Halloween events was the large LAE shown in Fig. 15.4. Plotted as a function of time are the 1-h densities of the bulk solar

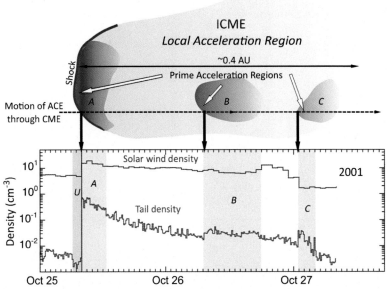

FIGURE 15.4
Top panel: Schematic representation of the local acceleration region (LAR) that produced the October 25, 2001, local acceleration event. In prime acceleration regions (PARs), tail densities are further enhanced, as shown in the shaded regions *A*, *B*, and *C* in the bottom panel. Bottom panel: Time profiles of the 1-h averaged density of bulk solar wind protons (*blue*) and the 12-min averaged tail density (*red*).

wind protons (blue trace) and the 12-min density of the tail particles computed from the observed solar wind frame phase space between $3V_{SW}$ and $9V_{SW}$. Also shown on the top of the figure is the relationship of the various features of these profiles to the passage past ACE of the LAR, most likely an interplanetary coronal mass ejection (ICME). We envision the LAR to be a 3-D region in space moving away from the Sun at the speed ($\sim 400-450$ km/s in the present case) of the strong shock at its nose. Compared to its surroundings, the LAR is a region of enhanced solar wind density and is very likely of high compressive turbulence. The density and turbulence levels are by no means uniform throughout the LAR. Rather, there are "hot" spots (red regions) where presumably the turbulence is higher compared to its immediate surroundings. These hot spots are the PARs where the intensities of locally accelerated particles are so high that contributions from remotely accelerated particles are unimportant.

In the frame of the LAR, ACE moved to the right, as indicted (for simplicity) by the dashed line. The strong shock ($R \approx 4$, $M_A \approx 3.6$, $\theta_{Bn} = 30° \pm 22°$) passed ACE at 8:01 on October 25, 2001 (see the bottom panel of Fig. 15.4). Almost immediately downstream of the shock, ACE entered the first PAR, where it remained for ~ 5 h. Then, about a day after the shock passage, ACE entered a second PAR and spent the next 11 h in it.

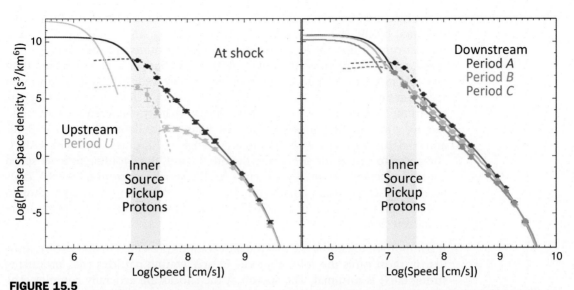

FIGURE 15.5

Left panel: 12-min averaged, sector-averaged solar wind frame velocity distribution functions of protons during the passage of a strong shock (a compression ratio of $R = 3.5 \pm 0.5$) on October 25, 2001 (DOY 298) at 08:01 UT (*red symbols and curve*) and during period *U* (see Fig. 15.4, *shaded area U*) upstream of the shock (*gray symbols and curves*). Right panel: sector-averaged solar wind frame velocity distribution functions of protons averaged over periods *A*, *B*, and *C* (see Fig. 15.4, *shaded areas A, B,* and *C*), respectively. Except for period *B*, hot inner source pickup protons (*shaded area*) dominate the spectra between ~100 and 300 km/s and are most exposed upstream of the shock.

Fig. 15.5 shows the sector-averaged phase space density of protons as a function of proton speed in the solar wind frame for the October 25, 2001, large LAE during the time intervals indicated by the shaded regions in the bottom panel of Fig. 15.4 and during the time of shock passage. The spectrum averaged over a 12-min time period that included the time of shock passage (*red symbols* and *curve* in the left panel of Fig. 15.5) is the common spectrum above $\sim 3 \cdot 10^7$ cm/s, a -5 power law with a gentle exponential rollover at an e-folding speed of $\sim 7 \cdot 10^8$ cm/s. Between $\sim 10^7$ and $\sim 3 \cdot 10^7$ cm/s is the characteristic spectrum of PickUp Ions (PUIs), and below $\sim 10^7$ cm/s is the Maxwellian spectrum of the bulk solar wind (density of 15.8 cm^{-3}, thermal speed equal to 44.6 km/s). The pickup proton spectrum has a broad rollover (rather than a sharp cutoff) at ~ 160 km/s, suggesting that these are Inner Source pickup protons (rather than Interstellar pickup protons) produced by an electron impact ionization close to the Sun (at ~ 0.3 AU). Furthermore, the maximum phase space density of Interstellar protons near their cutoff would be only $\sim 10^4$ (s^3/km^6), far below the observed densities. Our observations of Inner Source pickup protons suggest that large fluxes of energetic electrons existed upstream of the October 25 shock when it was at ~ 0.3 AU. These electrons were further heated and compressed by the shock and accelerated by the pump mechanism in the strong PAR downstream of the strong shock, creating even more pickup protons.

In the right-hand panel of Fig. 15.5 are shown spectra in each of the three PARs indicated by the shaded regions *A*, *B*, and *C* of Fig. 15.4. In the PAR *A* right downstream of the shock, Inner Source pickup protons are still visible in the speed range between $\sim 10^7$ and $\sim 3 \cdot 10^7$ cm/s, but at a speed higher than $\sim 3 \cdot 10^7$ cm/s the common spectrum is clearly observed. In PAR *B* the common spectrum starts at $\sim 10^7$ cm/s, concealing entirely any PUIs. Finally, in PAR *C* the spectrum is similar, but the phase space density between $\sim 3 \cdot 10^7$ and $\sim 3 \cdot 10^8$ cm/s is visibly lower. In fact, the phase space density between $\sim 3 \cdot 10^7$ and $\sim 3 \cdot 10^8$ cm/s decreases gradually, being ~ 20 times lower in PAR *C* compared to that at the shock. On the other hand, the e-folding rollover speed gradually increases from $\sim 6 \cdot 10^8$ cm/s for the shock spectrum to $\sim 3 \cdot 10^9$ cm/s for the PAR *C* spectrum. Finally, the density at $\sim 3 \cdot 10^9$ cm/s (corresponding to ~ 4 MeV protons) remains the same in all spectra shown, i.e., 4 MeV protons are not accelerated.

It is important to emphasize that the observations presented in this section show that in most cases the role of shocks in accelerating particles (i.e., producing strong tails) is minimal. The spectra at the shocks are generally complex. The acceleration is shown to take place in PARs often removed from the shocks. Indeed, in many instances no shocks are observed in acceleration regions. In PARs, the observed high-density spectra (in all of the cases that we have examined so far) are the common spectrum: −5 power law tails with rollovers at characteristic speeds of 2 to 5000 km/s. These tails often begin at particle speeds of several times that of the bulk solar wind flow speed.

15.3.2 Common Spectrum of Anomalous Cosmic Rays Accelerated in the Nose of the Heliosheath

The Voyager spacecraft are currently exploring the nose region of the heliosheath and are finding it to be a region of extensive particle acceleration, in effect one large LAR. Unlike LARs in the inner heliosphere that are in motion and whose size, shape, and properties may change rapidly as they travel away from the Sun, the nose of the heliosheath is a stationary region of approximately constant size and shape, extending from the termination shock whose distance from the Sun, typically 80 to 100 AU, is controlled by the supersonic solar wind, to the heliopause, some 40 AU beyond the termination shock (Fisk and Gloeckler, 2014; Gloeckler and Fisk, 2014). The solar wind moves through the stationary heliosheath with a constantly diminishing radial speed that at ~ 115 AU (at Voyager 1) reaches its lowest value of only about 5−6 km/s and continues at this speed until it reaches the heliopause. The PUIs produced inside the termination shock are heated by the termination shock and become the dominant pressure in the immediate downstream region of the termination shock. The magnetic field is mostly azimuthal in the nose of the heliosheath and exhibits regions with strong compressions and expansions.

Within a short distance downstream from the termination shock, both Voyager 1 and Voyager 2 observed the common spectrum (with a spectral index of −1.5) for particles accelerated to about a few MeV/nucleon (Decker et al., 2006; Gloeckler et al., 2008). The termination shock thus behaves in the same manner as shocks in the inner heliosphere discussed in Section 15.3.1. The common spectrum is not observed at the shock but rather downstream. Moreover, the common spectrum seen by Voyagers 1 and 2 had the same absolute intensity, even though the two Voyagers were more than 100 AU apart, and these common spectra remained essentially unchanged for years after the termination shock crossings.

Further into the heliosheath, the Voyager 1 observations of Stone et al. (2013) and Krimigis et al. (2013) reveal that ACRs, which are accelerated out of the Interstellar PUIs, attain their highest energies at ~ 117 AU. This is then the prime acceleration region for ACRs. Shown in Fig. 15.6 is the ACR O spectrum observed by Krimigis et al. (2013) in the PAR for ACRs. The spectrum is well fit by the common spectrum, an intensity spectrum that is a power law with a spectral index-1.5 and an exponential cutoff at ~ 100 MeV.

As is discussed in detail in Fisk and Gloeckler (2013), the acceleration of the ACRs must be in the heliosheath and cannot occur at the termination shock. The pressure in the ACRs in the PAR is observed to be about half the total particle pressure in the heliosheath (Gloeckler and Fisk, 2010). If ACRs originate at the termination shock on the flanks, e.g., as suggested by Schwadron and McComas (2006), the ACR pressure would be added to the pressure in the PUIs that are convected radially downstream from the termination shock with the solar wind. The pressure would then not be constant, as is required in the subsonic

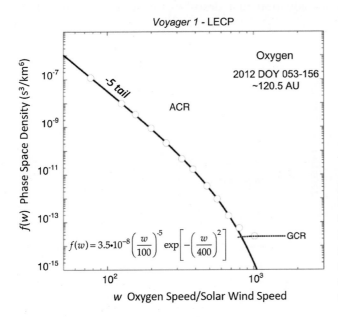

Voyager 1 - LECP

Oxygen

2012 DOY 053-156
~ 120.5 AU

ACR

-5 tail

GCR

$$f(w) = 3.5{\cdot}10^{-8}\left(\frac{w}{100}\right)^{-5}\exp\left[-\left(\frac{w}{400}\right)^{2}\right]$$

f(w) Phase Space Density (s³/km⁶)

w Oxygen Speed/Solar Wind Speed

FIGURE 15.6

Phase space density of anomalous cosmic ray oxygen as a function of oxygen speed/solar wind speed in the prime acceleration region of the heliosheath at ~ 120 AU. The solar wind speed is taken to be 10 km/s. The equation shown in the figure is the formula for the common spectrum and provides an excellent fit to the observations.

heliosheath, unless the PUI pressure could be reduced, which seems unlikely. Rather, it is necessary to accelerate the ACRs out of the PUIs in the PAR.

15.3.3 Common Spectrum in the Fast, Polar Coronal Holes Solar Wind

Ulysses pioneered the exploration of the high-latitude heliosphere. During a solar minimum it studied the properties of the fast solar wind emanating from the large polar coronal holes. This solar wind was found to be remarkably quiet, with no shocks or LAEs recorded during the high-latitude passes at or near the solar minimum.

Using data from the SWICS instrument on Ulysses it is possible to characterize the proton velocity distribution in the fast solar wind both in the northern and southern coronal holes. The proton spectrum, combining data from the southern and northern high-latitude passes around the solar minimum, is shown in Fig. 15.7. One of the most interesting features of this spectrum is the discovery of a suprathermal tail. These were months and months of super-quiet times, no shocks, no LAEs. Yet not only is there a tail, but this tail is the common spectrum. The tail, however, is very weak compared to the tails discussed in Section 15.3.1 for acceleration in the slow solar wind. At $u_{\text{proton}} \sim 10^8$ cm/s the extrapolated phase space density is only $f_0 \approx 0.01$ s^3/km^6.

15.4 Acceleration Mechanism That Yields the Common Spectrum

In this section, we develop an acceleration mechanism that yields the common spectrum and derive an equation that describes the time evolution of the

FIGURE 15.7

Phase space density versus proton speed in the solar wind frame, as measured by the Ulysses Solar Wind Ion Composition Spectrometer (SWICS) in the fast, high-latitude solar wind from the large polar coronal holes. Data from the southern/northern polar passes were combined to improve counting statistics at the highest SWICS energies. The tail of the bulk solar wind, interstellar pickup protons with a sharp cutoff at ~800 km/s, and the −5 suprathermal tail are shown.

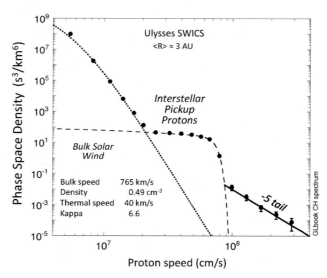

common spectrum. We begin by specifying the conditions in which the acceleration mechanism must operate, as revealed by the observations discussed in Section 15.3. We then describe an acceleration mechanism that will accelerate particles under these conditions and yield the common spectrum. We present a derivation of the equation that governs this acceleration, discuss the subtleties and implications of both the derivation and the governing equation, show how the pump acceleration mechanism can accelerate particles directly out of the thermal distribution, and determine the spectrum that links the thermal distribution to the common spectrum. It is this governing equation that can be used to fit observations of the common spectrum, in whatever astrophysical setting it occurs. In Section 15.5, we apply the governing equation to acceleration downstream from shocks in the solar wind, acceleration of galactic cosmic rays (GCRs) in the interstellar medium, and acceleration in the solar corona.

15.4.1 Conditions in Which the Acceleration Must Operate

As discussed in Section 15.3.2, the most pronounced common spectrum is the ACR spectrum in the heliosheath, which is a subsonic region, with extensive and relatively large-scale compressions and expansions of the solar wind. There is also a ready source of particles to accelerate: interstellar PUIs. The common spectrum is also pronounced downstream from shocks, which is also a subsonic region with compression and expansion regions, with a ready source of particles to accelerate the solar wind that is heated when crossing the shock (Section 15.3.1). The common spectrum is also strongest in the slow solar wind, which tends to have compressive turbulence, as opposed to the fast solar wind from the polar coronal holes, where the turbulence is more Alfvénic (Section 15.3.3).

We can thus assume that a region containing large-scale compressions and expansions is a requirement for the common spectrum, and presumably a source of particles to accelerate is required. The regions where the common spectrum occurs also do not exhibit strong average spatial gradients. The spatial variations in the average ACR intensity in the heliosheath, or in the accelerated particles downstream from shocks, are small relative to the scale sizes of the turbulence. The absence of large-scale spatial gradients across the volume where particles are accelerated has the consequence that the volume can be considered to be thermally isolated. The flows across any spatial boundaries are uncorrelated, with the result that particles are being accelerated in the volume, not elsewhere and flowing into the volume, nor are particles being lost from the volume.

15.4.2 Illustration of How the Acceleration Mechanism Works

To understand how particles can be accelerated in the conditions described in the previous section, it is useful to consider the illustration in Fig. 15.8. We have a volume of plasma containing compressions and expansions. The particle speed is plotted on the vertical axis and the position is plotted on the horizontal axis. There are three particle populations in the volume: (1) the

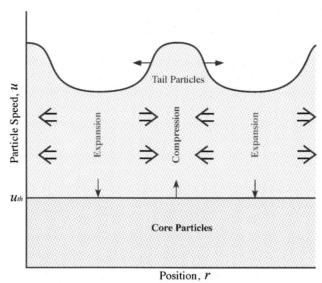

FIGURE 15.8
A schematic illustration of the principles underlying the pump acceleration mechanism.

thermal plasma, e.g., the thermal solar wind, which contains the mass and is responsible for the random compressions and expansions (not shown in the figure); (2) a particle population with speeds greater than the thermal speed of the bulk plasma and less than an upper threshold speed, $u \leq u_{th}$, which is the source particle population; the source particles also undergo random compressions and expansions but are not mobile and do not readily escape by spatial diffusion from a compression/expansion region, e.g., interstellar PUIs in the solar wind make an ideal source particle population or heated solar wind downstream from shocks; and (3) a particle population with speeds above u_{th} that is being accelerated out of the source particle population; the distinction between the source and the accelerated particles is that the accelerated particles can spatially diffuse and escape from a compression or diffuse into an expansion region.

Consider then what happens in the compression shown in the center of Fig. 15.8. The source particles are compressed adiabatically, and energy and particles flow across the threshold boundary at $u = u_{th}$. The accelerated particles are also compressed adiabatically and raised in energy, as noted by the extension in the compression region to higher particle speeds.

The opposite behavior occurs in the two expansion regions on either side of the compression region. In the expansion regions, particles and energy flow from the accelerated particle population back into the source particle population, and the energy of the accelerated particles is reduced. Note the large spatial gradients that result at higher particle speeds between the compression and the surrounding expansion regions. Accelerated particles are able to spatially diffuse, and so at higher particle speeds, particles will diffuse in response to these gradients out of the compression region and into the surrounding expansion regions.

Subsequently, the compression region will become an expansion region, and the process will be reversed. Particles and energy will flow back into the source from the accelerated particles. However, since particles have escaped from the compression region by spatial diffusion, there are fewer particles and less energy to return to the source.

If the process of compressions and expansions is repeated sequentially, then the accelerated particles and the energy they contain will systematically increase in time. This is a classic pump mechanism. The combination of adiabatic compressions and expansions and spatial diffusion of the accelerated particles will pump particles out of the source population and extend the distribution of the accelerated particles to higher and higher energies.

15.4.3 Parker Equation: Describing the Behavior of Accelerated Particles

The appropriate equation that describes the behavior of accelerated particles is the standard Parker transport equation, which is used in cosmic ray modulation studies. The compression and expansion regions in Fig. 15.8 have cross-sectional dimensions that are large compared to the gyro-radii of the particles that are being accelerated, and thus each is like a small modulation region. The distribution function of the particles being accelerated, $f(\vec{r}, \vec{u}, t)$, thus behaves as

$$\frac{\partial f}{\partial t} + \delta \vec{u} \cdot \vec{\nabla} f = \frac{1}{3} \left(\vec{\nabla} \cdot \delta \vec{u} \right) u \frac{\partial f}{\partial u} + \vec{\nabla} \cdot \left(\mathrm{K} \cdot \vec{\nabla} f \right). \tag{15.2}$$

Here, $\delta \vec{u}$ is the convective velocity of the compressions and expansions, and K is the spatial diffusion tensor. Note that implicit in the Parker equation is the assumption that there is a magnetic field that will couple the particle behavior within the compression and expansion regions.

It is important to note that the compression and expansion regions in Eq. (15.2) are not random and are correlated with each other. We have a fixed volume in which the acceleration occurs, which requires that each compression region must be surrounded by expansion regions. This is a fundamentally different assumption than assuming that the compressions and expansions are random. The former assumption of correlated expansions/compressions will lead to the pump acceleration mechanism illustrated in Fig. 15.8; the latter assumption of random compressions and expansions will lead to traditional stochastic acceleration.

We can see the role of our assumption of a fixed, thermally isolated volume by integrating Eq. (15.2) over all particle speeds above the threshold speed that separates the low-energy source of particles from the accelerated particles,

$u \geq u_{th}$, and over the fixed volume to form an equation for the behavior of the pressure, P, of the accelerated particles. The resulting equation contains divergences of flows of energy, which when integrated over the volume become flows across the spatial boundaries. The system is thermally isolated, in which case the boundary terms become zero, and we find that

$$\int_V d^3\vec{r}\left[\frac{\partial P}{\partial t} + \delta\vec{u}\cdot\vec{\nabla}P + \frac{5}{3}(\vec{\nabla}\cdot\delta\vec{u})P\right] = -\int_V d^3\vec{r}\,\frac{1}{3}\vec{\nabla}\cdot\delta\vec{u}\,\frac{4\pi}{3}m\,u^5f(u = u_{th}),$$

(15.3)

where m is the particle mass, and $d^3\vec{r}$ is the infinitesimal volume spanning the total volume V. The integrand on the left side of Eq. (15.3) is the time rate of change of the average pressure in the volume plus the time rate of change of the pressure due to the change in volume in a compression or expansion region. The expression on the right side of Eq. (15.3) is the net flow of energy across the threshold boundary due to the compressions and expansions. The total volume of the system is constant. The changes in the pressure due to changes in the volume in a compression or expansion region must thus integrate to zero. Eq. (15.3) then requires that the only source of energy to the average pressure of the accelerated particles is the net flow of energy from the low-energy source of particles due to the compressions and expansions.

15.4.4 Why Is the Spectral Index in the Velocity Space Equal to −5?

The acceleration illustrated in Fig. 15.8 is an irreversible process occurring in a thermally isolated volume. A thermally isolated system with an irreversible process will tend to a state of maximum entropy, i.e., the accelerated particles will approach a state of maximum entropy. Yet, in a state of maximum entropy, the compressions and expansions continue, and thus each compression and expansion must be isentropic, or equivalently for an ideal gas, adiabatic. At energies below where the rollovers occur in Fig. 15.8, the spatial gradients are relatively small, in which case spatial diffusion can be ignored. The behavior of the accelerated particles in this portion of the spectrum, from Eq. (15.2), is described by

$$\frac{\partial f}{\partial t} + \delta\vec{u}\cdot\vec{\nabla}f - \frac{1}{3}(\vec{\nabla}\cdot\delta\vec{u})u\frac{\partial f}{\partial u} = 0.$$

(15.4)

In order for each compression and expansion to be adiabatic (isentropic), the pressure P in the portion of the spectrum governed by Eq. (15.4) must satisfy the equation for an adiabatic or isentropic compression or expansion:

$$\frac{\partial P}{\partial t} + \delta\vec{u}\cdot\vec{\nabla}P + \frac{5}{3}(\vec{\nabla}\cdot\delta\vec{u})P = 0.$$

(15.5)

If we have a power law spectrum, $f \propto u^{-\alpha}$, α must be equal to 5 in order for the pressure associated with the power law portion of the spectrum to satisfy Eqs. (15.4) and (15.5).

15.4.5 Deriving an Equation for the Time Evolution of the Common Spectrum

In order to turn the Parker Eq. (15.2) into an equation that describes the time evolution of the common spectrum, we take the following steps: (1) we divide f into a mean term and a variable term, $f = f_0 + \delta f$, but unlike other approaches, we make nonconventional definitions of f_0 and δf that will facilitate obtaining our required equation; (2) we approximate the spatial diffusion term in Eq. (15.2) as a loss/gain term; and (3) given steps (1) and (2), we can derive the equation for the time evolution of the common spectrum simply by inspection of Eq. (15.2).

1. We define f_0 and δf in terms of the spatially-averaged pressure P_0 and the local deviation in the pressure in a compression or expansion region, δP, or

$$P_0 = \frac{4\pi}{3} m \int_{u_{th}}^{\infty} u^4 f_0 \, du \text{ and } \delta P = \frac{4\pi}{3} m \int_{u_{th}}^{\infty} u^4 \delta f \, du, \tag{15.6}$$

   where, by definition, $\langle \delta P \rangle = 0$; the angular brackets denote spatial average.

   It is important to note the f_0 and δf in Eq. (15.6) are defined in fundamentally different ways than in traditional stochastic acceleration, as considered, e.g., by Jokipii and Lee (2010). In traditional stochastic acceleration, f_0 is defined as the spatial average of f at each particle speed, in which case the spatial average of δf is $\langle \delta f \rangle = 0$. In Eq. (15.6), δf is integrated over particle speed to determine the pressure, which in turn averages to zero. Since the spectrum of δf varies between the compression and expansion regions, δf, as defined in Eq. (15.6), does not necessarily average to zero.

2. The diffusion term in Eq. (15.2) is difficult to deal with, since it involves local gradients. We can, however, approximate the effects of the diffusion with a loss/gain term, $\delta f / \tau$, where τ is the characteristic time for escape due to spatial diffusion. Note that in this approximation of spatial diffusion, τ is a quantity that needs to be solved for. The escape time τ depends upon the spatial diffusion coefficient, but it also depends upon the spatial gradients in the variable δf.

3. Then, substituting in $f = f_0 + \delta f$, approximating the diffusion term as a loss/gain term and rearranging terms, Eq. (15.1) becomes

$$\frac{\partial (f_0 + \delta f)}{\partial t} + \delta \vec{u} \cdot \vec{\nabla} (f_0 + \delta f) + \frac{5}{3} \left(\vec{\nabla} \cdot \delta \vec{u} \right) (f_0 + \delta f)$$

$$= \frac{\vec{\nabla} \cdot \delta \vec{u}}{3u^4} \frac{\partial}{\partial u} \left(u^5 \delta f \right) + \frac{\vec{\nabla} \cdot \delta \vec{u}}{3u^4} \frac{\partial}{\partial u} \left(u^5 f_0 \right) - \frac{\delta f}{\tau}. \tag{15.7}$$

We then determine the equation for f_0 by inspection by first integrating Eq. (15.7) to form an equation for the pressure and then over volume:

$$\int_V d^3\vec{r} \int_{u_{th}}^{\infty} u^4[\ldots]du = 0, \quad [\ldots] \equiv \frac{\partial(f_0 + \delta f)}{\partial t} + \delta\vec{u} \cdot \vec{\nabla}(f_0 + \delta f) +$$

$$+\frac{5}{3}(\vec{\nabla} \cdot \delta\vec{u})(f_0 + \delta f) - \frac{\vec{\nabla} \cdot \delta\vec{u}}{3u^4} \frac{\partial}{\partial u}(u^5 \delta f) - \frac{\vec{\nabla} \cdot \delta\vec{u}}{3u^4} \frac{\partial}{\partial u}(u^5 f_0) + \frac{\delta f}{\tau}.$$

(15.8)

We compare Eq. (15.8) with Eq. (15.3), which states that the only source of energy to the average pressure of the accelerated particles is the net flow of energy from the low-energy source of particles due to the compressions and expansions. Equivalently, f_0, which determines the average pressure, can vary in time only due to the second-order average flow of energy from the low-energy source, or

$$\frac{\partial f_0}{\partial t} = \frac{1}{u^4} \frac{\partial}{\partial u} \left(\frac{1}{3} u^5 \langle (\vec{\nabla} \cdot \delta\vec{u}) \delta f \rangle \right),$$

(15.9)

where the angular brackets denote average over volume. Then, we subtract Eq. (15.9) from Eq. (15.8), use the definitions of P_0 and δP in Eq. (15.6), and rearrange the terms as

$$\int_V d^3\vec{r} \left[\frac{\partial \delta P}{\partial t} + \delta\vec{u} \cdot \vec{\nabla} \delta P + \frac{5}{3}(\vec{\nabla} \cdot \delta\vec{u})(P_0 + \delta P) \right]$$

$$= \frac{4\pi}{3} m \int_V d^3\vec{r} \int_{u_{th}}^{\infty} u^4 \left[-\frac{\vec{\nabla} \cdot \delta\vec{u}}{3u^4} \frac{\partial}{\partial u}(u^5 f_0) + \frac{\delta f}{\tau} \right] du$$

$$= \frac{4\pi}{3} m \int_V d^3\vec{r} \left[-\frac{1}{3}(\vec{\nabla} \cdot \delta\vec{u})u^5 f_0(u = u_{th}) - \int_{u_{th}}^{\infty} u^4 \frac{\delta f}{\tau} du \right]$$

(15.10)

Variations with time in the pressure, δP, are the result only of variations in the volume, and since the total volume is constant and thermally isolated, the left side of Eq. (15.10) must integrate to zero, or

$$\frac{4\pi}{3} m \int_V d^3\vec{r} \left[-\frac{1}{3}(\vec{\nabla} \cdot \delta\vec{u})u^5 f_0(u = u_{th}) - \int_{u_{th}}^{\infty} u^4 \frac{\delta f}{\tau} du \right] = 0.$$

(15.11)

The first term on the right side of Eq. (15.11) is the net first-order flow of energy into the accelerated particles due to compressions and expansions. The second term is the net flow of energy into or out of expansion and compression regions due to spatial diffusion.

Consider the circulation of energy between compression and expansion regions. By definition, each compression region must be surrounded by an expansion region. Energy flows into a compression region from the low-energy source of particles, according to the first term in the integrand of Eq. (15.11), and then outward into the surrounding expansion region by spatial diffusion, according to the second term. In the surrounding expansion region, the energy that flows inward by spatial diffusion from the compression region flows outward to the low-energy source. To the first order, the energy that flows back to the source in the surrounding expansion region must equal the energy that flows from the source into the compression region. This can be realized only if (to the first order) all of the energy that flows into the compression region from the source escapes by spatial diffusion into the expansion region, and all of the energy that flows into the expansion region by spatial diffusion from the compression region flows back into the core. Thus using the formulation in the second line of Eq. (15.10), for each compression and expansion,

$$\frac{\vec{\nabla} \cdot \delta \vec{u}}{3u^4} \frac{\partial}{\partial u} \left(u^5 f_0 \right) = \frac{\delta f}{\tau}. \tag{15.12}$$

Substituting Eq. (15.12) into Eq. (15.9), we find the desired equation for f_0:

$$\frac{\partial f_0}{\partial t} = \frac{1}{u^4} \frac{\partial}{\partial u} \left[\frac{1}{9} \left\langle \left(\vec{\nabla} \cdot \delta \vec{u} \right)^2 \tau \right\rangle u \frac{\partial}{\partial u} \left(u^5 f_0 \right) \right]. \tag{15.13}$$

It is evident that the solution to Eq. (15.13) is a spectrum of u^{-5} with some form of rollover at higher speeds, where the location of the rollover in the particle speed increases with time, as required. To evaluate the location of the rollover, we need to specify τ, which depends upon the local spatial diffusion coefficient, K, and the spatial gradients in δf in the rollover regions of the spectrum, which have scale sizes on the order of the scale size of a compression or expansion region, which we take to be λ. It is thus reasonable to take

$$\left\langle \left(\vec{\nabla} \cdot \delta \vec{u} \right)^2 \tau \right\rangle \approx \frac{\delta u^2}{\lambda^2} \frac{\lambda^2}{K} = \frac{\delta u^2}{K}. \tag{15.14}$$

If we then take the local spatial diffusion coefficient for escape from a compression region to have a standard form of particle speed times a power law in particle rigidity, or $K \propto u^{(1+\alpha)}$, the solution to Eq. (15.13) is

$$f_0 \propto u^{-5} \exp \left[- \frac{9K}{(1 + \alpha)^2 \delta u^2 t} \right]. \tag{15.15}$$

This is the solution for the acceleration of the energetic particles due to compressions and expansions in a thermally isolated volume derived by Fisk and Gloeckler (2008, 2012a, 2014) and Fisk et al. (2010).

Finally, we can use our method of solving by inspection to determine the equation for δf, noting from Eqs. (15.3) and (15.9) that first-order changes in δP and thus δf are the result only of volume changes, or

$$\frac{\partial \delta f}{\partial t} + \delta \vec{u} \cdot \vec{\nabla} \delta f + \frac{5}{3} \left(\vec{\nabla} \cdot \delta \vec{u} \right) (f_0 + \delta f) \approx 0. \tag{15.16}$$

Noting that $\delta f << f_0$, Eq. (15.16) can be solved to yield

$$\delta f = -\frac{5}{3} \int_{t-\tau}^{t} (\vec{\nabla} \cdot \delta \vec{u}) f_0 dt',$$ (15.17)

where the integration is over the past time history as particles are convected with $\delta \vec{u}$. We perform the integration only starting at $t-\tau$, i.e., only for a time interval during which the particles have not escaped by local spatial diffusion.

Eq. (15.16) is valid in the -5 portion of the spectrum since here τ is longer than the coherence time of a compression or expansion. In the -5 portion of the spectrum particles undergo compressions and expansions, without escape within a compression time, as described in Eq. (15.17). Note that in this region of the spectrum, the spatial average of δf is zero, $\langle \delta f \rangle = 0$.

In the rollover region of the spectrum, τ is short compared to the coherence time of a compression region, and Eq. (15.12) determines δf. There is a balance between the flow of energy from the core and escape by spatial diffusion. Substituting Eq. (15.14) into Eq. (15.12), we find that

$$\delta f = \frac{\tau \vec{\nabla} \cdot \delta \vec{u}}{3u^4} \frac{\partial}{\partial u}(u^5 f_0) = \frac{3 \vec{\nabla} \cdot \delta \vec{u}}{(1+\alpha)^2} \left(\frac{K}{\delta u^2 t}\right) \tau f_0.$$ (15.18)

In the rollover region the fluctuations are also proportional to $\vec{\nabla} \cdot \delta \vec{u}$, as in the -5 portion of the spectrum, as seen in Eq. (15.17), but they are intrinsically larger since $K/(\delta u^2 t) > 1$. However, the particles escape within a compression or expansion time, or $\left(\vec{\nabla} \cdot \delta \vec{u}\right)\tau < 1$, with the result that the fluctuation δf, relative to f_0, is smaller in the rollover region than in the -5 portion of the spectrum. But if we spatially average Eq. (15.16) and substitute in Eq. (15.12), we find that

$$\frac{\partial \langle \delta f \rangle}{\partial t} = -\frac{2}{3}\langle (\vec{\nabla} \cdot \delta \vec{u}) \delta f \rangle = -\frac{2}{9}\langle (\vec{\nabla} \cdot \delta \vec{u})^2 \rangle \tau \frac{1}{u^4} \frac{\partial}{\partial u}(u^5 f_0).$$ (15.19)

Thus in the rollover region of the spectrum, $\langle \delta f \rangle \neq 0$, unlike in the -5 portion of the spectrum.

15.4.6 Comparing Solutions for the Time Evolution of the Common Spectrum With Observations

Observations of the distribution functions of the accelerated particles are usually spatial averages, where the mean distribution at each particle speed is determined, and deviations from the mean at each particle speed are assumed to average to zero. As is described in Eq. (15.6), we did not define f_0 and δf with the expectation that $\langle \delta f \rangle$ must be zero. However, as we discussed in the previous section, our solution for f_0 is a valid description of the observed mean distribution function of accelerated particles at all particle speeds. In the -5 portion of the spectrum there are fluctuations, δf, that depend upon the local value of $\vec{\nabla} \cdot \delta \vec{u}$. They have the same -5 spectrum and average to zero, as can be seen in

Eq. (15.16). In the rollover region of the spectrum the fluctuations δf do not have the same spectrum as f_0, as can be seen in Eq. (15.12), and do not average to zero, as in Eq. (15.19). The average of these fluctuations, however, is small compared to f_0, as can be seen in Eq. (15.18), and does not affect the comparison of f_0 with observations.

15.4.7 Why Is Pump Acceleration the Dominant Acceleration Mechanism?

We should expect that the pump acceleration mechanism illustrated in Fig. 15.8 and described by Eqs. (15.17) and (15.19) dominates over any second-order stochastic acceleration process. The pump acceleration includes a first-order acceleration, as can be seen in the schematic, Fig. 15.8. Particles escape from a compression region into an expansion region only after the expansion region has undergone most of its expansion; this is when the spatial gradients are the strongest. Thus in the cyclic process, particles do not gain and lose an equal amount of energy, as would occur if the particles undergo a stochastic diffusion in velocity space. Rather, the particles gain more energy in a compression region than they lose in the expansion regions into which they escape, resulting in a first-order acceleration.

We can see the first-order acceleration by rewriting the basic equation that describes the pump acceleration, Eq. (15.17), as

$$\frac{\partial}{\partial t}\left(u^5 f_0\right) + \frac{1}{3}\left\langle (\vec{\nabla}\cdot\delta\vec{u})^2\tau\right\rangle u \frac{\partial}{\partial u}\left(u^5 f_0\right) = \frac{1}{u^2}\frac{\partial}{\partial u}\left[u^4\frac{1}{9}\left\langle(\vec{\nabla}\cdot\delta\vec{u})^2\tau\right\rangle\frac{\partial}{\partial u}\left(u^5 f_0\right)\right].$$

(15.20)

Recall that the solutions to the governing equation of the pump acceleration are proportional to u^{-5}. Thus the quantity $u^5 f_0$ marks where f_0 deviates from u^{-5}, i.e., it marks where the rollover in the spectrum occurs. The term on the right side of Eq. (15.20) is in the form of a traditional stochastic diffusion. The location of the rollover thus diffuses to higher particle speeds. The first-order acceleration of the location of the rollover is contained in the second term on the left of Eq. (15.20) and the first-order acceleration is

$$\frac{du}{dt} = \frac{1}{3}\left\langle(\vec{\nabla}\cdot\delta\vec{u})^2\tau\right\rangle u.$$

(15.21)

Thus if there is a sufficient source particle population present on which the pump acceleration can act, the pump acceleration process will be the dominate determinant of the spectrum of the accelerated particles and yield the $f_0 \propto u^{-5}$ spectra that are most commonly observed.

We should also expect that, as discussed in Section 15.3.1, the pump acceleration mechanism operating in the downstream region dominates over diffusive shock acceleration, as discussed in more detail in Section 15.5.1. Particles are accelerated in standard diffusive shock acceleration by experiencing multiple times the strong compression that occurs at the shock front. In that sense, the compression at the shock front is just one of many compressions; other compressions used by the pump acceleration mechanism occur throughout the downstream region. In fact, the compression at the shock front may not be particularly effective. The extent to which particles are accelerated depends upon the time spent in the compression. Shocks are at only one location, whereas the compressions downstream, albeit weaker than the compressions at the shock, occur in a much larger volume.

15.4.8 Subtleties Associated With Solutions for Pump Acceleration

Finally, we discuss some of the subtleties associated with the derivation of the governing equation for pump acceleration and the solution to this equation, which appear to have confused some researchers considering this problem.

Jokipii and Lee (2010) argue that density is not properly conserved in Eq. (15.9). They integrate Eq. (15.9) over all tail particle speeds to find an equation for the time rate of change of the density n_0 associated with f_0:

$$\frac{dn_0}{dt} = -\frac{4\pi}{3}u^3 \langle \delta f \, \vec{\nabla} \cdot \delta \vec{u} \rangle |u_{th} + \frac{8\pi}{3} \int_{u_{th}}^{\infty} u^2 \langle \delta f \vec{\nabla} \cdot \delta \vec{u} \rangle du. \tag{15.22}$$

The first term on the right represents a flow of particles across the threshold boundary from the source particle population, and Jokipii and Lee (2010) state that the second term on the right is a spurious source term that appears to be creating particles.

The actual requirement for the conservation of density is not that the density n_o is conserved but rather that the total average density is conserved, which is determined by $f_0 + \langle \delta f \rangle$. If we combine Eqs. (15.19) and (15.9), we find

$$\frac{\partial f_0}{\partial t} + \frac{\partial \langle \delta f \rangle}{\partial t} = \frac{1}{u^4} \frac{\partial}{\partial u} \left(\frac{1}{3}u^5 \langle (\vec{\nabla} \cdot \delta \vec{u}) \delta f \rangle \right) - \frac{2}{3} \langle (\vec{\nabla} \cdot \delta \vec{u}) \delta f \rangle$$

$$= \frac{1}{u^2} \frac{\partial}{\partial u} \left(\frac{1}{3}u^3 \langle (\vec{\nabla} \cdot \delta \vec{u}) \delta f \rangle \right). \tag{15.23}$$

which when integrated to form the density demonstrates that all accelerated particles originate in the source particle population. There are no spurious source terms. The extra contribution to the density results as follows. There is a balance between energy flowing into and out of the source population and energy flowing outward and inward by spatial diffusion. However, the energy flowing in and out of the source is carried by low-speed particles, whereas the energy flowing in and out by spatial diffusion is carried by higher speed particles. Thus although the energy flows balance, the particle flows do not, yielding an apparent particle source.

There has also been some confusion about the thermodynamics of the pump acceleration process. The basic premise of the pump acceleration mechanism is that there is interplay between acceleration in a compression region and escape by spatial diffusion, versus deceleration in an expansion region and inflow by spatial diffusion. This interplay pumps particles out of a source particle population under circumstances in which no net work is being done on the accelerated particles when averaged over multiple compressions and expansions. The source of energy for the accelerated tail is the energy in the source particles. It might seem then that the pump mechanism is thermodynamically impossible. Particles are being accelerated when no work is being done on them. Energy is flowing from the colder source into the hotter tail.

Consider, however, the following illustrative example. Suppose you have a fixed, thermally isolated volume that contains a gas that is undergoing particle–particle collisions. Suppose also that initially the distribution of the gas is not Maxwellian in that there are too few high-speed particles to be a Maxwellian distribution. In time, the distribution of the gas will evolve into a Maxwellian distribution, which is the state of maximum entropy. The particle–particle collisions will accelerate particles to the high speeds required to satisfy a Maxwellian distribution. Since these are particle–particle collisions, the total energy of the gas is unchanged in the process of accelerating the particles to high speeds. This is thus an example of an irreversible process that transfers particles and energy from a colder portion of the distribution to a hotter high-speed portion, while maintaining the overall thermodynamic constraints that the total pressure and density are constants.

The irreversible transfer of energy from the colder core particles to the hotter tail particles for the purpose of maximizing entropy also occurs in the pump acceleration process. We have a fixed volume, thermally isolated system in which there are not particle–particle collisions but rather organized compressions and expansions of the gas. There is an embedded magnetic field whose sole purpose is to couple the behavior of particles at all speeds in compression and expansion regions. The compressions and expansions are simply volume changes of the gas, and since the overall volume is constant there is no net work done by the compressions and expansions when summed over the volume. When we add spatial diffusion at tail particle speeds, we find that there is an irreversible flow of particles and energy from the core into the tail and an irreversible expansion of the tail particles in velocity space (i.e., acceleration). The pump process thus increases the entropy of our system.

A thermally isolated, constant volume system conserves particles and energy. The evolution of the system is determined by the behavior of the entropy. Allowable evolutions must increase the entropy until the entropy is a maximum, at which point the system can be said to be in equilibrium. The pump acceleration mechanism operates in a thermally isolated, constant volume system; it conserves the total energy and number of particles; and it increases the entropy. It is an allowable evolution of the system, and it will form a tail on the particle distribution function at high particle speeds.

15.4.9 Source Particle Spectrum

The pump acceleration mechanism pumps particles out of a source particle distribution to form the −5 tail. The source particle distribution is an extension of the solar wind thermal plasma or is due to PUIs. In general, the transition from the source to the tail appears to occur when the particle speeds are of the order of the solar wind flow speed. However, we do not expect the pump mechanism to start abruptly at the threshold between the source and the tail. Rather, the pump mechanism can be expected to modify the source particle spectrum or perhaps even create the source particles, and thus consideration of where the source particles come from provides an essential link between the thermal particles and the tail. We will consider only the case where the source is an extension of the solar wind thermal distribution. Source particles that are dominated by PUIs should be more influenced by this mass loading process than by the pump acceleration mechanism.

The pump acceleration mechanism creates u^{-5} tails because particles can escape from a compression region in time τ_{es} less than the characteristic compression time $\tau_c \approx \lambda_\perp / \delta u$, where $\lambda_\perp$ is the cross-sectional dimension of the compression region. Escape faster than a compression time distinguishes the u^{-5} tail from the source particle population. The source particles can also be accelerated by the pump acceleration mechanism but less efficiently than the tail particles. If a particle cannot escape within τ_c, it will experience multiple compressions and expansions in the same location, and the average value of $\vec{\nabla} \cdot \delta \vec{u}$ will be reduced.

Consider the equation that describes the basic pump acceleration mechanism, that is,

$$\frac{\partial f}{\partial t} = \frac{1}{u^4} \frac{\partial}{\partial u} \left(\frac{1}{3} u^5 \langle (\vec{\nabla} \cdot \delta \vec{u}) \delta f \rangle \right). \tag{15.24}$$

The term in angular brackets in Eq. (15.9) is a spatial average of the correlation between the local value of $\vec{\nabla} \cdot \delta \vec{u}$ and δf, the deviation of the distribution function from its mean value due to the time history of the values of $\vec{\nabla} \cdot \delta \vec{u}$ that a particle has experienced. If a particle escapes a compression region in a time less than the compression time, the particle experiences only the local value of $\vec{\nabla} \cdot \delta \vec{u}$, which leads to the pump acceleration equation for the tail in Eq. (15.13). If a particle remains within a region for a time that is long compared to the compression time, then a particle experiences multiple compressions and expansions, and the time averaged value of $\vec{\nabla} \cdot \delta \vec{u}$ that it experiences is reduced. As an example, we take

$$\vec{\nabla} \cdot \delta \vec{u} = \frac{1}{\tau_c} \cos(t/\tau_c), \tag{15.25}$$

where τ_c is the characteristic compression/expansion time. Then, from Eq. (15.12), the average deviation of the distribution function from its mean value within an escape time, due to the time history of the compressions and expansions, is

$$\delta f|_{av} = \left[\frac{1}{\tau_{es}} \int_0^\infty e^{-t/\tau_{es}} (\vec{\nabla} \cdot \delta \vec{u}) dt \right] \cdot \left[\frac{\tau_{es}}{3u^4} \frac{\partial}{\partial u} (u^5 f_0) \right]$$

$$= \frac{\tau_c/\tau_{es}}{1 + (\tau_c/\tau_{es})^2} \cdot \frac{1}{3u^4} \frac{\partial}{\partial u} (u^5 f_0). \tag{15.26}$$

Thus

$$\langle (\vec{\nabla} \cdot \delta \vec{u}) \delta f \rangle \cong \frac{1/\tau_{es}}{1 + (\tau_c/\tau_{es})^2} \cdot \frac{1}{3u^4} \frac{\partial}{\partial u} (u^5 f_0), \tag{15.27}$$

where we have used the approximation for the local value of $\vec{\nabla} \cdot \delta \vec{u}$, i.e., $|\vec{\nabla} \cdot \delta \vec{u}| \approx 1/\tau_c$.

We define the diffusion coefficient, which determines the escape from a compression, to be $K = \lambda_c^2/\tau_{es}$, where λ_c is the cross-sectional dimension of the compression region. Note that K is a cross-field diffusion coefficient. Thus for the core, where $\tau_c << \tau_{es}$, the acceleration Eq. (15.24) becomes

$$\frac{\partial f}{\partial t} = \frac{1}{u^5} u \frac{\partial}{\partial u} \left[\frac{K}{9\lambda_c^2} u \frac{\partial}{\partial u} (u^5 f) \right], \tag{15.28}$$

where we have $K/\lambda_c^2 = 1/\tau_{es}$. For the tail, where $\tau_c >> \tau_{es}$, the acceleration equation becomes our usual acceleration equation:

$$\frac{\partial f}{\partial t} = \frac{1}{u^5} u \frac{\partial}{\partial u} \left[\frac{\delta u^2}{9K} u \frac{\partial}{\partial u} (u^5 f) \right], \tag{15.29}$$

where we have substituted $\delta u^2 = \lambda_c^2/\tau_c^2$ and $\delta u^2/K = \tau_{es}/\tau_c^2$.

We normally take the diffusion coefficient to be $K = K_0 \cdot u^{1+\alpha}$, where the mass-to-charge ratio (A/Z) is contained in K_0. With this form, the acceleration equation for the core, Eq. (15.28), becomes

$$\frac{\partial f}{\partial t} = \frac{1}{u^5} u \frac{\partial}{\partial u} \left[\frac{K_0}{9\lambda_c^2} u^{2+\alpha} \frac{\partial}{\partial u} (u^5 f) \right]. \tag{15.30}$$

There are two steady state solutions to Eq. (15.30), $f \propto u^{-\beta}$, with $\beta = 5$ or $\beta = 6 + \alpha$. Either solution is acceptable and contains finite energy.

With this form of the diffusion coefficient, the acceleration equation for the tail, Eq. (15.29), becomes

$$\frac{\partial f}{\partial t} = \frac{1}{u^5} u \frac{\partial}{\partial u} \left[\frac{\delta u^2}{9K_0} u^{-\alpha} \frac{\partial}{\partial u} (u^5 f) \right]. \tag{15.31}$$

The tail extends to infinity, and thus a pure power law solution of $\beta = 5$ or $\beta = 4 - \alpha$ contains infinite energy. The only acceptable solution is the one we always use:

$$f \propto u^{-5} \cdot \exp\left\{ - 9K_0 u^{1+\alpha} \Big/ \left[(1 + \alpha)^2 \delta u^2 t \right] \right\}. \tag{15.32}$$

The solution to Eq. (15.30) with $\beta = 6 + \alpha$ represents a constant flow of energy through the core. If we multiply Eq. (15.30) by u^4 and integrate to an upper threshold u_{th}, we find that the flow of energy through u_{th} is

$$S_E \propto \frac{(1 + \alpha) K_0 u_{th}^{\beta}}{9 \lambda_c^2}, \tag{15.33}$$

which is independent of u.

We can calculate the energy flow into the tail across the boundary at $\tau_c \sim \tau_{es}$ by integrating Eq. (15.31) as we did for Eq. (15.33), and with the solution in Eq. (15.32), we find that

$$S_E \propto \frac{u_{th}^5}{(1 + \alpha) t}. \tag{15.34}$$

The energy required by the tail thus goes down with time, as the tail develops to higher energies.

The energy that flows through the core must be deposited in the tail. Otherwise, it remains in the core, and presumably the core should be heated and evolve from the steeper $\beta = 6 + \alpha$ spectrum to the $\beta = 5$ spectrum, which is the same as the tail. Thus as the energy requirements of the tail decrease with time, more energy is deposited into the core, and the break between the steeper $\beta = 6 + \alpha$ spectrum and the $\beta = 5$ spectrum should move downward in particle speed.

We should expect then that the break between the $\beta = 6 + \alpha$ spectrum and the $\beta = 5$ spectrum occurs not where $\tau_c \sim \tau_{es}$ but rather within the core itself, roughly where the energy flowing through the $\beta = 6 + \alpha$ portion of the core is equal and opposite to the energy flowing downward from the $\beta = 5$ portion, which is heated by energy that cannot be absorbed by the tail. This result is consistent with the derivation of the pump acceleration equation in Section 15.4.5. In the -5 portion of the spectrum, particles undergo adiabatic compressions and expansions and do not readily escape from the compression regions. The escape within a compression time occurs in the rollover region of the spectrum.

The downward flowing energy occurs within the core, where $\tau_c \gg \tau_{es}$, and thus to find the downward flowing energy we need to integrate Eq. (15.30) and evaluate it with the solution to Eq. (15.30) that corresponds to the solution to Eq. (15.31) in Eq. (15.32). We find that the downward flowing energy is

$$S_E \propto - \frac{u_{th}^5}{(1 + \alpha) t}. \tag{15.35}$$

We expect then that the break between the $\beta = 6 + \alpha$ spectrum and the $\beta = 5$ spectrum occurs where

$$\frac{(1 + \alpha)K_0 u_{th}^\beta}{9\lambda_c^2} = \frac{2u_{th}^5}{(1 + \alpha)t}.$$

(15.36)

Half of this energy flows back from the heated core, as in Eq. (15.35), and the remaining energy provides the energy to the tail.

We can express t in terms of the maximum value of u_m obtained by the tail from Eq. (15.32) or $1/t = \frac{1}{9}(1 + \alpha')^2 \delta u^2 / (K_0' u_m^{1+\alpha'})$, where we have allowed with the primed-quantities for the fact that the diffusion coefficient at u_m is in the tail and will be different from the core. We find that the break between the $\beta = 6 + \alpha$ spectrum and the $\beta = 5$ spectrum has to occur at

$$\frac{(1 + \alpha)^2 K_0 u_{th}^{1+\alpha}}{9\lambda_c^2} = \frac{2(1 + \alpha')^2 \delta u^2}{9K_0' u_m^{1+\alpha'}}.$$

(15.37)

Note that the threshold speed goes down as the maximum speed increases. In principle this is something that could be checked against observations.

The pump acceleration equation is based upon the Parker equation, which is strictly valid only when $u >> \delta u$. We are thus explicitly assuming that the thermal distribution of the solar wind has speeds of at least δu and is available to flow through the core and into the tail. This is the case in any of the applications we have discussed. For example, for the solar wind at 1 AU, the thermal speed and δu are comparable, and for acceleration in the heliosheath, the PUI pressure is dominant.

15.5 Applications of the Pump Acceleration Mechanism

The pump acceleration mechanism described in Section 15.4 has been applied to explain several different observations. In Fisk and Gloeckler (2008), the mechanism has been applied to account for suprathermal tails observed in the quiet solar wind near 1 AU. In this case, the acceleration mechanism competes against adiabatic deceleration in the expanding solar wind. For reasonable choices of the governing parameters, the acceleration and deceleration balance and particles are not accelerated to energies above a few MeV/nucleon in the supersonic solar wind, consistent with ACE measurements at 1 AU and observations of Voyager in the heliosphere at ~40 AU. In Fisk and Gloeckler (2009), the mechanism has been applied to accelerate ACRs in the heliosheath. The pump mechanism is particularly appropriate for this application; it creates the ACRs by pumping energy out of the PUIs, which are the dominant energy source in the heliosheath. In Fisk and Gloeckler (2012b), the pump acceleration mechanism is applied to the acceleration of GCRs in the interstellar medium.

In this section, we cite two key applications of the pump acceleration mechanism: the acceleration of energetic particles at shocks and the acceleration of particles in the solar corona.

15.5.1 Acceleration of Energetic Particles at Shocks

The most common mechanism invoked for accelerating energetic particles at shocks in astrophysical plasmas is diffusive shock acceleration, in which particles gain energy by making repeated crossings of the shock and thus experience repeated compressions between the upstream and downstream flows. This mechanism was discussed in Fisk (1971) and then perfected in a series of papers in the late 1970s by Axford et al. (1977), Krymsky (1977), Bell (1978), and Blandford and Ostriker (1978) for application primarily to the acceleration of GCRs in supernovae. It is now widely applied to the acceleration of particles in the solar corona and in the solar wind.

In the case of the solar wind, however, there is relatively little observational evidence that diffusive shock acceleration, at least in its simplest form, is the dominant acceleration mechanism at shocks. In simple diffusive shock acceleration, the spectral index of the accelerated particles is related to the compression ratio of the shock. However, the correlation between the Mach number of the shock, and thus the compression ratio, and the spectral index of the accelerated particles is not clearly seen. For example, van Ness et al. (1984) found a correlation between the spectral index and the compression ratio for 75% of the shocks observed but only when fairly generous brackets were put upon the compression ratio. The study by Desai et al. (2004) showed little or no correlation for the spectral index of oxygen accelerated at shocks.

In Section 15.3, we found that the most prevalent acceleration mechanism at shocks observed by ACE at 1 AU in 2001 is the pump acceleration mechanism that yields the common spectral shape of a power law with a spectral index of -5 when expressed as a velocity distribution function. It is thus appropriate to develop a more detailed model for acceleration by the pump mechanism downstream from shocks.

15.5.1.1 Model for the Shock

The model for the shock that we will use is illustrated in Fig. 15.9. There is a shock front and an upstream and downstream region. There is a magnetic field that downstream makes an angle ψ relative to the shock normal. There is compressive turbulence downstream (random compressions and expansions of the plasma). The scale lengths of the compressions/expansions are λ normal to the magnetic field and ℓ along the field. We assume that $\ell \gg \lambda$. Particles are assumed to move freely along the magnetic field, i.e., the mean free path, L_m, parallel to the magnetic field is $> \ell$. The principal means of escape of particles from a compression region or flow into an expansion

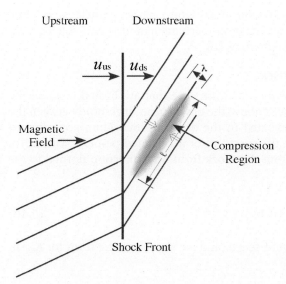

Upstream   Downstream

u_{us} u_{ds} λ

Magnetic
Field   Compression
Region

Shock Front

FIGURE 15.9
Schematic of the shock model used in Section 15.5.1. Source: Fisk
and Gloeckler (2012).

region is by following random walking field lines along which they move with parallel speed, $u/\sqrt{3}$, where we have taken the distribution of particles to be quasiisotropic. We assume that the mean square spread in the magnetic field due to the random walk is λ^2 in a distance ℓ, in which case the cross-field diffusion coefficient for escape is

$$K_{esc} = u\,\lambda^2 \big/ \left(\sqrt{3}\ell\right). \tag{15.38}$$

Large-scale spatial diffusion is also possible due to the mobility of particles along the field. The large-scale diffusion coefficient in the direction normal to the shock is then

$$K_{ls} = u\,L_m\cos^2\psi \big/ \sqrt{3}. \tag{15.39}$$

We assume that the particles are accelerated in the compressive turbulence downstream by the pump acceleration mechanism, which pumps particles out of a distribution, known as the core distribution, which lies between the thermal distribution of the solar wind and the suprathermal tail of accelerated particles, as explained in this chapter (see also Fisk and Gloeckler, 2012a, 2014).

Upstream of the shock is the solar wind thermal plasma, a core distribution, and a suprathermal tail, which we take to have the common spectral shape of -5. Crossing the shock, the thermal distribution and the core distribution are nonadiabatically heated, according to the Rankine-Hugoniot relationships. The upstream suprathermal tail experiences only an adiabatic compression in crossing the shock.

The nonadiabatic heating of the core distribution places particles above the threshold of the downstream acceleration mechanism and is available to form an enhanced downstream suprathermal tail.

15.5.1.2 Spectrum at the Shock

In our model, the particles in the upstream core that are heated crossing the shock and raised in particle speed above the downstream threshold between the core and the tail will be accelerated in the downstream region, according to Eq. (15.13). In a steady state, the spectrum of accelerated particles will evolve with distance r downstream from the shock front, in a direction normal to the shock front, as

$$u_{ds}\frac{\partial f}{\partial r} = \frac{1}{u^4}\frac{\partial}{\partial u}\left(\frac{\delta u^2}{9\lambda^2}\frac{\sqrt{3}\,\ell}{u}u\frac{\partial}{\partial u}\left(u^5 f\right)\right). \tag{15.40}$$

u_{ds} is the downstream solar wind speed, and we have used the form for K_{esc} in Eq. (15.40).

Eq. (15.40) has a ready solution:

$$f = f_{th}\left(\frac{u}{u_{th}}\right)^{-5}\cdot\exp\left[-\frac{9u\,u_{ds}}{\sqrt{3}\,\delta u^2}\cdot\frac{\lambda^2}{\ell\,r}\right], \tag{15.41}$$

where u_{th} is the threshold speed between the core and the tail. Note that as required, the spectrum has a spectral index of -5, with an exponential rollover that depends upon the random speed and cross-sectional dimension of the turbulence, the downstream solar wind speed, and the distance from the shock. The value of r is set so that at the shock $r = r_s$, the rollover in the spectrum approximates the rollover of the downstream spectrum of the upstream core that is heated and acquires speeds $u \geq u_{th}$ crossing the shock.

The solution in Eq. (15.41) results in spatial gradients in the rollover region; the particle speed at which the rollover occurs increases with r. In response to these spatial gradients, particles diffuse inward, back toward the shock front. We have a 1-D system; thus the distribution function at the shock front of particles diffusing inward from the appropriate rollover region is reduced by

$$f \propto \exp\left[-\frac{\sqrt{3}u_{ds}\,r}{u\,L_{mfp}\cos^2\psi}\right], \tag{15.42}$$

where we have assumed that the large-scale spatial diffusion coefficient, K_{ls}, is given in Eq. (15.39). It should be noted that for particle speeds at the rollover that are large compared to u_{th}, as is the case for particles diffusing back to the shock from downstream rollover locations, $r_s \ll r$ and has been ignored in Eq. (15.42). The spectrum at the shock front of particles diffusing inward from a rollover region at r is then

$$f = f_{th}\left(\frac{u}{u_{th}}\right)^{-5} \cdot \exp\left[-\frac{\sqrt{3}\,u_{ds}r}{u\,L_m\cos^2\psi} - \frac{9u\,u_{ds}}{\sqrt{3}\,\delta u^2}\cdot\frac{\lambda^2}{\ell\,r}\right].$$ (15.43)

We can find the location of the rollover region that contributes most to the spectrum at the shock front by finding the value of r, which for a given u yields the minimum value of the exponential and thus the largest value of f at the shock front. The result is

$$\frac{r}{u} = \frac{\sqrt{3}\,\lambda}{\delta u}\sqrt{\frac{L_m}{\ell}}\cos\psi.$$ (15.44)

Note that for simplicity we have ignored the velocity dependence of L_m. The spectrum at the shock front due to particles diffusing inward from rollover regions downstream is then

$$f = f_{th}\left(\frac{u}{u_{th}}\right)^{-5} \cdot \exp\left[-\frac{6u_{ds}}{\delta u\cos\psi}\cdot\frac{\lambda}{\sqrt{\ell\,L_m}}\right].$$ (15.45)

Note that the spectrum in Eq. (15.45) is u^{-5}; the reduction in f due to particles diffusing inward against the downstream flow is independent of particle speed. At the shock front, then, we expect at low particle speeds a u^{-5} spectrum with a rollover due to particles accelerated immediately downstream from the shock. There is then a second u^{-5} spectrum with a reduced value due to particles diffusing in from rollover locations further downstream. The reduction in Eq. (15.45), i.e., the difference between the two u^{-5} spectra, can be quite small. Recall that we assumed that $\lambda^2 \ll \ell\,L_m$.

An example of a low particle speed spectrum with a spectral index of -5 and a second reduced spectrum at higher particle speeds with the same spectral index of -5 is shown in Fig. 15.10 for a shock observed by the ACE/SWICS and the ACE/Ultra Low Energy Isotope Spectrometer in 2001.

The spectrum in Eq. (15.45) does not contain a rollover. This is due to the simplicity of our assumption in Eq. (15.43) that the rollover particle speed increases indefinitely with r. In fact it will not, for several reasons. We form the downstream tail from the energy in the core distribution that is heated crossing the shock. As this energy is pumped into the tail to increasingly higher particle speeds f_{th} will decrease with r. Thus the low particle speed portion of the downstream suprathermal tail peaks close to the shock and then decreases with distance behind the shock as the energy is pumped to higher energies. The higher particle speed portion will continue to rise as energy is pumped to these higher particle speeds, but then it will decrease in response to the reduction in the low particle speed portion of the spectrum. It is also possible that the acceleration will be limited when the particle gyroradius exceeds λ since now the dwell time of particles in a compression or expansion is greatly reduced.

We expect then a rollover in the second u^{-5} spectrum at the shock front, reflecting limits on the rollovers at larger values of r. The simplest way to

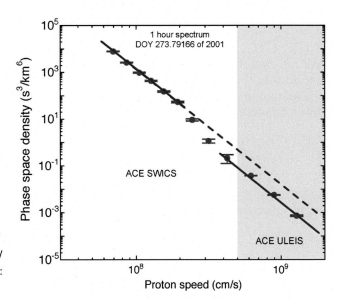

FIGURE 15.10
An example of two u^{-5} spectra observed by the ACE/
Solar Wind Ion Composition Spectrometer and the ACE/
Ultra Low Energy Isotope Spectrometer in 2001. Source:
Fisk and Gloeckler (2012).

determine the rollover spectrum at the shock is to determine the probability that
a particle accelerated at r reaches the shock front. We assume that the probability
depends upon r, with the particles accelerated at large r less likely to reach the
shock front. We follow the approach in Schwadron et al. (2010), assuming that
the probability is given by a Poisson distribution, in which the rate of a given
contribution to the spectrum at the shock front is proportional to $1/r$:

$$P(1/r) = \frac{\bar{r}^2}{r}\exp\left[-\frac{\bar{r}}{r}\right], \tag{15.46}$$

where $\bar{r}$ is the inverse of the average value of $1/r$. We then assume that the value of
f_{th} in the region where the acceleration is becoming limited deceases with r as
$r^{-\varsigma}$, as the energy is pumped to higher particle speeds. The resulting spectrum at
the shock front is then

$$f \propto u^{-5}\int_0^\infty \left(\frac{1}{r}\right)^{\varsigma+1}\exp\left[-\frac{\bar{r}}{r} - \frac{9u\,u_{ds}\lambda^2}{\sqrt{3}\,\delta u^2\ell\,r}\right]d\left(\frac{1}{r}\right)$$

$$= \Gamma(\varsigma+2)u^{-5}\left(\bar{r} + \frac{9\,u_{ds}\lambda^2}{\sqrt{3}\,\delta u^2\ell}u\right)^{-\varsigma-2}. \tag{15.47}$$

Note that for the low-speed particles that originate near the shock, for which
$u \ll \sqrt{3}\delta u^2\ell\,\bar{r}/(9\,u_{ds}\lambda^2)$, the spectrum at the shock is the u^{-5} spectrum found
above. However, for $u \gg \sqrt{3}\delta u^2\ell\,\bar{r}/(9\,u_{ds}\lambda^2)$, i.e., for particles that originate
from the region where there is a limit on the acceleration, the spectrum rolls over

into a power law with a spectral index of $u^{-5 \ -(\varsigma+2)}$. It is possible that the spectrum could be harder than u^{-5} if $\varsigma < -2$, but then there needs to be an additional limit since the total energy will diverge. Also, note that the spectral index of the power law rollover in Eq. (15.47) is independent of the shock compression ratio. It is thus not surprising that in observations no correlation between the spectral index and the compression ratio is found (e.g., Desai et al., 2004).

The technique of determining the spectrum outside of the immediate acceleration region in Eq. (15.47) can also be applied to regions further downstream. If the acceleration is confined to a region near the shock, $\bar{r}$ is small, and the particles escape by diffusion downstream, the rollover further downstream should also be a power law. This result appears to be consistent with observations by Voyager downstream from the termination shock (Decker et al., 2006; Gloeckler et al., 2008). The compressive turbulence appears to be strongest immediately downstream from the termination shock. Beyond this point, particles can escape by diffusion. As Voyager penetrated into the heliosheath, the spectra of particles accelerated downstream of the termination shock evolved to a u^{-5} spectrum (Voyager results are expressed as differential intensity, with the equivalent spectral index of -1.5) and a power law rollover.

In summary, we predict that the spectrum at low particle speeds at the shock will have a spectral index of -5. There can be a second, reduced spectrum with a spectral index of -5 at higher particle speeds due to particles diffusing inward from downstream. The second spectrum can be indistinguishable from the -5 spectrum at lower particle speeds, depending upon the parameters that control the propagation and acceleration. At particle speeds above the second -5 spectrum, the spectrum rolls over into a steeper power law. The three distinct spectra occur at the shock front. Downstream we expect a single -5 spectrum with an exponential rollover. Thus whether a power law or exponential rollover is observed depends upon the time period over which the data is averaged.

The spectra determined by the pump acceleration mechanism applied downstream from shocks provide excellent fits to the observations discussed in Section 15.3.

15.5.1.3 Role of Diffusive Shock Acceleration

We may ask the following: What is the role of standard diffusive shock acceleration versus acceleration by the pump acceleration mechanism? Particles are accelerated in standard diffusive shock acceleration by experiencing multiple times the strong compression that exists at the shock front. In that sense, the compression at the shock front is just one of many compressions; other compressions used by the pump acceleration mechanism occur throughout the downstream region. In fact, the compression at the shock front may not be particularly effective. The extent to which particles are accelerated depends upon the time spent in the compression. Shocks are at only one location, whereas the compressions downstream, albeit weaker than the compression at the shock, occur in a much larger volume.

Drury's (1983) equation, Eq. (15.2), provides the standard formula for the mean acceleration time from diffusive shock acceleration for the acceleration of particles with initial speed u_0 to speed u:

$$t = \frac{3}{u_{us} - u_{ds}} \cdot \int_{u_0}^{u} \left(\frac{K_{us}}{u_{us}} + \frac{K_{ds}}{u_{ds}} \right) \frac{du}{u}, \tag{15.48}$$

where the subscripts *us* and *ds* refer to upstream and downstream, respectively. The downstream flow speed is less than the upstream flow speed, and we take the downstream diffusion coefficient $K_{ds} = uL_m\cos^2 \psi/\sqrt{3}$ to be larger than the upstream diffusion coefficient, in which case we keep only the second term in the integrand in Eq. (15.48). We then differentiate both sides of Eq. (15.48) to find that the acceleration of particles at a shock is

$$\frac{du}{dt} \cong \frac{(u_{us} - u_{ds})u_{ds}}{\sqrt{3}L_m \cos^2 \psi}. \tag{15.49}$$

We found the equivalent first-order acceleration in the pump acceleration mechanism in Eq. (15.21) when expressed in terms of the diffusion coefficient:

$$\frac{du}{dt} = \frac{\delta u^2 \ell}{\sqrt{3} \, \lambda^2}. \tag{15.50}$$

Contrasting the acceleration in the pump mechanism in Eq. (15.49) with the acceleration in diffusive shock acceleration in Eq. (15.48), we see that

$$\frac{du/dt|_{pump}}{du/dt|_{dif \cdot shock}} = \frac{\delta u^2}{(u_{us} - u_{ds})u_{ds}} \cdot \frac{\ell \, L_m \cos^2 \psi}{\lambda^2} \gg 1. \tag{15.51}$$

The jump in the flow speed across the shock and the downstream flow speed is undoubtedly larger than δu^2. However, in our model we have assumed that $\ell \gg \lambda$ and $L_m > \ell$, with the result that the length scales are very much in favor of the pump mechanism and yield a larger acceleration rate for the pump mechanism.

Particles propagate with ease along the magnetic field in our model, with a resulting large spatial diffusion coefficient. As a result, particles do not spend much time at the shock, nor do they receive much net acceleration. The pump mechanism, in contrast, occurs throughout the downstream region. Particles are always experiencing downstream compressions, and thus for this choice of the propagation parameters the pump mechanism is the dominant acceleration mechanism at the shocks.

There could be circumstances where the length scales favor diffusive shock acceleration over the pump mechanism. In general, for this to be the case, a small mean free path is required. Even in the case where $\ell \sim \lambda$, Eq. (15.51) can hold for most shock angles, provided that the mean free path is large compared to ℓ. As noted in Section 15.3, there are very few shocks where diffusive shock acceleration can account for the observations.

15.5.2 Acceleration in the Solar Corona

There are many interesting applications of the pump acceleration mechanism in the solar corona: impulsive SEP events; large SEP events, associated with coronal mass ejections; and the creation of seed particles in the solar corona. The pump acceleration mechanism is applicable to all of these. As an illustration of these applications, we consider here the compositional variations that occur in impulsive SEP events. To begin, we first need to define the likely properties of the diffusion coefficient that controls the escape across the magnetic field from compression regions, in coronal conditions, where the magnetic field is strong and thus gyroradii are small and other spatial scales are also small.

15.5.2.1 Choice of the Diffusion Coefficient

For the pump acceleration mechanism to be successful in producing a suprathermal tail, there must be numerous compressions and expansions. We assume that the magnetic field in the solar corona consists of intertwined and independent flux tubes of length dimension λ_c, essentially "fibrils." The motions of the fibrils are due to magnetic forces and thus should be compressed and expanded at about the Alfvén speed. If there is an overall volumetric compression on a scale that is large compared to λ_c, with a compression speed at the outer edge of the compression of about the Alfvén speed, then the speed inside the compression must be proportional to the distance from the center of the compression or small compared to the Alfvén speed. If the individual fibrils also have motions at speeds of about the Alfvén speed, they must be both positive and negative, i.e., compressions and expansions, so that the average motion within the compression is less than the Alfvén speed. These are the individual compressions and expansions that we will use to generate the suprathermal tails in the solar corona.

The most likely way for particles to escape from a fibril is by particle drift. The drift velocity is

$$\overrightarrow{u}_D = \frac{1}{3} u\, r_g \frac{\overrightarrow{\nabla} \times \delta \overrightarrow{B}}{B_0}. \tag{15.52}$$

The gyroradius is r_g, B_0 is the magnitude of the mean magnetic field, and $\delta \overrightarrow{B}$ is the turbulent magnetic field responsible for the curls that yield the particle drifts.

The drift speeds are not fast for low-speed particles. Moreover, the curls in the magnetic field can be expected to be coherent only for about time $\tau_c = \lambda_c/\delta u$. Thus for escape times that are long compared with the compression time, particles can execute a coherent drift only for a distance $u_D \tau_c$. The diffusion coefficient that describes escape by drifts is then

$$K = \frac{u^2 r_g^2}{9 B_0^2} \left\langle (\vec{\nabla} \times \delta \vec{B})^2 \right\rangle \frac{\lambda_c}{\delta u},$$ (15.53)

where the angular brackets denote a spatial average.

To determine the spatial average in Eq. (15.53), we need to specify the turbulence in which the particles are drifting. This is the turbulence within a fibril for which there are no measurements. We use the form from Fisk (1976) in which the turbulent power spectrum is

$$\left\langle \delta B^2 \right\rangle = C \eta^2 B_0^2 \lambda_c^2 \cdot \frac{\lambda_c^2 k_\perp^2}{\left(1 + \lambda_c^2 k_\perp^2\right)^{\frac{1}{2}\gamma + 2}},$$ (15.54)

where C is a dimensionless normalization constant with a value $\sim 8/3$, and $\eta^2 \equiv \delta B^2 / B_0^2$. We have ignored the dependence on $k_\perp$ under the assumption that the turbulence is elongated along the fibril. This form of the power spectrum yields a power spectrum at large values of wavenumber k that deceases as γ, i.e., for Kolomogorev turbulence $\gamma = 5/3$.

Thus noting that the Fourier transform of $\vec{\nabla} \times \delta \vec{B} = \vec{k} \times \delta \vec{B}$, we find that

$$\left\langle (\vec{\nabla} \times \delta \vec{B})^2 \right\rangle = C \eta^2 B_0^2 \lambda_c^2 \int_0^{1/r_g} \frac{\lambda_c^2 k_\perp^5}{\left(1 + \lambda_c^2 k_\perp^2\right)^{\frac{1}{2}\gamma + 2}} dk_\perp.$$ (15.55)

For particles in the core, $r_g \ll \lambda_c$ and thus for turbulent spectra with $\gamma < 2$, Eq. (15.55) becomes

$$\left\langle (\vec{\nabla} \times \delta \vec{B})^2 \right\rangle \approx \frac{C \eta^2 B_0^2}{\lambda_c^2} \left(\frac{\lambda_c}{r_g} \right)^{2 - \gamma},$$ (15.56)

and the diffusion coefficient in Eq. (15.53) becomes

$$K = \frac{u^2}{9} \frac{\lambda_c}{\delta u} \left(\frac{r_g}{\lambda_c} \right)^{\gamma},$$ (15.57)

where we have taken $C \eta^2 \approx 1$.

15.5.2.2 Impulsive Solar Energetic Particle Events

One of the interesting features of impulsive solar flare events is their compositional variation. The events are usually He^3-rich. They are also, according to Mason et al. (2004), events that exhibit enhancements in proportion to $(A/Z)^{3.26}$, where A is the atomic number, and Z is the charge. We consider in this section these latter compositional enhancements.

The model we propose is a coronal loop that is disrupted by reconnecting with open field lines. The loop is made up of fibrils of individual field lines, which, following the reconnection, are highly disturbed and contain ample compressions and expansions to accelerate particles out of the thermal plasma. The accelerated particles can escape onto the open field lines.

We found in Section 15.4.9 that there are two solutions to the pump acceleration mechanism: a solution for the source particle distribution and for the tail, and these should map continuously from the thermal distribution to the source particle distribution to the tail.

The source particle solution, from Eq. (15.30), is

$$f = f_0 \left(\frac{u_0}{u}\right)^{6+\alpha}, \tag{15.58}$$

where u_0 is the speed where the thermal distribution merges into the core, the diffusion coefficient is taken to be of the form $K = K_0 u^{1+\alpha}$, and the A/Z dependence is contained in K_0.

The tail solution, from Eq. (15.31), is

$$f_t = f_0 \left(\frac{u_0}{u_1}\right)^{6+\alpha} \left(\frac{u_1}{u}\right)^5, \tag{15.59}$$

where u_1 is the speed where the source particle distribution merges into the tail. We may assume that we have a Maxwell distribution:

$$f_{th} = f_{th,0} \exp\left(-\frac{u^2}{u_{th}^2}\right), \tag{15.60}$$

where u_{th} is the thermal speed of the plasma, which we take to be independent of the particle mass or mass-to-charge ratio. We assume that f_{th} merges smoothly into f_c, i.e., the slopes of the two distributions are the same:

$$2\frac{u_0^2}{u_{th}^2} = 6 + \alpha. \tag{15.61}$$

Thus combining Eqs. (15.58)–(15.61), we find that the tail distribution is

$$f_t = f_{th,0} \exp\left(-\frac{6+\alpha}{2}\right)\left(\frac{6+\alpha}{2}\right)^{\frac{1}{2}(6+\alpha)} \left(\frac{u_{th}}{u_1}\right)^{6+\alpha} \left(\frac{u_1}{u}\right)^5. \tag{15.62}$$

For impulsive SEP events we use the diffusion coefficient given in Eq. (15.57), where $\gamma = \alpha - 1$, corresponding to the spectral index of the turbulence, i.e., $\gamma = 5/3$ for Kolmogorov turbulence.

We saw in Eq. (15.37) that u_1 occurs when

$$\frac{(1+\alpha)^2}{9\lambda_c^2} \frac{u_1^2}{9} \frac{\lambda_c}{\delta u} \left(\frac{r_{g1}}{\lambda_c}\right)^\gamma = \frac{2}{9}(1+\alpha')^2 \delta u^2 \cdot \frac{9\delta u}{u_{max}^2 \lambda_c} \left(\frac{\lambda_c}{r_{g,max}}\right)^\gamma, \tag{15.63}$$

where u_{max} is the maximum speed a particle can obtain. Since particles escape along open field lines, we assume that u_{max} is the same for all particles, in which case

$$u_1 \propto (Z/A)^{2\gamma/(\gamma+2)}. \tag{15.64}$$

Thus the mass-to-charge dependence of the higher speed particles on the flare loop flare should be

$$f_t \propto \left(\frac{A}{Z}\right)^{(2\gamma/(\gamma+2))(\gamma+2)} = \left(\frac{A}{Z}\right)^{2\gamma}. \tag{15.65}$$

Note that if $\gamma = 5/3$, as in Kolmogorov turbulence, then the A/Z exponent becomes 10/3, which is close to the observed value of ~ 3.26.

15.6 Concluding Remarks

In this chapter, we have presented relevant observations of and a theoretical explanation for the common spectrum: a distribution function that is a power law with a spectral index of -5, with an exponential rollover at higher particle speeds or, equivalently, a differential intensity spectrum that is a power law with a spectral index of 1.5 and an exponential rollover at higher energies. We have cited those observations that are most revealing of the conditions in which the common spectrum occurs, and we have presented a detailed and thorough description of how a pump acceleration process operates and yields the time evolution of the common spectrum.

It is important to emphasize that the common spectrum is a real, observed phenomenon in the heliosphere, observed at shocks in the inner heliosphere and in the ACRs throughout the heliosheath. We encourage every theorist and observer who is currently studying particle acceleration in the heliosphere or in other astrophysical settings, by using diffusive shock acceleration or some form of traditional stochastic acceleration, to consider whether the pump acceleration mechanism that yields the common spectrum might provide a better description of the acceleration that he or she is studying. We should all recognize that there are new phenomena in heliospheric physics, which when understood will prove the relevance of our discipline to astrophysics.

15.7 Science Questions for Future Research

Future analyses and observations need to address the following questions:

1. What are the plasma/environment limitations of the pump mechanism?
2. Is this acceleration mechanism applied to other plasma processes?
3. Can a modified pump mechanism lead to kappa distributions with $\kappa > 1.5$?

Acknowledgments

This work was supported in part by NASA Grants NNX10AF23G, NNX13AE056, and NNX11AP01G and by NSF Grant AGS-1043012.

G.P. Zank
University of Alabama in Huntsville, Huntsville, AL, United States

Chapter Outline

16.1 Summary   609
16.2 Introduction   610
16.3 Upstream Distributions and Their
　　　Transmission Through
　　　Quasiperpendicular
　　　Shocks   614
16.4 Velocity Distribution Function
　　　Downstream of a
　　　Quasiperpendicular Shock   619
16.5 Simulations   623

16.6 Observational Tests   625
16.7 Dissipation and Particle
　　　Acceleration at
　　　Quasiperpendicular
　　　Shocks   629
16.8 Concluding Remarks   631
16.9 Science Questions for Future
　　　Research   632
Acknowledgments   632

16.1 Summary

Although wave—particle interactions, once formed, may maintain a preexisting kappa distribution throughout the solar wind, an important question is to identify the origin of the kappa distributions. Throughout much of the heliosphere beyond some 2 to 3 AU, PickUp Ions (PUIs) form a suprathermal proton component in the solar wind. The PUI distribution does not equilibrate with the thermal solar wind protons and forms a filled-shell distribution about the colder solar wind protons' Maxwellian core. In this chapter, we show that the suprathermal component is reflected preferentially by the cross-shock electrostatic potential at perpendicular shocks, unlike the colder solar wind proton distribution, which is transmitted through the shock. We show how the velocity distribution function downstream of quasiperpendicular shocks is constructed,

Kappa Distributions. http://dx.doi.org/10.1016/B978-0-12-804638-8.00016-4

and we find that the reflected suprathermal ions form an extended tail to the velocity distribution function. The total proton distribution function closely resembles a kappa distribution. Simulation results are presented that support the general theory for the formation of downstream kappa distributions at quasiperpendicular shocks in the presence of PUIs. Various observational tests of the theory for the formation of kappa distributions by quasiperpendicular shocks are presented. Finally, we discuss the close relationship between the dissipation process that forms kappa distributions at quasiperpendicular shocks and the diffusive shock acceleration (DSA) of protons at quasiperpendicular shocks, even at 1 AU.

Science Question: How does the plasma shock interaction lead to a kappa distribution?

Keywords: Cross-shock potential; PickUp Ions; Shock waves; Solar wind; Suprathermal particles.

16.2 Introduction

A fairly well-developed theory exists that describes the basic structure and dissipation mechanism at quasiperpendicular shocks in a plasma beta $\beta_p \sim O(1)$ collisionless plasma. Illustrated in Fig. 16.1 is an example of a classically structured quasiperpendicular shock. This is the so-called TS-3 shock observed by Voyager 2 during the crossing of the heliospheric termination shock (HTS). The structure of TS-3 (Fig. 16.1A) is remarkably consistent with the "classical" structure exhibited by quasiperpendicular shocks. An extended "foot" exists ahead of a much steeper "ramp." The ramp is followed by "overshoots" in the magnetic field strength B, after which the downstream state settles into its Rankine-Hugoniot downstream asymptotic state. The structure of TS-3 is consistent with our current understanding of quasiperpendicular shock structure (e.g., Goodrich, 1985; Lembège et al., 2009) where dissipation is provided by ion reflection at the cross-shock electrostatic potential. Recall that the cross-shock electrostatic potential forms as a result of charge separation induced by the overshoot at the shock of the more massive ions. Such an overshoot effect is observed to be present at both quasiperpendicular and quasiparallel shocks. The ions or protons reflected by the cross-shock electrostatic potential gain energy, being accelerated in the upstream motional electric field until they gain sufficient energy to overcome the cross-shock electrostatic potential. The reflected particles are therefore effectively heated, and the incident flow energy is dissipated in the form of heat. Although of interest, the exact profile of the entire cross-shock electric field is unimportant to our purposes here, since the electrostatic cross-shock potential is only needed to act as a barrier for incident low-velocity PUIs or other energetic particles. The nature of the proton reflection process is also the subject of some debate, although specular ion reflection is generally assumed. Incident ions can also be reflected in the foot preceding the electrostatic cross-shock potential, but the net effect is roughly consistent with specular reflection. Nonetheless, as

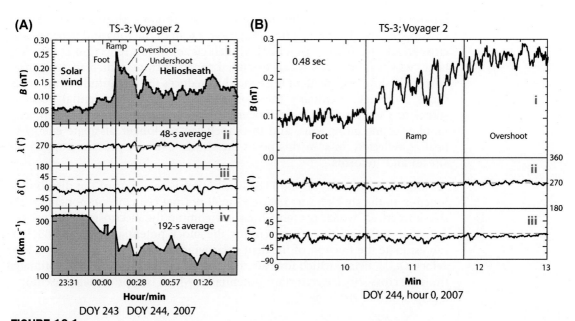

FIGURE 16.1

(A) (i) Magnetic field strength *B* measured using 48-s averages, (ii) its direction λ, (iii) elevation angle δ, and (iv) 192-s averages of the solar wind speed across TS-3 (noted with *V*). Clearly visible are an extended foot, a ramp, the magnetic field overshoot, and trailing oscillations. (B) (i) The internal structure of the ramp of TS-3, based on observations of the magnetic field strength *B*; (ii) azimuthal angle λ; and (iii) elevation angle δ at 0.48-s intervals. *DOY*, Day of the year. Source: Burlaga et al. (2008).

we discuss in the following sections, the steepest gradient in the electrostatic shock potential profile is important in determining an upper limit to the reflected ion energy gain.

The TS-3 shock compression ratio, as measured by the ratio of the upstream to downstream *B*, is 1.7 ± 0.1, which is consistent with the corresponding density ratio of 1.4 ± 0.2 (Burlaga et al., 2008; Richardson et al., 2008). The HTS is therefore comparatively weak. Fig. 16.1B shows that the HTS ramp possesses a considerably fine structure, exhibiting quasiperiodic oscillations. These have been observed before in other quasiperpendicular shocks. Burlaga et al. (2008) attempted to estimate the scales of TS-3, including the fine structure. However, the TS-3 structure observed by Voyager 2 is a single spacecraft measurement that does not include the PUI component because Voyager 2 has no instrument for this energy range. As a result, determining the shock speed and shock normal is difficult. Besides TS-3, Voyager 2 observed three separate crossings of the HTS (from an inferred five total crossings of the HTS). The different structures exhibited by the three observed crossings may possibly be attributed to these being different periods of shock reformation (Burlaga et al., 2008; Lembège et al., 2009). Shock reformation of quasiperpendicular shocks is a consequence of the inherent unsteady process of ion reflection by the cross-shock electrostatic potential. If the foot ahead of the ramp and cross-shock potential is small, then the ions are not appreciably slowed down before crossing the shock. This implies a large overshoot of the ions and hence a large cross-shock electrostatic potential. The efficiency of ion reflection is therefore large, which then leads to an extended foot structure, which in turn more effectively slows the incident ions and thus reduces the magnitude of the cross-shock potential. Eventually the cross-shock electrostatic potential becomes sufficiently weak so that it no longer provides a barrier to a sufficiently large number of incoming ions, and the shock can no longer dissipate energy by heating reflected ions. At this point the shock loses the protons trapped upstream and "reforms."

An important property of the PUI distribution is that part of the distribution function in the shock frame has very small normal velocity components at the shock interface, which prevents their overcoming the electrostatic cross-shock potential. By estimating or prescribing the shock potential (Leroy, 1983) one can estimate the fraction of incident solar wind or PUIs that are reflected. The velocity u_{spec}, below which ions are reflected at a given shock, can be estimated from the cross-shock electrostatic potential Φ, which is approximated usefully at a perpendicular shock as

$$e\Phi \cdot \eta \frac{1}{M_{A1}^2} \frac{\Delta B}{B_1} m_i u_{b1}^2, \tag{16.1}$$

where $\Delta B \equiv B_2 - B_1$ is the difference between the downstream B_2 and upstream B_1 magnetic fields, u_{b1} is the upstream flow speed normal to the shock, m_i is the proton mass, and M_{A1} is the upstream Alfvénic Mach number. Eq. (16.1) is derived from the momentum equation for electrons (which are assumed massless), and the parameter η has been introduced to approximate the contribution to Φ of the deflected bulk velocity in the y-direction (which arises through ion reflection at the shock ramp) and the jump in the electron pressure (Leroy, 1983). Here, we assume a coordinate system in which x is normal to the shock, z is parallel to $\vec{B}$ and orthogonal to x, and y completes the coordinate system. Empirically, η is found to be about 2 or greater. For a particle (of mass m and charge Z) to be reflected specularly at the shock, its velocity component in the shock normal direction u_x must satisfy $Ze\Phi_{sh} \geq \frac{1}{2} m u_x^2$. This of course neglects the particles' Lorentz force, but for the low particle speeds under consideration and the assumed structure of the shock potential, this is a reasonable simplification for which some level of analytic description can be retained. It then follows that those particles satisfying $u_x \leq u_{\text{spec}}$ are reflected at the shock, where

$$u_{\text{spec}}^2 = 2\eta \frac{Zm_i}{m} \frac{u_{b1}^2}{M_{A1}^2} \frac{\Delta B}{B_1}. \tag{16.2}$$

Here, m_i refers to the proton mass, and m and Z refer to the mass and charge of the particle of interest (pickup H^+, He^+, and so on). If we assume that the PUI distribution ahead of the shock is a simple shell, then the fraction of the distribution R_{ref} that is incapable of surmounting the cross-shock potential barrier is found to be

$$R_{\text{ref}} = \left[\frac{Zm_i}{m}\frac{\eta}{2M_{A1}^2}\frac{\Delta B}{B_1}\right]^{\frac{1}{2}} = \left[\frac{Zm_i}{m}\frac{\eta}{2M_{A1}^2}(r-1)\right]^{\frac{1}{2}}, \tag{16.3}$$

where, for a perpendicular shock, $B_2/B_1 = r$ is the shock compression ratio. As discussed by Zank et al. (1996a) and Burrows et al. (2010), these reflected ions are capable of being accelerated to large energies. If we accept this for the present, a couple of points are immediately apparent if we interpret the reflection efficiency as the injection efficiency for ion energization at shock waves: (1) heavier PUI species ($m>m_i$) are less efficiently injected, and (2) injection efficiency increases with an increasing particle charge state. Thus pickup H^+ should be twice as efficiently injected as pickup He^+, and if we assume that the alpha particle distribution is shell-like, then the injection efficiency of He^{++} should be intermediate to the aforementioned pickup species. This appears to be precisely what is observed by Ulysses at a forward shock (Gloeckler et al., 1994). Of course, the same reasoning applies to the solar wind ions if, instead, we assume the 1-D Maxwellian distribution:

$$f(\vec{r},u,t) = n\,\pi^{-\frac{1}{2}}\theta^{-1}\exp\left[-\frac{(u-u_b)^2}{\theta^2}\right] \tag{16.4}$$

where $\theta \equiv \sqrt{2k_BT/m}$ is the characteristic thermal speed of the particle with mass m and temperature T. The fraction of incident solar wind ions reflected by the electrostatic cross-shock potential, i.e., the injection efficiency, corresponds to

$$R_{\text{ref}}^{\text{SW}} = \frac{1}{n}\int_{u_{\text{cut}}}^{\infty} f(\vec{r},u,t)du = \frac{1}{2}\text{erfc}\left(\frac{u_{\text{cut}}-u_b}{\theta}\right) \tag{16.5}$$

where $\text{erfc}(x)$ is the complementary error function, and u_{cut} denotes the reflected solar wind ion cutoff velocity, approximated by u_{spec}. Under typical solar wind conditions in the outer heliosphere, $R_{\text{ref}}^{\text{SW}}$ is exceedingly small and many times smaller than that corresponding to a shell distribution such as PUIs. This is consistent with the observations made at Ulysses (Gloeckler et al., 1994).

The role of shock strength in determining the injection efficiency is difficult to precisely assess. Naively, since $R_{\text{ref}}^2 \propto (r-1)$, one might expect the injection efficiency to increase with the shock strength, but this increase might well be offset by a corresponding increase in M_{A1}^2. Furthermore, the presence of hot reflected PUIs may reduce the effective magnetosonic Mach number of the shock, thus leading to some form of self-regulation of the shock strength. To properly address these questions requires the development of a fully nonlinear theory.

In this chapter, we consider the transmission of thermal solar wind protons and suprathermal particles through quasiperpendicular collisionless shock waves throughout the solar wind (Zank et al., 1996a; Lee et al., 1996; Zilbersher and Gedalin, 1997; Lipatov and Zank, 1999; Le Roux et al., 2000; Burrows et al., 2010;

Zank et al., 2010, 2014). We show that the presence of a suprathermal charged particle distribution upstream of a quasiperpendicular shock leads to a distribution that closely resembles a kappa distribution function downstream of the shock. In the context of outer heliospheric physics, this work is extensively reviewed in Zank (1999, 2015). In Section 16.3, we address the transmission of upstream distributions at quasi = perpendicular shocks, and in Section 16.4 we address the transmitted distribution Section 16.5 presents an overview of supporting kinetic simulations. In Section 16.6, we discuss observational tests of the theory, while in Section 16.7, we discuss the dissipation and particle acceleration at quasi-perpendicular shocks. Finally, the concluding remarks are given in Section 16.8, while three general science questions for future analyses are posed in Section 16.9.

16.3 Upstream Distributions and Their Transmission Through Quasiperpendicular Shocks

In the supersonic solar wind, interstellar neutral H (and other species) is ionized by charge exchange, creating a beam of hot (~ 1 keV) ions. The beam is unstable to the generation of Alfvén waves, on which the PUIs subsequently scatter. In this way, a bispherical distribution (well approximated by a shell distribution) for PUIs results (Lee and Ip, 1987; Williams and Zank, 1994). The PUI shell convects at the background solar wind speed, experiencing adiabatic cooling with an increasing heliocentric distance. Since shell distributions are created at all radial distances beyond the ionization cavity, the complete PUI distribution in the supersonic solar wind is a filled-shell, expressed in the solar wind flow frame as (Vasyliunas and Siscoe, 1976)

$$f_p(c) = \frac{3n_p}{8\pi\, u_b^{\frac{3}{2}}} c^{-\frac{3}{2}} \qquad (16.6)$$

where c is the particle velocity in the solar wind frame, $c = |\vec{c}\,|$, n_p is the PUI number density, and $u_b = |\vec{u}_b|$ is the solar wind bulk speed. The pressure contributed by PUIs is therefore

$$P_p = \frac{1}{7} m_i n_p u_b^2. \qquad (16.7)$$

Because the energy gained in the pickup process is proportional to u^2, the PUI pressure dominates in the outer heliosphere, and this is reflected in the pressure/temperature distributions in the global heliospheric models (Pogorelov et al., 2008; Zank et al., 2013; Heerikhuisen et al., 2014).

The observed proton distribution in the solar wind is illustrated in Fig. 16.2. The total distribution is a superposition of the background or bulk solar wind thermal proton distribution, well fitted by a Maxwell—Boltzmann velocity distribution, and a filled-shell PUI distribution function (Vasyliunas and Siscoe, 1976). The shell distribution, when plotted in the spacecraft or stationary frame, exhibits a flat-topped form with a cutoff at roughly twice the solar wind speed;

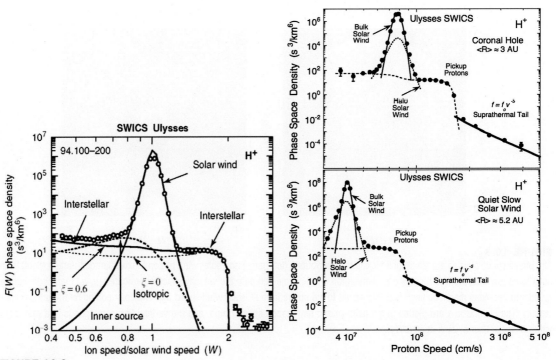

FIGURE 16.2

The observed phase space density as a function of proton speed. The left figure shows the superposition of the background thermal solar wind proton distribution (essentially a Maxwell—Boltzmann distribution) and the filled-shell pickup proton distribution, plotted in the spacecraft frame. The flat-top PickUp Ion (PUI) distribution exhibits a cutoff at twice the solar wind speed, consistent with a shell distribution. A small PUI contribution from inner source neutral atoms is also present. The right panel (*top*) shows the velocity distribution in the fast, high-latitude solar wind during solar minimum, and the bottom panel shows the proton velocity distribution observed in the in-ecliptic slow solar wind at ~5 AU. All observations were made by the Solar Wind Ion Composition (SWICS) instrument on Ulysses. The *solid, dotted, dashed,* and *solid bold* curves are model fits to the bulk solar wind, pickup protons, and u^{-5} power law tails, respectively. Source: Gloeckler et al. (1994, 2008).

see, for example, Zank (1999) for an extended discussion about the formation of the PUI distribution function. The PUIs form a dominant suprathermal halo population of PUIs in the solar wind that dominates the tails, both at low and high energies, of the thermal Maxwellian proton velocity distribution function. The right panels illustrate this clearly, where observations were made at ~3 AU in the high-latitude fast solar minimum wind and at ~5 AU in the slow ecliptic solar wind.

As discussed by Zank et al. (1996a) and Lee et al. (1996), protons with velocity u_x in the shock frame such that $\frac{1}{2}m_i u_x^2 \leq e\Phi$ are reflected by the cross-shock potential Φ (where $\vec{E} = -\vec{\nabla}\Phi$; $\hat{x}$ is the direction normal to the perpendicular shock front, and $\vec{B} = B\hat{z}$ is in the plane of the shock front; note that hereafter we consider $Z = 1$). As discussed in Section 16.2, we can define a threshold

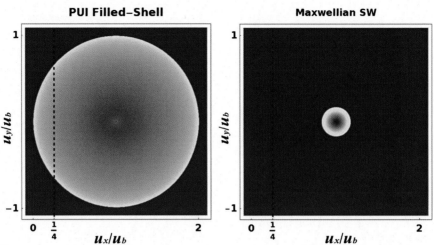

FIGURE 16.3

The phase space plots of the filled-shell PickUp Ion distribution (left panel) and solar wind proton Maxwellian distribution (right) in the solar wind just ahead of the heliospheric termination shock, projected onto the velocity space coordinates (u_x, u_z) (perpendicular and parallel to the mean magnetic field) and normalized to the upstream bulk solar wind speed u_{b1}. The color coding refers to the number density. The distributions are plotted in the heliospheric termination shock frame, and the vertical line corresponds to a reasonable value of u_{spec}. Source: Zank et al. (2010).

velocity $u_{spec}\widehat{x}$ such that protons with $u_x < u_{spec}$ in the shock frame will experience specular reflection at the cross-shock potential. PUIs, by virtue of their filled-shell distribution with a radius in the phase space of u (the solar wind speed), occupy a large part of the phase space that satisfies this condition compared to the cold solar wind proton Maxwellian distribution. This is illustrated in Fig. 16.3 where a projection of the phase space distribution is plotted in the 2-D (u_x, u_z) plane, normalized to the upstream flow speed u_{b1}. Here u_x and u_z are particle velocities perpendicular and parallel to the upstream mean magnetic field in the shock frame. The vertical line identifies a reasonable value of $(u_{spec}/u_{b1})\widehat{x}$, showing that, unlike the PUI distribution, virtually no solar wind protons can be reflected.

It was widely expected that the solar wind would undergo a transition from supersonic to subsonic flow at the HTS, with an accompanying heating of the solar wind plasma. However, the Voyager 2 plasma instrument during the 2007/242-244 TS crossing(s) measured both less solar wind heating than expected and a downstream flow speed that remained supersonic with respect to the heated solar wind sound speed. The observed temperature of less than ~200,000 K indicated that ~80% of the incident solar wind ram energy had to be heating some particle population other than that of the solar wind, possibly PUIs (Fig. 16.4) (Richardson, 2008; Richardson et al., 2008). In their investigation of the interaction of PUIs and solar wind ions with the HTS, Zank et al. (1996a) predicted that PUIs provide the HTS dissipation and heated

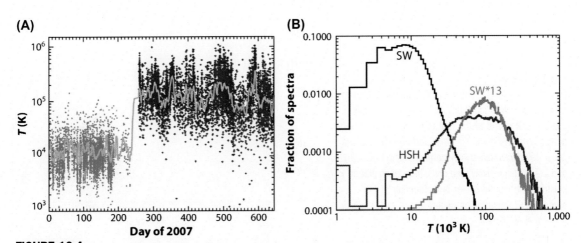

FIGURE 16.4

(A) Solar wind thermal proton temperature observed by Voyager 2 across the heliospheric termination shock: (*points*) high-time resolution data, (*lines*) daily averages. (B) Histograms of the temperature distributions in the solar wind and inner heliosheath: (*black*) solar wind temperature distribution (SW), (*red*) inner heliosheath temperature distribution (HSH), and (*blue*) the solar wind temperature distribution (SW*13) multiplied by 13, that is, the ratio between the upstream solar wind and downstream inner heliosheath temperatures. No reflected solar wind protons can be identified from the distribution function. Source: Richardson et al. (2008).

downstream plasma. They concluded that "P[U]Is may therefore provide the primary dissipation mechanism for a perpendicular [H]TS with solar wind ions playing very much a secondary role" (Zank et al., 1996a, p. 472). Thus solar wind ions were expected to be heated modestly.

Let us consider the transmission of an upstream filled-shell distribution through a quasiperpendicular shock. Burrows et al. (2010) extended the model of Zank et al. (1996a) by using test particle simulations of the transmission of a filled-shell PUI distribution through the TS-3 quasiperpendicular HTS structure observed by Voyager 2 (Burlaga et al., 2008). For a PUI distribution injected just ahead of the foot of a TS-3 model, the downstream (just behind the ramp) energy spectrum dN/N is plotted in Fig. 16.5, and the corresponding velocity phase space projections are plotted in Fig. 16.6. Several features of Fig. 16.5 are noteworthy. At low energies of $u^2/u_b^2 < 0.5$, the spectrum is very much like the initial filled-shell distribution but is less in intensity by about a factor of ~ 2. The energized spectrum merges smoothly into a power law distribution for $0.5 \leq u^2/u_b^2 \leq \sim 4$, with a slope of -2.25. (We use the notation of the normalized energy $\varepsilon \equiv u^2/u_b^2$.) An $\varepsilon^{-2.25}$ power law corresponds to $f \propto u^{-\alpha}$, $\alpha = 5.5$. The PUIs in this interval gained energy as a result of shock drift acceleration. At $u^2/u_b^2 \sim 4$, there is a clear break in the spectrum with a drop in dN/N of about 1.8×10^{-4}, and a harder power law emerges over the energy range $4 \leq u^2/u_b^2 \leq 65$ with a slope of ~ -1.5, corresponding to $f(u) \propto u^{-\alpha}$, $\alpha = 4$. The spectrum dN/N is already integrated over the energy, thus at high energies has the asymptotic behavior of

FIGURE 16.5
Downstream (just behind the ramp) energy spectrum dN/N for a PickUp Ion (PUI) filled-shell distribution injected just ahead of the foot of a TS-3 model. Two power laws, $dN/N \propto \varepsilon^{-2.25}$ (*thin dashed line*) and $dN/N \propto \varepsilon^{-1.5}$ (*thick dashed line*) (corresponding to $\kappa \sim 2.75$ and $\kappa \sim 2$, respectively, *see text*), have been plotted to illustrate the double power law character of the energized PUI distribution (where $\varepsilon \equiv u^2/u_b^2$). Source: Burrows et al. (2010).

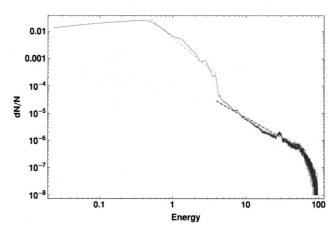

FIGURE 16.6
Velocity phase space portraits for the MRI-accelerated filled-shell distribution of Fig. 16.5. This is the distribution just behind the ramp of the TS-3 model. Source: Burrows et al. (2010).

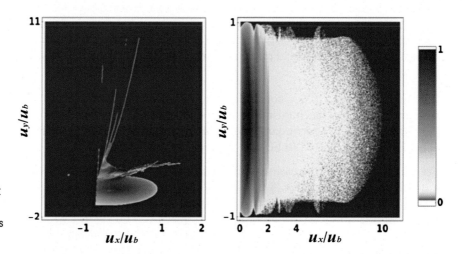

$\sim \varepsilon^{-\zeta}$, with $\zeta \equiv \kappa - \frac{1}{2}$. On the other hand, the distribution function of speeds, $f(u)$, has the asymptotic behavior of $\sim u^{-\alpha}$, with $\alpha \equiv 2\kappa$ (Livadiotis and McComas, 2009). In our example, we find $\zeta \sim 1.5$ and $\alpha \sim 4$, both values corresponding to $\kappa = 2$. In similar, the slope $\zeta \sim -2.25$ may correspond to another kappa $\kappa \sim 2.75$. A superposition, therefore, of at least two kappa distributions, $\kappa = 2$ and $\kappa \sim 2.75$, is possible. PUI reflection, often multiple reflection (Zank et al., 1996a), is responsible for this section of the energetic PUI spectrum. Finally, the rollover in the PUI spectrum near $u^2/u_b^2 \sim 100$ (i.e., about 50 keV, where $u_b = 300$ km s^{-1}) is a consequence of the particle's Lorentz force overcoming the gradient of the cross-shock potential and no longer being trapped ahead of the quasiperpendicular shock (Zank et al., 1996a). The precise value of this rollover is determined by the ramp thickness.

Fig. 16.6 shows the phase space plots of the downstream PUI distribution projected onto the (u_x, u_y) plane and represented in the $\left(u_\perp, u_\parallel\right)$ space in the right-hand panel. Here $u_\perp = u \sin\vartheta$ and $u_\parallel = u \cos\vartheta$ are the velocity components perpendicular and parallel to the magnetic field $\overrightarrow{B}$ (where ϑ is the usual polar angle for spherical coordinates). Fig. 16.6 exhibits a clear change in the downstream distribution function, indicating that incident particles satisfying $u_x < u_{\text{spec}}$ experience reflection $\left(\text{where} \quad u_{\text{spec}} = \sqrt{2e\Phi/m_i}\right)$, which is also reflected in the double power law character of the 1-D PUI distribution illustrated in Fig. 16.5.

An important point for subsequent analysis is that the transmitted PUIs remain approximately like a filled-shell, whereas the reflected PUIs form beams in the $\widehat{y}$ (motional electric field) direction with significant energy gains occurring for many particles. An interesting feature of the right-hand panel of Fig. 16.6 is that a fraction of the distribution not satisfying $u_x < u_{\text{spec}}$ (i.e., portions of the distribution with a relatively large value of $u_\parallel$) is significantly energized, indicating that some of these ions also experience reflection. The phase mixing of the distribution in the (u_x, u_y) plane, as it travels through the shock foot along with further phase mixing in the rather wide ramp, possessing multiple reflection sites, causes some particles not initially in $u_x < u_{\text{spec}}$ to migrate to a reflecting region of the phase space and thus to experience MRI acceleration.

The formation of beams in the motional electric field direction indicates some bending of the flow toward the motional electric field direction and the existence of an anisotropic pressure tensor just behind the ramp. The beams exhibit a considerable fine-scale structure, which is a direct consequence of PUI reflection by fine-scale fluctuations in the ramp. Thus, like Zilbersher and Gedalin (1997), because of the large gradients associated with their small structure, we find that fluctuations within the ramp are especially effective in reflecting PUIs.

Finally, the results of Burrows et al. (2010) show that an initial Maxwellian with a temperature of 20,000K injected just ahead of the ramp passes through all variations of the TS-3 model with essentially zero reflection, indicating that almost all thermal solar wind particles in the Maxwellian distribution continuously satisfy $u_x > u_{\text{spec}}$.

16.4 Velocity Distribution Function Downstream of a Quasiperpendicular Shock

To determine the heliosheath proton distribution function, we follow Zank et al. (2010) in that we consider separately the solar wind protons and the transmitted and reflected PUI distribution. For the solar wind protons, in view of Fig. 16.3 and the simulations of Burrows et al. (2010) discussed above, we shall assume that all are transmitted downstream and that they do not contribute to the reflected proton distribution. We further assume that the solar wind proton distribution is a Maxwellian. Since the number of PUIs reflected is comparatively small, we can make the simplifying assumption that the nonreflected PUI distribution can be approximated by the filled-shell distribution (Eq. 16.6) with a number density of $n_{p,1}^t$ (this is probably quite reasonable in view of phase mixing). Again, this assumption appears to be supported by the simulations of Burrows et al. (2010), discussed in the context of Fig. 16.5. On integrating the stationary transport equation for PUIs in the flow frame

$$u\frac{\partial f}{\partial x} + \frac{du}{dx}\frac{c}{3}\frac{\partial f}{\partial c} = 0,$$

across the shock, we have $u_{b1}f_1(c) = u_{b2}f_2(c)$, or equivalently the downstream distribution is simply

$$f_2(c) = r \cdot f_1(c) = \frac{3}{8\pi}\frac{m n_{p1}^t}{u_{b1}^{\frac{3}{2}}}c^{-\frac{3}{2}}, c \in [0, u_{b1}], \tag{16.8}$$

where $r \equiv u_{b1}/u_{b2}$ denotes here the shock compression ratio. Eq. (16.8), with the appropriate $n_{p,1}^t$, yields part of the downstream proton distribution, which will be a superposition of the transmitted solar wind, the transmitted PUI, and the downstream reflected PUI distributions.

To estimate the energy gained by a particle transmitted across the HTS, we must account for the proton deceleration by the cross-shock potential, thus

$$u_2(u_1) = \sqrt{u_1^2 - \frac{1}{2}e\Phi/m_i} \tag{16.9}$$

We Taylor expand this expression about the upstream flow speed u_1 and use

$$u_{b2} = \sqrt{u_{b1}^2 - \frac{1}{2}e\Phi \Big/ m_i} \text{ to find}$$

$$v_2(v_1) - u_2 = r(v_1 - u_1), \quad c_2 = rc_1 \tag{16.10}$$

Taking the scalar pressure moment of the transmitted PUIs and using Eq. (16.10) yields $P_{p2}^t = r^3 P_{p1}^t$, or

$$T_{p2}^t = r^2 T_{p1}^t \tag{16.11}$$

This result holds for arbitrary isotropic distributions at an idealized shock discontinuity. For $r = 2.5$, as observed at TS-3, and $T_{p1}^t = 1.56 \times 10^6$ K, we have $T_{p2}^t = 9.75 \times 10^6$ K for the transmitted PUIs. Eq. (16.11) also yields a downstream solar wind proton temperature $T_{SW2} = 125,000$K, which is quite close to the observed downstream solar wind temperature (see Fig. 16.4, which shows a temperature plateau for <180,000K).

To estimate the temperature of the downstream reflected PUIs, we note (Zank et al., 1996a) that reflected PUIs are trapped at the perpendicular shock front by a balance of the Lorentz force eu_yB_z and the cross-shock potential gradient $-eE_x$.

Then, approximating $e\Phi \sim \frac{1}{2}m_i u_{b1}^2$, we estimate the velocity u_y gained by a PUI in the upstream motional electric field as

$$u_y \cdot \frac{1}{B_{z1}} \frac{\partial \Phi}{\partial x} \cdot \frac{1}{B_{z1}} \frac{\Phi}{L_{\text{ramp}}} \cdot \frac{1}{2} \frac{r_{g1}}{L_{\text{ramp}}} u_{b1} \qquad (16.12)$$

where r_{g1} is the gyroradius of an upstream PUI. Evidently, the energy gain is large provided $L_{\text{ramp}} \ll r_{g1}$. On using values observed by Voyager 2, $r_{g1} \sim 63,000$ km, which is much greater than the observed ramp thickness of $\sim 5,500$ km. Thus, we find that

$$\langle u^2 \rangle_2 \cdot u_{b1}^2 \cdot \left[1 + \frac{1}{4}(r_{g1}/L_{\text{ramp}})^2 \right],$$

which yields an estimate for the downstream temperature of reflected PUIs as

$$T_{p,2}^{ref} \cdot \frac{m_i}{3k_B} \left[1 + \frac{1}{4}(r_{g1}/L_{\text{ramp}})^2 \right] \cdot u_{b1}^2 \qquad (16.13)$$

Thus, Voyager 2 observations and the estimates above yield $T_{p,2}^{ref} \sim 7.7 \times 10^7$ K for the reflected PUI temperature transmitted downstream.

Finally, to estimate the number densities of the transmitted and reflected PUIs, we assume a filled-shell distribution and slice off that section corresponding to particles with $u_x \leq u_{\text{spec}}$. This procedure yields

$$n_p^t = n_p \left(1 - \frac{3}{2} \frac{u_{\text{spec}}^2}{u_{b1}^2} \right), \qquad (16.14)$$

$$n_p^{\text{ref}} = \frac{3}{2} n_p \frac{u_{\text{spec}}^2}{u_{b1}^2}, \qquad (16.15)$$

We may, therefore, estimate the total temperature downstream of the quasi-perpendicular TS using Eq. (16.11) and Eqs. (16.13)−(16.15) on the assumption that $n_{p1} = 0.2 n_{i1}$, finding that $T_2 = 3.4 \times 10^6$ K. Such a total temperature is consistent with the postshock temperatures obtained from global magnetohydrodynamic (MHD)-kinetic simulations of heliospheric structures (i.e., models that incorporate charge exchange source terms in the conservation equations) (Pogorelov et al., 2008; Heerikhuisen et al., 2008; Zank et al., 2013), giving us confidence that the detailed shock microphysics is translating accurately into the MHD conservation laws.

The partitioning of downstream thermal energy into transmitted solar wind protons, transmitted PUIs, and reflected PUIs can be estimated using the above temperatures and number densities for the various transmitted and reflected components. By way of illustration, on using the estimated value $T_2 = 3.4 \times 10^6$ K and introducing $\Gamma_{\text{SW}} = T_{\text{SW}}/T_2$, etc., we find $\Gamma_{\text{SW}} \sim 6 \times 10^{-2}$, $\Gamma_p^t \sim 2.9$, and $\Gamma_p^{ref} \sim 22.6$, giving a downstream temperature/energy partition of $\sim 4.8\%$ in solar wind protons, 52% in transmitted PUIs, and 45% in reflected PUIs. Clearly, PUIs, both transmitted and reflected, dominate the thermal energy content in the heliosheath, just as PUIs do upstream of the HTS. Further details can be

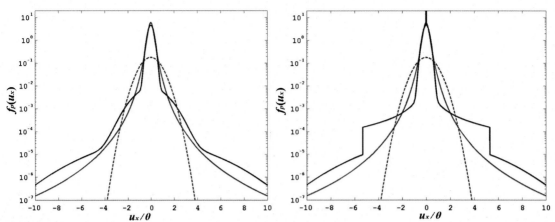

FIGURE 16.7

Plots of the local proton distribution function in the heliosheath. The *blue curve* shows the kappa distribution used by Heerikhuisen et al. (2008) with a value of $\kappa = 1.63$ (see: Chapter 2; Livadiotis and McComas, 2010a). The black curves depict the distribution constructed from a superposition of the transmitted solar wind protons, the transmitted PickUp Ions (PUIs), and the downstream reflected PUIs. The *red curve* illustrates a Maxwellian distribution with the downstream density and temperature. (Left) The heliosheath-constructed proton distribution (*black curve*) assuming that the transmitted PUIs evolve into a Maxwellian distribution. (Right) The heliosheath-constructed proton distribution (*black curve*) assuming that the transmitted PUIs possess a filled-shell distribution. Here, the particle velocity u_x is normalized to the thermal speed θ. Source: Zank et al. (2010).

found in Zank et al. (2010), who use this decomposition procedure as an algorithm to determine the partitioning of energy downstream of the HTS in a global model of the solar wind—interstellar medium interaction. From such a decomposition, a downstream proton distribution can be constructed (Zank et al., 2010).

Shown in Fig. 16.7 are examples of the heliosheath proton distribution, using, on the left, the assumption that the transmitted PUIs relax to a Maxwellian distribution, and, on the right, the assumption that the transmitted PUIs remain a filled-shell distribution. A corresponding kappa distribution is plotted with index $\kappa = 1.63$ (Chapter 2; Livadiotis and McComas, 2010a), together with a downstream total Maxwellian. Evidently, the kappa distribution resembles a smoothed form of both constructed heliosheath proton distributions. Moreover, both forms of the constructed distributions exhibit the important property that a significant number of protons reside in the wings of the distribution function, quite unlike the Maxwellian distribution. As seen in the simulations by Burrows et al. (2010), phase mixing smooths out the distribution. There are several noteworthy points about the distribution functions plotted in Fig. 16.7. The first is the narrowness of the transmitted solar wind proton distribution, the second is the broadening of the distribution function by the transmitted PUIs out to about $\sim 5\theta$, and finally the downstream reflected PUIs considerably extend the outermost wings of the total proton distribution function. The filled-shell assumption for the transmitted PUIs probably overestimates the hardness of the spectrum, and we would expect the spectrum to be intermediate to the two cases illustrated in Fig. 16.7. Nonetheless, the close correspondence between the constructed distributions and the kappa distribution with kappa index $\kappa \sim 1.63$ is quite remarkable (Chapters 2 and 10; Decker, et al., 2005; Livadiotis and McComas, 2010a). However, it is important to recognize that the constructed heliosheath proton distribution, under both assumptions for the transmitted PUIs, possesses some structure that may manifest itself in Energetic Neutral Atom (ENA) spectra observed at 1 AU by Interstellar Boundary EXplorer (IBEX). Zank et al. (2010) therefore suggested that the microphysics of the HTS plays a key role in determining the form of the total downstream or heliosheath proton distribution.

In Section 16.5, we discuss several kinetic simulations that provide support for the theory described here for the generation of downstream kappa distributions.

16.5 Simulations

Following the initial work of Lipatov and Zank (1999), Oka et al. (2011) used 1-D particle-in-cell (PIC) simulations to investigate energy dissipation and particle acceleration at the heliospheric termination shock, with a view to understanding the downstream proton distribution function. This work has been further extended by Yang et al. (2015). Oka et al. (2011) utilize a 1-D relativistic PIC code that treats both ions and electrons as particles and assumes an ion to electron mass ratio of $m_i/m_e \approx 40$. Oka et al. (2011) restrict their attention to a perpendicular shock. Three particle species are used in the simulation: solar wind ions, electrons, and PUIs. The former two species have Maxwellian distributions, and the upstream temperatures are chosen so that the solar wind ion and electron beta plasma values are $\beta_i \approx 0.2$ and $\beta_e \approx 0.5$, respectively. The PUIs have a spherical shell distribution with a radius equal to the upstream flow speed.

Oka et al. (2011) performed three runs with different values of PUI number density N_{PUI}. Run A has a large PUI number density of 30%. This is somewhat larger than was estimated from Voyager observations, as Richardson et al. (2008) estimated 20%, but this emphasizes the effect of PUIs on the shock structure. Run B, on the other hand, has a very small PUI number density, 1%, meaning that the PUIs behave like test particles, and the electromagnetic field variations should be very similar to a case with no PUIs. Lastly, Run C has an intermediate value of PUI number density, 15%.

Fig. 16.8 is an overview of Runs A (PUI 30%) and B (PUI 1%). The y-component of the magnetic field B_y is sliced along the x direction and is shown with respect to time. The shock front develops within a gyro period and shows cyclic behavior or "reformation" during the entire simulation. Typical structures of a perpendicular shock such as foot, ramp, overshoot, etc. are also evident in both cases. However, these features are more pronounced for the lower PUI density case (Run B, Fig. 16.8B) than the higher PUI density case (Run A, Fig. 16.8B).

Fig. 16.9 shows the upstream and downstream velocity distribution functions. The number of particles as a function of u_z is shown for the upstream (upper panels)

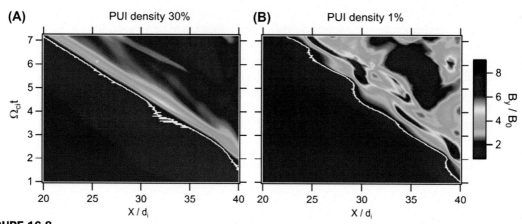

FIGURE 16.8

Overview of magnetic field magnitude B_y during Runs A, panel (A), and B, panel (B). The *white curves* indicate shock front positions used to calculate shock speed. $\Omega_{ci} = eB/m_i$ is the ion gyro frequency, and d_i is the ion inertial length scale. Source: Oka et al. (2011).

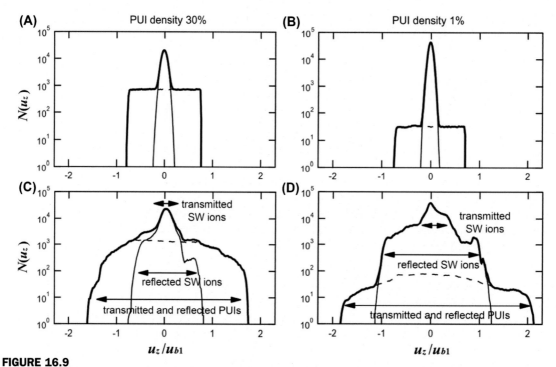

FIGURE 16.9

1-D cut of the velocity distribution function in the upstream (A,B) and the downstream (C,D) taken from Run A (A,C) and Run B (B,D). The particle velocities are normalized by the upstream flow speed u_{b1} in the shock rest frame. The solar wind (SW) and the PickUp Ions (PUIs) are denoted by the *solid and dashed curves*, respectively, and the *thick black curves* are their sum. Source: Oka et al. (2011).

and the downstream (lower panels) from Runs A (left panels) and B (right panels). The particle velocities are normalized by the upstream flow speed u_{b1} in the shock rest frame. The solar wind and the PUIs are denoted by the solid and dashed curves, respectively, and the thick black curves are their sum. It is evident that both the solar wind and PUIs are heated significantly when they are reflected at the shock, although the boundary between the reflected and transmitted PUIs is unclear. The highest energy PUI is nearly ~ 4 times the energy of the upstream bulk flow energy, consistent with previous simulations (e.g., Lee et al., 2005; Matsukiyo et al., 2007). Since the Voyager 2 plasma instrument does not cover the entire energy of the distribution, certainly in the range of $|u_z|/u_{b1} < 0.6$, the observations miss the PUI contribution to the number density in both cases and underestimate the down-stream temperature. It should also be noted that, because of the larger number density of PUIs, the integrated distribution function does not show a clear boundary between the solar wind ions and the PUIs in the PUI 30% case (Run A) whereas the PUIs are easily identified by the broadened wing of the distribution function in the PUI 1% case (Run B).

The Zank et al. (2010) model of the downstream velocity distribution functions were for a higher PUI density case (20%). The Zank et al. (2010) models agree rather well with the Oka et al. (2011) simulation results. That Zank et al. (2010) did not consider reflected solar wind ions is justified because, in the Oka et al. (2011) simulations, the solar wind ions were less heated (Fig. 16.9C) compared to the lower PUI density case (Fig. 16.9D) so that the distribution function (the thick black curve) does not show a clear bump for the reflected solar wind component. The major difference between the Zank et al. (2010) model and the simulations is that the boundary between reflected and transmitted PUIs was not clear in the simulation whereas it is in the model. The model assumed either a complete Maxwellian distribution or a shell distribution for the transmitted PUIs, but the Oka et al. (2011) simulations indicate a heated component with a smooth cutoff. These results appear to be supported by the more detailed and extensive simulations presented in Yang et al. (2015).

16.6 Observational Tests

The direct observation of PUIs accelerated by interplanetary shock waves has been discussed in several important papers by Gloeckler and his colleagues (Gloeckler et al., 1994, 2001). This work serves as a useful test of some of the ideas described previously. Illustrated in Fig. 16.10 is an example of the H^+, He^+, and He^{++} velocity distributions observed by the SWICS instrument upstream and downstream of a corotating interaction region (CIR) reverse shock (Gloeckler et al., 2001). This figure is related closely to observations reported in Gloeckler et al. (1994), in which solar wind protons, alpha particles, and interstellar pickup H^+ and He^+ were observed by Ulysses during the passage of a CIR at ~ 4.5 AU. Of particular note, Gloeckler et al. (1994, 2001) find that interstellar H^+ and He^+ are injected into the acceleration process almost 100 times more efficiently than solar wind protons and alpha particles.

FIGURE 16.10

Velocity distribution functions derived from the Ulysses Solar Wind Ion CompoSition (SWICS) instrument measurements of H^+ (*open squares*), He^+ (*solid triangles*), and alpha particles He^{++} (*solid diamonds*) in the spacecraft frame. Upstream and downstream distribution functions are shown for each population. The upstream pickup H^+ and He^+ exhibit cutoffs at twice the solar wind speed, consistent with a filled-shell distribution, and the upstream thermal solar wind proton distribution is a Maxwellian. Downstream of the corotating interaction region (CIR) reverse shock, the total H^+ spectrum is smoothed out, and a non-Maxwellian tail is formed. The downstream proton distribution appears to be consistent with dissipation due primarily to reflected pickup protons. Source: Gloeckler et al. (2001).

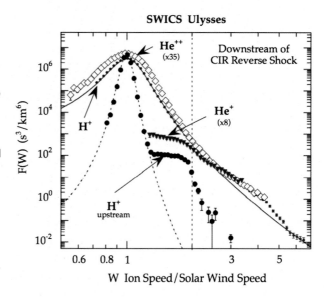

SWICS Ulysses

Consequently, interstellar H^+ and He^+ are the two most abundant suprathermal ion species in the solar wind beyond ~5 AU. We would therefore expect that quasiperpendicular interplanetary shocks as well the HTS primarily reflect pickup H^+ rather than solar wind protons at the cross-shock electro-static potential of quasiperpendicular shocks in the outer heliosphere. PUIs will therefore modify the downstream proton velocity distribution function in the distant supersonic solar wind as well as in the inner heliosheath, creating proton kappa distributions throughout the heliosphere.

IBEX observations, despite being remote, offer an opportunity to test the basic model of Zank et al. (1996a) that describes the microstructure of the HTS and the creation of a downstream kappa velocity distribution for the protons. Zank et al. (2010) developed a basic model of a quasiperpendicular HTS, mediated by PUIs, to derive the complete downstream proton distribution function in the inner heliosheath (IHS), determine the partitioning of energy between solar wind protons and PUIs, and infer the implications of the constructed IHS proton distribution function for the ENA spectral flux observed by IBEX.

Zank et al. (2010) introduced a multicomponent distribution approximation of the IHS plasma, comprising core solar wind protons, transmitted (without reflection) PUIs, and reflected (and then transmitted) PUIs. Zank et al. (2010) predicted that the constructed heliosheath proton distribution should possess a structure that would manifest itself in the ENA spectra observed at 1 AU by IBEX.

To test the possibility that the microphysics of the HTS would manifest itself in the IBEX ENA spectra observed at 1 AU, Desai et al. (2012), in an initial study, found that the fluxes, energy spectra, and energy dependence of the spectral indices of ~0.5–6 keV ENAs measured by IBEX-Hi along Voyager 1 and 2 lines of sight were consistent within a factor of ~2 with the model results of Zank et al. (2010). The observed ENA spectra do not exhibit sharp cutoffs at approximately twice the solar wind speed, as is typically found for shell-like PUI distributions in the heliosphere. Desai et al. (2012) concluded that the ENAs measured by IBEX-Hi are generated by at least two types of PUI populations whose relative contributions depend on the ENA energy: transmitted PUIs in the ~0.05–0.5 keV energy range and reflected PUIs above ~5 keV energy. The ~0.05–0.5 keV PUI distribution is probably a superposition of Maxwellian or kappa distributions and partially filled-shell distributions in velocity space (Desai et al., 2012; also see: Appendix A in Livadiotis and McComas, 2013a; and the theory of spectral statistics in Tsallis, 2009b).

The observed lower energy ENAs (below ~0.5 keV) are not well described by the Zank et al. (2010) theory, and most existing models underestimate the ENA fluxes between ~0.05–0.5 keV by an order of magnitude or more (Fuselier et al., 2012). To address the lower energies, Zirnstein et al. (2014) extended the Zank et al. (2010) model in two ways. First, they accounted for the extinction of solar wind protons and transmitted and reflected PUIs by charge-exchange in the composite proton distribution. The extinction process alters the distribution of energy in the IHS, compared to assuming that the relative energy densities of the core solar wind protons and transmitted and reflected PUIs remain constant. Determining an accurate partitioning of the energy is essential for under-standing the role that PUIs play in the heliosphere and its effect on H ENA flux.

The second extension introduced by Zirnstein et al. (2014) was to include ENAs created by PUIs from the Very Local InterStellar Medium (VLISM); see Zank (2015) for a further discussion and definition. Although ENAs are created everywhere in the solar wind–Local InterStellar Medium (LISM) interaction region, ENAs produced in the IHS easily propagate into the VLISM before charge-exchange occurs, creating a population of PUIs there. ENAs produced in the VLISM, however, do not easily charge-exchange in the IHS and therefore permeate the inner heliosphere and can be detected at 1 AU. One can similarly partition the VLISM energy into various proton populations (Zirnstein et al., 2014). The VLISM plasma consists mostly of protons, initially ~6300 K in the pristine LISM (Möbius et al., 2012; Bzowski et al., 2012; McComas et al., 2012), that are partially heated by charge-exchange near the H wall and by crossing a bow wave (McComas et al., 2012; Zank et al., 2013). However, the increase in thermal energy of the VLISM plasma near the HP is also due to energetic PUIs, which are created from charge-exchange between LISM protons and ENAs from the IHS (Zank et al., 1996b). The majority of PUIs are in close proximity to the HelioPause (HP) and drop off exponentially at larger distances due to the mean free path of their parent ENAs and due to advection with the LISM flow toward the HP (Zirnstein et al., 2014). As with the IHS, Zirnstein et al. (2014) determine the VLISM PUI properties by partitioning the total energy from the plasma-neutral results between LISM protons and PUIs. Since ENAs from IHS protons may propagate into the VLISM and charge-exchange to become PUIs, they treat the VLISM plasma as a five-component distribution, including protons from the

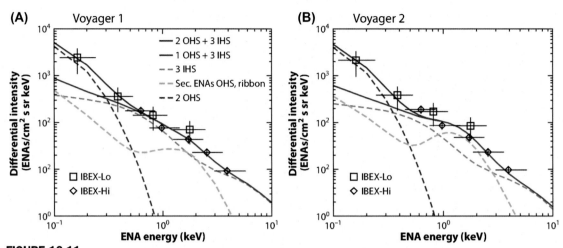

FIGURE 16.11

Interstellar Boundary EXplorer (IBEX)-Hi and IBEX-Lo ENA spectra compared with the simulations of Zirnstein et al. (2014), in the (A) Voyager 1 and (B) Voyager 2 direction. *Dashed green:* ENAs from a secondary Very Local InterStellar Medium (VLISM) population in the Outer HelioSheath (OHS), forming the ribbon; *dashed blue:* three inner heliosheath (IHS) populations with Maxwellian distributions; *dashed red:* ENAs from a hot, PickUp Ion VLISM population; *solid purple:* superposition of all three IHSs, the secondary ENAs from the ribbon, and a single completely thermalized VLISM population (not visible); *solid red:* superposition of all three IHSs, the secondary ENAs from the ribbon, and the two proton VLISM populations. Source: Desai et al. (2014).

core (and compressed) VLISM plasma and PUIs created by charge-exchange from IHS ENAs.

Fig. 16.11 shows various sources of the H spectrum in the (A) Voyager 1 and (B) Voyager 2 direction, based on an extended model (Zirnstein et al., 2014) with a comparison to the corrected IBEX data (Desai et al., 2014). Below ~0.5 keV, the flux is dominated by ENAs from VLISM secondary PUIs, while ENAs from HTS transmitted and reflected PUIs dominate above 0.5 keV. Although a small fraction of ENAs from core solar wind protons are visible at 1 AU, most exit the HP and become PUIs in the VLISM, producing significant flux near ~0.1 keV. Zirnstein et al. (2014) predict that a significant part of the ENA flux seen at 1e AU comes from the VLISM. ENAs created from solar wind PUIs in the VLISM dominate the flux below ~0.2 keV, while secondary injected, transmitted, and reflected PUIs contribute a significant flux up to keV energies, comparable to the flux from the IHS. Our current detailed model (Zirnstein et al., 2014) therefore exploits the properties of PUIs that contribute to heating the VLISM plasma, thereby establishing that not only the low but also the high energy flux is a result of the coupling between the IHS and VLISM plasmas through charge-exchange. PUIs from the IHS are the source of multiple PUI species in the VLISM. Our simulation results (Zirnstein et al., 2014) compare favorably with the IBEX data, although perhaps somewhat low at high energies compared to those observed by IBEX since VLISM PUIs created from supersonic solar wind ENAs, or time-dependent solar wind boundary conditions, were not included.

Nonetheless, these results suggest a strong coupling between the IHS and VLISM plasmas through ENA charge-exchange, and VLISM PUIs up to ~ 10 keV may dominate the globally distributed ENA flux visible at 1 AU.

The results from the theoretical models (Zank et al., 2010; Zirnstein et al., 2014) describing the interaction of the solar wind and the partially ionized LISM and the observational results (Desai et al., 2012, 2014) confirm that indeed the proton distribution downstream of the HTS is in the form of a kappa distribution and that quasiperpendicular shocks play a fundamental role in generating kappa distributions.

16.7 Dissipation and Particle Acceleration at Quasiperpendicular Shocks

In this final section, we discuss a possible relationship between the dissipative ion reflection process at a quasiperpendicular shock and the subsequent acceleration of protons via the DSA mechanism. The question of particle injection into the DSA mechanism has yet to be fully explained. Although fairly well addressed in the context of quasiparallel shocks (see Neergaard Parker and Zank, 2012; Giacalone, 2012, and references therein), less is understood about the injection and acceleration mechanism for quasiperpendicular shocks. There is an important difference in the physical injection process between quasiperpendicular and quasiparallel shocks. Charged particles are essentially tied to magnetic field lines, making crossing from upstream to downstream field lines and back again at a quasiperpendicular shock difficult. Since charged particles cannot easily diffuse back and forth across a quasiperpendicular shock, DSA is difficult. However, stochastic field line meandering can ensure that if a particle is already of sufficient energy, it can experience multiple crossings (Zank et al., 2006). Consequently, if the dissipation mechanism at a quasiperpendicular shock generates a kappa distribution that leads to an elevated number of energetic particles in the wings of the proton velocity distribution function compared to a Maxwellian distribution, then the DSA mechanism can work at a perpendicular shock when stochastic magnetic field line meandering is present.

Neergaard Parker et al. (2014) investigated the injection problem at quasi-perpendicular shocks, following related work for quasiparallel shocks (Neergaard Parker and Zank, 2012). Neergaard Parker et al. (2014) used the nonlinear guiding center theory (Matthaeus et al., 2003) to describe the random walk of magnetic field lines in terms of a perpendicular diffusion coefficient, which allows them to describe the diffusive acceleration of particles at a quasi-perpendicular shock front (Zank et al., 2006). Plasma observations at 1 AU ahead of observed shocks were used to construct upstream Maxwellian and kappa distributions describing the incident flow. This allowed Neergaard Parker et al. (2014) to compare the theoretically accelerated upstream power law distributions predicted by DSA, assuming either a Maxwellian or kappa distribution at the shock, with observations from the Electron, Proton, and Alpha

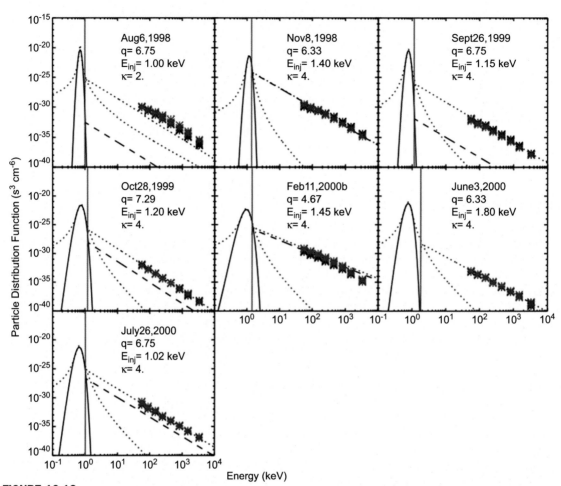

FIGURE 16.12

Plotted on each panel is an upstream Maxwellian (*solid line*) and a kappa distribution (*dotted line*) with the observed accelerated distributions overlaid. All best-fit downstream accelerated distributions shown correspond to accelerating an upstream $\kappa = 4$ particle distribution, except for 1998 August 6, which required $\kappa = 2$. The (Electron, Proton, and Alpha Monitor) EPAM observations for 10 min immediately preceding (*black points*) and immediately following (*red points*) the shock are overplotted. The injection energy is denoted by the vertical line, while the spectral index q is given in the figure labels. Source: Neergaard Parker et al. (2014).

Monitor (EPAM) instrument on Advanced Composition Explorer (ACE). As illustrated in Fig. 16.12, Neergaard Parker et al. (2014) find that in general, the assumed upstream particle distribution that best fits the energetic particle observations at quasiperpendicular shocks is best represented by a kappa distribution, typically with $\kappa = 4$. One case required a value of $\kappa = 2$. The key point here is that an upstream Maxwellian proton distribution could not account for the observed accelerated particle distribution. This is quite unlike the case for quasiparallel shocks, for which Neergaard Parker & Zank (2012) found that all the observed accelerated proton spectra could be explained by assuming an injection from an upstream Maxwellian distribution. We would therefore conclude that quasiperpendicular shocks, through their dissipation process, generate a kappa distribution that then provides a sufficiently large number of suprathermal particles that can be diffusively accelerated at a quasiperpendicular interplanetary shock provided that magnetic field line meandering occurs.

16.8 Concluding Remarks

In summarizing the underlying theory that we have developed, our conclusions are as follows.

1. We have developed a model describing the basic plasma kinetic processes and microphysics of the quasiperpendicular HTS in the presence of an energetic PUI population. We have shown that solar wind protons do not experience reflection at the cross-shock potential of the HTS and are transmitted directly into the heliosheath. PUIs, by contrast, can be either transmitted or reflected at the HTS, provide the primary dissipation mechanism at the shock, and dominate the downstream temperature distribution. We have derived an IHS proton distribution function that is consistent with Voyager 2 solar wind plasma observations and is similar to a kappa distribution with an index of 1.63. The latter result holds if we assume that the transmitted PUI distribution relaxes to either a Maxwellian or a filled-shell distribution.

2. The composite heliosheath proton distribution function is a superposition of cold transmitted solar wind protons, a hot transmitted PUI population, and a very hot PUI population that was reflected by the cross-shock electrostatic potential at least once before being transmitted downstream. The composite spectrum possesses more structure than the kappa distribution, but both distributions have approximately the same number of protons in the wings of the distribution (and therefore many more than a corresponding Maxwellian distribution). Phase mixing will smooth the distribution, yielding a kappa distribution.

3. Kinetic simulations of the quasiperpendicular HTS show that PUIs are indeed reflected preferentially at the cross-shock electrostatic potential and that the downstream velocity distribution function for the protons resembles a kappa distribution.

4. Direct observations of the acceleration of pickup H^+ and He^+ at forward and reverse interplanetary shocks show that the upstream superimposed Maxwellian and flat-topped PUI distribution becomes a smooth kappa-like proton distribution downstream of the shock.

5. IBEX observations of the ENA flux at 1 AU show that the spectrum from 1 to 6 keV possesses a structure that can be modeled on the basics of a kappa-like proton distribution function behind the heliospheric termination shock.

6. Finally, the proton reflection dissipation process at quasiperpendicular shocks generates a kappa distribution that then provides an enhanced number of suprathermal particles in the wings. These protons can be diffusively accelerated more easily at a quasiperpendicular interplanetary shock at which magnetic field line meandering occurs.

We note that we have assumed a filled-shell PUI distribution upstream of a quasiperpendicular HTS, but these results hold for any interplanetary shock in the outer heliosphere. Indeed, as illustrated in Fig. 16.2, we can expect that PUIs will be a primary dissipation mechanism for any quasiperpendicular shock from beyond 2 to 3 AU. We note that in neither the supersonic solar wind nor the IHS will PUIs ever thermally equilibrate with the thermal solar wind protons (Zank et al., 2014). Furthermore, if a suprathermal halo population is present in the solar wind ahead of an interplanetary shock, the same analysis given above will carry through. The net result is that a downstream kappa distribution almost inevitably emerges from the interaction of a suprathermal halo distribution interacting with a quasiperpendicular shock.

16.9 Science Questions for Future Research

Future analyses and observations need to address the following questions:

1. What is the role of the shock direction?
2. What shock parameters determine the downstream kappa indices?
3. How does the turbulence affect the generated kappa distribution?

Acknowledgments

We acknowledge the partial support of NASA grants NNX08AJ33G, Subaward 37102-2,NNX14AC08G, NNX14AJ53G, A99132BT, RR185-447/4944336, and NNX12AB30G.

D.C. Nicholls [1], M.A. Dopita [1,2,3], R.S. Sutherland [1], L.J. Kewley [1,3]
[1]Australian National University, Canberra, ACT, Australia;
[2]King Abdulaziz University, Jeddah, Saudi Arabia; [3]University of Hawaii
at Manoa, Honolulu, HI, United States

Chapter Outline

17.1 Summary 634
17.2 Introduction 634
  17.2.1 HII Regions 635
  17.2.2 Planetary Nebulae 636
  17.2.3 Kappa Distributions 637
17.3 Are Energy Kappa Distributions Present in Astrophysical Nebulae? 638
17.4 Ionization Structures in an HII Region 639
17.5 Magnetic Structures in HII Regions 641
17.6 Nebular Spectral Lines 642
  17.6.1 Collisional Excitation 643
  17.6.2 Recombination Lines 643
  17.6.3 Abundance Discrepancy Problem 644
17.7 Atomic Energy Levels and Kappa Distribution 646
  17.7.1 Calculating Electron Temperatures From Collisionally Excited Lines 647

17.7.2 Explaining the Optical Recombination Line/Collisonally Excited Line Abundance Discrepancy 650
17.7.3 Discrepancies in Collisionally Excited Line Temperatures 650
17.8 Diagnostics for the Kappa Index 651
17.9 Modeling of Photoionized Nebulae 652
  17.9.1 Modeling Programs 652
  17.9.2 Complexities in HII Region Structures 653
17.10 Other Applications of Kappa Distributions in Astrophysical Nebulae 653
17.11 Alternative Explanations of Abundance Discrepancy 654
17.12 Concluding Remarks 655
17.13 Science Questions for Future Research 655

Kappa Distributions. http://dx.doi.org/10.1016/B978-0-12-804638-8.00017-6

17.1 Summary

Until recently, it had been assumed that the electrons in astrophysical nebulae (HII regions and planetary nebulae) are in thermal equilibrium, with energies described by the Maxwell–Boltzmann distribution. This may not always be true. Based on satellite and space probe measurements of electron energies in the solar system, where nonequilibrium energy distributions are regularly encountered, it appears to be entirely plausible that such distributions also occur in nebulae, under similar physical conditions. If we adopt kappa distributions for the nebulae electron energies, we can resolve a long-standing problem where measurements of electron temperatures and chemical abundances using different methods yield discrepant results. This has been a major concern, as measurements of nebular abundances are widely used to study galaxy evolution and dynamics. By assuming nonequilibrium electron energies, we can resolve the measurement discrepancies and gain a deeper understanding of the physics in these regions.

Science Question: What are the features of kappa distributions in astrophysical nebulae?

Keywords: Atomic data; Atomic processes; Physical data and processes; Radiative transfer.

17.2 Introduction

Since the earliest spectra were observed from optical emissions in nebulae in the Milky Way, we have endeavored to understand the physics operating in these regions (e.g., Bowen, 1928; Baker et al., 1938; Hebb and Menzel, 1940). More recent analytical efforts began in the late 1960s, with the work of Peimbert and others (e.g., see Peimbert, 1967). Analytical methods and model simulations based on emission line intensities have developed considerably since then to estimate electron temperatures and abundances of the heavier elements. However, different methods to estimate abundances have always given conflicting results, known as the "abundance discrepancy problem," and none of the suggested explanations have proved to be fully satisfactory (Stasińska, 2002, 2004). Nicholls et al. (2012) suggested that a better explanation for the discrepancies might be due to electron energies in the plasmas not being in thermal equilibrium. This chapter addresses the application of the electron energy kappa distributions in HII regions and planetary nebulae (PNe) and its implications for understanding nebular physics, and, in particular, the abundance discrepancy problem.

In the following subsections we introduce the HII regions, PNe, and the formulation of kappa distributions used in this chapter. In Section 17.3, we make the case that kappa electron energy distributions can occur in astrophysical nebulae, which is still a matter of some controversy. In Section 17.4, we examine the ionization structure of a theoretical spherically symmetric HII region to show how the physical conditions vary throughout the nebula. Section 17.5

presents evidence for the presence of magnetic fields in these objects, a likely source of high-energy electrons. In Section 17.6 we detail the spectral emission lines observed in astrophysical nebulae and the physical processes giving rise to these lines. We describe a long-standing conundrum arising from the analysis of these emissions, the "abundance discrepancy problem," whereby different analysis methods give rise to substantially different results for the physical conditions in nebulae, particularly temperatures and element abundances. Section 17.7 is the core of our argument. It explains, in terms of the atomic energy levels giving rise to the observed emission lines, how a kappa electron energy distribution can provide a natural explanation of the abundance discrepancy problem. In Section 17.8 we examine possible observable emission line diagnostics for the presence of kappa electron energies. In Section 17.9 we discuss the complexities of numerical modeling of ionized nebulae, an essential part of understanding the physics of these regions. In Section 17.10 we describe attempts to search for evidence of nonequilibrium electron energies. In Section 17.11, we look at previous attempts to explain the abundance discrepancy problem. Finally, the concluding remarks are given in Section 17.12, while three general science questions for future analyses are posed in Section 17.13.

17.2.1 HII Regions

HII regions are volumes of ionized gas (on scales of a few tens to a few hundred light years) present in virtually all gas-rich galaxies. These gas regions consist principally of hydrogen but with ~ 10% helium and smaller amounts of oxygen, carbon, nitrogen, sulfur, neon, argon, and other trace elements. They arise in concentrations of gas and dust in the interstellar medium, where gravitational contraction and random turbulence create regions of higher density. At the core of these concentrations, stars form, singly, in small groups or in large clusters. A few of the stars will be massive, extremely hot stars that ionize the neutral gas surrounding them, allowing us to observe the star forming regions that might otherwise remain hidden. HII regions are the visible signposts of where massive stars are forming. They are important subjects of study, as they are intimately associated with star formation and are thus indicators of galaxy evolution.

Nebulae are heated by the photoionization of hydrogen by ultraviolet (UV) photons with energies >13.6 eV from the central star or cluster. The photoionization process generates free electrons, which heat the gas through elastic collisions. The relationship between the stellar photoionization heating rate and the electron temperature is described by a power law. Heating is balanced by radiative cooling, predominantly from the heavier ions collisionally excited by the hot electrons. The electron temperature reaches a balance when the heating rate equals the cooling rate. If the abundance of the heavier species increases, the cooling rate increases, and the heating/cooling balance is achieved at a lower electron temperature (see Fig. 9.5 in Dopita and Sutherland, 2003). This is the standard description of the energy balance in an HII region or PNe. However, other processes can add energy to the electrons, as we shall see.

The spectral lines emitted by the ionized gas in the radiative cooling process are characteristic of the atoms making up the cloud. Most prominent in the optical and near infrared spectrum are the recombination lines of the hydrogen (Balmer and Paschen series) and helium and prominent collisionally excited lines (CELs) of oxygen (O, O^+, and O^{++}); nitrogen (N and N^+); sulfur (S^+ and S^{++}); neon (Ne^{++}); and argon (Ar^{+++}). Because of the low densities of HII regions ($n_H \sim 10{-}10^3$ cm^{-3}), these heavier ionic species emit in quantum-mechanically forbidden lines such as [OIII] 5007 Å. Fainter emission lines from many other elements are present, such as Fe, and other prominent lines occur in the UV and mid to far infrared. Measuring the emission line fluxes, and knowing the energy levels and transition probabilities between these levels, allow us to understand the structure and physical conditions present within the HII region. Fig. 17.1 shows a typical HII region, called the "Rosette Nebula," where the different (synthetic) colors illustrate the variation of ionic concentrations through the nebula.

17.2.2 Planetary Nebulae

PNe arise from medium-size stars toward the end of their lives. These are stars that are not large enough to generate core-collapse supernovae nor are they able to accrete mass from a close binary companion to reach the critical Type Ia supernova detonation point. Instead, the stars shed their atmospheres in episodes of strong stellar winds in the red giant stage and are surrounded by shells of gas, enriched by the carbon, nitrogen, and oxygen produced by the star as it evolved. The remnant of the central star evolves to the white dwarf stage, a small very hot star ($T \sim 10^5$ K) with a strong UV emission, which ionizes the gaseous

FIGURE 17.1

The Rosette Nebula, NGC 2237, a nearly symmetric HII region, showing the central star cluster and the region evacuated by stellar winds. The false color image uses the 6717,31 Å [SII] forbidden lines (*red channel*), Hα (*green channel*), and the 5007Å [OIII] forbidden line (*blue channel*). Source: https://commons.wikimedia.org/wiki/File:Rosette_nebula_Lanoue.png.

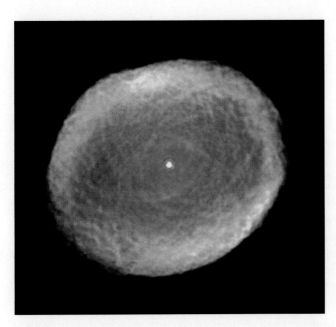

FIGURE 17.2
Planetary nebula IC418 (Hubble Space Telescope image), showing the central white dwarf star. The false color image uses the 6584Å [NII] forbidden line (*red channel*), Hα (*green channel*), and the 5007Å [OIII] forbidden line (*blue channel*), similar to Fig. 17.1.
Source: NASA and the Hubble Heritage Team, STScI/AURA.

shells. In some cases, these nebulae have a nearly circular cross-section, and, when first observed through small telescopes in the 18th century, were thought to resemble planetary disks, giving rise to the name. Fig. 17.2 illustrates a typical planetary nebula, the "Spirograph Nebula" (IC418), showing the central white dwarf star.

Both HII regions and PNe are analyzed using the same methods to determine their chemical abundances, densities, temperatures, and dynamics. A detailed discussion of these objects and the analysis methods for electron temperatures is given in Stasińska (2002).

17.2.3 Kappa Distributions

As previously mentioned, the energy kappa distribution resembles the Boltzmann distribution at lower energies but has a high energy power law tail. The electron density $n(\varepsilon)d\varepsilon$ with energies between ε and $\varepsilon + d\varepsilon$ may be expressed using the energy distribution (see Chapters 1 and 4; Livadiotis & McComas 2009; 2013a), i.e.,

$$f(\varepsilon)d\varepsilon = n \cdot \frac{2}{\sqrt{\pi}} \cdot \frac{\Gamma(\kappa+1)}{\left(\kappa - \frac{3}{2}\right)^{\frac{3}{2}} \Gamma\left(\kappa - \frac{1}{2}\right)} \cdot (k_B T)^{-\frac{3}{2}} \cdot \left(1 + \frac{1}{\kappa - \frac{3}{2}} \cdot \frac{\varepsilon}{k_B T}\right)^{-\kappa-1} \varepsilon^{\frac{1}{2}} d\varepsilon.$$

$$(17.1)$$

The values of the κ-index are in the range of $[3/2, \infty]$; at the limit $\kappa \to \infty$, the energy kappa distribution reduces to the Boltzmann distribution:

$$f(\varepsilon)d\varepsilon \ = \ n \cdot \frac{2}{\sqrt{\pi}} \cdot (k_B T)^{-\frac{3}{2}} \cdot \exp\left(-\frac{\varepsilon}{k_B T}\right) \varepsilon^{\frac{1}{2}} d\varepsilon. \tag{17.2}$$

In these equations, $n = \int_0^\infty f(\varepsilon)d\varepsilon$ is the number density, T is the temperature of electrons, Γ is the gamma function, and k_B is the Boltzmann constant.

17.3 Are Energy Kappa Distributions Present in Astrophysical Nebulae?

The suggestion that electrons in HII regions are not in thermal equilibrium runs counter to the conventional view in astrophysics, so it is important to state the rationale for suggesting that the accepted wisdom may be wrong. We argued this case in detail in earlier papers (Nicholls et al., 2012, 2013), and we refer the reader to those sources for a more extensive discussion. The following draws in part on that work.

Some of the earliest analytical calculations of electron velocity distributions in gaseous nebulae were presented by Bohm and Aller (1947). Their work led them to state that the velocity distribution is "very close to Maxwellian." Spitzer (1962, Chapter 5) also examined the thermalization process for electron energies in plasmas and found that electron energies equilibrate rapidly through collisions. This early work led authors to assume that the electrons in gaseous nebulae are always in thermal equilibrium. However, Spitzer's analysis showed that the "equilibration time" of an energetic electron (the timescale needed to be thermalized and described by the electron Boltzmann distribution) is proportional to the cube of its velocity (or $\varepsilon^{3/2}$), so high-energy electrons take much longer to equilibrate than cooler electrons. This aspect of the equilibration process does not appear to have been explored in detail since the initial arguments, but it suggests that in all space and astrophysical plasmas, high-energy electrons are less easily thermalized than cooler electrons (this may, in itself, contribute to the apparent stability of kappa distributions).

Space plasmas are commonly described by kappa distributions, while the Boltzmann distribution is rather a rarity (Livadiotis, 2015a,b). We are confronted with the fact that despite the early theoretical work suggesting that the electrons in such plasmas should be in thermal equilibrium, they are almost always not. This led us to consider that the same might also be true for the much lower density plasmas in HII regions and PNe. Therefore the case for nonequilibrium electron energy distributions in astrophysical nebulae can be summarized via the following statements: (1) kappa distributions occur in a variety of space plasmas, as confirmed by observations; (2) the thermodynamic conditions (density, temperature, pressure, magnetic field) characterizing the variety of space plasmas, where kappa distributions occur, span the conditions measured in astrophysical nebulae; and (3) several mechanisms that can generate kappa distributions in space plasmas (e.g., see Chapters 5, 6, 8, 15, and 16), can clearly also occur in the stellar winds surrounding HII region central star clusters and other astrophysical nebulae.

From a theoretical perspective, the kappa distribution has a sound basis in statistical mechanics (Chapter 1; Livadiotis and McComas, 2009). The requirement for this to occur is that there be long-range interactions affecting the particles, in addition to the shorter range Coulombic forces that give rise to Maxwell–Boltzmann equilibration.

Kappa distributions may arise when the energetic electrons are pumped with high-energy, nonequilibrium electrons on a timescale shorter than the collisional energy equilibration timescale, so that the system cannot relax to a classical equilibrium distribution. Collier (1993) has also shown that kappa-like energy distributions can arise as a consequence of normal power law variations of physical parameters such as density, temperature, and electric and magnetic fields.

There are several mechanisms likely to occur throughout HII regions and PNe that are capable of generating continuing sources of high-energy electrons. Magnetic fields play an important role in many of them. In particular, there are a number of possible energy injection mechanisms capable of pumping the energetic electron population (e.g., see Chapter 15). These may include magnetic reconnection followed by the migration of high-energy electrons along field lines, the development of inertial Alfvén waves, local shocks (driven either by the collision of bulk flows or by supersonic turbulence), and, most simply, the injection of high-energy electrons through the photo-ionization process itself. Energetic electrons can be generated by the photo-ionization of dust (Dopita and Sutherland, 2000), and X-ray ionization can produce highly energetic (~ 1 keV) inner-shell (Auger process) electrons (e.g., Shull and van Steenberg, 1985; Aldrovandi and Gruenwald, 1985; Petrini and da Silva, 1997; and references therein). If the processes giving rise to high-energy, nonequilibrium electrons occur spasmodically, interspersed with periods of equilibration, their signature will nonetheless be imprinted on the collisional excitation of atomic species and resulting spectra. An analogous effect occurs in the airglow, where nonequilibrium effects are observed in the spectrum of the hydroxyl molecule when it is formed in chemical reactions in the Earth's upper atmosphere (Nicholls et al., 1972).

17.4 Ionization Structures in an HII Region

Fig. 17.3 illustrates the ionization structure of an HII region, showing the typical radial ionic species distribution. The extent of the nebula is defined by the region in which there are unabsorbed UV photons from the central star cluster with energies greater than 13.598 eV, the ionization potential of neutral atomic hydrogen. It is also the region from which the hydrogen recombination lines are

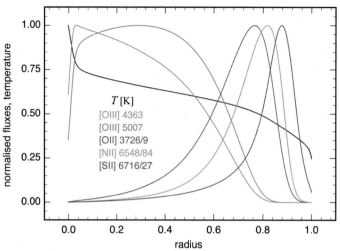

FIGURE 17.3

Simplified structure of a spherically symmetric HII region (Strömgren Sphere) showing the zones of (normalized) emission from different ions and the electron temperature T (*black curve*). At radius $r = 0$ is the outer edge of the evacuated region. The radial scale is nonlinear (that is, $-0.5 \cdot \log(1.01\text{-}R)$, where R is the physical radius in units of 10^{20} cm) to expand the outer regions. The temperature scale is normalized as a fraction of the maximum temperature at the inner edge, that is, $2.4195 \cdot 10^4$ K. Computed using the Mappings V photoionization code, e.g., see Nicholls et al. (2014). The model parameters used are as follows: ionization parameter $\sim 10^{8.5}$, constant pressure $P/k_B \sim 10^5$ with pressure in CGS units such that P/k_B is measured in 1 K cm^{-3}, solar heavy element abundances $= 0.1$, and for a Boltzmann electron energy distribution.

emitted. The core of the HII region is evacuated by the pressure of the stellar winds from the central star or star cluster.

Outside the evacuated region, different species are ionized by energetic radiation. Higher ionization potential species are collisionally excited (and emit radiation) in the inner regions and throughout the body of the nebula. Lower ionization potential species do the same in the outer regions (see Table 17.1, ionization potentials for several ions found in HII regions, from the NIST databases, e.g., Kramida et al., 2012).

In Fig. 17.3, the populations are normalized for comparison, and it is clear that [OIII] collisionally excited forbidden lines and HI recombination lines (here, H_β) are emitted throughout the body of the nebula, whereas spectra from lower ionization potential ions such as O^+, S^+, and N^+ are predominantly emitted from the outer, cooler, lower excitation regions. Neutral species such as O and N (not plotted) are present at the outermost edges. (Note the convention to refer to an ionic species, for example, doubly ionized oxygen, as O^{++}, but the spectral lines associated with radiative transitions from that ion as OIII, with square brackets indicating forbidden lines, as in [OIII], and the single "I" suffix indicating a line from the neutral atom.)

Table 17.1	Ionization Potentials for Some HII Region Ions
Ion	**Ionization Potential (eV)**
H^+	13.5984
He^+	24.5874
N^+	14.5341
O^+	13.6181
O^{++}	35.1211
Ne^{++}	40.9630
S^+	10.3600
S^{++}	23.3379
Ar^{++}	27.6297
Ar^{+++}	40.7350

Source: The NIST databases (Kramida et al., 2012).

It is also evident from Fig. 17.3 that the [OIII] 4363Å line, from the upper $^1S_0$ energy level (see Fig. 17.5), is preferentially excited in the inner regions of the nebula compared to the [OIII] 5007Å line, although both are generated thorough most of the body of the nebula.

17.5 Magnetic Structures in HII Regions

As many mechanisms in the solar system responsible for generating electron energy kappa distributions involve magnetic processes, evidence for their presence in astrophysical nebulae would support the case for similar energy distributions in those plasmas. There is evidence for the presence of magnetic fields in

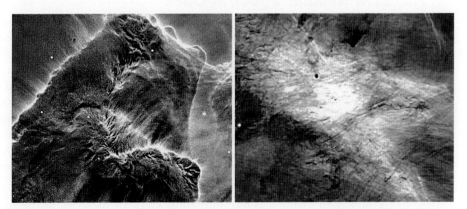

FIGURE 17.4
The fine hair-like structures suggest the presence of magnetically dominated structures, aligned with outflows from ionization fronts (the width of the "hairs" is ~2 AU). Source: Hubble Space Telescope images; see Nicholls et al. (2012).

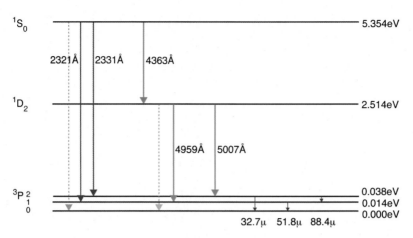

FIGURE 17.5

Lowest five energy levels for O^{++}, showing transitions with wavelengths in Å and microns (μm, abbreviated to "μ"). The *dotted* transitions are very weak, via electric quadrupole. Ultraviolet transitions are shown in *purple*, visible transitions in *green*, and far-infrared transitions in *red*. The 4363 Å transition is referred to as an "auroral" line, by analogy with the *green* 5577 Å line in neutral atomic oxygen auroral emissions, which arises from a similar transition.

HII regions of the sort capable of accelerating electrons to high energies. Fig. 17.4 shows Hubble Space Telescope images of the Eta Carina and M17 Nebulae, enhanced by unsharp mask techniques to show fine hair-like structures that correspond to magnetic field lines. The images suggest the existence of magnetically dominated microstructures aligned with outflows from ionization fronts. These bear a remarkable similarity to theoretical results from MHD simulations (see, for example, Figs. 4, 6, and 8 in Henney et al., 2009; Arthur et al., 2011).

17.6 Nebular Spectral Lines

In the concentrations of the heavier elements, HII regions retain the signature of the star formation history in the region of the gas cloud. In the primordial Universe, these clouds consisted solely of hydrogen, helium, and a trace amount of lithium. As the first stars formed, evolved, and exploded as core-collapse supernovae, the interstellar gas was enriched through the addition of heavy elements formed in the stars and expelled in the explosions. Generations of star formation and supernovae have resulted in the elemental concentrations we see today. Variations in the chemical abundances of heavy elements provide evidence of different star formation histories. This provides important data for studies of galaxy formation and the evolution of the Universe in general.

The spectra of HII regions and PNe include emission lines from two sources: collisional excitation and recombination. CELs are those where the atom or ion is excited to a higher level by collisions between the ion or atom in its ground state and electrons. Recombination emission lines, such as the hydrogen Balmer series, arise when an electron recombines with a hydrogen or helium ion or ions of other elements.

Higher abundances of the heavier elements provide more ways for the gas to cool by the emission of radiation: the more ions there are with available spectral emission lines, the greater the cooling. Higher chemical abundances therefore correspond to lower electron kinetic temperatures. Electron temperature can thus be used to estimate chemical abundance.

17.6.1 Collisional Excitation

In HII regions, the particle density is relatively low, typically $<1000 \text{ cm}^{-3}$. In planetary nebulae, it is $\sim 10^4 \text{ cm}^{-3}$, and the overwhelming majority of ions are in their ground electronic states. Electron collisions can excite them to levels above the ground state. Fig. 17.5 shows a simplified diagram of the lowest five energy levels of the O^{++} ion, one of the most important observed in astrophysical nebulae and key to measuring electron temperatures. The triplet-P ground state energies are separated by less than 0.04 eV, and the $^1D_2$ and $^1S_0$ levels are separated by a few eV. The kinetic energies of the electrons are easily capable of exciting the ions to these levels, although at typical HII region electron temperatures of $\sim 10^4$ K, the higher level ($^1S_0$) is substantially less populated by collisional excitation than the middle ($^1D_2$). These two levels emit photons as they decay back to ground or by cascade from the upper level through the middle level, primarily via forbidden magnetic dipole transitions. The intensities of the spectral lines are therefore very good indicators of the rate of excitation into these levels. They provide an elegant method for measuring the electron temperature, once we know the collision strengths for excitation to the levels, the statistical weights of the levels, and the radiative transition probabilities.

Similar energy and transition structures occur for several other important ions in HII regions and PNe, particularly N^+ and S^{++}, although the energy levels are different. Understanding how these energy levels and their differences relate to the electron energy distribution is key to explaining apparent abundance discrepancies, both between individual collisionally excited ions and compared to recombination line results.

17.6.2 Recombination Lines

Radiative recombination is the capture of a free electron by an ion and the radiation of the excess energy as photons. The electron is typically captured into an excited state, and the excess excitation energy is radiated away as several photons (see, for example, Section 5.2 in Dopita and Sutherland, 2003), giving rise to the recombination line spectra. Recombination lines occur throughout the spectra of HII regions and PNe. The most prominent are those of hydrogen (Lyman, Balmer, and Paschen series) and helium due to the high abundances of these species. Recombination lines from carbon, nitrogen, oxygen, and other species are also observed, though they are far fainter than the hydrogen and helium lines due to the relative scarcity of the species (typically less than 10^{-3} of the hydrogen abundance). However, they are all generated in the same way. Recombination can occur for ions at any level of ionization, with the resulting atom or ion at one level lower ionization than the precursor, e.g.,

$$O^{++} + e^- \rightarrow O^+ + \text{photon}$$

Observing planetary nebulae, Wyse (1942) found that permitted lines from O^+ originate from electron capture by O^{++} ions, just as the Balmer lines originate from electron capture by H^+ ions. The relative intensities of the OII and HI lines allow us to measure the O^{++}/H^+ abundance ratio. Similar measurements are possible for O^+, N^+, and C^+, among others. So, the measurement of permitted lines can give direct measurements of ionic and atomic abundances. They are effectively independent of the thermal and density structure of the nebula. However, the permitted lines being observed are very much weaker than the hydrogen Balmer lines and are also much weaker than the CELs from the same atomic and ionic species. Therefore identifying recombination lines that are not blended with stronger lines or are contaminated by other excitation mechanisms, such as Bowen Fluorescence, is critical to obtaining reliable abundance measurements.

In terms of electron kinetic energies, the recombination rate depends on the inverse square of the electron velocity. This means that ions are far more likely to recombine with slow electrons than faster ones, as one might expect intuitively. Recombination is therefore more likely to occur at lower rather than higher electron temperatures because of the increased low-energy population or with electron energy kappa distributions where the low-energy population is enhanced. These facts are key to the way in which a kappa distribution can explain the abundance discrepancy.

Directly related to recombination line measurements are the Balmer and Paschen jumps. Fig. 17.6 shows the change in continuum levels below and above the Balmer jump wavelength. This depends on the electron temperature, $\propto T^{-0.645}$ (Dopita and Sutherland, 2003), and can be used to measure the recombination temperature where there is a good signal to noise.

Photoionization models are also used to estimate abundances. Photoionization modeling allows us to compare predicted and observed recombination line fluxes and thus estimate abundances.

17.6.3 Abundance Discrepancy Problem

Since the 1940s, systematic discrepancies have plagued abundance measurements derived from observations of emission lines and emission continua in HII regions and PNe. Temperatures determined from hydrogen and helium bound-free continuum spectra (i.e., the Balmer jump) and recombination line flux measurements are consistently lower than those obtained from CELs. As a

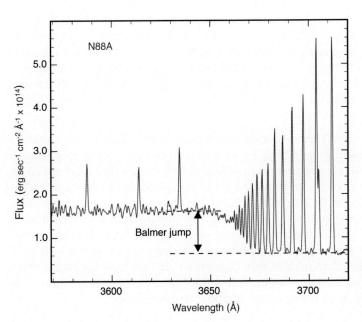

FIGURE 17.6
Balmer jump in the spectrum of the HII region N88A in the Small Magellanic Cloud.

consequence, chemical abundances determined from the optical recombination lines (ORLs) are systematically higher than those determined from CELs. In addition, abundances determined from CELs for different ions differ from one another. These discrepancies are often referred to as the "abundance discrepancy problem" and are sometimes even parameterized as an "abundance discrepancy factor." The problem was first observed over 70 years ago (e.g., see Wyse, 1942; Peimbert, 1967; Liu et al., 2000; Stasińska, 2004; García-Rojas and Esteban, 2007). Particularly in planetary nebulae, the abundance measurements may differ by factors of two or more and in some cases by as much as a factor of 20 (Liu, 2002).

A number of attempts have been made to explain these differences (see Section 17.10). The earliest attempt appears to be by Peimbert (1967), who proposed small temperature inhomogeneities through the emitting regions as the cause. Later, Liu et al. (2000) suggested the presence of a two-phase "bi-abundance" structure, where the emitting regions contain cool, metal-rich, hydrogen-poor inclusions. (Note that the terms "metal" and "metallicity" in astronomy refer to all elements heavier than hydrogen and helium and not just what are normally thought of as metallic elements.) However, neither explanation appears to be fully satisfactory: the temperature fluctuation model often requires large fluctuations to explain the observed discrepancies, without suggesting how these fluctuations could arise. The bi-abundance model requires proposing inhomogeneities where, in some cases, none are observed or where the physical processes militate against the stability of such inhomogeneities. The deficiencies in these explanations were discussed in detail by Stasińska (2004). Further, in neither of these mechanisms is the discrepancy between different CEL species explained. Binette et al. (2012) have suggested that shock waves may contribute to the apparent discrepancies, but they state that the mechanism needs to be explored further before it can be considered to be an explanation. A common feature of all of these approaches is that they assume that the electrons involved in collisional excitation and recombination processes are in thermal equilibrium.

Nicholls et al. (2012) showed that an electron energy kappa distribution is capable of explaining the ORL/CEL discrepancy and the differences between electron temperatures obtained using different CEL species. The mechanism has been shown, for example, to provide a possible explanation in the case of [OIII] and [SIII] CEL lines (Binette et al., 2012). It is interesting to note that extreme departures from an equilibrium electron energy distribution are not required to accomplish this, and if there is pumping of electron energies by mechanisms clearly likely to occur in gaseous nebulae, such distributions may not be difficult to achieve. In the remainder of this chapter, we discuss how the kappa distribution in electron energies can provide a third, and in many ways more satisfactory, explanation of the apparent abundance discrepancy.

17.7 Atomic Energy Levels and Kappa Distribution

Fig. 17.7 shows a plot versus energy of the population distribution ratios for different kappa distributions relative to the Boltzmann distribution for electron temperature $T \sim 10^4$ K. The figure also shows the excitation energies of different ions.

Two features should be noted. The first is that at low energies (<0.2 eV), the kappa distribution population is enhanced compared to the Boltzmann

FIGURE 17.7
Plot of the relative populations (logarithm) versus energy for various kappa indices and a fixed temperature of $T = 10^4$ K. The excitation energies for various ionic species are also shown.

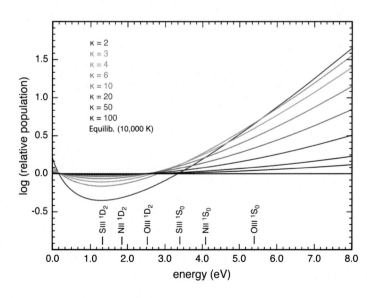

distribution, i.e., there are more low-energy electrons in a kappa distribution than in a Boltzmann distribution with the same temperature T (Fig. 10.1; see also Livadiotis, 2014a). This means that any radiative process that is sensitive to the low-energy electrons will be enhanced. If one assumes that one is looking at fluxes from an equilibrium distribution, the only way to explain the increased low energy population is to assume that the plasma is at a lower temperature. This immediately implies a higher abundance, following the reasoning presented earlier, and an explanation of the apparent higher abundances measured by recombination line methods. We will return to this shortly.

The second feature is that the populations of the $^1D_2$ level and the $^1S_0$ levels of O^{++} (and similarly for N^+ and S^{++}) are enhanced or depleted by a kappa distribution compared to a Boltzmann distribution. What is more, the relative enhancements and depletions differ for different ions. Inherent in this is the explanation of different electron temperatures for different ions. Considering the O^{++} energy levels, at $T \sim 10^4$ K, the $^1D_2$ is slightly depleted for a kappa distribution compared to a Boltzmann distribution, while the $^1S_0$ is substantially enhanced. In order to understand the effect of this latter difference, it is necessary to understand how the spectral fluxes originating from these levels are used to calculate the electron temperature T by what is known as the "direct method," which uses ratios of the spectral fluxes from the two levels.

17.7.1 Calculating Electron Temperatures From Collisionally Excited Lines

This method was developed by Menzel et al. (1941), very early in the history of nebular physics studies, and it is still used today as a standard method. It depends on the excitation through collisions from ^3P ground states to the $^1D_2$ and $^1S_0$, as shown in Fig. 17.5. Both levels can be collisionally populated by electrons at temperatures $\sim 10^4$ K, with the upper level population less than the lower, due to the relative electron energy populations at the two excitation energies. The ratio of the population rates depends on the electron temperature, and thus the flux of the emission lines from these levels gives a direct measure of the electron temperature (Nicholls et al., 2012).

Allowing for the branching ratio for the transitions from the upper state (direct to the ^3P ground states, via the $^1D_2$ level), Osterbrock and Ferland (2006) give a simple formula, for equilibrium electron energies, expressing the electron temperature T in terms of the fluxes of the 4363Å, 4959Å, and 5007Å CELs of [OIII]:

$$\frac{j_{5007} + j_{4959}}{j_{4363}} = 7.90 \exp\left(\frac{32900}{T}\right). \tag{17.3}$$

The formula is shown in graphical form as the black curve in Fig. 17.8. It is reliable for electron densities $<10^5$ cm^{-3}, above which the effects of collisional de-excitation reduce the population of the $^1D_2$ level, and for electron

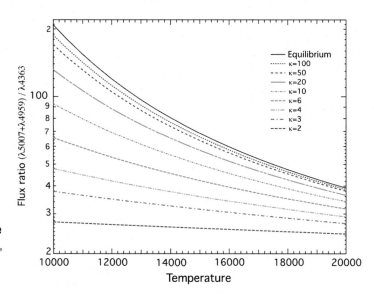

FIGURE 17.8

[OIII] line flux ratio versus temperature for the range of kappa indices from $\kappa = 2$ to $\kappa = 100$, compared to the thermal equilibrium curve (*black line*). Source: Nicholls et al. (2012).

temperatures up to $\sim 2 \cdot 10^4$ K. It is derived from a more complex equation for the population rate for collisional excitation from level 1 to level 2 (Nicholls et al., 2012):

$$R_{12}(\kappa \to \infty) = n_e n_i \frac{h^2}{4\pi^{\frac{3}{2}} m_e g_1} (k_B T)^{-\frac{3}{2}} \int_{E_{12}}^{\infty} \Omega_{12}(\varepsilon) \exp\left(-\frac{\varepsilon}{k_B T}\right) d\varepsilon, \qquad (17.4)$$

where n_e and n_i are the electron and ion densities, respectively, h is the Planck Constant, m_e is the electron mass, g_1 is the statistical weight of the lower energy state, T is the electron temperature, $\Omega_{12}(\varepsilon)$ is the collision strength, and E_{12} is the energy gap between levels 1 and 2.

The term in the integral is the collision strength averaged over the electron energy distribution at thermal equilibrium, usually expressed as U_{12} and available for a range of temperatures in published computations of O^{++} collision strengths. If one has a value for this parameter for transitions from level 1 to 2 and from level 2 to 3, the relative population rates, and therefore the line fluxes, can be computed, resulting in Eq. (17.3). There are some caveats. The values of Ω for the collisional excitation of O^{++} are critical to the accuracy of this method. They were first calculated by Seaton (1953) but have been very substantially improved since then, with the increase in computing power. They have been computed on five occasions since 1993, but the complexity of the quantum mechanical calculations is such that the studies do not present consistent results. The most recent computations are by Palay et al. (2012) and Storey et al. (2014). This is still a matter of some concern. The results of computed electron temperatures using different collision strengths can differ by several hundred degrees or $\sim 5\%$ (see Section 4.5.2 in Nicholls et al., 2013).

It is relatively simple to use Eq. (17.3) graphically to measure the electron temperature in a nebula when one has observed fluxes for the three [OIII] lines (see, for example, Fig. 5.1 in Osterbrock and Ferland, 2006). Similar equations and curves are available for other ions.

The situation is more complex with the kappa distribution, but an equation similar to Eq. (17.4) may be derived (Nicholls et al., 2012):

$$R_{12}(\kappa) = n_e n_i \frac{h^2}{4\pi^{\frac{3}{2}} m_e g_1}(k_B T)^{-\frac{3}{2}} \frac{\Gamma(\kappa+1)}{\left(\kappa-\frac{3}{2}\right)^{\frac{3}{2}}\Gamma\left(\kappa-\frac{1}{2}\right)} \cdot \int_{E_{12}}^{\infty}\left(1+\frac{1}{\kappa-\frac{3}{2}}\cdot\frac{\varepsilon}{k_B T}\right)^{-\kappa-1}\cdot\Omega_{12}(\varepsilon)dE.$$

(17.5)

In this case, the term under the integral is the collision strength averaged over the electron energy kappa distribution (out of thermal equilibrium). This needs to be computed numerically for each excitation from the tabulated values of Ω as a function of energy, where these are available.

Fig. 17.7 gives a rough idea of the effect of kappa distributions on the emitted spectra, but it is necessary to use Eqs. (17.4) and (17.6) to obtain a quantitative result. As a first approximation, we can assume that Ω_{12} is constant. Dividing Eq. (17.4) by Eq. (17.6), following Nicholls et al. (2012), we get the ratio of the excitation rates between levels 1 and 2:

$$\frac{R_{12}(\kappa)}{R_{12}(\kappa\to\infty)} = \frac{\Gamma(\kappa)}{\left(\kappa-\frac{3}{2}\right)^{\frac{1}{2}}\Gamma\left(\kappa-\frac{1}{2}\right)}\exp\left(\frac{E_{12}}{k_B T}\right)\cdot\left(1+\frac{1}{\kappa-\frac{3}{2}}\cdot\frac{E_{12}}{k_B T}\right)^{-\kappa}. \qquad (17.6)$$

In practice, numerical integration of the ratio of Eqs. (17.4) and (17.6) with nonaveraged values for Ω_{12} is necessary to get precise results. Fig. 17.8 shows the graph from Osterbrock and Ferland (2006; see, for example, Fig. 5.1 therein), recomputed using the values of Ω from Lennon and Burke (1994), for a range of values of kappa.

This figure itself has radical implications for measurements using the ratio method (Eq. (17.3)) to determine electron temperatures from [OIII] 4363 Å. If we are observing a region where electron energy kappa distributions are present, for values of $\kappa \lesssim 10$, the effect is to exaggerate the apparent electron temperature, as enhanced flux from [OIII] 4363 Å, conventionally interpreted, means a higher temperature. For smaller values of κ, however, the process breaks down altogether. As [OIII] 4363 Å is greatly enhanced, the method becomes largely insensitive to variations in T, and relatively modest true temperatures could be read as very high apparent temperatures. It is essential to discover if there is evidence for kappa-distributed energies in order to detect whether the ratio method gives reliable values for T.

However, even without detailed numerical computation, it is clear from Fig. 17.7 that some levels, at lower energies, have lower-kappa distributed populations compared to equilibrium, and others at higher energies have enhanced populations. As the population ratios of the lower and higher energy levels are used to estimate electron temperatures, assuming equilibrium, it is clear that different ions will generate different "apparent" temperatures when electron energy kappa distributions are present. The relative effects of the kappa distribution on the population rates for the upper state, compared to the lower state, will differ for different ions. This means that apparent electron temperatures calculated from spectra from these ions will differ under the influence of kappa distributions.

Another point is worth noting, in passing. Conventionally, observers wishing to measure electron temperatures using the [OIII] line ratio method (Eq. (17.3)) make the tacit assumption that emission lines arise from the same regions of the nebula. Fig. 17.3 shows that this is not strictly true. The 4363 Å line, from the higher energy $^1S_0$ level, is preferentially excited in higher excitation zones than the 5007 Å and 4959 Å lines. Similarly, it is weaker in the lower excitation regions. So, the observations inevitably average out the ratio over the emission zones. Fig. 17.3 also shows how the electron temperature varies through a uniform symmetric nebula. This temperature curve is calculated by photoionization modeling, but the observed spectral ratio can only calculate an average value. Given that HII regions are seldom regular or symmetric, compensating for this process is difficult, if possible at all, resulting in additional uncertainties in the values determined for T, even in the absence of kappa-distributed electron energies.

17.7.2 Explaining the Optical Recombination Line/ Collisonally Excited Line Abundance Discrepancy

Bringing both of these ideas together, it is apparent that the presence of kappa-distributed electrons makes the thermal equilibrium recombination estimates of abundance too high and the equilibrium direct method estimates of abundances calculated using the [OIII] CELs too low. On the other hand, calculating abundances (or temperatures) assuming a kappa distribution brings the values closer together and can potentially remove the discrepancy. Assuming that we have accurate values for the collision strengths, the only parameter necessary to adjust to achieve this is the κ-index.

17.7.3 Discrepancies in Collisionally Excited Line Temperatures

As noted earlier, the "abundance discrepancy" also applies to values of abundance and electron temperature calculated using CEL line ratios for different ionic species. While the observational evidence is limited due to noise in low signal to noise observations (a problem in all species except O^{++}), there appears to be consistent differences in the electron temperatures (and thus, abundances) between different elements. A kappa distribution can also be expected to contribute to such differences, as shown in Fig. 17.7. The best illustration is with S^{++}. The excitation of the $^1D_2$ level is somewhat diminished in kappa distributions (except for very low values of kappa) compared to the Boltzmann

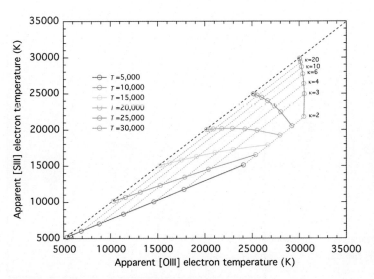

FIGURE 17.9
Locus of "apparent" electron temperatures measured using [SIII] and [OIII] flux ratios, as functions of the κ-index and the true temperature T. (*The dashed line shows the Boltzmannian case.*) With electron temperatures calculated assuming thermal equilibrium conditions from low-noise spectra for the two species, it should be possible to estimate both the κ-index and the true temperature. Source: Nicholls et al. (2013).

distribution, and the excitation to the $^1S_0$ level is slightly enhanced at $T = 10^4$ K. This means the line ratio will be only slightly higher, in contrast to the effect on O^{++}, where the ratio can be considerably enhanced. Nicholls et al. (2013) analyzed the effect this would have for different temperatures and κ indices, resulting in Fig. 17.9, which plots the apparent (equilibrium assumption) electron temperatures for [SIII] and [OIII] lines.

With sufficiently good observations and sufficiently well-known collision strengths, this might provide a diagnostic for the κ-index. However, it should be noted that the [SIII] and [OIII] emissions do not originate from exactly the same regions of a nebula, so we may not be making a useful comparison. An initial attempt to compare such observations was made by Binette et al. (2012) with mixed results, due in part to low signal to noise for the [SIII] observations. A better choice for comparison with O^{++} electron temperatures would be Ne^{++} or Ar^{+++}, which have ionization potentials closer to O^{++} and are therefore ionized in similar parts of the nebula (see Table 17.1). The abundances of Ne and Ar are lower than S, however, so the auroral lines are fainter.

17.8 Diagnostics for the Kappa Index

One of the questions asked in Section 17.3 is as follows: "Are there spectral parameters that can be measured to demonstrate the presence of kappa-distributed electron energies?" This is perhaps the most important question concerning the possibility of kappa distributions in astrophysical nebulae. Fig. 17.9 is an initial attempt to identify such observables, but the work is only beginning. If such features can be found, then the knowledge that such measurements gives us allows us to incorporate the value of kappa in photoionization models to simulate nebular physics.

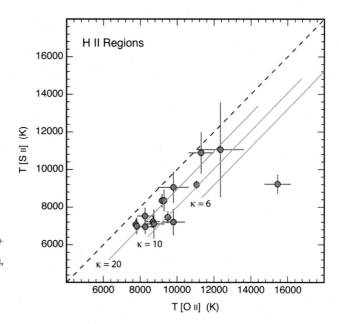

FIGURE 17.10

Plot of apparent electron temperatures for O^+ versus S^+ ions in HII regions, from published [OII] and [SII] spectra, with computed loci for thermal equilibrium ($\kappa \rightarrow \infty$), $\kappa = 20$, 10, and 6 electron energies. Source: Nicholls et al. (2012).

Another approach, using published data, is shown in Fig. 17.10, a plot of the electron temperatures derived from the [SII] versus [OII] collisionally excited forbidden lines. These ionic species have similar ionization energies (see Table 17.1) and therefore are excited in the same parts of the nebula, so they should experience similar collisional excitation. If the electron temperatures were the same for both ions, the points should lie along the dashed line. This would happen if the electrons giving rise to the collisional excitation of the lines were in thermal equilibrium. If instead a $\kappa = 20$ electron energy distribution were present, we would expect the points to lie along the blue line and likewise for $\kappa = 10$ (green) and $\kappa = 6$ (orange). As the figure suggests, the results indicate that the [SII] and [OII] "temperatures" differ, as we would expect in a nebula exhibiting characteristics of an electron energy kappa distribution. Nicholls et al. (2012, Figs. 9 and 10 of their paper) also examined a variety of electron temperature plots from data for both HII regions and PNe. Their analyses suggested that values of κ between 6 and 50 could explain the majority of the temperature discrepancies observed. The comparison between electron temperatures from different ionic species and with recombination temperatures is a particularly interesting area for investigation, given good (low noise) spectral flux measurements.

17.9 Modeling of Photoionized Nebulae

17.9.1 Modeling Programs

In this chapter we mention this topic only briefly, as it is a complex field. Much effort has been devoted to modeling HII regions. Early models considered only simplified toy model nebulae, as the sheer detail of the radiative transfer interactions makes this a very complex process. However, computational models have been developed that now take into account the complex physical and radiative interactions that occur in nebulae. Perhaps the best known of these is Cloudy (Ferland et al., 2013), a general astrophysical plasma modeling system. A second modeling application is MOCASSIN, a photoionization and dust radiative transfer code that employs a Monte Carlo approach to the transfer of radiation through the media of arbitrary geometry and density distribution, originally developed for the modeling of photoionized regions like HII regions and PNe (Ercolano et al., 2008). The application developed here is Mappings, now at version V. Mappings is designed to model ionized nebulae and includes detailed dust radiative transfer and shock wave physics. Most relevant to this chapter, it is unique in allowing the user to include the effects of kappa-distributed electron energies. It is designed to be very computationally efficient.

17.9.2 Complexities in HII Region Structures

Any search for images of HII regions will demonstrate that they are seldom symmetric or well organized. A few, centered around single or a few stars, do resemble the model "Strömgren Sphere" (Osterbrock, 1989), but this is not common. Some, like the Rosette Nebula in Fig. 17.1, are roughly symmetric, but others, like M11 and the Orion Nebula, are closer to chaotic. Two types of simple models can be used to approximate real nebulae: the "single slab" and the spherical model. In the latter, radiation from the far side of the sphere also affects the physics on the near side, whereas in the single slab model, we only need to look at radiation in one direction. As a reasonable approximation to real nebulae (such as the Orion Nebula), these two types of models can be combined.

The great advantage of models like Cloudy, MOCASSIN, and Mappings that emulate the detailed radiation and ionization structure of nebulae is that they can take into account the vast detail of atomic energy levels, transition probabilities, radiative transfer, ionization, recombination, excitation, and radiative cascade processes that defy simple analysis. A next step will be to take into account the fractal nature of ionized nebulae, and this work is planned. It will use the "Fyris" 3-D hydrodynamic code (Sutherland, 2010) and will extend the work by Bland-Hawthorn et al. (2011), which looked at the effects of supernova explosions in small fractal gas clouds.

17.10 Other Applications of Kappa Distributions in Astrophysical Nebulae

Following the initial work by Nicholls et al. (2012), several other authors have investigated the implications of electron energy kappa distributions in astrophysical nebulae. Noteworthy is the work by Storey and Sochi (2015), who have recalculated the emission and recombination coefficients of hydrogen in the presence of kappa-distributed electron energies. The emission and recombination coefficients for hydrogen are critical inputs to photoionization models of astrophysical nebulae, and therefore, it is also important to know how they vary when the electron energies are kappa distributed.

The free–free emission continuum (see, for example, Section 6.1 and Fig. 6.1 in Dopita and Sutherland, 2003) forms an essential part of the "nebular continuum" in the optical spectrum of HII regions and planetary nebular, which underlies the emission lines in observed nebular spectra and in the ionizing continuum up to X-ray energies in hot ($\gg 10^6$ K) plasmas such as in supernovae and in galaxy and galaxy-cluster atmospheres.

Modeling this accurately is important when trying to subtract the embedded stellar continuum, for example, and compensating for dust reddening. The spectral distribution of the free–free emission of nebulae depends on the plasma electron temperature and is computed for equilibrium conditions using temperature-averaged free–free Gaunt factors. These are the corrections to the simple classical model of electron collisions to account for quantum mechanical considerations. De Avillez and Breidtschwerdt (2015) recomputed the Gaunt factors for a range of values of κ. These allow the effects of the electron energy kappa distributions to be calculated.

17.11 Alternative Explanations of Abundance Discrepancy

As discussed by Stasińska (2004), there have been a number of attempts to explain the abundance discrepancy. Of these, the most important are the temperature fluctuation ("t^2") process, the cold high-metallicity inclusion process, and atomic data deficiencies.

The t^2 method was first proposed by Peimbert (1967) and has since been used extensively to explain abundance discrepancies. It proposes that temperature fluctuations within the nebular plasma explain the discrepant results. It suffers from the problem that in order to explain some discrepancies, unreasonably large fluctuations are required, and it is not clear how these might arise.

The cold metal-rich, hydrogen-poor inclusions idea was proposed by Liu et al. (2000) and has been used particularly in explaining the results obtained in planetary nebula observations. While it seems quite probable that such inclusions exist, particularly in planetary nebulae, the theory probably does not explain all observed discrepancies, as there is evidence that the discrepancies occur where no such metal-rich inclusions are observed (Stasińska, 2004).

Finally, we come to the question of unreliable atomic data. While our knowledge of atomic energy levels, transition probabilities, collision strengths, and recombination coefficients has improved markedly since the early days, the atomic data is not of a uniform quality or even consistent. An example is the collision strengths calculated for the O^{++} ion, mentioned earlier. Another notable problem is the paucity of computed recombination coefficients. As recombination events preferentially generate an atom or ion in an excited state, the recombination lines are generated in a radiative cascade from the initial level, through numerous permitted transitions. The sheer complexity of the cascade process makes computing all of the parameters necessary to describe the process extremely difficult. Different computations for the atoms and ions whose recombination lines are observed in nebulae give different results. We do not yet have a completely reliable picture of the process. How much this contributes to the abundance discrepancy remains to be seen, although if we select recombination lines whose processes are the best understood, we can minimize the problem.

17.12 Concluding Remarks

If we adopt kappa distributions for the nebulae electron energies, we can resolve a long-standing problem where measurements of electron temperatures and chemical abundances using different methods yield discrepant results. While there are alternative explanations of the abundance discrepancies, there are concerns that the idea runs counter to what they see as the findings of such authorities. It is likely that the true explanation of abundance measurement differences will have a variety of causes. However, kappa electron energy distributions are undoubtedly present in HII regions and PNe and contribute to the abundance and temperature discrepancies. Electron temperature errors consequent from ignoring this possibility may amount to 1000 or 2000 K. This is a new field of research, and we need to undertake observations to look for unambiguous signatures of the kappa distribution to measure kappa and explore its physical effects.

17.13 Science Questions for Future Research

Future analyses and observations need to address the following questions:

1. What are the key diagnostics of kappa distributions?
2. What kind of target is most likely to display these diagnostics?
3. Over what physical scales do they occur?

Appendix A: Abbreviations

Abundance discrepancy factor (ADF)
Active Magnetospheric Particle Tracer Explorer (AMPTE)
Advanced Composition Explorer (ACE)
Alfvén-Cyclotron (AC)
Anomalous cosmic ray (ACR)
Association of Universities for Research in Astronomy (AURA)
Astronomical unit (AU)
Atmospheric Imaging Assembly (AIA)
Auroral Electroject Index (AE)
Block-Adaptive-Tree-Solarwind-Roe-Upwind-Scheme (BATS-R-US)
Boltzmann—Gibbs (BG)
Cassini Plasma Spectrometer (CAPS)
Centimetre—Gram—Second (CGS)
Central limit theorem (CLT)
Charge Composition Explorer (CCE)
Collisionally excited line (CEL)
Coronal mass ejection (CME)
Corotating interaction region (CIR)
Cosmic Ray Subsystem (CRS)
Differential emission measure (DEM)
Diffusive shock acceleration (DSA)
Double layer (DL)
Electrical Power Subsystem (EPS)
Electromagnetic (EM)
Electron beam (eb)
Electron core (ec)
Electron firehose (EF)
Electron spectrometer (ELS)
Electron, Proton, and Alpha Monitor (EPAM)
Electrostatic (ES)
Electrostatic solitary wave (ESW)
Energetic ion mass spectrometer (EIMS)
Energetic neutral atom (ENA)
Energetic particle spectrometer (EPS)
Energetic particles detector (EPD)
European Space Agency (ESA)
Extreme ultraviolet (EUV)
Extreme-ultraviolet imaging spectrograph (EIS)
Fast Auroral SnapshoT explorer (FAST)
Galactic cosmic ray (GCR)

Heliopause (HP)
Heliosphere Instrument for Spectrum, Composition, and Anisotropy at Low Energies (HISCALE)
Heliospheric termination shock (HTS)
High-Energy (HE)
Index catalogue (IC)
Inner heliosheath (IH)
Inner heliosheath (IHS)
Interface Region Imaging Spacecraft (IRIS)
International Sun-Earth Explorer (ISEE)
Interplanetary (IP)
Interplanetary coronal mass ejection (ICME)
Interplanetary magnetic field (IMF)
Interplanetary Monitoring Platform (IMP)
Interstellar Boundary Explorer (IBEX)
Ion and Neutral Camera (INCA)
Ion Release Module (IRM)
Ion/electron acoustic (IA)
Korteweg-de Vries (KdV)
Local acceleration event (LAE)
Local acceleration region (LAR)
Local interstellar medium (LISM)
Los Alamos National Labs (LANL)
Low-Energy (LE)
Low energy charged particle detector (LECP)
Low Energy Magnetospheric Measurement System (LEMMS)
Low Energy Proton and Electron Differential Energy Analyzer (LEPEDEA)
Magnetohydrodynamics (MHD)
Magnetometer (MAG)
Magnetosheath (MSh)
Magnetosphere (MSp)
Magnetospheric imaging instrument (MIMI)
Massachusetts Institute of Technology (MIT)
Medium energy particle instrument (MEPI)
MErcury Surface, Space ENvironment, GEochemistry, and Ranging (MESSENGER)
modified Korteweg-de Vries (mKdV)
Murchison Widefield Array (MWA)
National Aeronautics and Space Administration (NASA)
National Oceanic and Atmospheric Administration (NOAA)
New general catalogue (NGC)
Nonlinear guiding center (NLGC)
Operating Missions as a Node on the Internet (OMNI)
Optical recombination line (ORL)
Orbiting Geophysical Observatory (OGO)

Particle-in-cell (PIC)
PickUp Ion (PUI)
Pioneer venus orbiter (PVO)
Planetary nebula (PNe)
Plasma sheet boundary layer (PSBL)
Polar cap boundary layer (PCBL)
Prime acceleration region (PAR)
Proton beam (pb)
Proton core (pc)
Proton Firehose (PF)
Research with Adaptive Particle Imaging Detectors (RAPID)
Reuven-Ramaty High-Energy Solar Spectroscopic Imager (RHESSI)
Saturn-Thermosphere-Ionosphere Model (STIM)
Solar and Heliospheric Observatory (SOHO)
Solar dynamics observatory (SDO)
Solar energetic particle (SEP)
Solar flux unit (SFU)
Solar TErrestrial RElations Observatory (STEREO)
Solar Ultraviolet Measurements of Emitted Radiation (SUMER)
Solar Wind Around Pluto (SWAP)
Solar Wind Electron, Proton, and Alpha Monitor (SWEPAM)
Solar Wind Ion Composition instrument (SWICS)
Solitary wave (SW)
Space Telescope Science Institute (STScI)
Spacecraft (S/C)
Statistical mechanics (SM)
Stochastic growth theory (SGT)
Suprathermal (ST)
Synchronous Orbit Particle Analyzer (SOPA)
Time History of Events and Macroscale Interactions during Substorms
 (THEMIS)
Ultra-Low Energy Isotope Spectrometer (ULEIS)
Ultraviolet (UV)
Ultraviolet Coronograph Spectrometer (UVCS)
Unified RAdio and Plasma Wave Experiment (URAP)
Universal time (UT)
Velocity distribution function (VDF)
Very Local InterStellar Medium (VLISM)
Weak double layer (WDL)
Whistler cyclotron (WC)
Zakharov–Kuznetsov (ZK)

Appendix B: Main Symbols

Several symbols may have been used in multiple symbolisms. Here, we list the primary usage of the most important symbols.

Description	Symbol
Adiabatic index	γ
Alfvén speed	V_A
Angular momentum	$\vec{L}$
Argument, entropic	φ_q
Atomic number	Z
Azimuthal spherical angle	φ
Bessel function	J
Beta function	B
Beta plasma	β
Boltzmann constant	k_B
Chi-square	χ^2
Collision frequency, electron−ion	ν_{ei}
Collision strength argument	U_{ji}
Compression ratio	R
Correlation	R, ρ, r
Correlation length	ℓ_C
Cross section, dimensionless	Ω_{ji}
Curie constant	C
Current density	J
Cyclotron frequency	Ω
Debye length	λ_D
Debye particle numbers	N_D
Degrees of freedom, kinetic	d_K
Degrees of freedom, potential	d_Φ
Degrees of freedom, total	d, f
Density of indices	D
Density of states, energy	g_E
Density of states, speed	g_V
Density, number	n
Dielectric tensor	$\overleftrightarrow{\varepsilon}$
Dispersion tensor	$\overleftrightarrow{D}$
Distribution function	f

(Continued)

Description	Symbol
Einstein coefficient	A_{ij}
Electric charge	E, q, Q
Electric field	$\vec{E}$
Electrical conductivity	σ
Energy	ε, E
Entropic index	q
Entropy	S
Flow bulk velocity	$\vec{u}_b$
Force	$\vec{F}$
Frequency, angular	ω
Gamma function	Γ
Gauss hypergeometric function	${}_2F_1$
Gravitational acceleration	g
Gravitational constant	G
Growth/damping rate	Γ
Gyroradius	r_g
Hamiltonian	H
Intensity	I
Internal energy	U
Interparticle distance	b
Inverse temperature	β
Ion acoustic, sound speed	V_S, c_S, C_S
Jeans radius	r_J
Kappa index	κ
Kappa index, invariant	κ_0
Kinetic energy	ε_K, E_K
Kurtosis	K
Landau length	x_L
Langevin function	L
Large-scale constant	h_*
Large-scale constant, reduced	$\hbar_*$
Mach number	M
Magnetic field	$\vec{B}$
Magnetic field angle to shock normal	ψ
Magnetic moments	$\vec{\mu}$
Magnetization	$\vec{M}$
Magnetosonic speed, fast	V_{ms}
Mass	m
Mass number	A
Mean free path	L_m
Metastability measure	M, M_q
Microstates number, N particles,	Ω_N
Momentum	$\vec{p}$

(Continued)

Description	Symbol
Nonlinear monotonic superposition	φ
Norms	L_p
Particle intensity/flux, differential	J, j
Particle numbers	N
Particle numbers, correlated	N_C
Particle velocity	$\vec{u}$
Partition function	Z
Permeability, vacuum	μ_0
Permittivity, vacuum	ε_0
Pitch angle	α
Planck constant, reduced	$\hbar$
Planck quantization constant	h
Plasma frequency	ω_{pl}
Polar spherical angle	ϑ
Polytropic index	a, ν
Position	$\vec{r}$
Positional dimensions	d_r
Potential energy	Φ
Power-law potential exponent	b
Pressure, thermal	p, P
Probability, escort	P
Probability, ordinary	p
PUI fluctuation number	Θ_d
Radius	r, R
Random walk distance	ℓ
Scale, length	σ_r
Scale, speed	σ_u
Spatial diffusion tensor	K
Spectral index	γ
Speed	u, w
Speed, thermal	θ
Statistical moment	μ
Susceptibility, dielectric/paramagnetic	χ
Temperature	T
Temperature anisotropy	Γ_s
Unit sphere surface, f-D	B_f
Variance	σ^2
Volume	V
Wavelength	λ
Wavevector	$\vec{k}$

References

Abe, S., 1999. Correlation induced by Tsallis' nonextensivity. Physica A 269, 403−409.

Abe, S., 2000. Axioms and uniqueness theorem for Tsallis entropy. Physics Letters A 271, 74−79.

Abe, S., 2001. Heat and entropy in nonextensive thermodynamics: transmutation from Tsallis theory to Rényi-entropy-based theory. Physica A 300, 417−423.

Abe, S., 2002. Stability of Tsallis entropy and instabilities of Rényi and normalized Tsallis entropies: a basis for q-exponential distributions. Physical Review E 66, 046134.

Abe, S., Suzuki, N., 2003. Itineration of the Internet over nonequilibrium stationary states in Tsallis statistics. Physical Review E 67, 016106.

Abe, S., Thurner, S., 2005. Complex networks arising from fluctuating random graphs. Physical Review E 72, 036102.

Abe, S., Okamoto, Y. (Eds.), 2001. Nonextensive Statistical Mechanics and Its Applications. Springer, Berlin.

Abe, S., Martínez, S., Pennini, F., Plastino, A., 2001. Nonextensive thermodynamic relations. Physics Letters A 281, 126−130.

Abraham-Shrauner, B., Feldman, W.C., 1977a. Electromagnetic ion-cyclotron wave growth rates and their variation with velocity distribution shape. Journal of Plasma Physics 17, 123−131.

Abraham-Shrauner, B., Feldman, W.C., 1977b. Whistler heat flux instability in the solar wind with bi-Lorentzian velocity distribution function. Journal of Geophysical Research 82, 1889−1892.

Abul-Magd, A.Y., 2006. Superstatistics in random matrix theory. Physica A 361, 41.

Achilleos, N., Guio, P., Arridge, C.S., 2010a. A model of force balance in Saturn's magnetodisc. Monthly Notices of the Royal Astronomical Society 401, 2349−2371.

Achilleos, N., Guio, P., Arridge, C.S., Sergis, N., Wilson, R.J., Thomsen, M.F., Coates, A.J., 2010b. Influence of hot plasma pressure on the global structure of Saturn's magnetodisk. Geophysical Research Letters 37, L20201.

Adnan, M., Mahmood, S., Qamar, A., 2014. Small amplitude ion acoustic solitons in a weakly magnetized plasma with anisotropic ion pressure and kappa distributed electrons. Advances in Space Research 53, 845−852.

Aharonian, F., et al., 2006. A low level of extragalactic background light as revealed by γ-rays from blazars. Nature 440, 1018−1021.

Ajello, J.M., Shemansky, D.E., Pryor, W.R., Stewart, A.I., Simmons, K.E., Majeed, T., Waite, J.H., Gladstone, G.R., 2001. Spectroscopic evidence for high-latitude aurora at Jupiter from Galileo extreme ultraviolet spectrometer and Hopkins ultraviolet telescope observations. Icarus 152, 151−171.

Akhiezer, A.I., Akhiezer, I.A., Polovin, R.V., G Sitenko, A., Stepanov, K.N., 1975. Plasma Electrodynamics. Pergamon Press, New York.

Aldrovandi, S.M.V., Gruenwald, R.B., June 1985. Electron temperatures of astrophysical plasmas from the forbidden O III line ratio. Astronomy and Astrophysics 147, 331−334.

André, N., Ferriére, K.M., 2008. Stratification-driven instabilities with bi-kappa distribution functions in the Io plasma torus. Journal of Geophysical Research 113, A09202.

Andrews, D.J., Bunce, E.J., Cowley, S.W.H., Dougherty, M.K., Provan, G., Southwood, D.J., 2008. Planetary period oscillations in Saturn's magnetosphere: phase relation of equatorial magnetic field oscillations and Saturn kilometric radiation modulation. Journal of Geophysical Research 113, A09205.

Andricioaei, I., Straub, J.E., 1996. Generalized simulated annealing algorithms using Tsallis statistics: application to conformational optimization of a tetrapeptide. Physical Review E 53, R3055–R3058.

Angelopoulos, V., Kennel, C.F., Coroniti, F.V., Pellat, R., Kivelson, M.G., Walker, R.J., Russell, C.T., Baumjohann, W., Feldman, W.C., Gosling, J.T., 1994. Statistical characteristics of bursty bulk flow events. Journal of Geophysical Research 99, 21257–21271.

Annou, K., 2015. Ion-acoustic solitons in plasma: an application to Saturn's magnetosphere. Astrophysics and Space Science 357, 163, 9.

Armstrong, T.P., Paonessa, M.T., Bell II, E.V., Krimigis, S.M., 1983. Voyager observations of Saturnian ion and electron phase space densities. Journal of Geophysical Research 88, 8893–8904.

Armstrong, T.P., Krimigis, S.M., Lanzerotti, L.J., 1975. A reinterpretation of the reported energetic particle fluxes in the vicinity of Mercury. Journal of Geophysical Research 80, 4015–4017.

Arnaud, M., Raymond, J., 1992. Iron ionization and recombination rates and ionization equilibrium. The Astrophysical Journal 398, 394–406.

Arnaud, M., Rothenflug, R., 1985. An updated evaluation of recombination and ionization rates. Astronomy and Astrophysics Supplement Series 60, 425–457.

Arnaud, K.A., et al., 1985. EXOSAT observations of a strong soft X-ray excess in MKN 841. Monthly Notices of the Royal Astronomical Society 217, 105–113.

Arridge, C.A., et al., 2009. Plasma electrons in Saturn's magnetotail: structure, distribution, and energisation. Planetary and Space Science 57, 2032–2047.

Artemyev, A.V., Petrukovich, A.A., Nakamura, R., Zelenyi, L.M., 2013. Profiles of electron temperature and Bz along Earth's magnetotail. Annales Geophysicae 31, 1109–1114.

Artemyef, A.V., Vasko, I.Y., Lutsenko, V.N., Petrukovich, A.A., 2014. Formation of the high energy ion population in the Earth's magnetotail: spacecraft observations and theoretical models. Annales Geophysicae 32, 1233–1246.

Arthur, S.J., Henney, W.J., Mellema, G., de Colle, F., Vázquez-Semadeni, E., June 2011. Radiation-magnetohydrodynamic simulations of H II regions and their associated PDRs in turbulent molecular clouds. Monthly Notices of the Royal Astronomical Society 414, 1747–1768.

Aschwanden, M.J., et al., 2000. Time variability of the "quiet" sun observed with *trace*. II. Physical parameters, Temperature evolution, and energetics of extreme-ultraviolet nanoflares. The Astrophysical Journal 535, 1047–1065.

Astfalk, P., Görler, T., Jenko, F., 2015. DSHARK: a dispersion relation solver for obliquely propagating waves in bi-kappa-distributed plasmas. Journal of Geophysical Research 120, 7107–7120.

Ausloos, M., Ivanova, K., 2003. Dynamical model and nonextensive statistical mechanics of a market index on large time windows. Physical Review 68E, 046122.

Axford, W.I., Leer, E., Skadron, G., 1977. The acceleration of cosmic rays by shock waves. In: 15th International Cosmic Ray Conference, vol. 11. B'Igorska Academiia na Naukite, Sofia, pp. 132–137.

Baboolal, S., Bharuthram, R., Hellberg, M.A., 1990. Cut-off conditions and existence domains for largeamplitude ion-acoustic solitons and double layers in fluid plasmas. Journal of Plasma Physics 44, 1.

Badnell, N., O'Mullane, M.G., Summers, H.P., Altun, Z., Bautista, M.A., Colgan, J., et al., 2003. Dielectronic recombination data for dynamic finite-density plasmas. I. Goals and methodology. Astronomy and Astrophysics 406, 1151–1165.

Baiesi, M., Paczuski, M., Stella, A.L., 2006. Intensity thresholds and the statistics of temporal occurence of solar flares. Physical Review Letters 96, 051103.

Baker Jr., G.A., 1975. Essentials of Padé Approximants in Theoretical Physics. Academic Press, New York, pp. 27–38.

Baker, D., Van Allen, J., 1976. Energetic electrons in the Jovian magnetosphere. Journal of Geophysical Research 81, 617–632.

Baker, J.G., Menzel, D.H., Aller, L.H., November 1938. Physical processes in gaseous nebulae. V. Electron temperatures. The Astrophysical Journal 88, 422.

Baluku, T.K., Hellberg, M.A., Kourakis, I., Saini, N.S., 2010. Dust ion acoustic solitons in a plasma with kappa-distributed electrons. Physics of Plasmas 17, 053702.

Baluku, T.K., Hellberg, M.A., Mace, R.L., 2011. Electron acoustic waves in double-kappa plasmas: application to Saturn's magnetosphere. Journal of Geophysical Research 116, A04227.

Baranyai, A., 2000a. Numerical temperature measurement in far from equilibrium model systems. Physical Review E 61, R3306–R3309.

Baranyai, A., 2000b. Temperature of nonequilibrium steady-state systems. Physical Review E 62, 5989–5997.

Barghouthi, I., Pierrard, V., Barakat, A.R., Lemaire, J., 2001. A Monte Carlo simulation of the H+ polar wind: effect of velocity distributions with Kappa suprathermal tails. Astrophysics and Space Science 277, 427–436.

Bastian, T.S., Benz, A.O., Gary, D.E., 1998. Radio emission from solar flares. Annual Review of Astronomy and Astrophysics 36, 131–188.

Basu, B., 2009. Hydromagnetic waves and instabilities in kappa distribution plasma. Physics of Plasmas 16, 052106.

Baumjohann, W., Paschmann, G., 1989. Determination of the polytropic index in the plasma sheet. Geophysical Research Letters 16, 295–298.

Baumjohann, W., Treumann, R.A., 1997a. Basic Space Plasma Physics. Imperial College Press, London.

Baumjohann, W., Treumann, R.A., 1997b. Advanced Space Plasma Physics. Imperial College Press, London.

Bazarghan, M., Safari, H., Innes, D.E., Karami, E., Solanki, S.K., 2008. Astronomy and Astrophysics 492, L13.

Beck, C., Lewis, G.S., Swinney, H.L., 2001. Measuring nonextensitivity parameters in a turbulent Couette-Taylor flow. Physical Review E 63, 035303.

Beck, C., Miah, S., 2013. Statistics of Lagrangian quantum turbulence. Physical Review E 87, 031002(R).

Beck, C., Schlögl, F., 1993. Thermodynamics of Chaotic Systems. Cambridge University Press.

Beck, C., Cohen, E.G.D., 2003. Superstatistics. Physica A 322, 267–275.

Beck, C., Cohen, E.G.D., 2004. Superstatistical generalization of the work fluctuation theorem. Physica A 344, 393–402.

Beck, C., 2009. Generalised information and entropy measures in physics. Contemporary Physics 50, 495–510.

Beck, C., Cohen, E.G.D., Swinney, H.L., 2005. From time series to superstatistics. Physical Review E 72, 026304.

Beck, C., 2003. Lagrangian acceleration statistics in turbulent flows. Europhysics Letters 64, 151.

Beck, C., 2001. Dynamical Foundations of Nonextensive Statistical Mechanics. Physical Review Letters 87, 180601.

Beck, C., 2007. Statistics of Three-Dimensional Lagrangian Turbulence. Physical Review Letters 98, 064502.

Beck, C., 2004a. Superstatistics in hydrodynamic turbulence. Physica 193D, 195–207.

Beck, C., 2004b. Generalized statistical mechanics of cosmic rays. Physica 331A, 173–181.

Beck, C., 2000. Non-extensive statistical mechanics and particle spectra in elementary interactions. Physica A 286, 164–180.

Beck, C., 2011. Philosophical Transactions of the Royal Society of London A 369, 453–465.

Bediaga, I., Curado, E.M.F., de Miranda, J.M., 2000. A nonextensive thermodynamical equilibrium approach in $e^+e^- \rightarrow$ hadrons. Physica A 286, 156–163.

Belcher, J.W., et al., 1989. Plasma observations near Neptune: initial results from Voyager 2. Science 246, 1478–1482.

Belcher, J.W., McNutt Jr., R.L., Richardson, J.D., Selesnick, R.S., Sittler Jr., E.C., Bagenal, F., 1991. The plasma environment of Uranus. In: Bergstralh, J.T., et al. (Eds.), Uranus. Univ. of Arizona Press, Tucson, Ariz, pp. 780–829.

Bell, A.R., 1978. The acceleration of cosmic rays in shock fronts – I. Monthly Notices of the Royal Astronomical Society 182, 147–156.

Benz, A.O., 2004. In: Dupree, A.K., Benz, A.O. (Eds.), Proc. IAU Symp. No. 219, Stars as Suns: Activity, Evolution, and Planets, p. 461.

Bernstein, I.B., Engelmann, F., 1966. Quasi-linear theory of plasma waves. Physics of Fluids 9, 937–952.

Berthomier, M., Pottelette, R., Malingre, M., 1998. Solitary waves and weak double layers in a two-electron temperature auroral plasma. Journal of Geophysical Research 103, 4261–4270.

Berthomier, M., Pottelette, R., Malingre, M., Khotyaintsev, Y., 2000. Electron acoustic solitons in an electron-beam plasma system. Physics of Plasmas 7, 2987–2994.

Bian, N., Emslie, G.A., Stackhouse, D.J., Kontar, E.P., 2014. The formation of a kappa-distribution accelerated electron populations in solar flares. The Astrophysical Journal 796, 142.

Binette, L., Matadamas, R., Hägele, G.F., Nicholls, D.C., Magris, C.G., Peña-Guerrero, M.Á., Morisset, C., Rodríguez-González, A., November 2012. Discrepancies between the [O III] and [S III] temperatures in H II regions. Astronomy and Astrophysics 547, A29.

Binsack, J.H., 1966. Plasma Studies with the IMP-2 Satellite (Ph.D. thesis). MIT.

Blandford, R.D., Ostriker, J.P., 1978. Particle acceleration by astrophysical shocks. The Astrophysical Journal 221, L29.

Bland-Hawthorn, J., Sutherland, R., Karlsson, T., July 2011. The minimum mass for a Dwarf galaxy. In: Koleva, M., Prugniel, P., Vauglin, I. (Eds.), EAS Publications Series, vol. 48. EAS Publications Series. EDP Sciences, Les Ulis, France, pp. 397–404.

Boerner, P., Edwards, C., Lemen, J., Rausch, A., Schrijver, C., Shine, R., et al., 2012. Initial calibration of the atmospheric imaging assembly (AIA) on the solar dynamics observatory (SDO). Solar Physics 275, 41–66.

Bohm, D., Aller, L.H., 1947. The electron velocity distribution in gaseous nebulae and stellar envelopes. The Astrophysical Journal 105, 131.

Bondfond, B., Grodent, D., Gérard, J.-C., Radioti, A., Dols, V., Delamere, P.A., Clarke, J.T., 2009. The Io UV footprint: location, inter-spot distances and tail vertical extent. Journal of Geophysical Research 114, A07224.

Borges, E.P., Tsallis, C., Anãnõs, G.F.J., de Oliveira, P.M.C., 2002. Nonequilibrium probabilistic dynamics at the logistic map edge of chaos. Physical Review Letters 89, 254103.

Borges, E.P., 1998. On a q-generalization of circular and hyperbolic functions. Journal of Physics A 31, 5281–5288.

Borland, L., 2002. Option pricing formulas based on a non-Gaussian stock price model. Physical Review Letters 89, 098701.

Bornatici, M., Cano, R., Barbieri, O.D., Engelmann, F., 1983. Electron cyclotron emission and absorption in fusion plasmas. Nuclear Fusion 23 (9), 1153.

Borovsky, J.E., Cayton, T.E., 2011. Entropy mapping of the outer electron radiation belt between the magnetotail and geosynchronous orbit. Journal of Geophysical Research 116, A06216.

Borovsky, J.E., Thomsen, M.F., Elphic, R.C., Cayton, T.E., McComas, D.J., 1998. The transport of plasma sheet material from the distant tail to geosynchronous orbit. Journal of Geophysical Research 103, 20297–20331.

Borovsky, J.E., Gary, S.P., 2014. How important are the alpha–proton relative drift and the electron heat flux for the proton heating of the solar wind in the inner heliosphere? Journal of Geophysical Research 119, 5210–5219.

Boström, R., Gustafsson, G., Holback, B., Holmgreen, G., Koskinen, H., Kintner, P., 1988. Characteristics of solitary waves and weak double layers in the magnetospheric plasma. Physical Review Letters 61, 82.

Bouchard, J.-P., Potters, M., 2003. Theory of Financial Risk and Derivative Pricing. Cambridge University Press, Cambridge.

Bounds, S.R., Pfaff, R.F., Knowlton, S.F., Mozer, F.S., Temerin, M.A., Kletzing, C.A., 1999. Solitary potential structures associated with ion and electron beams near 1 R_E altitude. Journal of Geophysical Research 104, 28709–28717.

Bourouaine, S., Marsch, E., Neubauer, F.M., 2011. On the relative speed and temperature ratio of solar wind alpha particles and protons: collisions versus wave effects. The Astrophysical Journal Letters 728, L3.

Bowen, I.S., January 1928. The origin of the nebular lines and the structure of the planetary nebulae. The Astrophysical Journal 67, 1–15.

Bradshaw, S.J., Mason, H.E., 2003a. A self-consistent treatment of radiation in coronal loop modeling. Astronomy and Astrophysics 401, 699–709.

Bradshaw, S.J., Mason, H.E., 2003b. The radiative response of solar loop plasma subject to transient heating. Astronomy and Astrophysics 407, 1127–1138.

Bradshaw, S.J., Klimchuk, J.A., 2011. What dominates the coronal emission spectrum during the cycle of impulsive heating and cooling? The Astrophysical Journal Supplement Series 194, 26.

Bradshaw, S.J., Klimchuk, J.A., Reep, J.W., 2012. Diagnosing the time-dependence of active region core heating from the emission measure. I. Low-frequency nanoflares. The Astrophysical Journal 758, 53.

Brandt, P.C., Khurana, K.K., Mitchell, D.G., Sergis, N., Dialynas, K., Carbary, J.F., Roelof, E.C., Paranicas, C.P., Krimigis, S.M., Mauk, B.H., 2010. Saturn's periodic magnetic field perturbations caused by a rotating partial ring current. Geophysical Research Letters 37, L22103.

Bridge, H.S., et al., 1986. Plasma observations near Uranus: initial results from Voyager 2. Science 233, 89.

Broiles, T.W., Livadiotis, G., Burch, J.L., Chae, K., Clark, G., Cravens, T.E., Davidson, R., Eriksson, A., Frahm, R.A., Fuselier, S.A., Goldstein, J., Goldstein, R., Henri, P., Madanian, H., Mandt, K.E., Mokashi, P., Pollock, C., Rahmati, A., Samara, M., Schwartz, S.J., 2016a. Characterizing cometary electrons with kappa distributions. Journal of Geophysical Research 121, 7407–7422.

Broiles, T.W., Burch, J.L., Chae, K., Clark, G., Cravens, T.E., Eriksson, A., Fuselier, S.A., Frahm, R.A., Gasc, S., Goldstein, R., Henri, P., Koenders, C., Livadiotis, G., Mandt, K.E., Mokashi, P., Nemeth, Z., Rubin, M., Samara, M., 2016b. Statistical analysis of suprathermal electron drivers at 67P/Churyumov-Gerasimenko. Monthly Notices of the Royal Astronomical Society 462, S312–S322.

Brown, J.C., 1971. The deduction of energy spectra of non-thermal electrons in flares from the observed dynamic spectra of hard X-ray bursts. Solar Physics 18, 489–502.

Bryans, P., 2005. On the Spectral Emission of Non-Maxwellian Plasmas (Ph.D. thesis). University of Strathclyde, Glasgow, UK.

Bryant, D.A., 1996. Debye length in a kappa distribution plasma. Journal of Plasma Physics 56, 87–93.

Burlaga, L.F., Viñas, A.F., 2005. Triangle for the entropic index q of non-extensive statistical mechanics observed by Voyager 1 in the distant heliosphere. Physica A 356, 375—384.

Burlaga, L.F., Ness, N.F., Acuña, M.H., Lepping, R.P., Connerney, J.E.P., Richardson, J.D., 2008. Magnetic fields at the solar wind termination shock. Nature 454, 75—77.

Burrows, R.H., Zank, G.P., Webb, G.M., Burlaga, L.F., Ness, N.F., 2010. Pickup ion dynamics at the heliospheric termination shock observed by voyager 2. The Astrophysical Journal 715, 1109—1116.

Buti, B., 1980. Ion-acoustic holes in a two-electron-temperature plasma. Physics Letters A 76, 251—254.

Bykov, A.M., 2001. Particle acceleration and nonthermal phenomena in superbubbles. Space Science Reviews 99, 317—326.

Bykov, A.M., 2014. Nonthermal particles and photons in starburst regions and superbubbles. Astronomy and Astrophysics Review 22, 77.

Bzowski, M., et al., 2012. Neutral interstellar helium parameters based on IBEX-Lo observations and test particle calculations. The Astrophysical Journal Supplement Series 198, 12.

Caan, M.N., McPherron, R.L., Russell, C.T., 1977. Characteristics of the association between the interplanetary magnetic field and substorms. Journal of Geophysical Research 82, 4837—4842.

Cairns, I.H., 1987. The electron distribution function upstream from the Earth's bow shock. Journal of Geophysical Research 92, 2315—2327.

Cairns, I.H., 2011. Coherent radio emissions associated with solar system shocks. In: Miralles, M.P., Sanchez Almeida, J. (Eds.), The Sun, the Solar Wind, and the Heliosphere. Springer, p. 267.

Cairns, I.H., Schmidt, J.M., 2015. Testing a theory for type II radio bursts from the Sun to near 0.5 au. In: Zank, G.P., (Ed.), Journal of Physics: Conference Series (JPCS), vol. 642, 012004.

Cairns, I.H., Robinson, P.A., 1999. Strong evidence for stochastic growth of Langmuir-like waves in Earth's foreshock. Physical Review Letters 82, 3066.

Cairns, I.H., Knock, S.A., Robinson, P.A., Kuncic, Z., 2003. Type II solar radio bursts: theory and space weather implications. Space Science Reviews 107, 27—34.

Carbary, J.F., Mitchell, D.G., 2013. Periodicities in Saturn's magnetosphere. Reviews of Geophysics 51, 1—30.

Carbary, J.F., Mauk, B.H., Krimigis, S.M., 1983. Corotation anisotropies in Saturn's magnetosphere. Journal of Geophysical Research 81, 8937—8946.

Carbary, J.F., Mitchell, D.G., Brandt, P., Roelof, E.C., Krimigis, S.M., 2008. Statistical morphology of ENA emissions at Saturn. Geophysical Research Letters 113, A05210.

Carbary, J.F., Kane, M., Mauk, B.H., Krimigis, S.M., 2014. Using the kappa function to investigate hot plasma in the magnetospheres of the giant planets. Journal of Geophysical Research 119, 8426—8447.

Carbary, J.F., Paranicas, C., Mitchell, D.G., Krimigis, S.M., Krupp, N., 2011. Energetic electron spectra in Saturn's plasma sheet. Journal of Geophysical Research 116, A07210.

Cargill, P.J., 2014. Active region emission measure distributions and implications for nanoflare heating. The Astrophysical Journal 784, 49.

Castaing, B., Gagne, Y., Hopfinger, E.J., 1990. Velocity probability density functions of high Reynolds number turbulence. Physica 46D, 177—200.

Cattaert, T., Hellberg, M.A., Mace, R.L., 2007. Oblique propagation of electromagnetic waves in a kappa Maxwellian plasma. Physics of Plasmas 14, 082111.

Cattaert, T., Verheest, F., Hellberg, M.A., 2005. Potential hill electron acoustic solitons and double layers in plasmas with two electron species. Physics of Plasmas 12, 042901.

Cattell, C.A., Dombeck, J., Wygant, J.R., Hudson, M.K., Mozer, F.S., Temerin, M.A., Peterson, W.K., Kletzing, C.A., Russell, C.T., Pfaff, R.F., 1999. Comparisons of polar satellite observations of solitary wave velocities in the plasma sheet boundary and the high altitude cusp to those in the auroral zone. Geophysical Research Letters 26, 425.

Chakrabarti, R., Chandrashekar, R., 2010. Extended Curie-Weiss law via Tsallis statistics. Journal of Statistical Mechanics 09, 09007.

Chakrabarti, S.K., Titarchuk, L.G., 1995. Spectral Properties of Accretion Disks around Galactic and Extragalactic Black Holes. The Astrophysical Journal 455, 623.

Champion, K.S.W., Minzner, R.A., Pond, H.L., 1961. Review of the ARDC model atmosphere, 1959. Planetary and Space Science 7, 454–462.

Chapman, S., Cowling, T.G., 1990. The Mathematical Theory of Non-uniform Gases. Cambridge University Press, Cambridge.

Chapman, D.A., Gericke, D.O., 2011. Analysis of Thomson Scattering from Nonequilibrium Plasmas. Physical Review Letters 107, 165004.

Charbonneau, P., McIntosh, S.W., Liu, H.L., Bogdan, T.J., 2001. Avalanche models for solar flares. Solar Physics 203, 321–353.

Chavanis, P.-H., 2006. Coarse-grained distributions and superstatistics. Physica A 359, 177–212.

Che, H., Goldstein, M.L., 2014. The origin of non-Maxwellian solar wind electron velocity distribution function: connection to nanoflares in the solar corona. The Astrophysical Journal Letters 795, L38.

Chen, F.F., 1974. Introduction to Plasma Physics. Plenum, New York.

Chen, L.-J., Pickett, J.S., Kinter, P., Franz, J., Gurnett, D.A., 2005. On the width amplitude inequality of electron phase space holes. Journal of Geophysical Research 110, A09211.

Cheng, A.F., 1990a. Global magnetic anomaly and aurora of Neptune. Geophysical Research Letters 17, 1697.

Cheng, A.F., 1990b. Triton torus and Neptune aurora. Geophysical Research Letters 17, 1669–1672.

Cheng, A.F., Johnson, R.E., Krimigis, S.M., Lanzerotti, L.J., 1987. Magnetosphere, exosphere, and surface of Mercury. Icarus 71, 430–440.

Chotoo, K., et al., 2000. The suprathermal seed population for corotaing interaction region ions at 1 AU deduced from composition and spectra of H+, He++, and He+ observed by Wind. Journal of Geophysical Research 105, 23107–23122.

Choudhuri, A.R., 1998. The Physics of Fluids and Plasmas: An Introduction for Astrophysicists. Cambridge University Press, Cambridge.

Christon, S.P., 1987. A comparison of the Mercury and Earth magnetospheres: electron measurements and substorm time scales. Icarus 71, 448–471.

Christon, S.P., Mitchell, D.G., Williams, D.J., Frank, L.A., Huang, C.Y., Eastman, T.E., 1988. Energy spectra of plasma sheet ions and electrons from $\sim$50 eV/e to $\sim$1 MeV during plasma temperature transitions. Journal of Geophysical Research 93, 2562–2572.

Christon, S.P., Williams, D.J., Mitchell, D.G., Huang, C.Y., Frank, L.A., 1991. Spectral characteristics of plasma sheet ion and electron populations during disturbed geomagnetic conditions. Journal of Geophysical Research 96, 1–22.

Christon, S.P., Williams, D.J., Mitchell, D.G., Frank, L.A., Huang, C.Y., 1989. Spectral characteristics of plasma sheet ion and electron populations during undisturbed geomagnetic conditions. Journal of Geophysical Research 94, 13409–13424.

Christon, S.P., Feynman, J., Slavin, J.A., 1987. Dynamic substorm injections: similar magnetospheric phenomena at Earth and Mercury. In: Lui, A.T.Y. (Ed.), Magnetotail Physics. Johns Hopkins Press, Baltimore.

Clausius, R.J.E., 1862. Sixth Memoir: On the Application of the Theorem of the Equivalence of Transformations to Interior Work, p. 215.

Clausius, R.J.E., 1870. On a mechanical theorem applicable to heat. Philosophical Magazine Series 4, 122–127.

Cohen, E.G.D., 2004. Superstatistics. Physica 193D, 35–52.

Cohen, O., Sokolov, I.V., Roussev, I.I., Gombosi, T.I., 2008. Validation of a synoptic solar wind model. Journal of Geophysical Research 113, A03104.

Cohen, O., Sokolov, I.V., Roussev, I.I., Arge, C.N., Manchester, W.B., et al., 2007. A semiempirical magnetohydrodynamical model of the solar wind. The Astrophysical Journal 654, L163.

Collier, M.R., 1993. On generating kappa-like distribution functions using velocity space Levy flights. Geophysical Research Letters 20, 1531–1534.

Collier, M.R., 1999. Evolution of kappa distributions under velocity space diffusion: a model for the observed relationship between their spectral parameters. Journal of Geophysical Research 104, 28559–28564.

Collier, M.R., Hamilton, D.C., 1995. The relationship between kappa and temperature in the energetic ion spectra at Jupiter. Geophysical Research Letters 22, 303–306.

Collier, M.R., Hamilton, D.C., Gloeckler, G., Bochsler, P., Sheldon, R.B., 1996. Neon-20, oxygen-16, and helium-4 densities, temperatures, and suprathermal tails in the solar wind determined with WIND/MASS. Geophysical Research Letters 23, 1191–1194.

Collins II, G.W., 1978. The virial theorem in stellar astrophysics. In: Astronomy and Astrophysics Series, vol. 7. Pachart Publishing House, Tucson, 143 p.

Connerney, J.E.P., Acuna, M.H., Ness, N.F., 1983. Currents in Saturn's magnetosphere. Journal of Geophysical Research 88, 8779–8789.

Copper, J.F., Stone, E.C., 1991. Electron signatures of satellite sweeping in the magnetosphere of Uranus. Journal of Geophysical Research 96, 7803–7821.

Cranmer, S.R., 2014. Suprathermal electrons in the solar corona: can nonlocal transport explain heliospheric charge states? The Astrophysical Journal Letters 791, L31.

Crosby, N.B., Aschwanden, M.J., Dennis, B.R., 1993. Frequency distributions and correlations of solar X-ray flare parameters. Solar Physics 143, 275–299.

Culhane, J.L., Harra, L.K., James, A.M., Al-Janabi, K., Bradley, L.J., Chaudry, R.A., et al., 2007. The EUV imaging spectrometer for Hinode. Solar Physics 243, 19–61.

Daglis, I.A., Sarris, E.T., Kremser, G., 1990. Indications for ionospheric participation in the substorm process from AMPTE/CCE observations. Geophysical Research Letters 17, 57–60.

Daglis, I.A., Sarris, E.T., Axford, W.I., Karagevrekis, G., Kasotakis, G., Livi, S., Wilken, B., 1997. Influence of interplanetary disturbances on the terrestrial ionospheric outflow. Physics and Chemistry of the Earth 24, 61–65.

Daglis, I.A., Kasotakis, G., Sarris, E.T., Kamide, Y., Livi, S., Wilken, B., 1999a. Variations of the ion composition during a large magnetic storm and their consequences. Physics and Chemistry of the Earth 24, 229–232.

Daglis, I., Thorne, R.M., Baumjohann, W., Orsini, S., November 4, 1999b. The terrestrial ring current: origin, formation, and decay. Reviews of Geophysics 37, 407–438.

Daniels, K.E., Beck, C., Bodenschatz, E., 2004. Defect turbulence and generalized statistical mechanics. Physica D 193, 208–217.

Daróczy, Z., 1970. Generalized information functions. Information & Control 16, 36–51.

Dash, S.K., Khuntia, S.R., 2010. Fundamentals of Electromagnetic Theory. PHI Learning Private Limited, New Delhi, p. 123.

Davidson, R.C., 1972. Methods in Nonlinear Plasma Theory. Academic Press, New York.

Dauxois, T., Ruffo, S., Arimondo, E., Wilkens, M., 2002. Dynamics and thermodynamics of systems with long-range interactions: an introduction. Lecture Notes in Physics 602, 1–19.

Dayeh, M.A., Desai, M.I., Dwyer, J.R., Rassoul, H.K., Mason, G.M., Mazur, J.E., 2009. The Astrophysical Journal 693, 1588.

De Avillez, M.A., Breitschwerdt, D., 2015. Temperature averaged and total free-free Gaunt factors for kappa and Maxwellian distributions of electrons. Astronomy and Astrophysics 580, A124.

De Pontieu, B., Title, A.M., Lemen, J.R., Kushner, G.D., Akin, D.J., Allard, B., Berger, T., et al., 2014. The interface region imaging spectrograph (IRIS). Solar Physics 289, 2733–2779.

Decker, R.B., Krimigis, S.M., 2003. Voyager observations of low-energy ions during solar cycle 23. Advances in Space Research 32, 597–602.

Decker, R.B., et al., 2005. Voyager 1 in the foreshock, termination shock, and heliosheath. Science 309, 2020–2024.

Decker, R.B., Roelof, E.C., Krimigis, S.M., Hill, M.E., 2006. Low-energy ions near the termination shock. In: Physics of the Inner Heliosheath, AIP Conference Proceedings, vol. 858. AIP, Melville, NY, pp. 73–78.

Del Zanna, G., 2011. Benchmarking atomic data for astrophysics: Fe XVII X-ray lines. Astronomy and Astrophysics 536, A59.

Del Zanna, G., 2013a. A revised radiometric calibration for the Hinode/EIS instrument. Astronomy and Astrophysics 555, A47.

Del Zanna, G., 2013b. The multi-thermal emission in solar active regions. Astronomy and Astrophysics 558, A73.

Del Zanna, G., Storey, P.J., 2012. Atomic data for astrophysics: Fe XIII soft X-ray lines. Astronomy and Astrophysics 543, A144.

Del Zanna, G., Storey, P.J., 2013. Atomic data for astrophysics: Fe XI soft X-ray lines. Astronomy and Astrophysics 549, A42.

Del Zanna, G., Dere, K.P., Young, P.R., Landi, E., Mason, H.E., 2015a. CHIANTI – an atomic database for emission lines. Version 8. Astronomy and Astrophysics 582, A56.

Del Zanna, G., Fernández-Menchero, L., Badnell, N.R., 2015b. Benchmarking atomic data for astrophysics: Si III. Astronomy and Astrophysics 574, A99.

Del Zanna, G., Storey, P.J., Mason, H.E., 2010. Atomic data for the IRON project. LXVIII. Electron impact excitation of Fe XI. Astronomy and Astrophysics 514, A40.

Del Zanna, G., Storey, P.J., Badnell, N.R., Mason, H.E., 2012a. Atomic data for astrophysics: Fe XII soft X-ray lines. Astronomy and Astrophysics 543, A139.

Del Zanna, G., Storey, P.J., Badnell, N.R., Mason, H.E., 2012b. Atomic data for astrophysics: Fe X soft X-ray lines. Astronomy and Astrophysics 541, A90.

Del Zanna, G., Storey, P.J., Badnell, N.R., Mason, H.E., 2014. Atomic data for astrophysics: Fe IX. Astronomy and Astrophysics 565, A77.

Delcourt, D.C., Pedersen, A., Sauvaud, J.A., 1990. Dynamics of single-particle orbits during substorm expansion phase. Journal of Geophysical Research 95, 20853–20865.

Deppman, A., 2012. Self-consistency in non-extensive thermodynamics of highly excited hadronic states. Physica A 391, 6380–6385.

Delcourt, D.C., 2002. Particle acceleration by inductive electric fields in the inner magnetosphere. Journal of Atmospheric and Solar Terrestrial Physics 64, 551–559.

Dere, K.P., 2007. Ionization rate coefficients for the elements hydrogen through zinc. Astronomy and Astrophysics 466, 771–792.

Dere, K.P., Landi, E., Mason, H.E., Monsignori-Fossi, B.C., Young, P.R., 1997. CHIANTI – an atomic database for emission lines. Astronomy and Astrophysics Supplement Series 125, 149–173.

Desai, M.I., et al., 2012. Spectral Properties of ∼0.5–6 kev Energetic Neutral Atoms Measured by the Interstellar Boundary Explorer (IBEX) along the Lines of Sight of Voyager. The Astrophysical Journal Letters 749, L30.

Desai, M.I., et al., 2014. Energetic Neutral Atoms Measured by the Interstellar Boundary Explorer (IBEX): Evidence for Multiple Heliosheath Populations. The Astrophysical Journal 780, 98.

Desai, M.J., Mason, G., Wiedenbeck, M., Cohen, C., Mazur, J., Dwyer, J., Gold, R.E., Krimigis, S.M., Hu, Q., Smith, C., Skoug, R., 2004. Spectral properties of heavy ions associated with the passage of interplanetary shocks at 1 AU. The Astrophysical Journal 611, 1156–1174.

Devanandhan, S., Singh, S.V., Lakhina, G.S., Bharuthram, R., 2012. Electron acoustic waves in a magnetized plasma with kappa distributed ions. Physics of Plasmas 19, 082314.

Devanandhan, S., Singh, S.V., Lakhina, G.S., 2011a. Electron acoustic solitary waves with kappa-distributed electrons. Physica Scripta 84, 025507.

Devanandhan, S., Singh, S.V., Lakhina, G.S., Bharuthram, R., 2011b. Electron acoustic solitons in the presence of an electron beam and superthermal electrons. Nonlinear Processes in Geophysics 18, 627–634.

Devanandhan, S., Singh, S.V., Lakhina, G.S., Bharuthram, R., 2015. Small amplitude electron acoustic solitary waves in a magnetized superthermal plasma. Communications in Nonlinear Science and Numerical Simulation 22, 1322.

Dial, G., 1982. Nonadditive entropies of order 1 and type β and of order α and type β; cascaded channels and the equivocation inequality. Information Sciences 27, l–8.

Dialynas, K., Brandt, P.C., Krimigis, S.M., Mitchell, D.G., Hamilton, D.C., Krupp, N., Rymer, A.M., 2013. The extended Saturnian neutral cloud as revealed by global ENA simulations using Cassini/MIMI measurements. Journal of Geophysical Research 118, 3027–3041.

Dialynas, K., Krimigis, S.M., Mitchell, D.G., Hamilton, D.C., Krupp, N., Brandt, P.C., 2009. Energetic ion spectral characteristics in the Saturnian magnetosphere using Cassini/MIMI measurements. Journal of Geophysical Research 114, A01212.

Diamond, P.H., Itoh, S.-I., Itoh, K., 2010. Modern Plasma Physics. Cambridge University Press, Cambridge.

Dopita, M.A., Sutherland, R.S., 2003. Astrophysics of the Diffuse Universe. Springer, New York.

Dopita, M.A., Sutherland, R.S., 2000. The importance of photoelectric heating by dust in planetary nebulae. The Astrophysical Journal 539, 742–750.

Dorelli, J.C., Scudder, J.D., 1999. Electron heat flow carried by Kappa distributions in the solar corona. Geophysical Research Letters 26, 3537–3540.

Dors, E.E., Kletzing, C.A., 1999. Effects of suprathermal tails on auroral electrodynamics. Journal of Geophysical Research 104, 6783–6796.

Dos Santos, R.J.V., 1997. Generalization of Shannon's theorem for Tsallis entropy. Journal of Mathematical Physics 38, 4104–4107.

Dos Santos, M.S., Ziebell, L.F., Gaelzer, R., 2014. Ion firehose instability in plasmas with plasma particles described by product bi-kappa distributions. Physics of Plasmas 21, 112102.

Dos Santos, M.S., Ziebell, L.F., Gaelzer, R., 2015. Ion-cyclotron instability in plasmas described by product-by-kappa distributions. Physics of Plasmas 22, 122107.

Dos Santos, M.S., Ziebell, L.F., Gaelzer, R., 2016. Ion firehose instability in a dusty plasma considering product-bi-kappa distributions for the plasma particles. Physics of Plasmas 23, 013705.

Dougherty, M.K., et al., 2005. Cassini magnetometer observations during Saturn orbit insertion. Science 307, 1266–1270.

Dougherty, M.K., Khurana, K.K., Neubauer, F.M., Russel, C.T., Saur, J., Leisner, J.S., Burton, M.E., 2006. Identification of a dynamic atmosphere at Enceladus with the Cassini magnetometer. Science 311, 1406–1409.

Downs, C., Roussev, I.I., van der Holst, B., Lugaz, N., Sokolov, I.V., Gombosi, T.I., 2010. Toward a realistic thermodynamic magnetohydrodynamic model of the global solar corona. The Astrophysical Journal 712, 1219.

Doyle, J.G., Giunta, A., Madjarska, M.S., Summers, H., O'Mullane, M., Singh, A., 2013. Diagnosing transient ionization in dynamic events. Astronomy and Astrophysics 557, L9.

Drude, P., 1900. Zur Elektronentheorie der Metalle. Annalen der Physik 306, 566.

Drury, L.O.'C., 1983. An introduction to the theory of diffusive shock acceleration of energetic particles in tenuous plasmas. Reports on Progress in Physics 46, 973−1027.

Du, J., 2004. The nonextensive parameter and Tsallis distribution for self-gravitating systems. Europhysics Letters 67, 893−899.

Dubouloz, N., Treumann, R.A., Pottelette, R., Malingre, M., 1993. Turbulence generated by a gas of electron acoustic solitons. Journal of Geophysical Research 98, 17415−17422.

Dubouloz, N., Pottelette, R., Malingre, M., Treumann, R.A., 1991. Generation of broadband electrostatic noise by electron acoustic solitons. Geophysical Research Letters 18, 155−158.

Dudík, J., Del Zanna, G., Dzifčáková, E., Mason, H.E., Golub, L., 2014b. Solar transition region lines observed by the interface region imaging spectrograph: diagnostics for the O IV and Si IV lines. The Astrophysical Journal Letters 780, L12.

Dudík, J., Del Zanna, G., Mason, H.E., Dzifčáková, E., 2014a. Signatures of the non-Maxwellian κ-distributions in optically thin line spectra. I. Theory and synthetic Fe IX-XIII spectra. Astronomy and Astrophysics 570, A124.

Dudík, J., Kašparová, J., Dzifčáková, E., Karlický, M., Mackovjak, Š., 2012. The non-Maxwellian continuum in the X-ray, UV, and radio range. Astronomy and Astrophysics 539, A107.

Dudík, J., Mackovjak, Š., Dzifčáková, E., Del Zanna, G., Williams, D.R., Karlický, M., Mason, H.E., Lörinčík, J., Kotrč, P., Fárník, F., Zemanová, A., 2015. Imaging and spectroscopic observations of a transient coronal loop: evidence for the non-maxwellian κ distributions. The Astrophysical Journal 807, 123.

Dufton, P.L., Kingson, A.E., Keenan, F.P., 1984. Observational evidence for non-Maxwellian electron energy distributions in the solar transition region. The Astrophysical Journal 280, L35−L37.

Dzifčáková, E., 1992. The ionization balance of the Fe in the solar corona for a non-Maxwellian electron distribution function. Solar Physics 140, 247−267.

Dzifčáková, E., 2002. The updated Fe ionization equilibrium for the electron κ-distributions. Solar Physics 208, 91−111.

Dzifčáková, E., 2006. The influence of the electron κ-distribution in the solar corona on the Fe VIII − Fe XV line intensities. Solar Physics 234, 243−256.

Dzifčáková, E., Kulinová, A., 2011. Diagnostics of the κ-distribution using Si III lines in the solar transition region. Astronomy and Astrophysics 531, A122.

Dzifčáková, E., Del Zanna, G., 2012. Fe XVII and the κ-distributions: diagnostics with Hinode EIS. In: Sekii, T., Watanabe, T., Sakurai, T. (Eds.), Proc. Hinode-3: The 3rd Hinode Science Meeting, ASP Conference Series, vol. 454, pp. 167−170.

Dzifčáková, E., Dudík, J., 2013. H to Zn ionization equilibrium for the non-Maxwellian electron κ-distributions: updated calculations. The Astrophysical Journal Supplement Series 206 (6), 9.

Dzifčáková, E., Kulinová, A., 2010. The diagnostics of the κ-distributions from EUV spectra. Solar Physics 263, 25−41.

Dzifčáková, E., Dudík, J., Kotrč, P., Fárník, F., Zemanová, A., 2015. KAPPA: a package for synthesis of optically thin spectra for the non-Maxwellian κ-distributions based on the Chianti database. The Astrophysical Journal Supplement Series 217, 14.

Elaydi, S., 2005. An Introduction to Difference Equations. Springer Science + Business Media, Inc, New York.

Elliott, H., McComas, D.J., Valek, P., Nicolaou, G., Weidner, S., Livadiotis, G., 2016. New horizons solar wind around Pluto observations of the solar wind from 11−33 au. The Astrophysical Journal Supplement Series 223, 19, pp. 21.

Eraker, J.H., Simpson, J.A., 1986. Acceleration of charged particles in Mercury's magnetosphere. Journal of Geophysical Research 91, 9973−9993.

Ercolano, B., Young, P.R., Drake, J.J., Raymond, J.C., April 2008. X-ray enabled MOCASSIN: a three-dimensional code for photoionized media. The Astrophysical Journal Supplement Series 175, 534—542.

Ergun, R.E., Carlson, C.W., McFadden, J.P., Mozer, F.S., Delory, G.T., Peria, W., Chaston, C.C., Temerin, M., Roth, I., Muschietti, L., Elphic, R., Strangeway, R., Pfaff, R., Cattell, C.A., Klumpar, D., Shelley, E., Peterson, W., Moebius, E., Kistler, L., 1998a. Fast satellite observations of large-amplitude solitary structures. Geophysical Research Letters 25, 2041.

Ergun, R.E., Calrson, C.W., McFadden, J.P., Mozer, F.S., Muschietti, L., Roth, I., Strangeway, R.J., 1998b. Debye-scale plasma structures associated with magnetic-field-aligned electric fields. Physical Review Letters 81, 826.

Eslami, P., Mottaghizadeh, M., Pakzad, H.R., 2011. Nonplanar dust acoustic solitary waves in dusty plasmas with ions and electrons following a q-nonextensive distribution. Physics of Plasmas 18, 102303.

Esser, R., Edgar, R.J., 2000. Reconciling spectroscopic electron temperature measurements in the solar corona with in-situ charge state observations. The Astrophysical Journal Letters 532, L71—L74.

Fahr, H.-J., Siewert, M., 2013. The multi-fluid pressures downstream of the solar wind termination shock. Astronomy and Astrophysics 558, A41.

Fan, C.Y., Gloeckler, G., Hovestadt, D., 1975. Energy spectra and charge states of H, He, and heavy ions observed in the Earth's magnetosheath and magnetotail. Physical Review Letters 34 (8), 495—498.

Feldman, W.C., Asbridge, J.R., Bame, S.J., Montgomery, M.D., Gary, S.P., 1975. Solar wind electrons. Journal of Geophysical Research 80, 4181—4196.

Ferland, G.J., Porter, R.L., van Hoof, P.A.M., Williams, R.J.R., Abel, N.P., Lykins, M.L., Shaw, G., Henney, W.J., Stancil, P.C., April 2013. The 2013 release of cloudy. Revista Mexicana deicana Astronomia y Astrofisica 49, 137—163.

Filbert, P.C., Kellogg, P.J., 1979. Electrostatic noise at the plasma frequency beyond the Earth's bow shock. Journal of Geophysical Research 84, 1369—1381.

Fillius, R.W., McIlwain, C.E., 1974. Measurements of the Jovian radiation belts. Journal of Geophysical Research 79, 3589—3599.

Fisk, L.A., 1971. Increases in the low-energy cosmic ray intensity at the front of propagating interplanetary shock waves. Journal of Geophysical Research 76, 1662—1672.

Fisk, L.A., 1976. The acceleration of energetic particles in the interplanetary medium by transit time damping. Journal of Geophysical Research 81, 4633—4640.

Fisk, L.A., Gloeckler, G., 2006. The common spectrum for accelerated ions in the quiet-time solar wind. The Astrophysical Journal 640, L79—L82.

Fisk, L.A., Gloeckler, G., 2007. Thermodynamic constraints on stochastic acceleration in compressional turbulence. Proceedings of the National Academy of Sciences of the United States of America 104, 5749—5754.

Fisk, L.A., Gloeckler, G., 2008. Acceleration of suprathermal tails in the solar wind. The Astrophysical Journal 686, 1466—1473.

Fisk, L.A., Gloeckler, G., 2009. The acceleration of anomalous cosmic rays by stochastic acceleration in the heliosheath. Advances in Space Research 43, 1471—1478.

Fisk, L.A., Gloeckler, G., 2012a. Particle acceleration in the heliosphere: implications for astrophysics. Space Science Reviews 173, 433—458.

Fisk, L.A., Gloeckler, G., 2012b. Acceleration of galactic cosmic rays in the interstellar medium. The Astrophysical Journal 744, 127.

Fisk, L.A., Gloeckler, G., 2013. The global configuration of the heliosheath inferred from recent Voyager 1 observations. The Astrophysical Journal 776, 79.

Fisk, L.A., Gloeckler, G., 2014. The case for a common spectrum of particles accelerated in the heliosphere: observations and theory. Journal of Geophysical Research 119, 8733–8749.

Fisk, L.A., Gloeckler, G., Schwadron, N.A., 2010. On theories for stochastic acceleration in the solar wind. The Astrophysical Journal 720, 533–540.

Fletcher, L., Dennis, B.R., Hudson, H.S., Krucker, S., Phillips, K., Veronig, A., et al., 2011. An observational overview of solar flares. Space Science Reviews 159, 19–106.

Font, J.A., 2007. An introduction to relativistic hydrodynamics. Journal of Physics: Conference Series 91, 012002.

Formisano, V., Moreno, G., Palmiotto, F., Hedgecock, P.C., 1973. Solar wind interaction with the Earth's magnetic field 1. Magnetosheath. Journal of Geophysical Research 78, 3714–3730.

Fort, J., Jou, D., Llebot, J.E., 1999. Temperature and measurement: comparison between two models of nonequilibrium radiation. Physica A 269, 439–454.

Frank, L.A., Ackerson, K.L., Wolfe, J.H., Mihalov, J.D., 1976. Observations of plasmas in the Jovian magnetosphere. Journal of Geophysical Research 81, 457–468.

Frank, L.A., Burek, B.G., Ackerson, K.L., Wolfe, J.H., Mihalov, J.D., 1980. Plasmas in Saturn's magnetosphere. Journal of Geophysical Research 85, 5695–5708.

Franz, J.R., Kintner, P.M., Pickett, J.S., 1998. Polar observations of coherent electric field structures. Geophysical Research Letters 25, 1277–1280.

Franz, J.R., Kintner, P.M., Pickett, J.S., Chen, L.-J., 2005. Properties of small-amplitude electron phase-space holes observed by polar. Journal of Geophysical Research 110, A09212.

Fried, B.D., Conte, S.D., 1961. The Plasma Dispersion Function. Academic, San Diego, California.

Frisch, P.C., Bzowski, M., Livadiotis, G., McComas, D.J., Möbius, E., Mueller, H.-R., Pryor, W.R., Schwadron, N.A., Sokól, J.M., Vallerga, J.V., Ajello, J.M., 2013. Decades-long changes of the interstellar wind through our solar system. Science 341, 1080.

Funsten, H., et al., 2009. The interstellar boundary explorer high energy (IBEX-Hi) neutral atom imager. Space Science Reviews 146, 75–103.

Fuselier, S.A., Allegrini, F., Bzowski, M., Dayeh, M.A., Desai, M., Funsten, H.O., Galli, A., Heirtzler, D., Janzen, P., Kubiak, M.A., Kucharek, H., Lewis, W., Livadiotis, G., McComas, D.J., Möbius, E., Petrinec, S.M., Quinn, M., Schwadron, N., Sokól, J.M., Trattner, K.J., Wood, B.E., Wurz, P., 2014. Low energy neutral atoms from the heliosheath. The Astrophysical Journal 784, 14.

Fuselier, S.A., et al., 2012. Heliospheric neutral atom spectra between 0.01 and 6 keV from IBEX. The Astrophysical Journal 754, 14.

Gabriel, A.H., Phillips, K.J.H., 1979. Dielectronic satellite spectra for highly charged helium-like ions. IV — iron satellite lines as a measure of non-thermal electron energy distributions. Monthly Notices of the Royal Astronomical Society 189, 319–327.

Gaelzer, R., Ziebell, L.F., 2014. The dispersion relation of dispersive Alfvén waves in superthermal plasmas. Journal of Geophysical Research 119, 9334–9356.

Gaelzer, R., Ziebell, L.F., 2016. Obliquely propagating electromagnetic waves in magnetized kappa plasmas. Physics of Plasmas 23, 022110.

Galand, M., Lilensten, J., Maurice, S., 1999. The ionosphere of Titan: ideal diurnal and nocturnal cases. Icarus 140, 92–105.

García-Rojas, J., Esteban, C., November 2007. On the abundance discrepancy problem in H II regions. The Astrophysical Journal 670, 457–470.

Gary, S.P., 1985. Electrostatic instabilities in plasmas with two electron components. Journal of Geophysical Research 90, 8213–8221.

Gary, S.P., 1993. Theory of Space Plasma Microinstabilities. Cambridge University Press.

Gell-Mann, M., Tsallis, C., 2004. Nonextensive Entropy: Interdisciplinary Applications. Oxford University Press.

Gershman, D.J., Slavin, J.A., Raines, J.M., Zurbuchen, T.H., Anderson, B.J., Korth, H., Baker, D.N., Solomon, S.C., 2014. Ion kinetic properties in Mercury's premidnight plasma sheet. Geophysical Research Letters 41, 5740–5747.

Ghosh, S.S., Iyengar, A.N.S., 1997. Anomalous width variations forion acoustic rarefactive solitary waves in a warm ion plasma with two electron temperatures. Physics of Plasmas 5, 3204.

Ghosh, S.S., Lakhina, G.S., 2004. Anomalous width variation of rarefactive ion acoustic solitary waves in the context of auroral plasmas. Nonlinear Processes in Geophysics 11, 219.

Ghosh, S.S., Ghosh, K.K., Iyengar, A.N.S., 1996. Large mach number ion acoustic rarefactive solitary waves for a two electron temperature warm ion plasma. Physics of Plasmas 4, 3939.

Ghosh, S.S., Pickett, J.S., Lakhina, G.S., Winningham, J.D., Lavraud, B., Decreau, P.M.E., 2008. Parametric analysis of positive amplitude electron acoustic solitary waves in a magnetized plasma and its application to boundary layers. Journal of Geophysical Research 113, A06218.

Giacalone, J., 2012. Energetic charged particles associated with strong interplanetary shocks. The Astrophysical Journal 761, 28.

Gibbs, J.W., 1902. Elementary Principles in Statistical Mechanics. Scribner's sons, New York.

Gloeckler, G., Geiss, J., 1998. Interstellar and inner source pickup ions observed with Swics on Ulysses. Space Science Reviews 86, 127–159.

Gloeckler, G., Hamilton, D.C., 1987. AMPTE ion composition results. Physica Scripta T18, 73–84.

Gloeckler, G., Fisk, L.A., 2010. Proton velocity distributions in the inner heliosheath derived from energetic hydrogen atoms measured with Cassini and IBEX. In: Proceedings of the 9th Annual International Astrophysics Conference, AIP, vol. 1302, p. 110.

Gloeckler, G., Fisk, L.A., 2014. A test for whether or not Voyager 1 has crossed the heliopause. Geophysical Research Letters 41, 5325–5330.

Gloeckler, G., et al., 1994. Acceleration of interstellar pickup ions in the disturbed solar wind observed on Ulysses. Journal of Geophysical Research 99, 17637–17643.

Gloeckler, G., Geiss, J., Balsiger, H., Bedini, P., Cain, J.C., Fischer, J., Fisk, L.A., Galvin, A.B., Gliem, F., Hamilton, D.C., Hollweg, J.V., Ipavich, F.M., Joss, R., Livi, S., Lundgren, R., Mall, U., McKenzie, J.F., Ogilvie, K.W., Ottens, F., Rieck, W., Tums, E.O., von Steiger, R., Weiss, W., Wilken, B., 1992. The solar wind ion composition spectrometer. Astronomy and Astrophysics Supplement Series 92, 267–289.

Gloeckler, G., Geiss, J., Fish, L.A., 2001. In: Balogh, A., Marsden, R.G., Smith, E.J. (Eds.), Heliospheric and Interstellar Phenomena Revealed from Observations of Pickup Ions, p. 287.

Gloeckler, G., Fisk, L.A., Mason, G.M., Hill, M.E., 2008. Formation of power law tail with spectral index 5 inside and beyond the heliosphere. In: Hu, G.L.Q., Verkhoglyadova, O., Zank, G.P., Lin, R.P., Luhmann, J. (Eds.), Particle Acceleration and Transport in the Heliosheath and beyond, AIP Conf. Proc, vol. 1039. AIP, Melville, NY, pp. 367–374.

Göğüş, E., et al., 1999. Statistical Properties of SGR 1900+14 Bursts. The Astrophysical Journal 526, L93–L96.

Goldman, M.V., Oppenheim, M.M., Newman, D.L., 1999. Nonlinear two-stream instabilities as an explanation for the auroral bipolar wave structures. Geophysical Research Letters 26, 1821–1824.

Goldman, M.V., Newman, D.L., Mangeney, A., 2007. Physical Review Letters 99, 145002.

Goldston, R.J., Rutherford, P.H., 1995. Introduction to Plasma Physics. Institute of Physics: Bristol.

Goodrich, C.C., 1985. Geophysical Monograph Series, vol. 35. American Geophysical Union, Washington, DC, p. 153.

Gopalswamy, N., Yashiro, S., 2011. The strength and radial profile of the coronal magnetic field from the standoff distance of a coronal mass ejection-driven shock. The Astrophysical Journal 736, L17.

Gopalswamy, N., Yashiro, S., Kaiser, M.L., Howard, R.A., Bougeret, J.-L., 2001. Radio signatures of coronal mass ejection interaction: coronal mass ejection cannibalism? The Astrophysical Journal Letters 548, L91.

Gosling, J.T., 1998. The solar wind. In: The Encyclopedia of the Solar System. Academic Press, San Diego, CA, USA.

Gougam, L.A., Tribeche, M., 2011. Debye shielding in a nonextensive plasma. Physics of Plasmas 18, 062102.

Grabbe, C., 2000. Generation of broadband electrostatic waves in Earth's magnetotail. Physical Review Letters 84, 3614–3617.

Graham, D.B., Cairns, I.H., 2015. The Langmuir waves associated with the 1 December 2013 type II burst. Journal of Geophysical Research 120, 4126.

Grigorenko, E.E., Hoshino, M., Hirai, M., Mukai, T., Zelenyi, L.M., 2009. "Geography" of ion acceleration in the magnetotail: X-line versus current sheet effects. Journal of Geophysical Research 114, A03203.

Gupta, G.R., Tripathi, D., Mason, H.E., 2015. Spectroscopic observations of a coronal loop: basic physical plasma parameters along the full loop length. The Astrophysical Journal 800, 140.

Gurnett, D.A., Bhattacharjee, A., 2005. Introduction to Plasma Physics with Space and Laboratory Applications. Cambridge University Press.

Gurnett, D.A., Frank, L.A., Lepping, R.P., 1976. Plasma waves in the distant magnetotail. Journal of Geophysical Research 81, 6059.

Gurnett, D.A., Scarf, F.L., Kurth, W.S., Poynter, R.L., 1986. First plasma wave observation at Uranus. Science 233, 106.

Gurnett, D.A., Persoon, A.M., Kurth, W.S., Groene, J.B., Averkamp, T.F., Dougherty, M.K., Southwood, D.J., 2007. The variable rotation period of the inner region of Saturn's plasma disk. Science 316, 442–445.

Haaland, S., Kronberg, E.A., Daly, P.W., Franz, M., Degener, L., Georgescu, E., Dandouras, I., 2010. Spectral characteristics of protons in the Earth's plasmasheet: statistical results from Cluster CIS and RAPID. Annales Geophysicae 28, 1483–1498.

Habeck, M., Nilges, M., Rieping, W., 2005. Replica-exchange Monte Carlo scheme for Bayesian data analysis. Physical Review Letters 94, 018105.

Hagedorn, R., 1965. Statistical thermodynamics of strong interactions at high energies. Nuovo. Cimento. Suppl. 3, 147–186.

Hahn, M., Savin, D.W., 2015. A simple method for modeling collision processes in plasmas with a kappa energy distribution. The Astrophysical Journal 809, 178.

Hamilton, D.C., Gloeckler, G., Krimigis, S.M., Lanzerotti, L.J., 1981. Composition of non-thermal ions in the Jovian magnetosphere. Journal of Geophysical Research 86 (A10), 8301–8313.

Hammond, C.M., Feldman, W.C., Phillips, J.L., Goldstein, B.E., Balogh, A., 1996. Solar wind double ion beams and the heliospheric current sheet. Journal of Geophysical Research 100, 7881–7889.

Hannah, I.G., Kontar, E., 2012. Differential emission measures from the regularized inversion of Hinode and SDO data. Astronomy and Astrophysics 539, A146.

Hannah, I.G., Kontar, E., 2013. Multi-thermal dynamics and energetics of a coronal mass ejection in the low solar atmosphere. Astronomy and Astrophysics 553, A10.

Hannah, I.G., Hudson, H.S., Hurford, G.J., Lin, R.P., 2010. Constraining the hard X-ray properties of the quiet Sun with new RHESSI observations. The Astrophysical Journal 724, 487–492.

Hannah, I.G., Hudson, H.S., Battaglia, M., Christe, S., Kašparová, J., Krucker, S., Kundu, M.R., Veronig, A., 2011. Microflares and the statistics of X-ray flares. Space Science Reviews 159, 263–300.

Hansen, C.J., et al., 2011. The composition and structure of the Enceladus plume. Geophysical Research Letters 38, L11202.

Hansen, C.J., Esposito, L., Stewart, A.I.F., Colwell, J., Hendrix, A., Pryor, W., Shemansky, D., West, R., 2006. Enceladus water vapour plume. Science 311, 1422–1425.

Hapgood, M.A., Bryant, D.A., 1992. Exploring the magnetospheric boundary layer. Planetary and Space Science 40, 1431–1459.

Hapgood, M., Perry, C., Davies, J., Denton, M., 2011. The role of suprathermal particle measurements in CrossScale studies of collisionless plasma processes. Planetary and Space Science 59, 618–629.

Hartquist, T.W., Dyson, J.E., Ruffle, D.P., 2004. Blowing Bubbles in the Cosmos. Oxford University Press, New York.

Hasegawa, A., Mima, K., Duong-van, M., 1985. Plasma distribution function in a superthermal radiation field. Physical Review Letters 54, 2608.

Hasegawa, H., 2005. Nonextensive thermodynamics of the two-site Hubbard model. Physica A 351, 273–285.

Hawkins, S.E., Cheng, A.F., Lanzerotti, L.J., 1998. Bulk Iows of hot plasma in the Jovian magnetosphere: a model of anisotropic Iuxes of energetic ions. Journal of Geophysical Research 103, 20031–20054.

Hebb, M.H., Menzel, D.H., November 1940. Physical processes in gaseous nebulae. X. Collisional excitation of Nebulium. The Astrophysical Journal 92, 408.

Heerikhuisen, J., Pogorelov, N.V., Florinski, V., Zank, G.P., le Roux, J.A., 2008. The effects of a κ-distribution in the heliosheath on the global heliosphere and ENA flux at 1 AU. The Astrophysical Journal 682, 679.

Heerikhuisen, et al., 2010. Pick-Up Ions in the Outer Heliosheath: A Possible Mechanism for the Interstellar Boundary EXplorer Ribbon. The Astrophysical Journal 708, L126–L130.

Heerikhuisen, J., Zirnstein, E.J., Funsten, H.O., Pogorelov, N.V., Zank, G.P., 2014. The effect of new interstellar medium parameters on the heliosphere and energetic neutral atoms from the interstellar boundary. The Astrophysical Journal 784, 1–14.

Heerikhuisen, J., Zirnstein, E., Pogorelov, N., 2015. Kappa-distributed protons in the solar wind and their charge-exchange coupling to energetic hydrogen. Journal of Geophysical Research 120, 1516–1525.

Hellberg, M.A., Mace, R.L., 2002. Generalized plasma dispersion function for a plasma with a kappa-Maxwellian velocity distribution. Physics of Plasmas 9, 1495–1504.

Hellberg, M.A., Mace, R.L., Baluku, T.K., Kourakis, I., Saini, N.S., 2009. Comment on "Mathematical and physical aspects of Kappa velocity distribution" [Physics of Plasmas 14, 110702 (2007)]. Physics of Plasmas 16, 094701.

Henney, W.J., Arthur, S.J., de Colle, F., Mellema, G., September 2009. Radiation-magnetohydrodynamic simulations of the photoionization of magnetized globules. Monthly Notices of the Royal Astronomical Society 398, 157–175.

Henning, F.D., Mace, R.L., Pillay, S.R., 2011. Electrostatic Bernstein waves in plasmas whose electrons have a dual kappa distribution: applications to the Saturnian magnetosphere. Journal of Geophysical Research 116, A12.

Hill, T.W., 1984. Magnetospheric structures: Uranus and Neptune. In: Bergstrahl, J.T. (Ed.), Uranus and Neptune. NASA Conf. Publ, p. 2330.

Hillan, D.S., Cairns, I.H., Robinson, P.A., 2012a. Type II solar radio bursts: modeling and extraction of shock parameters. Journal of Geophysical Research 117, A03104.

Hillan, D.S., Cairns, I.H., Robinson, P.A., 2012b. Type II solar radio bursts: 2. Detailed comparison of theory with observations. Journal of Geophysical Research 117, A06105.

Ho, G., et al., 2011. Observations of suprathermal electrons in Mercury's magnetosphere during the three MESSENGER flybys. Planetary and Space Science 59, 2016–2025.

Ho, G., Krimigis, C.S.M., Gold, R.E., Baker, D.N., Anderson, B.J., Korth, H., Slavin, J.A., McNutt, R.L., Winslow, R.M., Solomon, S.C., 2012. Spatial distribution and spectral characteristics of energetic electrons in Mercury's magnetosphere. Journal of Geophysical Research 117, A9.

Hollweg, J.V., 1974. On electron heat conduction in the solar wind. Journal of Geophysical Research 79, 3845–3850.

Holman, G.D., Sui, L., Schwartz, R.A., Emslie, A.G., 2003. Electron Bremsstrahlung hard X-ray spectra, electron distributions, and energetics in the 2002 july 23 solar flare. The Astrophysical Journal 595, L97–L101.

Hoover, Wm G., 2001. Time Reversibility, Computer Simulation, and Chaos. World Scientific, Singapore, p. 154.

Hoover, W.G., Hoover, C.G., 2008. Nonequilibrium temperature and thermometry in heat-conducting models. Physical Review E 77, 041104.

Hoover, Wm G., Aoki, K., Hoover, C.G., De Groot, S.V., 2004. Time-reversible deterministic thermostats. Physica D 187, 253–267.

Huang, C.Y., Goertz, C.K., Frank, L.A., Rostoker, G., 1989. Observational determination of the adiabatic index in the quiet time plasma sheet. Geophysical Research Letters 16, 563.

Ichimaru, S., 1973. Basic Principles of Plasma Physics: A Statistical Approach. Benjamin/Cummings Publishing Co., Reading, Massachussetts.

Ilovaisky, S.A., Chevalier, C., Motch, C., Chiappetti, L., 1986. GX 339-4 - EXOSAT observations in the off and soft states. Astronomy and Astrophysics 164, 67–72.

Issautier, K., Meyer-Vernet, N., Pierrard, V., Lemaire, J., 2001. Electron temperature in the solar wind from a kinetic collisionless model: application to high-latitude Ulysses observations. Astrophysics and Space Science 277, 189–193.

Javidan, K., Pakzad, H.R., 2013. Obliquely propagating electron acoustic solitons in a magnetized plasma with superthermal electrons. Indian Journal of Physics 87, 83–87.

Jeans, J.H., 1902. The stability of a spherical nebula. Philosophical Transactions of the Royal Society of London A 199, 1–53.

Jeffrey, N.L.S., Fletcher, L., Labrosse, N., 2016. First evidence of non-Gaussian solar flare EUV spectral line profiles and accelerated non-thermal ion motion. Astronomy and Astrophysics 590, A99.

Jokipii, J.R., Lee, M.A., 2010. Compression acceleration in astrophysical plasmas and the production of f(v) proportion to v-5 spectra. The Astrophysical Journal 713, 475–483.

Jonkers, J., Van Der Mullen, J.A.M., 1999. The excitation temperature in (helium) plasmas. Journal of Quantitative Spectroscopy and Radiative Transfer 61, 703–709.

Jurac, S., Richardson, J.D., 2005. A self-consistent model of plasma and neutrals at Saturn: neutral cloud morphology. Journal of Geophysical Research 110, A09220.

Jurac, S., McGrath, M.A., Johnson, R.E., Richardson, J.D., Vasyliunas, V.M., Eviatar, A., 2002. Saturn: search for a missing water source. Geophysical Research Letters 29, 2172.

Kadomtsev, B.B., 1965. Plasma Turbulence. Academic Press, New York.

Kaeppler, S.R., Nicolls, M.J., Strømme, A., Kletzing, C.A., Bounds, S.R., 2014. Observations in the E region ionosphere of kappa distribution functions associated with precipitating auroral electrons and discrete aurorae. Journal of Geophysical Research 119, 10164–10183.

Kakad, A.P., Singh, S.V., Reddy, R.V., Lakhina, G.S., Tagare, S.G., Verheest, F., 2007. Generation mechanism for electron acoustic solitary waves. Physics of Plasmas 14, 052305.

Kakad, A.P., Singh, S.V., Reddy, R.V., Lakhina, G.S., Tagare, S.G., 2009. Electron acoustic solitary waves in the Earth's magnetotail region. Advances in Space Research 43, 1945–1949.

Kakad, A.P., Lotekar, A., Kakad, B., 2016. First-ever model simulation of the new subclass of solitons "Supersolitons" in plasma. Physics of Plasmas 23, 110702.

Kallenrode, M.B., 1998. Space Physics an Introduction to Plasmas and Particles in the Heliosphere and Magnetospheres. Springer-Verlag, Berlin, Heielberg.

Kallenrode, M.-B., 2004. Space Physics, an Introduction to Plasmas and Particles in the Heliosphere and Magnetospheres. Springer-Verlag, Berlin, Heidellberg.

Kamei, T., Maeda, H., 1981. Auroral Electrojet Indices (AE) for January–June 1978, Data Book 3, World Data Center for Geomagn. Kyoto Univ., Kyoto, Japan.

Kamei, T., Maeda, H., 1982. Auroral Electrojet Indices (AE) for January—June 1979, Data Book 5, World Data Center for Geomagn. Kyoto Univ., Kyoto, Japan.

Kane, M., Mauk, B.H., Keath, E.P., Krimigis, S.M., 1992. A convected K distribution model for hot ions in the Jovian magnetodisc. Geophysical Research Letters 19, 1435—1438.

Kane, M., Mauk, B.H., Keath, E.P., Krimigis, S.M., 1994. Structure and dynamics of the uranian magnetotail: results from hot plasma and magnetic field observations. Journal of Geophysical Research 96, 11485—11499.

Kane, M., Mauk, B.H., Keath, E.P., Krimigis, S.M., 1995. Hot ions in Jupiter's magnetodisk: a model for Voyager 2 low-energy charged particle measurements. Journal of Geophysical Research 100, 19473—19486.

Kane, M., Williams, D.J., Mauk, B.H., McEntire, R.W., Roelof, E.C., 1999. Galileo energetic particles detector measurements of hot ions in the neutral sheet region of Jupiter's magnetodisk. Journal of Geophysical Research 26, 5—8.

Kane, M., Mitchell, D.G., Carbary, J.F., Krimigis, S.M., Crary, F.J., 2008. Plasma convection in Saturn's outer magnetosphere determined from ions detected by the Cassini INCA experiment. Geophysical Research Letters 35, L04102.

Kane, M., Mitchell, D.G., Carbary, J.F., Krimigis, S.M., 2014. Plasma convection in the nightside magnetosphere of Saturn determined from energetic ion anisotropies. Planetary and Space Science 91, 1—13.

Kartalev, M., Dryer, M., Grigorov, K., Stoimenova, E., 2006. Solar wind polytropic index estimates based on single spacecraft plasma and interplanetary magnetic field measurements. Geophysical Research Letters 111, A10107.

Kashyap, V.L., Drake, J.J., Güdel, M., Audard, M., 2002. Flare Heating in Stellar Coronae. The Astrophysical Journal 580, 1118—1132.

Kašparová, J., Karlický, M., 2009. Kappa distribution and hard X-ray emission of solar flares. Astronomy and Astrophysics 497, L13—L16.

Kaufmann, R.L., Paterson, W.R., Frank, L.A., 2004. Magnetization of the plasma sheet. Journal of Geophysical Research 109, A09212.

Kennel, C.F., Petschek, H.E., 1966. Limit on stably trapped particle fluxes. Journal of Geophysical Research 71, 1—28.

Kennewell, J., 2016. Monitoring the Radio Sun, http://www.spaceacademy.net.au/env/sol/solradp/solradp.htm.

Khazanov, G.V., Liemohn, M.W., Kozyra, J.U., Moore, T.E., 1998. Inner magnetospheric superthermal electron transport: photoelectron and plasma sheet electron sources. Journal of Geophysical Research 103, 23485.

Khurana, K.K., Mitchell, D.G., Arridge, C.S., Dougherty, M.K., Russell, C.T., Paranicas, C., Krupp, N., Coates, A.J., 2009. Sources of rotational signals in Saturn's magnetosphere. Journal of Geophysical Research 114, A02211.

Kifune, T., et al., 1995. Very high energy gamma rays from PSR 1706-44. The Astrophysical Journal 438, L91—L94.

Kivelson, M.G., Russell, C.T., 1995. Introduction to Space Physics. Cambridge University Press.

Klages, R., et al. (Eds.), 2008. Anomalous Transport: Foundations and Applications. Wiley-VCH.

Kletzing, C.A., Scudder, J.D., 1999. Auroral-plasma sheet electron anisotropy. Geophysical Research Letters 26, 971.

Kletzing, C.A., Scudder, J.D., Dors, E.E., Curto, C., 2003. Auroral source region: plasma properties of the high latitude plasma sheet. Journal of Geophysical Research 108, 1360.

Klimchuk, J.A., Karpen, J.T., Antiochos, S.K., 2010. Can thermal nonequilibrium explain coronal loops? The Astrophysical Journal 714, 1239—1248.

Knock, S.A., Cairns, I.H., 2005. Type II radio emission predictions: sources of coronal and interplanetary spectral structure. Journal of Geophysical Research 110, A01101.

Knock, S., Cairns, I.H., Robinson, P.A., Kuncic, Z., 2001. Theory of type II solar radio emission from the foreshock region of an interplanetary shock. Journal of Geophysical Research 106, 25041.

Knock, S.A., Cairns, I.H., Robinson, P.A., Kuncic, Z., 2003. Theoretically predicted properties of type II radio emission from an interplanetary foreshock. Journal of Geophysical Research 108, 1126.

Ko, Y.-K., Fisk, L., Gloeckler, G., Geiss, J., 1996. Limitations on suprathermal tails of electrons in the lower solar corona. Geophysical Research Letters 23, 2785−2788.

Koen, E.J., Collier, A.B., Maharaj, S.K., 2012. A simulation approach of high-frequency electrostatic waves found in Saturn's magnetosphere. Physics of Plasmas 19, 042102.

Kojima, H., Matsumoto, H., Chikuba, S., Horiyama, S., Ashour-Abdalla, M., Anderson, R.R., 1997. Geotail waveform observations of broadband/narrowband electrostatic noise in the distant tail. Journal of Geophysical Research 102, 14439.

Kokubun, S., McPherron, R.L., 1981. Substorm signatures at synchronous altitude. Journal of Geophysical Research 86, 11265−11277.

Kontar, E., Pecseli, H.L., 2002. Nonlinear development of electron-beam-driven weak turbulence in an inhomogeneous plasma. Physical Review E 65, 066408.

Koskinen, H.E.J., Lundin, R., Holback, B., 1990. On the plasma environment of solitary waves and weak double layers. Journal of Geophysical Research 95, 5921−5929.

Kourakis, I., Sultana, S., Hellberg, M.A., 2012. Dynamical characteristics of solitary waves, shocks and envelope modes in kappa-distributed non-thermal plasmas: an overview. Plasma Physics and Controlled Fusion 54, 124001.

Krall, N.A., Trivelpiece, A.W., 1973. Principles of Plasma Physics. Mc-Graw Hill Book Co, New York.

Kramida, A., Ralchenko, Y., Reader, J., June 2012. Current status of atomic spectroscopy databases at NIST. In: APS Division of Atomic, Molecular and Optical Physics Meeting Abstracts, p. D1004.

Krennrich, F., et al., 1999. Measurement of the Multi-TEV Gamma-Ray Flare Spectra of Markarian 421 and Markarian 501. The Astrophysical Journal 511, 149−156.

Krimigis, S.M., Roelof, E.C., 1983. Low-energy particle population. In: Dessler, A.J. (Ed.), Physics of the Jovian Magnetosphere. Cambridge University Press, New York, pp. 106−156.

Krimigis, S.M., Armstrong, T.P., Axford, W.I., Bostrom, C.O., Fan, C.Y., Gloeckler, G., Lanzerotti, L.J., 1977. The low energy charged particle (LECP) experiment on the Voyager spacecraft. Space Science Reviews 21, 329−354.

Krimigis, S.M., Armstrong, T.P., Axford, W.I., Bostrom, C.O., Fan, C.Y., Gloeckler, G., Lanzerotti, L.J., Keath, E.P., Zwickl, R.D., Carbary, J.F., Hamilton, D.C., 1979. Low-energy charged particle environment at Jupiter-A first look. Science 204, 998−1003.

Krimigis, S.M., Armstrong, T.P., Axford, W.I., Bostrom, C.O., Gloeckler, G., Keath, E.P., Lanzerotti, L.J., Carbary, J.F., Hamilton, D.C., Roelof, E.C., 1981a. Low-energy hot plasma and particles in Saturn's magnetosphere: results from Voyager 1. Science 212, 225−231.

Krimigis, S.M., Carbary, J.F., Keath, E.P., Bostrom, C.O., Axford, W.I., Gloeckler, G., Lanzerotti, L.J., Armstrong, T.P., 1981b. Characteristics of hot plasma in the Jovian magnetosphere: results from the Voyager spacecraft. Journal of Geophysical Research 86, 8227−8257.

Krimigis, S.M., Armstrong, T.P., Axford, W.I., Bostrom, C.O., Gloeckler, G., Keath, E.P., Lanzerotti, L.J., Carbary, J.F., Hamilton, D.C., Roelof, E.C., 1982. Low-energy hot plasma and particles in Saturn's magnetosphere. Science 215, 571−577.

Krimigis, S.M., Armstrong, T.P., Axford, W.I., Cheng, A.F., Gloeckler, G., Hamilton, D.C., Keath, E.P., Lanzerotti, L.J., Mauk, B.H., 1986. The magnetosphere of Uranus: hot plasma and radiation environment. Science 233, 97−102.

Krimigis, S.M., Carbary, J.F., Keath, E.P., Armstrong, T.P., Lanzerotti, L.J., Gloeckler, G., 1983. General characteristics of hot plasma and energetic particles in the Saturnian magnetosphere: results from the Voyager spacecraft. Journal of Geophysical Research 88, 8871−8892.

Krimigis, S.M., et al., 1989. Hot plasma and energetic particles in Neptune's magnetosphere. Science 246, 1483.

Krimigis, S.M., Mauk, B.H., Cheng, A.F., Keath, E.P., Kane, M., Armstrong, T.P., Gloeckler, G., Lanzerotti, L.J., 1990. Hot plasma parameters in Neptune's mangetosphere. Geophysical Research Letters 17, 1685–1688.

Krimigis, S.M., Bostrom, C.O., Armstrong, T.P., Axford, W.I., Fan, C.Y., Gloeckler, G., Lanzerotti, L.J., 1997. The low energy charged particle/LECP/experiment on the Voyager spacecraft. Space Science Reviews 21, 329–354.

Krimigis, S.M., et al., 2004. Magnetospheric imaging instrument (MIMI) on the Cassini mission to Saturn/Titan. Space Science Reviews 114, 233–329.

Krimigis, S.M., Sergis, N., Dialynas, K., Mitchell, D.G., Hamilton, D.C., Krupp, N., Dougherty, M., Sarris, E.T., 2009. Analysis of a sequence of energetic ion and magnetic field events upstream from the Saturnian magnetosphere. Planetary and Space Science 57, 1785–1794.

Krimigis, S.M., Decker, R.B., Roelof, E.C., Hill, M.E., Armstrong, T.P., Gloeckler, G., Hamilton, D.C., Lanzerotti, L.J., 2013. Search for the exit: Voyager 1 at heliosphere's border with the galaxy. Science 341, 144–147.

Krommes, J.A., 1984. Statistical descriptions and plasma physics. In: Handbook of Plasma Physics. Basic Plasma Physics II, vol. 2 (North-Holland, New York).

Kronberg, E.A., Daly, P.W., Dandouras, I., Haaland, S., Georgescu, E., 2010. Generation and validation of ion energy spectra based on Cluster RAPID and CIS measurements. In: Laasko, H., Taylor, M., Escoubet, C.P. (Eds.), The Cluster Active Archive. Studying the Earth's Space Plasma Environment. Springer.

Krupp, N., Lagg, A., Livi, S., Wilken, B., Woch, J., Roelof, E.C., Williams, D.J., 2001. Global flows of energetic ions in Jupiter's equatorial plane: first-order approximation. Journal of Geophysical Research 106, 26017–26032.

Krymsky, 1977. A regular mechanism for the acceleration of charged particles on the front of a shock wave. Doklady Akademii Nauk SSSR 234, 1306–1308.

Kurth, W.S., 1992. Comparative observations of plasma waves at the outer planets. Advances in Space Research 12, 83–90.

La Mantia, M., Skrbek, L., 2014. Quantum, or classical turbulence? Europhysics Letters 105, 46002.

La Porta, A., Voth, G.A., Crawford, A.M., Alexander, J., Bodenschatz, E., 2001. Fluid particle accelerations in fully developed turbulence. Nature 409, 1017–1019.

Lacombe, C., Salem, C., Mangeney, A., Hubert, D., Perche, C., Bougeret, J.-L., Kellogg, P.J., Bosqued, J.-M., 2002. Evidence for the interplanetary electric potential? Wind observations of electrostatic fluctuations. Annales Geophysicae 20, 609.

Lakhina, G.S., Singh, S.V., 2015. Generation of weak double layers and low-frequency electrostatic waves in the solar wind. Solar Physics 290, 3033–3049.

Lakhina, G.S., Kakad, A.P., Singh, S.V., Verheest, F., 2008a. Ion- and electron-acoustic solitons in two electron temperature space plasmas. Physics of Plasmas 15, 062903.

Lakhina, G.S., Singh, S.V., Kakad, A.P., Verheest, F., Bharuthram, R., 2008b. Study of nonlinear ion- and electron-acoustic waves in multi-component space plasmas. Nonlinear Processes in Geophysics 15, 903.

Lakhina, G.S., Singh, S.V., Kakad, A.P., Goldstein, M.L., Vinas, A.F., Pickett, J.S., 2009. A mechanism for electrostatic solitary structures in the Earth's magnetosheath. Journal of Geophysical Research 114, A09212.

Lakhina, G.S., Singh, S.V., Kakad, A.P., 2011a. Ion- and electron- acoustic solitons and double layers in multicomponent space plasmas. Advances in Space Research 47, 1558.

Lakhina, G.S., Singh, S.V., Kakad, A.P., Pickett, J.S., 2011b. Generation of electrostatic solitary waves in the plasma sheet boundary layer. Journal of Geophysical Research 116, A10218.

Lakhina, G.S., Singh, S.V., Kakad, A.P., 2014. Ion acoustic solitons/double layers in two-ion plasma revisited. Physics of Plasmas 21, 062311.

Laming, J.M., Moses, J.D., Ko, Y.-K., Ng, C.K., Rakowski, C.E., Tylka, A.J., 2013. On the remote detection of suprathermal ions in the solar corona and Their role as seeds for solar energetic particle production. The Astrophysical Journal 770, 73.

Lamy, H., Pierrard, V., Maksimovic, M., Lemaire, J., 2003. A kinetic exospheric model of the solar wind with a non monotonic potential energy for the protons. Journal of Geophysical Research 108, 1047−1057.

Landi, E., Young, P.R., 2010. The relative intensity calibration of Hinode/EIS and SOHO/SUMER. The Astrophysical Journal 714, 636−643.

Landi, S., Pantellini, F.G.E., 2001. On the temperature profile and heat flux in the solar corona: kinetic simulations. Astronomy and Astrophysics 372, 686.

Landi, E., Young, P.R., Dere, K.P., Del Zanna, G., Mason, H.E., 2013. CHIANTI − an atomic database for emission lines. XIII. Soft X-ray improvements and other changes. The Astrophysical Journal 763, 86.

Landi, S., Matteini, L., Pantellini, F., 2014. Electron heat flux in the solar wind: are we observing the collisional limit in the 1 au data? The Astrophysical Journal 790, L12.

Lane, J.H., 1870. On the theoretical temperature of the Sun under the hypothesis of a gaseous mass maintaining its volume by its internal heat and depending on the laws of gases known to terrestrial experiment. American Journal of Science and Arts 2, 57−74.

Lanzerotti, L.J., et al., 1992. Heliosphere instrument for spectra, composition, and anisotropy at low energies. Astronomy and Astrophysics Supplement Series 92, 349−363.

Lario, D., Hu, Q., Ho, G.C., Decker, R.B., Roelof, E.C., Smith, C.W., 2005. Statistical properties of fast forward transient interplanetary shocks and associated energetic particle events: ACE observations. In: Fleck, B., Zurbuchen, T.H., Lacoste, H. (Eds.), Proc. Solar Wind 11−SOHO 16 "Connecting Sun and Heliosphere". ESA SP-592, Noordwijk, the Netherlands, pp. 81−86.

Lawrence, D.J., et al., 2015. Comprehensive survey of energetic electron events in Mercury's magnetosphere with data from the MESSENGER Gamma-Ray and Neutron Spectrometer. Journal of Geophysical Research 120, 2851−2876.

Layden, B., Percival, D.J., Cairns, I.H., Robinson, P.A., 2011. First-order thermal correction to the quadratic response tensor and rate for second harmonic plasma emission. Physics of Plasmas 18, 022309.

Layden, A., Cairns, I.H., Li, B., Robinson, P.A., 2013. Electrostatic decay in a weakly magnetized plasma. Physical Review Letters 110, 185001.

Lazar, M., Poedts, S., 2009. Firehose instability in space plasmas with bi-kappa distributions. Astronomy and Astrophysics 494, 311−315.

Lazar, M., Poedts, S., 2014. Instability of the parallel electromagnetic modes in Kappa distributed plasmas − II. Electromagnetic ion-cyclotron modes. Monthly Notices of the Royal Astronomical Society 437, 641−648.

Lazar, M., Poedts, S., Schlickeiser, R., 2011a. Instability of the parallel electromagnetic modes in Kappa distributed plasmas − I. Electron whistler-cyclotron modes. Monthly Notices of the Royal Astronomical Society 410, 663.

Lazar, M., Poedts, S., Schlickeiser, R., 2011b. Proton firehose instability in bi-Kappa distributed plasmas. Astronomy and Astrophysics 534, A116.

Lazar, M., Yoon, P.H., Schlickeiser, R., 2012a. Spontaneous electromagnetic fluctuations in unmagnetized plasmas. III. Generalized kappa distributions. Physics of Plasmas 19, 122108.

Lazar, M., Schlickeiser, R., Poedts, S., 2012b. Suprathermal particle populations in the solar wind and corona. In: Lazar, M. (Ed.), Exploring the Solar Wind. InTech, ISBN 978-953-51-0339-4.

Lazar, M., Pierrard, V., Schlickeiser, R., Poedts, S., 2012c. Modeling space plasma dynamics with anisotropic Kappa distributions. Astrophysics and Space Science 33, 97−107.

Lazar, M., 2012. The electromagnetic ion-cyclotron instability in bi-Kappa distributed plasmas. Astronomy and Astrophysics 547, A94.

Le Chat, G., Issautier, K., Meyer-Vernet, N., Zouganelis, I., Maksimovic, M., Montcuquet, M., 2009. Quasi-thermal noise in space plasma: kappa distributions. Physics of Plasmas 16, 102903.

Le Chat, G., Issautier, K., Meyer-Vernet, N., Hoang, S., 2011. Large-scale variation of solar wind electron properties from quasi-thermal noise spectroscopy: Ulysses measurements. Solar Physics 271, 141–148.

Le Roux, J.A., Fichtner, H., Zank, G.P., 2000. Self-consistent acceleration of multiply reflected pickup ions at a quasi-perpendicular solar wind termination shock: a fluid approach. Journal of Geophysical Research 105, 12557–12578.

Le Roux, J.A., Webb, G.M., Shalchi, A., Zank, G.P., 2010. A generalized nonlinear guiding center theory for the collisionless anomalous perpendicular diffusion of cosmic rays. The Astrophysical Journal 716, 671–692.

Lee, M.A., Ip, W.-H., 1987. Hydromagnetic wave excitation by ionised interstellar hydrogen and helium in the solar wind. Journal of Geophysical Research 92, 11041–11052.

Lee, M.A., Shapiro, V.D., Sagdeev, R.Z., 1996. Pickup Ion Energization by Shock Surfing. Journal of Geophysical Research 101, 4777–4789.

Lee, R.E., Chapman, S.C., Dendy, R.O., 2005. Reforming perpendicular shocks in the presence of pickup protons: initial ion acceleration. Annales Geophysicae 23, 643–650.

Lee, E., Williams, D.R., Lapenta, G., 2013. Spectroscopic Indication of Suprathermal Ions in the Solar Corona arXiv: 1305.2939v.

Lemaire, J., 2010. Half a century of solar wind kinetic models. In: Maksimovic, et al. (Eds.), Twelfth International Solar Wind Conference, AIP CP, vol. 1216, p. 8.

Lemaire, J., Scherer, M., 1971. Kinetic models of the solar wind. Journal of Geophysical Research 76, 7479–7490.

Lemaire, J., Pierrard, V., 2001. Kinetic models of solar and polar winds. Astrophysics and Space Science 277, 169–180.

Lembège, B., Savoini, P., Hellinger, P., Trávníček, P.M., 2009. Nonstationarity of a two-dimensional perpendicular shock: Competing mechanisms. Journal of Geophysical Research 114, 3217.

Lemen, J.R., Title, A.M., Akin, D.J., Boerner, P.F., Chou, C., Drake, J.F., et al., 2012. The atmospheric imaging assembly (AIA) on the solar dynamics observatory (SDO). Solar Physics 275, 17–40.

Lennartsson, W., Shelley, E.G., 1986. Survey of 0.1- to 16-keV plasma sheet ion composition. Journal of Geophysical Research 91, 3061–3076.

Lennon, D.J., Burke, V.M., 1994. Atomic data from the IRON project. II. Effective collision strength S for infrared transitions in carbon-like ions. Astronomy and Astrophysics 103, 273–277.

Leroy, M.M., 1983. Structure of perpendicular shocks in collisionless plasma. Physics of Fluids 26, 2742–2753.

Leubner, M.P., 1982. On Jupiter's Whistler emission. Journal of Geophysical Research 87, 6335–6338.

Leubner, M.P., 2002. A nonextensive entropy approach to kappa distributions. Astrophysics and Space Science 282, 573–579.

Leubner, M.P., 2004a. Core-Halo distribution functions: a natural equilibrium state in generalized thermostatistics. The Astrophysical Journal 604, 469–478.

Leubner, M.P., 2004b. Fundamental issues on kappa-distributions in space plasmas and interplanetary proton distributions. Physics of Plasmas 11, 1308–1316.

Leubner, M.P., Vörös, Z., 2005. A nonextensive entropy approach to solar wind intermittency. The Astrophysical Journal 618, 547–555.

Li, B., Willes, A.J., Robinson, P.A., Cairns, I.H., 2003. Dynamics of beam driven Langmuir and ion-acoustic waves including electrostatic decay. Physics of Plasmas 10, 2748.

Li, B., Cairns, I.H., 2013a. Type III radio bursts in coronal plasmas with kappa particle distributions. The Astrophysical Journal 763, L34.

Li, B., Cairns, I.H., 2013b. Type III bursts produced by power-law injected electrons in Maxwellian background coronal plasmas. Journal of Geophysical Research 118, 4748–4759.

Li, B., Cairns, I.H., 2014. Fundamental emission of type III bursts produced in non-Maxwellian coronal plasmas with kappa-distributed background particles. Solar Physics 289, 951.

Li, B., Cairns, I.II., Robinson, P.A., 2008a. Simulations of coronal type III solar radio bursts: 1. Simulation model. Journal of Geophysical Research 113, A06104.

Li, B., Cairns, I.H., Robinson, P.A., 2008b. Simulations of coronal type III solar radio bursts: 2. Dynamic spectrum for typical parameters. Journal of Geophysical Research 113, A06105.

Li, B., Robinson, P.A., Cairns, I.H., 2006a. Numerical simulation of type III solar radio bursts. Physical Review Letters 96, 145005.

Li, B., Robinson, P.A., Cairns, I.H., 2006b. Numerical modelling of type III solar radio bursts in the inhomogeneous solar corona and interplanetary medium. Physics of Plasmas 13, 092902.

Li, B., Cairns, I.H., Robinson, P.A., 2011a. Effects of spatial variations in coronal temperatures on type III bursts. I. Variations in electron temperature. The Astrophysical Journal 730, 20.

Li, B., Cairns, I.H., Robinson, P.A., 2011b. Effects of spatial variations in coronal temperatures on type III bursts. I. Variations in ion temperature. The Astrophysical Journal 730, 21.

Li, B., Cairns, I.H., Robinson, P.A., 2012. Frequency fine structures of type III bursts due to localized medium-scale density structures along paths of type III beams. Solar Physics 279, 173.

Li, B., Robinson, P.A., Cairns, I.H., 2002. Multiple electron beam propagation and Langmuir wave generation in plasmas. Physics of Plasmas 9, 2976.

Lie-Svendsen, O., Hansteen, V.H., Leer, E., 1997. Kinetic electrons in high-speed solar wind streams: formation of high-energy tails. Journal of Geophysical Research 102, 4701–4718.

Lin, R.P., 1998. Wind observations of suprathermal electrons in interplanetary medium. Space Science Reviews 86, 61.

Lin, R.P., Hudson, H.S., 1971. 10–100 keV electron acceleration and emission from solar flares. Solar Physics 17, 412–435.

Lin, R.P., et al., 1996. Observation of an impulsive solar electron event extending down to 0.5 keV energy. Geophysical Research Letters 23, 1211–2013.

Lin, R.P., et al., 1997. Observations of the solar wind, the bow shock, and upstream particles with the wind 3D plasma instrument. Advances in Space Research 20, 645–654.

Lin, R.P., Dennis, B.R., Hurford, G.J., Smith, D.M., Zehnder, A., Harvey, P.R., et al., 2002. The Reuven Ramaty high-energy solar spectroscopic imager (RHESSI). Solar Physics 210, 3–32.

Lipatov, A.S., Zank, G.P., 1999. Pickup ion acceleration at low-βp perpendicular shocks. Physical Review Letters 82, 3609.

Liu, W.W., Rostoker, G., 1995. Energetic ring current particles generated by recurring substorm cycles. Journal of Geophysical Research 100, 21897–21910.

Liu, X.-W., Storey, P.J., Barlow, M.J., Danziger, I.J., Cohen, M., Bryce, M., 2000. NGC 6153: a super-metal-rich planetary nebula? Monthly Notices of the Royal Astronomical Society 312, 585–628.

Liu, Y., Liu, S.Q., Dai, B., Xue, T.L., 2014. Dispersion and damping rates of dispersive Alfvén wave in a nonextensive plasma. Physics of Plasmas 21, 032125.

Liu, H.-F., Tanga, C.-J., Zhanga, X., Zhua, L.-M., Zhao, Y., 2015. Pickup of thermal non-equilibrium ions by a low-frequency Alfvén wave via nonresonant and stochastic interaction. Advances in Space Research 56, 2298–2304.

Liu, X.-W., 2002. Optical recombination lines and temperature fluctuations. In: Henney, W.J., Franco, J., Martos, M. (Eds.), Revista Mexicana de Astronomia y Astrofisica Conference Series, vol. 12, pp. 70–76.

Livadiotis, G., 2005. Numerical approximation of the percentage of order for one-dimensional maps. Advances in Complex Systems 8, 15–32.

Livadiotis, G., 2007. Approach to general methods for fitting and their sensitivity. Physica A 375, 518–536.

Livadiotis, G., 2008. Approach to the block entropy modeling and optimization. Physica A 387, 2471–2494.

Livadiotis, G., 2009. Approach on Tsallis statistical interpretation of hydrogen-atom by adopting the generalized radial distribution function. Journal of Mathematical Chemistry 45, 930–939.

Livadiotis, G., 2012. Expectation value & variance based on Lp norms. Entropy 14, 2375–2396.

Livadiotis, G., 2014a. Lagrangian temperature: derivation and physical meaning for systems described by kappa distributions. Entropy 16, 4290–4308.

Livadiotis, G., 2014b. Chi-p distribution: characterization of the goodness of the fitting using Lp norms. Journal of Statistical Distributions and Applications 1, 4.

Livadiotis, G., 2015a. Statistical background and properties of kappa distributions in space plasmas. Journal of Geophysical Research 120, 1607–1619.

Livadiotis, G., 2015b. Kappa distribution in the presence of a potential energy. Journal of Geophysical Research 120, 880–903.

Livadiotis, G., 2015c. Kappa and q indices: dependence on the degrees of freedom. Entropy 17, 2062–2081.

Livadiotis, G., 2015d. Application of the theory of Large-Scale Quantization to the inner heliosheath. Journal of Physics: Conference Series 577, 012018, pp. 7.

Livadiotis, G., 2015e. Shock strength in space and astrophysical plasmas. The Astrophysical Journal 809, 111, pp. 21.

Livadiotis, G., 2016a. Curie law for systems described by kappa distributions. Europhysics Letters 113, 10003, pp. 6.

Livadiotis, G., 2016b. Non-Euclidean-normed statistical mechanics. Physica A 445, 240–255.

Livadiotis, G., 2016c. Superposition of polytropa in the inner heliosheath. The Astrophysical Journal Supplement Series 223, 13, pp. 13.

Livadiotis, G., 2016d. Modeling anisotropic Maxwell–Jüttner distributions: derivation and properties. Annales Geophysicae 34, 1–14.

Livadiotis, G., 2017a. Evidence for a large-scale Compton wavelength in space plasmas (submitted for publication).

Livadiotis, G., 2017b. Law of large numbers for non-euclidean Lp means (submitted for publication).

Livadiotis, G., Desai, M.I., 2016. Plasma-field coupling at small length scales in solar wind near 1 au. The Astrophysical Journal 829, 88, pp. 14.

Livadiotis, G., McComas, D.J., 2009. Beyond kappa distributions: exploiting Tsallis statistical mechanics in space plasmas. Journal of Geophysical Research 114, A11105, pp. 21.

Livadiotis, G., McComas, D.J., 2010a. Exploring transitions of space plasmas out of equilibrium. The Astrophysical Journal 714, 971–987.

Livadiotis, G., McComas, D.J., 2010b. Measure of the departure of the q-metastable stationary states from equilibrium. Physica Scripta 82, 035003, pp. 9.

Livadiotis, G., McComas, D.J., 2010c. Non-equilibrium stationary states in the Heliosphere: the influence of pick-up ions. AIP Conference Proceedings 1302, 70–76.

Livadiotis, G., McComas, D.J., 2011a. The influence of pick-up ions on space plasma distributions. The Astrophysical Journal 738, 64, pp. 13.

Livadiotis, G., McComas, D.J., 2011b. Invariant kappa distribution in space plasmas out of equilibrium. The Astrophysical Journal 741, 88, pp. 28.

Livadiotis, G., McComas, D.J., 2012. Non-equilibrium thermodynamic processes: space plasmas and the inner heliosheath. The Astrophysical Journal 749, 11, pp. 4.

Livadiotis, G., McComas, D.J., 2013a. Understanding kappa distributions: a toolbox for space science and astrophysics. Space Science Reviews 75, 183–214.

Livadiotis, G., McComas, D.J., 2013b. Evidence of large scale phase space quantization in plasmas. Entropy 15, 1118–1132.

Livadiotis, G., McComas, D.J., 2013c. Fitting method based on correlation maximization: applications in Astrophysics. Journal of Geophysical Research 118, 2863–2875.

Livadiotis, G., McComas, D.J., 2013d. Near-equilibrium heliosphere – far-equilibrium heliosheath. AIP Conference Proceedings 1539, 344–350.

Livadiotis, G., McComas, D.J., 2014a. Electrostatic shielding in plasmas and the physical meaning of the Debye length. Journal of Plasma Physics 80, 341–378.

Livadiotis, G., McComas, D.J., 2014b. Large-scale quantization from local correlations in space plasmas. Journal of Geophysical Research 119, 3247–3258.

Livadiotis, G., McComas, D.J., 2014c. Large-scale quantization in space plasmas: Summary and applications. ASP Conference Series 484, 130, pp. 6.

Livadiotis, G., Moussas, X., 2007. The sunspot as an autonomous dynamical system: a model for the growth and decay phases of sunspots. Physica A 379, 436–458.

Livadiotis, G., McComas, D.J., Dayeh, M.A., Funsten, H.O., Schwadron, N.A., 2011. First sky map of the inner heliosheath temperature using IBEX spectra. The Astrophysical Journal 734, 1, pp. 19.

Livadiotis, G., McComas, D.J., Randol, B., Möbius, E., Dayeh, M.A., Frisch, P.C., Funsten, H.O., Schwadron, N.A., Zank, G.P., 2012. Pick-up ion distributions and their influence on ENA spectral curvature. The Astrophysical Journal 751, 64, pp. 21.

Livadiotis, G., McComas, D.J., Schwadron, N.A., Funsten, H.O., Fuselier, S.A., 2013. Pressure of the proton plasma in the inner heliosheath. The Astrophysical Journal 762, 134, pp. 19.

Livadiotis, G., Assas, L., Dennis, B., Elaydi, S., Kwessi, E., 2015. A discrete time host-parasitoid model with an Allee effect. Journal of Biological Dynamics 9, 34–51.

Livadiotis, G., Assas, L., Dennis, B., Elaydi, S., Kwessi, E., 2016. Kappa function as a unifying framework for discrete population modeling. Natural Resource Modeling 29, 130–144.

Livi, R., Goldstein, J., Burch, J.L., Crary, F., Rymer, A.M., Mitchell, D.G., Persoon, A.M., 2014. Multi-instrument analysis of plasma parameters in Saturn's equatorial, inner magnetosphere using corrections for spacecraft potential and penetrating background radiation. Journal of Geophysical Research 119, 3683–3707.

Ljepojevic, N.N., MacNiece, P., 1988. Non-Maxwellian distribution functions in flaring coronal loops – comparison of Landau-Fokker-Planck and BGK solutions. Solar Physics 117, 123–133.

Lopez, R.E., Sibeck, D.G., McEntire, R.W., Krimigis, S.M., 1990. The energetic ion substorm injection boundary. Journal of Geophysical Research 95, 109–117.

Luhn, A., Hovestadt, D., 1985. Calculations of heavy ion charge state distributions for non-equilibrium conditions. In: Jones, F.C., Adams, J., Mason, G.M. (Eds.), 19th Int. Cosmic-Ray Conf. (La Jolla). NASA, Washington, pp. 245–248.

Lund, E.J., LaBelle, J., Treumann, R.A., 1994. On quasi-thermal fluctuations near the plasma frequency in the outer plasmasphere: A case study. Journal of Geophysical Research 99, 23651–23660.

Luzzi, R., Vasconcellos, A.R., Casas-Vázquez, J., Jou, D., 1997. Characterization and measurement of a non-equilibrium temperature-like variable in irreversible thermodynamics. Physica A 34, 699–714.

Mace, R.L., Hellberg, M.A., 1995. A dispersion function for plasmas containing superthermal particles. Physics of Plasmas 2, 2098–2109.

Mace, R.L., Hellberg, M.A., 2009. A new formulation and simplified derivation of the dispersion function for a plasma with a kappa velocity distribution. Physics of Plasmas 16, 072113.

Mace, R.L., Sydora, R.D., 2010. Parallel whistler instability in a plasma with an anisotropic bi-kappa distribution. Journal of Geophysical Research 115, A07206.

Mace, R.L., Hellberg, M.A., Treumann, R.A., 1998. Electrostatic fluctuations in plasmas containing suprathermal particles. Journal of Plasma Physics 59, 393.

Mace, R.L., Amery, G., Hellberg, M.A., 1999. The electron-acoustic mode in a plasma with hot suprathermal and cool Maxwellian electrons. Physics of Plasmas 6, 44.

Mace, R.L., Sydora, R.D., Silin, I., 2011. Effects of superthermal ring current ion tails on the electromagnetic ion cyclotron instability in multi-ion magnetospheric plasmas. Journal of Geophysical Research 116, A05206.

Mace, R.L., 1996. A dielectric tensor for a uniform magnetoplasma with a generalized Lorentzian distribution. Journal of Plasma Physics 55, 415–429.

Mace, R.L., 1998. Whistler instability enhanced by suprathermal electrons within the Earth's foreshock. Journal of Geophysical Research 103, 14643–14654.

Mace, R.L., 2003. A Gordeyev integral for electrostatic waves in a mag-netized plasma with a kappa velocity distribution. Physics of Plasmas 10, 2181–2193.

Mace, R.L., 2004. Generalized electron Bernstein modes in a plasma with a kappa velocity distribution. Physics of Plasmas 11, 507–522.

Mackovjak, Š., Dzifčáková, E., Dudík, J., 2013. On the possibility to diagnose the non-maxwellian κ-distributions from the Hinode/EIS EUV spectra. Solar Physics 282, 263–281.

Mackovjak, Š., Dzifčáková, E., Dudík, J., 2014. Differential emission measure analysis of active region cores and quiet Sun for the non-Maxwellian κ-distributions. Astronomy and Astrophysics 564, A130.

Maharaj, S.K., Bharuthram, R., Singh, S.V., Lakhina, G.S., 2012a. Existence domains of arbitrary amplitude nonlinear structures in two-electron temperature space plasmas. I. Low-frequency ion-acoustic solitons. Physics of Plasmas 19, 072320.

Maharaj, S.K., Bharuthram, R., Singh, S.V., Lakhina, G.S., 2012b. Existence domains of arbitrary amplitude nonlinear structures in two-electron temperature space plasmas. II. High-frequency electron-acoustic solitons. Physics of Plasmas 19, 122301.

Maksimovic, M., Pierrard, V., Lemaire, J., 1997a. A kinetic model of the solar wind with Kappa distributions in the corona. Astronomy and Astrophysics 324, 725–734.

Maksimovic, M., Pierrard, V., Riley, P., 1997b. Ulysses electron distributions fitted with Kappa functions. Geophysical Research Letters 24, 1151–1154.

Maksimovic, M., Zouganelis, I., Chaufray, J.-Y., Issautier, K., Scime, E.E., Littleton, J., Marsch, E., McComas, D.J., Salem, C., Lin, R.P., Elliot, H., 2005. Radial evolution of the electron distribution functions in the fast solar wind between 0.3 and 1.5 au. Journal of Geophysical Research 110, A09104.

Malacarne, L.C., Mendes, R.S., Lenzi, E.K., 2001. Average entropy of a subsystem from its average Tsallis entropy. Physical Review E 65, 017106.

Mangeney, A., Salem, C., Lacombe, C., Bougeret, J.-L., Perche, C., Manning, R., Kellogg, P.J., Goetz, K., Monson, S.J., Bosqued, J.-M., 1999. Wind observations of coherent electrostatic waves in the solar wind. Annales Geophysicae 17, 307–320.

Mann, G., Classen, H.T., Keppler, E., Roelof, E.C., 2002. On electron acceleration at CIR related shock waves. Astronomy and Astrophysics 391, 749–756.

Mann, G., Warmuth, A., Aurass, H., 2009. Generation of highly energetic electrons at reconnection outflow shocks during solar flares. Astronomy and Astrophysics 494, 669–675.

Maor, E., 1994. "e": The story of a Number. Princeton University Press, Princeton, New Jersey.

Marsch, E., 2006. Kinetic physics of the solar corona and solar wind. Living Reviews in Solar Physics 3, 1–100.

Marsch, E., Muhlhauser, K.-H., Rosenbauer, H., Schwenn, R., Neubauer, F.M., 1982. Solar wind helium ions: observations of the Helios solar probes between 0.3 and 1 au. Journal of Geophysical Research 87, 3552–3572.

Mason, G.M., Mazur, J.E., Dwyer, J.R., Jokipii, J.R., Gold, R.E., Krimigis, S.M., 2004. Abundances of heavy and ultraheavy ions in 3He-rich solar flares. The Astrophysical Journal 606, 555–564.

Mason, H.E., Mosignori-Fossi, B.C., 1994. Spectroscopic diagnostics in the VUV for solar and stellar plasmas. Astronomy and Astrophysics Review 6, 123–179.

Mason, G.M., et al., 2008. Abundances and Energy Spectra of Corotating Interaction Region Heavy Ions Observed during Solar Cycle 23. The Astrophysical Journal 678, 1458.

Matsukiyo, S., Scholer, M., Burgess, D., 2007. Pickup protons at quasi-perpendicular shocks: full particle electrodynamic simulations. Annales Geophysicae 25, 283–291.

Matsumoto, H., Kojima, H., Miyatake, T., Omura, Y., Okada, Y.M., Nagano, I., Tsutui, M., 1994. Electrostatic solitary waves (ESW) in the magnetotail: Ben wave forms observed by geotail. Geophysical Research Letters 21, 2915.

Matthaeus, W.H., Qin, G., Bieber, J.W., Zank, G.P., 2003. Nonlinear Collisionless Perpendicular Diffusion of Charged Particles. The Astrophysical Journal Letters 590, L53–L56.

Mauk, B.H., 1986. Quantitative modeling of the "convection surge" mechanism of ion acceleration. Journal of Geophysical Research 91, 13423.

Mauk, B.H., 2014. Comparative investigation of the energetic ion spectra comprising the magnetospheric ring currents of the solar system. Journal of Geophysical Research 119, 9729–9746.

Mauk, B.H., Meng, C.I., 1983. Characterizations of geostationary particle signatures based on the injection boundary model. Journal of Geophysical Research 88, 3055–3071.

Mauk, B.H., Fox, N.J., 2010. Electron radiation belts of the solar system. Journal of Geophysical Research 115, A12220.

Mauk, B.H., Krimigis, S.M., Keath, E.P., Cheng, A.F., Armstrong, T.P., Lanzerotti, L.J., Gloeckler, G., Hamilton, D.C., 1987. The hot plasma and radiation environment of the Uranian magnetosphere. Journal of Geophysical Research 92, 15283.

Mauk, B.H., Kane, M., Keath, E.P., Cheng, A.F., Krimigis, S.M., Armstrong, T.P., Ness, N.F., 1990. Energetic charged particle angular distributions near ($r < 2R_N$) and over the pole of Neptune. Geophysical Research Letters 17, 1701.

Mauk, B.H., Keath, E.P., Kane, M., Krimigis, S.M., Cheng, A.F., Acuna, M.A., Armstrong, T.P., Ness, N.F., 1991. The magnetosphere of Neptune: hot plasmas and energetic particles. Journal of Geophysical Research 96, 19061–19084.

Mauk, B., Keath, E., Krimigis, S., 1994. Unusual Satellite-Electron signature within the uranian magnetosphere and its implications regarding whistler electron loss processes. Journal of Geophysical Research 99, 19441–19450.

Mauk, B.H., Gary, S.A., Kane, M., Keath, E.P., Krimigis, S.M., Armstrong, T.P., 1996. Hot plasma parameters of Jupiter's inner magnetosphere. Journal of Geophysical Research 101, 7685–7695.

Mauk, B.H., McIntire, R.W., Williams, D.J., Lagg, A., Roelof, E.C., Krimigis, S.M., Armstrong, T.P., Fritz, T.A., Lanzerotti, L.J., Roederer, J.G., Wilken, B., 1998. Galieo-measured depletion of near-Io hot ring current plasmas since the Voyager epoch. Journal of Geophysical Research 103, 4715–4722.

Mauk, B.H., Mitchell, D.G., McEntire, R.W., Paranicas, C.P., Roelof, E.C., Williams, D.J., Krimigis, S.M., Lagg, A., 2004. Energetic ion characteristics and neutral gas interactions in Jupiter's magnetosphere. Journal of Geophysical Research 109, A09S12.

Mauk, B.H., et al., 2005. Energetic particle injections in Saturn's magnetosphere. Geophysical Research Letters 32, L14S05.

Maxwell, J.C., 1866. On the dynamical theory of gases. Philosophical Magazine 32, 390–393.

Mazzotta, P., Mazzitelli, G., Colafrancesco, S., Vittorio, N., 1998. Ionization balance for optically thin plasmas: rate coefficients for all atoms and ions of the elements H to Ni. Astronomy and Astrophysics Supplement Series 133, 403–409.

Mbuli, L.N., Maharaj, S.K., Bharuthram, R., Singh, S.V., Lakhina, G.S., 2015. Physics of Plasmas 22, 062307.

McComas, D., et al., 2008. The solar wind around Pluto (SWAP) instrument aboard New Horizons. Space Science Reviews 140, 261–313.

McComas, D.J., Allegrini, F., Bochsler, P., Bzowski, M., Christian, E.R., Crew, G.B., DeMajistre, R., Fahr, H., Fichtner, H., Frisch, P.C., Funsten, H.O., Fuselier, S.A., Glöckler, G., Gruntman, M., Heerikhuisen, J., Izmodenov, V., Janzen, P., Knappenberger, P., Krimigis, S., Kucharek, H., Lee, M., Livadiotis, G., Livi, S., MacDowall, R.J., Mitchell, D., Möbius, E., Moore, T., Pogorelov, N.V., Reisenfeld, D., Rölof, E., Saul, L., Schwadron, N.A., Valek, P.W., Vanderspek, R., Wurz, P., Zank, G.P., 2009. Global obser¬vations of the interstellar interaction from the Interstellar Boundary Explorer. Science 326, 959–962.

McComas, D.J., Bzowski, M., Frisch, P., Crew, G.B., Dayeh, M.A., DeMajistre, R., Funsten, H.O., Fuselier, S.A., Gruntman, M., Janzen, P., Kubiak, M.A., Livadiotis, G., Möbius, E., Reisenfeld, D., Schwadron, N.A., 2010. Evolving outer heliosphere: large-scale stability and time variations observed by IBEX. Journal of Geophysical Research 115, A09113, pp. 18.

McComas, D.J., Dayeh, M.A., Allegrini, F., Bzowski, M., DeMajistre, R., Fujiki, K., Funsten, H.O., Fuselier, S.A., Gruntman, M., Janzen, P.H., Kubiak, M.A., Livadiotis, G., Möbius, E., Reisenfeld, D.B., Reno, M., Schwadron, N.A., Sokol, J.M., Tokumaru, M., 2012a. The first 3 years of IBEX observations & our evolving heliosphere. The Astrophysical Journal Supplement Series 203, 1, pp. 36.

McComas, D.J., et al., 2012b. The Heliosphere's Interstellar Interaction: No Bow Shock. Science 336, 1291–1293.

McComas, D.J., et al., 2014. IBEX: The first five years (2009–2013). The Astrophysical Journal 213, 20.

McIlwain, C.E., Fillius, R.W., 1975. Differential spectra and phase space densities of trapped electrons at Jupiter. Journal of Geophysical Research 80, 1341–1345.

McIntosh, S.W., Charbonneau, P., 2001. Geometric Effects in Avalanche Models of Solar Flares: Implications for Coronal Heating. The Astrophysical Journal 563, L165.

McIntosh, S.W., et al., 2002. Geometrical properties of avalanches in self-organized critical models of solar flares. Physical Review E 65, 046125.

McNutt Jr., R.L., et al., 1987a. The low energy plasma in the Uranian magnetosphere. Advances in Space Research 7, 237–241.

McNutt Jr., R.L., Selesnick, S.S., Richardson, J.D., 1987b. Low-energy plasma observations in the magnetosphere of Uranus. Journal of Geophysical Research 92, 4399–4410.

Melrose, D.B., 1980. Plasma Astrophysics. Gordon and Breach, New York.

Melrose, D.B., 1986. Instabilities in Space and Laboratory Plasmas. Cambridge, Cambridge.

Menzel, D.H., Aller, L.H., Hebb, M.H., 1941. Physical processes in gaseous nebulae. XIII. The Astrophysical Journal 93, 230.

Metzner, C., et al., 2015. Superstatistical analysis and modelling of heterogeneous random walks. Nature Communications 6, 7516.

Mewe, R., 1972. Interpolation formulae for the electron impact excitation of ions in the H-, He-, Li-, and Ne- sequences. Astronomy and Astrophysics 20, 215.

Meyer-Vernet, N., 1999. How does the solar wind blow? A simple kinetic model. European Journal of Physics 20, 167–176.

Meyer-Vernet, N., 2001. Large scale structure of planetary environments: the importance of not being Maxwellian. Planetary and Space Science 49, 247–260.

Meyer-Vernet, N., 2007. Basics of the Solar Wind, Atmos. Space Sci. Ser., Cambridge, 463 p.

Meyer-Vernet, N., Perche, C., 1989. Toolkit for antennae and thermal noise near the plasma frequency. Journal of Geophysical Research 94, 2405–2415.

Meyer-Vernet, N., Issautier, K., 1998. Electron temperature in the solar wind: generic radial variation from a kinetic collisionless model. Journal of Geophysical Research 103, 29705—29717.

Meyer-Vernet, N., Moncuquot, M., Hoang, S., 1995. Temperature inversion in the Io plasma torus. Icarus 116, 202—213.

Miah, S., Beck, C., 2014. Lagrangian quantum turbulence model based on alternating superfluid/normal fluid stochastic dynamics. Europhysics Letters 108, 40004.

Mihalov, J.D., Wolfe, J.II., Frank, L.A., 1976. Survey for non-Maxwellian plasma in Jupiter's magnetosheath. Journal of Geophysical Research 81, 3412—3416.

Milovanov, A.V., Zelenyi, L.M., 2000. Functional background of the Tsallis entropy: coarse-grained systems and kappa distribution functions. Nonlinear Processes in Geophysics 7, 211—221.

Mitchell, D.G., et al., 1998. The imaging neutral camera for the cassini mission to Saturn and Titan, measurements techniques in space plasmas: fields. Geophysical Monograph 103, 281—287.

Mitchell, D.G., et al., 2005. Energetic ion acceleration in Saturn's magetosphere: substorms on Saturn? Geophysical Research Letters 32, L20S01.

Mitchell, D.G., et al., 2009. Recurrent energization of plasma in the midnight-to-dawn quadrant of Saturn's magnetosphere, and its relationship to auroral UV and radio emissions. Planetary and Space Science 57, 1732—1742.

Mitchell, D.G., Brandt, P.C., Roelof, E.C., Hamilton, D.C., Retterer, K.C., Mende, S., 2003. Global imaging of O+ from IMAGE/HENA. Space Science Reviews 109, 63—75.

Miyamoto, S., Kimura, K., Kitamoto, S., Dotani, T., Ebisawa, K., 1991. The Astrophysical Journal 383, 784.

Möbius, E., et al., 2012. Interstellar gas flow parameters derived from interstellar boundary explorer-Lo observations in 2009 and 2010: analytical analysis. The Astrophysical Journal Supplement Series 198, 11.

Moncuquet, M., Bagenal, F., Meyer-Vernet, N., 2002. Latitudinal structure of the outer Io plasma torus. Journal of Geophysical Research 108, 1260.

Montemurro, A., 2001. Beyond the Zipf-Mandelbrot law in quantitative linguistics. Physica A 300, 567—578.

Montgomery, D.C., Tidman, D.A., 1964. Plasma Kinetic Theory. McGraw-Hill Inc, New York.

Montgomery, M.D., Bame, S.J., Hundhausen, A.J., 1968. Solar wind electrons: Vela 4 measurements. Journal of Geophysical Research 73, 4999—5003.

Moore, L., Mendillo, M., 2005. Ionospheric contribution to Saturn's inner plasmasphere. Journal of Geophysical Research 110, A05310.

Moyano, L.G., Tsallis, C., Gell-Mann, M., 2006. Numerical indications of a q-generalised central limit theorem. Europhysics Letters 73, 813.

Mozer, F.S., Ergun, R.E., Temerin, M., Cattell, C., Dombeck, J., Wygant, J., 1997. New features of time domain electric field structures in the auroral acceleration region. Physical Review Letters 79, 1281.

Muschietti, L., Ergun, R.E., Roth, I., Carlson, C.W., 1999. Phase- space electron holes along magnetic field lines. Geophysical Research Letters 26, 1093—1096.

Navarro, R.E., Muñoz, V., Araneda, J., Viñas, A.F., Moya, P.S., Valdivia, J.A., 2015. Magnetic Alfvén-cyclotron fluctuations of anisotropic non-thermal plasmas. Journal of Geophysical Research 120, 2382—2396.

Neergaard Parker, L., Zank, G.P., 2012. Particle acceleration at quasi-parallel shock waves: theory and observations at 1 au. The Astrophysical Journal 757, 97.

Neergaard Parker, L., Zank, G.P., Hu, Q., 2014. Particle acceleration at quasi-perpendicular shock waves: theory and observations at 1 au. The Astrophysical Journal 782 (52).

Nelson, G.J., Melrose, D.B., 1985. Type II bursts. In: McLean, D.J., Labrum, N.R. (Eds.), Solar Radiophysics. Cambridge, p. 333.

Ness, N.F., Acuna, M.H., Behannon, K.W., Burlaga, L.F., Connerney, J.E.P., Lepping, R.P., Neubauer, F.M., 1986. Magnetic fields at Uranus. Science 233, 85.

Ness, N., Acufia, M.H., Burlaga, L.F., Connerney, J.E.P., Lepping, R.P., M Neubauer, F., 1989. Magnetic fields at Neptune. Science 246, 1473.

Newbury, J.A., Russell, C.T., Lindsay, G.M., 1997. Solar wind index in the vicinity of stream interactions. Geophysical Research Letters 24, 1431−1434.

Nicholls, D.C., Evans, W.F.J., Llewellyn, E.J., 1972. Collisional relaxation and rotational intensity distributions in spectra of aeronomic interest. Journal of Quantitative Spectroscopy and Radiative Transfer 12, 549−558.

Nicholls, D.C., Dopita, M.A., Sutherland, R.S., 2012. Resolving the electron temperature Discrepancies in H II regions and planetary nebulae: κ-distributed electrons. The Astrophysical Journal 752, 148.

Nicholls, D.C., Dopita, M.A., Sutherland, R.S., Kewley, L.J., Palay, E., 2013. Measuring nebular temperatures: the effect of new collision strengths with equilibrium and κ-distributed electron energies. The Astrophysical Journal Supplement Series 207, 21.

Nicholls, D.C., Dopita, M.A., Sutherland, R.S., Jerjen, H., Kewley, L.J., 2014. Metal-poor dwarf galaxies in the SIGRID galaxy sample. II. The electron temperature-abundance calibration and the parameters that affect it. The Astrophysical Journal 790, 75.

Nicolaou, G., Livadiotis, G., 2016. Misestimation of temperature when applying Maxwellian distributions to space plasmas described by kappa distributions. Astrophysics and Space Science 361, 359, pp. 11.

Nicolaou, G., Livadiotis, G., 2017. Modeling the plasma flow in the inner heliosheath with a spatially varying compression ratio. The Astrophysical Journal 838, 7, pp. 7.

Nicolaou, G., McComas, D.J., Bagenal, F., Elliott, H.A., 2013. Properties of plasma ions in the distance Jovian magnetosheath using solar wind around Pluto data on new horizons. Journal of Geophysical Research 119, 3463−3479.

Nicolaou, G., Livadiotis, G., Moussas, X., 2014. Long term variability of the polytropic Index of solar wind protons at ∼1 au. Solar Physics 289, 1371−1378.

Nicolaou, G., McComas, D.J., Bagenal, F., Elliott, H.A., Ebert, R.W., 2015. Jupiter's deep magnetotail boundary layer. Planetary and Space Science 111, 116−125.

Nieves-Chinchilla, T., Viñas, A.F., 2008. Solar wind electron distribution functions inside magnetic clouds. Journal of Geophysical Research 113, A02105.

Niven, R.K., Suyari, H., 2009. The q-gamma and (q,q)-polygamma functions of Tsallis statistics. Physica A 388, 4045−4060.

Norgren, C., André, M., Vaivads, A., Khotyaintsev, Y.V., 2015. Slow electron phase space holes: Magnetotail observations. Geophysical Research Letters 42, 1654−1661.

Norman, J.P., Charbonneau, P., McIntosh, S.W., Liu, H.-L., 2001. Waiting-Time Distributions in Lattice Models of Solar Flares. The Astrophysical Journal 557, 891.

Nose, M., Lui, A.T.Y., Ohtani, S., Mauk, B.H., McEntire, R.W., Williams, D.J., Mukai, T., Yumoto, K., 2000. Acceleration of oxygen ions of ionospheric origin in the near-Earth magnetotail during substorms. Journal of Geophysical Research 105, 7669−7678.

Nose, M., Koshiishi, H., Matsumoto, H., C:son Brandt, P., Keika, K., Koga, K., Goka, T., Obara, T., 2010. Magnetic field dipolarization in the deep inner magnetosphere and its role in development of O+ rich ring current. Journal of Geophysical Research 115, A00J03.

O'Dwyer, B., Del Zanna, G., Mason, H.E., Weber, M.A., Tripathi, D., 2010. SDO/AIA response to coronal hole, quiet Sun, active region, and flare plasma. Astronomy and Astrophysics 521, A21.

O'Connor, J.J., Robertson, E.F., 2016. The Number e. St Andrews University.

Ogasawara, K., Angelopoulos, V., Dayeh, M.A., Fuselier, S.A., Livadiotis, G., McComas, D.J., McFadden, J.P., 2013. Characterizing the dayside magnetosheath using ENAs: IBEX and THEMIS observations. Journal of Geophysical Research 118, 3126−3137.

Ogasawara, K., Dayeh, M.A., Funsten, H.O., Fuselier, S.A., Livadiotis, G., McComas, D.J., 2015. Interplanetary magnetic field dependence of the suprathermal energetic neutral atoms originated in subsolar magnetopause. Journal of Geophysical Research 120, 964−972.

Ohtaki, Y., Hasegawa, H.H., 2003. Superstatistics in Econophysics arXiv: cond-mat/0312568.

Oka, M., Zank, G.P., Burrows, R.H., Shinohara, I., 2011. In: Florinski, V., Heerikhuisen, J., Zank, G.P., Gallagher, D.L. (Eds.), American Institute of Physics Conference Series, vol. 1366, p. 53.

Oka, M., Ishikawa, S., Saint-Hilaire, P., Krucker, S., Lin, R.P., 2013. Kappa distribution model for hard X-ray coronal sources of solar flares. The Astrophysical Journal 764, 6.

Oka, M., Krucker, S., Hudson, H.S., Saint-Hilaire, P., 2015. Electron energy partition in the above-the-looptop solar hard X-ray sources. The Astrophysical Journal 799, 129.

Olbert, S., 1968. Summary of experimental results from M.I.T. detector on IMP-1. In: Carovillano, R.L., McClay, J.F., Radoski, H.R. (Eds.), Physics of the Magnetosphere. Springer, New York, pp. 641−659.

Olluri, K., Gudiksen, B.V., Hansteen, V.H., 2013a. Non-equilibrium ionization in the Bifrost stellar atmosphere code. Astronomical Journal 145, 72.

Olluri, K., Gudiksen, B.V., Hansteen, V.H., 2013b. Non-equilibrium ionization effects on the density line ratio diagnostics of O IV. The Astrophysical Journal 767, 43.

Omura, Y., Matsumoto, H., Miyake, T., Kojima, H., 1996. Electron beam instabilities as generation mechanism of electrostatic solitary waves in the magnetotail. Journal of Geophysical Research 101, 2685−2697.

Omura, Y., Kojima, H., Miki, N., Mukai, T., Matsumoto, H., Anderson, R., 1999. Electrostatic solitary waves carried by diffused electron beams observed by the Geotail spacecraft. Journal of Geophysical Research 104, 14627−14637.

Osherovich, V.A., Farrugia, C.J., Burlaga, L.F., 1993. Dynamics of aging magnetic clouds. Advances in Space Research 13, 57−62.

Osterbrock, D.E., Ferland, G.J., 2006. Astrophysics of Gaseous Nebulae and Active Galactic Nuclei, second ed. University Science Books.

Osterbrock, D.E., 1989. Astrophysics of Gaseous Nebulae and Active Galactic Nuclei. University Science Books, Sausalito, CA.

Ourabah, K., Ait Gougam, L., Tribeche, M., 2015. Nonthermal and suprathermal distributions as a consequence of superstatistics. Physical Review E 91, 012133.

Owocki, S.P., Scudder, J.D., 1983. The effect of a non-Maxwellian electron distribution on oxygen and iron ionization balances in the solar corona. The Astrophysical Journal 270, 758−768.

Palay, E., Nahar, S.N., Pradhan, A.K., Eissner, W., 2012. Improved collision strengths and line ratios for forbidden [O III] far-infrared and optical lines. Monthly Notices of the Royal Astronomical Society 423, L35−L39.

Pandey, R.S., Kaur, R., 2015a. Oblique electromagnetic cyclotron waves for kappa distribution with AC field in planetary magnetospheres. Advances in Space Research 56, 714−724.

Pandey, R.S., Kaur, R., 2015b. Theoretical study of electromagnetic electron cyclotron waves in the presence of AC field in Uranian magnetosphere. New Astronomy 40, 41−48.

Pandey, R.S., Pandey, R.P., Srivastava, A.K., Karim, S.M., Hariom, 2008. The electromagnetic ion-cylclotron instability in the presence of a.c electric field for lorentzian kappa. Progress in Electromagnetic Research M 1, 207−217.

Pang, X.X., Cao, J.B., Liu, W.L., et al., 2015a. Polytropic index of central plasma sheet ions based on MHD Bernoulli integral. Journal of Geophysical Research: Space Physics 120, 4736−4747.

Pang, X.X., Cao, J.B., Liu, W.L., et al., 2015b. Case study of small scale polytropic index in the central plasma sheet. Sci. China Earth Sci. 58, 1993−2001.

Paoletti, M., Fisher, M.E., Sreenivasan, K.R., Lathrop, D.P., 2008. Velocity Statistics Distinguish Quantum Turbulence from Classical Turbulence. Physical Review Letters 101, 154501.

Paranicas, C.P., Paterson, W.R., Cheng, A.F., Mauk, B.H., McEntire, R.W., Frank, L.A., Williams, D.J., 1999. Energetic particle observations near Ganymede. Journal of Geophysical Research 104, 17459–17469.

Paranicas, C.P., Decker, R.B., Williams, D.J., Mitchell, D.G., Brandt, P.C., Mauk, B.H., 2005. Recent research highlights from planetary magnetospheres and the heliosphere. Johns Hopkins APL Technical Digest 26, 156–163.

Paranicas, C.P., et al., 2014. The lens feature on the inner saturnian satellites. Icarus 234, 155–161.

Parnell, E.N., Jupp, P.E., 2000. Statistical analysis of the energy distribution of nanoflares in the quiet sun. The Astrophysical Journal 529, 554–569.

Pauluhn, A., Solanki, S.K., 2007. A nanoflare model of quiet Sun EUV emission. Astronomy and Astrophysics 462, 311.

Pavlos, G.P., Malandraki, O.E., Pavlos, E.G., Iliopoulos, A.C., Karakatsanis, L.P., 2016. Non-extensive statistical analysis of magnetic field during the March 2012 ICME event using a multi-spacecraft approach. Physica A 464, 149–181.

Peimbert, M., December 1967. Temperature determinations of H II regions. The Astrophysical Journal 150, 825.

Pesnell, W.D., Thompson, B.J., Chamberlin, P.C., 2012. The solar dynamics observatory (SDO). Solar Physics 275, 3–15.

Peterson, W.K., Sharp, R.D., Shelley, E.G., Johnson, R.G., Balsiger, H., 1981. Energetic ion composition of the plasma sheet. Journal of Geophysical Research 86, 761–767.

Petrini, D., da Silva, E.P., January 1997. C III and C IV line emission following K-shell photoionization. Astronomy and Astrophysics 317, 262–264.

Phillips, K.J.H., Feldman, U., Landi, E., 2008. Ultraviolet and X-ray Spectroscopy of the Solar Atmosphere. Cambridge Univ. Press, Cambridge, UK.

Pickett, J.S., Chen, L.-J., Kahler, S.W., Santolik, O., Gurnett, D.A., Tsurutani, B.T., Balogh, A., 2004. Isolated electrostatic structures observed throughout the cluster orbit: relationship to magnetic field strength. Annales Geophysicae 22, 2515–2523.

Pickett, J.S., Chen, L.-J., Kahler, S.W., Santolík, O., Goldstein, M.L., Lavraud, B., Décréau, P.M.E., Kessel, R., Lucek, E., Lakhina, G.S., Tsurutani, B.T., Gurnett, D.A., Cornilleau-Wehrlin, N., Fazakerley, A., Réme, H., Balogh, A., 2005. On the generation of solitary waves observed by cluster in the near-Earth magnetosheath. Nonlinear Processes in Geophysics 12, 181–193.

Pickett, J.S., Chen, L.-J., Mutel, R.L., Christopher, I.H., Santolik, O., Lakhina, G.S., Singh, S.V., Reddy, R.V., Gurnett, D.A., Tsurutani, B.T., Lucek, E., 2008. Furthering our understanding of electrostatic solitary waves through cluster multi-spacecraft observations and theory. Advances in Space Research 41, 1666–1676.

Pierrard, V., Lazar, M., Poedts, S., Stverak, S., Maksimovic, M., Tranicek, P.M., 2016. The electron temperature and anisotropy in the solar wind. 1. Comparison of the core and halo populations. Solar Physics 291, 2165–2179.

Pierrard, V., 1996. New model of magnetospheric current-voltage relationship. Journal of Geophysical Research 101, 2669–2675.

Pierrard, V., 2009. Kinetic models for the exospheres of Jupiter and Saturn. Planetary and Space Science 57, 1260–1267.

Pierrard, V., 2012a. Solar wind electron transport: interplanetary electric field and heat conduction. Space Science Reviews 172, 315–324.

Pierrard, V., 2012b. Kinetic models for solar wind electrons, protons and ions, chapter. In: Lazar, M. (Ed.), Exploring the Solar Wind. Intech, pp. 221–240.

Pierrard, V., 2012c. Effects of suprathermal particles in space plasmas. In: Physics of the Heliosphere: A 10 Year Retrospective, ICNS Annual International Astrophysics Conference Proc., American Institute of Physics, 1436, pp. 61–66.

Pierrard, V., Borremans, K., 2012a. The ionosphere coupled to the plasmasphere and polar wind models, numerical modeling of space plasma flows: astronum-2011. In: Pogorelov, N.V., Font, J.A., Audit, E., Zank, G.P. (Eds.), ASP Conference Series, vol. 459, pp. 234–239.

Pierrard, V., Borremans, K., 2012b. Fitting the AP8 spectra to determine the proton momentum distribution functions in space radiations. Radiation Measurements 47, 401–405.

Pierrard, V., Lamy, H., 2003. The effects of the velocity filtration mechanism on the minor ions of the corona. Solar Physics 216, 47–58.

Pierrard, V., Lazar, M., 2010. Kappa distributions: theory and applications in space plasmas. Solar Physics 267, 153–174.

Pierrard, V., Lemaire, J., 1996a. Lorentzian ion exosphere model. Journal of Geophysical Research 101, 7923–7934.

Pierrard, V., Lemaire, J., 1996b. Fitting the AE-8 energy spectra with two maxwellian functions. Radiation Measurements 26, 333–337.

Pierrard, V., Pieters, M., 2015. Coronal heating and solar wind acceleration for electrons, protons, and minor ions, obtained from kinetic models based on kappa distributions. Journal of Geophysical Research 119, 9441–9455.

Pierrard, V., Stegen, K., 2008. A three dimensional dynamic kinetic model of the plasmasphere. Journal of Geophysical Research 113, A10209.

Pierrard, V., Lamy, H., Lemaire, J., 2004. Exospheric distributions of minor ions in the solar wind. Journal of Geophysical Research 109, A02118.

Pierrard, V., Maksimovic, M., Lemaire, J., 1999. Electron velocity distribution function from the solar wind to the corona. Journal of Geophysical Research 104, 17021–17032.

Pierrard, V., Issautier, K., Meyer-Vernet, N., Lemaire, J., 2001a. Collisionless solar wind in a spiral magnetic field. Geophysical Research Letters 28, 223–226.

Pierrard, V., Maksimovic, M., Lemaire, J., 2001b. Core, halo and strahl electrons in the solar wind. Astrophysics and Space Science 277, 195–200.

Pierrard, V., Maksimovic, M., Lemaire, J., 2001c. Self-consistent kinetic model of solar wind electrons. Journal of Geophysical Research 107, 29305–29312.

Pierrard, V., Khazanov, G.V., and Lemaire, J., 2007. Current-voltage relationship. In: Tam, Pierrard and Schunk (Guest Eds.), Recent Advances in the Polar Wind Theories and Observations. Journal of Atmospheric and Solar Terrestial Physics 69, 2048–2057.

Pierrard, V., Borremans, K., Stegen, K., Lemaire, J., 2014. Coronal temperature profiles obtained from kinetic models and from coronal brightness measurements obtained during solar eclipses. Solar Physics 289, 183–192.

Pierrard, V., Lazar, M., Schlickeiser, R., 2011. Evolution of the electron distribution function in the wave turbulence of the solar wind. Solar Physics 269, 421–438.

Pilipp, W.G., Miggenrieder, H., Montgomery, M.D., Mühlhäuser, K.-H., Rosenbauer, H., Schwenn, R., 1987. Characteristics of electron velocity distribution functions in the solar wind derived from the helios plasma experiment. Journal of Geophysical Research 92, 1075–1092.

Pinfield, D.J., Keenan, F.P., Mathioudakis, M., Phillips, K.J.H., Curdt, W., Wilhelm, K., 1999. Evidence for non-Maxwellian electron energy distributions in the solar transition region: Si III line ratios from SUMER. The Astrophysical Journal 527, 1000–1008.

Pisarenko, N.F., Budnik, E.Yu, Ermolaev, YuI., Kirpichev, I.P., Lutsenko, V.N., Morozova, E.I., Antonova, E.E., 2002a. The ion differential spectra in outer boundary of the ring current: November 17, 1995 case study. Journal of Atmospheric and Solar Terrestrial Physics 64, 573–583.

Pisarenko, N.F., Kirpichev, I.P., Lutsenko, V.N., Budnik, E.Yu, Morozova, E.I., Antonova, E.E., 2002b. The structure of the ring current: the November 13, 1995 event. Cosmic Research 40, 15–24.

Plastino, A., Rocca, M.C., 2017. Tsallis' q-non Extensivity Strongly Limits the System's Number of Particles, eprint: arXiv: 1702.04806.

Pogorelov, N.V., Heerikhuisen, J., Zank, G.P., 2008. Probing Heliospheric Asymmetries with an MHD-Kinetic model. The Astrophysical Journal Letters 675, L41.

Polito, V., Reep, J.W., Reeves, K.K., Simões, P.J.A., Dudík, J., Del Zanna, G., Mason, H.E., Golub, L., 2016. Simultaneous IRIS and Hinode/EIS observations and modelling of the 2014 October X2.0-class flare. The Astrophysical Journal 816, 89.

Poquerusse, M., 1994. Relativistic type 3 solar radio bursts. Astronomy and Astrophysics 286, 661.

Porco, C.C., et al., 2006. Cassini observes the active south pole of Enceladus. Science 311, 1393–1401.

Porporato, A., Vico, G., Fay, P.A., 2006. Superstatistics in hydro-climatic fluctuations and interannual ecosystem productivity. Geophysical Research Letters 33, L15402.

Pottelette, R., Malingre, M., Dubouloz, N., Aparicio, B., Lundin, R., Holmgren, G., Marklund, G., 1990. High frequency waves in the Cusp/cleft regions. Journal of Geophysical Research 95, 5957–5971.

Pottelette, R., Ergun, R.E., Treumann, R.A., Berthomier, M., Carlson, C.W., McFadden, J.P., Roth, I., 1999. Modulated electron-acoustic waves in auroral density cavities: FAST observations. Geophysical Research Letters 26, 2629–2632.

Powell, K.G., Roe, P.L., Linde, T.J., Gombosi, T.I., De Zeeuw, D.L., 1999. A solution-adaptive upwind scheme for ideal magnetohydrodynamics. Journal of Computational Physics 154, 284.

Prato, D., Tsallis, C., 1999. Nonextensive foundation of Lévy distributions. Physical Review E 60, 2398.

Prested, C., et al., 2008. Implications of solar wind suprathermal tails for IBEX ENA images of the heliosheath. Journal of Geophysical Research 113, A06102.

Provan, G., Andrews, D.J., Arridge, C.S., Coates, A.J., Cowley, S.W.H., Milan, S.E., Dougherty, M.K., Wright, D.M., 2009. Polarization and phase of planetary-period magnetic field oscillations on high-latitude field lines in Saturn's magnetosphere. Journal of Geophysical Research 114, A02225.

Pudovkin, I.M., Meister, C.-V., Besser, B.P., Biernat, H.K., 1997. The effective polytropic index in a magnetized plasma. Journal of Geophysical Research 102, 27145–27149.

Qureshi, M.N.S., et al., 2003. Solar Wind Particle Distribution Function Fitted via the Generalized Kappa Distribution Function: Cluster Observations. In: Velli, M., Vruno, R., Malara, F. (Eds.), Proceedings of the 10th Solar Wind Conference, AIP Conference Proceedings, Vol. 679, American Institute of Physics, Melville, NY, pp. 489–492.

Qureshi, M.N.S., Nasir, W., Masood, W., Yoon, P.H., Shah, H.A., Schwartz, S.J., 2015. Journal of Geophysical Research 119, 10059–10067.

Raadu, M.A., Shafiq, M., 2007. Test charge response for a dusty plasma with both grain size distribution and dynamical charging. Physics of Plasmas 14, 012105.

Rabassa, P., Beck, C., 2015. Superstatistical analysis of sea-level fluctuations. Physica A-Statistical Mechanics and Its Applications 417, 18–28.

Raines, J.M., et al., 2013. Distribution and compositional variations of plasma ions in Mercury's space environment: the first three Mercury years of MESSENGER observations. Journal of Geophysical Research 118, 1604–1619.

Rajagopal, A.K., 2006. Superstatistics − a quantum generalization. ArXiv: cond-mat/0608679.

Rama, S.K., 2000. Tsallis statistics: averages and a physical interpretation of the Lagrange multiplier/β. Physics Letters A 276, 103–108.

Randol, B.M., Christian, E.R., 2014. Simulations of plasma obeying Coulomb's law and the formation of suprathermal ion tails in the solar wind. Journal of Geophysical Research: Space Physics 119, 7025–7037.

Randol, B.M., Christian, E.R., 2016. Coupling of charged particles via Coulombic interactions: numerical simulations and resultant kappa-like velocity space distribution functions. Journal of Geophysical Research 121, 1907–1919.

Raymond, J.C., Winkler, P.F., Blair, W.P., Lee, J.-J., Park, S., 2010. Non-Maxwellian Hα profiles in Tycho's supernova remnant. The Astrophysical Journal 712, 901.

Reale, F., 2010. Coronal Loops: Observations and Modeling of Confined Plasma. ArXiv: 1010.5927v1.

Reddy, R.V., Lakhina, G.S., 1991. Ion acoustic double layers and solitons in auroral plasma. Planetary and Space Science 39, 1343.

Reddy, R.V., Lakhina, G.S., Verheest, F., 1992. Ion-acoustic double layers and solitons in multispecies auroral beam-plasmas. Planetary and Space Science 40, 1055.

Reid, H.A.S., Ratcliffe, H., 2014. A review of solar type III radio bursts. Annual Review of Astronomy and Astrophysics 14, 773.

Reynolds, A.M., 2003. Superstatistical Mechanics of Tracer-Particle Motions in Turbulence. Physical Review Letters 91, 084503.

Richardson, J.D., 1998. Thermal plasma and neutral gas in Saturn's magnetosphere. Reviews of Geophysics 36, 501–524.

Richardson, J.D., 2008. Plasma temperature distributions in the heliosheath. Geophysical Research Letters 35, 23104.

Richardson, J.D., Smith, C.W., 2013. The radial temperature profile of the solar wind. Geophysical Research Letters 30, 1206.

Richardson, J.D., Eviatar, A., Delitsky, M.L., 1990. The Triton torus revisited. Geophysical Research Letters 17, 1673–1676.

Richardson, J.D., Kasper, J.C., Wang, C., Belcher, J.W., Lazarus, A.J., 2008. Cool heliosheath plasma and deceleration of the upstream solar wind at the termination shock. Nature 454, 63–66.

Rizzo, S., Rapisarda, A., 2004. Proceedings of the 8th Experimental Chaos Conference, Florence, AIP Conf. Proc, vol. 742, p. 176 (cond-mat/0406684).

Robbins, H., 1955. A remark on Stirling's formula. American Mathematical Monthly 62, 26–29.

Robinson, P.A., Cairns, I.H., Gurnett, D.A., 1993. Clumpy Langmuir waves in type III radio sources: comparison of stochastic-growth theory with observations. The Astrophysical Journal 407, 790.

Robledo, A., 1999. Renormalization group, entropy optimization, and nonextensivity at criticality. Physical Review Letters 83, 2289–2292.

Roelof, E.C., Keath, E.P., Bostrom, C.O., Williams, D.J., 1976. Fluxes of >50-keV Protons and >30-keV electrons at ~35 RE 1. Velocity anisotropies and plasma flow in the magnetotail. Journal of Geophysical Research 81, 2304–2314.

Roussel-Dupré, R., 1980. Non-Maxwellian velocity distribution functions associated with steep temperature gradients in the solar transition region. Solar Physics 68, 243–263.

Roussev, I.I., Sokolov, I.V., Forbes, T.G., et al., 2004. A numerical model of a coronal mass ejection: shock development with implications for the acceleration of GeV protons. The Astrophysical Journal 605, L73.

Roussev, I.I., Gombosi, T.I., Sokolow, I.V., et al., 2003. A three-dimensional model of the solar wind incorporating solar magnetogram observations. The Astrophysical Journal 595, L57.

Rubab, N., Murtaza, G., 2006. Debye length in non-Maxwellian plasmas. Physica Scripta 74, 145.

Rufai, O.R., Bharuthram, R., Singh, S.V., Lakhina, G.S., 2012. Low frequency solitons and double layers in a magnetized plasma with two temperature electrons. Physics of Plasmas 19, 122308.

Rufai, O.R., Bharuthram, R., Singh, S.V., Lakhina, G.S., 2014. Effect of hot ion temperature on obliquely propagating ion-acoustic solitons and double layers in an auroral plasma. Communications in Nonlinear Science and Numerical Simulation 19, 1338.

Rutherford, K.D., Moos, H.W., Strobel, D.F., 2003. Io's auroral limb glow: hubble space telescope FUV observations. Journal of Geophysical Research 108, 1333.

Rymer, A.M., et al., 2007. Electron sources in Saturn's magnetosphere. Journal of Geophysical Research 112, A02201.

Sackur, O., 1911. Die Anwendung der kinetischen Theorie der Gase auf chemische Probleme (The application of the kinetic theory of gases to chemical problems). Annalen der Physik 36, 958–980.

Sagdeev, R.Z., Shapiro, V.D., Shevchenco, V.I., Szego, K., 1986. MHD Turbulence in the solar wind-comet interaction region. Geophysical Research Letters 13, 85–88.

Sagdeev, R.Z., 1966. Cooperative phenomena and shock waves in collisionless plasmas. In: Leontovich, M.A. (Ed.), Reviews of Plasma Physics, vol. 4. Consultants Bureau, New York.

Sahu, B., 2010. Electron acoustic solitary waves and double layers with superthermal hot electrons. Physics of Plasmas 17, 122305.

Saini, N.S., Kourakis, I., 2010. Electron beam–plasma interaction and ion-acoustic solitary waves in plasmas with a superthermal electron component. Plasma Physics and Controlled Fusion 52, 075009.

Saini, N.S., Kourakis, I., Hellberg, M.A., 2009. Arbitrary amplitude ion-acoustic solitary excitations in the presence of excess superthermal electrons. Physics of Plasmas 16, 062903.

Saito, S., Forme, F.R.E., Buchert, S.C., Nozawa, S., Fujii, R., 2000. Effects of a kappa distribution function of electrons on incoherent scatter spectra. Annales Geophysicae 18, 1216–1223.

Sakagami, M., Taruya, A., 2004. Self-gravitating stellar systems and non-extensive thermostatistics. Continuum Mechanics and Thermodynamics 16, 279–292.

Salem, C., Lacombe, C., Mangeney, A., Kellogg, P.J., Bougeret, J.-L., 2003. Weak double layers in the solar wind and their relation to the interplanetary electric field. In: Velli, M., Bruno, R., Malara, F. (Eds.), CP679, Solar Wind 10: Proceedings of the Tenth International Sola Wind Conference, Am. Inst. of Phys., New York, p. 513.

Sarris, E.T., Krimigis, S.M., Armstrong, T.P., 1976. Observations of magnetospheric bursts of High-Energy protons and electrons at 35 RE with Imp 7. Journal of Geophysical Research 81 (13), 2341–2355.

Sarris, E.T., Krimigis, S.M., Lui, A.T.Y., Ackerson, K.L., Frank, L.A., Williams, D.J., 1981. Relationships between energetic particles and plasmas in the distant plasma sheet. Geophysical Research Letters 8, 349–352.

Sato, Y., Shimada, E., Ohta, I., Tasaka, S., Tanaka, S., 1985. Journal of the Physical Society of Japan 54, 4502.

Sattin, F., Salasnich, L., 2002. Derivation of Tsallis statistics from dynamical equations for a granular gas. Physical Review E65, 035106(R).

Schippers, P., et al., 2008. Multi-instrument analysis of electron populations in Saturn's magnetosphere. Journal of Geophysical Research 113, A07208.

Schmelz, J.T., Nasraoui, K., Rightmire, L.A., Kimble, J.A., Del Zanna, G., Cirtain, J.W., DeLuca, E.E., Mason, H.E., 2009. Are coronal loops isothermal or multithermal? The Astrophysical Journal 691, 503–515.

Schmelz, J.T., Pathak, S., Brooks, D.H., Christian, G.M., Dhaliwal, R.S., 2014. Hot topic, warm loops, cooling plasma? Multithermal analysis of active region loops. The Astrophysical Journal 795, 171.

Schmidt, J.M., Gopalswamy, N., 2008. Synthetic radio maps of CME-driven shocks below 4 solar radii heliocentric distance. Journal of Geophysical Research 113, A08104.

Schmidt, J., Cairns, I.H., 2012a. Type II radio bursts: 1. New entirely analytic formalism for the electron beams, Langmuir waves, and radio emission. Journal of Geophysical Research 117, A04106.

Schmidt, J., Cairns, I.H., 2012b. Type II radio bursts: 2. Application of the new analytic formalism. Journal of Geophysical Research 117, A11104.

Schmidt, J.M., Cairns, I.H., 2014a. Type II solar radio bursts predicted by 3D MHD CME and kinetic radio emission simulations. Journal of Geophysical Research 119, 69–87.

Schmidt, J., Cairns, I.H., 2014b. The solar type II radio bursts of 7 March 2012: detailed simulation analyses. Journal of Geophysical Research 119, 6042–6061.

Schmidt, J.M., Cairns, I.H., 2016. Quantitative prediction of type II solar radio emission from the Sun to 1 AU. Geophysical Research Letters 43, 50.

Schmidt, J.M., Cairns, I.H., Hillan, D.S., 2013. Prediction of type II solar radio bursts by 3D MHD CME and kinetic radio emission simulations. The Astrophysical Journal Letters 773, L30.

Schmidt, J.M., Cairns, Iver, H., Lobzin, V.V., 2014. The solar type II radio bursts of 7 March 2012: Detailed simulation analyses. Journal of Geophysical Research 119, 6042–6061.

Schmidt, J.M., Cairns, I.H., Xie, H., St Cyr, O.C., Gopalswamy, N., 2016. CME flux rope and shock identifications and locations: comparison of white light data, graduated cylindrical shell model, and MHD simulations. Journal of Geophysical Research 121, 1886–1906.

Schreier, E., Gursky, H., Kellogg, E., Tananbaum, H., Giacconi, R., 1971. The Astrophysical Journal 170, L21–L27.

Schriver, D., et al., 2011. Quasi-trapped ion and electron populations at Mercury. Geophysical Research Letters 38, L23103.

Schroedter, M., et al., 2005. A Very High Energy Gamma-Ray Spectrum of 1ES 2344+514. The Astrophysical Journal 634, 947–954.

Schwadron, N.A., McComas, 2006. Modulation of anomalous and galactic cosmic rays beyond the termination shock. Geophysical Research Letters 34, L14105.

Schwadron, N.A., et al., 2009. Comparison of Interstellar Boundary Explorer Observations with 3D Global Heliospheric Models. Science 326 (5955), 966–968.

Schwadron, N.A., Dayeh, M.A., Desai, M., Fahr, H., Jokipii, J.R., Lee, M.A., 2010. Superposition of stochastic processes and the resulting particle distributions. The Astrophysical Journal 713, 1386–1392.

Sckopke, N., Paschmann, G., Haerendel, G., et al., 1981. Structure of the low-altitude boundary layer. Journal of Geophysical Research 86, 2099.

Scudder, J.D., 1992a. On the causes of temperature change in homogeneous low-density astrophysical plasmas. The Astrophysical Journal 398, 299–318.

Scudder, J.D., 1992b. Why all stars possess circumstellar temperature inversions. The Astrophysical Journal 398, 319–349.

Scudder, J.D., 1992c. The cause of the coronal temperature inversion of the solar atmosphere and the implications for the solar wind. In: Marsch, E., Schwenn, R. (Eds.), Solar Wind Seven. Pergamon Press, Oxford, pp. 103–112.

Scudder, J.D., Karimabadi, H., 2013. Ubiquitous non-thermals in astrophysical plasmas: restating the difficulty of maintaining Maxwellians. The Astrophysical Journal 770, 26.

Scudder, J.D., Olbert, S., 1979a. A theory of local and global processes which affect solar wind electrons: the origin of typical 1 AU velocity distribution functions – steady state theory. Journal of Geophysical Research 84, 2755–2772.

Scudder, J.D., Olbert, S., 1979b. A theory of local and global processes which affect solar wind electrons: the origin of typical 1 AU velocity distribution functions – 2. Experimental support. Journal of Geophysical Research 84, 6603–6620.

Scudder, J.D., Olbert, S., 1983. The collapse of the local, Spitzer-Härm formulation and a global-local generalization for heat flow in a homogeneous, fully ionized plasma. In: Negebauer (Ed.), Solar Wind Five, vol. 2280. NASA Conf. Pub. CP, p. 163.

Scudder, J.D., Sittler Jr., E.C., Bridge, H.S., 1981. A survey of the plasma electron environment of Jupiter: a view from Voyager. Journal of Geophysical Research 86, 8157–8179.

Seaton, M.J., July 1953. Electron excitation of forbidden lines occurring in gaseous nebulae. Proceedings of the Royal Society of London Series A 218, 400–416.

Selesnick, R.S., Stone, E.C., 1991. Neptune's cosmic ray cutoff. Geophysical Research Letters 18, 361.

Selesnick, R.S., McNutt Jr., R.L., 1987. Voyager 2 plasma ion observations in the magnetosphere of Uranus. Journal of Geophysical Research 92, 15249.

Selesnick, R.S., Richardson, J.D., 1986. Plasmasphere formation in arbitrarily oriented magnetospheres. Geophysical Research Letters 13, 624–627.

Sharma, B.D., Taneja, I.J., 1975. Entropy of type (α,β) and other generalized measures in information theory. Metrika, 22, 205–215, Physica-Verlag, Wien.

Shemansky, D.E., Matherson, P., Hall, D.T., Tripp, T.M., 1993. Detection of the hydroxyl radical in Saturn's magnetosphere. Nature 363, 329–332.

Shimizu, T., 1995. Energetics and occurrence rate of active-region transient brightenings and implications for the heating of the active-region corona. Publications of the Astronomical Society of Japan 47, 251–263.

Shizgal, B.D., December 2007. Suprathermal particle distributions in space physics: kappa distributions and entropy. Astrophysics and Space Science 312, 227–237.

Shoub, E.C., 1983. Invalidity of local thermodynamic equilibrium for electrons in the solar transition region. I. Fokker-Planck results. The Astrophysical Journal 266, 339–369.

Shull, J.M., van Steenberg, M., 1982. The ionization equilibrium of astrophysically abundant elements. The Astrophysical Journal Supplement Series 48, 95–107.

Shull, J.M., van Steenberg, M.E., November 1985. X-ray secondary heating and ionization in quasar emission-line clouds. The Astrophysical Journal 298, 268–274.

Silva, R., Plastino, A.R., Lima, J.A.S., 1998. A Maxwellian path to the q-nonextensive velocity distribution function. Physics Letters A 249, 401–408.

Silva, R., França, G., Vilar, C., Alcaniz, J., 2006. Nonextensive models for earthquakes. Physical Review E 73, 026102.

Simpson, J.A., Eraker, J.H., Lamport, J.E., Walpole, P.H., July 12, 1974. Electrons and protons accelerated in Mercury's magnetic field. Science 185, 160–166.

Singh, N., 2003. Space-time evolution of electron-beam driven electron holes and their effects on the plasma. Nonlinear Processes in Geophysics 10, 53–63.

Singh, S.V., Lakhina, G.S., 2001. Generation of electron-acoustic waves in the magnetosphere. Planetary and Space Science 49, 107–114.

Singh, S.V., Lakhina, G.S., 2004. Electron acoustic solitary waves with non-thermal distribution of electrons. Nonlinear Processes in Geophysics 11, 275–279.

Singh, S.V., Devanandhan, S., Lakhina, G.S., Bharuthram, R., 2013. Effect of ion temperature on ion-acoustic solitary waves in a magnetized plasma in presence of superthermal electrons. Physics of Plasmas 20, 012306.

Singh, S.V., Lakhina, G.S., Bharuthram, R., Pillay, S.R., 2011. Electrostatic solitary structures in presence of non-thermal electrons and a warm electron beam on the auroral field lines. Physics of Plasmas 18, 122306.

Singh, S.V., Reddy, R.V., Lakhina, G.S., 2001. Broadband electrostatic noise due to nonlinear electronacoustic waves. Advances in Space Research 28, 1643–1648.

Siscoe, G.L., Ness, N.F., Yeates, C.M., 1975. Substorms on Mercury? Journal of Geophysical Research 80, 4359–4363.

Sitenko, A.G., 1982. Fluctuations and Non-linear Wave Interactions in Plasmas. Pergamon Press, New York.

Sittler Jr., E.C., Ogilvie, K.W., Selesnick, R.S., 1987. Survey of electrons in the Uranian magnetospshere, Voyager 2 observations. Journal of Geophysical Research 92, 15263–15281.

Sittler Jr., E.C., Strobel, D.F., 1987. Io plasma torus electrons: Voyager 1. Journal of Geophysical Research 92, 5741–5762.

Slavin, J.A., et al., 2009. MESSENGER observations of magnetic reconnection in Mercury's magnetosphere. Science 324, 606–610.

Slavin, J.A., et al., 2010. MESSENGER observations of extreme loading and unloading of Mercury's magnetic tail. Science 329, 665–668.

Soker, N., Rahin, R., Behar, E., Kastner, J.H., 2010. Comparing shocks in planetary nebulae with the solar wind termination shock. The Astrophysical Journal 725, 1910–1917.

Sotolongo-Costa, O., Posadas, A., 2004. Fragment-asperity interaction model for earthquakes. Physical Review Letters 92, 048501.

Sotolongo-Costa, O., Rodriguez, A., Rodgers, G., 2000. Tsallis entropy and the transition to scaling in fragmentation. Entropy 2, 172–177.

Southwood, D.J., 2015. Formation of magnetotails: fast and slow rotators compared. In: Keiling, A., Jackman, C., Delamere, P. (Eds.), Magnetotails in the Solar System, Geophysical Monograph, vol. 207, pp. 424.

Southwood, D.J., Kivelson, M.G., 2007. Saturnian magnetospheric dynamics: elucidation of a camshaft model. Journal of Geophysical Research 112, A12222.

Speiser, T.W., 1965. Particle trajectories in model current sheets, 1, analytical solutions. Journal of Geophysical Research 70, 4219–4226.

Spence, E.H., Kivelson, M.G., 1990. the variation of the plasma sheet popytropic index along the midnight meridian in a finite width magnetotail. Geophysical Research Letters 17, 591–594.

Spitzer, L., 1962. Physics of Fully Ionized Gases, second ed. Interscience Publishers, New York.

Starr, R.D., Schriver, D., Nittler, L.R., Weider, S.Z., Byrne, P.K., Ho, G.C., Rhodes, E.A., Schlemm II, C.E., Solomon, S.C., Trávníček, P.M., 2012. MESSENGER detection of electron-induced X-ray fluorescence from Mercury's surface. Journal of Geophysical Research 117, E00L02.

Stasińska, G., 2002. The electron temperature in ionized nebulae. In: Henney, W.J., Franco, J., Martos, M. (Eds.), Revista Mexicana de Astronomia y Astrofisica Conference Series, vol. 12, pp. 62–69.

Stasińska, G., 2004. Abundance determinations in H II regions and planetary nebulae. In: Esteban, C., García López, R., Herrero, A., Sánchez, F. (Eds.), Cosmochemistry. The Melting Pot of the Elements, pp. 115–170.

Steffl, A.J., Bagenal, F., Stewart, I.A.F., 2004. Cassini UVIS observations of the Io plasma torus. II. Radial variations. Icarus 172, 91–103.

Stepanova, M., Antonova, E.E., 2015. Role of turbulent transport in the evolution of the κ distribution functions in the plasma sheet. Journal of Geophysical Research 120, 3702–3714.

Stix, T.H., 1992. Waves in Plasmas. American Institute of Physics, New York.

Stone, E.C., Cooper, J.F., Cummings, A.C., McDonald, F.B., Trainor, J.H., Lal, N., McGuire, R., Chenette, D.L., 1986. Energetic charged particles in the Uranian magnetosphere. Science 233, 93.

Stone, E.C., Vogt, R.E., McDonald, F.B., Teegarden, B.J., Trainor, J.H., Jokipii, J.R., Webber, W.R., 1977. Cosmic ray investigation for the Voyager missions: energetic particle studies in the outer heliosphere — and beyond. Space Science Reviews 21, 355–376.

Stone, E.C., Cummings, A.C., McDonald, F.B., Heikkila, B.C., Lal, N., Webber, W.R., 2013. Voyager 1 observes low-energy galactic cosmic rays in a region depleted of heliospheric ions. Science 341, 150–153.

Storey, P.J., Sochi, T., 2015. Emission and recombination coefficients for hydrogen with κ-distributed electron energies. Monthly Notices of the Royal Astronomical Society 446, 1864–1866.

Storey, P.J., Sochi, T., Badnell, N.R., July 2014. Collision strengths for nebular [O III] optical and infrared lines. Monthly Notices of the Royal Astronomical Society 441, 3028–3039.

Štverak, S., Travnicek, P.M., Maksimovic, M., Marsch, E., Fazakerley, A.N., Scime, E.E., 2008. Electron temperature anisotropy constraints in the solar wind. Journal of Geophysical Research 113, A03103.

Štverak, S., Maksimovic, M., Travnicek, P.M., Marsch, E., Fazakerley, A.N., Scime, E.E., 2009. Radial evolution of nonthermal electron populations in the low-latitude solar wind: Helios, Cluster, and Ulysses observations. Journal of Geophysical Research 114, A05104.

Sugiyama, H., Singh, S., Omura, Y., Shoji, M., Nunn, D., Summers, D., 2015. Electromagnetic ion cyclotron waves in the Earth's magnetosphere with a kappa-Maxwellian particle distribution. Journal of Geophysical Research 120, 8426–8439.

Sultana, S., Kourakis, I., Hellberg, M.A., 2012. Oblique propagation of arbitrary amplitude electron acoustic solitary waves in magnetized kappa-distributed plasmas. Plasma Physics and Controlled Fusion 54, 105016.

Summers, M.E., Strobel, D.F., 1991. Triton's atmosphere: a source of N and H for Neptune's magnetosphere. Geophysical Research Letters 18, 2309–2312.

Summers, D., Thorne, R.M., 1991. The modified plasma dispersion function. Physics of Fluids B 3, 1835–1836.

Summers, D., Thorne, R.M., 1992. A new tool for analyzing microinstabilities in space plasmas modeled by a generalized Lorentzian (kappa) distribution. Journal of Geophysical Research 97, 16827–16832.

Summers, D., Xue, S., Thorne, R.M., 1994. Calculation of the dielectric tensor for a generalized Lorentzian (kappa) distribution function. Physics of Plasmas 1, 2012–2025.

Sutherland, R.S., 2010. A new computational fluid dynamics code I: Fyris Alpha. Astrophysics and Space Science 327, 173–206.

Suyari, H., 2006. Mathematical structures derived from the q-multinomial coefficient in Tsallis statistics. Physica A: Statistical Mechanics and its Applications. 368, 63–82.

Suzuki, S., Dulk, G., 1985. Bursts of type III and V. In: McLean, D.J., Labrum, N.R. (Eds.), Solar Radiophysics. Cambridge, p. 289.

Swanson, D.G., 2003. Plasma Waves, second ed. American Institute of Physics Publishing-IOP, New York.

Swinney, H.L., Tsallis, C. (Eds.), 2004. Anomalous Distributions, Nonlinear Dynamics, and Non-extensivity, Physica D, vol. 193. Elsevier, Amsterdam.

Sylwester, J., Schrijver, J., Mewe, R., 1980. Multitemperature analysis of solar X-ray line emission. Solar Physics 67, 285–309.

Syunyaev, R.A., et al., 1994. Observations of X-ray novae in Vela (1993), Ophiuchus (1993), and Perseus (1992) using the instruments of the Mir-Kvant module. Astronomy Letters 20, 777–786.

Tagare, S.G., Singh, S.V., Reddy, R.V., Lakhina, G.S., 2004. Electron acoustic solitons in the Earths magnetotail. Nonlinear Processes in Geophysics 11, 215–218.

Tam, S.W.Y., Chang, T., Pierrard, V., 2007. Kinetic modeling of the polar wind. Journal of Atmospheric and Solar Terrestrial Physics 69, 1984–2027.

Tao, J.B., Ergun, R.E., Andersson, L., Bonnell, J.W., Roux, A., LeContel, O., Angelopoulos, V., McFadden, J.P., Larson, D.E., Cully, C.M., Auster, H.-U., Glassmeier, K.-H., Baumjohann, W., Newman, D.L., Goldman, M.V., 2011. A model of electromagnetic electron phase-space holes and its application. Journal of Geophysical Research 116, A11213.

Tatrallyay, M., Russell, C.T., Luhmann, J.G., Barnes, A., Mihalov, J.G., 1984. On the proper Mach number and ratio of specific heats for modeling the Venus bow shock. Journal of Geophysical Research 89, 7381–7392.

Temerin, M., Cerny, K., Lotko, W., Mozer, F.S., 1982. Observations of double layers and solitary waves in the auroral plasma. Physical Review Letters 48, 1175.

Testa, P., De Pontieu, B., Martínez-Sykora, J., DeLuca, E.E., Hansteen, V., Cirtain, J.W., et al., 2013. Observing coronal nanoflares in active region moss. The Astrophysical Journal 770, L1.

Testa, P., De Pontieu, B., Allred, J., Carlsson, M., Reale, F., Daw, A., et al., 2014. Evidence of nonthermal particles in coronal loops heated impulsively by nanoflares. Science 346, 315.

Tetrode, H., 1912. Die chemische Konstante der Gase und das elementare Wirkungsquantum (The chemical constant of gases and the elementary quantum of action). Annalen der Physik 38, 434–442.

Thomsen, M.F., Reisenfeld, D.B., Delapp, D.M., Tokar, R.L., Young, D.T., Crary, F.J., Sittler, E.C., McGraw, M.A., Williams, J.D., 2010. Survey of ion plasma parameters in Saturn's magnetosphere. Journal of Geophysical Research 115, A10220.

Thorne, R.M., Horne, R.B., 1994. Landau damping of magnetospherically reflected whistlers. Journal of Geophysical Research 99, 17249–17258.

Thorne, R.M., Summers, D., 1991a. Landau damping in space plasmas. Physics of Fluids B 3, 2117–2123.

Thorne, R.M., Summers, D., 1991b. Enhancement of wave growth for warm plasmas with a high-energy tail distribution. Journal of Geophysical Research 96, 217–223.

Tigik, S., Ziebell, L.F., Yoon, P.H., Kontar, E.P., 2016. Two-dimensional time evolution of beam-plasma instability in the presence of binary collisions. Astronomy and Astrophysics 586, A19.

Toral, R., 2003. On the definition of physical temperature and pressure for nonextensive thermo-statistics. Physica A 317, 209–212.

Toth, G., van der Holst, B., Sokolov, I.V., De Zeeuw, D.L., Gombosi, T.I., Fang, F., Manchester, W.B., Meng, X., Najib, D., Powell, K.G., Stout, Q.F., Glocer, A., Ma, Y.-J., Opher, M., 2012. Adaptive algorithms in space weather modeling. Journal of Computational Physics 231, 870.

Totten, T.L., Freeman, J.W., Arya, S., 1995. An empirical determination of the polytropic index for the free-streaming solar wind using Helios 1 data. Journal of Geophysical Research 100, 13–17.

Touchette, H., Beck, C., 2005. Asymptotics of superstatistics. Physical Review E 71, 016131.

Touchette, H., 2004. Temperature fluctuations and mixtures of equilibrium states in the canonical ensemble. In: Gell-Mann, M., Tsallis, C. (Eds.), Nonextensive Entropy – Interdisciplinary Applications. Oxford University Press.

Treumann, R.A., 1997. Theory of superdiffusion for the magnetopause. Geophysical Research Letters 24, 1727–1730.

Treumann, R.A., 1999a. Kinetic theoretical foundation of Lorentzian statistical mechanics. Physica Scripta 59, 19–26.

Treumann, R.A., 1999b. Generalized lorentzian thermodynamics. Physica Scripta 59, 204–214.

Treumann, R.A., Baumjohann, W., 2001. Advanced Space Plasma Physics. Imperial College Press, London, UK.

Treumann, R.A., Jaroschek, C.H., Scholer, M., 2004. Stationary plasma states far from equilibrium. Physics of Plasmas 11, 1317–1325.

Tribeche, M., Mayout, S., Amour, R., 2009. Effect of ion suprathermality on arbitrary amplitude dust acoustic waves in a charge varying dusty plasma. Physics of Plasmas 16, 043706.

Tripathi, A.K., Singhal, R.P., 2007. Electrostatic electron-cyclotron harmonic instability in outer planetary magnetosphere. Planetary and Space Science 55, 867–888.

Tripathi, D., Mason, H.E., Dwivedi, B.N., Del Zanna, G., Young, P.R., 2009. Active region loops: Hinode/extreme-ultraviolet imaging spectrometer observations. The Astrophysical Journal 694, 1256–1265.

Tsallis, C., 1988. Possible generalization of Boltzmann-Gibbs statistics. Journal of Statistical Physics 52, 479–487.

Tsallis, C., 1999. Non-extensive statistics: theoretical, experimental, and computational evidences and connections. Brazilian Journal of Physics 29, 1–35.

Tsallis, C., 2009a. Computational applications of nonextensive statistical mechanics. Journal of Computational and Applied Mathematics 227, 51–58.

Tsallis, C., 2009b. Introduction to Nonextensive Statistical Mechanics. Springer, New York.

Tsallis, C., Brigatti, E., 2004. Nonextensive statistical mechanics: a brief introduction. Continuum Mechanics and Thermodynamics 16, 223.

Tsallis, C., de Albuquerque, M.P., 2000. Are citations of scientific papers a case of nonextensivity. European Physical Journal B 13, 777–780.

Tsallis, C., Souza, A.M.C., 2003a. Stability of the entropy for superstatistics. Physics Letters A 319, 273–278.

Tsallis, C., Souza, A.M.C., 2003b. Constructing a statistical mechanics for Beck-Cohen super-statistics. Physical Review E 67, 026106.

Tsallis, C., Levy, S.V.F., Souza, A.M.C., Maynard, R., 1995. Statistical-mechanical foundation of the ubiquity of Lévy distributions in Nature. Physical Review Letters 75, 3589–3593.

Tsallis, C., Mendes, R.S., Plastino, A.R., 1998. The role of constraints within generalized non-extensive statistics. Physica A 261, 534–554.

Tsallis, C., Bemski, G., Mendes, R.S., 1999. Is reassociation in folded proteins a case of non-extensivity? Physics Letters A 257, 93–98.

Tsallis, C., Anjos, J.C., Borges, E.P., 2003. Fluxes of cosmic rays: a delicately balanced stationary state. Physics Letters A 310, 372–376.

Tsurutani, B.T., Arballo, J.K., Lakhina, G.S., Ho, C.M., Buti, B., Pickett, J.S., Gurnett, D.A., 1998. Plasma waves in the dayside polar cap boundary layer: bipolar and monopolar electric pulses and whistler mode waves. Geophysical Research Letters 25, 4117–4120.

Tsytovich, V.N., 1977. Theory of Turbulent Plasma. Consultants Bureau, New York.

Tu, C.-Y., Marsch, E., Wilhelm, K., Curdt, W., 1998. Ion temperatures in a solar polar coronal hole observed by SUMER on SOHO. The Astrophysical Journal 503, 475–482.

Umarov, S., Tsallis, C., Steinberg, S., 2008. A generalization of the central limit theorem consistent with nonextensive statistical mechanics. Milan Journal of Mathematics 76, 307–328.

Van Allen, J.A., Baker, D.N., Randall, B.A., Sentman, D.D., 1974. The magnetosphere of Jupiter as observed with Pioneer 10, 1. Instrument and principal findings. Journal of Geophysical Research 79, 3559–3577.

Van Ness, P., Reinhard, R., Sanderson, T.R., Wenzel, K.-P., Zwicki, R.D., 1984. The energy spectrum of 35 to 1600 keV protons associated with interplanetary shocks. Journal of Geophysical Research 89, 2122–2132.

Varotsos, P.A., Sarlis, N.V., Skordas, E.S., 2014. Journal of Geophysical Research 119, 9192–9206.

Vasyliūnas, V.M., 1968. A survey of low-energy electrons in the evening sector of the magnetosphere with OGO 1 and OGO 3. Journal of Geophysical Research 73, 2839–2884.

Vasyliunas, V.M., 1971. Deep space plasma measurements. In: Lovberg, R.H. (Ed.), Methods of Experimental Physics, vol. 9B. Academic Press, New York, pp. 49–88.

Vasyliunas, V.M., 1986. The convection-dominated magnetosphere of Uranus. Geophysical Research Letters 13, 621–623.

Vasyliunas, V.M., Siscoe, G.L., 1976. On the flux and the energy spectrum of interstellar ions in the solar system. Journal of Geophysical Research 81, 1247–1252.

Venters, T.M., Pavlidou, V., 2007. The Spectral Index Distribution of EGRET Blazars: Prospects for GLAST. The Astrophysical Journal 666, 128–138.

Verner, D.A., Ferland, G.J., 1996. Atomic data for astrophysics. I. Radiative recombination rates for H-like, He-like, Li-like, and Na-like ions over a broad range of temperatures. The Astrophysical Journal Supplement Series 103, 467.

Veronig, A.M., Rybák, J., Gömöry, P., Berkebile-Stoiser, S., Temmer, M., Otruba, W., Vršnak, B., Pötzi, W., Baumgartner, D., 2010. Multiwavelength imaging and spectroscopy of chromospheric evaporation in an M-class solar flare. The Astrophysical Journal 719, 655–670.

Viall, N., Klimchuk, J.A., 2011. Patterns of nanoflare storm heating exhibited by an active region observed with solar dynamics observatory/atmospheric imaging assembly. The Astrophysical Journal 738, 24.

Vignat, C., Plastino, A., Plastino, A.R., 2005. Superstatistics based on the microcanonical ensemble arXiv: cond-mat/0505580.

Viñas, A.F., Mace, R.L., Benson, R.F., 2005. Dispersion characteristics for plasma resonances of Maxwellian and Kappa distribution plasmas and their comparisons to the IMAGE/RPI observations. Journal of Geophysical Research 110, A06202.

Viñas, A.F., Moya, P.S., Navarro, R., Araneda, J.A., 2014. The role of higher-order modes on the electromagnetic whistler-cyclotron wave fluctuations of thermal and non-thermal plasmas. Physics of Plasmas 21, 012902.

Viñas, A.F., Moya, P.S., Navarro, R.E., Valdivia, J.A., Araneda, J.A., Muñoz, V., 2015. Electromagnetic fluctuations of the whistler-cyclotron and firehose instabilities in a Maxwellian and Tsallis-kappa-like plasma. Journal of Geophysical Research 120, 3307–3317.

Vocks, C., Mann, G., 2003. Generation of suprathermal electrons by resonant wave-particle interaction in the solar corona and wind. The Astrophysical Journal 593, 1134–1145.

Vocks, C., Mann, G., Rausche, G., 2008. Formation of suprathermal electron distributions in the quiet solar corona. Astronomy and Astrophysics 480, 527–536.

Von Montigny, C., et al., 1995. High-energy gamma-ray emission from active galaxies: EGRET observations and their implications. The Astrophysical Journal 440, 525.

Vourlidas, A., Wu, S.T., Wang, A.H., Subramanian, P., Howard, R.A., 2003. Direct detection of a coronal mass ejection-associated shock in large angle and spectrometric coronagraph experiment white-light images. The Astrophysical Journal 598, 1392.

Wang, Q.A., Nivanen, L., Le Méhauté, A., Pezeril, M., 2002. On the generalized entropy pseudoadditivity for complex systems. Journal of Physics A 35, 7003–7007.

Wang, C.-P., Lyons, L.R., Chen, M.W., Wolf, R.A., Toffoletto, F.R., 2003. Modeling the inner plasma sheet protons and magnetic field under enhanced convection. Journal of Geophysical Research 108, 1074.

Wang, L., Lin, R.P., Salem, C., Pulupa, M., Larson, D.E., Yoon, P.H., Luhmann, J., 2012. Quiet-time interplanetary $\sim 2-20$ keV superhalo electrons at solar minimum. The Astrophysical Journal Letters 753, L23.

Wannawichian, S., Ruffolo, D., Kartavykh, Yu Yu, 2003. Ionization fractions of slow ions in a plasma with kappa distributions for the electron velocity. The Astrophysical Journal Supplement Series 146, 443–457.

Warren, H.P., Winebarger, A.R., Brooks, D.H., 2012. A systematic survey of high-temperature emission in solar active regions. The Astrophysical Journal 759, 141.

Washimi, H., Taniuti, T., 1966. Propagation of ion-acoustic solitary waves of small amplitude. Physical Review Letters 17, 996.

Watanabe, T., Hara, H., Yamamoto, N., Kato, D., Sakaue, H.A., Murakami, I., Kato, T., Nakamura, N., Young, P.R., 2009. Fe XIII density diagnostics in the EIS observing wavelengths. The Astrophysical Journal 692, 1294–1304.

Wild, J.P., McCready, L.L., 1950. Observations of the spectrum of high-intensity solar radiation at metre wavelengths. I. The apparatus and spectral types of solar burst observed. Australian Journal of Scientific Research A3, 387.

Wild, J.P., Smerd, S.F., 1972. Radio bursts from the solar corona. Annual Review of Astronomy and Astrophysics 10, 159.

Wilhelm, K., Curdt, W., Marsch, E., Schüle, U., Lemaire, P., Gabriel, A., et al., 1995. SUMER – solar ultraviolet measurements of emitted radiation. Solar Physics 162, 189.

Williams, L.L., Zank, G.P., 1994. Effect of magnetic field geometry on the wave signature of the pickup of interstellar neutrals. Journal of Geophysical Research 99, 19229–19244.

Williams, D.J., McEntire, R.W., Jaskulek, S., Wilken, B., 1992. The Galileo energetic particles detector. Space Science Reviews 60, 385–412.

Williams, D.J., Mauk, B., McEntire, R.E., 1998. Properties of Ganymede's magnetosphere as revealed by energetic particle observations. Journal of Geophysical Research 103, 17523–17534.

Winterhalter, D., Kivelson, M.G., Walker, R.J., Russell, C.T., 1984. Advances in Space Research 4, 287.

Withbroe, G.L., 1975. The analysis of XUV emission lines. Solar Physics 45, 301–317.

Wong, C.-Y., et al., 2015. From QCD-based hard-scattering to nonextensive statistical mechanical descriptions of transverse momentum spectra in high-energy pp and pp^- collisions. Physical Review D 91, 114027.

Wu, C.S., 1984. A fast Fermi process – energetic electrons accelerated by a nearly perpendicular bow shock. Journal of Geophysical Research 89, 8857.

Wyse, A.B., May 1942. The spectra of ten gaseous nebulae. The Astrophysical Journal 95, 356.

Xiao, F., Shen, C., Wang, Y., Zheng, H., Whang, S., 2008. Energetic electron distributions fitted with a kappa-type function at geosynchronous orbit. Journal of Geophysical Research 113, A05203.

Xu, D., Beck, C., 2016. Transition from log-normal to chi-square superstatistics for financial time series. Physica A 453, 173–183.

Yalcin, C., Rabassa, P., Beck, C., 2016. Extreme event statistics of daily rainfall: Dynamical systems approach. Journal of Physics A 49, 154001.

Yamano, T., 2002. Some properties of q-logarithmic and q-exponential functions in Tsallis statistics. Physica A 305, 486–496.

Yang, L., Wang, L., Li, G., He, J., Salem, C., Yu, C., Wimmer-Schweingruber, R.F., Bale, S.D., 2015a. The angular distribution of solar wind superhalo electrons at quiet times. The Astrophysical Journal Letters 811, L8.

Yang, Z., Liu, Y.D., Richardson, J.D., Lu, Q., Huang, C., Wang, R., 2015b. Impact of pickup ions on the shock front nonstationarity and energy dissipation of the heliospheric termination shock: Two-dimensional full particle simulations and comparison with Voyager 2 observations. The Astrophysical Journal 809, 28.

Yashiro, S., Akiyama, S., Gopalswamy, N., Howard, R.A., 2006. Different Power-Law Indices in the Frequency Distributions of Flares with and without Coronal Mass Ejections. The Astrophysical Journal 650, L143–L146.

Ye, G., Hill, T.W., 1994. Solar-wind-driven convection in the Uranian magnetosphere. Journal of Geophysical Research 99, 17225–17235.

Yoon, P.H., 2014. Electron kappa distribution and quasi-thermal noise. Journal of Geophysical Research 119, 7074–7087.

Yoon, P.H., Rhee, T., Ryu, C.M., 2006. Self-consistent formation of electron κ distribution: 1. Theory. Journal of Geophysical Research 111, A09106.

Yoon, P.H., 2000. Generalized weak turbulence theory. Physics of Plasmas 7, 4858.

Yoon, P.H., 2005. Effects of spontaneous fluctuations on the generalized weak turbulence theory. Physics of Plasmas 12, 042306.

Yoon, P.H., Ryu, C.-M., Rhee, T., 2003. Particle kinetic equation including weakly turbulent mode coupling. Physics of Plasmas 10, 3881.

Yoon, P.H., Ziebell, L.F., Gaelzer, R., Lin, R.P., Wang, L., 2012. Langmuir turbulence and suprathermal electrons. Space Science Reviews 173, 459–489.

Young, D.T., et al., 2005. Composition and dynamics of plasma in Saturn's magnetosphere. Science 307, 1262–1266.

Young, D.T., et al., 2004. Cassini plasma spectrometer investigation. Space Science Reviews 114, 1–112.

Young, P.R., Watanabe, T., Hara, H., Mariska, J.T., 2009. High-precision density measurements in the solar corona. I. Analysis methods and results for Fe XII and XIII. Astronomy and Astrophysics 495, 587–606.

Younsi, S., Tribeche, M., 2010. Arbitrary amplitude electron-acoustic solitary waves in the presence of excess superthermal electrons. Astrophysics and Space Science 330, 295–300.

Zank, G.P., 1999. Interaction of the solar wind with the local interstellar medium: a theoretical perspective. Space Science Reviews 89, 413–688.

Zank, G.P., 2015. Faltering steps into the galaxy: the boundary regions of the heliosphere. Annual Review of Astronomy and Astrophysics 53, 449.

Zank, G.P., Heerikhuisen, J., Pogorelov, N.V., Burrows, R., McComas, D.J., 2010. Microstructure of the heliospheric termination shock: implications for energetic neutral atom observations. The Astrophysical Journal 708, 1092.

Zank, G.P., Heerikhuisen, J., Wood, B.E., Pogorelov, N.V., Zirnstein, E., McComas, D.J., 2013. Heliospheric structure: the bow wave and the hydrogen wall. The Astrophysical Journal 763, 20.

Zank, G.P., Hunana, P., Mostafavi, P., Goldstein, M.L., 2014. Pickup ion mediated plasmas. I. Basic model and linear waves in the solar wind and local interstellar medium. The Astrophysical Journal 797, 87.

Zank, G.P., Li, G., Florinski, V., Hu, Q., Lario, D., Smith, C.W., 2006. Particle acceleration at perpendicular shock waves: Model and observations. Journal of Geophysical Research 111, A06108.

Zank, G.P., Pauls, H.L., Cairns, I.H., Webb, G.M., 1996a. Journal of Geophysical Research 101, 457.

Zank, G.P., Pauls, H.L., Williams, L.L., Hall, D.T., 1996b. Journal of Geophysical Research 101, 21639.

Zank, G.P., Pogorelov, N.V., Heerikhuisen, J., Washimi, H., Florinski, V., Borovikov, S., Kryukov, I., Müller, H.R., 2009. Space Science Reviews 146, 295.

Zhang, Y., Liu, X.-W., Zhang, B., 2014. H-I free-bound emission of planetary nebulae with large abundance discrepancies: two-component models versus κ-distributed electrons. The Astrophysical Journal 780, 93.

Zhuang, H.C., Russell, C.T., 1981. An analytic treatment of the structure of the bow shock and magnetosheath. Journal of Geophysical Research 86, 2191–2205.

Ziebell, L.F., Gaelzer, R., Yoon, P.H., 2001. Nonlinear development of weak beam-plasma instability. Physics of Plasmas 8, 3982.

Ziebell, L.F., Yoon, P.H., Petruzzellis, L.T., Gaelzer, R., Pavan, J., 2015. Plasma emission by nonlinear electromagnetic processes. The Astrophysical Journal 806, 237.

Zilbersher, D., Gedalin, M., 1997. Pickup ion dynamics at the structured quasi-perpendicular shock. Planetary and Space Science 45, 693–703.

Zirnstein, E.J., McComas, D.J., 2015. Using kappa functions to characterize outer heliosphere proton distributions in the presence of charge-exchange. The Astrophysical Journal 815, 31.

Zirnstein, E.J., Heerikhuisen, J., Zank, G.P., Pogorelov, N.V., McComas, D.J., Desai, M.I., 2014. Charge-exchange Coupling between Pickup Ions across the Heliopause and its Effect on Energetic Neutral Hydrogen Flux. The Astrophysical Journal 783, 129.

Zouganelis, I., 2008. Measuring suprathermal electron parameters in space plasmas: implementation of the quasi-thermal noise spectroscopy with kappa distributions using in situ Ulysses/URAP radio measurements in the solar wind. Journal of Geophysical Research 113, A08111.

Zouganelis, I., Maksimovic, M., Meyer-Vernet, N., Lamy, H., Issautier, K., 2004. A transonic collisionless model of the solar wind. The Astrophysical Journal 606, 542–554.

Zouganelis, I., Meyer-Vernet, N., Landi, S., Maksimovic, M., Pantellini, F., 2005. Acceleration of weakly collisional solar-type winds. The Astrophysical Journal 626, L117–L120.

Zurbuchen, T.H., Raines, J.M., Slavin, J.A., et al., 2011. MESSENGER observations of the spatial distribution of planetary ions near Mercury. Science 333, 1862–1865.

Zwillinger, D., 1997. Handbook of Differential Equations, third ed. Academic Press, Boston, MA.

Further Reading

Burlaga, L.F., Vinas, A.F., 2004. Multiscale structure of the magnetic field and speed at 1 AU during the declining phase of solar cycle 23 described by a generalized Tsallis probability distribution function. Journal of Geophysical Research 109, A12107.

Cairns, I.H., 1986. The source of free energy for Type II solar radio bursts. Publications of the Astronomical Society of Australia 6, 444.

Cairns, I.H., 1988. A semi-quantitative theory for the $2f_p$ radiation observed upstream from the Earth's bow shock. Journal of Geophysical Research 93, 3958–3968.

Dopita, M.A., Sutherland, R.S., Nicholls, D.C., Kewley, L.J., Vogt, F.P.A., 2013. New strong-line abundance Diagnostics for H II regions: effects of κ-distributed electron energies and new atomic data. The Astrophysical Journal Supplement Series 208, 10.

Gaelzer, R., de Juli, M.C., Ziebell, L.F., 2010. Effect of superthermal electrons on Alfvén wave propagation in the dusty plasmas of solar and stellar winds. Journal of Geophysical Research 115, A09109.

Galvão, R.A., Ziebell, L.F., Gaelzer, R., de Juli, M.C., 2012. Alfvén waves in dusty plasmas with plasma particles described by anisotropic kappa distributions. Physics of Plasmas 19, 123705.

Gary, S.P., Karimabadi, H., 2006. Linear theory of electron temperature anisotropy instabilities: whistler, mirror, and Weibel. Journal of Geophysical Research 111, A11224.

Gary, S.P., Tokar, R.L., 1985. The second-order theory of electromagnetic hot ion beam instabilities. Journal of Geophysical Research 90, 65–72.

Holzer, T.E., Axford, W.I., 1970. Solar wind ion composition. Journal of Geophysical Research 75 (31), 6354.

Kuncic, Z., Cairns, I.H., Knock, S.A., 2002. Analytic model for the electrostatic potential jump across collisionless shocks, with applic-ation to Earth's bow shock. Journal of Geophysical Research 107. SSH 11.

Luczka, J., Zaborek, B., 2004. Brownian Motion: a Case of Temperature Fluctuations. Acta Physica Polonica B 35, 2151.

Mauk, B.H., Krimigis, S.M., Keath, E.P., Cheng, A.F., Armstrong, T.P., Lanzerotti, L.J., Gloeckler, G., Hamilton, D.C., 1987. The hot plasma and radiation environment of the Uranian magneto-sphere. Journal of Geophysical Research 92, 15283–15308.

Mauk, B.H., Krimigis, S.M., Cheng, A.F., Selesnick, R.S., 1995. Energetic particles and hot plasmas of Neptune. In: Cruikshank, D.P. (Ed.), Neptune and Triton. Univ. of Arizona Press, Tucson, Ariz, pp. 169–232.

Plastino, A.R., Plastino, A., 1995. Non-extensive statistical mechanics and generalized Fokker-Planck equation. Physics A 222, 347–354.

Sooklal, A., Mace, R.L., 2004. The magnetized electron-acoustic instability driven by a warm, field-aligned electron beam. Physics of Plasmas 11, 1996.

Titov, V.S., Demoulin, P., 1999. Basic topology of twisted magnetic configurations in solar flares. Astronomy and Astrophysics 351, 707.

Tsallis, C., Bukmann, D.J., 1996. Anomalous diffusion in the presence of external forces: Exact time-dependent solutions and their thermostatistical basis. Physical Review 54E, R2197.

Van Kampen, N.G., 1981. Stochastic Processes in Physics and Chemistry. North Holland, Amsterdam.

Index

A

Abundance discrepancy
  alternative explanations, 654–655
  factor, 644–645
  problem, 634, 650–651
Acceleration, 437, 438f, 571, 582–597
  classic pump mechanism, 585
  conditions, 583
  Parker equation, 585–586
  particle speed, 583–584
  pump acceleration mechanism, 584f, 591–592
  in solar corona, 605–608
    diffusion coefficient choice, 605–606
    impulsive solar energetic particle events, 606–608
  source particle spectrum, 594–597
  spectral index in velocity space, 586–587
  spectrum time evolution
    comparing solutions for, 590–591
    driving equation for, 587–590
  statistics, 325–326
  subtleties with solutions for pump acceleration, 592–593
ACRs. *See* Anomalous cosmic rays (ACRs)
Advanced Composition Explorer (ACE), 571, 629–631
AIA. *See* Atmospheric Imaging Assembly (AIA)
Alfvén cyclotron
  instability, 343–348, 344f
  plasma waves, 339–342
Alfvénic Mach number, 612
Algebraic equations, 336–337

Analytic theory, 563
Analytical methods, 634
  plasma methods, 66
Angular potentials, 231–232
Anisotropic distributions, 186–187, 196–217. *See also* Isotropic distributions
  correlated degrees of freedom, 196–199
  correlation between projection at certain direction and perpendicular plane, 200–201
  potentials forming anisotropic distributions of velocity, 174–175
  self-correlated
    degrees of freedom, 201–206
    projections, 204–205, 207–209
  self-correlation and intercorrelation between degrees of freedom, 210–217
Anisotropic kappa distributions, 331–332, 475
  function, 493
  of velocities, 430–432
Anisotropy, 174, 490–492, 495
Anomalous cosmic rays (ACRs), 571, 580–582
Anticorrelations between density and temperature, 452–455
Antiequilibrium, 402–403, 424, 432–439, 571–573
  divergent temperature at, 48–49
  double role of, 102
  intermediate transitions toward, 101
  state, 51, 445–446, 449–450
Argon (Ar), 636
Argument, 78–87, 244
Astronomical unit (AU), 571

Astrophysical nebulae, 634–635, 638–639
  alternative explanations of abundance discrepancy, 654–655
  applications of kappa distributions in, 653–654
  atomic energy levels and kappa distribution, 646–651
  diagnostics for kappa index, 651–652
  HII regions, 635–636
  ionization structures in HII region, 639–641
  kappa distributions, 637–638
  magnetic structures in HII regions, 641–642
  nebular spectral lines, 642–646
  photoionized nebulae modeling, 652–653
  PNe, 636–637, 637f
Astrophysics, 553
Asymptotic decay, 316–317
Atmospheres, 483
Atmospheric Imaging Assembly (AIA), 532–533
  observations, 535f
  response to emissions, 534–535
Atomic data deficiencies, 654
Atomic energy levels and kappa distribution, 646–651
  calculating electron temperatures from CEL, 647–650
  CEL abundance discrepancy, 650
  discrepancies in CEL temperatures, 650–651, 651f
  optical recombination line abundance discrepancy, 650
AU. *See* Astronomical unit (AU)
Aurora, 493–494
Auxiliary temperature, 38–39, 39f

B

Barometric formula, 164–165
BATS-R-US code, 564–565
Beam
  distributions, 553
  speed, 557
  velocity, 551
Bessel differential equation, 275
Bessel functions, 337–340
"Bi-abundance" model, 645–646
Bi-Maxwellian distribution, 343–344, 349–351
Blast wave shock, 561–563
Boltzmann constant, 332
Boltzmann distribution, 637
Boltzmann equation for dilute gas, 364
Boltzmann–Gibbs (BG)
  energy distribution, 291
  entropic formulation, 66–67
  entropy, 68–69
  statistical mechanics, 14–15, 15f, 84
Boltzmannian exponential distribution, 27
Bowen Fluorescence, 644
Bremsstrahlung spectrum, 532, 554
Brownian particle, 316, 318
Bulk plasma velocity, 489–490
Buneman instability, 358

C

Canonical distributions, 14–16
Canonical probability distribution, 34
Cascade random variable, 319
CELs. See Collisionally excited lines (CELs)
Central Limit Theorem (CLT), 318–319
Cerenkov emission, 554
Cerenkov resonance, 554
Charge excess, 272
CHIANTI 7.1 database, 533
CIR. See Corotating interaction region (CIR)
Circularly polarized waves, 343
Classical Lagrangian turbulence, 324–325
Clausius thermodynamic definition, 36
Cloudy (plasma modeling system), 652–653

CLT. See Central Limit Theorem (CLT)
CMEs. See Coronal mass ejections (CMEs)
Coherent wave process, 556
Cold high-metallicity inclusion process, 654
Cold protons, 405–407
Collision frequency, 292–295
Collisional excitation, 643
Collisionally excited lines (CELs), 636, 642–643
  abundance discrepancy, 650
  calculating electron temperatures from, 647–650
  discrepancies in CEL temperatures, 650–651, 651f
Collisionless plasmas, 296
Common spectrum of particles, 571
  acceleration mechanism, 582–597
  heliosphere, 573
  observations, 574–582
    anomalous cosmic rays accelerated in heliosheath, 580–582
    in fast, polar coronal holes solar wind, 582
    in inner heliosphere, 575–580
  pump acceleration mechanism applications, 597–608
  shock and particle acceleration results, 572t
Conditional distributions, 230
Continuous description, 28–29
Continuum intensities, 526–535
  continuum spectrum, 531–533
  excitation rates, 529–531
  ionization and recombination rates, 526–529, 527f
  KAPPA package, 533
  synthetic spectra and atmospheric imaging assembly, 534–535, 535f
Continuum spectrum, 531–533
Convected distribution, 485
"Convective motion", 485
Core component, 352–353
Core distribution, 599
Coronal heating, 470–472, 473f
Coronal loop, 606
Coronal mass ejections (CMEs), 551, 553
Coronal radiation, 524

Corotating interaction region (CIR), 574, 625–626
Correlations, 60–61, 300–302
  correlated degrees of freedom, 196–199
  degeneracy, 77
  dependence of kappa index on number of correlated particles, 52–56
  length, 74, 77
  between projection at certain direction and perpendicular plane, 200–201
Cosmic ray, 483
Coulomb force, 330–331
Cross-shock electrostatic potential, 610–611
Curie constant, 295–300
Cyclotron resonances, 554
Cylindrical coordinates, 337–338

D

1-D
  Langmuir wavevector distribution, 558
  linear gravitational potential, 124–127
  Maxwellian distribution, 612–613
  PIC simulations, 623
  potential, 276–277
3-D
  kappa index, 188, 192–193, 198, 201–202, 205–206, 550
  paraboloid shock, 563–564
  potential, 278
d–D potential, 279
de Hoffman-Teller frame, 555–556, 563
Debye length, 74–75
  in equilibrium and nonequilibrium plasmas, 272–285
    charge/field dimensions and symmetry, 279t
    interpretations, 281–285, 283f, 285f, 285t
    large potential energy, 279–281
    Poisson equation for electrostatic potential, 272–276
    symmetric Poisson equation and solutions, 276–279
  misestimation of, 49–50
Debye spheres, 83, 281

Debye—Hückel potential, 275
Decay processes, induced and
    spontaneous, 386—388
Deceleration, 437, 438f
Degeneration of kappa index, 228,
    233
    1D linear gravitational potential,
        124—127
    positive attractive power law
        potential, 127—133
    rationale, 123—124
Degrees of freedom, 38—39
    correlated, 196—199
    per particle, 186—187
    self-correlated, 201—206
    self-correlation and
        intercorrelation between,
        210—217
DEM. *See* Differential emission
    measure (DEM)
Density
    diagnostics, 536—537, 537f
    of kappa indices, 195
    normalized distribution,
        337—338
Dependence of kappa index
    on degrees of freedom, 55f
    on number of correlated particles,
        13—14, 52—56
Dielectric tensor, 359—361
Dielectronic recombination
    process, 526—527
Differential emission measure
    (DEM), 536
    for kappa distributions, 544—546,
        545f
Diffusion coefficient, 604
    choice, 605—606
Diffusive shock acceleration (DSA),
    571, 603—605, 629
Discrete descriptions, 22—27
Discrete distributions, 245
Discrete dynamics, 96—97
    discrete map of stationary states
        transitions, 97—98
    numerical application of discrete
        transitions of stationary states,
        98—101
    stationary state transition stages,
        101—103
Discrete energy spectrum, 18—19
Discrete phase space of map,
    97—98
Discrete probability distribution,
    18—19

Dispersion
    relation, 336—339, 341
    tensor, 336—337
Dissipation
    dissipative ion reflection process,
        629
    at quasiperpendicular shocks,
        629—631
Distribution function, 571
Distributions with potential,
    217—233
    angular potentials, 231—232
    equivalent local distribution,
        222—223
    general Hamiltonian distribution,
        218
    magnetization potential,
        232—233
    marginal and conditional
        distributions, 230
    negative attractive potential,
        220—221
    negative power law central
        potential, 226—227
    positive attractive potential,
        218—219
    positive power law central
        potential, 224—225
    properties, 228—230
    small positive/negative attractive/
        repulsive potential,
        221—222
"Disturbed" plasma sheet spectra,
    510—511
d_K-dimensional kappa index,
    188—189, 197, 202,
    206
Double layers (DLs), 399—400,
    408—409. *See also*
    Suprathermal electrons
    in plasmas with suprathermal
        electrons, 402—414
    nonlinear electrostatic waves,
        403
    nonlinear evolution of
        ion-acoustic waves, 404
    three-component plasmas,
        407—411
    two-component magnetized
        plasmas, 413—414
    two-component plasmas,
        405—407
Downstream
    proton distribution,
        619—620

velocity distribution functions,
    619, 623—625, 624f
Drifting bi-kappa distribution.
    See Anisotropic kappa
    distribution
Drude model for electrical
    conduction, 289
DSA. *See* Diffusive shock
    acceleration (DSA)
Dual-kappa distributions,
    498—499
Duality relation, 33
Dynamical creation of kappa
    distributions, 314—317

E
Earth, 506—513
    atmospheres, 483
    "disturbed" plasma sheet spectra,
        510—511
    high-latitude spectra, 512—513
    magnetosheath spectra, 513
    middle magnetosphere and
        substorms, 511—512
    orbit, 551
    "quiet" plasma sheet spectra,
        507—510
eb component. *See* Electron beam
    component (eb component)
ec component. *See* Electron core
    component (ec component)
EF. *See* Electron firehose (EF)
EIS. *See* Extreme-Ultraviolet
    Imaging Spectrograph (EIS)
Electrical conductivity, 286—291
Electromagnetic (EM)
    Alfvén cyclotron plasma waves,
        low-frequency, 339—342
    disturbances, 334—335
    plasma waves, 341
Electron, Proton, and Alpha
    Monitor (EPAM), 629—631
Electron beam component (eb
    component), 352—353
Electron core component (ec
    component), 352—353
Electron distributions. *See also* Ion
    distributions
    coronal heating by velocity
        filtration to suprathermal
        electrons, 470—472
    heat flux, 472—474, 473f
    observations and origins, 466—470
    in space plasmas, 466

Electron distributions (*Continued*)
  suprathermal electrons on
    acceleration of escaping
    particles, 474–478
Electron firehose (EF), 349
  instability, 351–352, 353f
Electron(s), 345, 470–471,
    507–510, 509f, 623
  beams, 557–561, 560f, 562f
  capture, 526–527
  density, 473–474
  electron-acoustic solitons in
    plasmas, 415–417
  energy kappa distribution, 646
  kinetic energies, 644
  reflections, 561–565, 562f, 564f
  spectra, 498–500
  thermal anisotropy, 351
Electrostatic (ES)
  potential, 475, 555–556
    Poisson equation for
    electrostatic potential,
    272–276
  waves, 359–360
Electrostatic solitary waves (ESWs),
    400
Emission measure (EM), 535
Emissions, induced and
    spontaneous, 386, 393–394
ENA. *See* Energetic neutral atoms
    (ENA)
Energetic electrons, 639
  bursts, 515, 516f
Energetic ions, 504–506, 505f
  spectra, 488–490, 501–503
Energetic neutral atoms (ENA),
    487, 622–623
Energetic Particle Detector (EPD),
    485–486, 489–490, 491f
Energetic Particle Spectrometer
    (EPS), 515
Energetic particle(s), 487
  acceleration at shocks, 598–605
  moments, 492–493
Energetic solar particle (ESP), 574
Energy. *See also* Electron(s)
  distribution, 245
  energy-flux spectral index,
    571–573
  spectrum, 23, 571
Entropy, 14–15, 28–29, 86f,
    458
  discrete dynamics, 96–97
  entropic formula for velocity
    kappa distributions, 78–87

for isothermal transitions between
    stationary states, 87–96
  maximization, 314
    continuous description, 28–29
    discrete description, 22–27
  scale parameters in entropic
    formulation, 67–78
EPAM. *See* Electron, Proton, and
    Alpha Monitor (EPAM)
EPD. *See* Energetic Particle Detector
    (EPD)
EPS. *See* Energetic Particle
    Spectrometer (EPS)
Equilibrium
  density, 268
  equilibrium plasmas, Debye length
    in, 272–285.
  see also Space plasmas
  "equilibrium temperature",
    misleading, 46
  final transitions toward, 102
  statistical mechanics, 315
  temperature out of, 41–42
Equivalence, 33
Equivalent dielectric tensor,
    334–335
Equivalent local distribution,
    222–223
  with no potential energy, 223
  phase space distribution with
    potential, 223
ES. *See* Electrostatic (ES)
Escape state, 7, 448, 456
Escort distribution function,
    79–80
Escort probability distributions,
    18–20
ESP. *See* Energetic solar particle
    (ESP)
ESWs. *See* Electrostatic solitary
    waves (ESWs)
EUV. *See* Extreme-ultraviolet (EUV)
Excitation
  mechanisms, 644
  rates, 529–531
Exospheric model, 475, 477
Extreme-ultraviolet (EUV), 524,
    529–530, 536
Extreme-Ultraviolet Imaging
    Spectrograph (EIS), 536

F

F distributions, 320
Far equilibrium, 456–458

back to far-equilibrium region,
    102
  inner heliosheath, 451–452, 454f
Flare X-ray emission, 532–533
Flares, 552
Fluctuations, 322, 334–335
Fokker–Planck equation, 283
Fokker–Planck model, 467–468
Fourier transform, 606
Fourier–Laplace transformation,
    336–337, 368–369
Free energy, 335
Free-space permittivity, 337–338
Free-streaming Klimontovich
    function, 367
Free–free emission continuum,
    654
"Fundamental" stationary state, 91
"Fyris" 3-D hydrodynamic code,
    653

G

Galactic cosmic rays (GCRs), 582–583
Gamma distribution, 314–315
Gamma function, 331–332
Gas giant planets, kappa distribution
    in magnetospheres of,
    488–506
  Jupiter, 488–494
  Neptune, 503–506
  Saturn, 494–500
  Uranus, 500–503
Gaunt factors, 654
Gauss hypergeometric function,
    340
Gauss's law of electrodynamics,
    Poisson equation of,
    267–268
Gaussian behavior, 323
Gaussian distribution of polytropic
    indices, 255
Gaussian random variables,
    318–319
GCRs. *See* Galactic cosmic rays
    (GCRs)
General Hamiltonian distribution
  Hamiltonian function, 218
  phase space distribution, 218
Gibbs path, 22
Gordeyev integral, 359–360
Gravitational potentials, 109–110.
    See also Negative potentials
  linear, 164–165, 169f
  spherical, 166–170, 169f

Virial theorem and Jeans radius, 170–171
Gravitational pull, 162
Gravitational type potential, 153–162, 226–227
Gyro resonant process, 554
Gyroradius (r_g), 605

H

Hagedorn temperature parameter, 327
Halo
  components, 467–468
  population, 466–467
Hamiltonian distribution, 110–113, 116f. *See also* General Hamiltonian distribution
Hamiltonian function, 37, 218, 236, 291
Hamiltonian kappa distribution, 109–110, 186–187
Hamiltonian probability distribution, 142
Heat flux, 472–474, 473f–474f
Heavier ions, 407–411
HelioPause (HP), 627–628
Heliosheath, 571. *See also* Inner heliosheath
  anomalous cosmic rays accelerated in, 580–582
  proton distribution, 622–623
Heliosphere, 36. *See also* Inner heliosphere
  observations and measurements of kappa indices, 444–450, 447f, 450t–451t
Heliospheric termination shock (HTS), 610–611
Helium, 636
High-latitude spectra, 512–513
HII regions, 634–636, 636f
  complexities in structures, 653
  ionization structures in, 639–641
  magnetic structures in, 641–642
HP. *See* HelioPause (HP)
HTS. *See* Heliospheric termination shock (HTS)
Hubble space telescope, 641–642
Hybrid probability distribution, 92–93
Hybrid velocity distribution, 359–360

Hydrogen, 636
  Balmer series, 642–643
Hyperbolic functions, 20–21

I

IBEX. *See* Interstellar Boundary Explorer (IBEX)
ICME. *See* Interplanetary coronal mass ejection (ICME)
IHS. *See* Inner heliosheath (IHS)
IMF. *See* Interplanetary magnetic field (IMF)
Impulsive solar energetic particle events, 606–608
INCA. *See* Ion and Neutral Camera (INCA)
Incident ions, 610–611
Incoherent emission process, 554
Inertial reference frame, 189–193
Inner heliosheath (IHS), 106–107, 422–423, 626
  observations
    anticorrelation between density and temperature, 452–454
    far-equilibrium inner heliosheath, 451–452, 454f
    isobaric process, 454–456
Inner heliosphere, 106–107, 422–423, 571
  common spectrum in, 575–580
  LAEs, 575–580
Interface Region Imaging Spacecraft, 536
Internal energy, 245
International Sun Earth Explorer (ISEE), 517f
Interplanetary coronal mass ejection (ICME), 577–578
Interplanetary magnetic field (IMF), 511
Interplanetary shock waves, 625–626
Interpretations, Debye length, 281–285, 283f, 285f, 285t
Interstellar Boundary Explorer (IBEX), 451, 622–623, 626, 628f
Invariant kappa index (k_0), 13–14, 51, 82, 82f, 187–191, 193, 196–197, 200
Inverse temperature variable, 327
Ion and Neutral Camera (INCA), 487
Ion distributions. *See also* Electron distributions

antiequilibrium, 432–439
arrangement of stationary states, 439–444
interpreting observations
  in inner heliosheath, 451–456
  and measurements of kappa indices, 444–450, 447f, 450t–451t
  near *vs.* far equilibrium, 456–458
  PUI in transitions of kappa distributions, 458–462
  in space plasmas, 422–423
Ion kappa distribution
  formulations, 425–432
  anisotropic kappa distribution of velocities, 430–432
  "negative" kappa distribution of velocities, 427–429, 428f–429f
  phase space marginal kappa distribution of velocities, 432
  standard kappa distribution of velocities, 425–427
  superposition of kappa distribution of velocities, 429–430
Ion spectra, 494–497
Ion-acoustic plasma waves, 355–359, 359f
Ion-acoustic solitons with suprathermal electrons, 402–414
  nonlinear electrostatic waves, 403
  nonlinear evolution of ion-acoustic waves, 404
  three-component plasmas, 407–411
  two-component
    magnetized plasmas, 413–414
    plasmas, 405–407
Ionization
  diagnostics involving ionization equilibrium, 540–541, 541f–542f
  ionized metal ions, 524
  potential ions, 640
  rates, 526–529, 527f
  structures in HII region, 639–641, 641t
Ionospheres, 483
Irreversibility. *See* Dissipation
ISEE. *See* International Sun Earth Explorer (ISEE)
Isentropic switching, 95–96, 99, 101–102, 102f

Isobaric process, 454–456
Isothermal transitions, entropy for stationary states between
  derivation, 87–89
  spontaneous entropic procedures, 94–96
  survey, 89–94, 91f
Isotropic distributions, 186–187. *See also* Anisotropic distributions
  negative, multidimensional kappa distribution, 193–194
  standard multidimensional kappa distributions, 187–189
    in inertial reference frame, 189–193
  superposition of multidimensional kappa distributions, 194–196

J
Jovian ion spectra, 489
Jupiter, 488–494, 491f
  energetic ion spectra, 488–490
  energetic particle moments, 492–493
  particle energization processes and anisotropies, 490–492
  plasma sources and aurora, 493–494

K
κ-index, 332–333, 338–339, 344, 351, 520
  single-ion diagnostics of, 537–540, 538f
Kappa distribution(s), 5, 106–108, 250, 321–322, 333, 422–423, 434f, 436f, 470–471, 524, 529, 571–573, 622–623, 629, 637–638. *See also* Phase space kappa distributions
  applications in astrophysical nebulae, 653–654
  atomic energy levels and, 646–651
  base of, 37–39
  dynamical creation, 314–317
  of electrons, 550
  examples of space plasmas, 10f
  exponent of, 39–41
  fitting and correlation, 11f

formulae, 186
  anisotropic distributions, 196–217
  discrete distributions, 245
  distributions with potential, 217–233
  isotropic distributions, 187–196
  multiparticle distributions, 234–243
  non-Euclidean–normed distributions, 243–244
  function, 485–486, 486f
  in high-energy scattering processes, 326–327, 327f
  kappa index, 51–62
  kappa velocity distribution plasma waves
    at oblique propagation, 359–361
    at parallel propagation, 339–359
  in magnetospheres
    of gas giant planets, 488–506
    for magnetospheric research, 518–521
    of terrestrial planets, 506–518
  mathematical motivation, 14–15
  nonextensive statistical mechanics, 16–22, 29–37, 423
  PUI in kappa distribution transitions, 458–462
  solar spectra and, 525–526, 536
    DEM for, 544–546, 545f
    plasma diagnostics from emission line spectra, 535–543
    Reuven-Ramaty high-energy solar spectroscopic imager, 525f
    synthetic line and continuum intensities, 526–535
  in space plasmas, 487
  statistical background, 5
  temperature, 41–51
  weak turbulence theory for plasmas by, 380–389
Kappa electron distribution, 389–397
  steady-state Langmuir turbulence, 392–397
  steady-state particle distribution, 389–392
Kappa index (q), 7, 10f, 36, 51–62, 85–87, 193–194, 291, 424, 426, 439, 441, 467–468, 468f

connection with, 257–261
  degeneration, 233
    angular potentials, 232
    in potential energy, 123–133
  dependence on number of correlated particles, 52–56
  diagnostics for, 651–652
  in heliosphere, 444–450, 447f, 450t–451t
  misleading considerations
    correlation, 60–61
    dependence of temperature on, 44–45
    problem of divergence, 61–62
    sets upper limit on total number of particles, 59–60
  N-particle kappa distributions, 56–57
  negative, 57–59, 149–153, 154f
  positive, 145–149
Kappa ion–exosphere model, 477
Kappa plasma dispersion function, 339–342
Kappa spectrum, 52, 267f, 443–444, 444f–445f
Kappa velocity distribution plasma waves
  high-frequency
    electromagnetic Langmuir plasma waves, 352–355, 354f
    electromagnetic Whistler cyclotron plasma waves, 349
    instabilities, 349–352
  low-frequency
    electromagnetic Alfvén cyclotron plasma waves, 342–343
    electromagnetic ion-acoustic plasma waves, 355–359, 359f
    instabilities, 343–348
  at oblique propagation, 359–361
  at parallel propagation, 339–359
Kappa-shaped particle spectra
  kappa distribution
    in magnetospheres of gas giant planets, 488–506
    in magnetospheres of terrestrial planets, 506–518
    for magnetospheric research, 518–521
    measuring and interpreting in space plasmas, 487

planetary magnetospheres, 484f, 485—486
KdV-type solitons. *See* Korteweg-de Vries-type solitons (KdV-type solitons)
Kennel—Petschek limit, 518
Kinetic energy, 12, 113
   distribution, 232—233
   standard d_K—dimensional kappa distribution of, 244
Kinetic temperature (T_K), 21—22
Kinetic theory, 483
Klimontovich function, 365—366
Knudsen number, 469, 472, 474, 479
Kolmogorov K62 theory, 325
Kolmogorov turbulence, 608
Korteweg-de Vries-type solitons (KdV-type solitons), 401
Kurtosis function, 322

L

LAEs. *See* Local acceleration events (LAEs)
Lagrangian temperature (T_L), 21, 24
Langmuir mode, 551
Langmuir plasma waves, high-frequency electromagnetic, 352—355, 354f
Langmuir waves, 381—382, 557—561, 560f, 562f
Large-scale quantization constant correlations, 300—302
   estimation of h^* for space plasmas, 304—307
   fast solar wind, 306f
   missing plasma parameters, 308—310
   phase space portion, 305f
   smallest phase space parcel in plasmas, 302—303
LARs. *See* Local acceleration regions (LARs)
Learmonth radio spectrograph, 565
LECP. *See* Low Energy Charged Particle (LECP)
Legendre transform, 319—320
LEMMS. *See* Low Energy Magnetospheric Measurement System (LEMMS)
Length scale (σ_r), 72—75
Levy-type models, 316
Line radiation, 531

Line ratio method, 650
Linear charge perturbation, 277—278
Linear gravitational potential, 164—165, 169f
Linear kinetic waves in plasmas
   kappa velocity distribution plasma waves
      at oblique propagation, 359—361
      at parallel propagation, 339—359
   non-Maxwellian
      particle velocity distributions, 334
      plasma particle distributions, 331—332
   normalized kappa distribution, 333f
   parameter kappa, 332—333
   plasma dielectric tensor and dispersion relation, 336—339
   spectral properties of waves, 334—335
Linear superposition, 194—196, 430
Liouville's theorem, 471, 563
LISM. *See* Local InterStellar Medium (LISM)
Local acceleration events (LAEs), 574
Local acceleration regions (LARs), 574, 577—578
Local correlations, 298
Local InterStellar Medium (LISM), 627—628
Local kappa distribution, 134—141
Local parameters, 228—229
Lorentz force, 612
Lorentzian distributions, 5, 466
Low Energy Charged Particle (LECP), 487, 490f, 495f
Low Energy Magnetospheric Measurement System (LEMMS), 496

M

Macdonald function, 275
Magnetic field, 341, 477, 639
"Magnetic reconnection", 551—552
Magnetic structures in HII regions, 641—642
   fine hair-like structures, 641f
   lowest five energy levels, 642f

Magnetization, 295—300, 299f
   magnetic moment and velocity vectors, 295f
   positional distribution and, 298f
   potential, 232—233
Magnetohydrodynamic simulation (MHD simulation), 563—565, 621
Magnetosheath spectra, 513, 514f
Magnetosonic
   mode, 345
   waves, 342
Magnetospheres, 483
   kappa distributions
      of gas giant planets, 488—506
      of terrestrial planets, 506—518
Magnetospheric electron velocity distributions, 333—334
Magnetospheric Imaging Instrument (MIMI), 496, 496f
Magnetospheric research, kappa distributions for, 518—521
Magnetospheric systems, 483
Marginal distribution, 116—119, 230, 318
Mathematical motivation, 14—15
Maxwell distribution(s), 5—7, 107—108, 422—423
Maxwell velocity distribution, 69—70
Maxwell's equations, 331, 365
Maxwell's kinetic theory, 123
Maxwell—Boltzmann distributions, superposition of
   applications, 324—327
      classical Lagrangian turbulence, 324—325, 325f
      kappa distributions in high-energy scattering processes, 326—327, 327f
      quantum turbulence, 325—326, 326f
   asymptotic behavior, 319—321
   dynamical creation of kappa distributions, 314—317
   from measured time series to superstatistics, 322—324, 323f—324f
   timescale separation in nonequilibrium situations, 317—318
   universality, 321—322
      classes, 318—319

Maxwell—Boltzmann exponential formulation, 108
Maxwell—Boltzmann velocity distribution, 338—339, 614—615
Maxwellian decomposition technique, 531
Maxwellian distribution, 14, 355, 359—360, 593, 607, 623
Maxwellian limit, 356—357, 571—573
Maxwellian plasma, 351
Maxwellian temperature, 82
Maxwellian velocity distribution, 474, 476f
Maxwellian Z-function, 349
Mean
  energy, 26
  free path, 292—295
  kinetic energy, 35—36, 42, 109—110, 174—175
    defining temperature, 42—43
    in potential energy, 120—123
Mercury, 514—518
MESSENGER measurements, 515—518
"Metal", 645—646
"Metallicity", 645—646
Metastable distributions, 334
MHD simulation.
    See Magnetohydrodynamic simulation (MHD simulation)
Microinstabilities, 334—335
Microscopic random variables, 318—319
MIMI. *See* Magnetospheric Imaging Instrument (MIMI)
MOCASSIN application, 652—653
Monte Carlo approach, 652—653
Moon, 488
Multidimensional kappa distributions
  negative, 193—194
  superposition of, 194—196
Multiparticle distributions, 186
  multispecies distributions, 240—243
  N-particle (N·d)—dimensional kappa distributions with potential, 235—236
  negative N-particle (N·d)—dimensional kappa distribution, 235

standard N-particle (N·d)—dimensional kappa distributions, 234, 236—240
Multispecies distributions, 240—243
Murchison Widefield Array, 565

N

N-particle
  kappa distributions, 56—57
  phase space kappa distribution, 52
N-particle (N·d)—dimensional kappa distributions
  negative, 235
  with potential, 235—236
  standard, 234, 236—240
Nanoflares, 449
Near equilibrium, 456—458
Nebulae, 635
Nebular continuum, 654
Nebular spectral lines, 642—646
  abundance discrepancy problem, 644—646
  collisional excitation, 643
  recombination lines, 643—644
Negative attractive potential
  phase space distributions, 220
  positional distribution function, 221
  potential degrees of freedom, 221
Negative attractive power law potential, 153—162
Negative kappa index, 57—59
  negative potential, 149—153, 154f
  positive potential, 145—149
Negative potentials, 109—110.
    See also Gravitational potentials
  formulation of phase space distributions, 145—153
  Hamiltonian probability distribution, 142
  negative attractive power law potential, 153—162
  "negative" kappa distribution, 143—145
  with positive and negative values acquiring stable/unstable equilibrium points, 163—164
Negative power law central potential, 226—227
"Negative" kappa distribution, 143—145
  of velocities, 427—429, 428f—429f

Neon (Ne), 636
Neptune, 503—506
  energetic ions, 504—506
Neutral(s), 497—498
  species, 640
Nitrogen (N), 636
Non-Euclidean—normed distributions
  argument (x*), 244
  standard d_K—dimensional kappa distribution of kinetic energy, 244
  standard d_K—dimensional kappa distribution of velocity, 243
Non-Gaussian behavior, 316
Non-Maxwellian
  distributions, 524—525
  drifting bi-kappa distribution, 338—339
  particle velocity distributions, 334
  plasma
    particle distributions, 331—332
    waves, 551
  power law, 532—533
  tails, 341—342
Noncollisional value, 474
Nonequilibrium
  Debye length in nonequilibrium plasmas, 272—285
  electron energy distributions, 638—639
  situations, timescale separation in, 317—318, 317f
  system, 318—319
  "temperatures", 41
  misleading "nonequilibrium temperature", 45—46
  transitions, 461f—462f
Nonextensive statistical mechanics, 5, 16—22, 36f, 78, 96—97, 315
  connection of kappa distributions with
    derivation, 29—34
    historical comments, 34—37
  entropy maximization, 22—29
  ordinary and escort probability distributions, 18—20
  physical temperature, 21—22
  q-deformed functions, 17—18
  Tsallis entropy, 20—21
Nonlinear electrostatic waves, 403

Nonlinear guiding center theory, 629–631
Nonlinear spectral balance equation, 377–378
Nonlinear superposition, 196
Nonlinear wave–particle interaction
  kappa electron distribution, 389–397
  plasma weak turbulence theory, 365–389
  turbulent quasiequilibrium, 389–397
Nonmagnetized plasma. *See* Thermal plasma
Nonseparability, 339
Nonthermal particle distributions, 466
Normal modes, 331
Normalization of phase space kappa distribution, 113–116
Normalized ordinary distribution, 88

O

Oblique propagation, kappa velocity distribution plasma waves at, 359–361
Ohm's law, 287. *See also* Power law
Optical recombination line abundance discrepancy, 650
Optical recombination lines (ORLs), 644–645
Ordinary probability distribution(s), 18–20, 78–79, 82
Orthonormal frame, 336–337
Oscillation type potential, 224–225
Oscillator type, 127–133
Outer heliosphere, 106–107, 422–423
Oxygen (O), 636

P

Parallel propagation, kappa velocity distribution plasma waves at, 339–359
  high-frequency
    electromagnetic Langmuir plasma waves, 352–355, 354f
    electromagnetic Whistler cyclotron plasma waves, 349
    instabilities, 349–352
  low-frequency
    electromagnetic Alfvén cyclotron plasma waves, 342–343
    electromagnetic ion-acoustic plasma waves, 355–359, 359f
    instabilities, 343–348
Parker equation, 585–586
PARs. *See* Prime acceleration regions (PARs)
Particle(s)
  acceleration, 572t, 580
    description, 334
    at quasiperpendicular shocks, 629–631
  detectors, 487
  distributions, 483
  energies, correlation between, 267–272
  energization processes, 490–492
  flux, 440
  independent of total number of, 60–61
  velocity distribution, 67
Particle-in-cell (PIC), 623
Partition function, 244–245
pb. *See* Proton beam (pb)
Pearson's correlation coefficient, 7–10
Perpendicular diffusion coefficient, 629–631
Perpendicular shock, 612. *See also* Quasiperpendicular shock waves
PF instability. *See* Proton firehose instability (PF instability)
Phase space
  marginal kappa distribution of velocities, 432
  parcel in plasmas, 302–303
Phase space density. *See* Distribution function
Phase space distribution(s), 224, 226, 232. *See also* Electron distributions; Ion distributions
  angular potentials, 231
  formulation, 145–153
    negative potential, 149–153
    positive potential, 145–148
  general Hamiltonian distribution, 218
  negative attractive potential, 220
  positive attractive potential, 218–219
  with potential, 223
small positive/negative attractive/repulsive potential, 221–222
Phase space kappa distributions, 109. *See also* Kappa distribution(s)
  degeneration of kappa index, 123–133
  gravitational potentials, 164–171
  Hamiltonian distribution, 110–113, 116f
  local kappa distribution, 134–141
  marginal distributions, 116–119
  mean kinetic energy in potential energy, 120–123
  negative potentials, 142–164
  normalization of, 113–116
  potentials forming anisotropic distribution of velocity, 174–175
Photoionization
  models, 644, 650
  photoionized nebulae modeling complexities in HII region structures, 653
  programs, 652–653
  process, 635
Physical temperature, 21–22, 41–42
PIC. *See* Particle-in-cell (PIC)
Pickup ions (PUIs), 458–462, 610–612, 614–615, 623, 624f
Pickup process, 614
Planar charge perturbation, 276–277
Planck constant, 76, 526
Planck quantization constant, 298, 301f
Planet, 494
  planetary ions, 514–515
  planetary magnetospheres, 5, 106–107, 422–423
  "planetary wind", 492–493
Planetary nebulae (PNe), 634, 636–637, 637f
Plasma, 364. *See also* Space plasma(s)
  current density, 336–337
  dielectric tensor, 336–339
  energy spectra, 501
  environments, 107
  instability, 335
  particles, 45, 337–338

Plasma (*Continued*)
  physics, 553
  processes, 520
  sources, 493–494
  waves emissions, 551
    qualitative aspects for generation
      and damping of, 554–557
Plasma diagnostics
  density diagnostics, 536–537, 537f
  from emission line spectra,
    535–543
  in ionization equilibrium,
    540–541, 541f–542f
  single-ion diagnostics of κ-index,
    537–540, 538f
  from transition region lines,
    542–543
Plasma parameters
  collision frequency, 292–295
  correlation between particle
    energies, 267–272
  Debye length in equilibrium and
    nonequilibrium plasmas,
    272–285
  dependence on kappa index of,
    289–290
  electrical conductivity, 286–291
  infogram of, 311f
  kappa distribution, 250
  large-scale quantization constant,
    300–310
  magnetization, 295–300
  mean free path, 292–295
  plasma conductivity, 292f
  polytropes, 251–266
Plasma sheet boundary layer
  (PSBL), 400
Plasma weak turbulence theory,
  365–389
  general formulation, 365–380
  weak turbulence theory for
    plasmas, 380–389
PNe. *See* Planetary nebulae (PNe)
Point charge perturbation, 278
Poisson distribution, 601–602
Poisson equation
  for electrostatic potential,
    272–276
  of Gauss's law of electrodynamics,
    267–268
  symmetric Poisson equation and
    solutions, 276–279
Polar angle, 174
  distribution, 233
Polytrope(s), 162, 169–170, 251

generalized polytropic relations,
  254–257
kappa index
  connection with, 257–261
  relations between polytropic
    index and, 261–266
polytropic index, 45, 229,
  252–253, 254t
polytropic invariant pressure, 255
polytropic spectrum, 252f
Positional distribution, 225, 227
  function, 221
    positive attractive potential, 219
    small positive/negative
      attractive/repulsive potential,
      222
Positional kappa distribution, 126
Positive attractive potential
  phase space distributions,
    218–219
  positional distribution function,
    219
  potential degrees of freedom, 219
Positive attractive power law
  potential, 127–133
"Positive kappa distribution", 57
Positive kappa index
  negative potential, 149–153, 154f
  positive potential, 145–149
Positive potential, 145–148, 155t
Positive power law central
  potential, 224–225
Potential degrees of freedom, 221,
  224, 226
  angular potentials, 232
  positive attractive potential, 219
  small positive/negative attractive/
    repulsive potential, 222
Potential energy
  kappa index degeneration in
    1-D linear gravitational
    potential, 124–127
  positive attractive power law
    potential, 127–133
  rationale, 123–124
  mean kinetic energy in,
    120–123
Power law, 333–334, 571, 635
  Boltzmann factors, 314–315
  energy distribution, 448
Prime acceleration regions (PARs),
  574
Probability density, 317
Probability distribution, 16, 78,
  100f, 333–334

Proton beam (pb), 355–357
Proton firehose instability (PF
  instability), 345–348, 348f
Proton(s), 407–411, 413–414,
  470–471
  proton–electron plasma, 408
  reflection process, 610–611
PSBL. *See* Plasma sheet boundary
  layer (PSBL)
PUIs. *See* Pickup ions (PUIs)
Pump acceleration mechanism,
  573–574, 591–592
  applications, 597–608
    acceleration in solar corona,
      605–608
    energetic particles acceleration at
      shocks, 598–605
  subtleties with solutions for,
    592–593

Q

q-Boltzmannian distributions,
  27
q-deformed
  "exponential", 16
  functions, 17–18
    q-exponential, 17
    q-gamma, 18
    q-hyperbolic, 18
    q-logarithm, 17
    q-unity, 17
  gamma functions, 88–89
q-exponential
  distribution, 35
  function, 17
"q-frozen state", 51
q-gamma function, 18, 113
q-hyperbolic function, 18
q-logarithm function, 17
q-Maxwellian distributions, 27
q-Maxwellian probability
  distribution, 21
q-unity function, 17
Quantization constant, 76, 78
Quantum turbulence, 325–326,
  326f
Quasifluid theory, 346
Quasilinear equations, 558
Quasilinear relaxation, 557
"Quasinormal closure", 376
Quasiperpendicular shock waves,
  610–611, 629
  dissipation and particle
    acceleration at, 629–631

magnetic field strength, 611f
observational tests, 625–629
simulations, 623–625
upstream distributions and transmission, 614–619, 615f
velocity distribution function downstream of, 619–623, 622f
"Quiet" plasma sheet spectra, 506–507, 508f

R
R-Matrix calculations, 543
Radiation
belt–planetary atmosphere interaction, 503–504
radiative recombination, 643–644
radiative transfer, 652–653
Radio emissions, 551
qualitative aspects for generation and damping of, 554–557
"Ramp", 610–611
Rankine-Hugoniot conditions, 563–564
Rankine-Hugoniot downstream asymptotic state, 610–611
Rankine-Hugoniot relationships, 599
Ratio–ratio diagrams, 539–540
Recombination
emission lines, 642–643
lines, 643–644
rates, 526–529, 527f
Regularized inversion method, 544
Relativistic Maxwellian plasma, 359–360
"Renormalized kinetic theory", 371
Reuven-Ramaty high-energy solar spectroscopic imager, 532–533
"Rosette Nebula", 636, 636f

S
S/C. *See* Spacecraft (S/C)
Sackur–Tetrode equation, 76–77
Satellites, 483
Saturn, 494–500, 497f
electron spectra, 498–500
ion spectra, 494–497
neutrals, 497–498
Saturnian ionosphere, 477–478, 478f

Scale parameters in entropic formulation
impact on entropy, 76–78
length scale, 72–75
speed scale, 75–76
units' paradox, 67–72
Scattering processes
induced and spontaneous, 388–389, 395–397
kappa distributions in high-energy, 326–327, 327f
SDO. *See* Solar Dynamics Observatory (SDO)
Self-correlated
degrees of freedom, 201–206
projections, 204–205, 207–209
SEP events. *See* Solar energetic particle events (SEP events)
Shock(s), 551, 553
energetic particles acceleration at, 598–605
diffusive shock acceleration, 603–605
model for shock, 598–600, 599f
spectrum at shock, 600–603, 602f
reformation, 611–612
wave, 561–565, 562f, 564f, 612–613, 652–653
Short-range electric fields, 330–331
Si III spectrum, 542–543, 543f
Simulations, 623–625
magnetic field magnitude, 624f
1-D cut of velocity distribution function, 624f
Single-ion diagnostics of k-index, 537–540, 538f
Solar and Heliospheric Observatory (SOHO), 470–471, 542
Solar corona, 524, 526
acceleration in, 605–608
Solar Dynamics Observatory (SDO), 534–535
Solar energetic particle events (SEP events), 574
Solar flares, 449, 525, 551
Solar radio bursts
kappa distributions of electrons, 550
qualitative aspects for generation and damping of plasma waves and radio emissions, 554–557
solar radio emissions, 552f
type II bursts, shocks, and electron reflections, 561–565

type III bursts, electron beams, and Langmuir waves, 557–561
Solar spectra, 525–526, 536
kappa distributions and, 524
DEM for, 544–546, 545f
plasma diagnostics from emission line spectra, 535–543
Reuven-Ramaty high-energy solar spectroscopic imager, 525f
synthetic line and continuum intensities, 526–535
Solar System, 488
Solar wind, 402, 407–411, 483
common spectrum in fast, polar coronal holes, 582
electrons, 467–468
ions, 623
models, 475, 476f
plasma, 402
proton distribution, 619–620
protons, 460
Solar Wind Ion Composition Spectrometer (SWICS), 571
Solitary waves (SWs), 399–400. *See also* Suprathermal electrons
Source particle spectrum, 594–597
Space physics, 553
Space plasma(s), 297, 469, 638–639
electron distributions in, 467–468
coronal heating by velocity filtration, 470–472
heat flux, 472–474, 473f
observations and origins, 466–470
suprathermal electrons, 474–478
ion distributions in
antiequilibrium, 432–439
arrangement of stationary states, 439–444
formulations of ion kappa distributions, 425–432
interpreting observations, 444–462
measuring and interpreting kappa distribution in, 487
particle populations, 35
space plasma–related analyses, 106–107
Spacecraft (S/C), 400–401
ACE, 571
Cassini, 445–446, 498

Spacecraft (S/C) (*Continued*)
  frame, 485
  Ulysses, 467
  Voyager, 571, 580
  Voyager 2, 611–612
Spatial diffusion, 592–593
Spectrum
  at shock, 600–603, 602f
  spectral index in velocity space, 586–587
  spectral properties of waves, 334–335
Speed distribution, 46–48
Speed scale, 75–76
Spherical gravitational potential, 166–170, 169f
Spherical shell distribution, 92–93
"Spirograph Nebula", 636–637
"Split-bands", 565
Standard d_K–dimensional kappa distribution
  of kinetic energy, 244
  of velocity, 243
Standard kappa distribution of velocities, 425–427
Standard multidimensional kappa distributions, 187–189
  in inertial reference frame, 189–193
Stationary bi-kappa distribution, 342
Stationary distributions, 317
Stationary probability distribution, 96–97
Stationary state(s), 251
  arrangement, 448
    generalized measure, 441–443
    kappa spectrum, 443–444, 444f
    measure of thermodynamic distance from thermal equilibrium, 439–441
  entropy for isothermal transitions between
    derivation, 87–89
    spontaneous entropic procedures, 94–96
    survey, 89–94
  numerical application of discrete transitions of, 98–101
  transitions, 67
    back to far-equilibrium region, 102
    discrete map of, 97–98

double role of antiequilibrium, 102
  final transitions toward equilibrium, 102
  intermediate transitions toward antiequilibrium, 101
  isentropic switching and importance, 101–102
  stages of, 101–103
Statistical mechanics, 13–14
  principle for, 50–51
Steady theory, 96–97
Steady-state
  Langmuir turbulence, 392–397
    absence of steady-state decay processes, 394–395
    balance of induced and spontaneous emission, 393–394
    balance of induced and spontaneous scattering, 395–397
  particle distribution, 389–392
Stellar coronae, 524
Stochastic
  acceleration, 571
  growth theory, 563
Strahl, 466–468
"Strömgren Sphere" model, 653
Sulfur (S), 636
Sun, 551
Superhalo electrons, 467
Superposition
  of kappa distribution of velocities, 429–430
  of multidimensional kappa distributions, 194–196
  of polytropes, 256, 455–456
Supersonic solar wind, 614
Supersonic turbulence, 639
Superstatistics, 318–319
  approach, 315–316
  from measured time series to, 322–324, 323f–324f
  superstatistical model, 316, 325–326
  superstatistical probability densities, 319–320
  superstatistical system, 316–317
Suprathermal electrons, 405–411, 413–414, 571–573
  on acceleration of escaping particles, 474–478
  coronal heating by velocity filtration to, 470–472

model for electron-acoustic solitons in plasmas, 415–417
model of ion-acoustic solitons and double layers in plasmas, 402–414
  observations and origin of, 400–402, 466–470
Suprathermal particles, 401–402, 613–614
Suprathermal tails, 338–339, 582, 582f, 597, 599, 605
SWICS. *See* Solar Wind Ion Composition Spectrometer (SWICS)
SWs. *See* Solitary waves (SWs)
Symmetric Poisson equation and solutions, 276–279
Synchrotron radiation, 554
Synthetic line, 526–535
  continuum spectrum, 531–533
  excitation rates, 529–531
  ionization and recombination rates, 526–529, 527f
  KAPPA package, 533
  synthetic spectra and atmospheric imaging assembly, 534–535, 535f
Synthetic spectra, 534–535

T
"t^2" process, 654
Taylor expansion, 294
Temperature, 43f, 87–88
  out of equilibrium, 41–42
  fluctuation process, 654
  mean kinetic energy defining, 42–43
  misleading considerations, 43–50
    dependence of temperature on kappa index, 44–45
    divergent temperature at antiequilibrium, 48–49
    "equilibrium temperature", 46
    frequent speed, 46–48
    misestimation of Debye length, 49–50
    misestimation of thermal pressure, 49
    "nonequilibrium temperature", 45–46
    temperature-like parameter, 43–44

physical temperature, 41−42
principle for statistical mechanics, 50−51
temperature-averaged free−free Gaunt factors, 654
Terrestrial planets
  Earth, 506−513
  kappa distribution in magnetospheres of, 506−518
  Mercury, 514−518
Thermal anisotropy, 339−340
Thermal bi-Maxwellian distribution, 339
Thermal equilibrium, 7, 8t−9t, 13−14, 66, 257−258, 402−403, 432−439
  out of, 259−261
  measure of thermodynamic distance from, 439−441
Thermal plasma, 298−299
Thermal pressure, misestimation of, 49
Thermal solar wind protons, 613−614
"Thermalization", 426, 638
Thermodynamic(s), 89
  definition of Clausius, 42
  distance measure from thermal equilibrium, 439−441
  of far-equilibrium region, 457
  limits, 84−87
  variables, 49
Three-component plasmas, 407−411
Time series to superstatistics, measured, 322−324, 323f−324f
Timescale separation in nonequilibrium situations, 317−318, 317f
Tracer particle, 325−326
Transition region lines, diagnostics from, 542−543
Trigonometric functions, 20−21
TS-3 shock, 610−612
Tsallis entropy, 20−21
Tsallis nonextensive statistical mechanics, 333−334
Turbulence, 469
  power spectrum, 606
  quasiequilibrium steady-state Langmuir turbulence, 392−397

steady-state particle distribution, 389−392
Two-component
  magnetized plasmas, 413−414
  plasmas, 405−407
2-D potential, 277−278
Type II bursts wave, 561−565, 562f
Type II solar radio bursts, 551−552
Type III bursts wave, 557−561, 560f, 562f
Type III solar radio bursts, 552

U

Ultraviolet (UV), 551, 635
  emissions, 492
Units' paradox, 67−72
Universality, 321−322
  classes, 318−319
Upstream
  distributions and transmission, 614−619, 615f
    downstream energy spectrum, 618f
    phase space plots, 616f
    solar wind thermal proton temperature, 617f
    velocity phase space portraits, 618f
  of shock, 599
  temperatures, 623
  velocity distribution functions, 623−625, 624f
Uranus, 500−503, 502f
  energetic ion spectra, 501−503
  plasma energy spectra, 501
  whistler waves, 503
UV. See Ultraviolet (UV)

V

VDF. See Velocity distribution function (VDF)
Velocity, 12, 116
  anisotropic kappa distribution of, 430−432
  distributions, 483, 498f
  filtration to suprathermal electrons, coronal heating by, 470−472
  "negative" kappa distribution of, 427−429, 428f−429f
  phase space marginal kappa distribution of, 432

potentials forming anisotropic distribution of, 174−175
probability distribution, 68−69
spectral index in velocity space, 586−587
standard d_K−dimensional kappa distribution of, 243
standard kappa distribution of, 425−427
superposition of kappa distribution of, 429−430
variance of, 44−45
Velocity distribution function (VDF), 333−334, 337−338, 352−353, 355−356, 466, 467f
  downstream of quasiperpendicular shock, 619−623, 622f
Velocity kappa distributions. See also Electron distributions; Ion distributions
  argument, 78−82
  entropy
    formula, 82−84
    thermodynamic limits, 84−87
Very Local InterStellar Medium (VLISM), 627−628
Virial theorem and Jeans radius, 170−171
Vlasov equation, 331, 366
Vlasov−Maxwell differential equations, 96−97
VLISM. See Very Local InterStellar Medium (VLISM)
Voyager Low Energy Charged Particle and Cosmic Ray Subsystem instruments, 571
Voyager spacecraft, 571, 580
  Voyager 2, 611−612

W

Wave frequency−wave vector relation, 338−339
Wave−particle interactions, 334
Wave−wave resonance condition delta function, 395
WC. See Whistler cyclotron (WC)
Weak double layers (WDLs), 402
Weak turbulence theory for plasmas, 380−389
  induced and spontaneous decay processes, 386−388

Weak turbulence theory for plasmas
  (*Continued*)
  induced and spontaneous
      emissions, 386
  induced and spontaneous
      scattering processes, 388–389
Weakly coupled plasmas, 296
Whistler cyclotron (WC), 349
  instability, 349–351, 350f
  plasma waves, high-frequency
      electromagnetic, 349

Whistler waves, 503
Withbroe–Sylwester method, 544

X
X-ray, 524–525, 529–530,
    532–533, 536
  ionization, 639
  spectrum, 533

Y
Yukawa potential, 275

Z
Zero-th law of thermodynamics,
    44–45